THE TIMETABLES
OF ™ SCIENCE

A Chronology of the Most Important People
and Events in the History of Science

Alexander Hellemans and Bryan Bunch

A TOUCHSTONE BOOK

Published by Simon & Schuster Inc.

New York   London   Toronto   Sydney   Tokyo   Singapore

Touchstone

Rockefeller Center
1230 Avenue of the Americas
New York, New York 10020

A HUDSON GROUP BOOK

7   9   10   8   6

Library of Congress Cataloging in Publication Data
Hellemans, Alexander, date.
The timetables of science: a chronology of the most important
people and events in the history of science/Alexander Hellemans
and Bryan Bunch.
p.   cm.
"A Touchstone book."
Reprint. Originally published: New York: Simon and Schuster,
© 1988.
Includes index.
1. Science—History—Chronology—Tables.   I. Bunch, Bryan H.
II. Title.
Q125.H557   1991
509—dc20   90-48253
CIP
ISBN 0-671-62130-0
ISBN 0-671-73328-1 Pbk.

CONTENTS

PREFACE

Like the arts, science is an important part of our cultural heritage. Since earliest times people have tried to explain the universe, as in the myths that arose about the sun and the moon. By at least 600 BC a few scholars started to replace these myths with rational explantions, the beginning of science. Well before that time, the body of wisdom and skill we call technology had become essential to the human way of life.

This book tells what happened in science and technology, when it happened, and who made it happen. This information is largely in the timetables, a chronological, subject-by-subject chart.

We also want to explain the context in which science takes place as it changes from period to period. This is the purpose of the ten overviews. Each one covers a period characterized by particular trends in scientific activity. Sometimes such a period is roughly the same as a well-recognized cultural event, such as the Renaissance, or a standard time interval, such as the nineteenth century. But the scientific periods are bounded by events important to the development of science.

Historians of science have abandoned the idea that science develops linearly, according to simple rules. Instead, the growth of science is much like a stream, growing slowly from its source, meandering through plains, and fed by other small streams until it becomes a river. In the timetables, you can note that sometimes, as in the early part of the twentieth century, physics is dominant—the main part of the stream. You can also see where new streams merge into the main body of science.

Short individual entries cannot do justice to the most important or most interesting developments in science. We have selected more than a hundred significant topics for special attention. Each of these topics, ranging from the first scientists to the recent use of genetic engineering to correct in-born defects, is explained briefly in boxes within the overviews or within the timetables.

The Timetables of Science is based upon the notion that chronology is important: when something happened tells something about the event. If you know that Lucretius supported the atomic theory, it makes a difference in understanding what that means to know when Lucretius lived. The early supporters of atomism date from the fifth century BC and almost none of their work survives; we only have second hand reports of it. The people who made atomism a general belief, on the other hand, were chemists working near the start of the nineteenth century AD. The place of Lucretius—first century BC—tells that he was neither the philosophical innovator nor the practical chemist, but was important in transmitting the idea.

Unfortunately for this premise, it is not alway clear in retrospect exactly when something happened. We list about 10,000 separate events. Especially in the early years, the day, the year, or even the decade for many events are not known. We have relied upon the consensus of many historians of science whose works we consulted to help locate the events as closely as possible.

We particularly want to thank the historians of science who read *The Timetables of Science* in manuscript: Walter Purkert of the University of Leipzig, Steven J. Dick of the US Naval Observatory, and David E. Rowe of Pace University. Although their guidance was invaluable, they are not responsible for any errors in fact or interpretation that might appear in the book. We also want to thank the following people and organizations that made important contributions to the project at various stages: Bob Bender, our editor at Simon & Schuster; Mary Bunch, who aided in the research; Felice Levy of AEIOU, Inc., who copyedited the book and helped prepare the index; Laurie Harris Benjamin of JetSet Typography, who set the type; and Gerry Burstein of G&H SOHO, LTD., who designed the book and whose staff prepared the layouts.

Alexander Hellemans
Hastings, United Kingdom

Bryan Bunch
Briarcliff Manor, NY

OVERVIEW

Science before there were scientists

Science as an organized body of thought is generally considered to have begun with the Ionian school of Greek philosophers about 600 BC. Discoveries or inventions prior to 600 BC are nearly always examples of technology, since they are specific devices or techniques. For the purposes of the Timetable, however, it is useful to classify these discoveries or inventions into astronomy, mathematics, and life science where possible. For example, domestication of plants or animals requires understanding of the requirements of specific living things, but this knowledge is not organized as science—an attempt to systematically investigate the way the universe works. Instead, it is a specific technology, developed to solve a specific need. We classify domestication as life science, however, since such technologies are the seeds of the science that will develop in later times.

If science does not begin until 600 BC, when does technology begin? It is likely that wooden tools were used very early, but wood does not survive millions of years except in exceptional circumstances. Therefore, we must begin with the stone tools first found in the Olduvai Gorge in Tanzania by Louis and Mary Leakey and others, and since found elsewhere in Africa as well. It is customary to think that these tools were made by our ancestor, *Homo habilis*, although there is some evidence that they were made by early relatives that were not on the direct line to modern humans.

The next technology we know about came after *H. habilis* was replaced by a different species, *Homo erectus*. *H. erectus* greatly improved the stone tools and, more than a million years after the first stone tools, learned how to control fire. Technological change proceeded very slowly in those times. After *H. erectus* comes *Homo sapiens*—ourselves.

From very early on we know that *H. sapiens* developed a picture of the universe based on religion. Ancient burials provide the earliest evidence. In later times, statues, shrines, and, most likely, cave paintings attest to these early beliefs, even if evidence does not tell much about the specific beliefs.

The hunter-gatherers

Many anthropologists think that *H. habilis* was largely a scavenger of meat and gatherer of berries, nuts, and roots. Hunting became increasingly important to *H. erectus*. The early hunter-gatherers continued to make progress in several areas of technology and science.

The progress in technology is most clearly seen in the further refinement of stone tools. While *H. erectus* had used essentially the same tool kit for 1.5 million years, *H. sapiens* invented new forms of stone tools at a progressively faster rate before largely abandoning stone tools for metal ones some 7000 to 8000 years ago. Metal tools were preceded by bone tools, such as needles, that could not have been easily made from stone. Although remains are scarce in the archaeological record, there is no doubt that similar progress was being made in tools of more perishable materials, especially wood. Hunting weapons, such as the sling, the bow, the bolo, the fish hook, and the spear thrower, are among the technological innovations of this period. Although most early wooden tools have vanished, some of these inventions can be seen in early paintings.

There is also some evidence that mathematics and astronomy, which have been linked throughout most of history, were in use. Notches on artifacts have been interpreted as tally marks or counters, as calendars, and as records of the lunar cycle.

Other evidence of astronomy is less certain. Various early structures have been theorized to be observatories, but most of these interpretations are controversial. Traditions that seem to go very far back in Egypt and in Central America are associated with specific apparent motions of Sirius and Venus. Although we know of these traditions from a much later period, it seems likely that primitive hunter-gatherers had already detected patterns in the apparent motions of the stars and possibly even in the real motions of the planets through the night sky.

Hunter-gatherers living today have an extensive knowledge of wild animals and plants; there is no reason to expect that this was not also true of early hunter-gatherers. Early botanical taxonomy was undoubtedly accurate, for if it were not, people would not have known which plants were good for food. It is also likely that knowledge of plants whose chemical properties are useful as poisons, dyes, or medicines had its beginnings during the hunter-gatherer period.

Since our ancestors expanded rapidly, such knowledge had to be adjusted for it to be practical worldwide. *H. habilis* had been confined to Africa. However, *H. erectus* spread through Asia and Europe. Around 45,000 years ago, *H. sapiens* reached Australia and possibly even—although opinions differ sharply on this—the Americas. Each of these expansions must have presented the explorers with many taxonomic puzzles: Is this plant that looks like a lentil really a kind of lentil, or is it some poisonous or worthless plant? This led to a steadily expanding grasp of the beginnings of biology.

The agricultural revolution

Starting about 10,000 years ago, people made the major technological advance of domesticating animals and plants. This occurred independently in the Middle East, the Orient, and the Americas. At one time, historians assumed that the agricultural revolution was simply a form of progress. This interpretation is now in dispute. Historians now hypothesize that people knew how to raise crops and keep animals earlier, but were reluctant to do so until either rising population or reduced natural food supplies forced them into agriculture. This is partly supported by a rise in population preceding the adoption of agriculture, as indicated in the archaeological record. Another supporting element is that modern hunter-gatherers know about agriculture, but think it is too much work.

Another belief that has largely been discarded is that urban life began a result of the agricultural revolution. Towns were forming before farming became a way of life. The principal purpose of preagricultural settlements was trade. Towns arose at the juncture of trade routes or near supplies of goods that could be traded. Jericho, for example, was founded well before agriculture started.

It is hard to find any evidence of the physical sciences around the time of the agricultural revolution except for that implied by developing technology, such as the introduction of sun-dried bricks and mortar. Major developments of the period following the agricultural revolution were largely in astronomy, mathematics, and technology.

Civilization

Following the agricultural revolution, societies that we recognize as civilizations began to arise. It is thought that the need to maintain stability after the annual flood contributed to a strong central state in Egypt, while centralized control of irrigation projects was the stimulus in Mesopotamia. Less is known about the origins of the other early civilizations, the Indus culture, centered at Harappa and Mohenjo-Daro in the Punjab region of India, and the early Chinese empire. All were flourishing by about 3000 BC.

Our understanding of these cultures has been influenced by the way we became aware of them. Egypt was well known to the Greeks, archaeology began there under Napoleon, and we have known how to read hieroglyphics since the work of Jean Champollion and Thomas Young in 1822. For these reasons, Egyptian culture is well understood. Many of the Mesopotamian civilizations were familiar from the Bible's Old Testament. Other civilizations in the Americas were unknown to the Western world until after the fifteenth century. The early Maya in Central America, the Olmecs in Mexico, and various civilizations in South America domesticated plants and animals, used medical skills for mummifying and trephining (cutting holes in the skull), and built stone monuments. The Maya, especially, developed elaborate numerical and calendrical systems, along with a way of writing that is only now being deciphered.

A note on dating

In this period, only a few events can be exactly dated as to year. You should assume that unless a specific year is cited, the periods—whether fifty or five hundred years— are the best approximation historians or scientists have been able to locate.

The best true dates for the period, in general, are Egyptian. The Chinese tradition has many specific dates, but scholars think that these are based more upon legends than reality for this early part of Chinese history.

Major advances

Astronomy. The agricultural revolution led to further developments in astronomy. After the revolution, there is evidence of improved observatories, such as Stonehenge. The Great Pyramid is situated almost directly on north-south lines. People needed to have better information about the proper time to plant crops, as attested to by early calendars of 360 or 365 days. There is also some evidence of the prediction of eclipses. Sundials were invented and star catalogs began to be compiled, probably for use by astrologers.

Life science. Little is known about the biological knowledge of this period. The process of mummification was developed independently in South America and Egypt, suggesting considerable practical knowledge of biology. Surgery was performed in Egypt and presumably in Mesopotamia, since the Code of Hammurabi states that a fair price for successful surgery is between 2 and 10 shekels, while an unsuccessful surgeon should have his hands cut off. An Egyptian surgical manual of about the same time provides sensible procedures for many types of operations. Egyptian medicine, the most advanced of its time, also included various drugs, some of which are recognized as effective today.

Mathematics. A major development of the period after the agricultural revolution was the invention of numeration systems. Written numbers preceded any known form of written words. In fact, they seem to have led directly to writing in a series of steps that first took place in the Middle East. It is not clear how writing arose in other parts of the world, but numerals probably preceded words in the Orient and in the Americas as well. By 4000 years ago, positional notation was in use in Mesopotamia, with separate developments by the Chinese and the Maya some

---

Egyptian medicine

Sometime before 2000 BC, Egyptian priests began to develop the world's first sophisticated medical practice. Although it is believed that all early cultures learned some forms of herbal cures and methods of surgery, the Egyptian priests were the first to codify their knowledge in a way that scholars can interpret today. The fame of the Egyptian healers was so great that the rich and noble from all around the Middle East and, later, the Mediterranean would travel to Egypt to be treated.

One semilegendary figure, Imhotep, who flourished sometime around 2950 BC, is often considered to be the first scientist known by name, although he was not a scientist in the modern sense of the word. Famous as a physician, he is also credited with being the architect of the step pyramid for the pharaoh Zoser. After his death, Imhotep was assigned a godly ancestry and magical powers. Isaac Asimov has noted that Imhotep is the only scientist ever to have become a god.

It is widely assumed that Egyptian priests gained their understanding of the human body by preparing mummies. Priests removed for special treatment the internal organs of a cadaver being mummified. Although dissection presumably was the basis of their knowledge, the evidence is that most treatments for disease were based on trial-and-error experimentation. Moldy bread was put on wounds, an application of the mold that produces penicillin. Castor oil as a purgative and poppy juices to relieve pain were also commonly used. Radishes, garlic, and onions were eaten by the thousands who built the pyramids and temples, for the Egyp-

tian priests believed like modern herbalists that these vegetables prevent epidemic disease, a view partially confirmed by modern scientists. At the least, these vegetables contain ingredients that have antibiotic properties: Other ingredients were less likely to be efficacious, such as Nile mud, dung, and urine. There was also a tendency to mix medicines into wine or beer; this contributed to a feeling of well-being in the patient.

Treatment of diseases with unknown causes was done strictly in a religious context. Egyptians believed that gods watch over each body part and priests devoted to a particular god were also specialists in treating that body part. Surgery for wounds and broken bones was quite different. The priests knew that the cause of these injuries was not the gods. A papyrus known as the Edwin Smith Surgical Papyrus (because it was purchased by American Egyptologist Edwin Smith) does not mention the gods, even though it dates from about 1550 BC and is probably a copy of an earlier manuscript that may be as much as a thousand years older. The papyrus tells how to set bones, about the pumping function of the heart, and that the pulse can be used to determine how the heart is functioning. Another medical papyrus from about this time, the Ebers Papyrus, gives other medical practices, such as prescribing medications or diets.

Although Egyptian medicine went into decline about 1200 BC, its reputation as the best in the ancient world continued until well after Egypt became a Roman province in 30 BC.

As far as archaeologists can tell, mathematics preceded writing. Indeed, mathematics was the likely inspiration for writing. Bones inscribed with tally marks date from 30,000 BC, although there is considerable evidence that these tallies represented time rather than amounts. More definite information comes from clay containers of trade tokens used in the Middle East more than 5000 years ago. Apparently, the original purpose was to ensure that a shipment of goods was complete upon arrival. If, say, 123 sheep were sold, they were accompanied by a baked clay container containing 123 tokens. Later, 123 marks were made on the outside of the tablet as well. Gradually, a system evolved in which a number and the goods being traded could be represented by a few signs. The clay containers became clay tablets, the medium for the unique cuneiform system of writing both numerals and words.

Many unrelated peoples, ranging from the Sumerians to the Persians, used cuneiform. We often refer to the related cultures of the cuneiform users as Babylonian, after the city that was the center of many of the empires that occupied the region between the Tigris and Euphrates rivers, or Mesopotamia. It is more accurate, however, to call these cultures Mesopotamian.

Because baked clay tablets are easily preserved, especially in a dry climate, much is known about Mesopotamian mathematics. Some historians think that it is likely that a great deal of the mathematical knowledge of the ancient world, ranging from Rome to China, diffused from Mesopotamia. The Mesopotamian numeration system was based on 60 as well as 10, and scholars can trace this division through many different languages. The most notable reflections of this system today are in the divisions of hours, minutes, and seconds in time calculations and in the divisions of degrees, minutes, and seconds for angle measurements. The break at 10 in Mesopotamian notation was purely additive, as in many simple and inconvenient numeration systems, such as the Egyptian and Greek; but the break at 60 represented true place value, making the Mesopotamians one of the four cultures that developed place value (along with the Chinese, Indian, and Maya). Some evidence suggests, however, that the Chinese and Indian place-value systems were influenced by diffusion from the Mesopotamian system, although it is also possible that the Indian system developed from diffusion from China. In any case, the Mesopotamians lacked one essential for a modern place-value system: they did not have zero. A symbol for zero was probably invented either in Indochina or India about the seventh century AD. Zero was also invented independently by the Maya, probably a hundred or so years earlier than the Indian invention, but it did not have a chance to spread around the world.

Even without zero, the Mesopotamian place-value system provided many benefits, including simple algorithms for the basic arithmetic operations. Furthermore, the Mesopotamians made the logical step of extending the places to numbers smaller than one, just as we do with decimals. Sexagesimal fractions are just as convenient as decimal fractions. They contributed to Mesopotamia's developing a practical method for finding square roots, essentially the same method taught in elementary and secondary schools in the United States today.

Mesopotamian mathematicians were the most skilled algebraists of the ancient world. They were able to solve any quadratic equation and many cubic equations. It is possible that their methods diffused to India and from there back to the Arab world, from which algebra reached the West.

It was once fashionable to say that the Mesopotamians were good in algebra but weak in geometry. Later discoveries have forced a revision of this belief, since it is clear that the Mesopotamians were the earliest people to know the Pythagorean theorem (Pythagoras, who is known to have traveled in the East, may have learned his famous theorem there). The Mesopotamians also possessed all the theorems of plane geometry that the Greeks ascribed to Thales, including the theorem of Thales: an angle inscribed in a semicircle is a right angle. It seems unlikely, however, that they proved these theorems from first principles, as Thales is said to have done. Probably the criticism of Mesopotamian geometry began because some of their writings seemed to use a value of three for π, a value that made its way into the Bible (indirectly, a circular bowl is described as having a circumference three times its diameter). Later discoveries, however, have shown that at least some Mesopotamians used 3.125 for π, about as good a value as their contemporaries in Egypt had (*see* "The value of π," p 360).

hundreds or thousands of years later. In Mesopotamia, the base-60 system of numeration led to a mathematics capable of solving quadratic equations. Geometry also progressed in both Egypt and Mesopotamia, with improvements in the ability to measure area and volume, improvements in the value of π, and the discovery of the Pythagorean theorem. Toward the end of the period, symbols for zero as a placeholder were introduced.

Technology. The major advances in technology after the agricultural revolution include smelting, the use of metals, and the development of the wheel, used both for transportation and pottery making. In Egypt, papyrus—and later, parchment—was being used for writing. Oars replaced paddles and early sails began to appear. Building techniques improved, and large temples and palaces were constructed of stone or brick. Such structures as the pyramids in Egypt required a technology capable of quarrying, moving, and lifting very heavy stones. Standard weights, measures, and coins were introduced, and time was measured by water clocks as well as sundials.

2,400,000
BC

1,000,000
BC

100,000
BC

79,000
BC

60,000
BC   Based on the dating of bone fragments from Central
Australia early humans reach Australia

30,000
to
25,001
BC

25,000
to
20,001
BC   Music is produced by humans in what is now France;
archaeological evidence includes cave paintings, footprints
in caves that appear to be those of dancers, and carved
bones that appear to be wind and percussion instruments

LIFE SCIENCE	MATHEMATICS	TECHNOLOGY	
		Hominids in Africa manufacture stone tools	**2,400,000** BC
		Ancient hearths found in the Swart-krans cave in South Africa indicate that *Homo erectus,* the immediate predecessor of modern human beings, uses fire	**1,000,000** BC
		The earliest known ornament, a decorative amulet, is made from a piece of mammoth's tooth by a Neanderthal; it is found in what is now Hungary	**100,000** BC
		Simple forms of stone lamps, probably fueled with animal fat and using grass or moss for a wick, are in use	**79,000** BC
			60,000 BC
	Paleolithic peoples in central Europe and France use tallies on the bones of animals, ivory, and stone to record numbers; for example, a wolf bone from this period shows 55 cuts arranged in groups of five	Beads, bracelets, and pendants are worn by humans Fired ceramics appear in what is now Czechoslovakia, although ceramics are not used to make pots until 20,000 years later	**30,000 to 25,001** BC
	People make artifacts with primitive geometrical designs	The Venus figurines, small statues of faceless pregnant women with large breasts and buttocks, are made in Europe; they will continue to be manufactured for the next 2000 years People in what is now Poland are the first to use the boomerang, about 13,000 years before the first Australian boomerangs; the early Polish boomerang is made from mammoth tusk The sewing needle is in use in southwestern France and tailored clothing is known from what is now the Soviet Union The bow and arrow are invented, according to evidence from sites at Parpallo (Spain), and the Sahara; stone points from Parpallo appear to be tips of arrows; drawings are found at the North African site; other evidence, however, suggests a later origin, perhaps as late as 8000 BC	**25,000 to 20,001** BC

20,000
to
10,001
BC

10,000
to
9001
BC

9000
to
8001
BC

8000
to
7001
BC

7000
to
6001
BC

6000
to
5001
BC

Early metallurgy

The earliest metallic objects known, small jewels and tools, were directly carved or cut from pieces of metal found in the native state. Gold, copper, and some silver were the first metals used in this way, since they are the only metals commonly found in their elemental form. As early as 7000 years ago, especially in the Middle East and Afghanistan, people found these metals, conspicuous by their luster, in mountains and rivers.

The next stage in the development of metallurgy was when people found that metals would become softer and more malleable when heated and could thus be given different shapes. The first forged and cast objects looked like real metal objects, not like the earlier ones that resembled stone tools.

In the vicinity of native metals one often finds colored minerals, such as malachite, turquoise, and the deep-blue lapis lazuli. The next important discovery was that when heated in a charcoal fire, those minerals yield copper. These ores were much more abundant than native metals, and thus larger amounts of metals became available. The Egyptians exploited mines in the Sinai peninsula, and produced thousands of tons of copper during the predynastic period (about 3200 BC).

Bronze was the most important result of early attempts of rendering copper more resistant by adding different metals. Bronze, a mixture of 90 percent copper and 10 percent tin, can easily be cast, and its use had a large influence on several civilizations—it was the first metal technology that changed society.

The general use of iron appeared around 1000 BC, approximately 2000 years after the appearance of bronze, and was probably introduced by the Hittites in Anatolia. The processing of iron requires higher temperatures only achievable with bellows. Iron was occasionally used before the iron age in ornaments and ceremonial weapons. It was then very expensive; Homer mentions iron as a precious metal on a par with gold.

Quench hardening of iron became known early during the iron age. Again, the influence of this new technology on society was great. Weapons in hardened iron were superior to bronze ones. Tools and weapons were manufactured only by forging because temperatures high enough to melt iron were only achieved in Europe in the 14th century AD. In China cast iron, which has a higher carbon content and melts at a lower temperature, was known around the third century BC.

The Maya make astronomical inscriptions and constructions in Central America

A marked bone from this time or as late as 6500 BC found in Ishango (Zaire) is probably used as a record of months and lunar phases

LIFE SCIENCE	MATHEMATICS	TECHNOLOGY	
	People inhabiting caves in what are now Israel and Jordan use notches in bones to record sequences of numbers; the devices are thought to function primarily as lunar calendars	About this time or by 13,000 BC the spear thrower and the harpoon are invented

People make wall paintings in caves, for example, in the cave of Lascaux, France

The first known artifact with a map on it is made on bone at what is now Mezhirich, USSR; it appears to show the region immediately around the site at which it was found

Rope is in use, according to evidence at Lascaux, France | **20,000 to 10,001** BC |
The dog is domesticated in Mesopotamia (Iraq) and Canaan (Israel)		Houses of sun-dried brick held together without mortar are built in Jericho (Israeli-occupied Jordan)	**10,000 to 9001** BC
Goats and sheep are domesticated in Persia (Iran) and Afghanistan; emmer wheat and barley are cultivated in Canaan (Israel)			**9000 to 8001** BC
Potatoes are domesticated in Peru, pumpkins in middle America; beans are cultivated in Peru and rice in Indochina; floodwater agriculture is used in the Nile valley and in southwestern Asia	People in Mesopotamia use clay tokens to record numbers of animals and measures of grain; this practice gradually develops over the next 5000 years into the first system of numeration and the first writing		**8000 to 7001** BC
The pig and water buffalo are domesticated in eastern Asia and China; the chicken is domesticated in southern Asia

Einkorn wheat is cultivated in Syria and durum (macaroni) wheat in Anatolia (Turkey); sugar cane is being grown in New Guinea and yams, bananas, and coconuts in Indonesia

Domesticated cattle are found in southeastern Anatolia (Turkey)

Flax is cultivated in southwestern Asia

Farming based on corn (maize), squash, beans, and peppers is practiced in the Tehuacan valley of Mexico | | People at Jericho are now using mortar along with sun-dried brick to build houses

People in various sites in the Near East and what is now Turkey make pottery from clay

The oldest known woven mats are made in Beidha (Jordan); basketry probably began much earlier

Weaving of cloth is known in Anatolia (Turkey); the first known samples of cloth are from the early city Catal Hüyük | **7000 to 6001** BC |
| Modern-type domesticated bread wheat and lentils are cultivated in southwestern Asia; citrus fruit in Indochina; bulrush millet in southern Algeria; finger millet in Ethiopia; and squash in Mexico; irrigation is in use in Mesopotamia | | | **6000 to 5001** BC |

	GENERAL	ASTRONOMY
6000 to **5001** BC cont		
5000 to **4001** BC	The Sumerians enter Mesopotamia and start a civilization that will introduce cuneiform writing to the region	The Egyptian calendar, the first known based on 365 days (12 months of 30 days and 5 days of festival, starting with the day that Sirius, the Dog Star, rises in line with the sun in the morning, which coincides with the annual flood of the Nile) is possibly instituted (from astronomical evidence only) as early as 4241 BC, although it may be about 1500 years later
4000 to **3501** BC	Sumerian city of Ur is founded in Mesopotamia (Iraq) According to Marija Gimbutas, horsemen from the Kurgan culture of southern Russia migrate to east central Europe, bringing the Indo-European language and male gods and displacing the "Old Europeans," a matrilineal people whose religion revolves around the Venus figurines, statues of women with exaggerated female secondary-sexual characteristics	
3500 to **3001** BC	The Minoan civilization in Crete begins Menes unites the kingdoms of Upper Egypt and Lower Egypt, becoming the first pharaoh An early form of hieroglyphic writing is in use in Egypt; the first known hieroglyphs are those found on King Narmer's Palette	

LIFE SCIENCE	MATHEMATICS	TECHNOLOGY	
Chinchorro Indians on what is now the coastline of Chile and Peru produce human mummies that survive until today Foxtail millet and peaches are grown in central China			6000 to 500l BC cont
The llama and alpaca are domesticated in Peru; avocados are grown in Mexico Date palms are cultivated in India The horse is domesticated in the Ukraine (USSR) Cotton is grown in Mexico		Sailing ships are known in Mesopotamia Stone is used to construct buildings in Guernsey, an island in the English channel The Egyptians mine copper ores and smelt them	5000 to 400l BC
Wine and domesticated grapes are known in Turkestan; oil palm and sorghum are cultivated in Sudan Beer is known in Mesopotamia Zebu cattle are domesticated in Thailand Olives are cultivated in Crete		People in Mesopotamia fire bricks in kilns, although sun-dried brick continues to be used for ordinary purposes Pictograms with 2000 signs are used in Erech, Sumeria The ard, a primitive form of plow, is in use in China; plows pulled by cattle are known in northern Mesopotamia The Egyptians and Sumerians smelt silver and gold	4000 to 350l BC
	The Egyptians develop their number system to the point where they can record numbers as large as necessary, although they still have to introduce new symbols as the numbers grow larger Sumerians introduce a new set of clay tokens in addition to the traditional tokens used throughout the Middle East to represent numbers of animals or measures of grain; the new tokens stand for finished merchandise—clothing, jars of olive oil, loaves of bread, metal objects	The potter's wheel is introduced in Mesopotamia Wheeled vehicles are used in Sumeria The Egyptians use papyrus to write on The Egyptians mine and process iron, using it mostly for utensils Sailing ships are used in Egypt Candles are in use Metal mirrors are used in Egypt The Egyptians and Babylonians make extensive use of bronze, an alloy of copper and tin; bronze will be the dominant metal until about 1400 BC when the Hittites begin large-scale use of iron A ziggurat in Ur (Mesopotamia), 12 m (36 ft) high, shows that Sumerians are familiar with columns, domes, arches, and vaults	3500 to 300l BC

3000 to 2901 BC	Cuneiform writing is first developed by the Sumerians as an outgrowth of their method of recording numbers	The Babylonians predict eclipses

The calendar

The first quantity that people could measure with any degree of accuracy, and on which all people could agree, was time, although only fairly large amounts of time. Large amounts of time can be easily measured because the universe itself supplies "clockwork" in the daily and annual motion of Earth and the moon. Even so, the measurement of time was not easy to work out. A day is one revolution of Earth; a moon is from one new moon to the next; but it is not so easy to measure a year. Even the day is not as easy to measure as it seems. It took a while to learn to measure the day from one noon to the next (noon is when the sun reaches its highest point in the sky).

The ancient Egyptians were the first to establish a good length for the year, possibly because the Nile floods around the same time each year. This flooding generally coincides with the helical rising of Sirius: that is, when Sirius rises at about the same time as the sun. Although there are 365 days between such risings, the year is actually 365 days, 5 hours, 48 minutes, and 46 seconds, or about a quarter of a day longer than 365 days.

The Egyptian calendar is known to have accurately matched the seasons with dates in 139 AD. Since the year is not exactly 365 days, the Egyptian calendar gradually went into and out of alignment with the seasons with a period of about 1455 years. Knowing this, astronomers have speculated that the year of 365 days was instituted around 4228 BC or 2773 BC.

Hellenic astronomers added the missing ¼ day to the Egyptian calendar by adding an extra (leap) day every four years, but most people ignored it. The calendar with a leap day was finally adopted by the Romans under Julius Caesar in 46 BC. Since then, the calendar has had one major modification, when Pope Gregory, in 1582, on the advice of astronomers, dropped the leap day in years that end in two zeros.

2900 to 2801 BC	Sumerian writing progresses from simple pictographs to symbols that can represent syllables, greatly extending the scope of written language; a key element in the transition is the representation of proper names with symbols that stand for the parts of the name	
2800 to 2701 BC		
2700 to 2601 BC		
2600 to 2501 BC		The Chinese use a vertical pole to project the shadow of the sun for estimating time
2500 to 2401 BC	Units for length, weight, and capacity are legally fixed in Mesopotamia	

LIFE SCIENCE	MATHEMATICS	TECHNOLOGY	
Donkeys and mules are domesticated in what is now Israel; camels in what is now Iran and Arabia; and elephants in India Cotton is cultivated in India Tooth filling occurs in Sumer Imhotep, Egyptian physician and architect, b near Memphis (fl 2980-2950)	Impressions of clay tokens used in Sumeria for showing measures of grain become standardized as the first numerals; a small measure of grain becomes 1, while a larger measure becomes 10; around the same time symbols are introduced for 60 and 360		**3000 to 2901 BC**
		The Great Pyramid of Giza is built as a tomb for Egyptian pharaoh Cheops (Khu-fu); the base is a almost a perfect square, with the greatest deviation from a right angle only 0.05 percent; the orientation of its sides is exactly north-south and east-west The first version of Stonehenge is constructed on the Salisbury Plain (England); it consists of an earthen bank and a ditch, along with 56 pits known as the Aubrey Holes, after John Aubrey, who discovered them, and only three stones, including the Heel Stone	**2900 to 2801 BC**
		Astronomical evidence shows that by 2773 BC the Egyptians have instituted a 365-day calendar, although the evidence can also indicate that the calendar was introduced as early as 4228 BC Corbeled arches and domes are built in Mesopotamia; corbeling refers to building an arch or dome with layers of brick or stone set up so that each layer projects beyond the one beneath it, like an upside-down staircase	**2800 to 2701 BC**
Culture of silkworms is started in China; according to legend, the wife of Emperor Huang-ti is the first to unroll a cocoon and make silk			**2700 to 2601 BC**
			2600 to 2501 BC
Egyptian carvings from this time show a surgical operation in progress The yak is domesticated in Tibet and the cat in Egypt; cultivation of the yam is found in western Africa, while peanuts are domesticated in tropical America		Flint is mined in Europe A form of soldering to join sheets of gold is used by the Chaldeans in Ur (Mesopotamia)	**2500 to 2401 BC**

2500 to 2401 BC cont		
2400 to 2301 BC		The Chinese introduce a method of taking observations of the sky based on the equator of Earth and the poles; this method is not adopted in the West until Tycho Brahe in the sixteenth century AD, although it is now the standard way to record astronomical observations all over the world
2300 to 2201 BC	The Akkadians conquer the Sumerians	Chinese observers record a comet in 2296, the earliest known record of a comet sighting
2200 to 2101 BC		The Sumerians use a 360-day year, 12-month solar calendar along with a 354-day lunar calendar; the calendar has an extra month every eight years to keep it in step with the seasons
2100 to 2001 BC		
2000 to 1951 BC	Babylon becomes the capital of a small kingdom of Canaanites in 1894 BC	

LIFE SCIENCE	MATHEMATICS	TECHNOLOGY	
		Standard weights, used in trade, are developed by the Sumerians; they are based on the shekel of 8.36 g (129 grains or 0.29 oz) and the mina, 60 times as heavy	2500 to 2401 BC cont
		Civilizations in Crete, the Cyclades, and parts of Greece work metal and build tile-roofed houses with several rooms	
		Chariot wheels are used by Sumerian armies	
	Positional notation is developed in Mesopotamia by the Sumerians; unlike most other common numeration systems, the Sumerian system has a base of 60 instead of 10; this sexagesimal system using cuneiform symbols continues in use throughout Mesopotamia until Hellenic times, although even then astronomers continue to use it	Sargon of Akkad produces maps in Mesopotamia for land taxation purposes	2400 to 2301 BC
		The oldest preserved weight, found in the Mesopotamian city of Lagash, is 477 g (about 17 oz)	
		People in Central America make pottery	2300 to 2201 BC
		A map of the city of Lagash in Mesopotamia is carved in stone in the lap of a statue of a god; it is the oldest surviving map of a city	
		Eighty bluestones are set up at Stonehenge (England) in the form of two concentric circles	2200 to 2101 BC
		Queen Semiramis builds the first tunnel below a river (the Euphrates), linking the royal palace in Babylon with the Temple of Jupiter	
		The oldest preserved standard for length is the foot of the statue of the ruler Gudea of Lagash; it is divided in 16 parts and is 26.45 c (10.41 in.) long	2100 to 2001 BC
The guinea pig is domesticated in Peru	Mesopotamian cultures learn to solve quadratic equations; that is, equations in which the highest power is two	The Cretan palace of Minos introduces interior bathrooms with a water supply	2000 to 1951 BC
Contraceptives are introduced in Egypt		The Sumerian system of measures includes, besides the shekel and mina, units of capacity: the log (541 ml or 33 cu in.); the homer, 720 logs; and the cubit and the foot, which is two-thirds of a cubit; a cubit is the distance from the tip of the middle finger to the elbows, about 43 to 56 cm (17 to 22 in.)	
Alfalfa is cultivated in Iran			
The first zoo in China, the Park of Intelligence, is founded			
Physicians form an important professional group in Babylon and Syria; their medical practice is strongly based on astrology and belief in demons			

13

	GENERAL	ASTRONOMY
2000 to 1951 BC cont		
1950 to 1901 BC		
1900 to 1801 BC		
1800 to 1751 BC		Star catalogs and planetary records are compiled in Babylonia under Hammurabi
1750 to 1701 BC	The famous Code of Hammurabi, the first known set of laws, is written	
1700 to 1651 BC	The Hyksos dominate Egypt	
1650 to 1601 BC	The volcanic island of Thera, also known as Santorini, explodes in 1628 or 1645, according to sequences of tree-ring dates from the United States and Ireland (1628) or glacial cores found in Greenland (1645); this eruption, which probably destroys the nearby Cretan civilization, is more conventionally placed at 1550 BC; it may account for the unusually harsh winters from 1628 to 1626 BC as recorded in bristlecone pines	
1600 to 1551 BC	In Mesopotamia the Kasserites rule while the New Kingdom is started in Egypt	The zodiac is identified by Chaldean astrologers in Mesopotamia

Santorini and Atlantis

We know from geological and archaeological records that a major eruption of the volcano Santorini (also known as Thera) took place about 3500 years ago. Recently, tree-ring and ice-core data have shown that the date was probably 1628 or 1645 BC. This eruption has been blamed for everything from the destruction of the Minoan civilization to the legend of Atlantis to the parting of the Red Sea by Moses.

Santorini is in the Mediterranean Sea, only 70 miles from the island of Crete. The people of Crete who lived about 1645 BC are known today as the Minoans, after King Minos, their legendary ruler. Shortly after 1400 BC, the Minoan civilization vanished from the records. A current theory holds that earthquakes associated with the Santorini eruption caused extensive damage to the Minoan civilization, especially from fires from lamps that were knocked over.

Atlantis was supposed to have been a large island west of the strait of Gibraltar, in the Atlantic Ocean. Legend has it that Atlantis was the leading civilization of its time. Plato wrote that "in a single day and night of misfortune, it sank into the sea."

Some scientists think that Atlantis was an island where Santorini is today. When the great eruption happened, the island and everyone on it vanished. The volcano continues to be active. It erupted as recently as 1950.

Life Science	Mathematics	Technology	Date
		The ard, a primitive form of plow, is found in Uruk (Iraq), while the earliest iron plowshares are in use in Canaan (Israel); paddy culture of rice occurs in southeastern Asia People in Asia Minor and Persia develop wheels with spokes	2000 to 1951 BC cont
		A copper bar from Nippur weighing 41.5 kg (91.3 lb) and 110.35 cm (43.44 in.) long is an early standard measure; it is divided into 4 "feet" of 16 "inches" each	1950 to 1901 BC
	Mesopotamian mathematicians discover what is now known as the Pythagorean theorem		1900 to 1801 BC
	Multiplication tables appear in Mesopotamia		1800 to 1751 BC
	The Moscow papyrus, dating from the Twelfth dynasty, shows that the Egyptians possess considerable knowledge of geometry, including a formula for the volume of a truncated pyramid; the Moscow and Rhind papyruses (see also MATHEMATICS 1650-1601) are the two main sources of knowledge of Egyptian mathematics		1750 to 1701 BC
		The Phoenicians are writing with a 22-letter alphabet	1700 to 1651 BC
	Ahmes, an Egyptian scribe, affixes his name to a papyrus dealing with solutions of simple equations; this papyrus, known as the Rhind papyrus after A. Henry Rhind, who purchased it in 1858, is the main source of information about early Egyptian mathematics; Ahmes probably copied an earlier document from about 1825 BC		1650 to 1601 BC
		Bellows are used in the manufacture of glass and in metallurgy True plows made of bronze are in use in Vietnam	1600 to 1551 BC

	GENERAL	ASTRONOMY
1550 to **1501** BC		
1500 to **1451** BC	Pictograph writing appears in China during the Shang dynasty Archaeological evidence, such as distinctive axes, swords, and other manufactured items from China found in various Middle-Eastern sites, shows considerable contact between China and the Middle East	The gnomon, an L-shaped indicator used as a sundial, is known by the Egyptians Thutmosis III erects in Heliopolis the "Needle of Cleopatra"; its shadow is used to calculate the time, season, and solstices
1450 to **1401** BC		
1400 to **1351** BC	Warriors from the north—Homer calls them the Achaeans, but the most common term today is the Mycenaeans—enter Greece to form what later becomes the Greek civilization	
1350 to **1251** BC	The Assyrian supremacy in Mesopotamia begins Moses leads the Hebrews in their Exodus from Egypt	
1250 to **1201** BC	Ramses II reigns in Egypt	
1200 to **1151** BC	In Anatolia (Turkey), the Trojan War is fought	

LIFE SCIENCE	MATHEMATICS	TECHNOLOGY	
The Edwin Smith Surgical Papyrus is written, although it appears to be a copy of a manuscript written about 2500 BC; the papyrus is a scientific treatise on surgery The Papyrus Ebers gives a description of 700 medications; it also shows that physicians prescribe diets, fasts, and massage and that some practice hypnosis			**1550 to 1501 BC**
The dog is the only domesticated animal in North America The soybean is cultivated in Manchuria Bone inscriptions in China refer to the making of beer		Liquor is distilled in parts of Asia Light carts with two spoked wheels are used in warfare The Sumerians invent the single-tube seed drill	**1500 to 1451 BC**
		Egyptians build water clocks, or clepsydras Writing on parchment and the use of a balance with a pointer for weighing are developed in Egypt Stonehenge achieves the form in which it is known today	**1450 to 1401 BC**
Multiple cropping within the same year is in use in China Manioc (cassava) is grown in South America and sunflowers in North America		The Egyptians and Mesopotamians produce glass; a glassworks from at Tel-el-Amarna is discovered by Flinders Petrie in 1894	**1400 to 1351 BC**
	Decimal numerals are in use in China		**1350 to 1251 BC**
		The Egyptians build a canal from Lake Timsaeh (the Nile) to the Red Sea; this is the first of several canals linking the Nile and the Red Sea The Babylonians develop an instrument that can determine when a star or planet is due south	**1250 to 1201 BC**
		Bells cast in bronze appear in China	**1200 to 1151 BC**

1150 to 1101 BC	The Amorite king Nebuchadrezzar takes control of Babylon in 1124
1100 to 1051 BC	The Dorians complete their piece-by-piece invasion of the Mycenaeans, occupying the land that is now Greece
1050 to 1001 BC	The Shang dynasty in China is overcome by the Chou (from what are now Kansu and Shensi provinces) in 1027
1000 to 951 BC	
950 to 901 BC	
900 to 851 BC	
850 to 801 BC	
800 to 751 BC	The first Olympiad is celebrated in Greece in 776 According to tradition, Rome is founded in 753
750 to 701 BC	Homer and Hesiod compose their works
700 to 601 BC	The first standard coinage is in use in Lydia (western Turkey) Babylon under the Chaldean king Nebuchadnezzar II becomes the largest city on Earth, with an area of 10,000 hectares (25,000 acres)

Early measures

The ancients developed moderately accurate ways to measure four quantities: length, area, volume, and weight or mass (which were not distinguished between). They had no way to measure other quantities.

Ancient peoples used time to measure large lengths and areas. A journey was so many hours, days, or—for a very long journey—moons. An acre was the amount of land that a person driving a yoke of oxen could plow in a day, and the length of the furrow was the furlong. Since small amounts of time could not be measured accurately, other means for smaller measures were sought.

Many ancient measures came from body parts or easily obtainable materials. For example, in the customary US system, we still speak of the foot and the grain. Other measures based on body parts are less obvious. The inch was probably once the last joint of the thumb, since the word "inch" seems to come from *uncia*, the Latin word for thumb. The yard was the distance from the tip of the nose to the end of the fingers when the right arm was outstretched.

One problem with these measures is easy to see: bodies differ in size. Early attempts to solve this problem included defining the yard in terms of the king's nose and outstretched arm. The earliest preserved standard for length is the foot of a statue of Gudea, the governor of Lagash, a Mesopotamian city of about 4000 years ago.

From ancient times it was clear that different measures need to be related easily to make sense of them. It is simpler to have 10 or 12 or 16 inches in a foot, not 12.3583 inches or some other nonintegral multiple. These relationships were codified early on in most countries.

The Babylonians record a solar eclipse in 763 BC, the oldest eclipse recorded

Chinese solar eclipse records start in 720 BC

LIFE SCIENCE	MATHEMATICS	TECHNOLOGY	
			1150 to 1101 BC
Early measures (continued) Another rule for the inch was the length of three grains of barley laid end to end. The ancient Egyptians had decreed that there were four digits in a palm and seven palms in a cubit. It is apparent that the Egyptian cubit was standardized as well at 20.62 inches (52.3748 centimeters). War or other factors resulted in loss of the standard from time to time. One of the first sets of modern standard measures was enacted in England in 1215. But, when the British houses of Parliament burned on October 16, 1834, the standard yard and standard pound burned with them. In the United States, the British system was kept after the American Revolution, but lack of a central standard persisted until 1832. At that point, Congress averaged the different weights and measures from each of the main customs houses in the nation and declared these averages to be the standards.			1100 to 1051 BC
		The Duke of Chou in China builds either an early "south-pointing carriage" or magnetic compass; a south-pointing carriage uses a differential gear to keep a part of the carriage pointing in the same direction as the carriage turns	1050 to 1001 BC
The Etruscans use a form of false teeth, mostly for cosmetic reasons Oats are cultivated in central Europe	Chinese counting boards originate	Dyes made from the purple murex are introduced by the Phoenicians	1000 to 951 BC
		Darius I has a canal constructed from the Nile to the Red Sea, effectively linking the Mediterranean and the Indian Ocean, although, like other ancient canals, it eventually falls into disuse	950 to 901 BC
	The symbol for zero is used in an inscription in India in 876 BC, the first known reference to this symbol, although the concept may have originated earlier	Natural gas from wells is used in China	900 to 851 BC
		The first known arched bridge is built in Smyrna (Izmir, Turkey)	850 to 801 BC
		The Olmec build pyramids in La Venta (Tabasco, Mexico)	800 to 751 BC
			750 to 701 BC
Patients are treated at temples of the god Asklepios in Greece		Water clocks are used in Assyria Glaucus of Chios invents soldering with an alloy that melts easily	700 to 601 BC

OVERVIEW

Greek and Hellenistic science

Science that presents an organized view of the universe developed with the rise of the Greek civilization. The Greeks developed institutions such as the Academy, the Lyceum, and the Museum; these institutions carried on scientific research in somewhat the way that universities do today. When the Academy and Lyceum were closed in 529 AD, and the Museum was destroyed at about the same time, the Greek era in the history of science was over, although Greek writings continued to have great influence for another thousand years or more.

Greek ways and the origin of science

The origins of the people that made up the Greek civilization in antiquity are still not known. It is generally believed that waves of warriors came from the north or from western Asia and invaded Greece around 1400 BC. They soon developed into a seafaring and trading people, and it was perhaps that aspect that predestined them to be the first to practice science in the form we know it today. Navigation was important, and from it the Greeks developed a keen awareness of space and a sense of geometry. It was especially those Greeks who spoke the Ionian dialect and settled on the coast of Asia Minor who were responsible for the birth of science.

Around the sixth century BC, the first Ionian philosophers, Thales, Anaximander, and Anaximenes, started speculating about nature. Although Greek science may have been a continuation of ideas and practices developed by the Egyptians and Babylonians, the Greeks were the first to look for general principles beyond observations. Science before the Greeks, as practiced in Babylonia and Egypt, consisted mainly of the collection of observations and recipes for practical applications.

Speculative philosophy was the new element in Greek thinking. The Greeks detached themselves from observations and tried to formulate general theories that would explain the universe. The Greek philosopher's search for understanding was not inspired by religion or practical application; it was based entirely on the wish to know and to understand. The Greeks were the first to introduce a scientific method, although it was based on reasoning and observation, not on systematic experimentation. (Several Greek scholars performed experiments; for example, Pythagoras is said to have experimented with strings, investigating changes in pitch for various lengths. Empedocles is claimed to have proved that air is material by immersing a tube that is closed at one end.)

There are several reasons why science first developed in Greece and why rational thinking was maintained in that country. First, the Greeks colonized other countries where they encountered myths that were different from their own for the explanation of natural phenomena, thus holding their own beliefs and myths up to question. Second, although popular religion was widespread, the Greeks did not have a strongly organized priesthood or monolithic religious hierarchy. Science, which in Babylonia and Egypt was mainly in the hands of priests, became in Greece a lay movement. The Greeks felt masters of their fate, unlike those who believed in the fatalism of Chaldean astrology. The Greeks were a seafaring people with a decentralized economy who lived in city-states that were largely controlled by upper-class citizens. This led to freedom of expression and thought; philosophical ideas could be discussed freely.

A theory of creation was absent in Greek religion. Science, in a way, played the role of religion in providing theories on the origins of phenomena. However, clashes between established religion and the natural philosophers occurred during the fifth century BC, when the irreligious attitude of the philosophers became an issue. It resulted in the condemnation of Anaxagoras, who had to leave Athens in exile, the death of Socrates, and even attacks on Aristotle.

Growth and decline of Greek science

Greek culture and scientific thinking first developed on the Ionian coast of Asia Minor and then moved to the Aegean isles and the Greek colonies in southern Italy. Early Ionian science was materialistic. The atomists, such as Leucippus and Democritus, believed reality to be embodied by matter. The Pythagoreans viewed the universe differently; to them it was to be found in form and number. Pythagorean ideas strongly influenced the school of Plato, and scientific thinking became more metaphysical in nature.

Around the fourth century BC, Athens became the center of Greek intellectual activity. Aristotle, who lived in Athens, where he led the Lyceum, was the most important scholar in Greek antiquity. He was also the first true philosopher of science, introducing the inductive-deductive method, a "scientific method" that still plays a role in scientific thinking today. Aristotle argued that the investigator of nature should deduce general principles from observations—the inductive phase—and then explain the observations by deducing them from the general principles—the deductive phase.

Aristotle was the tutor of Alexander the Great, who conquered the world from Greece to India before his death in 323 BC. Alexander's armies spread the Greek, or Hellenic, culture, producing a fusion that we now call Hellenistic. Hellenistic science was particularly strong in Egypt, especially in the city of Alexandria. Hellenistic mathematicians advanced the study of curved figures and algebra considerably, while Hellenistic astronomers developed a complex view of the universe that produced accurate observational results.

From 146 BC on, although the Greek traditions persisted, most of the Mediterranean, including Greece itself, was dominated by Rome, although Egypt was independent for another couple of centuries. While Roman rule did not explicitly attack science, science did not thrive under it either. Much of the Hellenistic science after the period of Roman domination began took place in Egypt. After the third century AD, however, Hellenistic science fell into further decline, and thereafter little that was new was produced. When most of the Hellenistic world became part of the Byzantine Empire in 395 AD, the situation became, if anything, worse.

The First Scientists

Scientific thinking originated in Greece with the Ionian philosophers Thales, Anaximander, and Anaximenes. All three of them were born in Miletus, a city-state on the coast of Turkey. Although the Egyptian Imhotep might be claimed as a scientist, the Ionian philosophers were the first to believe that people could understand the universe using reason alone rather than mythology and religion. They searched for a prime cause for all natural phenomena. No personal forces of gods were involved, only impersonal, natural processes.

Thales of Miletus (about 600 BC) is regarded as the founder of the Ionian school of natural philosophy. He probably studied in Egypt, where he was exposed to new ideas. It is likely that there he learned the craft of land surveying, from which he deduced geometry. In Mesopotamia he studied astronomy, and it is believed that he predicted a solar eclipse, a feat that gained him a great reputation in Miletus.

Thales searched for a unifying principle or essence underlying all phenomena and identified this essence as water. Matter appeared to him to exist in three forms: mist, water, and earth. He considered mist and earth as forms of water. In the field of astronomy, which he had learned from the Babylonians, Thales claimed that the substance of stars is water.

Anaximander was a pupil of Thales c 610-545 BC; he is believed to have written the earliest scientific book, now lost. His basic principle, the *apeiron*, can be compared with the concept of "ether" of the nineteenth century (see "Does the ether exist," p 366). Anaximander also formulated a theory of the origin and evolution of life. He believed that life originated in the sea from the "moist element," which was evaporated by the sun. The presence of shells and marine fossils was for him the proof that the sea covered much of Earth's surface. He postulated that humans must have originated in the sea and have resembled fish.

Anaximenes (c 570-500 BC), who may have been a pupil of Anaximander, is known for his opinion that the rainbow is a natural phenomenon rather than a divine one. He believed that air is the basic principle of the universe.

The rise of the Christian Church also played an important part in the decline of science. The teachings of the Church were unfavorable toward empirical knowledge for several reasons. During the first Christian millenium, many thought that the world was to come to an end very soon, an attitude hardly encouraging investigation of how the universe might work. Early Christian theology had also absorbed many of the ideas of Plato and Aristotle. Although the works of the Greek scholars were not known directly until after the first millenium, many Greek ideas had been passed on to the Christians through Alexandria, where many of the early church fathers lived. The metaphysical and idealistic (in the philosophical meaning of the word) Platonic view of the world, verging on mysticism and leaving little place for actual observation or experimentation, became a mode of thought imposed by Christian theology. Also the Christian outlook on life itself discouraged secular knowledge since it was believed that such knowledge would not help in a life after death. St. Augustine's teaching that all natural processes had a spiritual purpose strongly influenced this view of nature. Symbols and allegories were used to explain natural phenomena. Secular knowledge and science became associated with heathenism, an idea that lead to the destruction of the Library of the Temple of Serapis in Alexandria in 390 by Bishop Theophilus and the murder of the mathematician Hypatia, instigated in 415 by St. Cyril, Bishop of Alexandria.

A note on dates

Dates in Greek and Hellenistic times were not kept rigorously. For one thing, most dates were given in terms of Olympiads; however, to say that an event occurred during the 19th Olympiad only localized it to within a four-year period. Historians have worked out approximate dates for the lives of many of the noted philosophers and scientists of the time from scraps of evidence. Thus, you may read in some sources that Thales was born in 624 BC and died in 546 BC, but neither date has a definite source. Instead, historians know the year that Thales is said to have predicted an eclipse, which astronomical calculations put at 585 BC. When only one date in a person's life is known, it is the custom of historians to assume that the person must have been around 40 when the event occurred, so it is assumed that Thales may have been born around 624 BC. There is a tradition that he lived to be 70, so he may have died in 546 BC.

Given the uncertainty that applies to most dates of this period (Chinese dates of the time are somewhat more certain), we have listed events by decade. Even events for which no specific dates are given did not necessarily occur in the decade listed, but sometime near that decade. In particular, events listed during the first decade of a century may be known only to the century, and often even the century is uncertain.

Major advances

Astronomy. Unlike the Babylonians, who believed that celestial bodies were gods, the Greeks tried to find physical explanations for the celestial phenomena they observed. Because the Greeks considered celestial phenomena to be emanations or evaporations from Earth, early Greek astronomy was not distinguishable from meteorology.

Astronomy occupied a less central position in Greek culture than in Babylonia. The Greeks were less practical and their early astronomical observations were less accurate than those of the Mesopotamian astronomers; however, later Hellenic astronomers improved on Mesopotamian work. The lack of interest in accurate observations was reflected in the state of disarray of the calendar: every Greek city kept time differently.

The main interest of the early Greeks was cosmology, and Greek astronomers came up with a multitude of cosmological models. Thales assumed Earth to be floating in water. Anaximander believed that Earth was a circular disk suspended freely in space; he explained the daily motion of the stars by assuming that they were attached to a sphere that rotated around Earth. The Pythagoreans believed that Earth rotated around a central "fire" that they did not identify with the sun. Parmenides of Elea and Pythagoras of Samos believed Earth to be spherical. Later thinkers, such as Aristotle, believed Earth to be the center of the universe. Aristarchus of Samos challenged this idea and proposed that the sun is at the center of the universe, with Earth and the planets circling around it. Aristarchus' idea was not generally accepted, and Aristotle's geocentric universe, adopted by the Alexandrian astronomer Ptolemy, remained unchallenged until the Renaissance.

Unlike the early Greek astronomers, however, Hellenistic astronomers, such as Eratosthenes, Hipparchus, and Ptolemy, were keen observers. They were able to calculate correctly the size of Earth and the distance to the moon, as well as the positions that the planets would occupy at any given time. The system of planetary motions developed by Hipparchus and refined by Ptolemy became the dominant astronomical work for over a thousand years, until it was replaced by the system of Copernicus, Kepler, and Newton.

Life science. Aristotle is considered the father of the life sciences. Alexander the Great became his patron, funded his work, and arranged for Aristotle to receive samples of plants and animals from all corners of the Alexandrian empire. Aristotle undertook the classification of animals and plants on a large scale. He divided animals into those with and without blood systems, and divided those of the first group into fish, amphibians, reptiles, birds, and mammals. Aristotle believed strongly in a progressive design in the different forms of organisms, an early form of the idea of the Great Chain of Being. He classified organisms into a hierarchy ranging from the most imperfect (plants)

In 430 BC a great plague struck Athens, killing the Athenian leader Pericles in 429. The Athenians appealed to the oracle at Delos to provide a remedy. The oracle said that Apollo was angry because his cubical altar was too small. If it were doubled, the plague would end. The Athenians had a new altar built that was twice the original in length, breadth, and height. The plague became worse. Consulting the oracle again, the Athenians learned that Apollo was angrier than ever. The god wanted the volume of the cube doubled, and the Athenians had octupled it. The plague went on until 423. The problem of doubling the cube went on until the nineteenth century.

Or so the story goes. There are other stories (but not so good) about how the Delian problem, as doubling the cube came to be known, arose. Whatever its origin, the Delian problem together with the problems of squaring the circle and trisecting the angle became the three classic unsolved problems of Greek mathematics.

It is essential to understand that the method of solution for the classic problems was restricted. A solution would count only if it were accomplished by a geometric construction using an unmarked straightedge and a compass that collapsed when it was lifted from the paper. Tradition has it that Plato is responsible for setting this requirement. Although a great many other problems were easily solved under the restriction, the three classics stubbornly refused to yield.

It is likely that the first classic problem was squaring the circle. It is thought that Anaxagoras worked on solving it, about 450 BC, while he was in prison for having claimed that the sun is a giant red-hot stone and that the moon shines by its reflected light. The problem specifically was to construct a square that has the same area as a given circle.

The trisection problem also arose around the same time. In that case, the problem was to find an angle whose angle measure is exactly one third that of a given arbitrary angle. Trisection appears deceptively simple since among the easiest constructions are the trisection of a line and the bisection of an angle. Like the other classic problems, it stumped the ancient Greeks, but not completely.

The apparent progress toward squaring the circle by Hippocrates of Chios (not to be confused with Hippocrates of Cos, the father of medicine) around 430 probably encouraged other mathematicians to continue. Hippocrates succeeded in squaring a region bounded by two arcs of circles. This region looks like a crescent moon and is therefore called a lune. This was the first successful attempt to convert an area bounded by curves to one bounded by straight lines.

A number of mathematicians decided to stretch the rules and solve the problem ignoring Plato's restriction. Hippias of Elis is supposed to have found two different ways to square the circle, both explicitly rejected by Plato (according to Plutarch). One of the methods attributed to Hippias, based on the first curve other than the circle that was well defined and constructible, was used to square the circle at a later date. Similarly, while trying to solve the Delian problem, Menaechmus seems to have discovered the conic sections—the ellipse, the parabola, and the hyperbola. Use of these curves enabled him to duplicate the cube. Archimedes solved two of the classic problems, trisection of the angle and squaring the circle, with his famous Archimedian spiral (a curve invented by Conon of Alexandria). Archimedes also discovered that all that is needed to trisect an angle is a marked straightedge instead of an unmarked one. In a related development, Archimedes managed to square a region bounded by a parabola and a line.

Around 320 AD, Pappus declared that it was impossible to solve any of the classic problems under the Platonic restriction, although he did not offer a proof of this assertion. Nevertheless, many continued to work on the classic problems in the traditional manner. Finally, the nineteenth century produced the definitive "solutions" to all three problems. Each problem was shown to be incapable of any solution that met the Platonic requirement, just as Pappus had stated. In 1837, Pierre Wantzel supplied a rigorous proof that an angle cannot be trisected with an unmarked straightedge and collapsing compass. In 1882, Ferdinand Lindemann showed that π, the ratio of the circumference of a circle to its diameter, is a transcendental number, implying that the circle cannot be squared (*see* "The value of π," p 360). As for the Delian problem, it had been shown even earlier that it required construction of a line whose length is the cube root of two. With a straightedge and compass, it is only possible to construct square roots.

Although it is clear to mathematicians that the solution of the classic problems under Plato's restriction is impossible, amateurs continue to offer their "proofs" of success.

to the most perfect (man). He studied over 540 species and compared the anatomy of 48 species by dissecting them.

Of great importance was Aristotle's work in embryology. He observed and described the development of embryos and was the founder of comparative embryology. His main discovery in embryology was that the contribution of the mother is as important as that of the father in procreation. Before Aristotle, the common Greek notion was that man supplied the seed that grew into the new individual and that the role of woman in relation to the seed was similar to the role of the soil in relation to the seed of a plant.

Theophrastus, the successor of Aristotle at the Lyceum, continued Aristotle's method of observation and classification by applying it to plants. He described and classified a large number of plants, and coined numerous terms in botany.

Many of the Greek philosophers, including Anaxagoras, Empedocles, Democritus, and Philolaus were physicians or were interested in medicine. Alcmaeon is considered by many the founder of medicine; he knew that the brain is the central controlling organ of the body. He also discovered the optical nerve. Hippocrates of Cos became the best known physician, but during his time several medical schools already existed in Greece and its colonies.

In Homer's time, the map of the world showed the continental mass formed by parts of Asia, Africa, and Europe surrounded by a vast body of water, Oceanus. Herodotus, however, felt that the Homeric world had too much water and not enough land. To balance things out, he replaced Oceanus with a great desert.

Symmetry was an overriding concern of the Greeks. For example, they noted that a line drawn between the Nile and the Danube would give a somewhat symmetrical version of the known world. But they obtained even more symmetry by carrying Pythagorean ideas over; that is, that Earth must be a sphere. Plato accepted this on purely geometrical grounds, but Aristotle offered a variety of proofs of sphericity from observation. Later Greek geographers accepted the spherical Earth as a matter of course and worked from there.

The Greek belief in symmetry resulted in the accidental insight that there must be a great continent in the Southern Hemisphere. Since the Greeks knew there was a great land mass in the Northern Hemisphere, they believed it must be balanced by one to the south. This tradition, supposedly originating with Ptolemy, allowed considerable freedom in execution for the mapmakers. Pomponius Mela, writing about 43 AD, made Ceylon the northern tip of *terra australis nondum cognita* (the undiscovered southern continent), as the great continent came to be known. By the fifteenth century, maps showed that Terra Australis had attached itself to southern Africa, effectively barring ships from reaching the East by sailing around Africa. When Bartolomeu Dias rounded the Cape of Good Hope in 1487, however, Terra Australis began to shrink.

Some maps from early in the sixteenth century, notably a Turkish map from 1513 and the Orantes Pinnaeus world map of 1532, show Antarctica of about the right size and location. Even the bays and mountain ranges are close to where they are now known to be. Some people think that these are copied from truly ancient maps prepared by a seafaring people who had sailed the globe. It is more probable that among the many versions of Terra Australis, these came closest to the truth.

Another persistent tradition is that some maps before the time of Columbus and Magellan correctly located the Americas. The Portuguese are supposed to have seen a map in Java when they first reached the island that showed South America. The Vinland map, originally dated as 1430 AD, showed North America. Although it was thought to be a fake soon after its discovery, later research suggests that it may be a copy of a real Viking map.

The discovery of the north coast of Australia by Europeans (it may have been known to the Chinese as early as the thirteenth century) would seem to have justified the belief in a giant southern continent, but in the mid-seventeenth century, Abel Tasman circumnavigated Australia, shrinking the polar continent once again. Even in the eighteenth century, however, people continued to believe in the existence of the continent at the South Pole, arguing from the same premise of symmetry as had the Greeks. Alexander Dalrymple was the particular champion of the "Great Southern Continent," but he was passed over on the British government's expedition to find it. Instead, the expedition was put under the leadership of Captain James Cook. While the official purpose of Cook's voyage was to observe the transit of Venus from Tahiti, he also had secret orders to find the Great Southern Continent.

Cook's first voyage located New Zealand, but by circumnavigation showed that it was not the missing continent. In 1772 he started out again, this time with location of the continent as his major goal. Cook reached within 75 miles of Antarctica, but failed to find land. He concluded that the Great Southern Continent was either a myth or so close to the pole as to be beyond navigation. Finally, in the nineteenth century, various explorers reached land at various points and the true nature of Antarctica gradually became known. The world map was finally complete on the basis of exploration, not just on theoretical ideas.

Hippocrates and his followers explained health and disease by the balance of "humors," a theory that remained unchallenged for many centuries and eventually hampered the development of medicine. Yet Hippocrates freed medicine from religion and superstition and put it on a more scientific footing. Accurate observation of the course of an illness became important, as well as the healing power of nature. Herbs, massages, diets, and baths were prescribed for the treatment of disease, but psychosomatic treatments also had their place.

When Alexandria became the center of scientific activity, a number of scholars there made important discoveries. Herophilus of Chalcedon studied the functions of the brain and the nerves, and was the first to distinguish between arteries and veins. Greek medical teaching spread in Rome. Among the practitioners of medicine there, Galen became the best known. He experimented with animals and dissected them, formulating theories that were treated as dogma in medicine for the next 1500 years.

Mathematics. Mathematics occupied an important place in Greek science and was further developed than any of the other sciences. Mathematics is based entirely on reasoning—the scientific activity Greeks preferred most—and does not require observation or experimentation as physics does.

The Egyptians were concerned mainly with applied mathematics: sets of rules that can be used to solve specific problems. Thales, who according to tradition was familiar with Egyptian mathematics, was the first to formulate general mathematical laws for measuring and to prove theorems. He is considered to be the founder of geometry.

Pythagoras, in the fifth century BC, founded a school stressing high moral standards and the pursuit of science.

He ranked mathematics as the most important science. Pythagoras was convinced that the "natural order" could be expressed in numbers (*see* "Mathematics and Mysticism," p 36).

Mathematics also occupied an important position in the school of Plato. Plato himself, although not a mathematician, viewed geometry as the basis for the study of any science. Theaetetus and Eudoxus, who belonged to the school of Plato, developed the theory of incommensurable magnitudes and the theory of proportions, respectively. Menaechmus made the first study of conic sections—ellipses, parabolas, and hyperbolas—a class of curves that later proved to be of extreme importance to astronomers and physicists.

Geometry culminated with the Alexandrian mathematician Euclid. He was the author of the *Elements,* comprising 13 books that summarize and organize the geometric thought of the Greeks. The *Elements* is a coherent mathematical construction based on a small number of axioms from which are derived a large number of propositions by applying strictly logical rules. The *Elements* formed the basis for teaching geometry in schools well into the twentieth century.

Archimedes is often classed among the first-rank mathematicians in history. Although he is popularly known for his work in physical science (see below), he worked largely in the context of mathematics. Among his accomplishments are the demonstration that numbers as large as one likes can be written (which he showed by writing a numeral for the number of grains of sand it would take to fill the universe); finding the properties of spirals; the expert use of Eudoxus's method of exhaustion (mathematically the same as the integral calculus, although formally quite different); and finding the ratios of the volumes of various figures to one another, such as showing that the volume of a sphere is two-thirds that of a cylinder whose height is the diameter of the sphere and whose radius is the same as that of the sphere.

Like Euclid and Archimedes, Apollonius of Perga was Hellenistic rather than Greek. Many Hellenistic works have been lost, including works by Euclid and Archimedes, but Apollonius' works suffered more than most. His work on the conics—the circle, ellipse, parabola, and hyperbola—along with a treatise called *Cutting-off of a ratio* are the only survivors. Nevertheless, many of his works have been restored on the basis of comments written about them. Apollonius was the last great synthetic geometer until the end of the eighteenth century.

Diophantus, the "father of algebra," worked with mathematical ideas that are equivalent to solving equations in several variables for integral solutions. Such equations are still called diophantine today.

Physical Science. The fundamental nature of matter was one of the first problems of the early Greek natural philosophers. Thales believed that water is the basic constituent of all matter, while others believed it to be fire or air (*see* "The first scientists," p 21). From Democritus we have the idea that matter is built up of atoms (*see* "Early atomists," p 32), an idea that appears more promising in retrospect than it did at the time. Empedocles introduced the idea of elements (*see* "The elements," p 31). Although the Greek ideas on matter were based on philosophical speculation rather than observation and experiment, developments in physics and chemistry during the nineteenth and early twentieth centuries proved them to be close to the mark. Unfortunately, their ideas about motion were erroneous. Aristotle believed motion to be induced by the striving of each object to reach its natural place (*see* "Replacing Aristotle's physics," p 118).

Archimedes made a number of advances in physical science, starting the tradition of what is called mathematical physics today. Although levers and other simple machines were undoubtedly known earlier, Archimedes was the first to work out the mathematical law of the lever. Similarly, he developed the first applications of hydrostatics when he showed that a body immersed in a fluid would displace its own mass. There are famous legends connected with each of these discoveries. According to one, Archimedes used a system of levers to pull a fully loaded ship onto shore to demonstrate the idea expressed by the statement attributed to him "give me a lever long enough and a place to stand and I can move the Earth." The other has Archimedes discovering the laws of hydrostatics in the bath and running naked through the streets of Syracuse shouting *"Eureka"* ("I have found it").

Technology. Because society in antiquity was based on slavery, little incentive existed to develop technologies that would ease labor. The Greeks mastered the smelting of iron, and because iron ores were abundant, iron weapons and tools became widespread. In Alexandria, largely because of Egyptian influence, there was more interest in technology. Many consider Ctesibius to be the founder of the Alexandrian school of engineering. His younger contemporary, Philon of Byzantium, reported some of his technical achievements; among them were a force pump, a water organ, and a mechanically driven water clock. Hero of Alexandria described a series of automata and experimented with model steam engines. Archimedes applied scientific principles to technology. His most famous invention is the Archimedian screw, a device for raising water in irrigation systems.

On the other hand, the presence of slaves did nothing to discourage building aqueducts, bridges, and roads. While the Romans became the master of these, the first water-carrying tunnel through a mountain was a Greek achievement, being the work of Epalinus of Megara in the sixth century BC. The Romans also developed concrete and used it in many of their constructions.

	GENERAL	ASTRONOMY
600 to 591 BC	Phoenician sailors travel by ship around Africa Greek poets Sappho and Alcaeus flourish on the isle of Lesbos The archon Solon, one of the nine main magistrates, reforms the government of Athens in 594, making it more democratic	
590 to 581 BC	In 586 Nebuchadrezzar II removes several thousand Jews from their homeland and transports them to his capital of Babylon; this becomes known as the Babylonian captivity	
580 to 571 BC		Thales of Miletus in what is now Turkey (b about 624) correctly predicts a solar eclipse that occurs on May 28, 585; the Medes and Lydians, taking the eclipse to be an ill omen, call off their war
570 to 561 BC		
560 to 541 BC	The Greek playwrights Aeschylus, Sophocles, and Euripides flourish	Thales, Greek astronomer, mathematician, and philosopher, d Miletus (Turkey)
540 to 531 BC	The Persians under Cyrus the Great conquer Babylon in 538	

The first known date

The earliest history of humanity has no known dates. After writing began to be used, people started dating events, but these were usually in relation to other events that we can no longer date. The long histories of Egyptian and Chinese dynasties, however, provide fairly good year dates for those cultures back to 3000 BC.

The Mayans recorded specific day dates that go back in time tens of thousands of years. Since many of the events dated in this system occurred before the most optimistic early date for humans in North America, it is assumed that these very early dates are for mythical events that were invented much later. Very early Chinese dates also are thought to have been later inventions.

A strong candidate for the first real date in history—that is, the first specific day on which an event can be pinpointed as having occurred—is May 28, 585 BC. The event was a battle between the Medes and the Lydians that was suddenly called off when an eclipse of the sun frightened both armies. This eclipse, supposed to have been forecast correctly by Thales (although at best he would have had the year right, not the date), could only have been the one observable in the Middle East on May 28, 585. No other solar eclipse would have been visible in that region for many years on either side of 585.

A similar candidate for the first date, September 6, 775 BC, comes from Chinese astronomical records. It is the earliest date that records an astronomical event that we can say occurred at a particular time. However, it is not connected with any other event, such as the war of the Medes and Lydians.

MATHEMATICS	PHYSICAL SCIENCE	TECHNOLOGY	
The Chinese text *Arithmetic classic of the gnomon and the circular paths of heaven* contains the first known proof of a version of the Pythagorean theorem		Sundials are used in China A road of 7.4 km (4.6 mi) is built across the isthmus of Corinth to transport ships across land on wheels The Chinese develop the practice of fumigating houses to rid them of pests King Sennacherib of Assyria (Mesopotamia) builds an 80-km (50-mi) canal connecting Nineveh with Bavia In 598, the second year of the reign of King Chao of Yen, whale-oil lamps with asbestos wicks are used; however, this may have been in 308 BC, as there were two King Chaos of Yen Anarcharis the Scythian is believed to invent the anchor	**600** to **591** BC
			590 to **581** BC
Mathematician and philosopher Pythagoras b Samos (Greek island off the coast of Turkey) According to Proclus, Thales of Miletus (Turkey) is the first to prove general geometrical propositions on angles and triangles, an approach that will be at the basis of Euclid's work	Thales of Miletus (Turkey) gives his interpretation of matter: water is the basis of all things		**580** to **571** BC
	Xenophanes, Greek philosopher, b Colophon, Ionia (Turkey); he speculates that because fossil sea shells are found on the tops of mountains, the surface of Earth must have risen and fallen in the past, one of the earliest ideas of earth science		**570** to **561** BC
			560 to **541** BC
There is a productive period in mathematics, with the Pythagoreans developing their arithmetic and geometry, the Chinese introducing rod numerals, and the Indians developing a geometry based on stretching ropes (similar to an earlier Egyptian development)			**540** to **531** BC

	GENERAL	ASTRONOMY	LIFE SCIENCE
530 to 521 BC	Persian general Cambyses, the son of Cyrus, conquers Egypt in 525, making the Persian empire the largest to date in the Middle East		
520 to 511 BC		Anaximander, (b Miletus in what is now Turkey about 610) makes the first known attempt to model Earth according to scientific principles; his concept is that Earth is a cylinder with a north-south curvature; he prepares a map of the known Earth based on this idea Anaximander, Greek astronomer and philosopher, d Miletus (Turkey)	Anaximander's *On nature* introduces the idea of evolution, assuming that life starts out in slime and eventually moves to drier places
510 to 501 BC	Pythagoreans are killed and dispersed by a mob in Croton (Crotone, Italy), and Pythagoras flees to Tarentum (Taranto, Italy), because of their advocacy of government by an elite		
500 to 491 BC	Greek philosopher Anaximenes d	The Pythagoreans teach that Earth is a sphere and not in the shape of a disk	Reindeer are domesticated in central Asia The Greek Pythagorean physician Alcmaeon of Croton (Italy), b about 535, is the first person known to have dissected human cadavers for scientific purposes; he notes the optic nerve and the Eustachian tubes, and recognizes the brain as the seat of the intellect Navigator Hanno of Carthage (Tunisia), b about 530, describes the gorilla after a voyage down the African coast In India, Susrata performs the first cataract operations
490 to 481 BC	Greek city-states combine to defeat the Persian emperor Darius at Marathon, Sep 12, 490 Religious reformer Buddha (Gautama Siddhartha) d 483		

MATHEMATICS	PHYSICAL SCIENCE	TECHNOLOGY	
Pythagoras leaves Samos and founds a school in Croton in southern Italy; the society that is centered around this school, the Pythagorean Brotherhood, later makes many striking discoveries in mathematics; the Pythagorean theorem, however, was known much earlier	Greek philosopher Anaximenes of Miletus (Turkey), b about 570, suggests that air is the primary substance; it can be changed into other substances by thinning, forming fire, or thickening, forming wind, clouds, rain, hail, earth, and rock	Greek architect and engineer Eupalinus of Megara (Greece), b about 600, builds an aqueduct and water-supply system for Megara Theodorus of Samos (Greek island near Turkey) flourishes; he is credited with the invention of ore smelting and casting, the bubble level, locks and keys, the carpenter's square, and the lathe In 522 architect and engineer Eupalinus of Megara constructs a tunnel of 1100 m (3600 ft) under 300-m-high (1000-ft-high) Mount Castro on the island of Samos (Greece) to supply water	**530 to 521** BC
	Anaximander believes in a substance that is boundless and eternal but contains the qualities of hot, cold, dry, and wet Greek philosopher Parmenides b Elea (Italy); he holds that change is illusory, since being is one, and that nothing can be created or destroyed	Anaximander introduces the sundial to Greece; it had been known for centuries in Mesopotamia and Egypt	**520 to 511** BC
			510 to 501 BC
Pythagoras, Greek mathematician and philosopher, d Metapontum (Italy)		Steel is made in India Chinese farmers use practices, such as planting crops in rows, hoeing weeds, and applying manure, that will not be used in the West until the eighteenth century Greek traveler and historian Hecataeus of Miletus (Turkey), b about 550, develops a map of the world showing Europe and Asia as semicircles surrounded by ocean	**500 to 491** BC
			490 to 481 BC

	GENERAL	ASTRONOMY	LIFE SCIENCE
490 to 481 BC cont	Greek philosopher Gorgias b Leontini (Sicily); his nihilist philosophy says nothing exists; if it did exist, we could not know it; if we knew it, we could not tell anyone about it		
480 to 471 BC	In 480, 300 Spartans and about 2000 other Greeks fight a large army of Persians—about 1,700,000 according to Herodotus—at the battle of Thermopylae; the Greeks refuse to retreat and fight until the last man is killed Hecataeus, Greek traveler, map maker, and historian d Greek philosopher Xenophanes, known for the idea that the surface of Earth changes with time, d The Greek navy, mainly that of Athens, defeats the Persian navy at Salamis on Sep 20, 480, watched from the shore by the Persian king Xerxes; Xerxes returns to Persia; although the war continues for another year, the Greeks finally succeed in stopping the Persian westward expansion Athens forms the Delian League in 478, officially a confederacy of Greek city states, but actually a disguised Athenian empire	Greek philosopher Oenopides b Chios; he is believed to be the first to calculate the angle that Earth is tipped with respect to the plane of its orbit; his value of 24 degrees is only half a degree from the presently accepted value of about 23-1/2 degrees	Greek historian Thucydides b Athens (Greece), 471; he is known primarily for his account of the Peloponnesian War between Athens and Sparta, in the course of which he describes the plague of Athens in great detail, a mysterious, fatal disease that scientists today have failed to identify
470 to 461 BC	Socrates b Athens The Age of Pericles begins in 461, a period of peace and prosperity in Athens that allows the arts and philosophy to flourish		
460 to 451 BC			Physician Hippocrates b Cos (Aegean island)
450 to 441 BC	Greek historian Herodotus of Halicarnassus (Turkey, b 484) explores the Middle East, Babylon, and Persia and gives the first description of cotton, which he finds in India Parmenides, Greek philosopher, d Greek historian Herodotus moves to Thurium in 443, in what is now the south of Italy, and begins to write his *History*	Pythagorean philosopher Philolaus, b Tarentum or Croton (what is now Italy) about 480, suggests that there is a central fire around which the Earth, sun, moon, and planets revolve; he also believes that Earth rotates	Greek philosopher Empedocles of Akragas (Sicily), b about 492, recognizes that the heart is the center of the system of blood vessels, but wrongly agrees with the notion that this organ is the seat of the emotions, an idea that persists in folk tradition

		490 to 481 BC cont

The elements

The concept of element, as well as that of atom, is of Greek origin, although the Greeks did not think of elements or atoms as we do today. The term *element* was coined by Plato, but the idea existed before him. Empedocles believed that all matter was made up of four primary substances, or elements: earth, air, fire, and water. All the appearances of matter could be explained by the commingling and separation of these elements under two influences that acted upon them: love, comparable to our force of attraction; and hate, comparable to our concept of repulsion.

Plato adopted Empedocles' theory of elements. Because Plato believed that geometry is the best method for thinking about nature, he provided the elements with exact mathematical form. The smallest part (or atom) of fire was in the shape of a tetrahedron, air of an octahedron, water of an icosahedron, and earth of a cube.

Aristotle also adopted the concept of four elements. But contrary to Empedocles' theory, in which the elements are immutable and compounds differ only in their composition, the elements of Aristotle undergo changes when they combine. Aristotle believed that besides physical extension, elements have qualities that are based on how we experience matter: hot, cold, dry, and moist. He saw each element as endowed with two of the qualities, so earth was dry and cold, water moist and cold, air moist and hot, and fire hot and dry. An element could change into another one by changing one or two of its qualities. For example, earth could change into water by changing from dryness to moistness.

---

480 to 471 BC — Greek philosopher Protagoras b Abdera (Greece); he believes that sense perceptions are all that exist, so reality may be different from one person to another

Greek philosopher Heraclitus of Ephesus (Turkey), b about 540, d about 475, teaches that change is the essence of all being and that fire is the primary substance

Greek philosopher Anaxagoras of Clazomenae (Turkey), b about 500, postulates that a large number of "seeds" make up the properties of materials; he claims that heavenly bodies are made of the same materials as Earth and that the sun is a large, hot, glowing rock; creation came when Earth expelled all the heavenly bodies

---

470 to 461 BC — The Pythagorean Hippasus of Metapontum (f. 450) discovers the dodecahedron, a regular solid with 12 regular pentagons as its faces

---

460 to 451 BC

---

450 to 441 BC — The Greeks develop a method of writing numbers based on letters of the alphabet

Hippasus of Metapontum, a member of the Pythagorean Brotherhood, discovers that some magnitudes are not commensurable, a result that we interpret today as meaning that some numbers are not rational; for example, the diagonal and side of a square do not have a measure in common

Empedocles of Akragas (Sicily) introduces a system in which fire, air, earth, and water are the elemental substances; they can undergo changes through the actions of two opposing forces, love and strife

Greek philosopher Leucippus of Miletus (Turkey), b about 490, introduces the first idea of the atom, an indivisible unit of matter

	GENERAL	ASTRONOMY
450 to 441 BC cont		
440 to 431 BC	Empedocles, Greek philosopher, d Mount Etna (Sicily); legend states that he jumped into the volcano's crater after failing to be taken into the heavens to become a god, which he had predicted would happen Athens and Sparta start the Peloponnesian War in 431; it lasts until 404, when Athens surrenders	Meton of Athens (b about 440) develops the Metonic cycle, the approximately 19-year period in which the motions of the sun and moon seem to come together from our vantage point on Earth; this period can be used to predict eclipses and forms the basis of the Greek and Jewish calendars
430 to 421 BC	Plato b Athens Greek historian and traveler Herodotus d 425	Anaxagoras, Greek philosopher, d Lampsacus (Turkey)
420 to 411 BC	Protagoras, Greek philosopher, d	
410 to 401 BC	Xenophon writes the *Anabasis* in 404, an account of his military mission leading the Ten Thousand, a mercenary army fighting for one faction in Persia, in their retreat from Mesopotamia Greek historian Thucydides d 401	Horoscopes setting out the positions of the planets at the time of an individual's birth are available in Chaldea

Early atomists

The concept of the atom as the smallest, indivisible entity is one of the oldest ideas of science. It has its origin in a philosophical problem the Greeks tried to solve. Heraclitus believed that the basic nature of all things is *change*. Parmenides, however, disagreed with Heraclitus and stated that reality is unchangeable and that change is a mere illusion.

Democritus (and probably Leucippus, the father of atomic theory, but of whom little is known) viewed change as part of reality, but held that this change takes place within the philosophical framework of Parmenides. Change for Democritus was the local motions of parts that in themselves were unchangeable and invisible: the atoms. Atoms comprised all being; everything else was void.

Democritus tried to explain the different forms in which matter exists, and some of his ideas seem quite modern, although coming from a perspective that does not really resemble today's ideas about atoms. For example, Democritus thought that substances differ from each other because of the shapes, positions, and grouping of the constituting atoms. In solid bodies the atoms stick together. In fluids and gases atoms do not stick together, they rebound from one another in random directions. Denser bodies are made up of larger, but still indivisible, atoms. In his theory, there was no limit to the size of an atom; he believed that atoms as big as a world could exist somewhere.

Atomism had few followers before the 17th century. Aristotle and Plato rejected the theories of Democritus, and Dante allocated him a very low place in hell.

MATHEMATICS	PHYSICAL SCIENCE	TECHNOLOGY	
Zeno of Elea (Italy), b about 490, expresses paradoxes contrasting continuity with discreteness, raising questions that are yet to be answered completely; the best known is the paradox of Achilles, who tries to catch a tortoise that has a headstart; Zeno argues that Achilles cannot win the race			**450** to **441** BC cont
		The sculptor Phidias, working with architects Actinus and Callicrates, completes the Parthenon in Athens in 438	**440** to **431** BC
When the Athenians, suffering a plague, appeal to the oracle at Delos, they are told that they need to double the cubical altar of Apollo to rid themselves of the plague; their failure to do so leads to one of the three classic problems, the duplication of the cube Hippocrates of Chios writes his *Elements*, which predates Euclid's more famous *Elements* by more than a century; since the work has been lost, it is not certain what it contained Zeno, Greek philosopher and mathematician, d Elea (Italy)	Democritus of Abdera (Thrace, now part of Greece), b about 470, expands on the concept introduced by his teacher, Leucippus, of atoms as indivisible bodies; he shows how every form of matter can be explained by his version of atoms	An optical telegraph using torches to signal from hilltop to hilltop operates in Greece	**430** to **421** BC
Hippias of Elis (Greece) discovers the curve called the quadratrix, which can be used to trisect an angle (and to square a circle, although Hippias probably did not know this); the quadratrix is the first curve known that cannot be constructed with straightedge and compass Mathematician Theaetetus b; he is considered to be the original source for Euclid's Books X and XIII, which deal with commensurable and incommensurable numbers and with the five Platonic solids		Archytas of Tarentum (Italy) b; he builds a series of toys, among them a mechanical pigeon propelled by a steam jet, and develops the theory of the pulley	**420** to **411** BC
		The forerunner of the catapult is introduced by Dionysius the Elder, ruler of Syracuse, in 406; his workers also develop *quadriremes*, ships with four banks of rowers; the motive for this activity is an imminent invasion by forces from Carthage	**410** to **401** BC

	GENERAL	ASTRONOMY	LIFE SCIENCE
400 to 391 BC	Socrates, Greek philosopher, d Athens, 399, of his own hand following the sentence of the state for corrupting youth		Hippocrates founds the profession of physicians, develops the Hippocratic oath, and encourages the separation of medicine from religion at his school of medicine at Cos; this earns him the sobriquet "the Father of Medicine"
390 to 381 BC	Plato founds a school in a grove on the outskirts of Athens; since the grove once belonged to the hero Academos, the school is named the Academy; the Academy continues as a place of learning until 529 AD Aristotle b Stagira (Greece)	Greek astronomer Heracleides b Heraclea Pontus (Turkey); he is the first person known to suggest that Venus and Mercury may orbit the sun	
380 to 371 BC	Gorgias, Greek philosopher, d	Democritus, Greek philosopher d; he recognizes that the Milky Way consists of numerous stars, that the moon is similar to Earth, and that matter is composed of atoms	
370 to 361 BC		Greek astronomer Callippus b Cyzicus (Turkey); a student of Eudoxus, he is a careful observer who shows that at least 34 spheres are needed to account for the movements of the stars, moon, and planets	Hippocrates of Cos, Greek physician, d Larissa, Thessaly (Greece)
360 to 351 BC		Eudoxus of Cnidus, Greek astronomer and mathematician, d Cnidus (Turkey) Chinese observers report a supernova in 352, the earliest known record of such a sighting	
350 to 341 BC	Plato, Greek philosopher, d Athens In 343 Aristotle becomes the tutor of Alexander, the son of Philip II of Macedon, later known as Alexander the Great		Aristotle classifies animals; 500 known species are divided into eight classes
340 to 331 BC	Philip II of Macedon overpowers Greek forces at Charonea in 338 and establishes himself as the head of all Greek states except Sparta Alexander the Great succeeds his father in 336 and starts the conquests that will spread Greek culture from Egypt to India	Kiddinu b Babylonia; he works out an early and somewhat inaccurate version of the precession of the equinoxes, the apparent change in the position of the fixed stars over a period of 26,700 years that is caused by Earth's wobbling in its orbit	Praxagoras, Greek physician, b Cos (Aegean island); he distinguishes between veins and arteries, although he thinks that arteries are hollow tubes that carry air throughout the body

MATHEMATICS	PHYSICAL SCIENCE	TECHNOLOGY	
The Pythagorean Theodorus of Cyrene (Libya), Plato's teacher in mathematics, demonstrates that the square roots of 3, 5, 6, 7, 8, 10, 11, 12, 13, 14, 15, and 17 are irrational		Catapults operating on the principle of torsion are introduced by the forces of Philip of Macedon Semilegendary Chinese artisan Lu Pan produces the first known kite	400 to 391 BC
	Plato begins his trilogy, *Timaeus*, *Critias*, and *Hemocrates*, finishing only the *Timaeus*, in which he puts forth his theory of four elements—earth, water, air, and fire—and hints at a fifth element, the ether Plato conceives the idea that there must be a continent directly opposite Europe, on the other side of the globe, which he calls the antipodes		390 to 381 BC
		Plato is said to invent a water clock with an alarm Eudoxus of Cnidus (Turkey), b about 408, improves on the primitive map of Hecataeus	380 to 371 BC
Theaetetus, Greek mathematician, d Athens, 369	Aristotle discovers that free fall is an accelerated form of motion, but believes that heavier bodies fall faster than lighter bodies, a concept that will be corrected by Galileo		370 to 361 BC
			360 to 351 BC
Dinostratus finds a way to square the circle using the trisectrix of Hippias Archytas, Greek mathematician, d Menaechmus, the brother of Dinostratus, writes on conic sections, showing that they can be used to duplicate the cube	Aristotle's *De caelo* (On the heavens) defines chemical elements as constituents of bodies that cannot be decomposed into other constituents	Celtic chiefs begin building Maiden Castle in south Dorset, one of the great fortified castles of Britain	350 to 341 BC
Eudemus of Rhodes writes a *History of mathematics*; although his manuscript is lost, a later writer makes a summary of it, which is also lost; but Proclus in the fifth century AD has access to the summary, and it is from his account that we know much of the little known early history of Greek mathematics	Aristotle teaches that space is always filled with matter Strato, Greek physicist, b Lampsacus (Turkey); he conducts experiments that lead him to believe, correctly, that bodies accelerate when they fall and, incorrectly, that heavier bodies fall faster than light ones; he also studies the lever, but does not work out its law		340 to 331 BC

	GENERAL	ASTRONOMY	LIFE SCIENCE

340 to 331 BC cont

Aristotle founds the Lyceum in Athens in 334; it is also called the peripatetic school of philosophy because the philosophers were taught while walking about the neighborhood of the Lyceum, a temple to Apollo

In 332 Alexander the Great founds Alexandria, a city in which culture and science will soon flourish

In 331 Alexander the Great captures Babylon, Susa, and Persepolis, the principal cities of the Persian empire

Mathematics and mysticism

In the fifth century BC, Pythagoras of Samos founded a society that was both a mystical religion and one of the most productive schools of mathematics in history. A fundamental mystical belief of the Pythagorean society was that "all is number," which is to say that the entire universe can be explained in terms of numbers.

The Pythagoreans had other mystical beliefs as well, such as an aversion to beans and a belief in the transmigration of souls (that is, that people's souls after death occupy the bodies of animals and vice versa).

But the principal ideas of the Pythagoreans that have been handed down concern numbers. The Pythagoreans assigned gender to numbers, saying that odd numbers were male and even numbers female. They also worked with figurate numbers, numbers that can be identified by a shape when dots are used. Square numbers form squares, triangular numbers form triangles, and so forth.

Although the name Pythagoras is most closely associated with the Pythagorean theorem (the sum of the squares on the legs of a right triangle equals the square on the hypotenuse), this theorem was known to the Chinese and probably to the Babylonians before the time of Pythagoras.

330 to 321 BC

Alexander the Great's army invades India in 326

In 323 Alexander the Great d at the age of 33 after carrying Greek civilization to the limits of the known world

Aristotle, having fled Athens following the death of Alexander, d in Chalcis on the Aegean island of Euboea (now Evvoia), 322

320 to 311 BC

In 318 Prince Xuan establishes in the capital of the Qi state in China an academy of scholars

The Seleucid Era begins in Mesopotamia in 311 after Selecus I, one of Alexander's generals, is awarded Mesopotamia in the breakup of Alexander's empire after his death

Zeno of Citium (Cyprus) establishes the philosophical school of Stoics, teaching in the Painted Porch at Athens, the *stoa poecile*

Heracleides, Greek astronomer, d
Athens

310 to 301 BC

Ptolemy I (Soter), one of the generals of Alexander the Great, starts his reign in Egypt in 306

300 to 291 BC

The Museum is built in Alexandria, a home for scholars and artists of all types, but essentially the center of Hellenistic mathematics

Callippus, Greek astronomer, d

Chinese astronomers Shih Shen, Gan De, and Wu Xien independently compile star maps that will be used for the next several hundred years

The turkey is domesticated in Mexico

Greek physician Diocles (b Carystus, Greek island of Euboea, about 350), a student of Aristotle's, writes the first anatomy book and the first book of herbal remedies

Epicurus argues that organs develop through exercise and weaken when not used

Greek physicians Herophilus, b Chalcedon (Turkey) about 320, and Erasistratus, b Chios about 304, flourish in Alexandria, where they perform dissections in public; they describe the liver, spleen, retina,

340 to 331 BC cont

Mathematics and mysticism (continued)

When the Pythagoreans talked about numbers, they meant the natural numbers, 1, 2, 3, and so forth. Fractions are simply ratios of natural numbers, so fractions posed no problem to them. But the Pythagoreans proved that objects in the real world could not be measured completely using ratios of natural numbers. For example, if the unit chosen to be one is a side of a square, the diagonal of the square cannot be measured exactly using that unit.

330 to 321 BC

Pytheas, Greek geographer and explorer, b Massalia (Marseille, France); he sails into the North Atlantic and Baltic Sea, probably reaching Norway; observing the strong Atlantic tides, he correctly theorizes that they are caused by the moon

320 to 311 BC

The lost *Conics* of Aristaeus is written, as well as a similar work on conics—the curves formed when a plane intersects a cone—by Euclid (fl 300); both are presumed to have been superseded by the *Conics* of Apollonius, of which seven books of eight survive

The Via Appia, from Rome to Alba Longa (later extended to Capua and Brindisi), and the first Roman aqueduct, the Aqua Appia, bringing water to Rome from springs 16 km (10 mi) away, are built by Appius Claudius Caecus in 312

310 to 301 BC

Epicurus of Samos (Greek island near Turkey), b 341, founds a philosophical school in which atoms are a fundamental part of the philosophy

The Chinese invent a form of bellows, known as the double-acting piston bellows, that produces a continuous stream of air; such a bellows was not known in the West until the sixteenth century

300 to 291 BC

The *Chou pei suan ching*, among the oldest mathematical classics from China, is written

The Egyptians know that right triangles can be formed with sides in the ratios of 3:4:5, 5:12:13, and 20:21:29; they probably knew the 3:4:5 triangle much earlier

Conon, Greek mathematician, b Samos (Greek island near Turkey); he is best known for convincing Berenice, queen of Egypt, that her hair, which had been cut off and dedicated to the gods, had not been stolen but had become the constellation *Coma Berenices*

The Chinese concepts of yin and yang, paired opposites, are incorporated into the Chinese model of how the universe is organized

The Chinese *Book of the devil valley master* contains the first clear reference to a lodestone's alignment with Earth's magnetic field; the lodestone is called a "south-pointer"

The Chinese invent a form of harness that is pushed by a horse's chest instead of a horse's throat, a great improvement that was not adopted in the West until the eighth century AD

Mohist writings in China describe the first known use of poison gas in warfare; the Mohists were followers of Mo-Tzu, a fifth-century Chinese philosopher

	GENERAL	ASTRONOMY	LIFE SCIENCE
300 to 291 BC cont			duodenum, ovaries, Fallopian tubes, and prostate gland, and determine that the seat of reasoning is in the brain, not the heart Aristotle's student Theophrastus of Eresus (Lesbos, Greece), b 372, describes over 550 plants in his *Natural history of plants,* including Indian cane sugar and the coconut palm
290 to 281 BC			Theophrastus, Greek botanist, d Athens
280 to 271 BC			Greek physician Erasistratus comes very close to recognizing the circulation of the blood, especially by noting the relationship of the lungs to the circulatory system
270 to 261 BC	Epicurus of Samos, Greek philosopher, d First Punic War, between Rome and Carthage, starts in 264	Aristarchus of Samos (island near Turkey, b about 310) challenges Aristotle's teachings by asserting that the sun is the center of the solar system and that the planets revolve around the sun; he estimates the distance of the sun from Earth by observing the angle between the sun and the moon when it is exactly half full	
260 to 241 BC			Erasistratus, Greek physician, d Mycale (Turkey)
240 to 231 BC	King Asoka d, 232; he is called the "Buddhist Constantine" for his influence in converting much of northern India to Buddhism	Chinese astronomers in 240 observe Halley's Comet in its first known recorded visit Eratosthenes of Cyrene (Libya), b about 276, calculates the circumference of Earth from the difference in latitude between Alexandria and Syene (Aswan) and finds a figure (46,000 km or 28,500 mi) close to the present value; he also lays down the first lines of longitude on a map of Earth	

MATHEMATICS	PHYSICAL SCIENCE	TECHNOLOGY	
Euclid's *Elements* summarizes and organizes the mathematical knowledge developed in Greece in the three preceding centuries; it includes information on plane and solid geometry, the theory of incommensurables, and the theory of numbers; it will be the basic textbook in mathematics for the next 2000 years		Convex lenses are introduced in Carthage (Tunisia) The Chinese invent cast iron Dicaerchus of Messina (Sicily), b about 355, a student of Aristotle's, develops a map of Earth that is on a sphere; it has lines of latitude based correctly on the lines where the noonday sun is at a given angle on a particular day	300 to 291 BC cont
	Dicaearchus, Greek geographer, d	The Pharos lighthouse is built at Alexandria in 283	290 to 281 BC
		Under Ptolemy Philadelphus, a canal linking the Nile to the Red Sea is completed; it was begun by Pharaoh Necho about 300 years earlier	280 to 271 BC
	Strato, Greek physicist, d Athens Zou Yan, a leading member of Prince Xuan's Zhi-Xia Academy, systemizes the Chinese theory of five elements: water, metal, wood, fire, and earth		270 to 261 BC
The Maya develop their system of numeration, a place-value system based on 20 that is complete with zero; it is the most sophisticated way of writing numbers for the next 1000 years	The *Mo Ching*, a collection of writings by followers of Mo-tzu, contains a clear statement of the first law of motion later formulated by Newton	The Great Wall of China is started Archimedes (b Syracuse, Sicily, about 287) works out the principle of the lever and other simple machines; he demonstrates this by pulling a large ship onto land by himself Philon of Byzantium (Turkey), b about 200, designs a chain drive for use in repeat loading of a catapult, advocates the use of bronze springs in catapults, and experiments with the expansion of air by heat	260 to 241 BC
			240 to 231 BC

	GENERAL	ASTRONOMY	LIFE SCIENCE
230 to 221 BC	The Chinese state of Qi develops the federal bureaucratic system that will run China for the next 2000 years; it also leads the Chinese states in putting the Great Wall into place	Aristarchus, Greek astronomer, d Alexandria (Egypt)	
220 to 211 BC			
210 to 201 BC			
200 to 191 BC		Eratosthenes, Greek astronomer and mathematician, d Alexandria (Egypt), 196	
190 to 181 BC		Seleucus, Greek astronomer, b Seleucia (Iraq); he is the last known astronomer to champion the heliocentric theory of the solar system until Copernicus	
180 to 171 BC			

MATHEMATICS	TECHNOLOGY	
Eratosthenes of Cyrene (Libya) develops a method of locating all prime numbers: the sieve of Eratosthenes		**230 to 221 BC**
Conics by Apollonius of Perga (Turkey, b 262) summarizes and extends the theory of the class of curves called conic sections (the circle and ellipse, the parabola, and the hyperbola); they can be obtained by passing a plane through a cone		
Archimedes, Greek mathematician and engineer, d Syracuse, Sicily; according to legend, while contemplating geometric figures drawn in the sand, he is killed by a soldier whose commander had ordered Archimedes' life spared	In 219 Shih Lu constructs the Magic Canal linking two rivers, one flowing south and the other north; although only 32 km (20 mi) long, the canal enables a ship to sail from Canton (or anywhere else on the China Sea) to the latitude of present-day Beijing in inland China In 215 Ling Ch'u constructs a canal of 145 km (90 mi) from Ch'ang-an to the Yellow River	**220 to 211 BC**
	In 210 Emperor Ch'in Huang Ti has constructed for his tomb a relief map of his empire with the rivers formed from flowing mercury and the heavens depicted above	**210 to 201 BC**
China begins to develop place-value notation	The Chinese develop a malleable form of cast iron The development of gears leads to the ox-powered waterwheel for irrigation Concrete is used in the Roman town of Palestrina in Italy Alexandrian engineer Ctesibius, b about 250, improves the water clock, making it the most accurate timekeeping device available for nearly 2000 years	**200 to 191 BC**
Apollonius, Greek mathematician, d Alexandria (Egypt)		**190 to 181 BC**
It is believed that Hypsicles' *De ascensionibus*, a work on astronomy, introduces the 360-degree circle to Greek mathematics Diocles investigates the properties of the curve he names the cissoid (meaning *ivy*)		**180 to 171 BC**

	GENERAL	ASTRONOMY	LIFE SCIENCE
170 to 161 BC	Judas Maccabaeus, in revolt against Syria, succeeds in freeing the Temple at Jerusalem in 165	Chinese astronomers record sunspots in 165, probably the first accurately dated record	
160 to 141 BC	Rome occupies Greece and destroys Carthage in 146	Hipparchus of Nicea (Turkey), b about 190, draws up a listing of fixed stars and discovers the precession of the equinoxes	
140 to 131 BC			
130 to 121 BC	In 121 Roman reformer Caius Gracchus is killed in riots; after Gracchus fails to be reelected as tribune, his reforms are overthown, which results in the riots	Hipparchus of Nicea uses a total eclipse of the sun and parallax to determine correctly the distance to and the size of the moon; the same data are an order of magnitude too small for the distance and size of the sun	Asclepiades of Bithynia (region in Turkey) b; he believes that disease, which he treats with baths, diet, and exercise, is caused by a disturbance in the particles that make up the body
120 to 111 BC		Hipparchus, Greek astronomer and mathematician, d	
110 to 101 BC			
100 to 91 BC	The Old Silk Road, a route on which Chinese and European goods are exchanged, is established The Greek explorer Hippalus discovers the regularity of the monsoon between east Africa and south Asia	Babylonian astronomers work out "zigzag functions" to describe changes in speed of the sun and moon at different times of the year	The Chinese version of the circulation of the blood comes far closer to the modern view than anything in the West before William Harvey in 1628

MATHEMATICS	PHYSICAL SCIENCE	TECHNOLOGY	
		The first paved roads are built in Rome	170 to 161 BC
		Ctesibius of Alexandria builds a "wind gun," a type of air gun	
			160 to 141 BC
The astronomer Hipparchus of Nicaea (Turkey) compiles the first tables of chord lengths, a forerunner of trigonometric tables	Han Ying's *Moral discourses illustrating the Han text of the "book of songs"* contains the first known reference to the hexagonal nature of snowflakes	The Chinese make paper; it is used as a packing material, for clothing, and for personal hygiene, but not for writing	140 to 131 BC
	Poseidonius, Greek philosopher, b Apamea (Syria); his incorrect calculation of Earth's circumference encourages Columbus 1500 years later to believe that Asia is only about 3000 miles west of Europe		
		In 126 Chang Ch'ien brings wine grapes—*vinifera*—to China from the West	130 to 121 BC
		In 119 the Han dynasty in China nationalizes cast iron and salt production	120 to 111 BC
		The Chinese invent the collar harness for horses, the most efficient form of harness to this day; it is not used in the West until the Middle Ages	110 to 101 BC
The Chinese begin to use negative numbers		The Chinese invent the crank handle for turning wheels	100 to 91 BC
		The Great Wall of China is completed	
		The Romans discover that possolana, volcanic ash from Puteoli, near Naples, makes excellent concrete that will set and keep its integrity even under fresh or salt water	
		In Illyria (Yugoslavia and Albania), water-powered mills are used for grinding corn	

	GENERAL	ASTRONOMY	LIFE SCIENCE
90 to 81 BC		Sosigenes, Greek astronomer, b; his advice to Julius Caesar is the basis of the adoption of the Julian calendar	
80 to 71 BC			
70 to 61 BC	Rome occupies Mesopotamia	Geminus of Rhodes views astronomical systems as mathematical conveniences rather than physical realities	
60 to 51 BC	In 56 *De rerum natura* (On the nature of things) by Lucretius (b Rome about 95 BC) revives the philosophy of Epicurus, emphasizing the atomic nature of matter and that the laws of nature, not mysterious gods, control fate Lucretius, Roman philosopher, d Rome	In 52 Chinese astronomer Ken Shou-Ch'ang builds an armillary ring, a metal circle that represents the equator and is used in observing stars	
50 to 41 BC	Poseidonius, Greek philosopher, d Julius Caesar is assassinated on Mar 15, 44	Andronikos of Kyrrhestes builds in Athens the Tower of Winds, a water clock combined with solar clocks erected in the eight principal directions; it is the most famous timekeeping device of the Greeks In May and Jun, 44, Roman and Chinese observers report a red comet that is visible in daylight; its red color is due to the volcanic dust in the air caused by eruptions of Mount Etna; Roman citizens widely believe the comet to be Julius Caesar, who has become a god after his assassination	The *Ayurveda* is compiled; it becomes the basic Hindu medical treatise
40 to 31 BC	In 40 Marc Antony gives the 200,000 scrolls from the library at Pergamon to Cleopatra, who adds them to the library at Alexandria, making it by far the greatest library on Earth In 31 Octavian's general Agrippa wins the decisive sea battle of Actium, effectively defeating the forces of Antony and Cleopatra and signaling the final fall of Egypt to Rome; Cleopatra, killing herself with the bite of an asp, is the last independent ruler of Egypt until the modern democracy		Asclepiades of Bithynia (region in Turkey), Greek physician, d

MATHEMATICS	PHYSICAL SCIENCE	TECHNOLOGY	
		Ssuma Ch'ien's *Historical records* includes the first known reference to parachutes	**90** to **81** BC
Cicero locates and restores the tomb of Archimedes, on which a sphere inscribed in a cylinder is diagrammed, representing Archimedes' favorite among his own mathematical theorems; the grave has since been lost		Greek engineers invent the differential gear	**80** to **71** BC
			70 to **61** BC
			60 to **51** BC
	In 44 Mount Etna in Sicily begins a series of major eruptions between mid-Mar and May; dust from the eruptions darkens the skies; Chinese sources record three years of crop failures as a result	Glass blowing is introduced in Syria A book on architecture and engineering by Roman architect Vitruvius (Marcus Vitruvius Pollio), b about 70, also discusses astronomy, acoustics, sundials, and waterwheels; it will remain the main source of knowledge about Roman construction until the Renaissance Acting on the advice of Greek astronomer Sosigenes, the Julian calendar of three 365-day years followed by one of 366 days is introduced in Rome by Julius Caesar in 46; as a result of changes to make the seasons correct, the year 46 has 445 days, making it the longest year on record	**50** to **41** BC
			40 to **31** BC

	GENERAL	ASTRONOMY	LIFE SCIENCE
30 to 21 BC		In 28 the official imperial histories of China begin a sunspot record that continues until 1638 AD; the record of 28 BC mentions "a black vapor as large as a coin"	
20 to 11 BC			
10 to 1 BC	Roman philosopher, dramatist, and statesman Lucius Annaeus Seneca b Cordoba (Spain)		Celsus, Roman medical encyclopedist, b; Celsus wrote of many things, but only the medical books survive; they include descriptions of surgical procedures and dentistry, and of such disorders as cataracts of the eye
1 to 9 AD			
10 to 19 AD			Thaddeus of Florence describes the medical uses of alcohol in *De virtutibus aquae vitae* (On the virtues of alcohol)
20 to 39 AD			

The Almagest

Ptolemy was a profilic writer in science, but he is best known for his *Almagest*. Ptolemy was the last great astronomer of the Alexandrian school in Egypt; he wrote the book *Megale Syntaxis tes astronomias* (Great astronomical composition) during the second century. The book is a full description of all that was known in astronomy around Ptolemy's time and is a synthesis of Ptolemy's ideas and of those of other Greek scholars, such as Aristotle, Pythagoras, Apollonius of Perga, and Hipparchus. These scholars believed that Earth was the center of the Universe. Ptolemy also included a catalog of stars compiled by Hipparchus in 130 BC. The Arabs were so taken with this book that they began to call it *Al magiste,* (The greatest), later corrupted to *Almagest*. The most important part of the *Almagest* is its description of the Ptolemaic system, a model of planetary motion in which Earth is the center of the universe and the sun and moon move around Earth in perfect circles. Because the planets seem to move backwards some of the time, however, their observed motion cannot be explained by perfect circles. Ptolemy adopted a solution to this problem that had been devised by Apollonius: each planet moves on a small circle, called an *epicyle*. The epicycle has as its center a point called *deferent,* and the deferent itself moves on a circle around Earth.

MATHEMATICS	PHYSICAL SCIENCE	TECHNOLOGY	
		In 27 Vitruvius gives a description of gold amalgam, that is, gold dissolved in mercury Vitruvius, Roman architect, d	**30** to **21** BC
		In 19 Roman general Agrippa has the Aqua Virgo aqueduct built to supply public baths in Rome; also the aqueduct of Nîmes with its 86-story-high Pont du Gard bridge In 15 Yang Hsiung's *Dictionary of local expressions* indicates that the Chinese have invented the belt drive, which will not be known in Europe until 1430	**20** to **11** BC
	Greek geographer Strabo of Amasya (Pontus, a nation in what is now Turkey), b about 53 BC, divides the world into frigid, temperate, and tropic zones Roman geographer Mela b Tingentera (Spain); he believes that the torrid zone near the equator is so hot that people cannot cross it to reach the southern temperate zone, in which people totally unknown to the north live	Herod the Great has the first large harbor constructed in the open sea built to support his new town of Caesarea Palestinae (near present-day Haifa); the harbor is constructed of giant blocks of concrete poured into wooden forms The Chinese invent methods for drilling deep wells (to obtain salt water and natural gas); the wells reach 1460 m (4800 ft)	**10** to **1** BC
Liu Hsin is the first person known to have used a decimal fraction		The Chinese build suspension bridges of cast iron that are strong enough for vehicles to cross The semi-legendary Ko Yu invents the wheelbarrow The earliest known depiction of a ship's rudder is made in China In 1 AD the aqueduct Aqua Alsientina is built to supply water for a 360-m (1181-ft) by 540-m (1800-ft) artificial lake designed for mock sea battles to amuse the Romans An adjustable caliper is in use in China	**1** to **9** AD
			10 to **19** AD
	Strabo's *Geography* attempts to collect all known geographical information; it is correct about the size of Earth, and therefore posits that there must be continents occupying the unknown parts of Earth Strabo, Greek geographer, d	In 31 Tu Shih invents the water-powered bellows for use in working cast iron	**20** to **39** AD

	GENERAL	ASTRONOMY	LIFE SCIENCE
40 to 49 AD	Greek sailors learn how to use the monsoon winds to sail from the horn of Africa to Southern India in only 40 days, shifting the spice trade to a new route and increasing trade between India and the Roman Empire		*De materia medica* by Greek physician Pedanius Dioscorides of Anazarbus (Turkey), b about 20 AD, deals with the medical properties of about 600 plants and nearly a thousand drugs
50 to 59 AD	Seneca's *Quaestiones naturales* (Natural questions), written about this time, deals with physics, geography, and meteorological phenomena		Pliny the Elder (Gaius Plinius Secundus) b Novum Comum (Como, Italy), 23 AD, writes *Naturalis historia* (History of nature), a work of 37 volumes summarizing all that is known in his time about astronomy, geography, and zoology; taken entirely from secondary sources, many legendary monsters are treated as real
60 to 69 AD	Seneca, Roman philosopher, dramatist, and statesman, d 65 by opening his veins		
70 to 79 AD			Pliny the Elder d 79 of asphyxiation while observing an eruption of Mount Vesuvius in Italy
80 to 89 AD		In 84 Fu An and Chia Kuei improve on the armillary ring for locating stars by combining a ring that shows the equator with one that shows the sun's motion through the sky (the ecliptic)	
90 to 99 AD			
100 to 109 AD	The Greek merchant Alexander reaches the south of China by sea	Bhaskara measures the diameter of the sun	The Chinese discover that a powder of dried chrysanthemum flowers can be used to kill insects, making it the first insecticide; the active ingredient, pyrethrum, is widely used today, especially on vegetables, since it is biodegradable and virtually harmless to mammals

MATHEMATICS	PHYSICAL SCIENCE	TECHNOLOGY	
			40 to 49 AD
		In 50 Emperor Claudius of Rome has a 5.6-km (3.5-mi) tunnel constructed to drain Lake Fucino	50 to 59 AD
		The Alexandrian Greek Hero (also known as Heron, b about 20 AD) builds a toy that is powered by steam, using jets of steam to turn a kettle; there is apparently no idea of putting it to practical use	60 to 69 AD
	In 79 Pliny the Younger (Gaius Plinius Cecilius Secundus) writes the first detailed account of the eruption of a volcano; the eruption of Vesuvius that destroys Pompeii and Herculaneum also kills his uncle, Pliny the Elder	The Grand Canal of China, eventually 965 km (600 mi) long, is started In 75 Emperor Vespasian orders the building of the Colosseum (originally the amphitheater of Vespasia) in Rome; it will be the largest amphitheater in the world until the construction of the Yale Bowl in 1914	70 to 79 AD
	In 83 Wang Ch'ung's *Lun hêng* (Discourses weighed in the balance) mentions a "south-controlling spoon" made from lodestone that points south when placed on a polished bronze plate; although this is clearly a compass, it is used only for divination	Wang Ch'ung's *Discourses weighed in the balance* gives the first known reference to the chain pump, a method of raising water from rivers or lakes	80 to 89 AD
The Chinese develop the first magic squares		In 97 a two-volume work on Roman aqueducts by Frontinus (Sextus Julius Frontinus, b about 30 AD) summarizes major advances in construction since ancient times The Chinese invent a device that winnows grain using a rotating fan to separate the grain from the chaff	90 to 99 AD
Menelaus of Alexandria's *Spherics* establishes spherical trigonometry Nichomachus of Gerasa's *Introductio arithmetica* summarizes the existing knowledge of number theory; it lists the only four known perfect numbers: 6, 28, 496, and 8128 (a number is perfect if the sum of its factors is the number, as 1 + 2 + 3 = 6 and 1 + 2 + 4 + 7 + 14 = 28)	Hero of Alexandria describes experiments with air, including the expansion of air caused by heat; he also writes books about simple machines and about light and mirrors	The Chinese invent the multitube seed drill Sextus Julius Frontinus, Roman administrator, d 104 Chinese tradition has it that paper is invented by the eunuch Tsai Lun (b Kueiyang, Kweichow, about 50 AD) in 105; however, archaeological evidence suggests that paper is invented at least 250 years earlier, although used for packing and other purposes, not for writing	100 to 109 AD

	GENERAL	ASTRONOMY	LIFE SCIENCE
110 to 119 AD	In 116 Trajan (Marcus Ulpius Trajanus) extends the Roman empire by conquering Parthia, a rival empire centered in northeastern Iran		
120 to 129 AD	Roman engineers and soldiers build Hadrian's Wall in 122 to protect Britain from northern tribes	In 125 Zhang Heng adds a third ring to the armillary, bringing it to its fully developed form	
130 to 139 AD		In 132 Zhang Heng combines a water clock with an armillary to produce a device that, somewhat like the modern planetarium, keeps track of where stars are expected to be in the sky	
140 to 149 AD		Ptolemy (Claudius Ptolemaeus), b about 100 AD, writes *Megale syntaxis tes astronomias* (Arab translation: Almagest); it becomes the most important text on astronomy during the Middle Ages	
150 to 169 AD			
170 to 179 AD		Ptolemy, Greek astronomer, d	Roman physician and anatomist Galen of Pergamum (Turkey), b about 130, becomes the first to use the pulse as a diagnostic aid
180 to 189 AD	Roman emperor and stoic philosopher Marcus Aurelius d	The Chinese observe a "guest star" in 185, most likely a supernova, in the constellation Centaurus, which remains visible for 20 months	Galen compiles in his writings all medical knowledge that exists at the time in one systematic treatment; this compilation will be used until the end of the Middle Ages
190 to 209 AD			Galen, Greek physician, d Sicily
210 to 259 AD			

MATHEMATICS	PHYSICAL SCIENCE	TECHNOLOGY	
		The oldest known piece of paper used for writing dates from 110	**110 to 119 AD**
		Zhang Heng develops the method of using a grid to locate points on a map	
		Tsai Lun, Chinese inventor, d	
Theon of Smyrna's *Expositio* expounds on Plato's theories of music			**120 to 129 AD**
	In 132 Zhang Heng invents the world's first seismograph, a device that indicates the direction of an earthquake by dropping a ball from the mouth of a bronze dragon into the mouth of a bronze frog		**130 to 139 AD**
			140 to 149 AD
	Ptolemy's *Geographia* (Geography) includes an atlas of the known world based on the travels of Roman legions		**150 to 169 AD**
			170 to 179 AD
	The first writings on alchemy appear in Egypt		**180 to 189 AD**
Chinese mathematicians use powers of 10 to express numbers Liu Hui uses polygons of up to 3072 sides to calculate π to 3.14159 Liu Hui develops the method of approximation known as Horner's method, since it was rediscovered by W. G. Horner in 1819		The Chinese invent the whippletree, a device that allows two oxen to pull a single cart together The Chinese develop porcelain	**190 to 209 AD**
Arithmetica by Diophantus (b about 210) is the earliest treatise known devoted to algebra, although many problems involve solutions in the integers only (diophantine equations); Diophantus obtains negative numbers as solutions to equations, but dismisses them as "absurd"			**210 to 259 AD**

	GENERAL	ASTRONOMY	LIFE SCIENCE
260 to 269 AD	In 269 the library at Alexandria is partly burned when Septima Zenobia, queen of Palmyra, captures Egypt		
270 to 279 AD			
280 to 289 AD	In 286 the Roman empire is divided by Diocletian into eastern and western halves for administrative purposes		
290 to 299 AD			
300 to 309 AD		Chinese astronomer Chen Zhuo combines the star maps produced by Shih Shen, Gan De and Wu Xien in the fourth century BC into a single map	In 304 the first recorded example of biological control of pests is made by Hsi Han's *Record of plants and trees of the southern regions,* in which he reports on how specially selected ants are used to protect mandarin oranges from other insects that might attack the plants
310 to 329 AD			
330 to 339 AD	In 330 Constantinople is founded by Constantine I		
340 to 369 AD		The Chinese report a new star in 369, possibly a supernova that remains visible for five months	
370 to 379 AD	The Visigoths defeat Roman emperor Valens at the battle of Adrianople (Edirne, Turkey) in 378		

MATHEMATICS	PHYSICAL SCIENCE	TECHNOLOGY	
	Mount Ilopango in what is now El Salvador erupts, sending the incipient Maya civilization hundreds of miles away; their characteristic culture does not revive for about 200 years	Ma Chün builds a "south-pointing carriage" using differential gears	260 to 269 AD
		In China, the first form of a compass is probably in use for finding south in 271	270 to 279 AD
			280 to 289 AD
Diophantus, Greek mathematician, d			290 to 299 AD
The Chinese may have begun to develop the abacus, although the first printed reference to one comes in 1593 AD		The Maya develop the day-count calendar, which combines the 365-day Olmec calendar that has a 52-year cycle with the 260-day cycle known as the tzolkin, which has 13 cycles of 20 days each; this calendar dates events back to 3000 BC The Chinese learn to use coal instead of wood as the fuel in making cast iron In China the earliest known depiction of a stirrup is made in 302	300 to 309 AD
The *Mathematical collections* of Pappus of Alexandria (Egypt, b about 260) summarizes geometric knowledge and includes ingenious proofs ascribed to Pappus himself; the book is the best account we have of various missing books of earlier Hellenistic mathematicians such as Euclid and Apollonius			310 to 329 AD
			330 to 339 AD
			340 to 369 AD
Hypatia, Greek philosopher and mathematician, and the only woman with such credentials known to history, b Alexandria (Egypt)			370 to 379 AD

	GENERAL	ASTRONOMY
380 to 389 AD	The library at Alexandria is destroyed during a riot of Christians against pagans in 389	The Chinese report a new star in 386, possibly a nova, that remains visible for three months
390 to 399 AD	The Roman emperor Theodosius closes the pagan temples in 392, effectively making Christianity the official religion of the Roman empire On the death of Roman emperor Theodosius I in 395, the Roman empire becomes completely divided into two parts, each ruled by a separate emperor	The Chinese observe a "guest star" in 393, most likely a supernova, in the constellation Scorpio; it remains visible for eight months
400 to 409 AD		
410 to 419 AD	The Visigothic king Alaric invades and sacks Rome in 410	
420 to 459 AD	The Vandals sack Rome in 455	
460 to 469 AD		
470 to 479 AD	The Roman empire, long in decline, ends completely in 476 when the German invaders elect one of their own generals to replace the ruling emperor and to reign as king of the Germans in Italy	

Classical volcanoes

The classical world of Greece and Rome was well-placed geographically to know about volcanoes. Although not in the Pacific "Ring of Fire," the Mediterranean, stressed by Africa pushing into Europe, is home to a number of volcanoes that were active in classical times, some of which are still active today. The oldest known historical record of a volcano erupting refers to an eruption of Mount Etna in 479 BC.

Classical mythology explained volcanoes as the work of a god, Vulcan, who lived underground. Vulcan was a blacksmith, and the smoke and fire of the volcano supposedly came from his forge. Vulcan's forge was located on the island of Stromboli, which was almost continuously active in classical times.

Greek scientists also tried to explain volcanoes. Aristotle noticed that the volcanoes with which he was familiar were near the sea (actually, most, but not all, volcanoes *are* near the sea). He suggested that waves on a beach push air into the ground. When that air reaches a mineral that can burn, such as coal or sulfur, the mineral deposits catch fire. These fires then produce volcanoes at the earth's surface.

The Roman people who lived at the base of Mount Vesuvius in later classical times did not recognize that it was a volcano (although Strabo identified Vesuvius's volcanic nature about the first decade AD). Even when earthquakes rocked the Bay of Naples, near the mountain, for 15 years, the citizens did not connect the tremors with the mountain. By 79 AD, the quakes were becoming frequent.

The coastline at the base of Vesuvius was a popular resort region for the Romans. Many had summer homes in the town of Pompeii, nearby along the shore, or on offshore islands. At the very base of the mountain was the town of Herculaneum. When an eruption began on August 24, the area was filled with summer visitors. Also, part of the Roman navy was anchored offshore.

The admiral of this fleet was Pliny the Elder, the well-known scientist. When the eruption began, he headed toward the volcano out of curiosity. It soon became apparent that his ships were needed for a rescue effort.

It is largely from the accounts of Pliny's nephew, Pliny the Younger, who was also there, that people today know the details of the eruption. Excavations by archaeologists since 1800 have completed the story.

Pompeii was destroyed twice by the eruption. As the residents and visitors were packing their most precious possessions to escape, clouds of poison gas rolled into the city. Those who were out of their houses and who had not been killed by falling rocks died instantly from the gas. People who were still in their houses died from lack of oxygen.

| | | | 380 to 389 AD |

| Theon of Alexandria flourishes; he is the father of Hypatia and the editor of an important edition of Euclid | | | 390 to 399 AD |

| | The term *chemistry* is used for the first time by Alexandrian scholars for the activity of changing matter | The Chinese learn to make steel by forging together cast and wrought iron

The Chinese invent the umbrella | 400 to 409 AD |

Hypatia, Greek mathematician and philosopher, d Alexandria, 415; she is murdered by a band of monks for teaching pagan ideas

Playfair's postulate, equivalent to Euclid's fifth postulate, named after a nineteenth-century rediscovery by John Playfair, is discovered by Greek mathematician Proclus, b Constantinople (Istanbul, Turkey), 410; it states that through a given point only one line parallel to a given line can be drawn

Tsu Ch'ung-Chih and Tsu Keng-Chih, father and son, calculate π as 3.1415929203 using a circle 3 m (10 ft) across (the correct value to the same number of decimal places is 3.1415926535)

410 to 419 AD
420 to 459 AD
460 to 469 AD
470 to 479 AD

Classical volcanoes, continued

Pliny had reached the coast at the town of Stabiae, where a friend needed rescue. By the time Pliny landed, high seas and strong winds prevented his ships from leaving again. As a volcano heats the air around it, the hot air rises very fast. Cooler air rushes toward the volcano, causing strong winds in that direction.

Pliny and his friends waited for the winds to die down, but the falling ash was slowly covering the exits from his friend's house. Escape by sea was impossible, not only because of the winds, but also because the sea level was falling, which sometimes happens during eruptions or earthquakes. The sailors tied pillows to their heads and tried to escape on foot. Pliny, however, was overcome by the poisonous gases from the volcano and died, although others with him survived. The town of Stabiae was totally destroyed.

In the meantime, the second destruction of Pompeii was progressing. As the ash rained down, it gradually covered the entire town, so that only the tops of taller buildings showed above the surface. Rain was also falling, which turned the ash into a kind of cement that completely enclosed the bodies of the dead who were not in houses. The bodies remained so well preserved that today archaeologists are able to make plaster casts that show the expressions and even the clothing the victims wore on their last day on Earth.

The town of Herculaneum, at the base of Vesuvius, was also buried. Volcanic eruptions usually produce large amounts of rain. In this case, the rain was so intense that it caused giant mud slides. One of them buried Herculaneum in 60 feet of mud. Most of the people in the town had left by the time of the mudslide.

Until 1982, it was believed that the people who left Herculaneum had escaped safely. That year, excavations on the beach revealed that many of them had been killed and buried there.

	GENERAL	ASTRONOMY
480 to **489** AD	Philosopher Boethius (Anicius Manlius Severinus) b Rome, 480; although best known for his *On the consolation of philosophy*, he also translates Aristotle into Latin, making Aristotle the only source of Greek thought during the Middle Ages Clovis becomes king of the Franks in 481	
490 to **499** AD	The East-Gothic empire is founded in 493 Frankish king Clovis I adopts Christianity in 496 along with 3000 of his followers	In 497 Aryabhata I (b India, 476) recalculates Greek measurements of the solar system; although he mostly accepts Ptolemy's scheme of the universe, he also puts forward the idea that Earth rotates
500 to **509** AD		
510 to **529** AD	In 525 Dionysius Exiguus introduces a calendar based on the birth of Jesus Christ The Greek merchant Kosmas reaches India and Ceylon in 525 Boethius d 524 Ostrogothic king Theodoric the Great d 526 The monastery at Monte Cassino is founded in 529	

Proclus, Greek mathematician, d Apr 17, 485, Athens			**480** to **489** AD
Indian mathematician Aryabhata's *Aryabhatiya* of 499 includes a good value of 3.1416 for the ratio of a circle's circumference to its diameter, uses the decimal, place-value numeration system, and supplies a variety of rules for algebra and trigonometry, not all of them correct			**490** to **499** AD
The abacus is used in Europe, although counting boards based on the same principle were used by the ancient Greeks and Romans as much as a thousand years earlier			**500** to **509** AD
The Academy and the Lyceum, the schools started by Plato and Aristotle at Athens, are closed in 529 by the emperor Justinian		The first European paddle-wheel boats are designed, to be powered by oxen walking in circles, as in a mill; it is unlikely that these were built	**510** to **529** AD

OVERVIEW

Science in many lands and medieval science

With the disappearance of the last great centers of learning from antiquity—the Academy and Lyceum in Athens were closed down by the Byzantine emperor Justinian in 529 and the Museum of Alexandria, only a shadow of its former self, was destroyed by the Arabs in 641—scientific activity almost ceased in Europe. Although there was a revival among church scholars in the twelfth century, this was snuffed out in the general disarray following the Black Death in the fourteenth century. A true revival of science can be dated from the middle of the fifteenth century, when the fall of Constantinople to the Turks resulted in scholars bringing many Greek manuscripts into Europe in 1453. Printing also began about this time, changing the way that scientific ideas could be communicated. Thus, we use 530 and 1452 as the boundary dates for this period.

The decline of science in Europe

Several reasons for the decline of science in Europe between 530 and 1000 have been put forward by historians. For one, European culture was still strongly influenced by the Romans, who were notoriously little interested in theoretical science. With the disintegration of the Roman empire, society underwent several important changes; for example, the large cities disappeared, roads and aqueducts disintegrated, and trade became more limited. The general infrastructure in which learning and science could develop had ceased to exist.

Medicine was an exception. To an extent, it survived the Dark Ages because the Christians felt it to be their duty to take care of the sick. During the sixth century, the works of Hippocrates and Galen were studied at monasteries. In addition, the high level of teaching of medicine at Salerno during the ninth century heralded the revival of science during the later Middle Ages.

Arab science

The Islamic culture, which flourished from about 700 to 1300, played a key role in bridging the gap between the Hellenistic period and the Renaissance. During that period, the Islamic empire became the most advanced civilization in the Western world. The role of Islamic culture in the development of science is demonstrated by the large number of scientific terms with Arabian origins, such as *azimuth* and *algebra*.

There are several factors that contributed to the flourishing of science in the Islamic civilization. Because of intense commercial activity, the Arabs came into contact with a large number of cultures, such as the Indian and Chinese cultures (see "Indian science," p. 65, and "Science in China" p. 59). Several different cultures became part of the Islamic world: the Iranian, Turkish, Jewish, Orthodox and Nestorian Christians, and Gnostic, known as Sabian or Mandaean. Each of these civilizations had new ideas to contribute to Arab thinking.

A strong unifying factor in the Islamic empire, besides the Islamic religion, was the Arabic language, although various parts of the empire also used Persian, Syriac, and other languages. Many of the works of the ancients have been preserved because they were translated into Arabic. During the sixth century, many of the important Greek works were also translated into Syriac; when the Arabs conquered Syria, these works were translated into Arabic. Several Hindu works were also translated into Arabic.

A number of centers of learning appeared throughout the Islamic empire. Baghdad developed during the eighth century into one of the most important ones: Al-Ma'mun (786-833) founded a "House of Wisdom" there that contained an astronomical observatory. Many of the important Greek treatises, such as Galen's medical writings and Ptolemy's astronomy, were translated at the House of Wisdom. During the tenth century, Islamic Cordoba, in Spain, which became the richest and largest city in Europe, contained a library of 400,000 volumes.

The Arabs not only translated and preserved the scientific works of the ancients and Indians, they also contributed themselves to several scientific fields. In astronomy, they made a large number of accurate observations with instruments superior to those of the Greeks. In addition, they compiled astronomical tables that remained in use until the Renaissance. The Arabs revived astronomy by studying the ancients and repeating their observations. The Toledan tables, compiled by al-Zarkali (also known as Arzachel) about 1029-1087, became the basis of the Alfonsine tables that were published in Spain by order of Alfonso el Sabio (Alfonso X of Castile-León, 1252-1284). This compilation of planetary and stellar positions remained in use for the next three centuries.

Arabian mathematics blossomed in part because it combined the mathematical knowledge of the Greeks with that of the Indians. The introduction of Indian numerals, often called Arabic numerals, with their decimal place-value system, simplified calculations. However, Arab mathematicians were not content merely to preserve the Greek and Indian traditions; they added new ideas of their own, especially in solving equations, trigonometry, and numerical calculation. The term *algebra* is based on

Muhammad ibn Musa Al-Khowarizmi's book *Al jabr w'al muqābalah* (Restoring and balancing, or possibly Equations and reduction). Also, the modern term *algorithm* comes from the first sentence of the Latin translation from that book: "*Algorismi dicit . . .*" (Al-Khowarizmi says . . .).

Chemistry developed into an experimental science among the Arabs; substances unknown to the Greeks, such as borax and sal-ammoniac, were prepared by Arab chemists.

In physics, as in astronomy, Arab scholars excelled in the craft of instrument making. Alhazen's *Treasury of optics* remained influential until the sixteenth century.

Medicine was highly developed, and a complete infrastructure for health care existed. For example, in Baghdad, 1000 government-licensed physicians practiced medicine. There were numerous hospitals and even mental institutions where humane treatment of patients was

Science in China

Chinese society has been known throughout history for the stability of its traditions and its bureaucracy. Yet, until about the fifteenth century, China was more successful than western Europe in applying scientific knowledge.

The Chinese discovered paper as early as 105 AD. The first printed text (in block printing) appeared in 868 AD. Around 100 BC the Chinese discovered that a magnet orients itself towards the North Pole, but they did not use magnets for navigation at sea until the tenth century. In 969 AD rockets were used in warfare. The Venetian traveler Marco Polo reported that the Chinese used firearms in 1237.

The escapement, the most important part of the mechanical clock, was invented in 725 AD. The Chinese were familiar with many other mechanical devices, such as the eccentric and the connecting rod, as well as the piston rod, long before they became known in Europe. Furnace bellows driven by waterwheels contributed to the astonishing developments in iron and steel production during the Sung period (960-1279 AD).

Besides their technological progress, Chinese civilization had gone far in developing theories on matter and living be-

ings. The Chinese seem to have always believed that the blood circulated, for example. Isaac Newton's first law of motion, that a body will not stop moving unless stopped by an opposing force, was stated by Chinese philosophers almost 2000 years before Newton.

Large amounts of empirical data on a number of subjects were accumulated. Most notably, Chinese astronomy reached a high level during the eighth century BC; it was outstanding for the quality and quantity of observations. For example, the precise Chinese records of the supernova that was visible in 1054 AD allowed present-day astronomers to establish that the supernova was the origin of the Crab nebula.

The attitude of Chinese society toward nature was quite different from the attitude that developed in Europe during the Renaissance. The Chinese never separated the material from the sacred world, and did not have the conviction that people can dominate nature. They were not interested in developing a scientific method; thus their theories often remained divorced from observation and experimentation.

offered. However, the science of anatomy did not progress because dissection of corpses was not allowed by Islamic law.

The knowledge gathered by the Arabs found its way to Christian Europe through translations made in the twelfth century from Arabic into Latin.

The revival of science in Europe

Science returned to western Europe to some degree during the twelfth century. Although the Christian church had played an inhibiting role in the development of science in earlier centuries, it was the church that caused much of ancient learning to be preserved. The members of the clergy and of monastic orders were practically the only people who were literate. The emperor Charlemagne, who learned as an adult to read and tried unsuccessfully to learn to write, decreed in 787 that every monastery must establish a school. These cathedral schools became the forerunners of the first universities. Technology, especially crafts and farming, was actively pursued at monasteries. However, the strongest factor in the revival of science during that time in Europe was the establishment of contact with the Islamic culture.

Islamic civilization was at its height during the cultural low point in western Europe that followed Charlemagne. As cities with a more educated population began to develop in Europe, Christian scholars were eager to absorb the knowledge amassed by the Muslims. Toledo (Spain), which had remained a Christian bishopric throughout the Muslim occupation, was one of the important centers of Islamic learning. The city was conquered by Alfonso VI of Castile in 1085, and many scholars went there to study with the Arabs. The Christians mostly learned about the Islamic culture, however, through the Crusades.

A large number of Greek works became available to European scholars from 1150 to 1270. Some scholars, such as Adelard of Bath and Gerard of Cremona, learned Arabic. Gerard of Cremona translated Ptolemy's *Almagest* from Arabic into Latin as well as some of Aristotle's works that had been translated into Arabic, Euclid's *Elements*, and the works of Galen and Hippocrates. Adelard of Bath translated Al-Khowarizmi's astronomical tables. Robert of Chester translated Al-Khowarizmi's *Algebra*. By 1270 the whole corpus of Aristotle's work, a large part of it translated from the original Greek by William of Moerbeke, became known to medieval scholars.

The Scholastics were Christian philosophers of the thirteenth century. They set out to absorb the newly gained knowledge of the ancients and to reconcile it with the teachings of the church. St. Thomas Aquinas, one of the founders of the Scholastic school, argued that knowledge can be obtained through both religious faith and natural reason. He believed that Plato and Aristotle, especially Aristotle, were compatible with Christian religion. Some other Scholastics, such as Siger of Brabant and Boethius of Dacia (not the well-known Boethius), disagreed. Throughout the early thirteenth century, it became more and more apparent that Aristotle and the church were on a collision course. As early as 1210, the local synod decreed that certain of Aristotle's works could not be taught at the University of Paris. After all, they said, Aristotle's works imply that God did not create the world, that there can be no transubstantiation of the host and wine during communion, that miracles cannot occur, and that the soul does not survive the body. In 1277 the pope issued a condemnation of 219 propositions, including many from Thomas Aquinas, that were tainted with Aristotle's views.

During the fourteenth century, a number of scholars developed a new style of philosophy that was less tainted by Aristotelianism and therefore not banned under the condemnation of 1277. Roger Bacon wrote in his *Opus maius* that it is wrong to rely only on the authority of past scholars and that experiment is an important means of gaining knowledge. William of Ockham and Jean Buridan criticized Aristotle's theory of motion. Nicholas of Cusa (a cardinal in the church) challenged Aristotle's concept of the fixity of the Earth.

Major advances

Astronomy. Ptolemy's *Almagest* became the standard text for astronomy throughout the later Middle Ages. Astronomy was still based on Plato's principle that all observed motions of heavenly bodies had to be explained in terms of uniform circular motions. Ptolemy's and Aristotle's views became incorporated into church dogma, mainly through the efforts of St. Thomas Aquinas. Aristotle's cosmology required a "prime mover" that keeps the planets and stars moving. The existence of a prime mover became for St. Thomas his "first proof" of the existence of God.

Although the Arabs accepted Ptolemy's system of cycles and epicycles, they greatly improved observational techniques and developed trigonometry as a part of their astronomical work.

Life science. Later medieval biology was strongly influenced by Aristotle's method of investigation: try to find the function and purpose of organic structures. Botany and zoology became separate disciplines. The main interest in botany was medical, while zoology often played a moral and didactic role, emphasizing fables more than facts. Both sciences incorporated tales and fables of all kinds; famous were the so-called bestiaries, stories describing incredible animals and monsters, usually believed literally by the population. The most accomplished biologist was Albertus Magnus, who studied plants all over Germany not just for medical or agricultural reasons, but also for scientific purposes as well. He also studied a large number of animals, birds, and insects, and ridiculed the fantastic stories from the bestiaries.

The monasteries generally preserved the medical

heritage of Greece and Rome, and medicine became the most successful practical art in the Middle Ages. During the eleventh century, medicine underwent a revival at Salerno (Italy), where it reached a high level because of contact with the Arabs in nearby Sicily. During the twelfth century, Montpellier (France) became a medical center, and during the thirteenth century, medical schools appeared in Bologna, Padua, and Paris.

Medical practice relied on the Hippocratic method combined with the use of herbs and drugs, although astrological influences also were believed to play a role in the course of an illness. Drugs based on arsenic, sulfur, and mercury were used, especially mercury ointments for the treatment of skin disease. Opium was used as an anesthetic during surgery. The art of ophthalmology, acquired from Arab physicians, reached a high level and operations removing cataracts were performed successfully.

Mathematics. Mathematics was at a very low level during the early Middle Ages. Most calculations were done on the abacus because, before the advent of Hindu-Arabic numerals, mathematical operations were difficult to perform. Leonardo of Pisa, also known as Leonardo Fibonacci, introduced Hindu-Arabic numerals in Europe, although they were known previously to some mathematicians. He also is known for the Fibonacci series, a mathematical series in which each term is the sum of the two preceding terms (1, 1, 2, 3, 5, 8, . . .).

Nicolas Oresme introduced notions that correspond to the idea of rational (fractional) exponents as well as a concept similar to that of a function. He also introduced a graphical system for studying mathematical curves. While his graphs for uniform motion were not based on a coordinate system in the Cartesian sense, they implied (for the specific examples used) the fundamental law of the calculus: the way a function changes determines the area under a curve describing the function; however, such an interpretation is possible only if one already knows calculus.

Physical sciences. The physical sciences were dominated by the views of the ancients, especially by Aristotle's ideas on motion. It was believed that motion is possible only if something continuously pushes the object being moved. William of Ockham was one of the first to introduce the concept of *impetus* and to reject the idea of a prime mover and thus the validity of St. Thomas's first proof of the existence of God. He argued that God put the celestial bodies in motion during the Creation, and that they keep moving because they retain the impetus conferred upon them. Jean Buridan perfected Ockham's theory and Albert of Saxony distinguished between uniform and irregular motion.

Although magnetism was known in antiquity, one of the earliest descriptions of the behavior of magnets is from Petrus Peregrinus. In his *Epistola de magnete* he describes an experimental method of studying magnetism; for this,

he is viewed by many as the first experimental physicist. Robert Grosseteste and Roger Bacon performed optical experiments; Bacon viewed experimentation as an important means of increasing knowledge.

Chemistry was dominated by alchemy, taken over from Arab sources. The alchemists searched for a method of making gold, and a large number of manuscripts appeared on that subject. The church vigorously opposed alchemy, but there were practitioners of alchemy within its ranks. Alchemy led to experimentation; several new chemicals, such as mineral acids, and some practical applications of alchemic techniques appeared. Several types of distillation methods were improved; for example, water cooling of the condensing tube was introduced. The distillation of alcohol appeared during the twelfth century, and *aqua vitae*, 96 percent alcohol or 192 proof, was obtained during the thirteenth century by redistillation. Toward the end of the thirteenth century, the first oil-based paints appeared.

Technology. Because society in the Middle Ages was not based on slavery, and because of the emergence of trade and artisans, technology made striking progress. Many of the devices used, such as waterwheels, geared wheels, and windmills, were known in antiquity—although windmills were first described about 600 AD—but it is only during the Middle Ages that these devices came into use on a large scale. In 1086 there were some 5000 water mills in England. Waterwheels were used to drive trip hammers for crushing bark or for tanning; later waterwheels drove forge hammers and bellows in forges. The size of furnaces to reduce iron ores increased, and several types of carbon steel and cast iron were produced, although the exact role of carbon content in iron was not understood.

Paper-making technology, introduced from China by the Arabs, reached Europe during the twelfth century, and paper mills were established in several cities. This paved the way for printing in Europe, which began at the very end of this period, around 1440. The spinning wheel also appeared around that time.

Agriculture was improved by the introduction of the iron plowshare, about 1000 AD, and the horse collar, in which the pull is placed on the shoulders of the horse instead of the windpipe. The more efficient agriculture and food production resulted in an increased wealth: cities blossomed and commerce became more intensive.

Virtually all of these technological advances were known earlier in China, but in most cases it is impossible to determine which inventions diffused from China.

The empirical technology of engineering and architecture underwent its greatest growth since the time of the Romans. Medieval cathedrals developed over the centuries as builders found new ways to solve the problems of building very large structures from stone. Similar technology could be applied to other large structures, such as town halls or bridges.

530 to **539**	*The rise of time (part one)* In ancient time-measurement systems, daylight and nighttime were each given 12 hours. This was convenient for use with sundials, which were used in China in a primitive form as early as 2600 BC. Because the length of daylight and nighttime varies with the season, so did the length of the Chinese hour. When water clocks came into use, about a thousand years after sundials, a conflict between the two forms of measurement was apparent. A water clock works because water from a container flows through an opening at a steady rate. The amount of water in another container is used to move an indicator of some kind—in simplest form, the level of water in a container. Since the water flows at an almost steady rate, as the indicator moves along a marked face it shows the hours. But when the length of the hour changed from season to season, a different water clock was	
540 to **559**	needed for each month. Ancient peoples solved this problem in various ways, such as by having different marked faces for each month. In that way the water clock was never far out of line with the sundial, which also remained in use. Later, instead of modifying water clocks to change with the seasons, sundials were constructed to show hours of the same length all year. In the eighth century AD, the Chinese began to fashion water clocks with primitive escapements. The escapement is a ratchet that causes a wheel to move only so far and then stop. Continuous motion is replaced with	In 541 a pandemic of the bubonic plague strikes Europe and the empire of Justinian, killing at its peak 10,000 persons a day in Constantinople; the pandemic continues until 544
560 to **569**	discrete "ticks." By the beginning of the fourteenth century, the concept of an escapement was known in Europe, where the escapement was used to slow down the motion of a falling weight attached to it by a cord or chain. This motion can then be converted with gears to turn the dial of a clock. Mechanical clocks using escapements and weights were gradually improved and put in towers all over Europe.	
570 to **579**	Mohammed b Mecca (Saudi Arabia)	
580 to **589**		
590 to **599**	Gregory the Great is elected pope in 590	
600 to **609**		Korean priests introduce the calendar and astronomy into Japan in 602

MATHEMATICS	PHYSICAL SCIENCE	TECHNOLOGY	
Koreans introduce Chinese mathematics into Japan in 534		The water-powered flour sifting and shaking machine invented in China is mentioned for the first time; this device works like a steam engine in reverse, changing rotary motion into the back-and-forth motion needed for sifting; it is the first known machine capable of doing this In 532 Eastern emperor Justinian orders the building of Santa Sophia, designed by Isidore of Miletus (Turkey), the first building with a dome this large, 37 m (120 ft) across and 14 m (46 ft) high	**530** to **539**
		The Liang emperor Luan's *Book of the Golden Hall master* describes wind-driven land vehicles used in China; a wind-driven carriage could carry 30 people and travel hundreds of kilometers in a single day; sails were also applied to wheelbarrows, and wheelbarrows with sails became a common symbol of Chinese culture in eighteenth-century Europe	**540** to **559**
Eutocius writes commentaries on the works of Archimedes and Apollonius; they remain the best source of certain of their mathematics		The Avars, an Eastern people who eventually settled in the Caucasus, invade Hungary in 568, bringing with them the trace harness and the stirrup for use with horses; both were originally developed in China	**560** to **569**
		In 577 women in Northern Ch'i under siege from neighboring kingdoms in China invent matches so that they can start fires for cooking and heating	**570** to **579**
		The oldest known reference to paper for use as toilet paper is made in China in 589	**580** to **589**
			590 to **599**
Zu Chong-Zhi and his son Zu Geng-Shi calculate π to be between 3.1415926 and 3.1415927 Decimal notation is in use in India		The Chinese print whole pages with woodblocks, although the earliest known surviving pages are from the 700s The earliest known windmills are built in Persia (Iran); they use a vertical shaft and horizontal sails, partially protected from the wind, to grind grain	**600** to **609**

GENERAL	ASTRONOMY	LIFE SCIENCE
610 to 619	The Visigoths take Spain from the Roman empire in 616		
620 to 629	Mohammed flees Mecca for Medina in 622, a flight known as the Hegira; Islamic timekeeping begins with this year		
630 to 639	*Etymologia sive origine* (Etymologies) by Isidore of Seville (Spain, b about 560) is a first attempt at a general synthesis of pre-Christian learning; drawing heavily on Pliny the Elder, it is filled with legends; it is immensely popular in its time and about 1000 manuscripts survive Isidore of Seville, scholar who was influential in preserving some Greek writings and in keeping astrology popular despite prohibitions of astrology in the Bible, d Seville (Spain), Apr 4, 636	A clear statement of the rule that the tail of a comet always points away from the sun is written in China in 635	
640 to 649	Islamic armies conquer Alexandria (Egypt) and complete the destruction of the Library and the Museum, formerly home to the works of such mathematicians as Eratosthenes, Euclid, Apollonius, and Pappus		The Chinese physician Chen Ch'üan d 643; he is the first person to have noted the symptoms of diabetes mellitus, including thirst and sweet urine
650 to 659			
660 to 669		Brahmagupta, Hindu astronomer and mathematician, d	
670 to 679	The Venerable Bede, English scholar, b Jarrow, Durham (England) 673 or 672; although most noted as a historian, Bede also writes about the calendar, the shape of Earth, and the tides, giving a correct account in each case	The first sundial in England is built in Newcastle in 675	
680 to 689			

MATHEMATICS

Brahmagupta (b central India, about 598) writes *Brahmasphuta siddhānta*; it concerns mensuration, trigonometry, and algebra, including excellent work on diophantine equations; Brahmagupta also uses negative numbers

The value of π is given as 3.1415927 (using decimals) in the official history of the Sui dynasty in 635

In 662 Bishop Severus Sebokht of Syria mentions Hindu numerals as using only nine digits

The first uses of a "goose-egg" sign for zero appear in Cambodia and Sumatra, although the Chinese had longed used an empty space as a placeholder and later Mesopotamian numeration also used a sign as a placeholder; it is not clear when 0 comes to be understood as a number and not just a placeholder

PHYSICAL SCIENCE

Indian science

Because of the religious character of Indian civilization, the development of science was less pronounced in India than in other areas of the world. However, Indian science excelled in three fields: linguistics, mathematics, and astronomy.

Buddhist philosophy taught that the universe is periodically destroyed and recreated. Indian mathematics was influenced by the idea of these cosmic cycles and related Hindu ideas. As a result, Indian mathematicians were interested early in very large numbers. Mathematics was mainly numerical and was also influenced by the algebraic tradition of the Babylonians. There was little interest in geometry. The Indians used the decimal system, including zero, to express numbers, making it the most successful system for extensive calculations.

Ancient Indian astronomy was mainly concerned with calendar computation. Calendar computations were required early for religious ceremonies; they were based on the motions of the moon and sun. The Indians showed little interest in the motion of the planets. Later, Indian astronomy was influenced by the theories from Babylonia and ancient Greece.

Medicine developed very early. It was based on preventing disease and on hygiene as much as on curing disease. Hospitals and herbal gardens existed in the third century BC. Health was viewed to be the result of the balance of three elements: air, fire, and water. In relation to the body, these elements became wind, bile, and phlegm, comparable to the three of the four "humors" in Western medicine at that time.

TECHNOLOGY

Li Ch'un builds the Great Stone Bridge over the Chiao Shui River in China; it is the first example of a segmental arch bridge and it survives intact	**610 to 619**
	620 to 629
	630 to 639
	640 to 649
Callincus (b Heliopolis, Egypt, about 620) invents a substance that will burn in water and thus be a weapon against wooden ships; it is known as Greek fire	**650 to 659**
	660 to 669
	670 to 679
In 688 the empress Wu Tse of China has a pagoda 90 m (294 ft) tall built of cast iron	**680 to 689**

	GENERAL	ASTRONOMY	LIFE SCIENCE
690 to 699			
700 to 709			
710 to 719	The Arabs invade Spain in 711		
720 to 729			
730 to 739	Charles Martel's defeat of the Islamic armies at Poitiers in 732 stops the westward expansion of Islam that began with the death of Mohammed in 632 Venerable Bede, English scholar, d Jarrow, Durham (England)		
740 to 749	Charlemagne, Frankish emperor, b Aachen (Germany)		
750 to 759			
760 to 769	Charlemagne succeeds Pepin III, king of the Franks, in 768		
770 to 779			
780 to 789	In 782 Charlemagne calls on Alcuin of York (b England, about 732) to organize education in his kingdom of the Franks and Lombards; Alcuin establishes schools associated with important cathedrals and also teaches Charlemagne to read		

MATHEMATICS	PHYSICAL SCIENCE	TECHNOLOGY	
		In 695 the empress Wu Tse has a cast-iron column made from about 1325 tons of iron to commemorate the Chou dynasty	690 to 699
The first printed reference to "ball arithmetic," which may have been an early form of the abacus, appears in Chinese texts; it implies that the abacus goes back to the second century AD		Between 704 and 751 the earliest printed text known, a Buddhist charm scroll, is created using woodblocks to produce images	700 to 709
			710 to 719
		In 725 Buddhist monk I-Xing and Chinese engineer Liang Ling-Zan build a water clock that has an escapement, the device that causes a clock to tick; the clockwork was used to power various astronomical devices rather than to indicate the hour	720 to 729
			730 to 739
		The first printed newspaper appears in Beijing, China, in 748	740 to 749
	Arabic alchemist Abu Musa Jabir ibn Hayyan, b possibly in Al-Kufah (Iraq) about 721, known as Geber, describes how to prepare aluminum chloride, white lead, nitric acid, and acetic acid		750 to 759
			760 to 769
Various Hindu works on mathematics are translated into Arabic			770 to 779
Muhammed ibn Musa Al-Khowarizmi, Arabian mathematician, b Khwarizm (Uzbek Republic, Soviet Union); his works, when translated, introduce Hindu-Arabic notation to Europe; from one comes the origin of the word *algebra*; Al-Khowarizmi's own name is the source of the words *algorism* and *algorithm*		In 789 Charlemagne introduces the royal foot as the unit of length and the "Karlspfund" (about 365 g, or 13 oz) as a unit of weight	780 to 789

	GENERAL	ASTRONOMY	LIFE SCIENCE
790 to 809	The ancestor of paper money is in use in China, essentially as bank drafts on money deposited; it can be exchanged for hard cash at a later date or at a different bank On Christmas day 800, the pope crowns Charlemagne, king of the Franks, Roman emperor, reviving a form of the Roman empire for a time; almost 200 years later, the same idea is revived in the Holy Roman empire (962 AD) Alcuin, English scholar, d Tours, France, May 19, 804 Al Kindi, the author of books that cover most of the sciences, b 801	In 807 Einhard's *Life of Charlemagne* provides the first Western reference to sunspots	
810 to 819	Charlemagne, Frankish emperor, d Aachen (Germany), Jan 28, 814 Bayt al-Hikma, a science library in Baghdad (Iraq), is founded in 815		
820 to 829		Afragamus (Abu-al-'Abbas al Farghani), Egyptian astronomer and geographer who presents Ptolemy clearly in Arabic, b A complete translation of Ptolemy's *Megale syntaxis*, better known from its Arabic title as the *Almagest*, is made into Arabic in 827	
830 to 839	The "House of Wisdom" is founded in Baghdad in 832		
840 to 849	Suleiman travels to China; the account of his trip is the first reporting about China in the Arab world In 843, the Treaty of Verdun divides the empire of Charlemagne among his three sons Alfred the Great, English monarch, b Wantage, Berkshire, England, 849	Abū al Fadl Ja'far ibn al-Muqtafī is the first Arab known to record sunspots	Rhazes (Arabic: Abu-Bakr Muhammad ibn Zakariyya Ar-Razi), Persian physician and alchemist, b Rhages (Iran)

MATHEMATICS	PHYSICAL SCIENCE	TECHNOLOGY	
Chinese mathematicians use the method of finite differences to solve equations A complete translation of Euclid's *Elements* is made into Arabic		Blast furnaces for making cast iron are built in Scandinavia	**790** to **809**
	Geber, Arabian alchemist, d, possibly at Al-Kufah (Iraq)	The Chinese government takes over the issuing of paper bank drafts, the ancestor of paper money, in 812	**810** to **819**
Al-Khowarizmi's *De numero indorum* (Concerning the Hindu art of reckoning) gives a set of rules for computing with Hindu-Arabic numerals			**820** to **829**
Al-Khowarizmi's *Al-jabr wa'l muqābalah*, known in the West as *Algebra*, gives methods for solving all equations of the first and second degree with positive roots Arabian mathematician Thabit ibn Qurra b Harran (Turkey), 836; he translates Greek works and tries to solve the problem of Euclid's fifth postulate		The *Utrecht psalter* provides the oldest evidence of a crank handle for turning wheels in the West, nearly a thousand years after the Chinese use of such handles	**830** to **839**
			840 to **849**

69

	GENERAL	ASTRONOMY	LIFE SCIENCE
850 to 859		Albategnius (Arabic: Abu-'Abdullah Muhammad ibn Jabir Al-Battani), Arabian astronomer, b Haran (Turkey), 858; he refines Ptolemy's work by making more careful measurements and introduces the use of trigonometry to Arabic astronomy	
860 to 869		Afragamus, (Al Farghani), Egyptian astronomer, d 861	
870 to 879	Al-Farabi (Alfarabicus), Persian scholar who reconciles Aristotle and Islam, b Al Kindi d 873		
880 to 909	Alfred the Great, English monarch, d Winchester, England, Oct 29, 900	Albategnius (Al-Battani) calculates the length of the year; he also determines more accurately than his predecessors the precession of the equinoxes	
910 to 919	The Benedictine abbey at Cluny is founded in 910; it will be the largest of its order in Europe		
920 to 929		Albategnius (Arabic: Abu-'Abdullah Muhammad ibn Jabir Al-Battani), Arabian astronomer, d Samarra (Iraq), 929	
930 to 939			Rhazes d Rhages (Iran)
940 to 949		The Dunhuang star map is produced in China; it uses a Mercator projection, the first known use of this kind of map projection (Mercator's first use is in 1568)	
950 to 959	Al Farabi (Alfarabicus) d in Damascus (Syria)		
960 to 969	Otto is crowned the first Holy Roman emperor in 962		

MATHEMATICS	PHYSICAL SCIENCE	TECHNOLOGY	
Al-Khowarizmi, Arabian mathematician, d		In China, *Classified essentials of the mysterious Tao of the true origin of things,* attributed to Cheng Yin, describes a primitive form of gunpowder; it warns against mixing it because of the danger of burning both the experimenter and the house in which the experiment is taking place	850 to 859
		The *Diamond sutra* is printed in 868, the earliest complete printed book (actually a scroll) extant	860 to 869
Al Kindi d 873			870 to 879
Thabit ibn Qurra, Arabian mathematician, d Baghdad (Iraq), Feb 18, 901	Arab chemists and physicians prepare alcohol by distilling wine Ar-razi, known as Rhazes, is the first to classify chemical substances into mineral, vegetable, animal, and derivative; he further classifies minerals into metals, spirits, salts, and stones; and describes how to make plaster of Paris and metallic antimony	Real paper money—that is, printed paper used as a medium of exchange—is in use in Szechuan Province in China	880 to 909
			910 to 919
			920 to 929
			930 to 939
Abu'l Wefa b; he introduces the tangent ratio to Arabic mathematics and further develops spherical trigonometry, basing his work more on Indian trigonometry than on Greek			940 to 949
		In 954 the emperor Shih Tsung has the largest single piece of cast iron in ancient China made to celebrate his campaign against the Tartars; known as the Great Lion of Tsang-chou, it weighs about 40 tons	950 to 959
The churchman Gerbert, b Aurillac, Auvergne (France) about 945, introduces the abacus and Hindu-Arabic numerals to Europe, although the new method for writing numbers does not catch on; he seems to have been unaware of 0			960 to 969

	GENERAL	ASTRONOMY	LIFE SCIENCE
970 to 979			A hospital in Baghdad is founded in 977; it employs 24 physicians and contains a surgery and a department for eye disorders
980 to 989	Vikings, led by Eric the Red, set up a camp on Greenland in 982, establishing a larger colony three years later Hugh Capet succeeds to the throne of France in 987		
990 to 999	Gerbert becomes Pope Sylvester II in 999		
1000 to 1009	The Vikings, led by Leif Ericson (Leif, son of Eric the Red), reach North America Gerbert (Pope Sylvester II), French scholar, d Rome, May 12, 1003 Dar al-ilm, a science library in Cairo, Egypt, is founded in 1005	A calendar with a year of 360 days divided into 12 months of 27 or 28 days is introduced in India; since this system falls short of an actual year, the Indians add an extra month at regular intervals; it is also possible that they use months of 30 days, still falling short of the actual length of the year The seven-day week is introduced to China by Persians or by merchants from central Asia; before this, the most common Chinese week was ten days A supernova or "guest star" is reported in China, Japan, Europe, and the Arab lands in 1006; it remains visible for several years Ibn Yunus's *The large astronomical tables of al-Hakim* (the "Hakimitic tables," named after Caliph al-Hakim, who built an important observatory completed in 1008) contains accurate astronomical and mathematical tables based on the observations of the past 200 years; the tables are used in later Arab astronomy	The Arabs introduce the lemon plant to Sicily and Spain The *Canon of medicine*, written about this time by Avicenna (Arabic: ibn Sina, b near Bukhara in present-day USSR, 980), is a five-volume treatment of Greek and Arabic medicine that dominates the teaching of medicine in Europe until the seventeenth century
1010 to 1029			
1030 to 1039	Al-Biruni writes *History of India*, giving a general account of India's history based on a selection of sources		Avicenna, Persian physician, d Hamadan (Iran), Jun, 1037

MATHEMATICS	PHYSICAL SCIENCE	TECHNOLOGY	
Mathematician, traveler, and physicist Al-Biruni b 973; his *History of India* becomes the principal introduction to Hindu numeration for the Arabs		Chang Ssu-Hsün invents in 976 the chain drive for use in a mechanical clock	**970 to 979**
		In 984 Ch'iao Wei-Yo, concerned about the amount of stealing taking place as boats are hauled over spillways, invents the lock for canals; previously, boats frequently broke up as they were being hauled; people waiting for just such a moment would dash into the wreckage and steal as much cargo as possible	**980 to 989**
			990 to 999
	The Chinese burn coal for fuel		**1000 to 1009**

The other Omar Khayyám

Most people who speak English know about Omar Khayyám. They think he wrote about "a jug of wine, a loaf of bread, and Thou." In fact, he did not. That was Edward FitzGerald, a nineteenth-century English poet whose translation of Omar was so loose that most scholars consider the FitzGerald poetry as a separate work.

Omar was a great poet. But he is also the Persian mathematician who solved the general cubic equation of the third degree hundreds of years before Tartaglia, the sixteenth-century mathematician generally given credit for this feat (*see* "A great scoundrel," p 107). Omar's method for solving the cubic did have some limitations, however. It was completely geometrical, so it produced only positive roots (there cannot be a line segment with a negative length).

Omar's work was also a step toward the unification of algebra and geometry that came with Descartes and Fermat. Omar pointed out that algebra is not just a collection of tricks for obtaining an answer, but a science deeply related to geometry. Despite this, he believed that it would be impossible to solve the cubic with purely algebraic means. Because of his commitment to geometrical methods, Omar also believed that equations of higher degrees than the third did not describe reality in any way, since space has only three dimensions.

In addition to his mathematical work, Omar also contributed to astronomy. His greatest feat as an astronomer was the reform of the Muslim calendar so that it would keep good time with the heavens.

MATHEMATICS	PHYSICAL SCIENCE	TECHNOLOGY	
	The Arabian physicist Abu 'Ali Al-Hasan ibn Al-Haytham (b Basra in what is now Iraq about 965), known as Alhazen, correctly explains how lenses work and develops parabolic mirrors, like those used in today's reflecting telescopes		**1010 to 1029**
	Alhazen, Arabian physicist, d Cairo, Egypt, 1039	A painting in 1035 shows a spinning wheel in use in China	**1030 to 1039**

	GENERAL	ASTRONOMY
1040 to 1049		
1050 to 1059		The supernova that now forms the Crab nebula is observed in China, Japan, and the Arab lands on Jul 4, 1054; it is visible for 22 months
1060 to 1069	William the Conqueror defeats Harold at the battle of Hastings in 1066, beginning the Norman domination of England	A large comet is sighted in 1066; it is connected in England with the invasion by William of Normandy; today it is known as Halley's comet
1070 to 1089	Alfonso VI of Castile takes Toledo (Spain), an important center of learning, from the Arabs in 1085	Arzachel suggests that planetary orbits are elliptical
1090 to 1099	Sicily (Italy), after almost 200 years of Arabic rule, is conquered by the Normans in 1091 In 1096 Pope Urban II, Peter the Hermit, and Walter the Penniless mount the First Crusade to take Jerusalem from the Arabs	
1100 to 1109	Henry I is crowned king of England in 1100 In 1101 Henry I of England introduces a measure of length equal to the length of his arm—the yard	Chinese astronomers build a stone planisphere of the heavens that correctly demonstrates the cause of solar and lunar eclipses
1110 to 1119	*Questiones naturales* by Adelard of Bath (England, b about 1090), written in 1111, is an early attempt at scientific method; it covers some of the scientific topics learned from the Arabs—meteorology, optics, acoustics, and botany; about this time Adelard also writes *Rules of the abacus* and *Usage of the astrolabium*	

MATHEMATICS	PHYSICAL SCIENCE	TECHNOLOGY	
Mathematician, traveler, and physicist Al-Biruni d 1048		In China Tseng Kung-Liang publishes the first recipes for three varieties of gunpowder	**1040 to 1049**
		Sometime between 1041 and 1048 Pi Sheng, an obscure commoner, invents movable type	
The decimal system is introduced by the Arabs into Spain		Some Chinese books are printed with movable type	**1050 to 1059**
			1060 to 1069
	Shen Kua's *Dream pool essays* of 1086 outlines the principles of erosion, uplift, and sedimentation that are the foundation of earth science	Tseng Kung-Liang's *Wu ching tsung yao* (Compendium of important military techniques) in 1084 describes magnetized iron "fish" that float in water and can be used for finding south; about this time the Chinese begin to use the compass for navigation, most likely using the iron "fish"	**1070 to 1089**
		The Domesday Book in 1086 lists 5624 waterwheel-driven mills in England south of Trent, or about one mill for each 400 persons	
		In 1086 Chinese scientist Shen Kua's *Dream pool essays* contains the first known reference to the use of a magnetic compass for navigation	
		In China, Su Sung builds a giant water clock and mechanical armillary sphere in 1092; it is considered the finest mechanical achievement of the time	**1090 to 1099**
Jia Xien states the method of forming the Pascal triangle; the triangle was probably known before his time	Alcohol is distilled at Salerno (Italy)	Italians learn to distill wine to make brandy	**1100 to 1109**
The Persian mathematician and poet Omar Khayyàm, b Nisipar (Iran), May 15, 1048, becomes the first to solve some types of equations with a degree of three, also known as cubics	Abu 'l Fath Al-Chuzini writes a book on the history of science; it includes tables of specific densities and a general formulation of the laws of gravity	The Chinese build in 1105 a pagoda 28 m (78 ft) high from cast iron, each level cast as a single piece	
		In 1107 the Chinese invent multicolor printing, mainly to make paper money harder to counterfeit	
		In 1117 Chu Yü's *P'ingchow table talk* contains the first mention in Chinese literature of a compass used for navigation at sea	**1110 to 1119**

	GENERAL	ASTRONOMY	LIFE SCIENCE
1120 to 1129	*Sic et non* (Yes and no) by Peter Abelard (b Le Pallet, Brittany—now part of France—1079), written in 1122, is a collection of contradictory statements found in the writings of various authorities; it demonstrates that authority is of no use and that reason and logic are needed to solve problems Arabian philosopher Averroës (Arabic: Abu-Al-Walid Muhammad ibn Ahmad ibn Rushd), b Cordoba (Spain), 1126; his lengthy commentaries on the works of Aristotle are to be a major influence in Europe in the Middle Ages	Adelard of Bath translates *Astronomical tables* by Al-Khowarizmi from the Arabic in 1126; about this time he also translates Al-Khowarizmi's *Liber ysagogarum alchorismi,* a work about arithmetic	Stephen of Antioch translates Haly Abbas's *Liber regalis,* a medical encyclopedia, in 1127
1130 to 1139	Jewish philosopher and physician Maimonides (Hebrew: Moses ben Maimon) b Cordoba (Spain), Mar 30, 1135; his *Guide for the perplexed* attempts to reconcile Aristotle's ideas with the Old Testament	Omar Khayyám, Persian poet, astronomer, and mathematician, d Nishapur (Iran), Dec 4, 1131	
1140 to 1149	Peter Abelard, French scholar, d Chalon-sur-Saône, Apr 21, 1142		The Norman king Roger II decrees in 1140 that only physicians with a license from the government may practice medicine
1150 to 1159	Adelard of Bath, English scholar, d Solomon Jarchus publishes an almanac The University of Bologna (Italy) is founded in 1158		
1160 to 1169	Oxford University is founded in England in 1167 or 1168, although there may have been lectures at Oxford as early as 1133 Scholar Robert Grosseteste b Stradbroke, Suffolk, England; the teacher of Roger Bacon, he arranges to have Aristotle translated from the original Greek and also experiments with mirrors and lenses in an effort to understand light and such phenomena as the rainbow The University of Paris (France) is founded as a definite entity before 1170, although its seeds go back to the early 1100s		

MATHEMATICS	PHYSICAL SCIENCE	TECHNOLOGY	
		In 1129 Abbé Suger starts construction on the abbey church of St. Denis, the first Gothic church with flying buttresses	**1120 to 1129**
			1130 to 1139
Adelard of Bath translates Euclid's *Elements* (15 books, of which 13 are now believed to be genuine) from the Arabic in 1142 Robert of Chester translates Al-Khowarizmi's *Algebra* from Arabic into Latin in 1145			**1140 to 1149**
Siddhanta siromani (Head jewel of an astronomical system) by Bhaskara (also known as Bhaskara II, b 1114) summarizes the arithmetic and algebraic knowledge in India of the time, focusing on solving diophantine equations	Bhaskara (also known as Bhaskara II) describes a wheel that he claims will run indefinitely, one of the first descriptions of a supposed perpetual motion machine Eugenius of Palermo translates Ptolemy's *Optics* from Arabic into Latin in 1154 Henricus Aristippus translates Aristotle's *Meteorologica* from Greek into Latin in 1156	The Chinese develop the first rockets A map of western China is printed in China in 1155; this is the oldest known printed map	**1150 to 1159**
			1160 to 1169

	GENERAL	ASTRONOMY	LIFE SCIENCE
1170 to 1179	Thomas à Becket is murdered	Gerard of Cremona (Italy, b about 1114) translates the *Almagest* from the Arabic into Latin in 1175; around this time he also translates from Arabic works by al-Kindi, Thabit ibn Qurra, Rhazes, al-Farabi, pseudo-Aristotle, Avicenna, Hippocrates, Aristotle, Euclid, Archimedes, Diocles, and Alexander of Aphrodisias	
1180 to 1189	Gerard of Cremona, Italian scholar and translator of Ptolemy's *Almagest*, along with many other Greek works, d Toledo (Spain), 1187	A supernova is reported in China and Japan in 1181; it remains visible for 183 days	Burgundio of Pisa translates various treatises by Galen from Greek into Latin
1190 to 1199	Frederick II, German emperor, b Iesi (Italy), Dec 26, 1194

Averroës (Arabic: Abu-Al-Walid Muhammad ibn Ahmad ibn Rushd), Arabian philosopher, d Marrakesh, (Morocco), Dec 10, 1198 | | |
| **1200 to 1209** | The Crusaders sack Constantinople in 1204

Maimonides (Hebrew: Moses ben Maimon), Jewish philosopher, d Cairo, Egypt, Dec 13, 1204 | | |

Perpetual motion

The idea of building a machine that can deliver power indefinitely without the need of wind, water, or muscle power has existed for a long time. The first description of such a device can be found in a Sanskrit manuscript from the fifth century. Around 1150 the Indian mathematician Bhaskara (Bhaskara II) described a similar wheel that was supposed to run forever.

Another machine, proposed by the French architect Villard de Honnecourt around 1235, used wheels on which weights would move in such a fashion that one side of the wheel would contain more weights than the other; the imbalance was supposed to cause the wheel to turn. As water mills and windmills became common, a machine that would deliver power without the need of a river or wind began to interest engineers. Several concepts, such as a waterwheel connected to a pump that runs the wheel, or a windmill that actuates giant bellows that drive the windmill, were put forward.

The first to formulate precisely why such perpetual-motion machines could not work was Gottfried Wilhelm Leibniz, who in his *Essay on dynamics* stated that energy could not be created out of nothing. Although Leibniz's idea was not paid much attention, by the end of the eighteenth century most scientists had concluded that perpetual-motion machines could not work. The Paris Academy decided to reject any proposals for perpetual-motion machines. Notwithstanding, inventors continue to propose perpetual-motion machines right up to present.

	GENERAL	ASTRONOMY	LIFE SCIENCE
1210 to 1219	The teaching of Aristotle's works is forbidden at the University of Paris in 1210 as a threat to Christianity

King John is forced by English barons to sign the Magna Carta in 1215; among other provisions, a unified system of measures is introduced

Alexander Neckam, English scholar, d Kempsey, Worcestershire, England, 1217 | Michael Scot translates *Liber astronomiae* (Book of astronomy) by Abu Ishaq al-Bitruji al-Ishbilt (Alpetragius) in 1217, introducing into Europe the Aristotelian model of astronomy; about this time he also translates works by Averroës and Aristotle | |
| **1220 to 1229** | Alfonso X of Castile, Spanish monarch, b Burgos (Spain), Nov 23, 1221

The University of Padua (Italy) is founded in 1222 | | Physician Taddeo Alderotti b Florence (Italy), 1223; he will bridge the gap between Greek and European medicine, urging other physicians to read Galen, Hippocrates, and Avicenna |

MATHEMATICS	PHYSICAL SCIENCE	TECHNOLOGY	
			1170 to **1179**
Bhaskara d	The oldest Western evidence for the use of a rudder is in carvings made about this time The Marco Polo Bridge across the Yung-ting River is built in 1189; 213 m (700 ft) long, the bridge is still heavily used by bus and truck traffic A paper mill is established in Herault, France, in 1189; it is probably the first paper mill in Europe		**1180** to **1189**
	De naturis rerum (On natural things) by Alexander Neckam (b St. Albans, England, Sep 8, 1157) contains the first known Western reference to the magnetic compass The oldest known depiction of a fishing reel is made in China		**1190** to **1199**
The Chinese devise a sign for zero In 1202 *Liber abaci* (Book of the abacus) by Leonardo Fibonacci (b Pisa Italy about 1180) introduces 0 to Europe and, in a problem about the reproduction of rabbits, the sequence of numbers known as the Fibonacci sequence: 1, 1, 2, 3, 5, 8, 13, 21, . . .; each new number is found by adding the two preceding numbers in the sequence			**1200** to **1209**
			1210 to **1219**
The *Arithmetica* of Jordanus Nemorarius uses letters as variables instead of generalizing from specific numerical cases; his other works include *Algoriismus demonstratus* (Algorithm demonstrated) and *De*	Jordanus Nemorarius' *Mechanica* contains a law of the lever and the law of composition of movements; his *Elementa Jordani super demonstrationem ponderis* (Elements for the demonstration of weights) contains an	A window at Chartres Cathedral shows the earliest known Western wheelbarrow	**1220** to **1229**

	GENERAL	ASTRONOMY	LIFE SCIENCE
1220 to 1229 cont	The University of Naples (Italy) is founded in 1224 Thomas Aquinas, theologian, b Roccasecca (Italy) The University of Toulouse (France) is founded in 1229		
1230 to 1239	Cambridge University in England is founded in 1231		
1240 to 1249	The University of Rome (Italy) is founded in 1244 Giovanni de Carpini, commissioned by Pope Innocent IV in 1245, explores south Russia and reaches the Mongol empire at Karakorum (Mongolia)		
1250 to 1259	Frederick II, German emperor, d Lucera (Italy), Dec 13, 1250 The Sorbonne University in Paris (France) is founded in 1253 Commissioned by Louis IX, Willhelm von Rubruck explores the road to Karakorum (Mongolia)	In 1250 Alfonso X of Castile orders the compilation of astronomical tables, the *Alfonsine tables:* they will be printed in 1483 In 1259 Nasir al-Din al Tusi starts construction of the observatory at Maragha (Iran), where Zij-i ilkhani's observations will lead to the completion of his astronomical tables in 1272	*On animals* by Albertus Magnus, b Lauingen (Germany), 1193, describes his observations and dissections of a number of animals and insects Pietro D'Abano, Italian physician, b near Padua, 1257; his *Conciliator* attempts to reconcile Greek and Arabic ideas about medicine
1260 to 1269	Bartholomew of Messina translates *Problemata* by pseudo-Aristotle from Greek into Latin Nicolo and Maffeo Polo travel to China in 1260 and return in 1269 William of Moerbeke starts the translation of the works of Aristotle from Greek into Latin; he also translates Hippocrates, Hero of Alexandria, Alexander of Aphrodisias, and Simplicius Simon de Monfort summons the "first" Parliament in England in 1265; previous groups, also called parliaments, consisted only of lords; government by Parliament was far in the future, although some governmental rights were given Parliament in 1340 Johannes Duns Scotus, Scottish theologian and philosopher, b Berwick, Scotland; he opposes many of the ideas of Thomas Aquinas and is thought to have influenced the thinking of such English scholars as Roger Bacon and William of Ockham		

MATHEMATICS	PHYSICAL SCIENCE	TECHNOLOGY	
numeris datis (On given numbers), collections of rules for solving problems Fibonacci's *Liber quadratorum* (1225) deals with diophantine equations of the second degree	early form of the principle of virtual displacements applied to the lever Alchemist Arnold of Villanova b near Valencia (Spain); his contributions to science include the near discovery of carbon monoxide and the first preparation of pure alcohol	In 1221 the Chinese use bombs that produce shrapnel and cause considerable damage; previous uses of gunpowder relied more on the frightening power of loud explosions	**1220 to 1229 cont**
		During a Mongol siege in 1232, the Chinese use kites to send messages behind enemy lines	**1230 to 1239**
In 1245 Ch'in Chiu-Shao's *Mathematical treatise in nine sections* treats equations of a degree higher than the third			**1240 to 1249**
Leonardo Fibonacci (Leonardo of Pisa), Italian mathematician, d The decimal system is introduced into England in 1253 by John of Halifax (b about 1200), better known as Sacrobosco	Robert Grosseteste, English scholar and optical experimenter, d Buckden, Huntingdonshire, England, Oct 9, 1253	The goose feather (quill) is used for writing	**1250 to 1259**
Johannes Campanus's translation of Euclid includes a method for the trisection of an angle that is probably owed to Jordanus Nemorarius; in any case, it produces only an approximation	In 1269 *Epistola Petri Peregrini de Maricourt ad Syerum de Foucaucourt, militem, de magnete* (Letter on the magnet of Peter the Pilgrim of Maricourt to Sygerus of Foucaucourt, Soldier), by Petrus de Maricourt (Petrus Peregrinus, b about 1240), is the first Western account of the forces between the poles of a magnet and of a compass dial	Roger Bacon (b Ilchester, Somerset, England, about 1220) discusses in *Opus majus*, written in 1267 and 1268 but not published until 1733, spectacles for the farsighted; he is also the first European to mention gunpowder	**1260 to 1269**

1270 to 1279

The Crusades effectively come to an end in 1271, although attempts to retake the Holy Land persist for several centuries

Marco Polo, b Venice (Italy) about 1254, begins his great voyage to the Far East in 1271, reaching Japan; he returns in 1295

Thomas Aquinas, Italian theologian, d Fossanuova (Italy), Mar 7, 1274

Roger Bacon is imprisoned for heresy in 1277

Kuo Shou-Ching builds the "simplified instrument," the first known astronomical device to use an equatorial mounting; it is a torquetum, an Arab improvement on the armillary sphere

The *Alfonsine tables,* a compilation of planetary positions that will be used extensively during the following three centuries, are completed in 1272

In 1276 Chinese astronomer Zhou Kung sets up a 12-meter (40 ft) gnomon for measuring the sun's shadow

Mondino de'Luzzi, Italian anatomist, b Bologna, 1275

De formatione corporis in utero (On the formation of the body in the uterus) by Giles of Rome (b about 1245) is a treatise on generation in the Aristotelian tradition; it discusses the contributions of both parents to procreation

William of Moerbeke translates various treatises by Galen from the Greek

Moses Farachi translates Rhazes's *Liber continens,* a medical encyclopedia, from Arabic into Latin in 1279 on the orders of Charles of Anjou, king of Sicily

1280 to 1289

Albertus Magnus, German scholar, d Cologne (Germany), Nov 15, 1280

Alfonso X of Castile, Spanish monarch, d Seville (Spain), Apr 24, 1284

In 1281 Qutb al-Din al Shirazi's *Nihayat al-idrak fi dirayat al-aflak* contains an alternative planetary model to that of Ptolemy, making more use of uniform circular motions

1290 to 1299

Roger Bacon, English scholar, d Oxford, England, Jun 11, 1292

William of Saint-Cloud determines the angle of the ecliptic from the sun's position at the solstice; he finds 23 arc degrees and 34 arc minutes for the value of this angle, only two arc minutes off from the presently accepted value

Taddeo Alderotti, Italian physician, d Bologna

1300 to 1309

The English statute acre is defined in 1305 as 4840 sq yds

Dante starts writing *La commedia* (Divine comedy) in 1307

Johannes Duns Scotus, Scottish theologian and philosopher, d Cologne (Germany), Nov 8, 1308

Eyeglasses become common

1310 to 1319

Pietro D'Abano, Italian physician, d Padua

Mondino de'Luzzi's *Anatomia* (Anatomy), written in 1316, is the first work devoted to human anatomy and the art of dissection to be written in the West; it is based on the dissection of corpses

MATHEMATICS	PHYSICAL SCIENCE	TECHNOLOGY	
William of Moerbeke completes his translation of almost the complete works of Archimedes from Greek into Latin	Witelo's *Perspectiva* (Perspectives, 1270), a treatise on optics, deals with refraction, reflection, and geometrical optics; Witelo rejects the notion that rays are emitted from the eye	A paper mill operates in Montefano (Italy) in 1276	**1270 to 1279**
		The earliest known Western reference to a spinning wheel is made in the statutes of a guild in Speyer (Germany) The first known gun is made in China in 1288; a small cannon, it probably has predecessors that go back 10 years or more	**1280 to 1289**
In 1299 Florence (Italy) bans the use of Arabic numerals	Marco Polo's 1298 book of his travels correctly describes such substances as coal and asbestos for the first time in Europe	South Americans build cable bridges across deep canyons in the Andes In 1295 the Castle of Beaumaris on the shores of Menai Straits is the last of the Welsh castles to be started; it will be completed in 1320	**1290 to 1299**
The first record of Pascal's Triangle is published in 1303 by the great Chinese algebraist Chu Shih-Chieh; it is believed that he obtained the information from the Arabs, who discovered the triangle of number relationships in the eleventh century	Alum is discovered at Rocca, Syria The False Geber (b about 1270), writing under the name of the alchemist Geber of five centuries earlier, describes sulfuric acid In 1304 Theodoric of Freibourg (Germany, b about 1250) takes the suggestion of the master general of his Dominican order that he investigate the rainbow; this leads to *De iride* (On the rainbow), in which Theodoric reports on his experiments with globes of water that correctly explain many aspects of rainbow formation		**1300 to 1309**
	Theodoric of Freibourg (Germany), early experimenter, d Arnold of Villanova, Spanish alchemist, d near Genoa (Italy), Sep 6, 1311	The first mechanical clocks appear in Europe, apparently in response to stories about the existence of mechanical clocks in China; the European clocks are driven by a weight whose descent is controlled by an escapement; the Chinese clocks also have escapements, but still use water as the power source	**1310 to 1319**

1310 to 1319 cont

1320 to 1329

Marco Polo, Italian explorer, d Venice (Italy), Jan 9, 1324

The Aztecs found Tenochtitlán, where they see an eagle sitting on a cactus with a snake in its beak, an omen they had been seeking; later the Spanish rebuild the city as Mexico, the present capital of the nation of Mexico

In 1326 Ibn Battuta from Tangier begins his exploration of India, Ceylon, and the Far East, returning in 1350; he becomes the most traveled person of his time

Henri de Mondeville's *Chirurgia* (Surgery) advocates sutures and cleansing of wounds

Mondino de'Luzzi, Italian anatomist, d Bologna, 1326

1330 to 1339

Summa totius logicae by William of Ockham (England, b about 1280) introduces the notion that when several explanations of a phenomenon are offered, the simplest must be taken; often expressed as "entities must not be needlessly multiplied," this concept, called Ockham's razor, has become one of the foundations of science

England and France begin the Hundred Years' War in 1338

A botanical garden is founded in Venice (Italy) in 1333; it is probably the first in Europe since antiquity

---

1340 to 1349

In 1346 or 1347 Italian ships bring rats carrying fleas infected with the Black Plague to Europe; the epidemic kills 25 million people, a third of the population, before 1351; during the next 80 years it recurs over and over, at least once every eight years, and kills three-fourths of the population of Europe

Boccaccio starts to write the *Decameron* in 1348

William of Ockham, English scholar, d Munich (Germany), 1349

Impetus

In antiquity, Greek scholars wondered what keeps the stars, which they viewed as being on a crystal sphere, in motion. Aristotle did not believe that motion without an agent of motion is possible and suggested that some agent must keep rotating the sphere that contains the fixed stars.

Many of Aristotle's ideas about motion were generally accepted until the seventeenth century. However, in the sixth century AD, the philosopher Johannas Philoponus had already rejected Aristotle's idea that a body would only move as long it is pushed. Philoponus believed that once a body is set in motion, it will keep moving in the absence of friction because it contains a certain quantity of motion. The sphere of the stars can move because nothing opposes it.

In the thirteenth century, Thomas Aquinas believed that it is God that keeps our sphere in rotation; the fact that the stars move was for him a proof of the existence of God. This view was made official in the condemnation of 1277, a papal decree intended to suppress speculation that might contradict the teachings of the church.

Nevertheless, the ideas of Philoponus were adopted by some scholars in the fourteenth century. William of Ockham and Jean Buridan called this quantity of motion *impetus*. These scholars avoided the problem of the condemnation of 1277 by invoking God as the originator of impetus and by saying that God could change motions if He pleased.

Galileo added experimentation to medieval speculation and showed that something like the impetus idea must be true. Galileo came very close to the modern idea of inertia. Finally, Newton replaced impetus with inertia. The main difference between impetus and inertia is that inertia applies both to bodies at rest and bodies in motion. Before Newton, none of the philosophers had thought that the same concept could describe stationary and moving bodies.

1350 to 1359

	The philosopher Raymond Lully of Majorca, b Palma (Spain), about 1236, the discoverer of ammonia gas, is stoned to death in 1315 by the inhabitants of Bougie in Africa for preaching against Islam In 1317 Pope John XXII issues a prohibition against the practice of alchemy	**1310** to **1319** cont
In 1321 Levi ben Gerson uses mathematical induction to establish formulas for the number of permutations of *n* objects and combinations of *n* objects taken *r* at a time In 1328 Thomas Brandwardine's *Tractatus de proportionibus* develops a theory of proportions that he uses to improve on Aristotle's laws of motion, although Brandwardine's laws also are not correct		China's Grand Canal, 1770 km (1100 mi) long, is completed in 1327; built over many centuries, starting in 70 AD, it connects Beijing to many parts of northern China and to the Yangtze	**1320** to **1329**
The University of Paris decrees in 1336 that no student can graduate without attending lectures on "some mathematical books"		**1330** to **1339**
		The first blast furnace is developed in or around Liège (Belgium) The first evidence of a gun in Europe, a drawing of a small cannon that fires arrows, dates from 1347	**1340** to **1349**
	Jean Buridan (b Béthune, France, about 1295) develops the idea of impetus, a concept close to the notion of inertia; he rejects the idea that God or angels propel the celestial bodies on their orbits continuously and he asserts that an initial impetus is sufficient to explain their motion Jean Buridan, French philosopher, d Paris	The mechanical clock at Strasbourg Cathedral is built in 1354	**1350** to **1359**

	GENERAL	ASTRONOMY	LIFE SCIENCE
1360 to 1369	In China, the Mongols are overthrown in 1368 and the Ming dynasty established		Guy de Chauliac's *Chirurgia magna* describes how to treat fractures and hernias
1370 to 1379	Italian poet Petrarch d 1374	Ibn ash-Shatir, who produced non-Ptolemaic theories of planetary motions, d	The first quarantine station is set up in the port of Ragusa (Dubrovnik, Yugoslavia) in 1377; those suspected of plague have to stay there for 40 days
1380 to 1389	Heidelberg University is founded in 1386, the oldest continuously existing university in Germany Geoffrey Chaucer starts writing *The Canterbury tales* in 1387		
1390 to 1399		Geoffrey Chaucer's *A treatise on the astrolabe* in 1391 shows how to construct an astrolabe and how to use it to compute the position of a star	Physician and mapmaker Paolo Toscanelli b Florence (Italy), 1397; it was his map, incorrectly showing Asia only 4830 km (3000 mi) west of Europe, that led Columbus to make his voyages of discovery
1400 to 1409	*Yung lo ta tien*, a Chinese encyclopedia in 22,937 volumes, is produced in 1403 in three copies John Bessarion, Greek scholar, b Trebizond (Turkey), Jan 2, 1403; he collects ancient Greek works and translates Aristotle into Latin Leipzig University (Germany) is founded by German refugees from Prague in 1409 St. Andrews University in Edinburgh, Scotland is founded in 1409	The Chinese know that the length of the solar year is about 365.25 days	
1410 to 1419	European expansion starts with the Portuguese capture of Ceuta (across from Gibraltar) in 1415, which Prince Henry the Navigator (b Oporto, Portugal, Mar 4, 1394) uses it as a base for further African operations	Hasdai ben Abraham Crescas's *Or Adonai* (The light of the Lord) contains a refutation of Aristotle's argument of the absence of other worlds	Benedetto Rinio's *Liber de simplicibus* describes and illustrates 440 plants that have medicinal uses Influenza is recorded for the first time in Paris in 1414

Nicolas Oresme's *Tractatus de figuratione potentiarum et ensurarum*, written about this time, describes his "latitude of forms," a precursor of analytic geometry, calculus, and even the fourth dimension; he also introduces rational and irrational powers in another work composed about this time

| | 1360 to 1369 |

Understanding fossils (part one)

From earliest times people must have seen fossils—remnants of organisms of the past preserved in various ways—but the first reports we have on the subject are from the ancient Greeks. Xenophanes of the early Ionian school is said to have noticed fossilized sea creatures high on mountains; he correctly interpreted this as meaning that these mountains had once been under water. Later, Herodotus reached the same conclusion regarding fossilized clam shells on mountains, but misinterpreted other fossils. For example, he equated the fossilized bones of large creatures with mythical animals or with giant humans. Theophrastus, Aristotle's successor at the Lyceum, is said to have written a book on fossils, but it is lost, although in one of his works that has been preserved he refers to fossil fishes. It appears that Theophrastus thought that the fossil fishes had become dried out by being trapped in sand. Later, Suetonius casually mentions that the emperor Augustus kept a collection of large fossil bones at his villa. These scattered references suggest that classical observers recognized that fossils were remains of organisms. At the same time they suspected nothing of the ancient origins of the fossils.

The Arab scholar Avicenna put forward another idea that was to confuse people about fossils for centuries. He argued against the concepts of alchemy, especially against the idea that one metal could be turned into another. But he extended the argument to cover bone turning into rock. Most ancient fossils are petrified: the original living tissue has been replaced by rock. Avicenna was right in saying that bone does not turn into rock.

The word *fossil* was coined by Georgius Agricola in the sixteenth century to mean anything dug up. Although Agricola was writing primarily about mining metals, he also described what we would still call fossils today. These were all fossil sea creatures, but Agricola lived most of his life far from the sea. He thought that the fossils were just stones that occur in particular shapes, rather as crystals do.

Writing about ten years later than Agricola, Konrad von Gesner picked up Agricola's new word and classified fossils into 15 different types. Although Gesner was acquainted with creatures of the sea, he failed to differentiate between real fossils and other "shaped rocks."

Ten years later, Bernard Palissy began to lecture on the natural world. He used fossils as props for his talks. Unlike Agricola and Gesner, Palissy recognized that fossils are remains of living organisms. Throughout the sixteenth century, scholars, including Jerome Cardan, Andrea Cesalpino, and Gabriel Fallipio, debated the nature of fossils.

In the seventeenth century, Nicolaus Steno and Robert Hooke argued persuasively that fossils were remains of living organisms, with Steno correctly identifying the origin of many common fossils, such as sharks' teeth. Of course, some fossils were of extinct creatures, but not even Steno realized this. Apparently Athanasius Kircher was the first to suggest the possibility of extinct species when he proposed that some animals failed to make it onto Noah's Ark.

The steel crossbow is introduced as a weapon of war

The first record of a canal lock is made in the West in 1373; such locks had been known in China for almost 400 years

| | 1370 to 1379 |

Rockets are used for the first time in Europe in the battle of Chioggia in 1380 between the Genoese and the Venetians

Cast iron becomes generally available in Europe

| | 1380 to 1389 |

A paper mill is established in Nuremberg (Germany) in 1391

| | 1390 to 1399 |

Oil is used as a base for paints

Coffee, which grows wild in Ethiopia, is made into a beverage there

The dulcimer is mentioned for the first time

Printer and moneylender Johann Fust, who later acquires Gutenberg's presses in repayment of debt, b; Fust becomes the first European to print in more than a single color and introduces type for Greek characters

In 1405 Konrad Kyeser's *Bellifortis* discusses military technology

A windmill is used in Holland in 1408 to carry water from inland areas out to sea

| | 1400 to 1409 |

Dutch fishers become the first to use drift nets

| | 1410 to 1419 |

	GENERAL	ASTRONOMY	LIFE SCIENCE
1420 to 1429	Louvain University (Belgium) is founded in 1426 Jeanne D'Arc (Joan of Arc) leads French armies against the English in 1428	Ulugh Beg (b Soltaniyeh (Iran), Mar 22, 1394), as Mongol astronomer Muhammad Taragay came to be known, builds an observatory at Samarkand (USSR)	
1430 to 1439	Joan of Arc is burned at the stake in Rouen, France, in 1431 Pachacutec founds the Inca empire in 1438	Ulugh Beg publishes new tables of star positions and a new star map; these are improvements on those of Ptolemy and Hipparchus	
1440 to 1449		Nicholas of Cusa's *De docta ignorantia* (On learned ignorance) contains the idea of an infinite universe and proposes that all heavenly bodies are essentially alike and that Earth revolves about the sun Ulugh Beg, Mongol astronomer, d Samarkand (USSR), Oct 27, 1449, assassinated by his son	
1450 to 1452			The first professional association for midwives is founded in 1452 in Regensburg (Germany)

The last alchemists

Alchemy was not merely misguided chemistry; it made many contributions to chemistry. Alchemy was, however, a magical or mystical way of looking at the world. Alchemy seems to have started with Taoists in China and the Pythagoreans in Greece, sometime after the sixth century BC. As the ideas of alchemy developed and moved westward, Taoist ideas about chemicals were combined with Pythagorean number mysticism. Another alchemical tradition came from the Egyptian embalmers, such as Zosimus, who wrote one of the first summaries of alchemical ideas extant.

In China, the early alchemists were searching for what came to be called the elixir of life, a way to provide immortality. Some of their accomplishments were remarkable, such as the embalming of the Lady of Tai. This woman was buried about 186 BC in a double coffin filled with a brown liquid containing mercuric sulfide and pressurized methane. Under these conditions there was no observable deterioration of her flesh when she was exhumed after more than 2000 years. She appeared more like someone who had died only a week or so before.

Chinese alchemy passed on to the Indians, who were more interested in using alchemical ideas to cure diseases. Eventually the Arabs put together the ideas from the East with the Alexandrian tradition of alchemy that had descended from the Pythagoreans. In this form of alchemy, astrological influences were important. Chemical reactions occurred because of the influences of the planets. Numerology and even the shapes of the vessels also helped determine reactions. In this tradition, the elixir of life became mingled with the concept of a philosopher's stone, an object whose presence would enable one to transmute other metals into gold. To some

MATHEMATICS	PHYSICAL SCIENCE	TECHNOLOGY	
Mathematician and astronomer Georg von Peurbach b Peurbach, near Vienna (Austria), May 30, 1423; he prepares tables of sines in Hindu-Arabic numerals and comments on Ptolemy's *Almagest*		Printer Peter Schoeffer b; working with Johann Fust, he is a pioneer in printing with colored inks and in the introduction of Greek characters	1420 to 1429
In 1434 a book on drawing by Leone Battista Alberti, b Genoa (Italy), Feb 18, 1404, includes the laws of perspective, a forerunner of projective geometry		The earliest representation of a drive belt in Europe shows it being used to turn a grindstone

Painter, sculptor, and engineer Andrea del Verrocchio b Italy, 1435 | 1430 to 1439 |
| | | Johann Gutenberg, b Mainz (Germany) about 1398, and Lauren Janszoon Koster invent printing with movable type, probably independently of the Chinese; however, the inventions of paper and printing with blocks, essential to printing with movable type, had both most likely been learned as a result of diffusion to Europe from China | 1440 to 1449 |
| The University of Paris requires in 1452 that students must have read the first six books of Euclid to get a master's degree | | Nicholas Krebs, known as Nicholas of Cusa (b Kues, Germany, 1401), constructs spectacles for the nearsighted | 1450 to 1452 |

The last alchemists, continued

degree, the mingling of the elixir with the philosopher's stone was due to Geber, who was the first to use the term "elixir." Geber's writings from the eighth century AD so dominated alchemy that a talented chemist of about 400 years later is known as "the False Geber," since, like many other alchemists, he signed all his works with the name of the master.

Despite mystical theories of matter, the alchemists managed to develop various practical tools, including the first strong acids and the distillation of alcohol.

During the Renaissance, the West absorbed Arabic alchemy along with more substantial Arabic science. By the sixteenth century, alchemy was being practiced mainly in Europe. Paracelsus was one of the alchemists who was also a successful physician and scientist. Some of his achievements include the first known description of zinc, the recognition that coal mining causes lung disease, and the use of opium to deaden pain. Paracelsus proclaimed that he had found the philosopher's stone and would live forever. Unfortunately, he drank a lot and died before he was 50 in a fall some attribute to drunkenness.

Libavius was a follower of Paracelsus who also was a successful scientist despite his belief in alchemy. His 1597 book *Alchemia* is really the first good book on chemistry. The tradition of alchemy persisted well into the eighteenth century. Newton spent much of his later life trying to find the philosopher's stone, and may have gone mad from mercury poisoning caused during his experiments. Finally, Lavoisier, in the later part of the eighteenth century, put together a scientific view of chemistry that effectively wiped out the alchemical tradition that had persisted for 2000 years.

OVERVIEW

The Renaissance and the Scientific Revolution

One way to date the beginning of the Renaissance is from May 29, 1453, the day the Turks captured the city of Constantinople and many Greek-speaking scholars escaped to the West. The scholars brought with them classical manuscripts in Greek along with the ability to translate the ancient writings into Latin, the common language of learning in Europe at the time. However, the roots of the Renaissance go back further. After the depletion of the population of Europe in the fourteenth century from the Black Death, cities and towns rebounded with vigor. A population that is suddenly much smaller must find new ways to function. Mechanical devices and more trade make up to some extent for missing people.

There was much change going on around 1453 for other reasons as well. Movable type was reinvented in Europe around 1440 and the Gutenberg 42-line Bible was printed just a year after the fall of Constantinople. The Moors were driven from Spain the year that Columbus reached America, 1492. In 1498 Vasco da Gama reached India by sailing around the Cape of Good Hope. And Luther nailed his 95 Theses to the church door in Wittenberg in 1517, starting the Protestant Reformation.

But if it barely possible to date the beginning of the Renaissance, it is almost impossible to say when the Renaissance ended. This is especially true for science. Gradually, scientists of the Renaissance began to perform experiments more frequently. The era of the Scientific Revolution that starts in the later Renaissance also has no specific ending. We conclude this chapter just before the founding of England's Royal Society, which inaugurates a new and better organized era in science dominated by Newton and Leibniz, an era in which the Scientific Revolution is completed.

The early Renaissance

The fifteenth century was primarily a time of the absorption of classical learning and the adoption of Arabic mathematics. It was much more a period of change in arts and letters than in science. When one thinks of the Renaissance, one thinks of Leonardo da Vinci and Michelangelo, not Regiomontanus, Paracelsus, and Tartaglia. Alchemy and astrology were more important than chemistry and astronomy during this time (see "The last alchemists," p 88). Physics had to absorb Aristotle's work before it could develop the modern approach to physical motion (see "Replacing Aristotle's physics," p 118).

Another major influence that had to be absorbed is often overlooked. The explorers of both the New World and the East discovered a wealth of previously unknown plants and animals. Although the short-term effect was a sudden change in the diets of the peoples of Europe and Asia, the long-term importance was that these discoveries led to the classification schemes of John Ray and Carolus Linnaeus.

In the continuation of a trend that predates the Renaissance, artists, especially painters, made significant contributions to science. The development of perspective drawing and the dissection of human bodies were valuable to both art and science. The artist Albrecht Dürer, for example, wrote papers on both geometry and anatomy. The most notable example of the artist-scientist during this period was Leonardo da Vinci, a remarkable inventor and engineer as well as an artist skilled in human anatomy. Even Michelangelo functioned as an engineer in his design for the dome of St. Peter's and other buildings, as well as being in charge for a time of the fortification of Florence (Italy). From these people and others like them the term *Renaissance man* came to mean one skilled in many fields.

The later Renaissance

The Renaissance, which began in Italy, reached northern Europe in the sixteenth century. There it joined with the new values of the Reformation. With the Renaissance came economic transformation: markets were opened on a world scale and trade boomed in the cities. By 1603 the English philosopher and scientist Francis Bacon had characterized the new century as an age distinguished by "the opening of the world by navigation and commerce, and the further discovery of knowledge." Bacon was among the first to view science as a powerful tool to conquer nature. In his fable *New Atlantis* he wrote that science's aim is "the enlarging of the bounds of human empire; to the effecting of all things possible."

The Scientific Revolution

Publication of Copernicus's heliocentric theory and Vesalius's anatomy in 1543 is a good beginning point for the Scientific Revolution (*see*, "1543: A remarkable year in publishing," p 108). Throughout the West, modern science began to take shape in many ways. The first scientific societies, such as the *Academia Secretorum Naturae* (Academy of the Secrets of Nature) and the *Accademia dei Lincei*, were formed in Italy during this period. The former society was founded in 1560 but was soon suppressed by the Inquisition. The *Accademia dei Lincei* (the exact meaning of *Lincei* is not known; the academy's symbol is a lynx), founded in 1603, is the oldest existing scientific society.

Both the Catholic church and the Protestant church of Martin Luther opposed the views of Copernicus, but the Catholic church persisted in its opposition and reaped most of the blame for unscientific attitudes. When the Inquisition burned Giordano Bruno in 1600 for his mystical heresies, it also appeared to condemn his Copernican views. By 1616, Copernicus's *De revolutionibus* was on the Catholic church's Index of prohibited books, and Galileo was warned not to promote Copernican ideas. Galileo's *Dialogues on two chief world systems* was viewed as possibly concealing some of Bruno's heresies and as breaking the ban on promoting Copernican views. The Catholic church succeeded in forcing Galileo to abjure his Copernican beliefs in 1633. More than 100 years passed before the church allowed the construction of a mausoleum for Galileo in 1734. Much later the church lifted its ban on the publication of works that defended the Copernican system. Several decrees to that effect were issued from 1757 to 1822.

Throughout the period, mathematicians introduced various symbols and conventions, including the cross as a multiplication sign, letters for constants and variables, and exponents. These greatly simplified the communication of mathematical ideas; as they gained acceptance, they made mathematics an almost universal language.

Not only were mathematical symbols beginning to make a difference, but the tools of mathematics expanded rapidly during the period. The new branches of the subject were symbolic algebra, analytic geometry, probability, and logarithms.

Mathematics is an intellectual tool of science, but the hardware of science also began to assume modern form during the period. The development of spectacles toward the end of the Middle Ages offered opportunities to experiment with lenses. In turn, this led to the invention of the telescope and the simple microscope around the beginning of the seventeenth century. Galileo's observations of the stability of the beat of a pendulum led to Huygens's invention of the pendulum clock (*see* "The rise of time," p 164). Galileo himself made one of the first thermometers (*see* "Galileo and measurement," p 137).

Galileo also introduced experimentation into science, thus laying the foundation of science as we know it today. His detailed study of motion and his method of expressing natural events mathematically opened the way to Newton's discovery of universal gravitation.

Major advances

Astronomy. With the translation of Ptolemy into Latin for the first time, astronomy became a science again in the West, but one founded on incorrect premises. Few Renaissance scholars questioned Ptolemy's scheme, although Nicholas of Cusa, whose thinking in general was far in advance of his time, presaged Copernicus in 1440 by claiming that Earth revolves about the sun. Georg von Peurbach, although he had met Nicholas of Cusa, accepted Ptolemy's scheme and revised it by correcting errors that had creeped in over the years.

Galileo's introduction of the astronomical telescope in 1609 changed astronomy forever (*see* "Galileo and his telescope," p 126). Kepler went beyond Copernicus to work out the detailed movements of the planets; he found empirical laws that governed these movements.

Life science. Most of the progress in life science during this period was in medicine, especially anatomy. The invention of printing led to books on surgery and medicine. Although these books often contained many errors, knowledge of these subjects was much more widely available. Anatomy was based primarily on the works of Galen, especially at the beginning of the period, but the practice of dissecting corpses in front of medical students in Italy led careful observers to disagree with Galen by the end of the period (see "1543: A great year in publishing, p 108).

Another factor in gaining a better understanding of anatomy was the study of musculature by painters and sculptors, most notably Leonardo da Vinci. With the discovery of the circulation of the blood by Harvey, a general picture of the body as a system of tubes, ducts, and valves grew during the course of the period. The mechanical view of the body became so prevalent that it acquired a name—iatromathematics, called iatrophysics today (iatro- means pertaining to medicine or healing). Also in the field of medicine, Paracelsus, heavily influenced by alchemy, became a strong advocate of the use of chemical medicines.

There were also notable gains in botany. Scientists such as Otto Brunfels and Leonhard Fuchs published works, influenced by Latin translations of Aristotle and Theophrastus, in which various plants were described or pictured (see "Peppers and a whole lot more," p 93). In the compendiums of Konrad Gesner, all known (and a few never to be known: Gesner still believed in mythical creatures) animals and plants were described. By the end of the sixteenth century, Gerard's *Herbal* stuck to real plants, but often ascribed almost magical properties to them.

Mathematics. As Hindu-Arabic numerals continued to replace the clumsy Roman system, various textbooks appeared that taught how to apply the new symbols. Standard algorithms and bookkeeping methods were introduced. Tables of trigonometric functions were printed and applied to new surveying techniques. Military uses of mathematics included ballistics and the improvement of fortifications. There was also gradual development of the symbolism of mathematics (see "Inventing signs," p 94). Calculation also advanced in other ways. Pascal invented an adding machine, and Napier discovered logarithms. All of these developments made mathematics easier and more a part of daily life.

On a more theoretical level, mathematicians were beginning to expand their understanding of what constitutes a number. Irrational numbers gained steady, if slow, acceptance as numbers, not merely magnitudes. Following the

irrationals were the negatives. Even writers who refused to accept negative solutions to equations discovered that algebra could be greatly simplified by accepting negative coefficients. Soon the best mathematicians were using negative numbers regularly, but still expressing skepticism about them. Imaginary numbers began to have some currency later in the period.

A major achievement of mathematics during this period was the algebraic solution of the general cubic and quartic equations—that is, polynomial equations whose highest power of the unknown variable is three or four (see "A great scoundrel," p 107).

The publication of a new edition of the works of Diophantus led Fermat and his circle to advance pure number theory. Much of their work was not published at the time. Similarly, probability theory was invented in an exchange of letters between Fermat and Pascal. Analytic geometry was developed by Fermat, who did not publish his discovery, and by Descartes, who published his finding in an appendix to a work of philosophy, the *Discourse on method.*

Physical science. As with mathematics, much of the development of physical science during the Renaissance had a strongly practical side to it. Thus, we find advances and summaries of knowledge in such fields as mining, assaying, distillation, and ballistics. In the early part of the period, discoveries outside of practical fields were isolated instances. Alchemists were probably making some progress in chemistry, but their interpretations of what they were doing and their tendency to keep results secret offered little to science.

A notable development took place when Galileo performed a series of experiments with moving bodies, making physics into an experimental science and founding the field of dynamics.

Technology. The greatest technological innovation of the period was printing from movable type; however, this may be characterized as a reinvention. Movable type had been invented in the eleventh century by the Chinese, who had earlier invented printing. Chinese movable type was not a major influence, however, because ideographic writing is not as conducive to movable type as is the Roman alphabet. Gutenberg probably did not get his idea from Chinese sources, even secondhand. His invention stems from about 1440, although the date is not certain. There is some evidence that movable type also was invented in Holland around the same time, which would suggest that the idea was generally known.

Printing and paper, however, almost assuredly did reach the West from China, although indirectly. Thus, Gutenberg's reinvention of movable type was built on a Chinese foundation.

Other technological innovations of the period furthered understanding of space and time. Maps, needed by sailors following the routes of the great explorers, were improved. Globes were introduced even before the Americas were known in Europe. Clocks based on suspended weights and toothed escapements had been introduced in Europe about a century before, primarily as tower clocks that rang the hours. During the Renaissance there was constant improvement of their mechanisms.

The most dramatic technological ideas of the early Renaissance come from Leonardo's notebooks. Written in mirror-writing and illustrated with numerous sketches and drawings, the notebooks provide an unparalleled record of the thoughts of a man far ahead of his time. Leonardo is given some credit for the invention of the helicopter, the parachute, and other devices. In many cases it is not clear whether Leonardo made working models of his devices. Remarkable as his technical ideas were, they were mostly secret and had little actual influence on the progress of science.

Advances in measuring devices and observational tools were another important technological feat of the period. Vacuum pumps, developed by Guericke and Boyle, also became tools of the new science (see "A lot out of nothing," p 142).

Commercial technology related to mining and smelting became important. This led to some of the early literature of technology, such as the works of Agricola and Biringuccio.

Peppers and a whole lot more

Columbus, from his first voyage onward, met Native Americans who were farmers. Columbus was looking for gold and spices; although he did not find much in the way of gold, he believed he had discovered spices, for he thought that a Caribbean shrub was cinnamon and that other plants were those reported from the East by Marco Polo. It is not clear whether he thought the *Capsicum* of the New World—the common green or red pepper—to be the same as black pepper, although he clearly refers to "very hot spices" that must have been *Capsicum*. By 1493, Matyr (Pietro Martire de Anghiera) was using the accounts of Columbus's first voyage to describe *Capsicum*, which he called peppers, although clearly differentiating the American peppers from black pepper.

Although archaeological remains in the Americas make it clear that *Capsicum* originated there, there are puzzling aspects to how it traveled from the Americas, which it did quite speedily. The padres that accompanied the Spanish explorers brought back seeds of many plants, including *Capsicum*, for their monastery gardens. After that, the pattern of distribution becomes unclear.

For example, most botanists before 1600 thought that the *Capsicum* had been imported from India, not the Americas. There was a reason for this belief. There is good evidence that *Capsicum* first reached Germany before 1542 from India, not from the Americas. Thus, in less than 50 years, *Capsicum* had circled three-quarters of the way around the globe, traveling in what would seem to be the wrong direction.

The conventional view is that Portuguese sailors took *Capsicum* to their colonies in India sometime between their first voyages in 1498 and 1513. There, peppers could have been introduced to the Indians, as well as to the Ottoman Turks, who besieged the Portuguese colonies in 1513 and 1538. Then the Turks carried peppers with them to the Balkan peninsula, which they occupied. Botanists in Europe got them from the Turks, who said the peppers were Indian. The tale becomes somewhat complicated because early writers attributed the origin of the name *Capsicum* to a thirteenth-century botanist, who, of course, wrote before Columbus's voyage.

The Indian connection remains somewhat of a mystery, especially because *Capsicum* is so much a part of Indian food. It should be noted, also, that traditional cooking from some regions of China is heavily dependent on *Capsicum*.

A further mystery is that *Capsicum* was being cultivated in Melanesia when the first Europeans arrived. Theories that explain this involve native traders bringing the plants by stages from India.

Of course, *Capsicum* quickly settled into Europe, becoming an essential part of the cuisine of Italy, for example, and Hungary (in the form of paprika). Other American food plants also quickly became naturalized, including the tomato, the potato, and the American bean. It is difficult to imagine any European cuisine before 1500 as a result.

The Americas also obtained European food plants from the early colonists. At the same time they acquired a multitude of weeds. Many of the common roadside weeds across the United States, such as Queen Anne's Lace (the wild carrot), are imports.

It is not surprising that the flora of the Old and New Worlds should be different after millions of years of separation, especially with regard to domesticated plants. The only domesticated plant found in both places before Columbus's voyages is cotton. How cotton was spread is an even greater mystery than how the *Capsicum* got to India.

The great plant interchange had great significance for science. About the same time as the new plants began to reach Europe, so did a rediscovery of the works of the first great botanist, Theophrastus, whose work had been stimulated in part by the Greek discovery of Indian flora in the conquests of Alexander. This rediscovery, along with the arrival of thousands of plants from the New World, contributed to the development of botany in the works of Leonard Fuchs, Konrad von Gesner, Charles Lécluse (Clusius), Caspar Bauhin, John Ray, and, ultimately, Linnaeus. The New World flora continued to have an impact well into the nineteenth century, forming one of the main influences on the ideas of Charles Darwin and those of Alfred Wallace, who had traveled to the Amazon basin and written about it before his sojourn in the East Indies.

1453 The Turks capture Constantinople on May 29

1454 Navigator Americus Vespucius, one of the first to recognize that North and South America are new continents, b Florence (Italy), Mar

1460 Navigator Juan Sebastián del Cano, who completes Magellan's voyage around Earth after Magellan is killed, b Guetaria (Spain) about this time

Prince Henry the Navigator d Sagres, Portugal, Nov 13, with his dream of sending an expedition to circumnavigate Africa unfulfilled

Physician and humanist Thomas Linacre, founder of the Royal College of Physicians in London and translator of Galen into Latin, b about this time

Heinrich von Pfolspeundt writes *Bündt-Ertzney* the first book on surgery published in Germany

1461

1464 Nicholas of Cusa, German scholar, d Lodi (Italy), Aug 11

Inventing signs

Today the language of mathematics is in many ways standardized around the world. A Japanese mathematics book can almost be read by an American mathematician who knows no Japanese. Furthermore, we have become used to the idea that mathematics is a kind of special language, one made up of letters and signs. This notion of mathematics as a language ignores the fact that much excellent mathematics dates to classical and Hellenistic times, before this language of signs and symbols was invented.

Although the signs + and − had been used in the fifteenth century to denote surpluses and deficits, they were not given their modern meaning until 1514. Our present day = was introduced in a slightly different form in 1557. In 1593 letters were used to represent variables. The year 1631 saw rival introductions of the · and × for multiplication, a rivalry that continues today. By 1659 the ÷ sign was present.

The importance of apt signs in mathematics is best illustrated with the calculus. Both Newton and Leibniz discovered the ideas of the calculus, and both invented a method to write down their calculations and results. Newton's way was to put dots and primes above the variables and to use *o* for an infinitesimal. Leibniz invented the notation *dy/dx* and the integral sign. In England, mathematicians used Newton's symbolism, while on the Continent, they followed Leibniz. For a century after Newton there was little English mathematics of consequence. Only after British mathematicians relented and adopted the Leibniz notation did British mathematics regain its position. This did not begin until after 1815, when the Analytical Society was formed at Cambridge with the express purpose of replacing the clumsy Newtonian notation with that of Leibniz.

1466 Afanassi Nikitin reaches India via Persia and returns in 1472; he publishes his experiences in *The voyage of Afanassi Nikitin over three seas*

1468

1469 Spain is united through the marriage of Ferdinand of Castile and Isabella of Seville

1470 John Bessarion, an influential figure in the rise of Platonism during the Italian Renaissance, d Ravenna (Italy), Nov 18

1471 Regiomontanus (b Johann Müller) builds an observatory in Nuremberg (Germany) and also establishes a printing press there

1472 Georg von Peurbach's *Theoricae novae planetarum* (New theory of the planets), published posthumously, is a version of Ptolemy's *Almagest*

Regiomontanus is the first to make a scientific study of a comet; the comet will later be known as Halley's Comet

MATHEMATICS	PHYSICAL SCIENCE	TECHNOLOGY	
			1453
		Johann Gutenberg prints the 42-line Bible at Mainz (Germany), inaugurating the era of movable type	1454
			1460
Georg von Peurbach, Austrian mathematician and astronomer, d Apr 8			1461
De triangulis omnimodis (On triangles of all kinds) by Regiomontanus, b Johann Müller, Königsberg (Germany), Jun 6, 1436, gives a major summary of trigonometry, including the law of sines; it will not be published until 1533			1464
		Johann Fust, German printer, d	1466
		Johann Gutenberg, German inventor, d Mainz, Feb 3	1468
			1469
	In Tuscany, alum is discovered		1470
			1471
Leone Alberti, Italian artist and geometer, d Rome (Italy), Apr 25			1472

	GENERAL	ASTRONOMY	LIFE SCIENCE
1473	Michelangelo paints the ceiling of the Sistine Chapel		The first complete edition of Avicenna's *Canon of medicine* is printed in Milan
1474		Regiomontanus's *Ephemerides astronomicae* appears, although without the tables of solar declination that later editions will have; the additional tables make the *Ephemerides* much more useful to navigators	
1475	Explorer Vasco Núñez de Balboa, European discoverer of the Pacific Ocean, b Badajoz, Spain		
1476		Regiomontanus, German astronomer, d Rome, Jul 8	

The nature of light (part one)

Throughout the history of physics, the nature of light has been a puzzle. In the twentieth century, the puzzle has been replaced with a paradox, but most scientists are finally in agreement that the paradox presents the truth.

One of the early atomists, Democritus, in the fifth century BC, proposed that all objects consist of atoms and that an object becomes visible because its atoms swarm into empty space to be received by the eye. This was the first known particle theory of light.

In Empedocles' visual-ray theory, an opposing view, vision is based on light rays emitted from the eye; objects become visible when touched by these rays. Plato modified this theory somewhat. He believed that eyes emit light rays that intercept the rays transmitted by objects.

The first person to state that light might be a wave phenomenon was Leonardo da Vinci. He compared light reflection to the reflection of sound waves in echoes.

	GENERAL	ASTRONOMY	LIFE SCIENCE
1482			Paolo Toscanelli, Italian physician, d Florence, May 15
1483		The Alfonsine tables, a revision of the Ptolemaic tables of the planetary positions created in Toledo about 1250, are printed in Toledo, Spain	
1485	Henry VII becomes the first Tudor ruler of England		
1488			

MATHEMATICS	PHYSICAL SCIENCE	TECHNOLOGY	
	Lucretius's *De rerum natura* (On the nature of things) is translated into Latin, making the atomic theory of Democritus known to scholars in the West		1473
		William Caxton prints the first book in English	1474
			1475
			1476
		Geographer and mathematician Johannes Schöner b Karlstadt (Germany), Jan 16; in 1515 he is the first to construct a globe showing the Americas	1477
The *Treviso arithmetic*, a list of rules for performing common calculations, becomes the first popular printed textbook in mathematics			1478
		Leonard da Vinci, b (Italy) Apr 15, 1452, describes a workable parachute	1480
		The first European lock on a canal is built by Pietro and Dionysius Domenico	1481
Johannes Campanus's translation of Euclid's *Elements* is the first mathematical book of significance to be printed			1482
			1483
Nicolas Chuquet's *Triparty en la sciences des nombres* is the first book on algebra to use negative numbers as exponents, but Chuquet does not accept them as solutions to equations			1484
			1485
		Artist and inventor Andrea del Verrocchio d	1488

	GENERAL	ASTRONOMY	LIFE SCIENCE
1489			
1490			An "anatomical theater" opens in Padua (Italy) for the demonstration of the dissection of corpses
1491			
1492	Christopher Columbus, b Genoa (Italy), 1451, reaches an island in the West Indies on Oct 12		
1493	Pope Alexander VI on May 4 draws a line on a map that gives Spain all the undiscovered lands to the west of the line and Portugal all the undiscovered lands to the east of it; subsequently, the line is moved in such a way as to give Portugal the right to claim Brazil		Christopher Columbus finds that Native Americans use tobacco as a medicine
1494			
1495			As the French army of Charles VIII takes Naples, camp followers of the defending Spaniards infect his troops with syphilis, most likely brought to Spain from the Americas by sailors on Columbus's voyages; the army then moves back to France, infecting the Italian peninsula and northern Europe
1496			

MATHEMATICS	PHYSICAL SCIENCE	TECHNOLOGY	
Johann Widmann's *Behend un hüpsch rechnung uff allen kauffmanschafften* (Mercantile arithmetic) is the first printed document to use the signs plus (+) and minus (-); however, Widmann uses them to mean a surplus or a deficit			1489
	Leonardo da Vinci notes that liquids in tubes with a small diameter tend to crawl up the tubes, taking the first notice of capillary action		1490
Filippo Calandri's arithmetic text introduces the algorithm for long division			1491
Mathematician and cartographer Pedro Nunes, reputedly the first to develop an instrument for measuring angles, b Portugal Francisco Pellos's *Compendio de lo abaco*, a commercial arithmetic, introduces a dot to represent division by 10, a precursor of the decimal point	Christopher Columbus discovers on his first voyage to the Americas that the magnetic compass somewhat changes the direction in which it points as the longitude changes	Graphite, known commonly as lead, is used for pencils around this time in England Martin Behaim makes the first globe map of Earth, omitting the about-to-be-discovered Americas and Pacific Ocean Leonardo da Vinci draws his conception of a flying machine	1492
			1493
Summa de arithmetica geometria proportioni et proportionalita by Luca Pacioli, b Sansepolcro (Italy), about 1445, is the most influential mathematics book of the time and the first printed algebra book; derived largely from Fibonacci's *Liber abaci*, published 300 years earlier, its popularity may result from its introduction of double-entry bookkeeping		The first paper mill appears in England Leonardo da Vinci makes a drawing of a clock with a pendulum	1494
			1495
		Leonardo da Vinci designs roller bearings and a rolling mill	1496

	GENERAL	ASTRONOMY	LIFE SCIENCE
1497		Nicolas Copernicus (b Toruń, Poland, Feb 19, 1473) observes and records the occultation of a star by the moon	Hieronymus Brunschwygk (b about 1452) publishes the first known book on the surgical treatment of gunshot wounds

The monk Romano Pane, who accompanied Columbus, gives a description of the tobacco plant and the smoking habits of the Indians |
| **1498** | Vasco da Gama reaches India by traveling around the Cape of Good Hope

Savonarola executed | | |
1499			Artist Jan Stefan van Calcar, probable illustrator of Vesalius's *De corporis humani fabric* (On the structure of the human body), b (Italy)
1500			Jakob Nufer of Switzerland performs the first recorded cesarean operation on a living woman
1501	*Laus stultitiae*, also called *Moriae encomium* (In praise of folly), by Desiderius Erasmus (b Holland, 1466) offers a satire on human nature		Botanist Leonhard Fuchs, for whom the shrub fuchsia and color of its blossoms is named, b Wemding, Bavaria (Germany), Jan 17
1502			
1503			
1504		Christopher Columbus, using an edition of Regiomontanus's *Ephemerides astronomicae*, frightens a group of Native Americans by correctly predicting a total eclipse of the moon on Feb 29	
1506	Christopher Columbus, Italian explorer, d Valladolid, Spain, May 20		

MATHEMATICS	PHYSICAL SCIENCE	TECHNOLOGY	
			1497
	Aldus Manutius completes a Latin translation of Aristotle	Venetian printer Ottaviano dei Petrucci (b 1466) invents a way of printing music using movable type	1498
			1499
	Hieronymus Brunschwygk's *Liber de arte distillandi de simplicibus*, known as the "Small Book," describes the construction of furnaces and stills, herbs usable for distillation, and medical applications of distillates; Brunschwygk publishes his "Big Book," dealing with the same subjects, in 1512	Leonardo da Vinci draws a wheel-lock musket, the first known appearance of a gun in the West About this time, Leonardo da Vinci designs the first helicopter; it is not built and probably would not have flown if built Chinese scientist Wan Hu ties 47 gunpowder rockets to the back of a chair about this time in an effort to build a flying machine; the device explodes during testing, killing Wan, who acted as pilot	1500
			1501
		Peter Henlein builds a spring-driven pocket watch, the first pocket watch ever made Printer Peter Schoeffer d about this time	1502
		Raw sugar is refined	1503
			1504
			1506

	GENERAL	ASTRONOMY	LIFE SCIENCE
1507	*Cosmographiae* by Martin Waldseemüller, b Baden (Germany), 1470, contains the first printed allusion to tobacco		
1509	Philosopher and scientist Bernardino Telesio, who urges the importance of knowledge based on experiment, b Cosenza, Calabria (Italy)		
1511			Physician Michael Servetus (Miguel Serveto) b Villanueva de Sixena, Spain, Sep 29
1512	Portuguese explorers de Abreu and Serrao reach the Moluccas, or Spice Islands Americus Vespucius, Italian navigator, d Seville, Spain, Feb 22		
1513	Machiavelli's *Il principe* (The prince) presents a study in how to rule and remain in power		
1514		In May Copernicus writes his first version of his heliocentric theory, which he does not publish until 1543	
1515			
1516	*Utopia* by Thomas More (b England, 1478) gives a classic account of a perfect state		
1517	Balboa, Spanish explorer, d Panama Martin Luther initiates the Protestant Reformation by posting 95 Theses about the Catholic church on the door of the *Schlosskirche* in Wittenberg (Germany)		Naturalist Pierre Belon, the first to notice homologies between specific bones in vertebrates from fish to mammals, b Soultière, France

MATHEMATICS	PHYSICAL SCIENCE	TECHNOLOGY	
		German cartographer Martin Waldseemüller publishes a thousand copies of a map on which the name America is first applied to the new continent discovered by Columbus and later explored by Americus Vespucius (between 1497 and 1504); unlike Columbus, Vespucius recognized that it was a new continent and not part of Asia	1507
			1509
			1511
French soldiers massacre Italians who have fled to the Brescia cathedral for sanctuary; among the dead is the father of the wounded Niccoló Fontana, b Brescia (Italy), 1499; as a result of his saber wound, young Niccoló develops a stammer that causes him—a mathematical genius—to be known to posterity as Tartaglia (Stutterer)	Hieronymus Brunschwygk d		1512
		Waldseemüller prepares an atlas containing 200 maps	1513
Dutch mathematician Vander Hoecke is the first to use the signs plus (+) and minus (−) with their modern meanings in an algebraic expression	Jodocus Trutfetter's *Summa in totam physicam, hoc est philosophiam naturalem* is an account of Theodoric of Freiburg's work on the rainbow; this book probably inspired Descartes in his research on the rainbow		1514
Scipione del Ferro (b 1465) discovers and keeps secret a method for solving one type of cubic equation			1515
			1516
Luca Pacioli, Italian mathematician, d Sansepolcro, Tuscany	Girolamo Fracastoro (b 1478) explains fossils as the remains of actual organisms; according to Fracastoro's calculations, while some fossils could have been buried during Noah's flood, they are at too many different geologic strata for all of them to have been caused by a flood that lasted only 150 days		1517

	GENERAL	ASTRONOMY	LIFE SCIENCE
1518			Smallpox reaches the Americas, causing a major epidemic among the Indians of the island of Hispaniola The Royal College of Physicians in London is established
1519	Portuguese explorer Ferdinand Magellan (b Sabrosa, Trás-os-Montes about 1480) starts the voyage that will take one of his ships around the world in the first circumnavigation		Physician and botanist Andreas Caesalpinus, who anticipated both William Harvey's theory of blood circulation and Linnaeus's scheme of plant classification, b (Italy)
1520			Turkeys are imported to Europe from America; the Portuguese import the orange tree from south China; maize (corn) is imported into Spain from the West Indies A smallpox epidemic among the Aztec demoralizes them and helps Hernando Cortes and a small groups of Spaniards, immune to the disease from childhood exposure, to take over the Aztec empire Physician and alchemist Philippus Aureolus Paracelsus (Theophrastus Bombast von Hohenheim, b Schwyz, Switzerland, May 1) introduces tincture of opium, which he names laudanum, into medicine about this time
1521	Magellan, Portuguese explorer and original leader of the first expedition to circumnavigate the globe, d Philippine Islands, Apr 27, killed by natives		
1522			Naturalist Ulisse Aldrovandi, influential in setting up the Bologna Botanical Garden in 1567, b (Italy) Sep 11
1524			Physician and humanist Thomas Linacre d Oct 20
1525			Smallpox reaches the Inca empire, killing Huayna Capac, the Inca ruler

MATHEMATICS	PHYSICAL SCIENCE	TECHNOLOGY	
		Martin Waldseemüller, German cartographer, d Alsace about this time	1518
		Leonardo da Vinci, Italian artist and inventor, d Amboise, France, May 2	1519
	The first publication of *Probierbüchlein* (Little book of assays) becomes an important guide to the assaying of metals	Peter Apian publishes a map that shows America	1520
			1521
Italian mathematician Luigi (Ludovico) Ferrari b; he will become the first to solve the general polynomial equation of the fourth degree (the quartic) Cuthbert Tunstall publishes the first book on arithmetic in England			1522
		Peter Apian publishes a book on cartographical methods	1524
Die Coss by German mathematician Christoff Rudolff (b about 1500) introduces a version of the modern symbol for the square root, $\sqrt{\ }$; it is one of the earliest books to use decimal fractions Artist Albrecht Dürer, b Nuremberg (Germany), May 21, 1471, writes a book on geometric constructions, including proofs of how to construct complicated curves			1525

	GENERAL	ASTRONOMY	LIFE SCIENCE
1526	Juan Sebastián del Cano, Spanish navigator, d Pacific Ocean, Aug 4, on his second voyage to the Pacific		
1527			
1528			
1530			Otto Brunfels's *Herbarum vivae eicones* (Living portraits of plants) describes 230 species of plants with detailed illustrations by Hans Weiditz Girolamo Fracastoro's *Syphilis sive de morbo gallico* (On syphilis, or the French disease) describes the symptoms, spread, and treatment of syphilis; he also coins the term *syphilis*, which is the name of a mythical young shepherd who develops the disease Paracelsus's *Paragranum* argues that medicine should be based on nature and its physical laws; he is the first to suggest the use of chemical substances, such as compounds of mercury and antimony, as remedies
1532	Franciso Pizarro captures the Inca Atahualpa, who rules over an empire devastated by smallpox and civil war; Pizarro puts Atahualpa to death and proceeds to conquer Peru		
1533			
1535	English statesman and author of *Utopia* Sir Thomas More d in the Tower of London, beheaded by order of Henry VIII for refusing to swear allegiance to the newly formed Church of England		Valerius Cordus's *Dispensatorium* describes drugs and medical preparations
1536	Desiderius Erasmus, Dutch philosopher and humanist, d Jul 12, Basel, Switzerland		

Scipione del Ferro d

1526

Pascal's triangle appears for the first time in Europe as the frontispiece of a book by German astronomer Peter Apian (Petrus Apianus, b Leisnig, Apr 16, 1495)

Paracelsus's *Archidoxis* (not published until 1570) reports that frozen wine will have a higher proof than liquid left unfrozen

1527

Albrecht Dürer, German artist and mathematician, d Nuremberg, Apr 6

1528

A great scoundrel

In the Italian Renaissance, the professions of scientist and mathematician were just beginning to be defined. One such scholar was Girolamo Cardano, known in English as Jerome Cardan. Cardan's basic source of income was his work as a physician, but he was also at various times a professor of mathematics at the universities of Milan, Pavia, and Bologna. Other sources of income included gambling and astrology; however, he was imprisoned for heresy after he cast Christ's horoscope.

Cardan's reputation as a mathematician is deservedly great, but marred by scandal. His 1545 work *Ars magna* (The great art) contains the first printed methods for solving all algebraic equations whose highest exponent is three (cubics) and all equations whose highest exponent is four (quartics).

At the end of the fifteenth century, it was believed that the cubic would never be solved. But Scipione del Ferro partially solved the cubic early in the sixteenth century, keeping his solution to himself. Before his death, however, del Ferro passed along his secret method to his pupil, Antonio Maria Fiore. Meanwhile, a self-taught mathematical genius known to posterity as Tartaglia (the Stutterer) discovered how to solve many kinds of cubics. He used his secret methods to best Fiore in a mathematical duel, solving all the problems Fiore posed while Fiore could solve none of Tartaglia's.

Cardan somehow persuaded Tartaglia to reveal his secret; in return, Cardan promised to keep the secret so Tartaglia could publish it later. Instead, however, Cardan put Tartaglia's secret method into his own book, *Ars magna*. He did, however, credit Tartaglia with the original work. Cardan also expanded on Tartaglia's work to cover all forms of the cubic.

Ars magna also contains the method for solving the quartic, the next higher-degree equation. But once again Cardan did not work out the solution on his own, Instead, he got it from a pupil, Luigi Ferrari, "who invented it at my request."

The spinning wheel is in general use in Europe

There are reports of matches being used in Europe, almost a thousand years after their invention in China

1530

1532

Geographer Reiner Gemma Frisius (b Dokkum, Friesland, Holland, 1508) is the first to point out in *De principis astronomiae et cosmographie* that knowing the correct time according to a mechanical clock and comparing it with sun time can be used to find the longitude

1533

Niccoló Tartaglia announces that he can solve certain types of cubic equations, types that were not previously solved by Scipione del Ferro

Diving bells are invented

1535

Hudalrichus Regius's *Utriusque arithmetices* includes the fifth known perfect number, 33,350,336

1536

	GENERAL	ASTRONOMY	LIFE SCIENCE
1537			
1538		Girolamo Fracastoro's *Homocentrica* introduces a model of the planetary system close to Eudoxus's model; it contains 79 spheres	
1539			
1540		*Astronomicon caesareum* by Peter Apian notes that the tails of comets always point away from the sun, a fact known to the Chinese as early as 635 AD and perhaps earlier, but not known in Europe Alessandro Piccolomini's *De le stelle fisse* (On fixed stars) is the first star atlas in which stars are labeled with letters *Narratio prima de libris revolutionum* by German mathematician Georg Joachim Rheticus (b Feldkirch, Austria, Feb 16, 1514) summarizes the heliocentric planetary model that Copernicus has developed but not published	
1541			Paracelsus, Swiss physician and alchemist, d Salzburg, Austria, Sep 24
1542			A botanical garden is started in Leipzig (Germany) A book on anatomy by Jean François Fernel (b Montdidier, Somme, France, 1497) is the first to describe appendicitis and peristalsis (the waves of contraction in the digestive system that move food through the alimentary canal) Leonhard Fuchs describes about 400 German and 100 foreign plants, including peppers, pumpkins, and maize from the New World
1543	The Portuguese arrive in Japan	Copernicus's *De revolutionibus orbium coelestium* (On the revolutions of celestial bodies) is a convincing recasting of planetary motions based on the assumption that Earth and other planets revolve around the sun	*De humani corporis fabrica* (On the structure of the human body) by Flemish anatomist Andreas Vesalius (b Brussels, Dec 31, 1514) is the first accurate work on human anatomy

1543: A great year in publishing

The year 1543 was marked by the publication of two books that revolutionized our view of humanity and the universe.

The more celebrated of the two, Nicholas Copernicus's *De revolutionibus orbium coelestium* (On the revolutions of celestial bodies), was actually completed in the 1530s, but Copernicus was reluctant to publish for fear of reprisals from the church. His work circulated among intellectuals rather as Russian *samizdat* manuscripts have circulated in the Soviet Union. Copernicus offered arguments in the book that Earth and other planets travel around the sun; however, his assumption of circular motion meant that he had to complicate the paths by adding 48 epicycles to them to accord with observation. Copernicus said that the stars do not appear to move as Earth moves in an orbit because the fixed stars are so far away, although he vastly underestimated how far away they actually are.

As Copernicus neared death, he was finally persuaded to publish by the mathematician Rheticus. The actual publication was overseen by a Lutheran minister, Andreas Osiander, who—presumably concerned because Luther did not believe Earth moves—added a preface that said in

		Niccoló Tartaglia's *Della nova scientai* (Of the new science) initiates the science of ballistics	1537
			1538
		Printer Ottaviano dei Petrucci d	1539
	De la pirotechnica (On pyrotechnics) by Italian mine superintendent Vannoccio Biringuccio (b 1480, d 1539) gives practical information on the treatment of ores, smelting of metals, distillation, and sublimation	Christoph Schurer uses cobalt in the production of blue glass	1540

1543: A great year in publishing, *continued*

effect that Earth does not move, but calculations would be easier if one assumes it does. A few hundred copies were printed a month or so before Copernicus's death; it is not clear whether he saw the published version.

From a publisher's point of view, the book was not a success. It was very high-priced and sold slowly. It went out of print, although a second edition was published in 1566 and a third in 1617. But most astronomers came to believe that Copernicus was right, notably Galileo, who spread the word more widely. The last great astronomer to hold a different view was Tycho Brahe, who died in 1601 still believing that Earth does not move, although he did think that the other planets revolve about the sun.

The other book from 1543 that changed the world was *De humani corporis fabrica* (On the structure of the human body) by Andreas Vesalius. Unlike Copernicus, Vesalius decided to publish when he was a young man, not quite 30. However, Copernicus might have had the right idea about keeping new ideas quiet. After Vesalius published his book, accusations of body snatching and heresy were directed against him by physicians loyal to older ideas. Whether because of this or for other reasons, Vesalius spent the rest of his life as a court physician, doing almost no more research.

De humani corporis fabrica is most important because it is a printed book with exceptional illustrations (usually attributed to the painter Stephen van Calcar). For the first time, the true anatomy of the human body as perceived by a great anatomist was reproduced in many accurate copies. Previous books on the subject generally lacked illustrations or had a few copied by hand, a method sure to introduce inaccuracies.

Before Vesalius, physicians were taught from Galen, even when Galen's authority clearly did not conform with what was observed. Although Vesalius had been educated in this tradition, he rejected it when confronted with the evidence of his own dissections. Although physicians still loyal to Galen denounced the book, later anatomists, like the astronomers after Copernicus, sought the truth instead of authority.

Mathematics			Year
Niccoló Tartaglia and Antonio Maria Fiore engage in a famous mathematical duel; challenged to solve each others' set of cubic equations, Tartaglia solves all of Fiore's, while Fiore fails to solve even one of Tartaglia's; Tartaglia has solved the general cubic; his solution is published by Cardan in 1545			1541
Robert Recorde's *The grounde of artes* is a popular arithmetic textbook that runs through 29 editions			1542
Peter Ramus (b Pierre de Ramée, Cuth, Picardy, France, 1515) attacks Aristotelian physics in his *Aristotelicae animadversiones* (Reproof of Aristotle)			1543

	GENERAL	ASTRONOMY	LIFE SCIENCE
1543 cont		Nicolas Copernicus, Polish astronomer, d Frombork, May 24	
1544			Physician, physicist, and earth scientist William Gilbert b Colchester, Essex, England, May 24
1545			A book on surgery by Ambroise Paré (b Mayenne, France, 1510) advocates abandoning the practice of treating wounds with boiling oil and using soothing ointments instead
1546			Girolamo Fracastoro's *De contagione* (On contagion) advances the idea that diseases are seedlike entities that are transferred from person to person
1547			
1550		Girolamo Cardano uses the measurement of parallax to establish that comets are not atmospheric phenomena	
1551		German astronomer and mathematician Erasmus Reinhold (b Thuringia, Oct 22, 1511) publishes astronomical tables based on Copernicus's theory; they are the first improvement over the Alfonsine tables of 1252 AD	The first volume of *Historia animalium* by Swiss naturalist Konrad von Gesner (b Zurich, Mar 26, 1516) is the beginning of the science of zoology; three more volumes are published in the period through 1558
1552		Peter Apian (Petrus Apianus), German astronomer, d Ingolstadt, Apr 21	Italian anatomist Bartolemeo Eustachio (b San Severino about 1510) describes the adrenal glands, the detailed structure of the teeth, and the Eustachian tubes, named after him; this work is not published until 1714
1553			Pedro de Cieza de Leon describes the potato in *Chronicle of Peru* Girolamo Fracastoro, Italian physician, d Aug 8, Incaffi (Italy) Michael Servetus's anonymously published book on theology contains

MATHEMATICS	PHYSICAL SCIENCE	TECHNOLOGY	
			1543 cont
Michael Stifel's *Arithmetica integra* summarizes the algebra and arithmetic known at the time	Georg Hartmann observes the magnetic dip (the magnetic needle, while pointing north, is not entirely horizontal)	Sebastian Münster's *Cosmographia*, published in Germany, is the first major compendium on world geography	1544
Ars magna (Great art) by Italian Girolamo Cardano—often known in English as Jerome Cardan—(b Sep 24, 1501, Pavia) is the first book of modern mathematics; it contains not only Cardano's stolen and purchased solutions to the cubic and quartic, but also the first acceptance of negative numbers and passing notice of complex numbers German Christoff Rudolff d about this time			1545
	De natura fossilium (About the nature of digging) by German metallurgist Georgius Agricola (b George Bauer in Glauchau, Sexony, Mar 24, 1494) introduces the word *fossil* for anything dug from the ground, including what Agricola believes to be odd rocks that look like bones or shells		1546
		Geographer Johannes Schöner d Nuremberg (Germany), Jan 16	1547
		Tobacco is cultivated in Spain	1550
Robert Recorde's *The pathewaie of knowledge* is a popular abridgment of the *Elements* of Euclid		Leonard Digges invents the theodolite, the telescope used in surveying; the invention is published by his son Thomas in 1571	1551
			1552
Erasmus Reinhold, German mathematician, d Thuringia, Feb 19			1553

1553 cont

his view, soon to be demonstrated by Colombo, that blood circulates from the heart to the lungs and back; his authorship discovered, his unorthodox theological views result in his being burned at the stake on Oct 27 in Geneva by John Calvin

1555

Pierre Belon's *L'histoire de la nature des oyseaux* (History of the nature of birds) contains a classification of 200 species and compares the bone anatomy of birds and humans

1556

Late in Jan or in early Feb, an earthquake strikes Shensi Province in China, killing perhaps 830,000 people—the worst earthquake in history

1557

1558

Elizabeth I becomes queen of England

Jean François Fernel, French physician, d Fontainebleau, Apr 26

1559

Italian anatomist Realdo Colombo (b Cremona about 1510), although a supporter of Galen against the new anatomy of Vesalius, demonstrates that Galen was wrong about the way the blood travels from the heart to the lungs and back; his *De re anatomica* (Of anatomical matters) claims that blood circulates from the right chamber of the heart to the lungs and then to the left chamber; Galen thought the blood passed directly between the two chambers

Realdo Colombo, Italian anatomist, d Rome

1560

Italian physicist Giambattista della Porta (b Vico Equense, Oct 1535) founds the *Academia Secretorum Naturae*, the first scientific society; it is suppressed by the Inquisition, and replaced by the *Accademia dei Lincei* in 1603

Bock's *Kräuterbuch* (Book on herbs) is published

Mathematics	Physical Science	Technology	
			1553 cont
	Agricola, German mineralogist, d Chemnitz, Saxony (Germany), Nov 21	Reiner Gemma Frisius, Dutch geographer, d Louvain, (Belgium)	**1555**
	Agricola's book *De re metallica*, published posthumously, offers a detailed discussion of where mineral veins are found; it remains an important reference until the end of the eighteenth century		**1556**
Robert Recorde's *The whetstone of witte* introduces an elongated version of the equal sign (=) into mathematics (asking what could be more equal than a pair of parallel lines); it also introduces the plus (+) and minus (−) signs for addition and subtraction into English Niccoló Tartaglia, Italian mathematician, d Venice, Dec 13	Julius Caesar Scaliger makes what seems to be the first known reference to platinum, discovered about this time		**1557**
			1558
			1559
			1560

	GENERAL	ASTRONOMY	LIFE SCIENCE
1561			*Observationes anatomicae* (Anatomical observations) by Italian anatomist Gabriel Fallopius (b about 1523, Modena) describes the organs of the inner ear and the female reproductive system, including the Fallopian tubes that are named for him
1562	Witchcraft is made a capital offense in England	Christian Severin (Longomontanus) b Longberg, Jutland (Germany), Oct 4; he becomes Tycho Brahe's assistant for eight years and then professor of mathematics at Copenhagen, Denmark; his plans for an astronomical observatory are completed after his death	Gabriel Fallopius, Italian anatomist, d Padua (Italy), Oct 9
1564	Shakespeare b; Michelangelo d		Pierre Belon, French naturalist, d Paris, Apr Andreas Vesalius, Flemish anatomist, d Zante, Greece, Oct 15
1565			The first potatoes arrive in Spain from America Konrad von Gesner, Swiss naturalist, d Zurich, Dec 13
1566			Leonhard Fuchs, German botanist, d Tübingen, May 10
1567			
1568			
1569			Sunflowers are imported into Spain from America
1570			
1572	Peter Ramus (Pierre de la Ramée), French philosopher and mathematician, d Aug 26 during the St. Bartholomew massacre in Paris	Tycho Brahe (b Knudstrop, Denmark, Dec 14, 1546) observes a new star, which he calls a *nova*, but which we now would call a *supernova*, in Cassiopeia; the star also is recorded by Chinese astronomers; as bright as Venus, it remains visible for 15 months	

The immutability of the heavens

The Greek concept of the universe was detailed by Aristotle. This Greek cosmology included the idea that something had to hold up the stars and planets, something that could not be seen. Since the only hard substance known that was also transparent was crystal, Greek logic insisted that spheres of crystal hold the stars, sun, moon, and planets in place.

The Greek tradition also included the idea that the heavens are very different from Earth. On Earth, everything is changing all the time. In the heavens, nothing ever changes. Comets, however, appear in the heavens, last a few days or weeks, and disappear. This constitutes change. Therefore, Aristotle argued, comets are earthly stuff, not heavenly stuff, and must be some kind of spontaneous fire in the upper atmosphere. Then the comet of 1577 showed that Aristotle could not be right about the heavens.

Tycho Brahe, Michael Mästlin, and other astronomers were able to plot the comet's path. Tycho Brahe also used parallax in an effort to determine the distance of the comet of 1577 from Earth. Parallax refers to the apparent angular difference in position of an object when seen from two locations. The amount of this difference can be used to calculate the distance to the object. These studies refuted much of the Greek view of the heavens. For one thing, the comet was farther away than the moon, so it had to be heavenly and not earthly. Secondly, the path of the comet was not part of a circle, so heavenly objects need not all move in circles. Finally, the path of the comet of 1577 was such that it must travel *through* the crystalline spheres. Hence, something other than crystalline spheres must be holding the sun, moon, planets, and stars in space.

MATHEMATICS	PHYSICAL SCIENCE	TECHNOLOGY	
			1561
			1562
		The horse-drawn coach is introduced into England from Holland	**1564**
	Konrad von Gesner's *De rerum fossilium* (On things dug up from the earth) contains the first drawings of fossils, although Gesner believed them to be rocks that just happen to look like bones or shells		**1565**
	Physicist Marko Antonije de Dominis b Rab, Yugoslavia; he is best known for his opposition to papal authority	Camillo Torello patents the first seed drill known in Europe, nearly 1700 years after its invention in China	**1566**
Michael Stifel, German mathematician and theologian, d Jena (Germany), Apr 19			**1567**
		Flemish geographer Gerardus Mercator, b Gerhard Kremer, Rupelmonde (Belgium), Mar 5, 1512, introduces the map projection that bears his name	**1568**
			1569
		About this time Giambattista della Porta investigates the *camera obscura*, or pinhole camera	**1570**
Rafael Bombelli's *Algebra* shows the first application of complex numbers to solve equations, including equations with real solutions; Bombelli also uses continued fractions to approximate roots			**1572**

	GENERAL	ASTRONOMY	LIFE SCIENCE
1573		Tycho Brahe publishes *De nova stella* (On the new star) giving an exact description of his observation of the supernova of 1572	
1574		Tycho Brahe's *Oratio de disciplinis mathematicis* presents an organistic view of the universe, comparing the parts of the universe to human organs	Bartolemeo Eustachio, Italian anatomist, d Urbino (Italy), Aug 27
1575	Leiden University in Holland is founded as a secular institution open to all faiths Philip II of Spain founds the *Academia de Ciencias Matematicas* in Madrid		
1576		Leonard Digges's *Prognostication euerlastingue* contains an appendix written by his son Thomas Digges in which the heliocentric system of Copernicus is placed in a system of stars occupying infinite space	
1577		Tycho Brahe tries to determine the distance of the great comet of 1577 from Earth using parallax; his crude observations are good enough to demonstrate that the comet has to be at least four times as distant as the moon, proving conclusively that a comet is not an atmospheric phenomenon	
1579			The first glass eyes are made about this time
1580		King Frederick II of Denmark builds for Tycho Brahe the first real astronomical observatory since that of Ulugh Beg on the island of Ven, between Denmark and Sweden	About this time Prospero Alpini, b Venice (Italy), Nov 23, 1553, becomes the first scientist to learn that plants, like animals, have two sexes (although this was probably known to the common farmer or gardener for certain plants centuries earlier)
1581			

MATHEMATICS	PHYSICAL SCIENCE	TECHNOLOGY	
		By this date Humphry Cole is supposed to have invented the ship's log for keeping track of the speed of a ship with respect to the water	1573
	Lazarus Ercker publishes *Beschreibung Allerfürnemisten mineralischen ertzt und bergwerksarten* (Description of mineral ores and mining techniques)		1574
Francesco Maurolico's *Arithmeticorum libri duo* includes the first known result proved by mathematical induction: that the sum of the first n odd numbers is n^2; this result was known as early as the Pythagoreans, but it is assumed that their proof was based on dot patterns used in figurate numerals		About this time Bernard Palissy (b Saintes, France, about 1510) rediscovers the enameling of pottery	1575
Girolamo Cardano, Italian mathematician and physician, d Rome, Sep 21 Georg Joachim Rheticus, German mathematician, d Cassovia (Hungary), Dec 4	Robert Norman (b Bristol, England, about 1550) shows that a compass needle when allowed freedom to swing in any direction will point below the horizon		1576
			1577
Canon-mathematicus by Franciscus Vieta (b François Viète, Pitou, France, 1540) argues for decimal representation of numbers and extends the trigonometric tables of Rheticus to each second of a degree			1579
	Bernard Palissy's *Discours admirables de l'art de terre, de son utilité, des esmaux et du feu* (Admirable discourse on pottery and its uses, on enamels, and on fire) covers a wide range of geological and chemical ideas		1580
	Galileo, b Pisa (Italy), Feb 15, 1554, studies the hanging lamps during a service at the cathedral of Pisa and concludes that the time of the swing of a pendulum is independent of the width of the swing; although this is not strictly correct, it directs attention to pendulums and eventually results in accurate clocks		1581

1581
cont

1582

Replacing Aristotle's physics

Aristotle was the first to formulate a unified system to explain physical phenomena. His thinking dominated philosophy and science for an enormously long time, almost 1900 years. His ideas about physics remained unchallenged until the Middle Ages, when such scholars as Jean Buridan, William of Okham, Nicolas Oresme, and Nicholas of Cusa started questioning and refining his assumptions.

Aristotle distinguished between "natural" and "unnatural" motion. Bodies moving naturally would move either up or down. Horizontal motion, as in a flying projectile, Aristotle termed "unnatural." Natural motion of a body was a consequence of the amount of the prime substances—earth, water, air, and fire—present in that body. Each of these elements had a

1583

The University of Edinburgh is founded in Scotland as a secular institution

Della causa, principio ed uno (On the cause, the principle, and the unity) by Italian philosopher Giordano Bruno (b Nola, Jan 1548) outlines his metaphysical ideas

Andrea Cesalpino's *De plantis* (On plants) gives a classification scheme for plants based on roots and fruits

1584

William of Orange is assassinated

Giordano Bruno's *Dell' infinito, universo e mondi* (Of infinity, the universe, and the world) gives the opinion that stars form planetary systems and that the universe is infinite

Giordano Bruno's *La cena delle ceneri* (The dinner of ashes) defends the Copernican view of the universe, although for mystical instead of astronomical reasons

1585

1586

Walter Raleigh imports the smoking habit into England from Virginia

1588

Sir Francis Drake defeats the Spanish Armada and the English become the rulers of the seas

Bernardino Telesio, Italian philosopher and scientist, d Cosenza, Calabria

Tycho Brahe's *De mundi aetherei recentioribus phaenomenis* rejects the idea of crystalline spheres holding the stars based on his observations of the comet of 1577

Robert Norman's *The new attractive* relates his discoveries about Earth's magnetism, especially the magnetic dip

1581
cont

Replacing Aristotle's physics, continued

natural place to which it moved if unimpeded. A stone falls because it contains earth and steam rises because it contains fire. The natural motion of celestial objects he viewed as different from motion on Earth; it was circular motion, with no beginning or end.

Horizontal motion resulted from a pull or push acting on the object. To explain the motion of a projectile, which apparently is not pushed when released, Aristotle assumed that the air flowing around it and filling up the empty space created behind it would push the projectile continuously. Aristotle had also noticed that bodies in a denser medium, such as water, fall at a slower speed than those in a less dense medium, such as air. From this he concluded that an object would fall at infinite speed in a vacuum; since he believed that infinite speed was not possible, an absolute vacuum could not exist.

In 1586 Simon Stevinus showed that the speed of falling bodies is not proportional to their weight, as Aristotle had assumed, but that all bodies fall equally fast. Galileo did similar experiments a few years later, often timing balls rolling down on inclined planes to slow down their "falling motion." Galileo deduced the following laws from his experiments: 1) Any body moving on a horizontal plane will continue at the same speed unless a force opposes it. 2) In a vacuum, all bodies fall at the same speed, no matter what their weight or constitution. 3) A body falling freely or rolling down an inclined plane undergoes uniform acceleration.

Newton generalized Galileo's ideas, showing that Aristotle's "unnatural motion" is in matter of fact quite useful. Interestingly, since Aristotle's ideas about motion correspond to what we observe in daily life, they remain as underlying assumptions for people unfamiliar with Newton's laws. Since we do not observe motion in a vacuum or without friction, objects on Earth continue to appear to obey Aristotle's laws of motion.

On the advice of astronomer Christoph Clavius, b Bamberg (Germany), 1537, Pope Gregory XIII reforms the calendar, dropping the 11 days between Oct 4 and Oct 15; in the new Gregorian calendar, century years that are not divisible by 400 will no longer be leap years (as they were in the Julian calendar); as a result, 1582 has 354 days, making it the shortest year on record

1582

Joseph Justus Scaligier (b Lot-et-Garonne, France, Aug 5, 1540) devises the Julian day count, still in use, which sets Jan 1, 4713 BC as day 1; he numbers all days subsequently from that, so Jan 1, 1990, is Julian day 2,447,527; it is called *Julian* in honor of his father, Julius Caesar Scaligier

1583

Chu Tsai-Yü invents equal temperament in music

1584

De thiende (The tithe), also known by its French translation as *La disme*, by Flemish mathematician Simon Stevinus (Stevin, b Bruges, 1548) presents a systematic account of how to use decimal fractions

Giovanni Battista Benedetti's *On mechanics* criticizes Aristotle's views on motion and discusses the "impetus" theory

1585

Simon Stevinus performs the key experiment in understanding gravity, dropping two different weights at the same time and noting that they strike the ground at the same time

Simon Stevinus shows that the pressure of a liquid on a given surface depends on the height of the liquid and on the area of the surface

A 327-ton Egyptian obelisk that the Romans had brought from Egypt in ancient times is raised to a vertical position by a team led by Domenico Fontana

1586

1588

1589

The first complete edition of Paracelsus's works is published in Basel, Switzerland, over a three-year period

1590

Giordano Bruno's *De minimo, magno et mensura* presents metaphysical views

Ambroise Paré, French surgeon, d Paris, Dec 20

1591

Philosophia sensibus demonstrata (Philosophy of the senses demonstrated) by Tommaso Campanella, b Stilo (Italy), Sep 5, 1568, opposes Aristotle's views

1592

Korean astronomers observe a nova in the constellation Cetus and track its changes over a 15-month period

David Fabricius, b Esens (Germany), Mar 9, 1564, discovers the star later named Mira, the first known variable star; he watches it gradually disappear

Mysteriumeriun cosmographicum (The mystery of the universe) by Johannes Kepler, b Weil der Stadt (Germany), Dec 27, 1571, includes the concept that the sphere of each planet is inscribed in or circumscribed about one of the five Platonic regular solids; this would explain why there are exactly six planets

1593

1594

1595

MATHEMATICS	PHYSICAL SCIENCE	TECHNOLOGY	
	Bernard Palissy, French ceramist, d Paris about this date	The Reverend William Lee of Cambridge invents the first knitting machine, the stocking frame Giambattista della Porta's *Natural magic* is the first Western book to mention kites and kite flying	1589
	Galileo's *De motu* (On motion) refutes Aristotelian physics and relates his experiments with falling bodies Zacherais Janssen probably invents the compound microscope	Robert Norman's *The safegarde of saylors*, a translation from the Dutch, is a navigational manual that contains maps of the appearance of the coast as seen from the sea Giacomo della Porta and Domenico Fontana complete the dome on St. Peters in Rome, still the largest church in the world; the church was designed in 1503 by Donato Bramante, but mostly completed by Michelangelo, who also designed the dome	1590
Isagoge in artem analyticam (Introduction to the analytic art) by Franciscus Vieta is the first work in mathematics to use letters for variables and constants (vowels for variables and consonants for constants)	English mathematician Thomas Harriot (b 1560) is the first European known to note that snowflakes are either six-pointed or six-sided, but he does not publish this observation; the hexagonal nature of snowflakes was well-known in China from at least the second century BC		1591
	Galileo's *Della scienza meccanica* (On the mechanical sciences) About this time Galileo develops a thermoscope, a primitive form of thermometer using air instead of liquid; it is grossly inaccurate, but forms the basis for measuring temperature for at least the next ten years		1592
The first description of a modern Chinese abacus appears in China			1593
John Napier (b Edinburgh, Scotland, 1550) first has the idea that will result in the invention of logarithms after 20 years of work		Gerardus Mercator, Flemish geographer, d Duisburg (Germany), Dec 2	1594
Bartholomaeus Pitiscus, b Grunberg (Germany), Aug 24, 1561, is probably the first to use the word *trigonometry* in print		Mercator's *Atlas sive cosmographicae*, published posthumously, contains a collection of detailed maps of Europe	1595

	GENERAL	ASTRONOMY	LIFE SCIENCE
1596			John Gerard's *Herbal* contains the most important survey of botanical knowledge at this time
			Li Shi-Chen's *Ben-zao gang-mu* contains descriptions of more than 1000 plants and 1000 animals along with 8000 medicinal uses for them
1597			
1598	Henry IV of France proclaims the Edict of Nantes, granting civil rights to French Protestants		
1599		Tycho Brahe moves to the court of Holy Roman Emperor Rudolph II in Prague (Czechoslovakia)	Ulisse Aldrovandi publishes his first volume of *Natural History* (3 volumes), the first serious work in zoology
1600	William Shakespeare's *Hamlet* is first performed	Johannes Kepler begins assisting Tycho Brahe at his Prague (Czechoslovakia) observatory	
		Giordano Bruno, Italian philosopher, accused of heresy because of his adherence to the theory that Earth revolves around the sun and other theories, is burned at the stake in Rome, Feb 17; his connection with the Copernican theory is said to have helped cause Galileo's persecution for writing in favor of a moving Earth	
1601		Tycho Brahe, Danish astronomer, d Benatky, near Prague (Czechoslovakia), Oct 24; Johannes Kepler is appointed his successor	Julius Casserius's *De vocis auditusque organis historia anatomica* (On the anatomy of the voice and hearing) is an illustrated work on the larynx and the ear
1602		Tycho Brahe's *Astronomiae instauratae progymnasmata* (Introduction to the new astronomy), published posthumously, contains detailed positions for 777 stars and a description of the 1572 supernova	

The *Opus palatinum de triangulis* by Rheticus is published posthumously; it contains trigonometric tables for the values of the six standard trigonometric functions		Li Shih-Chen's *The great pharmacopoeia* gives one of the clearest descriptions of how to distill wine into spirits, a technique known to the Chinese since about the seventh century	**1596**

Admiral Visunsin of Korea develops the first ironclad warship

---

1597

In correspondence with Kepler, Galileo admits that he has come to accept the Copernican scheme of the solar system

Alchemia (Alchemy) by Andreas Libavius (b Andreas Libau, Halle, Saxony, 1540) is one of the first important textbooks in chemistry, describing the preparation of hydrochloric acid, tin tetrachloride, and ammonium sulfate

---

1598

Philip III of Spain offers a prize and a lifetime pension to anyone who can discover a way to find the longitude at sea

---

1599

1600

Adriaen Anthoniszoon and his son calculate π as 3.1415929 (the correct value to that many places is 3.1415927)

Franciscus Vieta finds a solution to the "fourth Apollonian problem," constructing a circle touching three given circles

William Gilbert's *De magnete* (Concerning magnetism) suggests that Earth is a great spherical magnet and reports on various substances that can be used to produce static electricity; it is the first treatise on physical science based entirely on experimentation

---

1601

1602

Vincenzio Casarido discovers barium sulfide

	GENERAL	ASTRONOMY	LIFE SCIENCE
1603	The *Accademia dei Lincei* (Academy of Lynceus or Lynxes or Lynx-Eyed), a scientific society, is founded in Rome by Federigo Cesi; still active today, it may have been named for Lynceus, the sharp-eyed Argonaut with telescopic vision; the society adopted the lynx as their symbol Elizabeth I of England d	*Uranometria* by Johann Bayer, b Rain (Germany), 1572, introduces the method of describing the locations of stars and of naming them with Greek letters and by the constellation they are in; this system continues to be used today; it is the first attempt at a complete celestial atlas	Hieronymus Fabricius's *De venarum ostiolis* (On the valves in the veins) presents a detailed study of valves in veins; the functions of these valves will not be understood, however, until 1616 Sanctorius Sanctorius, b Justinopolis (Yugoslavia), Mar 29, 1561, describes his device that uses a pendulum for counting pulse beats Andrea Cesalpino, Italian doctor and botanist, d Feb 23 William Gilbert, English physician, physicist, and earth scientist, d London, Nov 30
1604		Kepler observes and describes a supernova in the constellation Ophiuchus, also seen by Korean and Chinese astronomers; it is not quite as bright as Venus and lasts for 12 months	*De formata foetu* (On the formation of the fetus) by Hieronymus Fabricius ab Aquapendente, b May 20, 1537 (Italy), is one of the first important studies of embryology; it contains a study of blood circulation in the umbilical cord
1605	*Advancement of learning* by Francis Bacon (b London, England, Jan 22, 1561) argues against magic and encourages the development of scientific methods		Ulisse Aldrovandi, Italian naturalist, d Bologna, May 10
1606		Kepler's *De stella nova* (On a new star) describes the nova first observed in 1604, including its astrological significance	
1607	Jamestown (Virginia) established		
1608			

Pietro A. Cataldi finds the sixth and seventh perfect numbers, which are 8,859,869,056 and 137,438,691,328		Hugh Platt discovers coke, a charcoal-like substance produced by heating coal	**1603**
Franciscus Vieta (François Viète), French mathematician, d Paris, Feb 23			

1604

Galileo announces correctly in a letter to Paolo Sarpi that a body falling freely increases its distance as the square of the time; he also states, incorrectly, that the velocity is proportional to distance; Galileo corrects his error in 1609

Johannes Kepler's *Ad vitellionem paralipomena quibus astronomiae pars optica traditor* (The optical part of astronomy) describes how the eye focuses light and shows that the intensity of light falls as the square of distance from the source

Johan Thölde edits or writes *Triumphwagen des antimonii* (Triumphal chariot of antimony), a detailed description of the metal antimony and the uses of its compounds; Thöde attributes the work to a sixteenth-century monk named Basil Valentine

---

1605

1606

1607

1608

Ein Kurser Tractat von der Natur der Elementum (An alchemical tract on the transmutation of elements) by Cornelius Drebbel (b Holland, 1572) anticipates the production of oxygen by heating saltpeter

Dutch scientist Hans Lippershey, b Wesel (Germany) about 1570, invents the telescope

	GENERAL	ASTRONOMY	LIFE SCIENCE
1609	Joseph Justus Scaliger, French scholar, d Leiden, Holland, Jan 21	Galileo builds his first telescope and, with modifications and improvements, eventually obtains a magnification of about 30 Kepler's *Astronomia nova* (New astronomy) contains his views that the planets revolve around the sun in elliptical orbits and that these orbits sweep out equal areas in equal time intervals, observations that later contribute to the confirmation of Newton's theory of gravity Thomas Harriot uses a simple telescope to sketch the moon	**Galileo and his telescope** In 1609 Galileo heard of a new invention shown at an exposition in Venice. It was an "optical tube" that would bring distant objects close. Galileo set out to reinvent such an instrument and quickly succeeded. During the next few months Galileo made some of the most important discoveries of astronomy. These observations also established the Copernican system as physical reality.
1610	French colony of Quebec established	Galileo turns his telescope to the night skies and sees the moons of Jupiter on Jan 7, Saturn's rings (which he does not understand), the individual stars of the Milky Way, and the phases of Venus; he reports this in *Sidereus nuncius* (Starry messenger), a series of newsletters that make him famous all over Europe	He observed the craters and mountains on the moon and was struck by the Earth-like appearance of the moon's landscape. He also noticed that the dark side of the moon was faintly illuminated by earth, a phenomenon known as Earthshine. With this he proved that Earth "shines" as much as the other planets and thus must reflect the light from the sun.
1611	The King James translation of the Bible is published	Thomas Harriot, Johannes Fabricius, Father Scheiner, and Galileo discover sunspots around the same time; Galileo claims to have seen a sunspot as early as 1607 but believed it to be the planet Mercury passing in front of the sun Nicolas Fabri de Peiresc (b Beaugensier, France, Dec 1, 1580) discovers the Orion nebula Christoph Scheiner, b Wald, Rhine Province (Germany), Jul 25, 1575, observes sunspots in Mar, drawing him into a controversy with Galileo as to who saw them first	Directing his telescope on the planets, Galileo discovered that, unlike the stars, which remained pointlike in his telescope, the planets appeared like small disks, or, as in the case of Venus, as a miniature crescent. This convinced him that the stars are much farther away from Earth than the planets. He also
1612		Simon Marius, b Gunzenhausen (Germany), Jan 20, 1573, is the first astronomer to mention the Andromeda galaxy Christoph Clavius, German astronomer, d Rome, Feb 6	John Gerard, English botanist, d London, Feb, 1612
1613		Galileo's *The sunspot letters* reports on his observations of sunspots; it is his first printed statement favoring the Copernican system and it contains his first formulation of the principle of inertia	Basilius Besler's *Hortus eychstettensis* is an important work in plant illustration Physician and architect Claude Perrault b Paris, France, Sep 25; among his designs is the facade for the Louvre known as the *Colonnade*

Galileo and his telescope, continued
discovered that the Milky Way consists of a large number of clusters of innumerable stars.

Observing Jupiter, Galileo discovered that it is surrounded by four "stars" (at that time the fixed stars, as well as the planets, were just called stars). He called these moons, as we now know them to be, "Medicean stars" in honor of the Medici, rulers of Florence. He started observing their motion around Jupiter and came to the conclusion that they must be circling the planet in the same way as the moon is circling Earth. This discovery was an important one because it showed that the Earth-moon system is not a unique anomaly, which many people thought an important objection to the Copernican system.

The strongest proof for the validity of the Copernican system came from Galileo's observations of the phases of Venus. Not only did it become clear that Venus also reflects light from the sun and does not shine by its own light, but Galileo also observed that the size of Venus changes with the changes of its phase: when the planet was visible as a complete disk, it was very small, and when it became a narrow crescent, its apparent diameter became large. This could only be explained by assuming that the planet circles the sun.

Finally, by observing the motion of sunspots, Galileo demonstrated the rotation of the sun.

1609

The first attempt is made to harness the tides in the Bay of Fundy as a source of power; small mills are successfully powered by this means

Hans Lippershey and separately Zacharias Janssen, invents the compound microscope; Janssen's invention may date to 1590

1610

Tyrocinium chymicum by Jean Beguin (b Lorraine, France, about 1550) is the first textbook of chemistry instead of alchemy

---

1611

Marco de Dominis publishes a scientific explanation of the rainbow

Johannes Kepler's *A new year's gift, or On the six-cornered snowflake* is the first published description of the hexagonal nature of snowflakes in the West, although this characteristic was known earlier in China

Kepler's *Dioptrice* describes the so-called astronomical telescope (inverting refractor)

The 57-m (186-ft) Tour de Condonan at the mouth of the Garonne River in France is the first lighthouse to have a revolving beacon

---

1612

Galileo's *Discurso intorno alle cose che stanno in su l'acqua* (Discourse on things that float in water) uses principles from Archimedes to develop elementary hydrostatics

Sanctorius Sanctorius's *Commentaria in artem medicinalem Galeni* contains the first printed mention of the thermoscope, a primitive thermometer invented by Galileo

---

1613

Pietro A. Cataldi develops methods of working with continued fractions; these are fractions whose denominators contain fractions, often in an infinite progression of fractions in the denominator

Bartholomaeus Pitiscus, German mathematician, d Heidelberg, Jul 2

1614

Sanctorius Sanctorius's *De statica medicina* offers the first study of metabolism; Sanctorius records measurements of changes in his own weight, pulse, and temperature

Physician Franciscus Sylvius (Franz de le Boë) b Hanau (Germany), Mar 15; he will be one of the first physicians to abandon the theory that illness is caused by an imbalance of the four humors (blood, black bile, yellow bile, and phlegm); instead, he thinks it is an imbalance between acids and bases

1615

William Baffin (b England about 1584) penetrates to within 1300 km (800 mi) of the North Pole, the closest anyone will be until the nineteeth century

1616

Pocahontas arrives in London

William Shakespeare and Miguel de Cervantes Saavedra both d Apr 23, Stratford-on-Avon, England, and Madrid, Spain, respectively

Galileo receives a warning from Cardinal Robert Bellarmine that he should not hold or defend the Copernican doctrine that Earth revolves about the sun; Copernicus's *De revolutionibus* is placed on the *Index Librorum Prohibitorum* of the church, from which it will not be removed until 1835

William Harvey (b Folkestone, England, Apr 1, 1578) lectures about the circulation of the blood to the Royal College of Physicians, England

Prospero Alpini, Italian botanist, d Padua, Nov 23

1617

Natural philosopher Elias Ashmole, founder of the Ashmolean Museum in Oxford, b Litchfield, England, May 23

David Fabricius, German astronomer, d Osteel, Ostfriesland, May 7

1614

John Napier's *Mirifici logarithmorum canonis descripto* (Description of the wonderful canon of logarithms) explains the nature of logarithms and gives tables and rules for their use

1615

Johannes Kepler's *Nova stereometria doliorum vinariorum* (New measurement of the volume of wine casks) uses infinitesimals to calculate the volume of various solids of revolution, such as wine casks, anticipating the integral calculus

Marin Mersenne (b Sarthe, France, Sep 8, 1588) calls attention to the cycloid, the curve formed by the motion of a point on a circle as the circle is rolled along a line; the cycloid is called "the Helen of geometers" because it provokes so many quarrels among seventeenth-century mathematicians; it is also the key to Huygens's first pendulum clocks

Giambattista della Porta, Italian physicist, d Naples, Feb 4

1616

Andreas Libavius, German alchemist, d Coburg, Bavaria, Jul 25

Galileo, hoping to win a prize and pension set up by Philip III of Spain for anyone who can find the longitude at sea, suggests using the satellites of Jupiter; although ignored by Philip, Galileo's method will later be used to find accurate longitudes on land

1617

Logarithmorum chilias prima (Logarithms of numbers from 1 to 1000) by Henry Briggs (b Yorkshire, England, Feb 1561) introduces common logarithms (logarithms based on powers of 10)

Eratosthenes Batavus by Willebrord van Roijen Snell (b Leiden, Holland, 1591) develops the method of determining distances by trigonometric triangulation

John Napier, Scottish mathematician, d Merchiston Castle, near Edinburgh, Apr 4

John Napier describes the device for multiplying that comes to be known as Napier's rods or Napier's bones

1618

The Thirty Years' War starts in Germany and spreads to other states in Europe; it is mainly a conflict between Protestant and Catholic states

Science administrator Henry Oldenburg b Bremen (Germany) about this time; as secretary of the Royal Society in the time of Newton and Leibniz, he is the recipient of many formal letters announcing developments in the calculus; he serves as a link between the two mathematicians

1619

René Descartes (b La Haye, France, Mar 31, 1596) has a famous dream on Nov 10 in which he is given the message that he should work out the unity of the sciences on a purely rational basis

Johann Cysat (b Lucerne, Switzerland, 1586) discovers the Orion nebula

Kepler's *Harmonice mundi* contains his third law of planetary motion (that the squares of times of revolution of any two planets are proportional to the cubes of their mean distances from the sun) and his theories of harmonies in nature

Kepler's *Epitome astronomiae Copernicanae* (Epitome of Copernican astronomy) is a defense of the Copernican system in three parts (the last published in 1621); it is immediately put in the *Index Librorum Prohibitorum* of the Roman Catholic Church

Kepler explains that a comet's tail points away from the sun as a result of what we now call solar wind—particles from the sun that push material from the comet out of the head and away from the sun

Hans Lippershey, German-Dutch optician, d Middelburg, Holland, about this time

Hieronymus Fabricius ab Aquapendente, Italian physician, d Padua, May 21

1620

Pilgrims land at Plymouth Rock (Massachusetts)

Francis Bacon's *Novum organum* (New organon, referring to Aristotle's book on logic, the *Organon*) recommends induction and experimentation as the bases of the scientific method

1621

Hieronymus Fabricius's *De formatione ovi et pulli* (On the development of eggs and chickens), published posthumously, elevates embryology to the level of a science

1618

Francesco Maria Grimaldi b Bologna (Italy), Apr 2; he discovers the interference pattern and diffraction of light waves, convincing evidence that light is a wave phenomenon, but little attention is paid to his work until Thomas Young rediscovers these phenomena in 1803

1619

Michelangelo Ricci b Rome (Italy), Jan 30; as a cardinal, he supports the work of such mathematicians as Stefano degli Angeli and becomes a close friend of Torricelli

Saturn's rings

In 1610, Galileo reported to Johannes Kepler: "SMAISMRMILMEPOETALEUMIBUNENUGTTAUIRAS."

This report of a major new discovery with the new telescope was an anagram, a scrambled message. Anagrams were used in the seventeenth century to establish the priority of a discovery without revealing what had been discovered. (Imagine Watson and Crick sending a colleague an anagram of "DNA IS DOUBLE HELIX.") In this case, Kepler correctly unscrambled the anagram as ALTISSIMUM PLANETAM TERGEMINUM OBSERVAI, or (loosely) "I have observed the most distant of planets to have a triple form." Kepler assumed that this meant that Galileo had found that Saturn has two satellites.

What Galileo saw was harder to interpret than that, however. In his notebook, the drawing of Saturn looks like a ball with handles. To make matters more complicated, three years later the "handles" had disappeared. Galileo asked, "Has Saturn perhaps devoured his own children?" But by 1616 the planet once again showed its odd triple shape.

Using a better telescope, Christiaan Huygens was able in 1656 to write an anagram announcing that Saturn has a ring. Huygens, however, assumed that the ring is a single, solid structure. Giovanni Cassini observed in 1675 that there were at least two rings separated by a gap and correctly guessed that the rings are made up of many small objects. But Cassini's view was not accepted until James Clerk Maxwell studied the problem mathematically in 1857, more than 200 years after Galileo had observed the rings for the first time.

1620

William, 2nd Viscount Brouncker, b England; he becomes the first president of the Royal Society and specializes in continued fractions

Arithmetische und geometrische progress-tabulen by Joost Bürgi (b Lichtensteig, Switzerland, Feb 28, 1552) present an independent discovery of logarithms

Jan (Johann) Baptista van Helmont (b Brussels, Belgium, Jan 12, 1579) coins the term "gas" to describe substances that are like air; the word *gas* is his own peculiar spelling of the Flemish word for chaos

Jean Beguin, French chemist, d Paris about this time

Flemish mathematician and physicist Simon Stevinus (Stevin) d The Hague, Holland, around Mar

Cornelius Drebbel builds a navigable submarine, powered by rowers, that can carry 24 persons; it cruises 5 m (15 ft) below the surface of the Thames in London on several occasions; the secret of the craft is probably the production of oxygen from saltpeter by a process Drebbel kept secret

1621

The famous edition of Diophantus's *Arithmetica* (Arithmetic) is published in Greek with a Latin translation, launching the revival of number theory; it is best known for having too narrow a margin to accommodate Fermat's "last theorem"

Willebrord Snell discovers that refraction of light is determined by the sine of the angle made by the incident ray (Snell's Law)

	GENERAL	ASTRONOMY	LIFE SCIENCE
1621 cont			
1622	The first German scientific academy, the *Societas Ereunitica*, is founded at Rostock William Baffin, English explorer, d Persian Gulf, Jan 23		
1623			*Pinax theatri botanici* by Gaspard Bauhin (b Basel, Switzerland, Jan 17, 1560) introduces the practice of using binomial names, one for the genus and a second for the species
1624		Simon Marius (b Mayr), German astronomer, d Anspach, Bavaria, Dec 26	Physician Thomas Sydenham, "the English Hippocrates," b Wynford Eagle, England, Sep 10; he is the first to describe measles and to identify scarlet fever; he advocates the use of opium to relieve pain, chinchona bark (quinine) to relieve malaria, and iron to relieve anemia Gaspard Bauhin, Swiss anatomist, d Basel, Dec 5
1625		Johann Bayer, German astronomer, d Augsburg, Bavaria, Mar 7	
1626	Francis Bacon, English philosopher, d London, Apr 9, a month after performing his first scientific experiment—stuffing a chicken with snow to see if it would decay less rapidly; the chill he caught during this experiment is thought to have led to his death	Godfried Wendilin, b Herken (Belgium), Jun 6, 1580, shows that Kepler's laws are also valid for the moons of Jupiter	Jan Baptista van Helmont proposes that diseases are caused by alien beings that he calls "archeae"
1627	Francis Bacon's *The new Atlantis*, published posthumously, is a utopian tale that predicts robots, telephones, tape recorders, and the electric motor, while emphasizing the importance of experimentation over deduction	Kepler's *Rudolphine tables* contain the locations of 1005 stars and calculation of the motions of the planets, 777 of which come from Tycho's previous catalog	The aurochs, the wild ancestor of domestic cattle, becomes extinct; the last known creature is seen in Europe, although the large animal previously also roamed over western Asia and northern Africa
1628			William Harvey's *Exercitatio anatomica de motu cordis et sanguinis in animalibus* (Anatomical treatise on the movement of the heart and blood in animals) describes his discovery of the circulation of blood

MATHEMATICS	PHYSICAL SCIENCE	TECHNOLOGY	
About this time William Oughtred (b Eton, England, Mar 5, 1574) invents the slide rule (according to Oughtred's statement in 1632) Mathematician, astronomer, and physicist Thomas Harriot d London, Jul 25			1621 cont
	Edmund Gunter (b England, 1581) discovers that magnetic declination varies with time		1622
Wilhelm Schickardt builds the first calculating machine based on the idea of Napier's bones; made of wood, it can add and subtract, and, with additional help from the operator, multiply and divide			1623
Henry Briggs's *Arithmetica logarithmica* (The arithmetic of logarithms) extends his tables of common logarithms to numbers from 1 to 20,000 and from 90,000 to 100,000; it also introduces Latin forms of the words *mantissa* and *characteristic*			1624
John Collins b Wood Eaton, England, Mar 5; he is primarily known because of his correspondence with Newton and other mathematicians concerning much of the early history of the calculus	Johann Rudolf Glauber, b Karlstadt (Germany), 1604, discovers Glauber's salt around this time		1625
Willebrord van Roijen Snell, Dutch mathematician and physicist, d Leiden, Oct 30 Edmund Gunter, English mathematician, d London, Dec 10		Marko Antonije Dominis, Yugoslav physicist and clergyman, d in prison, Rome, Sep 8	1626
			1627
			1628

1629

The church and science: Galileo

Galileo's first clash with members of the Catholic church was caused by a question of priority. Sunspots were observed by the Jesuit priest Christoph Scheiner, but in 1613 Galileo claimed to have discovered sunspots before Scheiner. His conflict with Scheiner had no immediate consequences, but caused problems later.

In letters Galileo argued that not everything in the Bible should be taken literally. Galileo further argued that the church had the burden of proving that the Copernican system is wrong. The Dominicans obtained copies of the letters and forwarded them to the Holy Office in Rome along with Galileo's book *Letters on solar spots*, in which he showed his support of Copernican views.

In December, 1615, Galileo traveled to Rome to defend his opinions. In February, 1616, the Inquisition declared that the opinion that the sun is the center of the world is heretical and decreed that the works of Copernicus be withdrawn until corrected. Galileo was admonished to cease defending the Copernican system.

Galileo lived in relative peace with the church until the publication in 1632 of his *Dialogue on the two chief world-systems, the Ptolemaic and the Copernican*. The book consists of a dialogue among a convinced Copernican, a stupid anti-Copernican, and a commentator. Scheiner, Galileo's enemy since the dispute over sunspots, convinced Pope Urban VIII that Galileo had intended to portray the pope as the stupid anti-Copernican. The pope summoned Galileo to appear before the Inquisition in 1633. Galileo was sentenced to abjure the Copernican doctrine, which he did, and was put under house arrest.

The church continued officially to deny the Copernican theory until 1922.

1630

Johannes Kepler, German astronomer, d Regensburg, Bavaria, Nov 15

1631

Pierre Gassendi (b France, Jan 22, 1592), using calculations made by Kepler, becomes the first to observe a transit of Mercury across the face of the sun

Michael Mästlin, German astronomer, d Tübingen, Oct 20

Physician Richard Lower b Tremeer, near Bodmin, England, about this time; he observes that contact with air turns dark blood from the veins bright and establishes that phlegm does not, as Galen claimed, originate in the brain

1632

Galileo's *Dialogo sopra i due massimi sistemi del mundo, Tolemaico e Copernico* (Dialogue concerning the two chief world systems, Ptolemaic and Copernican) puts him in disfavor with the pope and the book is banned; the "impartial" character Salviati in the dialog clearly represents Galileo's own, pro-Copernican position

1633

The Roman Catholic Inquisition forces Galileo to recant his Copernican view that Earth moves about the sun; tradition has it that at the end of his recantation, he was heard to mutter *"E pur se muove"* (Nevertheless, it moves)

MATHEMATICS	PHYSICAL SCIENCE	TECHNOLOGY	
Pierre de Fermat (b Beaumont-de-Lomagne, France, Aug 20, 1601) reconstructs the lost *Plane loci* of Apollonius and, in the process, develops the ideas that will lead to the invention of analytic geometry by the spring of 1636 Albert Girard's *L'invention nouvelle en l'algèbre* (The new science of algebra) asserts but does not prove a version of the fundamental theorem of algebra: every algebraic equation has as many solutions as the exponent of the highest term; this theorem is correctly proved for the first time by Karl Friederich Gauss in 1799		Giovanni Branca describes a steam turbine in which steam is directed at vanes on a wheel	**1629**
Henry Briggs, English mathematician, d Oxford, Jan 26	Cabaeus notices that electrically charged bodies first attract each other and then repel each other after contact		**1630**
Edward Cocker b England; his *Arithmetic*, published posthumously in 1678, exerts a major influence on English education Thomas Harriot's *Artis analyticae praxis* (Practice of the analytic art), printed posthumously, uses AAA to mean the cube of A and similar notations for other powers; however, Harriot symbolizes multiplication with a centered dot and introduces < and > for "is less than" and "is greater than," modern notations William Oughtred's *Clavis mathematicae* (Key to mathematics) introduces many new notations, of which only × for multiplication continues in use today		Pierre Vernier (b Ornans, France, Aug 19, 1580) describes his invention for precision measurement, known today as the Vernier scale	**1631**
In *Dialogue concerning the two chief world systems* Galileo notes that the number of perfect squares is as large as the number of roots of perfect squares, the first known matching of a set with a proper subset of itself Joost Bürgi, mathematician, d Kassel (Germany), Jan 31	Galileo's *Dialogue concerning the two chief world systems* introduces relativity to physics by pointing out that physics experiments made in a closed cabin on a ship cannot be used to tell whether the ship is moving or stationary		**1632**
	Dutch inventor, chemist, and instrument-maker Cornelius Drebbel d London, Nov 7	Military engineer Sebastian Le Prestre de Vauban b Saint-Léger, Nivernais, France, May 15; he develops a system of fortified fortresses to defend France against invasions	**1633**

	GENERAL	ASTRONOMY	LIFE SCIENCE
1634			
1635		Guillaume Schickard, German astronomer, d Tübingen, Oct 23	Naturalist Francis Willughby b Middleton, England, Nov 22; his systematic work on birds and fish helps pave the way for Linnaeus's system
1636	Harvard College (Massachusetts) is founded	Henry Gellibrand, English astronomer and mathematician, d London, Feb 16	Sanctorius Sanctorius (also known as Santorio Santorio), Italian physician, d Venice, Mar 6
1637	René Descartes' *Discours de la méthode pour bien conduire la raison et chercher la vérité dans les sciences* (Discourse on the method of rightly conducting reason and seeking truth in the sciences) argues for use of the deductive method in science, illustrated by three famous appendices	A permanent observatory is established in Copenhagen by King Christian IV of Denmark Nicolas-Claude Fabri de Peiresc, French astronomer, d Aix-en-Provence, Jun 24	
1638		Dutch astronomer Phocyclides Holwarda identifies the first variable star, Mira Ceta, previously observed in 1596 by David Fabricius, who saw Mira gradually disappear	

1634

Giles Personne de Roberval (b Roberval, France, Aug 1, 1602) proves that the area under a curve described by the motion of a point on a circle as the circle is rolled along a line for one circumference is exactly three times the area of the circle; the curve described by the point is the famous cycloid

---

Galileo and measurement

Before the seventeenth century, it was almost impossible to measure most things precisely. Although length and mass could be measured quite well, chemists did not realize the power of the balance, which was used primarily by assayers. Time could only be measured in large intervals. Temperature and fluid pressure could not be measured in numbers at all.

Galileo was at the heart of changing this situation. In 1581, when he was only 16, he noticed that the period of a pendulum appeared to be controlled solely by its length. This discovery led to the manufacture of good pendulum clocks by the end of the seventeenth century. In 1586, Galileo published his invention of a hydrostatic balance. In 1600, Galileo built the first primitive tool for measuring temperature. This was refined into a workable thermometer over the course of the seventeenth century and put into modern form by Fahrenheit in 1714. It was Galileo who suggested to Torricelli the investigation that led to the barometer. Galileo's telescopes inspired others to build astronomical telescopes, which in turn led to the micrometer. Telescopes used in surveying also inspired the vernier system for making accurate measurements of angles.

None of these tools for measurement was sufficiently accurate to contribute to the development of science in Galileo's lifetime. Shortly after his death in 1642, the new tools helped create modern science.

1635

Geometria indivisibilibus continuorum by Bonaventura Cavalieri, b Milan (Italy), 1598, develops his method of calculating volumes by using infinitely small sections, an important precursor of the integral calculus

Around this date, René Descartes discovers Euler's theorem that links the number of vertices, edges, and faces that a simple convex polyhedron must obey $(V - E + F = 2)$; since Descartes' discovery was not published until 1860, 108 years after Euler's rediscovery, the theorem is known as Euler's

---

1636

Pierre de Fermat writes to Marin Mersenne that he has discovered the first pair of amicable numbers (pairs such that the divisors of one number add to the other number) since the single pair of 220 and 284 known to the ancient Greeks; Fermat's pair is 17,296 and 18,416

Pierre de Fermat proposes in a letter to Mersenne that every natural number is the sum of at most three triangular numbers (stated publicly for the first time by Pascal in 1665 and finally proved in 1801 by Karl Friedrich Gauss)

Harmonie universelle (Universal harmony) by Marin Mersenne describes equal temperament in music for the first time in the West, although it was previously known in China

---

1637

An appendix to René Descartes' *Discours de la méthode* contains the first published development of analytic geometry (also developed in 1636 or earlier by Pierre de Fermat, but Fermat's work is not published until 1670)

René Descartes' *La Dioptrique*, an appendix to the *Discours de la méthode* contains his theory of refraction; in *Les météores* he explains the phenomenon of the rainbow and the formation of clouds

Isaac Beeckman, Dutch physicist, d Dordrecht, Holland, May 19

1638

Claude-Gaspar Bachet de Méziriac, French mathematician, d Bourg-en-Bresse, Feb 26

Galileo's *Discoursi e demonstrazione matematiche intorno à due nuove scienze* (Mathematical discourses and demonstrations on two new sciences) includes his mathematical discussion of the laws of motion and of friction, in which he corrects many of Aristotle's errors

Pierre Vernier, French engineer, d Ornans, Sep 14

	GENERAL	ASTRONOMY	LIFE SCIENCE
1639		Jeremiah Horrocks (b Toxteth Park, near Liverpool, England, 1618) observes the transit of Venus across the sun on Nov 24, which he had predicted; he is the first person known to have observed the transit	
1640			
1641		German astronomer Johannes Hevelius (b Danzig—now Gdańsk, Poland—Jan 28, 1611) builds an observatory in his house with an azimuthal quadrant of 1.5-m (5-ft) radius and a 1.8-m (6-ft) sextant Jeremiah Horrocks, English astronomer, d Toxteth Park, near Liverpool, Jan 13	Nicolaas Tulp describes the first living chimpanzee brought to Holland
1642	English Civil War starts	Galileo, Italian astronomer and physicist, d Arcetri, near Florence (Italy), Jan 8	
1643	Louis XIV becomes king of France		

1639

Florimond de Beaune (b Blois, France, Oct 7, 1601) writes Descartes to ask which curve has a constant subtangent and other questions about curves; Descartes' reply fails to solve the problem, but the exchange of letters becomes known to Leibniz, who shows in 1684 that the solution is a logarithmic curve

Brouillon projet d'une atteinte aux évenements des rencontres d'un cone avec un plan (Rough draft of an attempt to deal with the meeting of a cone with a plane) by Gerard Desargues (b Lyons, France, Mar 2, 1591) is the beginning of projective geometry as a systematic branch of mathematics

William Gascoigne (b Middleton, England about 1612) invents the micrometer about this time, placing it in the focus of a telescope for measuring the angular distance between stars

1640

Essai pour les coniques (Essay on conics) by Blaise Pascal (b Clermont-Ferrand, France, Jun 19, 1923) proves Pascal's theorem, also known as the mystic hexagram

Pierre de Fermat states in an Oct 19 letter to Bernard Frénicle de Bessy the "little theorem" of number theory; for any prime p and any natural number n, it is always true that p divides the difference of n raised to the power of p and n with no remainder

De motu gravium by Evangelista Torricelli, b Faenza (Italy), Oct 15, 1608, applies Galileo's laws of motion to fluids, making Torricelli the father of hydrodynamics

1641

The Grand Duke of Tuscany, Ferdinand II, b (Italy) Jul 14, 1610, invents a thermometer that uses liquid in a glass tube that has one end sealed; this is a slight improvement on Galileo's thermoscope

The son of Galileo designs a clock with pendulum, a device based on a concept of his father

1642

Pascal invents a machine that can add and subtract; after many experimental models, his machine will go on sale in 1645

The Briare Canal linking the Loire and the Seine in France is opened

1643

Paul Guldin, Swiss mathematician, d Graz, Austria, Nov 3

Torricelli makes the first barometer using mercury as a fluid in a glass column sealed at the top; when the tube is upended in a dish, the mercury sinks to about 76 cm (30 in.), leaving a partial vacuum at the top; this is also the first vacuum known to science

Benedetto Castelli, Italian physicist, d Rome Apr 19

	GENERAL	ASTRONOMY	LIFE SCIENCE
1644	Li Zi-Cheng overthrows the Ming dynasty, but the Manchus take over China	René Descartes' *Principia philosophiae* (Principles of philosophy) includes his vortex theory of the origin and state of the solar system, which he perceives in terms of whirling matter William Gascoigne, English astronomer, d Marston Moor, Yorkshire, Jul 2	Jan (Johannes) Baptista van Helmont, Flemish physician and alchemist, d Vilvoorde, near Brussels, Dec 30
1645	Meetings of The Invisible College, precursor of the Royal Society, begin in Gresham College, Oxford, England	Ismael Boulliau indicates in *Astronomia philolaoica* that the central force acting on planets must be proportional to the inverse square of the distance (later proven by Newton)	Daniel Whistler (b England, 1619) gives the first medical description of rickets in his thesis at the University of Leiden
1646			
1647		Johannes Hevelius's *Selenographia* is the first map of the side of the moon observable from Earth Christian Severin (Longomontanus), Danish astronomer, d Copenhagen, Oct 8	Yellow fever, brought along with slaves from Africa, strikes the island of Barbados in the Caribbean, presenting the first recorded cases in the Americas Jean Pecquet (b Dieppe, France, May 9, 1622) discovers the thoracic duct
1648	Thirty Years' War ends Semjon Deshnjov travels through the passage between America and Asia, proving that these two continents are not connected		Jan Baptista van Helmont's *Ortus medicinae* (On the development of medicine), published posthumously, tells of his famous experiment with growing a willow tree; it shows that the tree's increase in weight does not come from the soil
1649	Charles I of England is beheaded		

1644

1645

| | | Athanasius Kircher, b Geisa at the Ulster (Germany), May 2, 1601, invents the magic lantern | **1646** |

Bonaventura Cavalieri's *Exercitationes geometricae sex* (Six geometric exercises) demonstrates how to find the integral of x^n for $n = 3, 4, 5, 6, 7, 8, 9$

Claude C. Mydorge, French mathematician, d Paris, Jul

Bonaventura Cavalieri, Italian mathematician, d Bologna, Nov 30

Blaise Pascal's *Expériences nouvelles touchant le vide* (New experiments with a vacuum) describes his experiments with vacuums produced above a column of water or red wine

Evangelista Torricelli, Italian physicist, d Florence, Oct 25

1647

Desargues' Theorem is published by his friend Abraham Bosse; the theorem states that if two triangles are placed so that lines joining pairs of corresponding vertices pass through a single point, then the points of intersection of pairs of corresponding sides are on a single line, and vice versa

Mathematical magik by John Wilkins (b Fawsley, England, 1614) is an account of the fundamental principles of machines

Marin Mersenne, French mathematician, d Paris, Sep 1

Pascal arranges for his brother-in-law to demonstrate that the height of the mercury column in a barometer is balanced by the pressure of air by carrying a barometer up the mountain *Puy de Dôme*

1648

Geometria a Renato Des Cartes (Geometry by René Descartes) by Frans van Schooten (b Leiden, Holland, 1615) is translated into Latin; it expands on Descartes' *La géométrie*; from 1659 to 1661 a further expanded, 2-volume version will appear; this book introduces analytic geometry to scholars all over Europe

Pierre Gassendi's study of Epicurus, *Syntagma philosophiae Epicuri*, asserts that matter is made up of atoms

1649

1650 | Irish Bishop James Ussher sets the date of Creation at 4004 BC and the date of Noah's flood as 2349 BC, based on accounts in the Bible | Christoph Scheiner, German astronomer, d Neisse, Silesia (now Nysa, Poland), Jun 18 | *De rachitide* by Francis Glisson (b Dorset, England, 1596) gives a detailed clinical description of rickets

1651 | *Leviathan* by Thomas Hobbes (b England, 1588) proclaims man's life to be "solitary, poor, nasty, brutish, and short;" he argues that primitive people lived in anarchy and surrendered their rights to the state; he proposes a strong authoritarian government as an antidote; Hobbes wrote much on mathematics but nothing of lasting value | *Almagestum novum* by Giovanni Battista Riccioli (b Ferrara, Italy, Apr 17, 1598) compares Tycho Brahe's model of the planetary system with the Copernican system and favors the first because it is more in accordance with the Holy Scriptures; Tycho has a stationary Earth, but the other planets revolve about the sun | William Harvey's *Exercitationes de generatione animalium* (Exercises on the generation of animals) describes organ differentiation in the developing embryo

1652

De lacteis thoracicis by Thomas Bartholin (b Copenhagen, Denmark, Oct 20, 1616) offers a treatise on the lymphatic system, demonstrating it in humans

Olof Rudbeck (b Westerås, Sweden, Dec 12, 1630) demonstrates the lymphatic vessels to Queen Christiana of Sweden, using a dog

1654

Francis Glisson's *Anatomia hepatis* gives a detailed anatomical description of the liver

A lot out of nothing (part one)

Aristotle concluded that a vacuum could not exist. He believed that an object would travel with an infinite speed in a vacuum, but he did not believe an actual infinity could exist. Although scientists generally agreed that a vacuum could not exist in nature, artisans proceeded to devise pumps that used the vacuum anyway. These pumps consisted of a cylinder tightly fitted in a tube. When the cylinder was raised, water flowed upward in the tube. The explanation of how such pumps worked, based on Aristotle, was that water had to rush into a tube as a cylinder was lifted because "nature abhors a vacuum." If the water did not fill the empty space, the forbidden vacuum would be created.

Galileo learned from a workman that such vacuum pumps had a curious liability. They would not lift water more than about 10 m (30 ft). Galileo thought it odd that nature should abhor a vacuum for 10 m and then change its mind. He suggested that his assistant, Torricelli, investigate this matter. Torricelli soon realized that, contrary to Aristotle, a real vacuum might be created. Shortly after Galileo's death, Torricelli was able to produce a near vacuum with his mercury barometer. This was the first vacuum known to science, although it can be assumed that partial vacuums were produced in water pumps by people trying to exceed the 10-m limit.

Pumps could not exceed this limit because their action depends on air pressure. Air pressure is strong enough to lift water 10 m when it is not opposed by anything. Above 10 m, a partial vacuum can exist because air pressure will not lift the water to fill it.

Interest in the vacuum continued through the second half of the seventeenth century. Otto von Guericke developed an air pump that could produce an impressive vacuum; he delighted in exhibitions of its power,

synthesis," a procedure that included both an inductive and a deductive stage. Theories are formulated from observations; these theories are then used to predict other phenomena. However, in practice, science does not operate strictly according to scientific method. Newton's most important discoveries were probably the product of intuition, which he later backed up with experiment, reasoning, and mathematics.

Every natural phenomenon, according to Newton, could ultimately be explained by mathematical law, an approach to science not necessarily antagonistic to religion. In fact, Newton and Leibniz were both deeply religious. For example, Leibniz interpreted physical laws as nature's way of achieving a maximum or minimum of certain physical quantities. The presence of a direction in nature's laws implied for him the presence of a "Perfect Being" that created the universe.

It was during this time, however, that the separation of physics and metaphysics (or philosophy) took place. For example, Newton accepted the mathematical description of gravity, knowing full well that his laws of gravitation said nothing about the nature or cause of gravitation. Newton refused to frame hypotheses that were not verifiable, unlike René Descartes, who hypothesized the existence of vortices to explain gravitation.

Observation and experimentation became the pillars of scientific activity. The idea that real knowledge and wisdom could be found only in the writings of the ancients lost ground gradually. Scientists ceased to rely on the old masters, such as Aristotle, and started to study the results of their own observations and experiments when formulating theories. They began recording phenomena in terms of values such as weight, volume, or temperature. Scientific instruments became widely available in every major town. The development of a pendulum that kept perfect time by Christiaan Huygens and the invention of the balance spring by Robert Hooke made the clock a precision instrument.

Scientists, lacking the present-day concept of energy, tried to grasp such qualities as heat, electricity, magnetism, and chemical energy by describing them in terms of material, weightless fluids capable of flowing through solids. The wide experimentation of this period resulted in many observed phenomena that were not understood. Nevertheless, these experiments formed the basis for the theoretical advances in physics and chemistry at the end of the eighteenth and beginning of the nineteenth centuries.

Unity of the sciences

Newton's discovery that a single law gave an accurate description of phenomena on Earth as well as in celestial space broke down the long held belief that the heavenly bodies were of special and divine nature. Astronomy became an extension of Newton's mechanics and the return of the great comet of 1682 in 1758—at the time predicted by Edmund Halley—was viewed as the ultimate confirmation of the validity of Newton's theory of gravitation.

Science, still called "natural philosophy," was not yet divided into the disciplines we know today. Natural philosophy encompassed all phenomena of nature, including astronomy, optics, statics, hydraulics, and mathematics. Chemistry was considered to be closely allied to medicine.

Major advances

Astronomy. Newton's theory of gravitation provided a theoretical basis for the Copernican system and Kepler's laws. Newton's work was also the starting point of a new field, celestial mechanics, that was to dominate astronomy for the next 200 years (*see* "Newton's *Principia*," p 148).

Telescopes were greatly improved as both the reflecting telescope and the achromatic lens were introduced. They pushed further the limits of observational space. The idea that the sun is but one of myriads of stars led to the view that human beings are not the center of the world, as exemplified in Bernard Fontenelle's book, *De la pluralité des mondes.* The introduction of micrometers also improved the ability of astronomers to map stars.

Edmund Halley and Giovanni Domenico Cassini were

The greatest breakthroughs in physics occur when someone understands that apparently different phenomena have the same underlying cause. One of the great unifiers was Isaac Newton. Newton was the first to demonstrate that the falling of objects towards Earth's surface, the motion of the moon around Earth, the motion of planets around the sun, and the odd trajectories of the comets are all governed by one law: the law of universal gravitation.

However, the idea of universal gravitation only became clearly formed in Newton's mind in the 1680s, when he was preparing the manuscript for his *Principia*. He had had a first inkling of this idea some 20 years earlier. At the end of his life, Newton related that he started wondering whether one could consider the moon as falling toward Earth when he observed an apple falling from a tree.

In 1666 Newton studied circular motion. He found that the force acting on a circling body is inversely proportional to the square of the distance of the body from the point around which it rotates. This result was also found by Christiaan Huygens, who published it in his *Horologium oscillatorium* in 1673.

Newton tried to apply this concept to the Earth-moon system and to the planets. He realized that the motion of the moon could be viewed as having two components. One of these components is the moon's fall toward Earth; the other is its motion in a straight line. Both motions occur at once, resulting in the nearly circular path of the moon. In the absence of Earth, the moon would move on a straight line. Its path is curved toward Earth because of Earth's gravitational attraction.

Newton first determined how much the moon's motion deviates from a straight line. He then compared the pull acting on the moon with the pull acting on a body on the surface of Earth. Newton determined that a body dropped at Earth's surface will fall 490 cm (16 ft) in the first second. He then found that the moon deviates from a straight path—and thus "falls" toward Earth—over 366 cm (12 ft) during one hour. If the moon were at the surface of Earth it would fall 490×3600 cm (16×3600 ft) in an hour. Now Newton compared this with the inverse square law that he had derived for rotating bodies on Earth. Since the moon is about 60 Earth radii from Earth, it would be accelerated toward Earth 60^2 (3600) times slower, which is approximately the case based on observation. (He actually found this factor to be 4000; the error was caused because Newton used the incorrect value for the size of Earth.)

Newton came to the conclusion that a force inversely proportional to the square of the distance from a body to its center of rotation keeps the moon and planets in their orbits. In 1645 the French astronomer Ismael Boulliau had also suggested, without giving proof, that a central force inversely proportional to the square of their distance from the sun might keep the planets in their orbits.

For a period after 1666, however, Newton turned his attention to optics and chemistry. In 1679 he again became interested in planetary orbits when Robert Hooke asked him to demonstrate that a planet would move in an elliptical orbit when subjected to a central force inversely proportional to the square of its distance from the sun. Newton found a mathematical proof, but never forwarded it to Hooke.

Christopher Wren and Edmund Halley were also investigating the problem, but neither could solve it. In 1684 Halley, who was the clerk of the Royal Society, visited Newton and asked him what he thought the orbit of a planet might be under an inverse-square law. Newton immediately replied, "An ellipse, for I have calculated it." Halley asked him to produce his paper with the calculations. Newton replied that he had mislaid this paper, but that he would start over with his calculations.

A few months later Halley received a paper from Newton called *De motum corporum in gyrum* (On the motion of revolving bodies), which Halley presented to the Royal Society. Halley's interest spurred Newton to develop his ideas further. Over the next two and a half years he wrote the *Principia*. Halley edited the sections of the manuscript Newton sent him and saw it through the press, at one time employing two printers to speed up the process. Because the Royal Society had no funds available at this time, Halley paid for the printing of the *Principia* from his own pocket.

At an early stage of the writing of the *Principia*, Hooke discovered that Newton did not mention him in connection with the inverse-square law, which he believed he had discovered prior to Newton. Hooke, who never had given a mathematical proof, attacked Newton, and the dispute over the priority of the discovery furthered the bitter feud that had long existed between the two scientists.

The *Principia* was published late in 1687. Unlike Galileo, who wrote his books in Italian, Newton wrote the *Principia* entirely in Latin. Although he had developed calculus during earlier years, he drafted the book using classical geometry—as was the custom at that time—to tackle physical problems.

Book I of the *Principia* contains Newton's three laws of motion and a discussion of orbital motion. Book II deals with the motion of fluids and attempts to prove that the vortex model proposed by Descartes cannot explain the motion of celestial bodies. Book III, *The System of the World*, contains the proposition that gravity is proportional to mass. In this book, Newton discusses applications of his theory to planetary orbits. He shows how his theory accounts for the irregularities of the motion of the moon, as well as the trajectories of comets.

the leading astronomers of the day. Halley discovered periodicity of comets and proper motion of stars; he also started to map the sky as seen from the Southern Hemisphere. Cassini explored the solar system and discovered much about its planets and moons.

Biology. Better microscopes changed the view held by Descartes that living animals are relatively simple machines. His mechanistic interpretation of conception was displaced by the discovery of spermatozoa and by the discovery of the production of eggs in viviparous animals

The transmission of characteristics proper to each species formed a fundamental problem. Jan Swammerdam formulated his "preexistence theory," which assumed that the seeds of all living creatures are formed during the creation of the world, and that each generation is contained in the one before.

Anton van Leeuwenhoek, who persisted in using powerful simple microscopes (with only one lens), became the first to realize that the world is filled with creatures too small to be seen with the naked eye. His investigations of protists and sperm as well as his observations of bacteria revealed a new world. Hooke also investigated the invisible world, being the first to notice and name cells.

John Ray and others worked at classifying plants and animals, being the precursors to Carolus Linnaeus. Physiology improved, as Marcello Malpighi studied the lungs, nervous system, and the physiology of invertebrates, and Regnier de Graaf investigated the reproductive system.

Chemistry.
Until this period, the study of matter still relied almost completely on the traditional views of the alchemists, which incorporated a mixture of mystical and spiritualistic notions and empirical results. Jan Baptista van Helmont, and especially Robert Boyle, were instrumental in establishing chemistry as a rational and experimental science. Boyle introduced such concepts as elements, acids, and alkalis, developed Boyle's law of gases (although Edmé Mariotte independently discovered the same law), and discovered hydrogen and phosphorus (the latter independently discovered by Hennig Brand).

While little theoretical advance was made during the period in chemistry, a number of new substances were discovered. The major advances of the next period would not have been possible without the basic data that were developed during this one. One development, however, probably held chemistry back, the phlogiston theory (see "Phlogiston," p 182).

Mathematics.
In mathematics, everything led toward the development near the end of the seventeenth century of the differential and integral calculus by both Newton and Leibniz; however, there were many strides forward in side branches. The role of mathematics in the natural sciences also changed drastically. Mathematics became fully the language of science. Leibniz promoted mathematics to the level of a "universal language," hoping it could be applied in domains other than scientific ones that require logical reasoning.

Statistics got started with the work of John Graunt, Jan de Witt, John Arbuthnot, and Abraham De Moivre (see "The first statistician", p 150). At the same time, the Bernoulli family continued to develop probability theory.

New tools were developed that are still useful today. Mathematical induction, known earlier but popularized by the posthumous publication of a book by Blaise Pascal, is a method for proving the truth of a statement about integers. The calculus of variations, developed initially by Jacques Bernoulli, is a method for finding the maximum or minimum of a function.

Medicine.
Progress in biology, such as fundamental observations with microscopes, had an impact on medicine that would be felt when people started to apply their learning. Anatomists continued their description of the parts of the human body, such as the brain and the ear. Bernardino Rammazzini became the first doctor to note that certain kinds of cancer can be associated with environmental effects. The practice of inoculating children with smallpox spread from Turkey to England and its American colonies. Basic physiological measurements became easier with the availability of better thermometers and the first experiments in finding blood pressure.

Physics.
The major development in physics was Newton's formulation of his laws of motion, foreshadowed by Galileo in the seventeenth century. Galileo's work had shown the need for a force to keep planets in their orbits. Newton's law of universal gravitation provided the theoretical foundation for Johann Kepler's laws and Galileo's observations. Descartes' theory of vortices remained the only one accepted in France for a while, but it was discarded when sufficient proof for the validity of Newton's theory became available (see "Verifying Newton's theory of gravitation," p 178).

There were also many advances in the theoretical understanding of light, ranging from the first mathematical treatment of refraction to the discovery of the speed of light (see "The nature of light," p 172, and "The velocity of light," p 160). Both the wave and particle theories of light were advanced during this time, ultimately to be superseded in the twentieth century by the quantum theory, which combines wave and particle aspects.

After Otto von Guericke discovered how to create large amounts of static electricity, the move to investigate electric phenomena gained momentum, although it was to reach greater heights later in the eighteenth and early nineteenth centuries (see "Electricity and magnetism," p 270).

Technology.
The first steam engines, pumping water out of coal mines, appeared in England. More coal could be mined from deeper levels, clearing the way for Britain's Industrial Revolution. In addition, Abraham Darby discovered how to make iron with coal, making coal even more important to the new technology.

New inventions during this period, such as the flying shuttle and the spinning jenny, were not appreciated by workers whose jobs were displaced. These inventions, however, were forerunners of the Industrial Revolution.

1660

The Restoration of the kingdom in England begins

The Royal Society is founded in England, first as the Invisible College for the promoting of Physico-Mathematical Experimental Learning; it becomes The Royal Society for the Improvement of Natural Knowledge when Charles II seals its charter in 1663

Marcello Malpighi, b Crevalcore (Italy), Mar 10, 1628, shows that the lungs consist of many small air pockets and a complex system of blood vessels; by observing capillaries through a microscope he completes the work of Harvey in describing the circulation of the blood

Edmé Mariotte (b Dijon, France, 1620) discovers the blind spot in the eye

Robert Boyle (b Lismore Castle, Ireland, Jan 25, 1627) announces in *New experiments physico-mechanical touching the spring of air* that removing the air in a vacuum chamber extinguishes a flame and kills small animals, indicating that combustion and respiration are similar processes

1661

Louis XIV of France starts building Versailles

Boyle's *The sceptical chymist* introduces the modern concepts of element, alkali, and acid and refutes many of the ideas of Aristotle and Paracelsus on the chemical composition of matter

1662

Malvasia introduces the fixed-wire micrometer for use with telescopes

Boyle asserts that in an ideal gas under constant temperature, volume and pressure vary inversely (Boyle's law)

1663

René Descartes' works are placed on the Roman Catholic Church's *Index of Prohibited Books*, despite his efforts to avoid this fate

The first statistician

When the Royal Society was founded in 1660, the founding members were scientists, doctors, noblemen, lawyers, civil servants, or literary men. Two years later, the question arose as whether to admit a tradesman, a draper in fact. The members asked their patron, Charles II, what to do. Not only did he recommend admitting him, but added that if they could find any other tradesmen like him, the others should be admitted as well.

The draper in question was John Graunt, who wrote a remarkable book, published in 1662. Ever since 1563, the parish clerks of London had compiled records of the number of people buried. Starting with the barest information possible, these bills of mortality gradually began to include details about the causes of death. Beginning in 1625, the bills of mortality for each year were published. People who bought the published versions were mainly looking for unusual occurrences that would provide interesting gossip. John Graunt had the idea that there might be more to the bills than that. He obtained all of them from 1604 through 1661 and compiled the information. The compilation, with his analysis, was published as *Natural and political observations made upon the bills of mortality*.

Using his records, Graunt became the first to note such now familiar statistics as the following: more boys are born than girls; women live longer than men; and, except in epidemics, the number of people dying each year is relatively constant. Graunt also noted that male mortality was such that, by a suitable age for matrimony, the excess of boys over girls had vanished, so "every woman may have a husband, without the allowance of polygamy."

Graunt's most important innovation was the first mortality table, a table that could be interpreted to show how long a life might be expected after a given age. With this kind of information in hand, it was not long (37 years) before the first life insurance company was founded.

1660

Frans van Schooten d Leiden, Holland, May 29

William Oughtred, English mathematician, d Albury, Surrey, Jun 30

By this date England is producing 2 million tons of coal a year, more than 80 percent of all the coal produced in the world

Otto von Guericke is the first to use a barometer to forecast weather

1661

Gérard Desargues, French mathematician, d Lyons

Inventor Christopher Polhem b Visby, Gotland Island, Sweden, Dec 18; he specializes in new mining tools and also builds a factory to produce them

1662

Natural and political observations made upon the bills of mortality by John Graunt (b London, Apr 24, 1620) and William Petty (b Romsey, England, May 13, 1623) is the first book of statistics; it contains the London Life Table, the first table showing the ages at which people (in London) are likely to die

Blaise Pascal, French mathematician and physicist, d Paris, Aug 19

Johann Baptista van Helmont's *Oriatrike* or *Physic refined* is published in English translation and becomes very popular

1663

Isaac Barrow (b London, Oct 1630) becomes the first Lucasian professor of mathematics at Cambridge and inspires Isaac Newton to adopt an academic career

Liber de ludo aleae (Book on games of chance) by Girolamo Cardano (better known in English as Jerome Cardan) is published posthumously; it is the first known work on the theory of probability

John Wallis shows that the assumption that for every triangle there is a similar triangle of arbitrary size is equivalent to Euclid's fifth, or parallel, postulate

Optica promota by James Gregory (b Drumoak, Scotland, Oct, 1638) gives the first description of a reflecting telescope

Blaise Pascal's *Traité de l'équilibre des liqueurs* (On the equilibrium of liquids), published posthumously, suggests that in a fluid, pressure is transmitted equally in all directions (Pascal's law); Pascal probably discovered this around 1648

Francesco Maria Grimaldi, Italian physicist, d Bologna, Dec 28

The Marquis of Worcester claims to have discovered the power of steam to raise water from wells and to burst cannons

	GENERAL	ASTRONOMY	BIOLOGY	CHEMISTRY
1664	René Descartes' *Traité de l'homme et de la formation de foetus* (Treatise on man and the formation of the fetus), printed posthumously, describes animals as purely mechanical beings; that is, there is no "vital force" that makes animals different from other material objects	Giovanni Alfonso Borelli, b Naples (Italy), Jan 28, 1608, calculates the orbit of a comet and finds that it is a parabola (not a circle, ellipse, or line, as expected in various earlier theories) René Descartes' *Le monde* (The world), published posthumously, affirms the Copernican theory; Descartes had abandoned this project after learning of Galileo's problems with the Roman Catholic Church Robert Hooke discovers the Great Red Spot on Jupiter and Jupiter's rotation		
1665	The Royal Society begins to publish its *Philosophical Transactions* In France the *Journal des Savants* is founded	Giovanni Domenico Cassini (Jean-Dominique Cassini, b Perinaldo, (Italy), Jun 8, 1625) measures the rotational speed of Jupiter		Robert Hooke's *Micrographia* describes cells (viewed in sections of cork) for the first time Malpighi describes in *De cerebro* how the nervous system consists of bundles of fibers connected to the brain by the spinal cord
1666	London is ravaged by what comes to be known as the Great Fire Molière's *Misanthrope* is first performed	Cassini observes the polar ice caps of Mars		

MATHEMATICS	MEDICINE	PHYSICS	TECHNOLOGY	
	Thomas Willis's *Cerebri anatome* (Anatomy of the brain) is the most complete and accurate account of the brain and nervous system put forward so far	Christiaan Huygens proposes that the length of a pendulum with a period of one second be the standard unit for length	Cast iron is used for the pipes supplying water to the gardens at Versailles	**1664**
Isaac Newton (b Woolsthorpe, England, Dec 25, 1642) discovers the general binomial theorem in either this year or in 1664 Pascal's 1654 work *Traité du triangle arithmétique* (Treatise on figurative numbers), published posthumously, introduces mathematical induction to most mathematicians; although the first use of mathematical induction was in 1575, the method fails to have any impact until Pascal's use Pierre de Fermat, French mathematician, d Castres, near Toulouse, Jan 12		Francesco Maria Grimaldi's *Physico-mathesis de lumine coloribus et iride,* published posthumously, describes his experiments with diffraction of light and his theory that light is a wave phenomenon; it also notes his observation that muscles make sound as they expand and contract Robert Hooke's *Micrographia,* fundamentally the first book dealing with observations through a microscope, also compares light to waves in water The Great Plague in London kills 75,000 and closes Cambridge for two years; Isaac Newton retires to the country, where he invents the first form of calculus, discovers that white light is made from a mixture of colors, and develops the first version of his law of universal gravitation; at this time, according to Newton's later reconstruction, he observes an apple fall and realizes that the moon must also be attracted toward Earth (modern scholars, however, think Newton invented this story in his old age)	In his *Mettalum martis* Dud Dudley (b England, 1599) claims to have learned the secret of making iron with coal instead of charcoal	**1665**
Ars combinatoria by Gottfried Wilhelm Leibniz, b Leipzig (Germany), Jul 1, 1646, contains his suggestion that a mathematical language of reasoning can	Richard Lower demonstrates the direct transfusion of blood between two dogs	Robert Boyle's *Hydrostatical paradoxes* reports his experiments with fluids	Pierre-Paul Riquet builds the 290-km (180-mi) Canal du Midi, connecting the Mediterranean Sea and the Atlantic Ocean	**1666**

1666 cont

Robert Boyle's *The origine of formes and qualities* contains his view that everything is built up of atoms and reflects his mechanical view of nature

The *Académie Royale des Sciences* is founded in Paris; Christiaan Huygens, along with 19 other scientists, is elected as a founding member; after the Revolution, the *Royale* is dropped and the character of the academy changes; it later becomes the *Institut de France*

1667

The observatory of the French Academy is founded in Paris

Adrien Auzout's *Traité du micromètre ou manière exacte pour prendre le diamètre des planètes et la distance entre les petites étoiles* (Treatise on the micrometer or the exact method for measuring the diameter of the planets and the distance between the little stars) discusses his improvements of the telescope

Jean Picard (b La Flèche, France, Jul 21, 1620) introduces the micrometer for telescope use and discovers anomalies in the positions of stars, explained by James Bradley in 1728 as due to the aberration of light

Godefroy Wendelin (Vendelinus), Flemish astronomer, d Ghent

John Ray (b Black Notley, England, Nov 29, 1627) introduces a classification scheme for plants, dividing them into the monocots and dicots on the basis of the number of their seed leaves

Jean-Baptiste Denis (b Paris, 1640) transfers about 350 ml (12 oz) of lamb's blood into a sick boy, who recovers; later experiments are not so successful, and when two of his patients die after receiving blood from sheep, Denis is tried for murder; he is acquitted, but such transfusions are banned in France

1668

John Wilkins's *Essays toward a real character and a philosophical language* are presented to the Royal Society

Cassini's *Ephemerides bononienses mediceorum siderum* contains his computation of the configurations of the four satellites of Jupiter

Newton invents the reflecting telescope

Regnier de Graaf (b Schoonhoven, Holland, Jul 30, 1641) studies the testicles

Francesco Redi, b Aresso (Italy), Feb 18, 1626, disproves the idea that maggots arise spontaneously in rotten meat in one of the first controlled experiments in science

be devised; not until the nineteenth century will George Boole and others develop this idea into something useful

Newton describes his early work on the calculus in an untitled manuscript known to historians as the October 1666 Tract

1666 cont

Gregory of St. Vincent, German mathematician, d

In an experiment demonstrated before the Royal Society, Robert Boyle shows that an animal can be kept alive by artificial respiration

Thomas Willis's *Pathologiae cerebri* (Pathology of the brain) is the first account of the effects of the late stage of syphilis on the brain (although Willis did not recognize the cause of these symptoms)

1667

James Gregory's *Geometriae pars universalis* (The universal part of geometry) and *Exercitationes geometricae* (Geometrical exercises) include a geometrical version of the fundamental theorem of the calculus (that differentiation and integration are inverse operations) as well as the series named after him

Logarithmotechnia by Nicolaus Mercator (b Niklaus Kauffmann, Hols-

John Wallis is the first to suggest the law of conservation of momentum

1668

1668
cont

	GENERAL	ASTRONOMY	BIOLOGY	CHEMISTRY
1669	*De solido intra solidum naturaliter contento dissertationis prodromus* (Forerunner of a dissertation about a solid body enclosed by a process of nature within a solid) by Nicolaus Steno (b Niels Stenson, Cophenhagen, Denmark, Jan 11, 1638) gives a correct explanation of fossils and rock strata	Giovanni Domenico Cassini arrives in Paris to consider taking an appointment as the director of the Paris Observatory of the French Academy, a position he formally assumes in 1671; because he remains in France and becomes a French citizen, he is frequently cited in reference works as Jean-Dominique Geminiano Montanari discovers the variability of the star Beta Persei (Algol); he publishes his report in 1672	Richard Lower's *Tractatus de corde* describes the structure of the heart and its properties as a muscle, and notes that blood changes color in the lungs Malpighi's *Silkworms* is the first detailed description of the anatomy of an invertebrate Jan Swammerdam's general treatise on insects is published; he overthrows the idea of instant metamorphosis and details the reproductive parts of insects	*Physica subterranea* (Subterranean physics) by Johann Joachim Becher, b Speyer (Germany), May 6, 1635, contains his alchemistic ideas and experiments on the nature of minerals and other substances; his concept that a *terra pingus* (oily earth) causes fire later becomes the basis of the phlogiston theory Hennig Brand, b Hamburg (Germany), 1630, discovers phosphorus accidentally in an experiment with urine, but keeps the discovery secret
1670	Blaise Pascal's *Pensées* is published posthumously Grand Duke Ferdinand II of Tuscany, Italian ruler, d May 24	G. Mouton proposes the one-sixtieth part of a degree of the meridian of Earth as a unit for a decimal system of measure		Boyle discovers a flammable gas produced when certain metals are treated with acids; this gas is termed *inflammable air* by Cavendish in 1766 and is now known as hydrogen Johann Rudolf Glauber, German chemist, d Amsterdam, Mar 10
1671	Jean Picard's *Mesure de la terre* (Measure of the Earth) includes his determination of the length of a meridian of latitude, giving the most accurate figure since the ancient Greeks, very close to today's accepted values	Cassini discovers Iapetus, a satellite of Jupiter Cassini calculates the distance from Earth to Mars, which enables him to determine the distances of all the planets from the sun; his calculation is in near agreement with modern measurements Giovanni Battista Riccioli, Italian astronomer, d Bologna, Jun 25	Francesco Redi dissects the torpedo fish and studies the electric organ; he fails to recognize, however, that the shocking sensation is caused by electricity	
1672	John Wilkins, English scholar, d London, Nov 19	French physician (Guillaume) N. Cassegrain (b France, 1625) invents the reflecting telescope that is named for him Cassini discovers Rhea, a satellite of Saturn	Regnier de Graaf discovers and describes the ovarian follicle that is named after him Francis Willughby, English naturalist, d Middleton, Warwickshire, Jul 3	

				1668 cont

tein, Denmark, about 1619) includes various calculations of natural logarithms and methods of approximating them by series

---

1669

Newton's *De analysi per aequationes numero terminorm infintas* (On the analysis by equations unlimited in the number of their terms), circulated privately, contains his binomial theorem and methods for finding the areas under curves

Isaac Barrow's *Lectiones opticae* are published with the assistance of Newton

Experimentia crystalli Islandici disdiaclastici (A study of Iceland spar) by Erasmus Bartholin (b Roskilde, Denmark, Aug 13, 1625) describes the discovery of double refraction

Newton becomes a professor at Cambridge

1670

Isaac Barrow's *Lectiones geometricae* presents his theorems concerning drawing tangents to curves, finding lengths of curves, and finding the areas bounded by curves using methods similar to those of the calculus

William Neil, English mathematician, d Aug 24

Thomas Willis announces his rediscovery of the connection between sugar in the urine and diabetes mellitus (a connection also known to the Greeks, Chinese, and Indians)

Hans Hautsch, German mechanic, d

Francesco de Lana designs an airship—never built— that would be lifted by four copper spheres containing a near vacuum

---

1671

Jan De Witt's *Waerdye van lyf-renten naer proportie van los-renten* (A treatise on life annuities) introduces the concept of mathematical expectation

James Gregory discovers the series expansion for $\pi/4$ that is later rediscovered by Leibniz and is known as the Leibniz series

Leibniz designs a calculating machine that can multiply and divide; working models are built and demonstrated over the next four years

---

1672

Il problema della quadratura del circolo (The problem of squaring the circle) by Pietro Mengoli (b 1625) introduces many infinite series and products and contains many results, such as the divergence of

Francis Glisson gives a description of the "irritability" of living tissues, or the tendency of tissues to react to their environment, in *Tractatus de natura substantiae energetica*

Otto von Guericke's *Experimenta nova Magdeburgica de vacuo spatio* (New Magdeburg experiments concerning empty space) describes his experimental work with a vacuum

1672
cont

Mad Madge, the scientist

Throughout history men have erected barriers between women and scientific knowledge. Seventeenth-century England was no exception. Henry Oldenburg, the first secretary of the Royal Society, called science "masculine philosophy" in 1664. Thomas Spratt, the contemporary bishop of Rochester, also termed this new form of knowledge "masculine and durable."

In the beginning of modern science, Margaret Cavendish, duchess of Newcastle, fought against these prejudices. As a result of her interest in science, she came to be known as "Mad Madge." Despite this, she wrote prolifically about the nature of the new science.

Since the Royal Society was at the core of British science at the time, the duchess was determined to become a member. This campaign of hers was the talk of London, and Samuel Pepys recorded in his diary that "the town will be full of balades of it." In the face of opposition from Pepys and others, however, the duchess succeeded in becoming a member on May 30, 1667.

Contemporary descriptions of Duchess Margaret present a mixed picture. John Evelyn said he was "much pleased with the extraordinary fancifull habit, garb and discourse of the Duchess." His wife, on the other hand, found the duchess's conversation as "airy, empty, whimsical, and rambling as her books, aiming at science, difficulties, high notions, terminating commonly in nonsense, oaths, and obscenity."

No other woman was admitted to the Royal Society until 1945.

1673

Anton van Leeuwenhoek sends the English Royal Society letters on his discoveries with the simple microscope

Malpighi's *De formatione pulli* (On the formation of the chick in the egg) describes the development of the ovum

Regnier de Graaf, Dutch anatomist, d Delft, Aug 21

Thomas Wharton, English anatomist, d London, Nov 15

1674

Jean Pecquet, French anatomist, d Paris, Feb

Tractatus quineue medico-physici (Five medico-physical treatises) by John Mayow (b Bray, England, Dec, 1641) reports on experiments in measuring the air consumed by a mouse or a burning candle; Mayow is the first to note that the volume of air is reduced in respiration and that air must consist of two different gases

1675

Cassini discovers that the rings of Saturn are not a single flat disk surrounding the planet; the break in the rings he discovers is still known as the Cassini division

Greenwich Observatory is founded by King Charles II;

Malpighi's *Anatome plantarum* is the first important work in plant anatomy; it also contains the clearest description of the development of a chick up to that time

Nicolaus Steno demonstrates that eggs are

Cours de chymie (Chemistry course) by Nicolas Lémery (b Rouen, France, Nov 17, 1645) is a textbook that will be republished 31 times by 1756

MATHEMATICS	MEDICINE	PHYSICS	TECHNOLOGY	
the harmonic series and the sum of the reciprocals of the triangular numbers, often attributed to others

Euclides danicus by Georg Mohr (b Copenhagen, Denmark, Apr 1, 1640) contains his proof that all geometric figures constructible by straightedge and compass are also constructible with compass alone (this is known as Mascheroni's theorem, although Mascheroni was not to publish until 1797)

Dutch mathematician and politician Jan De Witt is assassinated by a mob, Aug 20, at The Hague | Thomas Willis's *De anima brutorum* is published

Franciscus Sylvius (b Franz de le Boë), Dutch physician, d Leiden, Holland, Nov 19 | Von Guericke develops a way to charge a ball of sulfur with static electricity, producing the greatest amount of electricity gathered in one place to this time

Jean Richer (b France, 1630) finds on an expedition to Cayenne that a pendulum of the same length has a longer period on the equator than in France; in 1687 Newton shows that this is because of a bulge at the equator and a flattening at the poles that is predicted by his theory of gravity | | 1672 cont |
| About this time Leibniz discovers the celebrated series for $\pi/4$ that bears his name; he was anticipated, however, by James Gregory in 1671 | | Christiaan Huygens's *Horologium oscillatorium sive de motu pendulorum* (The oscillation of pendulums) gives the calculation of equivalent pendulum lengths and the laws of centripetal force | Christiaan Huygens builds a motor driven by the explosion of gunpowder | 1673 |
| John Graunt, English statistician, d London, Apr 18 | | Robert Hooke's *Attempt to prove the motion of the Earth* puts forward a theory of planetary motion based on the balance between centrifugal force and gravitational attraction from the sun | | 1674 |
| Edward Cocker, English mathematician, d London

Gilles Personne de Roberval, French mathematician, d Paris

Bernard Frénicle de Bessy, French mathematician, d Paris, Jan 17 | Thomas Willis, English physician, d London, Nov 11 | Newton delivers his *Discourse on light and colour* to the Royal Society

Ole Römer (b Århus, Denmark, Sep 25, 1644) measures the speed of light by measuring time differences of the eclipses of satellites of Jupiter | | 1675 |

1675 cont

on Mar 4 he appoints John Flamsteed (b Denby, England, Aug 19, 1646) as the first astronomer royal

James Gregory, Scottish mathematician and astronomer, d Edinburgh, Oct

formed inside the dogfish before it gives birth to live offspring and leaps to the correct conclusion that mammals have eggs; somewhat earlier Regnier de Graaf independently reaches the same conclusion, also on suspect evidence

1676 Sir Hans Sloane starts the British Museum

Velocity of light

Most ancients, and even later scientists such as Johannes Kepler and René Descartes, believed the velocity of light to be infinite. Galileo thought otherwise and tried to measure light velocity with a method that had worked successfully for the measurement of the velocity of sound by Mersenne. Galileo and an assistant placed themselves on two hilltops some distance apart. Galileo flashed a light signal and the assistant had to respond by flashing a light signal in return as soon he could see Galileo's light flash. Galileo then measured the time elapsed between the emission of his light signal and the return of the light signal from his assistant. Galileo soon realized that the elapsed time measured was independent of the distance to the assistant. Instead, the elapsed time was the time needed for the assistant to react to Galileo's signal.

The Danish astronomer Ole Römer solved the problem by using a much longer distance over which to measure the velocity of light. He noticed that when Earth and Jupiter were on the same side of the sun, the eclipses occurred a few minutes earlier than when the sun was between Earth and Jupiter. Römer determined that it took 16 minutes longer for the light to travel from Jupiter when Earth was the farthest away from Jupiter than when Earth was closest. Römer deduced rightly that these 16 minutes were the time needed for the light to cover the distance between the closest and farthest positions of Jupiter. Römer calculated that the speed of light is 240,000 km (150,000 mi) per second.

Precise measurements of the velocity of light were made in the nineteenth century using specially designed apparatus. These relied on the wave phenomenon of interference, where the crests of light waves of the same frequency either enhance each other or cancel each other. With lasers, it is possible to have light waves with exact frequencies. As a result, the speed of light in a vacuum, which is always constant, is, along with the second, the basis of our system of measurement. Since 1983 the meter has been defined as the distance light travels in 1/299,792,458 second; the speed of light is exactly 299,792,458 meters per second. Furthermore, since the American inch is defined by Congress as 2.54 times the centimeter, the speed of light is exactly 186,292.03 miles per second.

1677 Henry Oldenburg, German/English science administrator, d London, Sep 5

Anton van Leeuwenhoek discovers protists (protozoa)

Anton van Leeuwenhoek in Nov confirms the discovery of sperm by Louis Dominicus Hamm; unlike Hamm, who thinks sperm are evidence of disease, Leeuwenhoek concludes that sperm are the source of reproduction, the larvas of humans (since Leeuwenhoek did not know of the mammalian egg, he did not propose fertilization)

Johann Kunckel, b Hutten (Germany), 1630, describes an aqueous solution of ammonia

On Nov 21, Leibniz becomes the first to use modern notation $\int f(x)\,dx$ for an integral, improving on his notation of less than a month before; he also correctly finds the product rule for differentiation

Traité des triangles rectangles en nombres by de Bessy (b about 1605), published posthumously, contains a proof of Fermat's "last theorem" for $n = 4$, using Fermat's special method of infinite descent, a form of indirect proof

Mathematician Jacopo Francesco Riccati b Venice (Italy), May 28; he is principally known for his study of a type of differential equation, now known as the Riccati equation ($dy/dx = A(x) + B(x)y + C(x)y^2$

On Jun 13 Newton writes Leibniz a letter, now known as the *Epistola prior*, describing his work on infinite series

On Oct 24, Newton writes Leibniz a second letter, now known as the *Epistola posteria*, that includes an anagram in Latin describing his method of fluxions

Leibniz discovers how to differentiate any integral or fractional power of x

Nehemiah Grew (b Mancetter Parish, England, Sep, 1641) coins the term "comparative anatomy"

Hooke uses an anagram to announce that he has found that the stretch of a spring varies directly with its tension, a rule that since has become known as "Hooke's Law"

Edmé Mariotte's *Essai sur la nature de l'air* (On the nature of air) states Boyle's law: for a gas, the product of volume and pressure is constant at a given temperature; in France, it is known as Mariotte's law

Robert Hooke invents the universal joint

Isaac Barrow, English mathematician, d London, May 4

Leibniz correctly determines the quotient rule for differentiation

Francis Glisson, English physiologist, d London, Oct 16

	GENERAL	ASTRONOMY	BIOLOGY	CHEMISTRY
1678			Chrysanthemums from Japan appear in Holland	
1679	The Writ of Habeas Corpus is instituted in England Thomas Hobbes, English philosopher, d Hardwick, Derbyshire, Dec 4	Cassini publishes *Carte de la lune,* a map of the moon that is based on the *Atlas de la lune* with 60 plates *Catalogus stellarum australium* by Edmund Halley (b Haggerston, England, Oct 29, 1656) gives the locations and descriptions of 341 southern stars, the first time that stars observable from south of the equator have been cataloged		
1680	Anton van Leeuwenhoek is elected a fellow of the Royal Society Athanasius Kircher, German scholar, d Rome (Italy), Nov 28	Francis Moore's *Old Moore's almanack* is started; it later becomes known as *Vox stellarum*	Giovanni Alfonso Borelli's posthumous *De motu animalum* (On the motions of animals) treats the movements and contractions of the muscles and includes his explanation of the power of a torpedo fish to produce shocks as a result of rapid contractions of a muscle now known as the electric organ (second volume published 1681) Jan Swammerdam, Dutch naturalist, d Amsterdam, Feb 17	Boyle's *The aerial noctiluca* describes phosphorus independently of Hennig Brand's discovery
1682	*Acta Eruditorum,* a learned journal, is founded; it is in Latin and continues to be published until 1776; Leibniz is a frequent contributor	Edmund Halley observes the "great comet," which will be named after him after he correctly predicts in 1705 that the comet will return in 1758 French astronomer Jean Picard d Paris, Jul 12	Nehemiah Grew's *The anatomy of plants* describes the different types of tissues making up stems and roots and identifies the male and female parts of flowering plants John Ray discusses the classification of plants in *Methodus plantarum nova* (New method of plants)	Johann Joachim Becher, German chemist, d London, Oct

MATHEMATICS	MEDICINE	PHYSICS	TECHNOLOGY	
Giovanni Ceva (b Milan, Italy, 1647) proves the theorem named after him on the division of sides of a triangle Edward Cocker's *Arithmetick, being a plain and easy method* becomes one of the most important textbooks; it continues to be popular for more than a century		Huygens's *Traité de la lumière* (Treatise on light) explains the wave theory of light; it is not published until 1690		1678
Leibniz introduces (in a letter to the Jesuit Joachim Bouvet) binary arithmetic by showing that every number can be represented by the symbols 0 and 1 only Giovanni Alfonso Borelli, Italian mathematician and physiologist, d Rome, Dec 31	John Mayow, English physiologist, d London, Sep	Hooke suggests in a letter to Newton that gravitational attraction varies inversely with distance from the sun and asks if this does not imply elliptical orbits for planets; Newton does not reply Jean Richer's *Observations astronomiques et physiques faites en l'ile de Cayenne* (Astronomical observations made at the Cayenne Isle) describes the change in the period of the pendulum in different locations on Earth because of changes of gravity	Johannes Kunckel invents the artificial ruby, a form of colored glass Denis Papin (b Coudraies, France, Aug 22, 1647) demonstrates his "steam digester," a pressure cooker with a safety valve used for cooking bones	1679
	Thomas Bartholin, Danish physiologist, d Copenhagen, Dec 4	Newton, spurred by boasts from Hooke that he has solved the riddle of planetary motion, calculates that an inverse-square law of attraction to the sun will produce an elliptical orbit	Clocks are equipped with hands to show minutes	1680
Pietro Mengoli, Italian mathematician, d Bologna Michelangelo Ricci, Italian mathematician, d Rome, May 12				1682

1683 The Siege of Vienna by the Turks leads indirectly to the development of the typical Viennese cafe and also to the introduction of the lilac into Europe

Anton van Leeuwenhoek observes bacteria, which will not be seen by other scientists for more than a century

1684 Cassini discovers Dione and Thetys, satellites of Saturn

The rise of time (part two)

In 1581 Galileo observed that a pendulum of a given length seemed to move through its cycle in a given amount of time. Many years later Galileo asked his son to build a clock using the pendulum as a regulator to improve the accuracy of a clock operated by a weight. But it was Christiaan Huygens a few years after that, in 1656, who made the pendulum clock practical. A simple pendulum does not swing at exact time intervals, but Huygens modified the pendulum to solve this problem. He built the first grandfather's clock. It was a sort of miniature tower clock made accurate enough to keep time to the minute through use of the pendulum.

Around the same time, Robert Hooke was experimenting with springs. His discoveries led to clocks powered with springs instead of weights, making clocks portable. By the eighteenth century, clocks that ticked the seconds were commonly available.

Mechanical clocks became more and more reliable and precise, but these clocks have limits. After World War II, scientists began to turn to other means to measure time. Atomic clocks use atoms that reliably vibrate millions of times a second. Even cheap silicon-chip wristwatches rely on hundreds or thousands of vibrations a second. As a result, time is regularly measured to almost unthinkable intervals today. A home computer uses a timer that may be anywhere from about 4000 to 20,000 cycles a second. Without such precise measurement of time, computers would be unable to operate with anything like the speed and efficiency we have come to expect of them.

1685 Louis XIV revokes the Edict of Nantes, taking religious freedom from French Protestants

1686 *Entretiens sur la pluralité des mondes* (Conversations on the plurality of worlds) by Bernard le Bovier de Fontenelle (b Rouen, France, Feb 11, 1657) popularizes the theories of Descartes

John Ray's *Historia plantarum* begins publication (it will be complete in three volumes in 1704); it describes 18,600 plant species, laying the groundwork for Linnaeus; it introduces the first definition of species, based on common descent, into science

MATHEMATICS	MEDICINE	PHYSICS	
Seki Kōwa produces a mathematical treatise with the first-known explanation of the concept of determinants, functions that assign a numerical value to a square array of symbols		John Théophile Desaguliers b La Rochelle, France, Mar 12; moves to England in 1685; later he repeats and extends Stephen Gray's experiments with electricity and introduces the words *conductor* and *insulator*	1683
Count Ehrenfried von Tschirnhous, b (Germany) 1651, reports in *Acta Eruditorum* on his transformations that simplify polynomial equations, providing new methods for solving equations of degrees three and four, the most important contribution to equation solving in the seventeenth century			
John Collins, English mathematician, d London, Nov 10			
Leibniz's "A new method for maxima and minima as well as tangents, which is impeded neither by fractional nor by irrational quantities, and a remarkable type of calculus for this" is the first account of differential calculus; its six pages are so dense, however, that few can understand them	Daniel Whistler, English physician, d	Hooke boasts to Christopher Wren and Edmund Halley that he has found the laws governing the movement of the planets; Wren correctly perceives that Hooke is wrong and offers a prize to anyone who can solve the problem; Halley visits Newton to ask what planetary orbits would be if the planets are attracted to the sun according to an inverse-square law; Newton immediately replies, "An ellipse, for I have calculated it;" this leads to Halley's encouraging Newton to write out his ideas, which become the *Principia*	1684
William, 2nd Viscount Brouncker, English mathematician, d Westminster, Apr 5			
		Edmé Mariotte, French physicist, d Paris, May 12	
John Wallis's *De algebra tractatus* (Treatise of algebra) contains the first publication of Newton's binomial theorem and an early way to represent complex numbers geometrically			1685
Leibniz describes the integral calculus in print for the first time in an issue of *Acta Eruditorum*		Edmé Mariotte's *Traité du mouvement des eaux et des autres corps fluides* (The motion of water and other fluids) is published posthumously	1686
		Newton presents the manuscript of the first volume of the *Principia, De*	

	GENERAL	ASTRONOMY	BIOLOGY	CHEMISTRY
1686 cont			Nicolaus Steno, Danish anatomist and geologist, d Schwerin (Germany), Nov 25 or Dec 5	
1687	Sir William Petty, English demographer, d London Dec 16	Kirch discovers the variability of the star Zeta Cygni

Johannes Hevelius, German, d Danzig (Gdańsk, Poland), Jan 28 | | |
| **1688** | The Glorious Revolution takes place in England; King James II is deposed and William and Mary of Holland are invited to become king and queen | | | |
| **1689** | Isaac Newton becomes a member of the House of Commons for Cambridge

Peter the Great becomes czar of Russia | | | |
| **1690** | *Essay concerning human understanding* by John Locke (b Somerset, England, Aug 29, 1632) argues that all knowledge of man comes from experience and sensations only

William III of England defeats former James II on Jul 2 at the battle of Boyne in Ireland, ensuring that England will stay Protestant | | | |
| **1691** | | | John Ray's *The wisdom of God manifested in the works of creation* suggests that fossils are the remains of animals from the distant past; it establishes Ray as the leader of the natural history movement in England | Robert Boyle, British physicist and chemist, d London, Dec 30 |
| **1692** | Elias Ashmole, English natural philosopher, d May 18 or 19 | | | |

MATHEMATICS	MEDICINE	PHYSICS	TECHNOLOGY	
		motu corporu (The motion of bodies), to the Royal Society		**1686 cont**
		Otto von Guericke, German physicist, d Hamburg, May 11		
Nicolaus Mercator, Danish mathematician and architect, d Paris, Jan 14		Guillaume Amontons (b Paris, Aug 31, 1633) invents an hygrometer		**1687**
		In Sep, Newton's *Philosophiae naturalis principia mathematica* (The mathematical principles of natural philosophy), known as the *Principia*, establishes Newton's three laws of motion and the law of universal gravitation		
Sometime before this date Lloyds of London is founded, showing that business people were already calculating probabilities	Claude Perrault, French physician and architect, d Paris, Oct 11			**1688**
	Thomas Sydenham, English physician, d London, Dec 29			**1689**
			Denis Papin is the first to use steam pressure to move a piston	**1690**
Méthode pour résoudre les égalités by Michel Rolle (b Ambert, France, Apr 21, 1652) contains, without proof or fanfare, the theorem of the calculus named after him	Clopton Havers (b Stambourne, England, about 1655) publishes the first complete textbook on the bones of the human body			**1691**
	Richard Lower, English physician, d London, Jan 17			
Leibniz introduces the terms *coordinate, abscissas,* and *ordinate*				**1692**

	GENERAL	ASTRONOMY	BIOLOGY	CHEMISTRY
1693			John Ray's *Synopsis animalium quadrupedem et serpentini* (A general view of four-legged animals and snakes) introduces the first important classification of animals; it follows Aristotle's divisions into "blooded" and "bloodless" and correctly classifies whales as mammals	
1694			*De sexu plantarum epistola* (Letter on the sex of plants) by Rudolph Jacob Camerarius, b Tübingen (Germany), Feb 12, 1665, clearly distinguishes the male and female reproductive organs in plants Marcello Malpighi, Italian biologist and physiologist, d Rome, Nov 29	
1695				Nehemiah Grew isolates magnesium sulfate—commonly known as Epsom salts—from spring water
1696		Jean Richer, French astronomer, d Paris William Whiston publishes *A new theory of the Earth*	*Arcana naturae* (Mysteries of nature) by Anton van Leeuwenhoek discusses his discovery of "animalculae" (microorganisms, chiefly those now known as protists)	
1697	The French Academy of Sciences is reorganized; Bernard le Bovier de Fontenelle becomes its secretary			Georg Ernst Stahl, b Ansbach (Germany), Oct 21, 1660, introduces the concept of phlogiston as the cause of burning and rusting, basing his idea on the 1669 idea of an "oily earth" that had the same role in the theories of Johann Joachim Becher

1693

Leibniz rediscovers the concept of determinants and communicates it to Antoine de L'Hospital in a series of letters, which will not be published until 1850

John Wallis's *Opera mathematica*, volume 2, contains the first complete, printed statement of Newton's "method of fluxions," essentially the calculus

1694

1695

Christiaan Huygens, Dutch physicist and astronomer, d The Hague, Jul 8

1696

Analyse des infiniment petits (Analysis of infinitesimals) by Marquis Antoine de l'Hospital (b Paris, 1661) is the first textbook on differential calculus; it contains "L'Hospital's rule" (bought by L'Hospital from Jean Bernoulli, who discovered it in 1694) and is influential for a century

1697

Johann (Jean) Bernoulli (b Basel, Switzerland, Aug 6, 1667) proposes the problem of the brachistochrone—finding the path of quickest descent, solved by Newton, l'Hospital, Leibniz, and Johann Bernoulli (incorrectly at first); the solution is a curve called the cycloid

Georg Mohr, Danish mathematician, d Kieslingswalde (Germany), Jan 26

Francesco Redi, Italian physician and poet, d Pisa, Mar 1

	GENERAL	ASTRONOMY	BIOLOGY	CHEMISTRY
1698				
1699	French playwright Jean-Baptiste Racine d		Otto Tachenius, German apothecary, d around this year	
1700		Guillaume Delisle produces a map in which coordinates of locations are determined by astronomical observations Edmund Halley's magnetic charts of the Atlantic and Pacific oceans show lines of equal magnetic variation Ole Römer invents the meridian telescope	Joseph de Tournefort's *Institutiones rei herbariae* is an illustrated work on plant life in the Mediterranean published in three volumes	
1701	Yale University is founded as the Collegiate School of America, becoming Yale College in 1718 Nehemiah Grew's *Cosmologia sacra* (Sacred cosmology) demonstrates the wonders of God's creation			
1702	The first daily newspaper, London's *Daily Courant*, is started		Naturalist Olof Rudbeck d Uppsala, Sweden, Sep 17	Wilhelm Homberg discovers boric acid

MATHEMATICS	MEDICINE	PHYSICS	TECHNOLOGY	
	Erasmus Bartholin, Danish physician and mathematician, d Copenhagen, Nov 4		Denis Papin builds a steam engine in which the piston is moved by the pressure of steam rather than atmospheric pressure	**1698**
			The Miner's Friend, invented by Thomas Savery (b Shilstone, England, about 1650), designed to pump water from coal mines, is patented; it will become the first practical machine powered by steam	
				1699
	Bernardino Rammazzini's *De morbis artificum* is the first systematic treatment of occupational disorders; he concludes that more nuns than married women develop breast cancer, possibly for reasons related to pregnancy and lactation			**1700**
	Giacomo Pylarini, considered by some the first immunologist, inoculates three children with smallpox in Constantinople in the hope of preventing their developing more serious cases of smallpox when they are older	Joseph Sauveur introduces the term "acoustics" in a work on the relation of tones of the musical scale	Jethro Tull (b Basildon, England, Mar, 1674) invents the machine drill for planting seeds	**1701**
	Physician Clopton Havers d Willingale, Essex, England, Apr	Guillaume Amontons invents a thermometer using air as the expandable fluid, but improves on Galileo's earlier design by using air pressure instead of volume	Thomas Savery's *The Miner's Friend* gives a description of his steam engine	**1702**
		David Gregory's *Astronomiae physicae et geometriae elementa* is the first astronomy text based on gravitational principles		
		Physico-mechanical experiments by Francis Hauksbee (b Colchester, England, about 1666) describes how air at low pressure glows during an electrical discharge		

1703

Isaac Newton is elected president of the Royal Society

1704

Lexicon technicum by John Harris (b Shropshire, England, about 1666) is a technical encyclopedia explaining 8000 scientific terms

Isaac Newton revives the almost defunct Royal Society, which had faltered because of lack of interest in natural science

Philosopher John Locke d High Laver, Essex, England, Oct 28

Giacomo Maraldi discovers the variability of the star R Hydrae

Römer constructs a private observatory with a meridian circle and a transit instrument

Anatomist Lorenzo Bellini d Florence, (Italy), Jan 8

1705

The Royal Observatory of Berlin is founded

George Cheyne's *Philosophical principles of natural religion* argues for the existence of a deity as a result of the observed phenomenon of attraction

Edmund Halley correctly predicts in *Synopsis astronomiae cometicae* (A synopsis of the astronomy of comets) the return in 1758 of the comet that appeared in 1682, which comes to be called Halley's comet; officially it is called Comet Halley today

Naturalist John Ray d Black Notley, England, Jan 17

The nature of light (part two)

Some seventeenth-century scientists promoted a wave theory of light, as had Leonardo da Vinci earlier. Robert Hooke and Robert Boyle had observed the colors appearing in oil on water. By 1665 both had explained this by suggesting interference of the rays reflected from the two surfaces of the thin film of oil.

Newton rejected the wave theory of light, although his theory allowed for wavelike constructs. But Newton did not believe that waves could cast sharp shadows. In principle, he was right. In reality, light does bend around sharp edges the way that waves do, but this could not be observed in Newton's time. The amount of bending is too small. As a result, Newton believed that light must consist of particles.

Particle theories were popular in England in the eighteenth century, but a series of experiments by Thomas Young, starting in 1801, demonstrated that light is a wave phenomenon. Young showed convincingly that rays of light could interfere with each other a phenomenon only possible for waves.

But Newton was not entirely wrong. In 1905 Albert Einstein showed that the photoelectric effect could be explained by assuming that light is composed of particles. In the photoelectric effect, electrons are knocked from a substance by light. Then Louis de Broglie suggested that entities recognized as particles could also be understood as waves. Experiments later confirmed this idea. Finally, physicists agreed that, whether light behaves as a wave or as a particle depends on the property being measured.

1706

Römer's *Observationum astronomicarum* contains his notes on astronomical observations

1707

Admiral Sir Cloudesley Shovel allows the British fleet to get lost in heavily overcast weather; thinking they are off the coast of Brittany, the ships run aground on the Scilly Islands off the southwest

MATHEMATICS	MEDICINE	PHYSICS	TECHNOLOGY	
Vincenzo Viviani d Florence (Italy), Sep 22 John Wallis d Oxford, England, Nov 8		De la Hautefeuille designs the first Western seismograph Robert Hooke d London, Mar 3		1703
Antoine de l'Hospital d Paris, Feb 2	*De aure humana tractatus* (Anatomy and diseases of the ear) by Antonio Maria Valsalva, b Imola (Italy), Jun 17, 1666, provides the first detailed description of the physiology of the ear Physician Jean-Baptiste Denis d Paris, Oct 3	Newton's *Optics* combines mathematics with experiment and accepts that light is particulate in nature, although these particles may create vibrations in the ether; the book becomes a standard text in experimental physics for the rest of the century; it closes with a series of famous unanswered questions posed by Newton	Nicolas Fatio de Duiller uses gems for bearings in clocks	1704
Jacques (Jakob) Bernoulli d Basel, Switzerland, Aug 16	Raymond Vieussens (b Vigan, France, about 1635) gives the first accurate description of the left ventricle of the heart, the valve of the large coronary vein, and the course of coronary blood vessels	Francis Hauksbee experiments with a clock in a vacuum to prove sound needs air to travel Richard Waller publishes posthumously Robert Hooke's lectures delivered at the Royal Society; one of them, "Lectures and discourses on earthquakes," explains how earthquakes might have changed the surface of Earth substantially since its creation Guillaume Amontons d Paris, Oct 11		1705
William Jones's *Synopsis palmariorum matheseos, or a new introduction to mathematics* is the first known text to use the Greek letter π to represent the ratio of the circumference of a circle to its diameter	*Adversaria anatomica prima* (Anatomical notes) by Giovanni Battista Morgagni, b Forlì (Italy), Feb 25, 1682, makes him known as an anatomist throughout Europe	Francis Hauksbee reports to the Royal Society that a glass tube that attracts bits of brass leaf after rubbing no longer attracts the brass after contact is made; instead the glass tube begins to repel the brass		1706
Newton's *Arithmetic universalis* (Universal arithmetic) includes a precise formulation of Descartes "rule of signs"	*The physician's pulse watch* by John Floyer (b Hintess, England, 1649) introduces pulse-rate counting in medical practice and puts forth a special watch for it		Denis Papin modifies Thomas Savery's high-pressure steam pump	1707

	GENERAL	ASTRONOMY	BIOLOGY
1707 cont	coast of England; four ships and 2000 men, including the admiral, are killed		
1708			Botanist Joseph de Tournefort d Paris, Dec 28
1709			*New theory of vision* by George Berkeley (b County Kilkenny, Ireland, Mar, 1685) presents the most significant contribution to psychology in the eighteenth century
1710	A private scientific society is founded at Uppsala, Sweden, by Erik Benzelius, Emanuel Swedenborg, and Eric Polhem; in 1728 it becomes the Kungliga Vetenskaps Societeten George Berkeley's *Principles of human knowledge* promotes his general doctrine that the "being" of things amounts to their being perceived	Ole Römer d Copenhagen, Denmark, Sep 19	
1711			Luigi Marsigli shows the animal nature of corals, formerly held to be plants
1712		John Flamsteed's first volume of his star catalog *Historia coelestis Britannica,* published without the author's permission, catalogs the position of nearly 3000 stars and	Botanist and physician Nehemiah Grew d London, Mar 25

MATHEMATICS	MEDICINE	PHYSICS	TECHNOLOGY	
	De subitaneis mortibus (On sudden death) by Giovanni Maria Lancisi, (b Rome, Oct 26, 1654) offers the first discussion of cardiac pathology			**1707** cont
	Georg Stahl's *Theoria medica vera* (The true story of medicine) expounds his animalistic theory associating the soul with the physiologic processes			
Scottish mathematician and astronomer David Gregory d Maidenhead, Berkshire, England, Oct 10	*Institutiones medicae* by Hermann Boerhaave (b Voorhout, Holland, Dec 31, 1668) explains his theory of inflammation; the book combines mechanical views of physiology with the idea that physiological processes are chemical fermentations		Count Ehrenfried von Tschirnhaus and J.F. Böttger discover a method of manufacturing true hard-paste porcelain	**1708**
	Hermann Boerhaave's *Aphorismi de cognoscendis et curandis* (Book of aphorisms), a text in medicine, remains in use for a century after its publication	Roger Cotes (b Burbage, England, Jul 10, 1682) corrects the minor mathematical errors for the second edition of Newton's *Principia* Gabriel Daniel Fahrenheit, b Danzig (Germany—now Gdańsk, Poland), May 24, 1686, constructs an alcohol thermometer	Abraham Darby introduces the use of coke for iron smelting at Coalbrookdale, England	**1709**
John Arbuthnot's short paper "An argument for Divine providence, taken from the constant regularity observ'd in the birth of both sexes" is one of the forerunners of modern statistical reasoning			Jacob Christoph Le Blon invents three-color printing	**1710**
				1711
Giovanni Ceva's *De re numeraria* (Concerning money matters) is the first clear application of mathematics to economics		French physicist Denis Papin d London	Thomas Newcomen (b Dartmouth, England, Feb 24, 1663) erects the first practical steam engine to use a piston and cylinder	**1712**

	GENERAL	ASTRONOMY	BIOLOGY	CHEMISTRY
1712 cont		replaces Kepler's catalog; an official, three-volume edition appears posthumously in 1725 Italian-French astronomer Giovanni Domenico Cassini d Paris, Sep 14		
1713			In *Physico-theology, or a demonstration of the being and attributes of God from His works of creation*, William Derham tries to show that this is the best of all possible worlds	
1714		William Derham publishes *Astro-theology*, an attempt to argue from astronomy to God		
1715				Nicolas Lémery d Paris, Jun 19
1716		The French government offers prizes for the solution of finding the longitude at sea with accuracy	Louis-Jean-Marie Daubenton b Montbard, France, May 29; he is active in many fields, but is best known as the author of *Advice to shepherds and owners of flocks on the care and management of sheep*	

1713

Nicholas Bernoulli (b St. Petersburg, Russia, 1695) poses the probability problem known as the Petersburg paradox in a letter to Pierre Rémond de Montmort; it involves the mathematical expectation of a simple coin-tossing game; mathematical reasoning gives a result that is clearly wrong according to practical experience

Jacques (Jakob) Bernoulli's *Ars conjectandi* (The conjectural arts), a treatise on probability published posthumously, contains the theorem named after him, a version of the law of large numbers; the theorem is also the first application of calculus to probability theory

Emanuel Timoni describes to the British Royal Society the Turkish practice of inoculating young children with smallpox to prevent more serious cases of the disease when they get older

The second revised edition of Newton's *Principia* is published with an introduction by Roger Cotes; it contains the famous *General scholium*

Francis Hauksbee d London, Apr

1714

Roger Cotes (b Burbage, England, Jul 10, 1682) gives one of the first forms of a theorem usually attributed to Euler: the natural logarithm of the sum of the cosine of an angle and i times the sine is equal to i times the angle; it is more familiar in the exponential form: $e^{ix} = \cos x + i \sin x$

Dominique Anel invents the fine-point syringe, still known by his name, for use in treating *fistula lacrymalis*

Gabriel Fahrenheit builds a mercury thermometer with a scale that will be named after him

The British Parliament passes a bill setting up a prize of £20,000 for the first person to develop a sufficiently accurate way to find the longitude at sea

1715

Brook Taylor develops in *Methodus incrementorum directa et inversa* the calculus of finite differences; the work also contains the series named after him; he also publishes *Linear perspective*

Anatomist Raymond Vieussens d Montpellier, France, Aug 16

John Harrison (b Foulby, England, Mar 24, 1693) constructs an eight-day clock

Engineer Thomas Savery d London, May

1716

Leibniz sends Newton a difficult problem (finding the orthogonal trajectories of a family of hyperbolas having the same vertex) to test his skills; Newton receives the problem about 5:00 PM on returning from a long day at

The French establish the first national highway department

Edmund Halley develops a diving bell with an air refreshment system and demonstrates its use

1716
cont

1717

1718

Halley discovers proper motion of fixed stars—that is, apparent motion with respect to other fixed stars—by comparing star positions with those given by Hipparchus and Ptolemy; Halley concludes that Sirius and other bright stars are in different positions than those recorded by the early Hellenic astronomers

1719   Encyclopedist John Harris d Norton Court, Kent, England, Sep 9

John Flamsteed d Greenwich, England, Dec 31

Verifying Newton's theory of gravitation

When Newton formulated his theory of gravitation he had one competitor, Descartes, with his theory of vortices. Newton's theory called for a universal force of gravitation that interacted between all bodies thoughout the universe. Descartes assumed that force could not be transmitted through empty space and that a vortex of matter moved Earth around the sun.

Descartes' theory predicted that Earth would be flattened at the equator and elongated at the poles. A Newtonian Earth should be flattened at the poles and fatter at the equator.

In 1718 Jacques Cassini published measurements of the curvature of Earth that he and his father, Jean-Dominique, had made. They found that the curvature of Earth was greater the closer one came to the North Pole, a result in agreement with Descartes' theory.

In 1733 Charles-Marie de la Condamine suggested an expedition to measure the curvature of Earth on the equator. La Condamine directed this expedition, which left France in 1736 and returned 10 years later. Another Frenchman, Pierre Maupertuis, directed an expedition to Lapland from 1736 to 1737. The measurements made by the two expeditions were compared, and it was shown convincingly that Earth is flattened at the poles and oblate at the equator, a result consistent with Newton's theory. The Cassinis' measurements had been in error.

A second challenge to Newton's theory during the eighteenth century concerned the problem that the theory could not explain the motion of the moon. To calculate the motion of the moon, one has to take into account not only the attraction between the moon and Earth, but also the force between the moon and the sun, a problem known as the "three bodies problem." Because it is impossible to solve the equations governing the motion of three bodies exactly, scientists, including Jean le Rond d'Alembert, Alexis Clairaut, and Leonhard Euler, developed methods for successive approximations to the solutions of these equations. All found that the calculated lunar apogee (the point in the moon's orbit farthest from Earth) was totally different from the one observed. Clairaut, however, who strongly believed that Newton's theory was correct, found a common error in all these approximations, which he announced in 1749. Clairaut was then able to use Newtonian mechanics to calculate the observed position.

One of the most convincing proofs that Newton's theory of gravitation was correct came in 1759. Edmund Halley had predicted that the great comet of 1682 would return in 1758 or 1759. Clairaut, using more precise calculations, predicted the return of the comet for 1759, with a margin of error of only 30 days. Halley's comet turned up at the predicted time.

the mint, where he is master, and solves it before going to bed

Roger Cotes d Cambridge, England, Jun 5

Philosopher and mathematician Gottfried Wilhelm Leibniz d Hanover (Germany), Nov 14

J.N. de la Hire invents the double-acting water pump, which produces a continuous stream of water

1716 cont

Abraham Sharp finds the value of π to 72 places

Giovanni Lancisi suggests in *De noxiis paludum effluviis* (On the noxious effluvia of marches) that malaria can be transmitted by a mosquito

Lady Mary Wortley Montagu brings back to England the Turkish practice of inoculation and has her own two children vaccinated against smallpox

Inventor Abraham Darby d Madley Court, Worcestershire, England, Mar 8

1717

Jacques (Jakob) Bernoulli's *Mémoires de l'Académie des Sciences* (Memoirs of the Academy of Sciences), published posthumously, contains the basic concepts of the calculus of variations, the mathematical treatment of functions that are the maximum or minimum function for some specific conditions

Doctrine of chances by Abraham De Moivre (b Vitry-le-François, France, May 26, 1667) is his first book on probability

Etienne Geoffroy presents *Table des différents rapports en chimie* (Table of different chemical affinities), the first table of affinities (attractions of substances to each other) to the French Academy

German physician Friedrich Hoffmann (b Halle, Feb 19, 1660) begins publication of the nine volumes of his *Medicina rationalis systematica*, in which he introduces the concept of muscle tone as a measure of health

Jacques Cassini (b Paris, Feb 8, 1677) publishes his and his father's (Giovanni Domenico Cassini) measurements supporting Descartes' prediction (later proven faulty) that Earth is elongated at the poles

1718

New principles of linear perspective by English mathematician Brook Taylor (b Edmonton, Middlesex, Aug 19, 1685) gives the first general statement of the principle of the vanishing point

1719

	GENERAL	ASTRONOMY	BIOLOGY	CHEMISTRY
1720	Stock in the South Sea Company collapses, ruining thousands of investors in England and damaging the English financial system, since the South Sea Company is a financial partner of the government as well as a private enterprise	Halley succeeds Flamsteed as astronomer royal and starts on an 18-year study of the moon; along the way he discovers its secular acceleration		
1721			Botanist Rudolph Camerarius d Tübingen (Germany), Sep 11	
1722	Dutch admiral Jakob Roggeveen discovers Easter Island			
1723		John Hadley (b Hertfordshire, England, Apr 16, 1682) builds a reflecting telescope of unusual high quality	Biologist and microscopist Anton van Leeuwenhoek d Delft, Holland, Aug 26	Georg Stahl's *Fundamenta chymiae dogmaticae et experimentalis* (Foundations of dogmatic and experimental chemistry) popularizes the views of Johann Becher

MATHEMATICS	MEDICINE	PHYSICS	TECHNOLOGY	
Daniel Bernoulli (b Groningen, Holland, Feb 9, 1700) solves one version of Riccati's differential equation, $y' = p(x)y^2 + q(x)y + r(x)$, although the general solution will not be found until 1760, when Euler solves the equation *Geometrica organica* by Colin Maclaurin (b Kilmodan, Scotland, Feb, 1698) discusses curves of the second, third, fourth, and general degree	Anatomist and pathologist Giovanni Lancisi d Rome, Jan 21		George Graham (b Cumberland, England, about 1674) invents the deadbeat escapement for clocks René Antoine Ferchault de Réaumur (b La Rochelle, France, Feb 28, 1683) builds the cupola furnace for melting iron	1720
	Zabdiel Boylston introduces inoculation against smallpox into America during the Boston epidemic Jean Palfyn introduces the use of forceps for facilitating birth	*Mathematical elements of natural philosophy confirmed by experiments, or an introduction to Sir Isaac Newton's philosophy* by Willem Jacob van 'sGravesande (b 'sHertogenbosch, Holland, Sep 26, 1688) offers strong Continental support for Newton's physics Christian Wolff, b Breslau (Germany), Jan 24, 1679, publishes *Allerhand nutzliche versuche, dadurch zu genauer erkenntnis der natur und kunst der weg gebahned wird* (Generally useful researches for attaining a more exact knowledge of nature and the arts)	George Graham develops the mercury compensating pendulum for clocks	1721
Roger Cotes's *Harmonia mensurarum*, published posthumously, is among the earliest works to recognize that the trigonometric functions are periodic			René de Réaumur's *L'art de convertir le fer forgé en acier* (The art of converting iron into steel) is the first technical treatise on iron	1722
	Anatomist Antonio Valsalva d Bologna (Italy), Feb 2	M.A. Capeller's *Prodomus crystallographiae* is the earliest treatise on crystallography	Jacob Leupold, b Planitz (Germany), Jul 25, 1674, publishes *Theatrum machinarum generale* (General theory of machines; nine volumes, 1723-1739), the first systematic treatment of mechanical engineering; the work includes the design of a noncondensing, high-pressure steam engine comparable to those built at the beginning of the nineteenth century Architect, mathematician, and astronomer Sir Christopher Wren d London, Feb 25	1723

1724   Paul Dudley discovers the
possibility of cross-
fertilizing corn

1725   John Flamsteed's three-
volume *Historia coelestis*,
published posthumously,
catalogs 2884 stars

Luigi Marsigli's *Histoire
physique de la mer*
(Physical history of the sea)
is the first treatise on
oceanography

Johann Scheuchzer's *Homo
diluvii testis* is an important
work on fossils

Phlogiston

The question of what fire really is has intrigued people since ancient times, but not until the Renaissance did scientists offer explanations for the phenomenon. Classical Greeks thought was that fire is an "element" no more explicable than earth, water, or air.

1726

Robert Boyle found, however, that fire depends on air. If one removes air in a vessel, combustion cannot take place. He also found that air is required for the respiration of small animals and for the production of oxides from metals by heating them. In this process, chemical changes caused by heating a substance to a temperature below the melting point (calcination) results in a compound called a *calx*.

1727

Around this time, however, chemists in Germany formulated a theory of combustion that did not require air. Johann Joachim Becher believed that combustible matter contains "oily earth," a substance released during combustion. George Stahl renamed this substance *phlogiston*. With the phlogiston theory these chemists could explain not only combustion but also the reverse of calcination, reduction. According to this theory, if calx of lead is brought into contact with charcoal and heated, the phlogiston flows from the charcoal into the calx of lead, changing it to metallic lead. Because charcoal burns so well, it was believed to contain a lot of phlogiston.

One of the first indications that something was amiss

Stephen Hales lays the foundations of plant physiology with *Vegetable staticks or Statistical essays on nutrition of plants and plant physiology*, discussing the ascent of liquids in plants and describing gases emanating from heated bodies as "air"

1728   Vitus Bering (b Horsens, Denmark, summer, 1681) discovers the Bering Strait

The first edition of Ephraim Chambers's *Universal dictionary of arts and sciences* appears in two volumes

In *Hesperi et phosphori nova phaenomena* Francesco Bianchini estimates the rotation period of Venus as 24-1/3 days

James Bradley (b Sherborne, England, Mar, 1693) explains the observed periodic shifts of stars' positions during the year by the abberation of light

1724

Peter the Great founds the Academy of Sciences at St. Petersburg, Russia; it attracts many of the leading mathematicians of Europe, including Nicholas and Daniel Bernoulli and Leonhard Euler

Hermann Boerhaave's *Elementae chemiae* (Elements of chemistry) argues that heat is a fluid

Gabriel Daniel Fahrenheit describes the supercooling of water

1725

Abraham De Moivre's *Annuities on lives* is the first important mathematical book on annuities

Nicolas Hartsoeker d Utrecht, Holland, Dec 10

Antoine Thiout (b near Vesoul, France, 1692) designs a clock showing solar time, called the equation clock

1726

Swiss mathematician Nicholas Bernoulli d St. Petersburg, Russia, Jul 31

Stephen Hales (b Bekesbourne, England, Sep 17, 1677) makes the first measurement of blood pressure of the horse

John Harrison constructs a gridiron compensating pendulum clock

1727

Leonhard Euler (b Basel, Switzerland, Apr 15, 1707) introduces *e* as the symbol for the base of the natural logarithms in correspondence; it will appear in print for the first time in Euler's *Mechanica, sive motus scientia analytice exposita* in 1736

Bernard de Fontenelle's *Eléments de la géométrie de l'infini* gives his theory of the calculus of infinity

Physicist and mathematician Sir Isaac Newton d London, Mar 20

Mechanical engineer Jacob Leupold d Leipzig (Germany), Jan 12

Phlogiston (continued)

with the theory was evidence that metals gain weight upon calcination. Clearly, this contradicted the idea that upon calcination, phlogiston leaves the metal. Chemists saved the phlogiston theory by assuming that the substance had "negative weight."

Around the same time Joseph Priestley experimented with heating calx of mercury and found that it produces "air" with special properties. If one introduces a candle into it, the candle burns vigorously. A mouse shut in a jar with this "air" lives longer than one shut in a jar filled with ordinary air. Priestly believed in phlogiston, so he tried to explain this result with the phlogiston theory. Priestley called this "air" *dephlogistonated air.* He believed that air has phlogiston added to it by fires and by breathing.

Antoine Lavoisier discovered in 1772 that phosphorus and sulfur absorb air when burning instead of expelling something. Lavoisier also discovered that the reduction of calx of lead releases large quantities of "fixed air," air that does not sustain life or combustion (now known as carbon dioxide). That is, reducing a calx to a metal expels something, it does not absorb something.

These results, and also the results of other experiments led Lavoisier to formulate a theory of combustion that superseded the phlogiston theory. Lavoisier had realized that "dephlogistonated air," a gas that he subsequently termed oxygen, was responsible for the combustion of materials.

1728

Pierre Fauchard's *Le chirurgien dentiste, ou traité des dents* (The surgeon dentist, or treatise of the teeth) puts dental treatment on a more scientific plane and describes how to fill a tooth infected with dental caries using tin, lead, or gold

Giovanni Lancisi's posthumously published *De motu cordis et aneurysmatibus* discusses heart dilatation

1729

Temperature

Some of the measurements we take for granted were not at all developed in the ancient world. Among them is temperature. Certainly ancient people knew that it was cold sometimes and hot sometimes, but no one had a way of putting a number to it. It was Galileo who made the first steps toward a quantitative means of measuring temperature. He noted that gases expand when heated. Using this principle, he developed a thermoscope, an inaccurate gas thermometer. Although Galileo did not know it, the gas he used—air—changed in volume according to outside air pressure as well as heating.

A really good thermometer was not made until 1714, when Gabriel Fahrenheit made the first mercury thermometer. Fahrenheit used changes in the expansion of mercury instead of air to measure temperature. He set 0° as the lowest temperature he could reach by freezing salted water, hoping to avoid negative temperatures.

Anders Celsius did not create a new type of thermometer, but he did create the scale that is most commonly used around the world today. In one of the great mistakes of science, Celsius first set his scale with 0° as the temperature of boiling water and 100° as the temperature of freezing water. Wiser heads prevailed, however, and the scale was reversed. Until 1948 the scale Celsius devised was known in the United States as *centigrade,* but that year scientists agreed to rename it after its inventor. Gradually the world has gotten used to the idea that 37°C means 37 degrees Celsius, not 37 degrees centigrade.

More accurate thermometers have been developed over the years. Instead of using the simple expansion of materials upon heating, other thermometers use differences in the coefficient of expansion of two metals,

Louis Bourget's *Lettres philosophiques sur la formation des sels et de cristaux et sur la génération et la mécanique organique* (Philosophical letters on the formation of salts and crystals and on generation and organic mechanisms) makes the distinction between organic and inorganic growth

1730

Otto Müller b Copenhagen, Denmark, Mar 2; he is among the first to see bacteria since Leeuwenhoek and the first to classify them into types

Anatomist Luigi Marsigli d Bologna (Italy), Nov 1

Conspectus chemiae theoretico-practicae (Conspectus of chemistry) by Johann Juncker, b Londorf (Germany), Dec 23, 1679, contains a systematic explanation of the ideas and experiments of Becher and Stahl concerning their phlogiston theory

Georg Brandt (b Riddarhyttan, Sweden, Jul 21, 1694) discovers cobalt

1731

The Royal Dublin Society is founded in Ireland

The Copley Medal is instituted by the Royal Society, London, following a 1709 bequest from Sir Godfrey Copley; it is awarded "to the living author of of such philosophical research ... that may appear to the council to be deserving of such honor"

John Hadley and Thomas Godfrey independently invent the reflecting quadrant, a precursor of the sextant

Volume I of Mark Catesby's *The natural history of Carolina, Florida, and the Bahama Islands* appears

Jethro Tull's *Horse-hoeing husbandry* advocates the use of manure, pulverization of the soil, growing crops in rows, and hoeing to remove weeds

Etienne Geoffroy d Paris, France, Jan 6

1732

First issue of *Poor Richard's Almanack* is published by Benjamin Franklin (b Boston, MA, Jan 17, 1706)

Nöel-Antoine Pluche publishes the first of eight volumes of *Le spectacle de la nature* (Spectacle of nature), popularizing "natural theology" in France

Isaac Greenwood's *Arithmetic vulgar and decimal* is the first mathematics textbook in America

---

Temperature (continued)

the amount of energy emitted at different temperatures by certain substances, or the changes in electrical resistance at different temperatures.

Fahrenheit's idea of a temperature scale with no negative numbers was achieved with the introduction of the Kelvin scale (sometimes called the absolute scale). Lord Kelvin observed in 1848 that Charles's law, which states that gases lose 1/273 of their volume for every degree below 0°C, implies that at −273°, the volume would become zero. He proposed that −273° must be the lowest temperature obtainable; this has since been verified and set more accurately as −273.18°C, or −497.7°F. On the Kelvin scale, then, the freezing point of water is 273.18°K and the boiling point is 373.18°K.

Essai d'optique sur la gradation de la lumière by Pierre Bouger (b Le Croisic, France, Feb 16, 1698) reports on some of the earliest measurements in photometry

Stephen Gray discovers that electricity can be transmitted from one object to another and over distances through conductors, and that static electric charges are on the surfaces of objects, not in the interiors

Andrew Motte publishes an English translation of Newton's *Principia*

Physicae experimentales et geometricae dissertationes by Pieter van Musschenbroek (b Leiden, Holland, Mar 14, 1692) is one of the first books where the term "physics" is used for "natural philosophy"

La science des ingénieurs by Bernard Forest de Belidor (b Catalonia, Spain, about 1693) is a popular manual of construction rules and tables that will be reprinted until 1830

Engineer Thomas Newcomen d London, Aug 5

1729

French surgeon Dominique Anel d

George Martine performs the first tracheotomy for treatment of diphtheria

Surgeon Jean Palfyn d Ghent (Belgium), Apr 21

René de Réaumur constructs an alcohol thermometer with a graduated scale from 0 to 80 to indicate freezing and boiling

1730

Brook Taylor d London, Dec 29

John Arbuthnot's *An essay concerning the nature of ailments* makes a suggestion for dieting

Stephen Gray demonstrates that anything can be charged with static electricity if it is isolated by nonconducting materials

The invention of the pyrometer by Pieter van Musschenbroek is reported in the Latin edition of *Saggi di naturali esperienze fatte nell' Accademia del Cimento*

1731

Discours sur la figure des astres (Discourse on the shape of the heavenly bodies) by Pierre Louis Moreau de Maupertuis (b St. Malo, France, Sep 28, 1698) predicts the shape of Earth using Newtonian mechanics

1732

1733

The Russian Great Nordic Expedition (1733-1743) starts; it looks for a northeastern sea passage

Anders Celsius (b Uppsala, Sweden, Nov 27, 1701) publishes a compilation of his observations of the aurora borealis

Chester Moor Hall invents the achromatic telescope, using an objective lens composed of two kinds of glass so that the chromatic abberation in one kind of glass is compensated for by the other kind of glass

Stephen Hales's *Statical essays, containing haemastatics, etc.,* two volumes, reports his investigations on blood flow and pressure in animals and the hydrostatics of sap in plants

Paleontologist, physician, and mathematician Johann Jakob Scheuchzer d Zurich Switzerland, Jun 23

1734

Lettres Anglaises ou philosophiques (English or philosophical letters) by Voltaire (François-Marie Arouet, b Paris, Nov 21, 1694), is the first introduction to Newtonian mechanics in French

René de Réaumur's *Mémoires pour servir à l'histoire des insectes* (History of insects) is one of the foundation works in entomology

Opera philosophica et mineralia by Emanuel Swedenborg (b Svedberg, Stockholm, Sweden, Jan 29, 1688) is a three-volume survey of the nature of matter and of the laws of motion

Georg Ernst Stahl d Berlin, (Germany), May 14

1733

Yale is the first American college to offer geometry, using Euclid's *Elements* as a textbook

Abraham De Moivre publishes the discovery of the normal curve of error, or the distribution curve

Euclid ab omni naevo vindicatus (Euclid cleared of every flaw) by Girolamo Saccheri, b San Remo (Italy), Sep 5, 1667, is an attempt to prove Euclid's parallel postulate that instead lays the groundwork for non-Euclidean geometry

Girolamo Saccheri d Milan (Italy), Oct 25

Charles François de Cisternay Du Fay (b Paris, France, Sep 14, 1698) discovers that there are two types of static electric charges and that like charges repel each other while unlike charges attract; this leads to Du Fay's two-fluid theory of electricity, later opposed by Benjamin Franklin's one-fluid theory; both have elements of truth since electric charge is transmitted by electrons (one fluid) but comes in two varieties, positive and negative

The Schemnitz Mining Academy founded at Schemnitz (Hungary) is the first technical college in the world

John Kay (b Bury, England, Jul 16, 1704) invents the flying shuttle loom

1734

Physician John Floyer d Lichfield, England, Feb 1

The French Academy of Sciences, suspicious of Newton's ideas for many years, awards a prize for the first time for a work based on Newton's theories

George Berkeley's *The analyst: or, a discourse addressed to an infidel mathematician*, a work addressed to Edmund Halley, attacks Newton's calculus, calling infinitesimals "neither finite quantities, nor quantities infinitely small, nor yet nothing. May we not call them the ghosts of departed quantities?"

Emanuel Swedenborg's *Regnum subterraneum* gives an account of mining and smelting techniques

OVERVIEW

The Enlightenment and the Industrial Revolution

The term *Enlightenment* was coined to reflect a change in the philosophical approach from that of the preceding ages in which faith in God or faith in classical authors was preeminent. During the Enlightenment, there was a critical examination of previous beliefs in the light of rationalism. The period of the Enlightenment is often equated with the latter part of the eighteenth century, starting with the works of Gotthold Ephraim Lessing (1729-1781) in Germany and Denis Diderot (1713-1784) in France, but this book is divided on the basis of scientific eras. Therefore, the next period begins with the Linnaean classification scheme, which represents scientifically the sense of order that characterized eighteenth-century thought, and terminates just before the publication of the discovery of electromagnetism by Hans Christian Oersted in 1820. In England, the period from about 1740 to 1780 is the time generally considered the Industrial Revolution (*see* "When was the Industrial Revolution?," p 192) which took place somewhat later in other Western nations. Thus, the Enlightenment and the Industrial Revolution overlap considerably.

Philosophy and science

One of the consequences of the revolution in science that took place during the seventeenth century was a profound change in philosophical thinking during the eighteenth century. The success of Isaac Newton's theories, the validity of which became well established during the first decades of the eighteenth century, not only had a profound impact on science itself, but also on philosophical outlook. Newton's success in explaining a large number of phenomena, such as falling bodies and the orbits of planets, by a simple set of laws encouraged people to view the physical sciences as a model for the other sciences. This resulted in the emergence of a "mechanical philosophy;" that is, the belief that not only physics, but also chemistry and biology, could be explained by sets of simple mechanical laws. Most attempts to apply Newtonian mechanics to problems in chemistry and physiology, however, were only partly successful during the eighteenth century.

The idea that mechanics would offer an explanation for all phenomena was taken the furthest by the materialistic philosophers of France such as Denis Diderot, Julien Offray de la Mettrie, and Baron d'Holbach. They denied the existence of a spiritual god, and viewed nature entirely as a mechanical system. In this system, they included humanity itself, body and spirit. They believed that all physiological processes could be reduced to physical processes, but also that spiritual life, social processes, and even the course of history, would fall under the same rules of nature.

Another factor that strongly influenced the philosophical outlook of the eighteenth century was the belief that nature embodied reason, and that the laws of nature would be reasonable laws.

Two approaches to philosophy in the seventeenth century were empiricism—the idea that knowledge comes from experience—and rationalism—the view that knowledge comes from reasoning. A leading philosopher of the eighteenth century was Immanuel Kant, who coined the term "Enlightenment." He reconciled empiricism and rationalism by saying that one can gain knowledge not only from experiences, but also from reasoning—both go hand in hand. The idea of causality—one event causes another to happen—a concept that David Hume rejected for many types of event, became for Kant an *a priori* concept, a general principle that cannot be proved but that has to be accepted as existing independently from our sense perceptions. Space and time, according to Kant, are also *a priori* concepts that are not accessible through the senses, but which condition our sense experiences of the physical world. Therefore, he argued, physics is not a science based only on sense experiences, but also on *a priori* concepts, making it possible to gain knowledge of the physical world using a rational system, such as mathematics.

Another philosophical idea that influenced science at this time had been revived by Gottfried Wilhelm Leibniz, although it had its roots in Plato and Aristotle. The Great Chain of Being, as this idea came to be known, envisioned all existence as continuous. For animals, for example, there was a continuous chain from the smallest protist to humanity, with no gaps. Some scientists of the Enlightenment, such as comte de Buffon and Charles Bonnet used this notion in their systems of classification, with Bonnet arguing that minerals graded into fossils which were continuous with plants that, somewhere in the vicinity of the green hydra, became animals. At the end of this century, Jean-Baptiste Lamarck used this concept in his theory of evolution. Not all scientists of the period accepted the Great Chain of Being. Georges Cuvier explicitly rejected it, for example.

Over the eighteenth century, this Great Chain of Being also was gradually modified to include the concept of progress—that as time went on, higher beings arose. Various early theories of evolution reflected this notion as well.

The *Encyclopédie*

The flagship publication of the Enlightenment was the *Encyclopédie*. Edited by Denis Diderot and Jean le Rond d'Alembert, the seventeen volumes of text and eleven of illustrations were published from 1751 to 1772. The *Encyclopédie* became important in the democratization of scientific knowledge. Technology was given equal importance to pure science or philosophy. The *Encyclopédie* allowed the philosophers of the Enlightenment to communicate their ideas to a broader public. In keeping with the trend toward systematizing knowledge, another aim of the editors was to lay the foundations of all the arts and sciences. Unlike earlier encyclopedias, such as Ephraim Chambers *Cyclopedia,* on which the *Encyclopédie* was based, the *Encyclopédie* adopted a strong editorial point of view, opposing both church and state.

The romantic reaction

The extreme materialism of some of the thinkers of the Enlightenment bothered many, and toward the end of the eighteenth century a new school of thought, often viewed as a romantic reaction to the materialism and rationalism of the philosophers of the Enlightenment, emerged in Germany: the school of *Naturphilosophie.* There was a similar reaction, generally called romanticism, in both France and England, The most prominent representatives of the *Naturphilosophie* school were the German poet and writer Johann Wolfgang von Goethe, who had also done considerable scientific work, Friedrich von Schiller, the philosopher Friedrich Wilhelm Joseph von Schelling, and the biologist Lorenz Oken. Instead of reason they preferred sentiment. Their view of nature was, to use a modern term, holistic; that is, they believed that all nature should be viewed as a single organism and imbued with spirit. In France, Jean-Jacques Rousseau had expressed similar ideas somewhat earlier, especially in his emphasis upon the importance of emotion.

In both France and the United States, political revolutions and the resulting new governments were heavily influenced by the philosophical trends of the time. Among the revolutionaries in the United States were the parttime scientists Benjamin Franklin and Thomas Jefferson; the French scientists Joseph-Louis Lagrange and Pierre-Simon Laplace held posts in the revolutionary French government. In France even more than in the United States, there was an effort to remake the culture along "scientific lines," including a short-lived reworking of the calendar and the creation of the metric system.

The Industrial Revolution

During the eighteenth century there was a general turn toward the development of machines that would make work faster or more efficient, especially in England. The motive for this surge of inventiveness is not completely clear. Some changes, such as the turn to coal for both heating and making iron, were caused by limitations in natural resources—in this case, wood. Trade certainly was a factor, for example the introduction of cotton from India and the United States. A new class, the capitalists, who were not really the same as the landowners, began to appear. They first bankrolled cottage industry, but soon found it more practical to set up factories. For these and probably other reasons, the Industrial Revolution started. Because of its success in England, it was quickly taken up in those nations that were similar to England; that is, in Europe, the United States, and Canada.

Influenced by technological developments, the Industrial Revolution in turn affected science. Interest in thermodynamics rose as a result of the steam engine. Such concepts as work and power began to be formalized, although this occurred mainly in the nineteenth century. A chemical industry based on chemical processes and not on such organic processes as fermentation slowly started. There was not much effect on life sciences, but industrialization did bring some new medical problems, mainly those connected with the workplace or with crowding.

Major advances

Astronomy. Astronomy profited enormously from Newton's theory of gravitation, from advances made in mathematics, and from the much higher accuracy of reflecting telescopes. Astronomers predicted many astronomical phenomena with great precision. Pierre-Simon Laplace, often called the "French Newton," solved many mechanical problems and demonstrated the stability of the solar system. Astronomers and philosophers abandoned the Aristotelian idea that everything in the universe is static; that is, that since its creation, the universe always has looked the way it looks now. Kant and Laplace introduced the concept of the evolution of the universe and independently formulated a theory, known as the nebular hypothesis, for the origin of the solar system.

The better and more accurate telescopes led to the improvement of positional astronomy, the study of double stars, and the discovery of Uranus and several new planetary satellites. William Herschel was the first to propose that the sun is a star belonging to a vast system of stars, now known as the Milky Way galaxy. By counting the stars in different directions, he tried to determine the shape of the galaxy and found that it must be shaped like a millstone with the sun at its center. Although Herschel was wrong about the details (we now think the Milky Way galaxy is shaped like a spiral with the sun far from the center), the basic idea of the sun's belonging to a vast system of stars was right. Kant suggested that the many patchy objects visible in the telescope may be systems similar to the sun's star system, an idea that was confirmed early in the twentieth century.

The taming of the longitude

The Phoenicians were the world's first great navigators. They determined the location of their ships in terms of the length and width of the Mediterranean, the sea on which they sailed. The Phoenicians taught their navigational methods to the Greeks, from whom the Romans learned. Eventually translated into Latin, the sailors' two perpendicular directions became the longitude (*longus*, or long) and latitude (*latus*, or wide). Eratosthenes drew a map of the known world around the third century BC that included lines of latitude and longitude, but his lines were determined by large cities; that is, they were not equally spaced. In the second century BC Hipparchus of Nicaea greatly improved on this by spacing the lines equally around the globe. By assigning 360° to the circumference of Earth, and by using Eratosthenes' nearly correct calculation for the size of Earth, Hipparchus was able to obtain good distances for a degree. Hipparchus's system continues in use today.

Sailors of classical times, however, used only half the system—the latitude. Early travelers observed the changes in the constellations as one travels north or south. One star, Polaris, is visible every clear night in the Northern Hemisphere, but its position dips closer to the horizon as one travels south. At the North Pole, Polaris is directly overhead, and it disappears below the horizon at the equator. Until sailors began to travel in the Southern Hemisphere, all that was needed to find the latitude was an instrument for measuring how high Polaris, the pole star, was above the horizon. The Portuguese sailors who rounded the Cape of Good Hope were terrified because they had lost their main navigational tool on the trip down the side of Africa.

Sailors did not use the longitude because they had no way to measure it. The historian of Magellan's circumnavigation reported that Magellan himself spent long hours trying to find ways to measure the longitude, but those under him were too proud of their navigational skills to speak of it. Common sailors and their pilots believed that they could navigate by a combination of charts, dead reckoning, and the latitude. Governments, however, realized that this was not good enough. The governmental view was particularly brought home to the English when in 1691 and again in 1707 large parts of the British Navy were lost because of navigational errors.

Much earlier, in 1598, Philip III of Spain offered the first of several prizes by sea-going nations for the person who could find the longitude. Among the schemes suggested to Philip was one from Galileo. Having discovered the four largest satellites of Jupiter, he proposed that they could be used to locate the longitude. Galileo's idea was based using charts that show the relative positions of the satellites at different times. By observing the positions, one can determine the time. From the time, one can find the latitude. By telling the time exactly, one can determine the longitude.

As a clock is carried from place to place, it will be off by one hour for each 15° of longitude. This is why there are four time zones in the lower 48 states of the United States. This part of the United States is approximately 4 × 15 or 60° of longitude wide. Comparing universal time with local time helps one to obtain the longitude. Local noon can be obtained easily, by determining when the sun reaches its daily zenith. It is universal time that is hard to obtain.

For example, if you get a long-distance phone call from someone while you are in New York City and you do not know where your friend is calling from, you can ask "What time is it?" If it is noon in New York, and your friend says 11 AM, then you know your friend's longitude is approximately between 90° W and 107° W, the approximate longitude of Central Time in the United States. To determine the *exact* longitude, however, you need to forget about time zones and compare sun times between two places. Specifically, compare local sun time with Greenwich Mean Time (GMT), the time at longitude 0°. If you know it is 1:00 PM GMT and you observe the sun time where you are to be noon, your longitude is exactly 15° W.

Astronomical events that repeat frequently are a good way to obtain universal time. Eclipses of the moon could be used, for example, if they were not too infrequent to be of much use to sailors. The positions of Jupiter's moons are more useful because one or another is frequently eclipsed.

Galileo's suggestion was a good one. While ignored by Philip, it was taken up in the seventeenth century by French astronomers, led by Giovanni Domenico (Jean-Dominique) Cassini. Using Jupiter's moons, the French were able to establish correctly the longitude of cities in Europe and of islands in the Atlantic. The observations required, however, were too difficult and too time-consuming to be done by sailors at sea.

In 1530, Gemma Frisius suggested that the easy way to solve the problem would be to carry a good clock, set to some universal time, with the ship. Others had the same idea, but the clocks were not good enough. Even after Huygens developed a pendulum clock that theoretically kept good time, it was too imprecise to find the longitude and too delicate to be used on ships.

Christopher Columbus noted on his first voyage that the compass needle changed its deviation from true north as he sailed across the Atlantic. Many expeditions were launched in the eighteenth century—notably that of Edmund Halley—to chart these magnetic deviations in the hopes that they would lead to the secret of the longitude. It was discovered, however, that the deviations varied in such an unpredictable manner that they were unsuitable for this purpose.

In 1714, the English Parliament provided a reward of £20,000 for anyone who could find the longitude and demonstrate the method on a voyage to the West Indies. Many of the leading scientists of the eighteenth century worked on the problem in competition for the prize, but the achievement was accomplished by a self-taught watchmaker, John Harrison. His clock—or marine chronometer, as it came to be called—Number Four made the West Indies trip in 1761 and passed the test with flying colors. All along the way Harrison's son, who was in charge of the chronometer, predicted landfalls with greater accuracy by far than the ship's pilots. Unfortunately for Harrison, the board governing the prize included a number of scientists who still hoped to get the money themselves. They gave him half the prize after four years, and then stalled him again for seven more years; they would probably have stalled longer if King George III had not intervened on Harrison's behalf.

Biology. During the eighteenth century the classification of plants and animals was still viewed as part of "natural history," a science that also included the study of minerals, for example. Plant and animal physiology belonged to physics. In the *Encyclopédie*, zoology, botany, and medicine were categorized as fields of physics and came under the general heading of "reason." The approach to physiology was consequently mechanistic.

Carolus Linnaeus began the modern system of classification of living organisms and introduced the binary notation system for naming species that is still in use today. Comte de Buffon disagreed with Linnaeus's system of classification because he did not believe that a system based on external characteristics could be a natural system of classification. Buffon proposed a system based on the reproductive history of organisms instead. He said, for example, that a fox is a different species from a dog if and only if it can be shown that matings between a fox and a dog either produced no offspring or (as in matings between a mare and a jackass) sterile offspring.

Toward the end of this period, Georges Cuvier began to classify extinct species. Jean-Baptiste Lamarck studied in detail the anatomy of invertebrates. Both of these men contributed to the concepts that would lead to the theory of evolution in the nineteenth century, although Cuvier did not believe in evolution and Lamarck accounted for it with a now-discredited mechanism (*see* "The theory of evolution," p 275).

Chemistry. During most of the eighteenth century, chemistry was dominated by the erroneous phlogiston theory (*see* "Phlogiston," page 182). A large number of chemists adhered to this theory, which may account for the fact that chemistry advanced at a much slower pace than physics. The scientific revolution that for physics took place during the seventeenth century, happened in chemistry only at the end of the eighteenth century with the work of Antoine-Laurent Lavoisier, Henry Cavendish, and John Dalton. Also, the number of technical applications was relatively small during the eighteenth century. A chemical industry did not flourish until the nineteenth century, when the mechanisms of chemical reactions were more fully understood.

Lavoisier proposed a correct theory of combustion at the end of the eighteenth century, opening the way to the rapid development of chemistry during the early years of the nineteenth century. At that time John Dalton put forth his atomic theory, which eventually became the foundation of the new chemistry.

When was the Industrial Revolution?

The French coined the term *Industrial revolution* in the nineteenth century as an analogy to their own political revolution. Although the phrase is in common use, it is not always clear what this phenomenon was or even when it occurred.

For example, it has been pointed out that humans underwent *an* industrial revolution at the beginning of the Bronze Age; yet, it is clear that *the* Industrial Revolution is something else. One theory holds that the Industrial Revolution happened when people stopped using human and animal power and began using inanimate power sources; this could place the Industrial Revolution as early as the first extensive use of wind and waterpower—medieval times in the West and somewhat earlier in China. A different suggestion is that it began when mills began to centralize the production of textiles. This might be interpreted as occurring when fulling mills (mills to work cloth so that it is fuller) began, in the thirteenth century. Perhaps the date ought to be tied to the large-scale production of iron in blast furnaces, which employed workers in something like a factory system. That postpones the revolution to the fifteenth and sixteenth centuries. Or maybe it started with the first factory, probably the six-story English silk-thread mill built in 1719 that employed 300 workers, mostly women and children.

A few years after that first factory, John Kay made one of the key inventions that started what we think of today as *the* Industrial Revolution: the flying shuttle, which made weaving much faster. And a few years before that factory, Abraham Darby discovered how to make steel using coal instead of wood—actually, coke instead of charcoal. This allowed for a greater output of cheap steel, since coal was more plentiful than wood at that time in England.

The problem with these early eighteenth-century inventions is that no one took them up. It was over 50 years before other steel makers started following Darby's example; and weavers afraid of losing their jobs destroyed Kay's loom and sent him packing to France.

About 1750, however, cotton workers, with less of a tradition behind them than the wool workers who attacked Kay's loom, started using the flying shuttle. This set into motion a chain of events that revolutionized the textile industry. With the flying shuttle, cotton workers were able to weave so much faster that they ran out of yarn. Seeing an opportunity, James Hargreaves invented a machine that multiplied the amount of yarn produced, the spinning jenny. This time the spinners were upset, and they destroyed some of Hargreaves's machines, but the cat was out of the bag. The spinning jenny could make only one of the two types of yarn needed for weaving. Richard Arkwright also saw opportunity knocking: he invented the water frame, a machine that produced the other type of yarn. Unlike the spinning jenny, the water frame was too large and too expensive to put in a cottage. Arkwright had to build a factory to house his machine; he is, in fact, considered the founder of the modern factory system. By 1769 the Industrial Revolution had definitely begun; many date the start back to the 1740s when the cotton weavers first adopted the flying shuttle.

However, there is another key invention that is not yet in place in 1769. Since at least 1629, when Giovanni Branca suggested using steam to propel a turbine, people had been experimenting with steam power. Branca was followed by the Marquis of Worcester in England, Denis Papin in France, and, in England again, Thomas Savery. Savery, at the very end of the seventeenth century, was the first to make a practical steam engine, known as "the Miner's Friend." It wasn't very efficient, however, and only a few were installed. Soon Thomas Newcomen recognized the opportunity, and developed a greatly improved steam engine that was more along the lines of the modern engine. Newcomen then got together with Savery, and the new engine was successfully manufactured and sold to drain mines. Over 100 Newcomen engines were installed during the eighteenth century. An engineer at Glasgow University, James Watt, in 1765 developed a new device, the steam condenser, that greatly improved the efficiency of the Newcomen engine. Ten years later, Watt teamed with a manufacturer of iron products, Matthew Boulton of Birmingham, to manufacture his new engine. Boulton had access to the technology needed to make finely machined parts that gave the Watt engine greater efficiency and durability. Boulton also convinced Watt in 1781 to convert the engine from a simple pump to a device producing rotary power—the first steam engine that could power other machinery. Four year later the first steam-powered cotton mill opened in Papplewick, Nottinghamshire. By this date, the revolution part of the Industrial Revolution in England had been completed. It would occur 30 years later in France and 20 years after that in Germany and the United States. Assigning the revolution a single date would be misleading. It took from the 1740s until the 1780s; after that, came consolidation of power by the revolutionaries.

Earth science. One of the results of the Industrial Revolution was an increase in prospecting for coal and ores. This stimulated interest in the study of fossils and rocks, giving birth to a new science, termed "geology" by Horace de Saussure. Until the middle of the eighteenth century, it was generally believed that fossils and rocks were deposited by Noah's flood, which, it was thought, had covered the whole planet. This was despite the fact that Robert Hooke had already shown that it was not possible that the thick layers of fossils could have been deposited in the 150 days the flood was believed to have lasted. The French naturalist Buffon believed that Earth was much older—80,000 years instead of 6000 years old. The figure of 6000 years was a value obtained from the Old Testament, however, and was widely accepted (*see* "The age of the Earth," p 404). Buffon also rejected the idea that Noah's flood was responsible for the existence of fossils, but he had to withdraw his writings because of opposition by the church.

Abraham Gottlob Werner, a professor of mineralogy at the Mining Academy of Freiberg, was the first to introduce systematic observation in the science of geology. He showed that rocks were formed layer by layer, with the oldest rocks at the bottom and the youngest ones on top. He argued that in the past Earth's surface was covered by a muddy sea and that Earth's crust was formed by deposition of suspended material. The crystalline rocks, such as granite or basalt, were, according to him, formed by the precipitation of minerals. His theory, called the *Neptunist* theory, was opposed by the *Vulcanists,* who argued that the layers of rock in Earth's surface were formed exclusively by volcanic action, and the *Plutonists,* who emphasized the importance of heat from within Earth. James Hutton, who believed that some rocks were formed by volcanic action, also accepted sedimentation as a rock-forming process. His major contribution to geology was the insight that the surface of Earth was formed by the same forces that still are active today—mainly erosion and volcanism (*see* "Neptunism vs. Plutonism," p 234).

Mathematics. The introduction of calculus by Newton and Gottfried Wilhelm Leibniz had caused mathematics to develop rapidly into a tool eminently suited for dealing with problems of theoretical mechanics. The notations of Leibniz and those used by Leonhard Euler were accepted in continental Europe. In England Newton's notation was used throughout this period (*see* "Inventing signs," page 94). The wholesale application of calculus to physical problems, initiated by the Bernoulli brothers early in the eighteenth century, was continued by Euler, d'Alembert, and Lagrange. Euler and Lagrange created the calculus of variations. Lagrange introduced the differential equations now called Lagrangians, which could be used to represent Newton's laws of motion in a more generalized form. Laplace applied Newton's theory of gravitation and calculus to astronomy in detail, furthering the field of celestial mechanics. A new mathematical field, statistics, initially developed by Jacques Bernoulli, was further refined by Abraham De Moivre and Laplace.

Medicine. Several important developments occurred in the medical sciences. A significant advance in physiology was the publication of Albrecht von Haller's medical encyclopedia, *Elementa physiologiae,* started in 1757. This was the beginning of modern physiology as an independent science. François-Xavier Bichat founded histology—the study of tissues of living things—around 1800. Edward Jenner introduced vaccination to prevent smallpox at the end of the eighteenth century, the first successful method in medicine for fighting disease on a large scale.

Physics. The developments in the science of mechanics were mainly the result of advances in mathematics. Newton's physics was finally accepted in France because it successfully predicted that Earth would be flattened at the poles and also correctly predicted the return of Halley's comet.

However, Newton's physics could only treat problems in which all the mass of a body is viewed as concentrated in one point, the center of gravity. Such a limitation is acceptable for the calculation of planetary orbits, but not useful in many other cases. Leonhard Euler first introduced the generalized coordinates of a body, a mathematical system that allows the whole motion of an extended system, such as a pair of scissors flying through space, to be analyzed. The individual motions of each half of the scissors could be accounted for instead of just the motion of the center of gravity of the scissors. Lagrange improved on Euler's coordinate system so that it could be applied to systems of several moving bodies taken all at once. He also gave a proof that the principle of least action (for example, that the kinetic energy takes on a minimum value when bodies are left moving freely) could be derived from Newton's laws of motion.

Count Rumford cast doubt on the commonly accepted solution to a problem that had intrigued scientists throughout the century: the nature of heat. Heat was generally viewed as a kind of fluid by most eighteenth-century scientists. Rumford measured the amount of heat produced by boring a cannon, which was enormous. He did not think that the cannon could contain so much caloric, the purported fluid. Instead, since heat was being produced by motion, he suggested heat is a form of motion itself (*see* "The nature of heat," p 320). Although he failed to convince many then, later studies showed that he was right.

The eighteenth century was also a period of extensive experimentation with electricity, although the nature of electricity remained largely a mystery.

Technology. During the eighteenth century technology in the modern sense of the word appeared: the direct application of science to machines. The introduction of the steam engine had an enormous influence on science as well as on manufacture and transportation. The improvement of steam engines was based on the science of gases that developed during the late seventeenth century and later on the development of the theory of heat. Engineers started measuring the efficiency of machines using scientific principles: the assessment of machines became a quantitative science.

At the same time, there was a lot of pure invention by amateurs. The people who remade the textile industry during this period were not scientists, but they certainly were inventors. Near the end of the period, the American inventor Eli Whitney introduced standardized parts into technology, providing a major change in the way products are manufactured.

It was generally understood that the teaching of science and technology is an important part of the technological development of a country, and several engineering schools were established throughout Europe. The Ecole Polytechnique in Paris, founded in 1794, became an important source of talent for the development of science in France during the nineteenth century.

	GENERAL	ASTRONOMY	BIOLOGY	CHEMISTRY
1735			*Systema naturae* (System of nature) by Carolus Linnaeus (b Råshult, Sweden, May 23, 1707) introduces the system of classification for organisms that is still in use today Swedish naturalist Petrus (Peter) Artedi d Amsterdam, Holland, Sep 27 Georges Louis Leclerc, comte de Buffon (b Montbard, France, Sep 7, 1707) translates *Vegetable statistics* of Stephen Hales into French	
1736				*Sur la base du sel marin* (On the composition of sea salt) by Henri Louis Duhamel du Monceau (b Paris, 1700) makes the first distinction between sodium and potassium salts
1737	Göttingen University (Germany) is founded	John Bevis, at Greenwich Observatory, observes the passage of Venus in front of Mercury	Linnaeus's *Genera plantorum* (Genera of plants) explains his method of systematic botany and classifies 18,000 species of plants Hermann Boerhaave prints Jan Swammerdam's *Biblia naturae* (Bible of nature), originally published to little notice in 1658; it contains	

EARTH SCIENCE	MATHEMATICS	MEDICINE	PHYSICS	TECHNOLOGY	
German explorer Johann Gmelin (b Tübingen, Aug 10, 1709) discovers permafrost on a voyage to Siberia George Hadley describes the Hadley cell, which models Earth's wind circulation	Leonhard Euler's Petersburg *Commentaries* introduces the notation $f(x)$ for functions	Physician John Arbuthnot d London, Feb 27	Francesco Algarotti's simplified version of Newton's optics, *Newtonianismo per le dame* (Newtonianism for the ladies) becomes one of the most popular explanations of Newton's physics Louis Castel's *L'optique des couleurs* (Optics of colors) contains a scheme for making colors and musical notes correspond Charles-Marie de la Condamine (b Paris, Jan 27, 1701) leads an expedition to Peru to measure the curvature of Earth at the equator and sends back samples of natural rubber and curare William Derham d Upminster, England, Apr 5	John Harrison, in response to the award put up by the British Board of Longitude in 1714, builds his first marine chronometer, known as Number One; each of his prototype chronometers is given a similar designation, with Number Four being the most famous	1735
Pierre de Maupertuis leads a French expedition to Lapland, accompanied by Alexis-Claude Clairaut and Anders Celsius, sponsored by the *Académie Française* to measure the length of a degree; he proves that Earth is flattened at the poles, which shows that Newton's theory of gravity is correct and Descartes' theory is false	Euler's *Mechanica sive motus scientia analytice exposita* is the first systematic textbook on mechanics based on differential equations John Colson's English version of Isaac Newton's *De methodis serierum at fluxionum* (The method of fluxions and infinite series) is the first published version of Newton's theory of fluxions (originally written in 1671)	Claudius Aymand performs the first succesful operation for appendicitis American physician William Douglass describes scarlet fever	German-Dutch physicist Gabriel Fahrenheit d The Hague, Holland, Sep 16	Electrical experimenter Stephen Gray d London, Feb 25	1736
An earthquake in Calcutta, India, kills about 300,000 people	Euler proves that the number *e*, known as the base for the natural logarithms, and its square are both irrational; that is, they cannot be represented by finite or repeating decimals		Pieter van Musschenbroek publishes *Essai de physique* (Essay on physics), one of the first books of his time using the term physics instead of natural or experimental philosophy; the term physics goes back at least to Aristotle	Bernard Forest de Bélidor publishes the first volume of his four-volume *Architecture hydraulique*, a manual that will influence building design and practice for more than a century; it will be complete in 1739	1737

	GENERAL	ASTRONOMY	BIOLOGY	CHEMISTRY
1737 cont			his reports on the dissection of insects under a microscope	
1738		Joseph-Nicolas Delisle (b Paris, Apr 4, 1688) tracks the location of sunspots using heliocentric coordinates	*Petri Artedi seuci, medici, ichthyologia sive opera omnia de piscibus* by Petrus (Peter) Artedi (b Angermanland, Sweden, Feb 27, 1705), edited by Linnaeus and published posthumously, establishes Artedi's reputation as the father of ichthyology for its taxonomy of fishes French natural philosopher Benoît de Maillet d	
1739	The Royal Society of Edinburgh is founded in Scotland A scientific society is founded in Stockholm, Sweden, ·by Linnaeus and others; in 1741 it becomes Kungliga Svenska Vetenskapsakademien David Hume (b Edinburgh, Scotland, Apr 26, 1711) publishes *Treatise of Human Nature*, an attempt to apply the experimental method to problems of psychology and human nature	*Notes on sunspots* by John Winthrop (b Boston, MA, Dec 19, 1714) contains the first set of such observations in the Massachusetts colony		
1740	Ephraim Chambers, English encyclopedist, d May 15	Anders Celsius is placed in charge of a new observatory in Uppsala, Sweden James Short builds telescopes to N. Cassegrain's design		

EARTH SCIENCE	MATHEMATICS	MEDICINE	PHYSICS	TECHNOLOGY	
				Pierre-Simon Fournier introduces the point system for measuring type sizes	1737 cont
	The second edition of Abraham De Moivre's *Doctrine of chances: or, a method of calculating the probability of events in play* introduces Stirling's formula for approximating factorials; the basic formula had been previously discovered by De Moivre, but James Stirling found the value of a missing constant	Physician Hermann Boerhaave d Leiden, Holland, Sep 23	Daniel Bernoulli's *Hydrodynamica* explains the relationship between the pressure and velocity of fluids and the idea behind the theorem named after him in terms of impact of atoms on the walls of the container Pierre de Maupertuis's *Sur la figure de la Terre* (On the shape of the Earth) reports his measurements made in Lapland, confirming that Earth is flattened at the poles Voltaire's *Eléments de la philosophie de Newton* (Elements of the philosophy of Newton), written with the help of Madame du Châtelet, introduces English empirical philosophy to the Continent	Charles Dangeau de Labelye develops the caisson, a device essential to building bridges and underwater tunnels, for a bridge over the Thames at Westminister	1738
			George Martine demonstrates that the amount of heat contained in an object is not proportional to its volume Charles Du Fay d Paris, Jul 16		1739
Antonio Moro's *Dei crostacei e degli altri corpi marini* is an important study of marine fossils			The French Academy of Sciences awards for the last time a prize for a memoir based on Descartes' ideas about physics		1740

	GENERAL	ASTRONOMY	BIOLOGY	CHEMISTRY
1741		James Bradley succeeds Edmund Halley as Britain's astronomer royal	Steller's sea cow is discovered living off the coast of the Kamchatka Peninsula (Soviet Union); 27 years later it will have been hunted to extinction Agriculturist Jethro Tull d Feb 21	
1742	The Royal Danish Academy of Sciences and Letters is founded at Copenhagen by Christian VI	Astronomer Edmund Halley d Greenwich, England, Jan 14	*Microscope made easy* by Henry Baker (b London, May 8, 1698) introduces the construction and use of the microscope to the layman Abraham Trembley (b Geneva, Switzerland, Sep 3, 1710) makes the first permanent graft of animal tissue using the hydra	
1743	Benjamin Franklin is instrumental in establishing the American Philosophical Society at Philadelphia, the United States' first scientific society Statesman and scholar Thomas Jefferson b Shadwell, VA, Apr 13; while his investigations of plant breeding, fossils, and Native American burial sites reveal no new scientific concepts of significance, he remains the closest the US has come to having a scientist-president	*Théorie de la figure de la terre, tirée des principes de l'hydrostatique* (Theory of the shape of the Earth based on the principles of hydrostatics) by Alexis-Claude Clairaut (b Paris, May 7, 1713) shows that Earth is flattened at the poles and demonstrates how to compute gravitational force at any latitude		

1741

Alexi Ilich Tschirikov and Vitus Jonassen Bering discover the Aleutian Islands and the coast of Alaska during the Russian Great Nordic Expedition

Danish-Russian navigator Vitus Jonassen Bering d Bering Island, east of Kamchatka, Dec 19

1742

Christian Goldbach, b Königsberg (Kaliningrad, USSR), Mar 18 1690, in a letter to Leonhard Euler, suggests that every even number greater than two is the sum of two prime numbers (alternatively, every even number greater than four is the sum of two odd primes); Goldbach's conjecture has resisted all attempts at proof or disproof

Treatise on fluxions by Colin Maclaurin treats calculus on the basis of Greek geometry

Benjamin Robins's *New principles of gunnery* describes his invention of the ballistic pendulum

Anders Celsius invents the Celsius scale (centigrade thermometer); in his original version, he uses 0 for the boiling point of water and 100 for the freezing point; Jean Pierre Christin switches these in 1743 to give the system we use today

Friedrich Hoffmann d Halle (Germany), Nov 12

Benjamin Huntsman (b Lincolnshire, England, 1704) introduces the crucible process for molten steel about this time

1743

Jean d'Anville produces his *Map of Italy*

Christopher Packe produces *A new philosophical chart of East Kent,* the first geological map

Marie-Jean-Antoine, Marquis de Condorcet b Ribemont, France, Sep 17

Traité de dynamique (Treatise on Dynamics) by Jean le Rond d'Alembert (b Paris, Nov 16, 1717) expands on Newton's laws of motion; d'Alembert's principle is that the actions and reactions in a closed system of moving bodies are in equilibrium, and the book applies the principle to the solution of problems in mechanics

1744

Jean Philippe Loys de Cheseaux restates the paradox known as Olbers' paradox (from Heinrich Olbers' 1823 version): Why is the sky dark?; if the number of stars is infinite, a stellar disk should cover every patch of sky; Cheseaux claims that a slight loss of light in space solves the problem

The Roman Catholic Church, while leaving Galileo's *Dialogue concerning the two chief world systems* on the Index of Prohibited Books, allows it to be printed as long as Galileo's recantation of Copernican theory is included in the same volume

Astronomer Anders Celsius d Uppsala, Sweden, Apr 25

Abraham Trembley's *Mémoires* summarizes his research on the hydra and his discovery of regeneration in polyps

1745

Traité d'insectologie (Treatise on insectology) by Charles Bonnet (b Geneva, Switzerland, Mar 13, 1720) describes his observations on parthenogenetic reproduction and metamorphosis of aphids

The lead-chamber process for making sulfuric acid is invented

1746 Princeton University is founded in NJ

Pensées philosophiques (Philosophical thoughts) by Denis Diderot (b Langres, France, Oct 5, 1713) demonstrates the existence of God through the order of nature

EARTH SCIENCE	MATHEMATICS	MEDICINE	PHYSICS	TECHNOLOGY	
César-François Cassini (b Thury, France, Jun 17, 1714) directs the triangulation of France, the first national geographic survey, resulting in the first map produced according to modern principles	Euler's *Methodus inveniendi lineas curvas maximi minimive proprietate gaudentes* (The art of finding curved lines that enjoy some maximum and minimum property) generalizes some basic methods of the calculus of variations, the mathematical study of maximum and minimum functions; it includes Euler's equation Euler's *Theorium motuum planetarium et cometarium* (Theory of the motion of planets and comets) is a precursor to the more exact calculations of orbits of Lagrange		Jean d'Alembert's *Traité de l'équilibre et du mouvement des fluides* (Treatise on the equilibrium and motion of fluids) uses his principle to describe fluid motion Mikhail Vasilievich Lomonosov (b Denisovka, Russia, Nov 8, 1711) publishes a paper on the causes of heat and cold; he correctly thinks that heat is a form of motion Pierre de Maupertuis originates the principle of least action; action is a quantity that is essentially force times distance times time; the principle states that nature will operate in such a way that action is at a minimum; modern physicists still view this rule as true French-English physicist John Théophile Desaguliers d London, Mar 10	Benjamin Franklin invents the Franklin stove about this time	1744
Comte de Buffon proposes that Earth was formed when a comet collided with the sun Mikhail Vasilievich Lomonosov publishes a catalog of 3030 minerals				Jacques de Vaucanson (b Grenoble, France, Feb 24, 1709) invents the self-acting loom for weaving silk	1745
Jean-Etienne Guettard (b Etampes, France, Sep 22, 1715) draws the first geological map of France	Colin Maclaurin d Edinburgh, Scotland, Jun 14		About this time Leonhard Euler works out the mathematics of the refraction of light by assuming that light is a wave phenomenon in which colors correspond to different wavelengths (as had been proposed by Christiaan Huygens in 1690) John Winthrop establishes the first laboratory for experimental physics in the United States at Harvard University	John Roebuck (b Sheffield, England, 1718) invents the lead-chamber process for the manufacture of sulfuric acid	1746

1746
cont

1747

Alexis-Claude Clairaut's *Théorie de la lune* (Theory of the moon), published this year, wins the 1750 prize of the Academy of Saint Petersburg; it is the first approximate resolution of the three-body problem, the intractable problem of finding how three different masses interact in space

Andreas Marggraf (b Berlin, Mar 3, 1709) discovers sugar in beets, laying the foundation for Europe's sugar-beet industry

---

1748

James Bradley announces his discovery of the nutation of Earth's axis, a small periodic change in the precession of the equinoxes caused by the gravitational forces of the sun and the moon

Benoît de Maillet's *Telliamed* advances an unorthodox evolutionary theory based on the action of a retreating sea; Maillet thinks that the universe is filled with "seeds" that grow into animals in the

EARTH SCIENCE	MATHEMATICS	MEDICINE	PHYSICS	TECHNOLOGY	
			In January the Leiden jar—the first practical way to store static electricity—is invented by both Pieter van Musschenbroek and by Ewald Georg von Kleist; Van Musschenbroek gets a shock (literally) when he first uses it, suggesting a connection between electricity and lightning William Watson's *Experiments on the nature of electricity*, published over a two-years period, describes his experiments with the Leiden jar		**1746** cont
Jean le Rond d'Alembert's *Réflexions sur la cause générale des vents* (On the general theory of the winds) presents the first general use of partial differential equations in mathematical physics	Bernhard Siegfried Albinus's *Tabula sceleti et musculorem corporis humani* (Plates of the skeleton and muscles of the human body) shows relative parts of bones and muscles in correct proportions *Primae lineae physiologiae* by Albrecht von Haller (b Berne, Switzerland, Oct 16, 1708) is the first textbook on physiology	Jean d'Alembert publishes his theory of vibrating strings, giving the general solution of the partial differential wave equation in two dimensions Benjamin Franklin describes in a letter his discovery that a pointed conductor can draw off electric charge from a charged body; this discovery is the basis for the lightning rod even before Franklin proves that lightning is a form of electricity Abbé Jean-Antoine Nollet (b Pimprez, France, Nov 19, 1700) constructs one of the first electrometers; it consists of a suspended pith ball William Watson tries to determine the velocity of electricity and incorrectly concludes that it is instantaneous	The first civil engineering school, the *Ecole des ponts et chaussées* (School for bridges and highways), is established in France	**1747**
Euler's *Introductio in analysin infinitorum* (Introduction to infinitesimal analysis) popularizes the concept of function, first introduced by Leib-	John Fothergill (b Car Eng, England, Mar 8, 1712) gives the first description of diphtheria in his *Account of the putrid sore throat*	Mikhail Vasilievich Lomonosov formulates the laws of the conservation of mass and energy	John Wilkinson (b Cumberland, England, 1728) establishes the first blast furnace in Bilston, England	**1748**	

1748
cont

sea; as the sea gradually diminishes, some of these creature evolve into land animals

L'homme machine (Man the machine) by Julien Offroy de la Mettrie (b St. Malo, France, Dec 25, 1709) depicts humans as machines without freedom or will

John Needham (b London, Sep 10, 1713) and comte de Buffon conduct a famous experiment that seems to prove spontaneous generation; they observe "little animals" that appear in boiled sealed flasks of broth

1749

Diderot's *Lettre sur les aveugles* (Letter on the blind) describes his materialist ideas, the dependence of humans on their senses, and his theory of variability and adaptation

Observations on man by David Hartley (b Yorkshire, England, Aug 30, 1705), the first work in English to use the term "psychology," is a systematic attempt to interpret the phenomena of mind by the theory of association

Jean le Rond d'Alembert's *Recherches sur la précession des équinoxes et sur la nutation de la terre* (Researches on the precession of the equinoxes and on the nutation of Earth) gives the first mathematical explanation of the regular changes in the orientation of Earth's axis

In *Protogaea*, published posthumously, Gottfried Wilhelm Leibniz explains that Earth probably passed from a gaseous through a molten state to its present form

Comte de Buffon gives the modern definition of a species, a group of organisms capable of breeding and producing fertile offspring

Comte de Buffon's *Histoire naturelle, générale et particulière* (General and particular natural history) begins a 55-year, 44-volume account of all that is known about animals and minerals; Buffon's popular style does much to encourage the study of nature among his readers

1748 cont

niz, develops the analytic interpretation of the trigonometic functions, and sums various infinite series with abandon, if not with rigor

Colin Maclaurin's posthumously published *A treatise of algebra* contains the method for solving systems of equations that is known as Cramer's rule, although Gabriel Cramer's version is not published until two years later

Johann (Jean) Bernoulli d Basel, Switzerland, Jan 1

Jean Nollet discovers and explains osmotic pressure

Physicist Ewald Georg von Kleist d Köslin, Pomerania (Poland), Dec 11

1749

John Canton (b Stroud, England, Jul 31, 1718) develops a method for making an artificial magnet

Gabrielle Emilie la Tonnelier de Breteuil, marquise du Châtelet, completes the only translation of Newton's *Principia* ever made into French, with the help of Alexis-Claude Clairaut; Madame du Châtelet's lover, Voltaire, encourages her in her work and writes the preface for the first (1759) edition

Jean-Jacques d'Ortous de Mairan describes the Chinese refrigerator in *Dissertation sur la glace* (Dissertation on ice), recognizing the cooling effect of evaporation

Benjamin Franklin installs a lightning rod on his home in Philadelphia

Writer Gabrielle Emilie la Tonnelier de Breteuil, marquise du Châtelet, d Luneville, France, Sep 10

Inventor Thomas Godfrey d Bristol Township, PA, Dec

Philip Vaughan patents radial ball bearings for carriage axles

Voltaire the scientist

In England, the *Principia* was generally accepted for the explanation of gravitation and the motion of celestial bodies. In France, however, Descartes' theory or vortices was still favored over Newton's theory of gravitation (*see* "Verifying Newton's theory of gravitation," p 178). One of the people instrumental in disseminating Newton's theory of gravitation to Europe was the French writer and philosopher Voltaire.

After having twice spent some time in the Bastille for writing witty verses directed against the French authorities, Voltaire went to England. There he found a society that he considered more rational than the French one. He also acquainted himself with the physics of Newton while he was there.

In 1734 Voltaire published an account of his stay in England, his *Lettres philosophiques* (Philosophical letters) in which he dedicated a large section to Newton's theory of gravitation. For many Frenchmen this was their first exposure to Newton's physics, although it was 50 years after the publication of the *Principia*. Because the *Lettres philosophiques* contained criticism of the French government, Voltaire was forced to leave Paris again. He settled with Madame du Châtelet in her chateau at Cirey and started writing his *Eléments de la philosophie de Newton* (Elements of Newton's philosophy), which was published in 1738. Madame du Châtelet started writing her own book on physics, *Institutions de physique* (Institutions of physics). Unlike Voltaire, she was more influenced by Leibniz than by Newton, and her book made Leibniz's work, which was in German, known to the French.

Madame du Châtelet started translating Newton's *Principia* in 1747. She died in childbirth as she was finishing the translation in 1749. Her translation still is the only one of the *Principia* ever attempted in French.

1750

ASTRONOMY

Nicolas de Lacaille (b Rumigny, France, May 15, 1713) makes the first extensive catalog of 2000 southern stars at the Cape of Good Hope; he also determines solar and lunar parallaxes

An original theory and new hypothesis of the universe by Thomas Wright (b Byers Green, England, Sep 22, 1711) explains the Milky Way as a collection of stars confined between two parallel planes and proposes that the sun and stars form a giant system rotating around a common center

1751

GENERAL

The calendar in Britain is altered, with Jan 1 becoming the beginning of year

The Göttinger wissenschaftliche Akademie is founded

Diderot and Jean le Rond d'Alembert publish the first volume of *Encyclopédie, ou dictionnaire raisonné des sciences, des arts, et des métiers* (Encyclopedia, or rational dictionary of science, art, and professions); known generally as the *Encyclopédie*, it will be complete in 1772 with 17 volumes of articles and 11 of illustrations

BIOLOGY

Carolus Linnaeus's *Philosophia botanica* (Botanical philosophy) rejects any notion of evolution and continues his work in classifying plants

Essay on the vital and other involuntary motions of animals by Robert Whytt (b Edinburgh, Scotland, Sep 6, 1714) refutes Georg Ernst Stahl's doctrine that the soul causes involuntary motion in animals and argues that the irritability of living tissues is caused by a stimulus

Pierre-Louis Moreau de Maupertuis's *Système de la nature* (System of nature) is a theoretical speculation on heredity and the origin of species by chance

CHEMISTRY

Axel Fredrik Cronstedt (b Turinge, Sweden, Dec 23, 1722) announces his discovery of nickel to the Swedish Academy of Sciences

Mikhail Vasilievich Lomonosov's *Discourse on the usefulness of chemistry* is published

Eléments de chymie théorique and *Eléments de chymie pratique* (Elements of theoretical chemistry and Elements of practical chemistry) published this year by Pierre-Joseph Macquer (b Paris, Oct 9, 1718) lead the field in chemistry textbooks for many years

1752

ASTRONOMY

Johann Tobias Mayer, b Marbach (Germany), Feb 17, 1723, publishes tables of the motion of the moon compared to the motion of the stars accurate enough for finding the longitude at sea; for this he obtains the 1755 prize offered by the British government

Great Britain and the American colonies adopt the Gregorian calendar by having Sep 14 directly follow Sep 3

BIOLOGY

René Antoine Ferchault de Réaumur discovers the role of gastric juices, showing that digestion is chemical and not merely mechanical; he feeds a hawk meat protected by a metal cylinder; upon recovery, the meat is partially digested; he also uses stomach juice from the hawk to digest meat outside the body

1750

Gabriel Cramer's *Introduction à l'analyse des lignes courbes algébriques* (Introduction to the analysis of algebraic curves) contains the famous Cramer's rule for solving systems of linear equations (actually discovered by Colin Maclaurin at least two years earlier)

Pierre-Louis Moreau de Maupertuis's *Essai de cosmologie* (Essay on cosmology) claims that the principle of least action proves the existence of God

A treatise on artificial magnets by John Michell (b Nottinghamshire, England, 1724) explains magnetic induction and reports his discovery of the inverse-square law for the repulsive forces of magnetism, now known as Coulomb's law from Charles-Augustin Coulomb's rediscovery about 30 years later

1751

The first mental institution is opened in London

Robert Whytt demonstrates that the contraction of the pupil in response to light is a reflex motion

French physician Julien de Lamettrie d Berlin, Nov 11

Benjamin Franklin describes electricity as a single fluid and distinguishes between positive and negative electricity in *Experiments and observations on electricity;* he shows that electricity can magnetize and demagnetize iron needles

The Ecole Supérieure de Guerre (High School of War) is founded in Paris

Benjamin Huntsman invents the crucible process for casting steel

Inventor Christopher Polhem d Tingstäde, Sweden, Aug 30

Clockmaker George Graham d England, Nov 20

1752

Nicolas Desmarest (b Soulaines, France, Sep 16, 1725) proposes that England and France were once connected by a land bridge that has since been washed away by sea currents

The first volume of Anton Friedrich Busching's *Neue erdbeschreibung* (New geography) is published; by 1792 it will

Euler states the theorem that in any simple convex polyhedron, the sum of the vertices and the faces equals two plus the number of edges; this is commonly known as Euler's formula $(V - E + F = 2)$, although it was discovered earlier by René Descartes

William Smellie's *Treatise on midwifery* is the first scientific approach to obstetrics

Surgeon William Cheselden d Bath, England, Apr 10

Jean le Rond d'Alembert's *Essai d'une nouvelle théorie de la résistance des fluides* (Essay on a new theory of fluid resistance) proposes his principles of hydrodynamics

In June Benjamin Franklin performs his famous kite experiment; the kite experiment shows that lightning is a form of

1752
cont

1753

The royal foundation charter is granted to the British Museum in London

Diderot publishes *Pensées sur l'interprétation de la nature* (Thoughts on the interpretation of nature), a philosophical essay inspired by Buffon's *Histoire naturelle*

Carolus Linnaeus's *Species plantarum* (Species of plants) completes his development of the use of binary nomenclature in botany; it remains the foundation for the present-day classification of species

---

1754

Claude d'Abbans' *"Discours préliminaire"* (Preliminary discourse) to Diderot's *Encyclopédie,* is a cardinal document of the Enlightenment

John Dollond (b London, Jun 10, 1706) invents the heliometer, a telescope that produces two images that can be manipulated to determine angular distances accurately, for finding the diameter of the sun (its intended use) or the distances between stars

Nicolas Louis de Lacaille indicates the existence of galactic star clusters

Charles Bonnet's *Recherches sur l'usage des feuilles des plantes* (Study of the use of plant leaves) details the nutritional value of plants

Etienne Bonnet de Condillac's *Traité de sensations* (Treatise on sensations) claims that knowledge reaches humans only through the senses

---

1755

The University of Moscow is founded

Explorer Johann Georg Gmelin d Tübingen, Württemberg (Germany), May 20

German philosopher Immanuel Kant, b Königsberg (Kalingrad, USSR), Apr 22, 1724, suggests in *Allgemeine Naturgeschichte und Theorie des Himmels* (General natural history and theory of the heavens) that the observed nebulas are large star systems like the Milky Way and that the solar system originated from a dust cloud

Sebastian Menghini experiments with the action of camphor on animals

Scottish chemist Joseph Black (b Bordeaux, France, Apr 16, 1728) shows that carbonates are compounds of a base and a gas; he terms carbon dioxide "fixed air"

EARTH SCIENCE	MATHEMATICS	MEDICINE	PHYSICS	TECHNOLOGY	
consist of 11 volumes, six of them describing European geography; it lays the foundation for modern statistical geography	Gabriel Cramer, Swiss mathematician, d Bagnoles, France, Jan 4		electricity, similar to the discharge from a Leiden jar		1752 cont
	Euler announces his solution to the problem of the bridges of Königsberg to the Russian Academy; he shows that with the configuration of seven bridges across the river to two islands, it is not possible to stroll the bridges without crossing one of them twice Philosopher George Berkeley, influential critic of the calculus, d Oxford, England, Jan 14	*Treatise on scurvy* by James Lind (b Edinburgh, Scotland, Oct 4, 1716) establishes the curative effect of lemon juice on scurvy	*Dell'elettricità* by Italian physicist Giovanni Beccaria (b Mondovi, Oct 3, 1716) supports Benjamin Franklin's theories about electricity George Wilhelm Richmann is killed performing a kite experiment similar to the one performed by Benjamin Franklin		1753
	Jacopo Riccati d Treviso (Italy), Apr 15 Christian Wolff d Halle (Germany), Apr 9 French-English mathematician Abraham De Moivre d London, Nov 27; he had declared that he would sleep 15 minutes longer each day until he slept through 24 hours, at which point he would die; this indeed is what occurs	The University of Halle (Germany) graduates the first female with a degree of medical doctor		The first iron-rolling mill is started at Fareham, Hampshire, England	1754
An earthquake in the ocean near Lisbon, Portugal, kills more than 60,000 people	Euler's *Institutiones calculi differentialis* is a textbook on the differential calculus that also contains several of Euler's own discoveries French mathematician Jean-Louis Lagrange, b Turin (Italy), Jan 25, 1736, writes to Euler about his new foundation for the calculus of variations and the differential equation of minimal surfaces				1755

	GENERAL	ASTRONOMY	BIOLOGY	CHEMISTRY
1756		John Canton makes the first observation of magnetic storms in Earth's magnetic field Jacques Cassini d Paris, Apr 18		Joseph Black's *Experiments upon magnesia, quicklime, and other alkaline substances* is the first detailed examination of chemical action and the first work on quantitative chemistry
1757	Science writer Bernard le Bovier de Fontenelle d Paris, Jan 9	Navy Captain John Campbell extends the arc of the reflecting quadrant used in navigation from 90 degrees to 120 degrees; the instrument becomes known as the sextant Alexis-Claude Clairaut obtains the best figures to date for the masses of the moon and the planet Venus		
1758	Claude Adrien Helvetius's *De l'esprit* (Essays on the mind) endorses the idea that the mind develops as a result of the sensations it perceives, rather than as a result of inborn tendencies; it is publicly burned in Paris and condemned by the English Parliament	John Dollond, independently of Chester Hall, who had accomplished the same feat in 1733 to little notice, constructs an achromatic telescope; that is, a telescope that avoids the defect of chromatic abberation, which Newton thought would be present in all lenses; Dollond's lens of flint glass and crown glass is presented to the Royal Society on Jun 8 Johann Georg Palitzsh locates Halley's comet on Dec 25; this is the first observation of a comet at a predicted return date; Alexis-Claude Clairaut had predicted earlier in the year that the comet would come closest to Earth within 30 days of Apr 1759; its closest approach is actually in Mar 1759	Henri Louis Duhamel du Monceau's *La physique des arbres* (The physics of trees) describes the structure and physiology of trees	Axel Fredrik Cronstedt's *Essay on the new mineralogy* distinguishes four classes of minerals: earths, bitumens, salts, and metals; it begins the classification of minerals by chemical structure as well as appearance

EARTH SCIENCE	MATHEMATICS	MEDICINE	PHYSICS	TECHNOLOGY	
Johann Gottlob Lehmann's study of the rocks of the Harz Mountains and the Erzgebirge is a pioneering work in local geology; soon other geologists begin to study specific sites	Philipp Pfaff's *Abhandlung von den Zähnen* gives the first description of casting models to make false teeth	William Cullen observes the cooling effect of evaporating liquids and publishes the results in *An essay on the cold produced by evaporating fluids and some other means of producing cold* Mikhail Vasilievich Lomonosov's *Theory of electricity* and *Origin of light and color* include his advocacy for the wave theory of light	The first cotton velvets are made at Bolton, Lancashire, England		1756
	Albrecht von Haller's *Elementa physiologiae corporis humani* (Elements of the physiology of the human body) is a survey of the physiological knowledge of his time; it details his research on the muscles and blood circulation of the human body Psychologist David Hartley d Bath, England, Aug 28	Louis Castel d Paris, Jan 11 Physicist and naturalist René Antoine Ferchault de Réaumur d La Bermondière, France, Oct 18	John Wilkinson patents a hydraulic blowing machine that uses waterpower to drive a bellows		1757
	Mathematician and hydrographer Pierre Bouguer d Paris, Aug 15			A commission in England sets standards for measures known as the "imperial standards" Jedediah Strutt (b Derbyshire, England, Jul 28, 1726) invents the ribbing machine for the manufacture of stockings	1758

	GENERAL	ASTRONOMY	BIOLOGY	CHEMISTRY
1759	The Bavarian Academy of Science is founded in Germany		*Theoria generationes* (Theory of generation) by Kaspar Wolff (b Berlin, Jan 18, 1734) claims the existence of a *vis essentialis* (essential force) that is at the heart of living matter; Wolff also describes the differentiation of tissues in a developing embryo, refuting the concept of miniature creatures in sperm	
1760		Louis Godin d	The Botanical Gardens in Kew, London, open Joseph Gottlieb Kölreuter's *Vorläufige nachricht von einigen das geschlecht der pflantzen betreffende versuche und beobachtungen* (Preliminary report on experiments and observations of certain species of plants) describes his research on heredity in plants	
1761	Jean-Jacques Rousseau publishes *La nouvelle Héloïse*	Joseph-Nicolas Delisle organizes a worldwide effort to observe the transit of Venus, which can be used to calculate solar parallax and the distance from Earth to the sun, an idea proposed by Jerimiah Horrocks and made popular by Edmund Halley Nicolas-Louis de Lacaille makes an accurate measurement of the distance of the moon, taking into account the nonsphericity of Earth	The first veterinary school is founded at Lyons, France Jean Baptiste Robinet's *De la nature* (On nature) (5 volumes) claims that organic species form a linear scale of progress without gaps Botanist and chemist Stephen Hales d Teddington, England, Jan 4	English chemical manufacturer Joshua Ward d Nov 21

EARTH SCIENCE	MATHEMATICS	MEDICINE	PHYSICS	TECHNOLOGY	
	Pierre-Louis Moreau de Maupertuis d Basel, Switzerland, Jul 27	Albrecht von Haller's *Experimenta physiologiae corporis humani* (Physiological experiments on the human body) (nine volumes, published from 1759 to 1776) presents a general overview of physiology	*Tentamen theoriae electricitatis et magnetismi* (Examination of a theory of electricity and magnetism) by German physicist Franz Ulrich Theodosius Aepinus (b Rostock, Dec 13, 1724) suggests that in the absence of electricity, ordinary matter repels itself; this idea supports Benjamin Franklin's one-fluid theory of electricity	Engineer James Brindley builds the first canal that crosses a river on a fixed aqueduct; the canal passes over the River Iswell in England instead of lowering to the level of the river with locks and then raising again

John Harrison completes Number Four, the marine chronometer that will eventually win the British Board of Longitude's prize for a practical way to find the longitude at sea

John Smeaton (b Leeds, England, Jun 8, 1724) builds a concrete lighthouse with mortar that sets underwater | 1759 |
| Mikhail Vasilievich Lomonosov explains the formation of icebergs

John Michell's "Essay on the causes and phenomena of earthquakes" proposes that earthquakes are waves produced when one layer of rock rubs against another; although he thinks that this is caused by a volcano's turning water into steam; he also notes that earthquake waves could be used to find where the earthquake starts | Euler uses his phi function to show that if two numbers are relatively prime, then one of them divides the difference found by subtracting one from the other raised to the phi function of the first; recently this theorem has become central to modern "open-key" codes | | In experiments with primitive apparatus, Daniel Bernoulli decides that the electric force obeys an inverse square law similar to that of gravity

Photometria by German physicist Johann Lambert (b Mulhause, Aug 26, 1728) is an investigation of light reflections from planets, introducing the term *albedo* ("whiteness") for the differing reflectivities of planetary bodies | Benjamin Franklin puts up lightning rods in Philadelphia | 1760 |
| | Johann Peter Süssmilch initiates the study of population statistics | Leopold Auenbrugger (b Graz, Austria, Nov 19, 1722) uses his musical knowledge to develop the technique of percussion for diagnosis of chest disorders; he publishes his findings in *Inventum Novum*

Dentist Pierre Fauchard d | Joseph Black discovers latent heat by finding that ice, when melting, absorbs heat without changing in temperature; later he measures the latent heat of steam—that is, the amount of heat that can keep water boiling without raising its temperature | French engineer and writer Bernard Forest de Belidor d Paris

Engineer John Rennie b Phantassie, Scotland, Jun 7; he builds the Waterloo and London bridges as well as many canals and harbors | 1761 |

1761 cont

ASTRONOMY

Johann Lambert's *Cosmologische briefe* (Cosmological letters) contains a theory for the structure of the Milky Way in which stars are contained in giant spherical clusters that are confined to a region of small thickness

Mikhail Vasilievich Lomonosov discovers the atmosphere of Venus while observing a transit of Venus across the sun

Optician John Dollond d London, Nov 30

The transit of Venus

Since orbits of the planets are in virtually the same plane, Mercury and Venus can be seen as dots passing in front of the sun. Such an event is a transit. Kepler was the first to recognize that transits must exist and to calculate when they would occur. With Kepler's calculations in hand, Pierre Gassendi was able to observe the transit of Mercury in 1631, a year after Kepler's death.

Transits of Venus have been especially important in the history of science. They occur in pairs, eight years apart, at intervals of 105.5 or 121.5 years. They can occur only near June or December. The next transit of Venus will be in June, 2004. Jeremiah Horrocks improved on Kepler's calculations and became the first to observe a transit of Venus, on November 24, 1639. It was Horrocks who realized that if a transit of Venus were observed simultaneously from several places on Earth, the information gained could be used to calculate the distance to Venus and the distance from Earth to the sun.

The first effort to accomplish this goal was organized by Joseph-Nicolas Delisle in 1761. Astronomers were dispatched to India, St. Helena, and other good viewing spots. War prevented some of the necessary observations from being made, and clouds prevented others. A few observations were made, however. It was during this transit that it was discovered that Venus has an atmosphere.

1762

James Bradley completes a new star catalog that contains the measured positions of 60,000 stars

Johann Tobias Mayer d Göttingen (Germany), Feb 20

Astronomer Nicolas Louis de Lacaille d Paris, Mar 21

James Bradley d Chalford, England, Jul 13

1763

The British mariner's guide by Nevil Maskelyne (b London, Oct 6, 1732) is a practical guide to methods of navigation

Joseph Gottlieb Kölreuter performs fertilization experiments on plants using animal pollinators

1764

Voltaire's *Dictionnaire philosophique* (Philosophical dictionary) argues against the Great Chain of Being, a concept that orders inanimate matter, plants, and animals into a hierarchy

Joseph-Louis Lagrange's *Libration de la lune* (Libration of the moon) explains the librations of the moon; that is, the periodic oscillation of the moon from side to side that permits us to see somewhat more than

Charles Bonnet's *Contemplation de la nature* (Contemplation of nature) gives his theory of "preformation;" he believes that each creature is already preformed in miniature in the egg, and that this tiny

David Macbride publishes *Experimental essays* in which he relates his discovery that van Helmont's gas sylvestre is identical to "fixed air" (now known as carbon dioxide)

The transit of Venus (continued)	Giovanni Morgagni's *De sedibus et causis morborum per anatomen indagatis* (On the causes of diseases) is the first important work in pathological anatomy	Pieter van Musschenbroek d Leiden, Holland, Sep 19	With John Harrison's marine chronometer Number Four aboard and his son William Harrison to take the readings, the H.M.S. *Deptford* sails off toward the West Indies to test whether Harrison's method can be used to find the longitude	**1761** cont

The June 3, 1769, transit of Venus became the most famous. It was to observe this transit from Tahiti that Captain James Cook undertook the first of his great voyages of discovery. Leaving England in the *Endeavor* on August 26, 1768, Cook made the necessary observations in Tahiti and returned to England on July 17, 1771.

Cook's was not the only trip of discovery based upon the 1769 transit. The Russians journeyed overland to Siberia to observe the event. Other observers were at various points around the globe.

In the mid-nineteenth century, Johann Encke used this transit data to calculate Earth's distance to the sun as 153,000,000 km (95,300,000 mi), the best calculation to that time.

For the transits of 1874 and 1882, George Airy organized vast expeditions to obtain data that could be used to improve on Encke's calculations. He failed, however, to recognize that the atmosphere of Venus causes enough error in the observations that no improvement can be made by this method. Later astronomers were able to arrive at more precise measurements by using the transits of asteroids, which have no atmosphere.

Lady Mary Wortley Montagu d London, Apr 29		Samuel Klingenstierna (b Linköping, Sweden, Aug, 18, 1697) wins the prize of the Russian Academy of Science for the best method of constructing optical instruments free of chromatic abberation John Roebuck converts cast iron into malleable iron by the action of pit fire and artificial blast at the Carron Ironworks in Stirlingshire, Scotland	**1762**

Jeremiah Dixon (b Durham, England, Jul 27, 1733) and Charles Mason (b England, 1730) begin their survey of the Pennsylvania-Maryland boundary that will end five years later with the Mason-Dixon line Antoine-Laurent Lavoisier (b Paris, Aug 6, 1743) helps the geologist Jean-Etienne Guettard with a mineralogical atlas of France	Gaspard Monge (b Beaune, France, May 10, 1746) develops descriptive geometry, which remains a French military secret until 1795	The first American medical society is founded in New London, CT	Josiah Wedgwood (b Burslem, England, Jul 12, 1730) patents the cream-colored earthenware that becomes the standard domestic pottery of England	**1763**

	German-Russian mathematician Christian Goldbach d Moscow, Nov 20	Robert Whytt's *Observations on nervous, hypochondriacal, or hysteric diseases* is one of the first important textbooks on neurology	Pierre Fournier's *Manuel typographique* is the first book on engraving and type founding	**1764**

1764 **cont**	half the lunar surface; it wins a prize from the *Académie des Sciences*	creature contains an egg with its descendant pre-formed within it, and so on *ad infinitum* Jacques Christophe Valmont de Bomare's *Dictionnaire raisonné universel d'histoire naturelle* (Universal dictionary of natural history) will be published in several editions until 1791	
1765	Nevil Maskelyne becomes director of the Greenwich Observatory (astronomer royal) John Winthrop's *Account of some fiery meteors* includes calculations for the mass of comets	Italian biologist Lazzaro Spallanzani (b Scandiano, Jan 12, 1729) suggests preserving food by sealing it in containers that do not permit air to penetrate	German chemist Karl Wilhelm Scheele (b Straslund, Dec 19, 1742) discovers prussic acid Chemist and writer Mikhail Vasilievich Lomonosov d St. Petersburg (Leningrad, USSR), Apr 15

The Lunar Society

Shortly after the arrival in Birmingham, England, of physicist and teacher William Small from Virginia in 1764 (where he had been one of Thomas Jefferson's most influential teachers), the Lunar Society was founded. Small carried a letter of introduction from Benjamin Franklin to Franklin's friend Matthew Boulton; Boulton, Small, and physician and poet Erasmus Darwin became the founding members of the society.

The Lunar Society occupies a special place in the history of the Industrial Revolution. It included physician John Roebuck who with another member, Samuel Garbett, founded the first sulfuric acid factory; James Watt, the inventor of many improvements on the steam engine; Joseph Priestley, one of the great chemists of the period; Josiah Wedgwood, the potter who made various advances in ceramics; William Withering, who introduced digitalis as a treatment for heart disease; William Murdock, who developed coal gas for lighting; John Wilkinson, who developed superior machinery for boring cylinders; and Benjamin Franklin, a corresponding member.

The Lunar Society received its name from its practice of holding its monthly meetings on the Monday night nearest the full moon so that members would have an easy time seeing their way home. The members, of course, were called Lunatics.

Interactions between the members of the society aided the development of both science and industry. For example, Boulton ran a water-powered mill that needed more power. He thought that a Newcomen steam engine could pump water from below the waterwheel to above it, increasing the flow. James Watt, who was already working on steam-engine improvements, visited Birmingham, and was introduced to Boulton, who bought a share of Watt's patents. Boulton then formed the Boulton & Watt firm to make steam engines, greatly improved by the excellent cylinders bored by Lunatic John Wilkinson.

1766	Matthew Boulton (b Birmingham, England, Sep 3, 1728) founds the Lunar Society, an institution to promote the arts and sciences; James Watt, Joseph Priestley, and Erasmus Darwin are among its members; the name comes from the practice of scheduling meetings so members can return home by moonlight after meetings	Joseph-Louis Lagrange's *Théorie des satellites de Jupiter* (Theory of Jupiter's moons) wins a prize from the *Académie des Sciences* German astronomer Johann Daniel Titius (b Konitz, Jan 2, 1729) proposes that the distances of the planets to the sun are proportional to the terms of the series 0, 3, 6, 12, 24, 48, 96; this law is published by German astronomer Johan Elert Bode (b Hamburg, Jan 19, 1747) and is commonly	Albrecht von Haller is the first to show that nerves stimulate muscles to contract and that all nerves lead to the spinal cord and brain	*On factitious airs* by English chemist Henry Cavendish (b Nice, France, Oct 10, 1731) announces his discovery of hydrogen, which he terms "inflammable air" Pierre-Joseph Macquer's *Dictionaire de chymie* is the first chemical dictionary organized on modern systematic lines

EARTH SCIENCE	MATHEMATICS	MEDICINE	PHYSICS	TECHNOLOGY	
				James Hargreaves (b Blackburn, England, 1720) introduces the spinning jenny, patented in 1770; his first model is able to spin eight threads at once, although later models reach 120 threads; the thread produced is especially suited for wefts	**1764** cont
Mineralogist Axel Fredrik Cronstedt d Stockholm, Sweden, Aug 19	Alexis-Claude Clairaut d Paris, May 17 Samuel Klingenstierna d Stockholm, Sweden, Oct 26	John Morgan founds the first medical school in America at the College of Pennsylvania	Leonhard Euler gives a general treatment of the motion of rigid bodies, including the precession and nutation of Earth, in *Theoria motus corporum solidorum seu rigidorum* (Theory of the motion of solid and rigid bodies)	A mining academy is established in Freiberg (Germany) John Harrison receives the first half the prize offered by the British Board of Longitude by building a chronometer accurate within one-tenth of a second per day and demonstrating that it could be used to find the longitude accurately on a voyage to the West Indies John Smeaton designs a cylinder boring machine James Watt (b Greenock, Scotland, Jan 19, 1736) builds a model of a steam engine in which the condenser is separated from the cylinder so that steam acts on the piston directly, resulting in a power source six times as effective as the Newcomen engine	**1765**
Giant bones, later named parts of a *Mosasaur* (Meuse lizard), are found near the Meuse River in the Netherlands; workers in the stone quarry where the bones are found do not know what they are; however, they are later identified as giant marine reptiles after the skull is found in 1780	Joseph-Louis Lagrange is summoned to Berlin by Frederick the Great, since, as Frederick puts it, "it is necessary that the greatest geometer of Europe should live near the greatest of kings;" Lagrange stays in Berlin for 20 years, completing most of his best work there	Physician Robert Whytt d Edinburgh, Scotland, Apr 15	Horace-Bénédict de Saussure (b Geneva, Switzerland, Feb 17, 1740) invents the electrometer, a device for measuring the electric potential by means of the attraction or repulsion of charged bodies German physicist Johan Carl Wilcke (b Mecklenburg, Sep 6, 1732) sets up the first chart of magnetic inclination	**1766**

1766 cont	known as Bode's law; the discovery of Neptune 70 years later proves the law is wrong		
1767	Nevil Maskelyne begins publication of an annual book that shows ephemeris, the position of heavenly bodies at specific times, the book is called *The British nautical almanac and astronomical ephemeris for the meridian of the Royal Observatory at Greenwich,* but is known simply as the *Nautical almanac*	Lazzaro Spallanzani, in a series of experiments, helps disprove John Needham's theory of spontaneous generation	*The history and present state of electricity* by Joseph Priestley (b Fieldhead, England, Mar 13, 1733) includes his explanation of the rings formed by an electric discharge on a metal; the rings are later named Priestley rings Priestley's *The history and present state of electricity* suggests that electrical forces follow an inverse square law, as gravitational forces do; Priestley is induced into writing this book after meeting Benjamin Franklin in London; the book contains the first detailed account of the famous kite experiment
1768	Publication of *Encyclopaedia Britannica* starts in weekly issues Antoine-Laurent Lavoisier is admitted to the French Academy at the unusually young age of 23	Joseph-Nicolas Delisle d Paris, Sep 11	Spallanzani's *Prodromo d'un ouvrage sur les reproductions animales* (Foreword to a work on animal reproduction) tells of his demonstration that spontaneous generation of animals does not take place in tightly closed bottles that have been boiled for more than 30 minutes Kaspar Friederich Wolff's *De formatione intestinarum* (On the formation of the intestine) establishes principles of the formation of organs in embryos	Georg Brandt d Stockholm, Sweden, Apr 29

EARTH SCIENCE	MATHEMATICS	MEDICINE	PHYSICS	TECHNOLOGY	
Louis-Antoine Bougainville (b Paris, Nov 11, 1729) begins his journey around the world, visiting Tahiti, Samoa, and the New Hebrides	*Theorie der Parallellinien* by Johann Heinrich Lambert is an attempt to prove Euclid's fifth, or parallel, postulate by the indirect method; that is, from the assumption that it does not hold, to find a contradiction; Lambert develops much of what we recognize as non-Euclidean geometry, finding no contradictions				**1766** cont
	Euler's *Vollständige Anleitung zur Algebra* (Complete instruction in algebra), one of the first books Euler dictates after his blindness, shapes elementary algebra in the form that it retains to this day				**1767**
Captain James Cook (b Marton, England, Oct 27, 1728) begins the first of his three voyages to the Pacific; he observes the transit of Venus in Tahiti and explores the coastline of New Zealand	Euler's *Institutiones calculi integralis*, published in three volumes between 1768 and 1770, is a textbook on the integral calculus that also contains many of Euler's own discoveries about differential equations Johann Lambert proves that π is an irrational number; thus, when π is expressed as an infinite decimal, it will not repeat any pattern, nor can π be exactly equal to the ratio of two natural numbers	Robert Whytt's *Observations on the dropsy of the brain*, published posthumously, gives the first description of tuberculous meningitis in children	Antoine Baumé (b Senlis, France, Feb 26, 1728) invents the graduated hydrometer; its scale has come to be known as the Baumé scale for specific gravities of liquids Jesse Ramsden invents an electrostatic machine with glass plates		**1768**

	GENERAL	ASTRONOMY	BIOLOGY	CHEMISTRY
1769	The American Philosophical Society Held at Philadelphia for Promoting Useful Knowledge, first proposed by Benjamin Franklin in 1743, becomes the first American society for advancing science	Observations of the transit of Venus are made from many stations all over the world, including Tahiti (from an expedition commanded by James Cook), the Russian-Chinese border (from an expedition commanded by Pierre-Simon Pallas), and Cavan, Ireland (by Charles Mason and colleagues)	Charles Bonnet's *Philosophical palingenesis, or ideas on the past and future states of living beings* contains his view that the females of every species contain the germs of all future generations Denis Diderot's *Le rêve d'Alembert* (D'Alembert's dream) deals with such thorny questions as the ultimate constitution of matter and the meaning of life	
1770	Paul, Baron d'Holbach's *Le système de la nature* (Nature's system), published under the pseudonym "Mirabaud," is an open attack on Christianity and an affirmation of materialism and atheism	Anders Jean Lexell is the first to observe a short-period comet, but Jupiter's gravity flings it out into space before it can return	Johann Gottleib Gahn (b Voxna, Sweden, Aug 19, 1745) and Karl Wilhelm Scheele discover that phosphorus is an essential component of bone	Karl Wilhelm Scheele discovers tartaric acid
1771	The first bound edition of the *Encyclopaedia Britannica* is published in three volumes The first issue of *Transactions* is published by the American Philosophical Society	Astronomer and surveyor Jeremiah Dixon d Durham, England		*Theoriam philosphiae naturalis redacta ad unicam legem virium in natural existentium* by Ruggiero Giuseppe Boscovich (b Dubrovnik, Yugoslavia, May 18, 1711) develops an unusual atomic theory based on point centers and their relations in space

EARTH SCIENCE	MATHEMATICS	MEDICINE	PHYSICS	TECHNOLOGY	
	Euler's *Dioptrica* (three volumes 1769-1771) lays the foundation for the calculation of optical systems		John Robison measures the repulsion between two charged bodies and shows that this force is inversely proportional to the distance between the two bodies	Richard Arkwright (b Preston, England, Dec 23, 1732) takes out a patent for the hydraulic, or water-frame, spinning machine, especially good for spinning warps; as it is too large and expensive to use in cottages, its development is one of the key changes in Britain's Industrial Revolution

Joseph Cugnot (b Poid, France, Sep 25, 1725), a military engineer, builds a steam carriage that can carry four people at speeds up to 3.6 km (2.25 mi) per hour, the first true automobile | 1769 |
| James Bruce discovers Lake Tana in the area of the sources of the Blue Nile | Euler's *Einleitung zur Algebra* (Introduction to algebra) treats determinate and indeterminate algebra; it also contains his proof that Fermat's "last theorem" $(x^n + y^n = z^n$ cannot be true for nonzero integers if n is greater than 2) is true for n = 3

Edward Waring's *Meditationes algebraicae* states without proof that every natural number is the sum of at most 9 cubes, 19 fourth powers, "and so on;" although this conjecture is widely believed to be true, it is not proved except for fourth powers, which is established some 216 years later | German anatomist Bernhard Siegfried Albinus d Leiden, Holland, Sep 9 | Abbé Jean Antoine Nollet d Paris, Apr 12 | Perrelet constructs a watch with automatic winding

Jacques de Vaucanson develops for use in silk reeling and throwing mills the first Western chain drive; this comes about 800 years after the development of the chain drive in China | 1770 |
| Louis-Antoine Bougainville's *Voyage autour du monde* (Voyage around the world) recounts his journey of 1766 to 1769 and argues that Earth has undergone great physical changes in the past | | *The natural history of the human teeth* by John Hunter (b East Kilbridge, Scotland, Feb 13, 1728) lays the foundations of dental anatomy and pathology | Henry Cavendish's work on electrical force leads to his mathematical single-fluid theory of electricity; this work anticipates many of the advances of the nineteenth century, but is not published until | The Smeatonian Club for engineers, named after John Smeaton, is founded in London

Inventor Chester Hall d Sutton, England, Mar 17 | 1771 |

1771
cont

1772

Swedish scientist and mystic Emanuel Swedenborg d London, Mar 29

A "law" that the distances of the planets to the sun are proportional to the terms of the series 0, 3, 6, 12, 24, 48, 96, proposed in 1766 by Johann Titius, is publicized by Johan Elert Bode; it becomes known as the Titius-Bode law or, commonly, Bode's law

Digressions académiques (Academic digressions) by Baron Louis Bernard Guyton de Morveau (b Dijon, France, Jan 4, 1737), published by the French Academy, contains the first demonstration that metals gain weight on calcination (the chemical change that occurs when some metals are heated below the melting point)

Antoine Lavoisier begins his experiments on combustion, proving that diamond can be burned, and that when sulfur or phosphorus burns, its gain in weight is due to its combination with atmospheric air

Joseph Priestley's *Observations on different kinds of air* reports that growing plants can restore air that has been made "lifeless" by animals or by fire; he notes that a dew forms when hydrogen burns in oxygen, eventually this series of observations reaches six volumes, with the last published in 1786

Priestley's *Directions for impregnating water with fixed air* gives instructions for making seltzer water from carbon dioxide and water

Daniel Rutherford (b Edinburgh, Scotland, Nov 3, 1749), Joseph Priestley, Karl Wilhelm Scheele, and

1771 cont

Anatomist Giovanni Battista Morgagni d Padua (Italy), Dec 5

after Maxwell's theory; thus it has no influence on the development of thinking about electricity

Italian anatomist Luigi Galvani (b Bologna, Sep 9, 1737) discovers accidentally the action of electricity on the muscles of a dissected frog: it causes a twitch; about 1780, while experimenting with this effect, he learns that certain metals can cause the same effect; he publishes his information in 1791

1772

James Cook's voyage (1772 to 1775) to the South Pacific proves there is no large southern continent except Australia

Denis Diderot's *Supplément aux voyages de Bougainville* (Supplement to the voyages of Bougainville) hints at the possibility of continental drift

Italian anatomist Antonio Scarpa (b Motta, Jun 13, 1747) discovers the labyrinth of the ear; that is, the semicircular canals, vestibule, and cochlea

Leonhard Euler's *Lettres à une princesse d'Allemagne* (Letters to a German princess), addressed to the niece of Frederick the Great, deals with mechanics, optics, acoustics, and astronomy at a popular level; it is successful with the general public and translated into seven languages

Nevil Maskelyne suggests to the Royal Society an experiment for measuring Earth's density

Jean Romé de Lisle's *Essai de cristallographie* describes the process of crystallization and identifies 110 crystal forms

Johan Carl Wilcke measures the latent heat of ice (the amount of heat absorbed when ice turns into water)

John Canton d London, Mar 22

Henry Clay patents his method for making papier-mâché

King George III intervenes to get the British Board of Longitude to give John Harrison the second half of the prize money he won by devising the marine chronometer

Engineer James Brindley d Turnhurst, England, Sep 30

	GENERAL	ASTRONOMY	BIOLOGY	CHEMISTRY
1772 cont				Henry Cavendish discover nitrogen independently, with Rutherford generally credited as the first discoverer Scheele discovers oxygen, which he calls "fire air," but the discovery is published in 1777, after Priestley's publication of his independent discovery of oxygen in 1774
1773		William Herschel, b Hanover (Germany), Nov 15, 1738, studies the proper motion of 13 stars and deduces that the sun moves toward the constellation Hercules Pierre-Simon Laplace (b Beaumond-en-Auge, France, Mar 29, 1749) contributes a paper to the French Academy of Science in which he proves the stability of the solar system: for example, if a planet is perturbed by another planet, its mean distance from the sun cannot change much in millenniums	Hilaire-Marin Rouelle investigates the compounds in vertebrate blood, finding sodium carbonate, potassium chloride, and sodium chloride	Antoine Baumé's *Chimie expérimentale et raisonnée* (Experimental and rational chemistry) distinguishes between eight different affinities (forces between constituents of chemical compounds)
1774	Henry Home, Lord Kames's *Sketches on the history of man* is published	Johann Elert Bode begins publication of the *Astronomisches jahrbuch* (Astronomical yearbook) of the Berlin Academy Nevil Maskelyne measures the mass of Earth by first determining the mass of the mountain of Schiehallion by how far it causes a plumb line to deviate from the horizontal; he concludes that Earth is 4.5 times as dense as water	*Exposition d'un nouvel ordre de plantes* (Explanation of a new order of plants) by Antoine-Laurent de Jussieu (b Lyon, France, Apr 12, 1748) proposes the principle of the relative value of traits Naturalist Henry Baker d London, Nov 25	Johann Gottlieb Gahn discovers the chemical element manganese Lavoisier's *Opuscules physiques et chymiques* (Small physical and chemical works) gives the results of his experiments with combustion, showing that the calx of a metal gains weight by absorbing something from the air during combustion Priestley independently discovers oxygen two years after Scheele, but is the first to publish his results Scheele discovers formic acid Scheele discovers chlorine, manganese, and barium, but is not assigned the credit for any of them; Scheele does not recognize that chlorine is an element, which is noted by Sir Humphry Davy about 25 years later

1773

Agostino Bassi, founder of the parasitic theory of infection, b Mairiago, Lombardy (Italy), Sep 25

Lazzaro Spallanzani discovers the digestive action of saliva

1774

Von den ausserlichen Kennzeichen der Fossilien (On the exterior characteristics of fossils) by German geologist Abraham Gottlob Werner (b Wehrau, Sep 25, 1750) introduces a method of classifying minerals by their physical characteristics, such as color, geometrical form, luster, transparency, and hardness

Geographer Charles-Marie de la Condamine d Paris, Jan 28

Anatomy of the human gravid uterus by William Hunter (b Lanarkshire, Scotland, May 23, 1718) is Hunter's greatest work in anatomy; it contains 24 masterpieces of anatomical illustration

German physician Franz Mesmer (b Baden, May 23, 1734) uses hypnotism to aid in curing disease

John Wilkinson patents a precision cannon borer

1775

Louis Claude, Marquis de Saint-Martin, known as *le philosophe inconnu* (the unknown philosopher), publishes *Des erreurs et de la vérité* (Of errors and truth), a mystical philosophy

Johann Fabricius's *Systema entomologiae* (Systematic entomology) classifies insects based on the structure of mouth organs rather than wings

Franz Anton Mesmer suggests that "animal magnetism" causes attractions between certain persons

Torbern Bergman's *Essay of electric attractions* is a study of chemical affinities

Priestley discovers hydrochloric and sulfuric acid

1776

The American colonies of Britain declare their independence

Albrecht von Haller publishes a bibliography of 52,000 scientific works

David Hume, Scottish philosopher, d

De generis humani varietate (On the natural varieties of mankind) by German anthropologist Johann Friederich Blumenbach (b Gotha, May 11, 1752) divides humanity into five different races—Caucasian, Mongolian, American Indian, Malayan, and Ethiopian

Tobias Mayer publishes a small but good lunar map

Scheele and Bergman independently discover uric acid

The French chemist F. De Lassone prepares carbon monoxide

1777

Scheele's *Chemische Abhandlung von der Luft und Feuer* (Chemical treatise on air and fire) describes his experiments for producing oxygen and claims that air consists of "fire air" (oxygen) and "foul air" (nitrogen)

EARTH SCIENCE	MATHEMATICS	MEDICINE	PHYSICS	TECHNOLOGY	
John Lorimer invents the dipping-needle compass		Sir Percival Potts suggests that chimney sweeps in London develop cancers of the scrotum and of the nasal cavity as a result of exposure to soot, the first indication that environmental factors can cause cancer William Withering (b Wellington, England, Mar, 1741) introduces digitalis to cure the dropsy associated with heart disease	Alessandro Volta describes his *eletrofore perpetuo* (electrophorus), a device for producing and storing a charge of static electricity; this device replaces the Leiden jar and eventually leads to modern electrical condensers	A mining academy is established in Clausthal (Germany) David Bushnell invents a hand-operated, one-man submarine, the *American Turtle* Pierre-Simon Girard invents a water turbine James Watt obtains a patent for his version of the steam engine John Wilkinson improves the cylinder boring machine, initially developed for boring cannons; by this date it is turned to its new main purpose of producing cylinders for James Watt's steam engines, which would not function unless good tolerances were maintained	1775
James Cook begins his last voyage of discovery, traveling to Kerguelen Island, New Zealand, Hawaii, and northeast of the Bering Strait James Keir suggests that such rock formations as the Giant's Causeway in Ireland may have been caused by molten rock crystallizing as it cooled		John Fothergill gives the first clinical description of trigeminal neuralgia (Fothergill's disease)	Pierre-Simon Laplace states that if all of the forces on all objects at any one time are known, then the future can be completely predicted	The first two of Watt's steam engines are installed John Wilkinson uses a steam engine to create the blast of air in a blast furnace, increasing its efficiency dramatically; by the end of the century there are 24 steam-driven blast furnaces in England Instrument-maker John Harrison d London, Mar 24	1776
Nicolas Desmarest correctly proposes that basalt is formed from lava emitted by volcanoes	Euler uses *i* to represent the square root of negative one in a manuscript that will not be published until 1794; Gauss actually popularizes the symbol with his *Disquisitiones arithmeticae* in 1801	Physiologist Albrecht von Haller d Berne, Switzerland, Dec 17	Charles-Augustin Coulomb (b Charente, France, Jun 14, 1736) invents the torsion balance	David Bushnell invents the torpedo	1777

GENERAL	ASTRONOMY	BIOLOGY	CHEMISTRY
1777 cont			German chemist Carl Wenzel (b 1740) determines the reaction rates of various chemicals, establishing, for example, that the amount of metal that dissolves in an acid is proportional to the concentration of acid in the solution
1778 Sir Joseph Banks (b London, Feb 13, 1743) is elected president of the Royal Society; although not a good administrator, his use of his fortune to advance scientific causes keeps him in office until 1821 James Cook discovers the Hawaiian islands on Jan 20 Philosopher Laura Bassi d Feb 20		*Flore française* (French flora) by Jean-Baptiste Lamarck (b Bazantin-le-Petit, France, Aug 1, 1744) discusses French plant life and leads to his position after 1781 as royal botanist Botanist Carolus (Carl) Linnaeus d Uppsala, Sweden, Jan 10	Torbern Bergman's *De analysi aquarum* (An analysis of water) gives the first comprehensive account of the analysis of mineral waters Peter Jacob Hjelm (b Sunnerbo Härad, Sweden, Oct 2, 1746) discovers molybdenum Lavoisier announces that air consists of two different gases, one that is respirable, which we now call oxygen, that is one quarter by volume, and another that is not respirable, which we now call nitrogen, that occupies the other three quarters Italian scientist Alessandro Volta (b Como, Feb 18, 1745) studies inflammable air from marshes and discovers methane gas
1779 English navigator James Cook d Kealakekua Bay, Hawaii, Feb 14	The comet Lexell passes as close to Jupiter as its inner satellites; it causes no disturbance in the path of the satellites; therefore, its mass must be smaller than 1/5000th of Earth John Winthrop d Cambridge, MA, May 3	Jan Ingenhousz (b Breda, Holland, Dec 8, 1730) discovers two distinct respiratory cycles in plants: at night oxygen is absorbed and carbon dioxide is exhaled; during the day carbon dioxide is absorbed and oxygen is exhaled (the exact nature of the exhaled gases becomes clear with the discoveries by Lavoisier) *Experiments on vegetables* by Jan Ingenhousz concludes that sunlight is essential for the production of oxygen by leaves Lazzaro Spallanzani studies the role of semen in fertilization and learns that the sperm must make physical contact with the egg for fertilization to take place Hilaire-Marin Rouelle d Apr 7	Adair Crawford's *Experiments and observations on animal heat and the inflammation of combustible bodies* develops methods of determining the specific heat of substances and treats heat generation in animals On Sep 5 Lavoisier proposes the name "oxygen" for the part of air that is respirable and responsible for combustion

Johann Henrich
Lambert d Berlin
(Germany), Sep 25

**1777
cont**

Jean-Etienne Guettard
and Antoine Grimoald
Monnet's *Atlas et
description minéralo-
gique de la France*
(Atlas and mineral
description of France),
completed in 1781, is
commissioned by the
French government

John Hunter's *A prac-
tical treatise on the
diseases of the teeth*
classifies teeth into
molars, bicuspids,
cuspids, and incisors

Joseph Bramah (b
Stainborough, Eng-
land, Apr 13, 1748)
builds an improved
toilet

John Wilkinson in-
vents the turning
lathe

Inventor James
Hargreaves d Apr 22

1778

Comte de Buffon
argues in *Epoques de
la nature* (Epochs of
nature) that 75,000
years have elapsed
since Creation; this is
the first modern
speculation that Earth
is older than biblical
evidence shows,
which is about 6000
years

Horace de Saussure
coins the term
"geology" in his work
*Voyages dans les
Alpes* (Travels in the
Alps), published bet-
ween 1779 and 1796;
he correctly analyzes
the movement of
glaciers, but thinks
that erratic boulders
are moved by streams
of water

Velocipedes, a type of
early bicycle, appear
in Paris

Samuel Crompton (b
Firwood, England,
Dec 3, 1753) develops
the spinning mule, a
cross between the
spinning jenny and
the water-frame spin-
ning machine, which
is capable of spinning
either warps or wefts

Abraham Darby III
builds the world's
first iron bridge,
which spans the
Severn River at
Coalbrookdale,
England

1779

GENERAL	ASTRONOMY	BIOLOGY	CHEMISTRY
1780 John Adams founds the American Academy of Arts and Sciences in Boston; George Washington, Benjamin Franklin, and Thomas Jefferson are among the first members		Spallanzani's *Dissertazioni de fisica animale e vegetale* interprets the process of digestion	Johann Wolfgang Döbereiner b Hof, Bavaria (Germany), Dec 13: he discovers that platinum causes organic gases to oxidize faster than they normally would, the first evidence that platinum acts as a catalyst; he also observes similarities between elements that suggest the periodic table Scheele discovers lactic acid John-Baptist-Michel Bucquet d Paris, Jan 24
1781	Charles Messier (b Badonviller, France, Jun 26, 1730) catalogs more than a hundred star clusters and nebulas that might be mistaken for comets; many are still known from his catalog numbers, such as M13, the giant star cluster in Hercules, and M31, also known as the Andromeda galaxy William Herschel discovers the planet Uranus on Mar 13, although at first he believes it to be a comet Anders Lexell discerns that the moving body discovered by William Herschel is a new planet—Uranus	English naturalist John Needham d Brussels, Belgium, Dec 30	Priestley ignites hydrogen in oxygen, obtaining water Scheele studies the composition of the mineral calcium tungstate, now called sheelite, but fails to pin down tungsten as a new element
1782	The 17-year-old, deaf-mute astronomer John Goodricke (b Groningen, Holland, Sep 17, 1764) becomes the first to suggest the correct explanation for variation in the light from the star Algol, which is caused by an invisible companion star; Goodricke is also the first to calculate the correct period for Algol's variations William Herschel is appointed astronomer royal by English king George III	Henri-Louis Duhamel du Monceau, French botanist, d Aug	Bergman's *Sciagraphia regni mineralis* does much to systematize chemical analyses of minerals Lavoisier observes that total weight does not change in chemical reactions, establishing the first version of the law of conservation of matter Austrian mineralogist Franz Joseph Müller (b Nagyszeben, Transylvania, Jul 1, 1740) discovers tellurium Scheele prepares hydrocyanic acid Andreas Marggraf d Berlin (Germany), Aug 7

EARTH SCIENCE	MATHEMATICS	MEDICINE	PHYSICS	TECHNOLOGY	
The huge skull of a creature later named a *Mosasaur* is found in a stone quarry near the Meuse River in the Netherlands; bones from the same creature had been found earlier; it will become the first remains of a giant, prehistoric reptile to be so identified when it is examined by Cuvier in 1795		Physician John Fothergill d London, Dec 26		James Pickard patents a steam engine Jacques-Germain Soufflot, French architect and designer of the Pantheon in Paris, d Aug 29	**1780**
René Haüy (b St.-Just-en-Choisée, France, Feb 28, 1743) drops a piece of calcite; his examination of the fragments leads to the geometrical law of crystallization	Immanuel Kant's *Critique of pure reason,* which tries to reconcile empiricism and rationalism, declares that Euclidean space is the pure product of thought and that no other possible geometry can exist, since there is only one way to think Joseph-Louis Lagrange expresses in a letter to his mentor Jean le Rond d'Alembert the fear that mathematics has reached its limit and that no further progress can be made		Coulomb's *Théorie des machines simples* (Theory of simple machines) is a study of friction Johan Carl Wilcke introduces the concept of specific heat Giovanni Battista Beccaria d Turin (Italy), May 27	Richard Arkwright builds a factory using his water frame for spinning, becoming the founder of the modern factory system The Marquis de Jouffroy designs (and later builds) the *Pyroschape,* a steam-engine-powered paddleboat; it is tested successfully on the River Saône near Lyon (France) in 1783 James Watt patents a way to change the power produced by a steam engine from a back-and-forth motion to rotary motion	**1781**
Cartographer Jean-Baptiste d'Anville d Paris, Jan 28	Daniel Bernoulli d Basel, Switzerland, Mar 17			James Watt patents a double-acting steam engine: steam is admitted alternatively on both sides of the piston, making the engine more efficient Josiah Wedgwood invents the pyrometer for checking the temperature in furnaces used to fire pottery Inventor Jacques de Vaucanson d Paris, Nov 21	**1782**

1783

William Herschel publishes his first list of double stars

Pierre-Simon Laplace's *Recherches sur le calcul intégral et sur le système du monde* (Investigations in integral calculus and the system of the world) contains a new method for solving the equations of motion of celestial bodies

Henry Cavendish observes that the combustion of hydrogen produces water, although he interprets this in terms of the now discredited phlogiston theory

Don Fausto d'Elhuyar (b Logroño, Spain, Oct, 11, 1755) and Juan José d'Elhuyar discover tungsten

Lavoisier's *Reflexions sur la phlogistique* (Reflections on phlogiston) depicts the contradictions and weaknesses of the phlogiston theory of combustion

Lavoisier repeats the experiment of Cavendish, who showed that hydrogen and oxygen combine to form water; unlike Cavendish, Lavoisier realizes that he is dealing with a new gas, which he names hydrogen

Scheele discovers glycerine

Chemist Nicolas Leblanc (b Issoudun, France, Dec 6, 1742) wins a prize set by the French government for finding a practical way of making sodium hydroxide and sodium carbonate from salt, although the money is never paid; among other benefits, the Leblanc process makes possible the large-scale manufacture of soap

Flight

Joseph and Etienne Montgolfier came from a family of papermakers, so it should come as no surprise that Joseph's idea of a hot-air balloon would be realized in paper. In their first successful experiment with a full-sized balloon (the balloon traveled a mile and a half), the fire suspended below the balloon lit the paper upon landing and the balloon burned. Word of this experiment, conducted in Annonay, quickly reached Paris. A public fund was set up in Paris to have a local scientist, Jacques Charles, duplicate the Montgolfier feat.

Charles, however, did not know how the Montgolfiers got their balloon to fly. He had been experimenting with hydrogen, which he knew to be lighter than air. Therefore, he began constructing a hydrogen balloon. He had to produce thousands of times as much free hydrogen as had ever been previously made—hydrogen had been known for less than 20 years. Charles and his assistants managed not to blow themselves up and succeeded in conducting the semipublic experiment—you had to have a ticket to get close to the 3.6 m (12-ft) balloon. When released, it shot straight up and quickly disappeared.

In the meantime, Etienne Montgolfier was also in Paris trying to arrange for a public demonstration of his own. He secured the support of Louis XVI and funding from the government. The second full-size hot-air balloon was 21 m (70 ft) tall, 12 m (40 ft) wide, and decorated like eighteenth-century wallpaper. This was because it was constructed with the help of another paper manufacturer, Jean-Baptiste Révillon, who made wallpaper. Unfortunately, the first attempt to fly the balloon ended in disaster when heavy rain dissolved the wallpaper. A third balloon was built, this time from taffeta coated with varnish. This balloon carried a sheep, a rooster, and a duck for about two miles in about eight minutes, landing gently enough not to harm the animals.

Competition now developed between the Montgolfier team and the Charles team to get the first human aloft. Although there are hints that the Montgolfier team might have achieved this even earlier, they definitely won with a public launch on the outskirts of Paris. François Pilatre de Rozier and the Marquis d'Arlandes flew across Paris for over 20 minutes, but then lost their nerve and let the fire die so that the vehicle would descend. Ten days later, Charles and a member of his team made the first hydrogen balloon flight. But it was too late. The Montgolfiers would always be known as the people who gave humanity flight.

1784

Denis Diderot, French encyclopedist, d Paris, Jul 31

César-François Cassini d Paris, Sep 4

Swedish astronomer Anders Lexell d St. Petersburg (Leningrad, USSR), Dec 11

Naturalist Abraham Trembley d Geneva, Switzerland, May 12

Biologist Otto Müller d Copenhagen, Denmark, Dec 26

Cavendish announces his discovery of the composition of water (probably made one or two years earlier) to the Royal Society

Gaspard Monge liquifies sulfur dioxide, the first to liquify a substance normally a gas

Scheele discovers citric acid

1783

EARTH SCIENCE

The volcano Laki in Iceland begins a disastrous eruption that eventually kills 10,000 Icelanders, one-fifth the population of the island; the eruption leads Benjamin Franklin to speculate that dust and gases from a volcanic eruption could result in lower temperatures by screening out some of the radiation from the sun

Abel Buell produces the first map of the United States as a new nation

J.B.L. Romé de Lisle's *Cristallographie* formulates the law of constancy of interfacial angles

MATHEMATICS

Leonhard Euler, Swiss mathematician, d St. Petersburg (Leningrad, USSR), Sep 18

Jean le Rond d'Alembert d Paris, Oct 29

MEDICINE

Horace de Saussure's *Essais sur l'hygrométrie* (Essay on measuring humidity) describes how to construct a hygrometer from human hair that can measure relative humidity

PHYSICS

Oxymuriatic acid is introduced for the bleaching of fabrics

TECHNOLOGY

Thomas Bell develops cylinder printing for fabrics

Jouffroy d'Abbans sails the Saône River with a paddlewheel steamboat

L.S. Lenormand, influenced by accounts from China, becomes the first Westerner to use a parachute, the name of which is his invention

The brothers Joseph-Michael (b Vidalon-les-Annonay, France, Aug 26, 1740) and Jacques-Etienne Montgolfier (b Vidalon-les-Annonay, France, Jan 6, 1745) demonstrate the hot-air balloon at Annonay on June 5

Physicist Jacques-Alexandre Charles (b Beaugency, France, Nov 12, 1746) builds the first hydrogen balloon on August 27; later in 1783 he begins a series of balloon ascents

Jean François Pilâtre de Rozier and François Laurent, Marquis d'Arlandes, make a 25-minute flight on Nov 21 in a hot-air balloon designed by the Montgolfier brothers, becoming the first human beings to fly

1784

EARTH SCIENCE

Mineralogist Torbern Olof Bergman d Medevi, Sweden, Jul 8

MATHEMATICS

In a paper on the planets, Adrien-Marie Legendre (b Toulouse, France, Sep 18, 1752) describes "Legendre polynomials," the polynomial solutions of an important differential equation

MEDICINE

German poet and natural philosopher Johann Wolfgang von Goethe (b Frankfurt-am-Main, Aug 28, 1749) discovers the human intermaxillary bone, a feature of the human upper jaw that is missing in most other mammals

PHYSICS

George Atwood accurately determines the acceleration of a free-falling body

TECHNOLOGY

Joseph Bramah invents an improved type of lock; a prize of 200 guineas is promised for anyone succeeding in picking the lock, which remains unpicked until 1851, when an American visitor spends 51 hours to achieve the task

Pierre-Joseph Macquer d
Feb 15

Neptunism vs. Plutonism

During the Industrial Revolution, it became apparent that geology was important to commerce. Placement of canals, railways, and highways depended on geology, as did the location of coal and iron deposits.

It is not surprising, then, that geology, especially in England, entered a period of growth. Many people devoted themselves to the study of specific types of formations or strata. Two theorists in particular, Abraham Gottlob Werner and James Hutton, put forward hypotheses that consolidated the newly discovered facts and offered explanations for Earth as it existed. Today, Hutton is called the father of modern geology, while Werner is hardly known at all outside of Germany. Toward the end of the eighteenth century, however, Werner's ideas were mostly ascendant.

Werner proposed that Earth was originally an irregular solid body covered completely with a heavily saturated fluid. Over time, solids precipitated from this fluid in a regular succession: first the primitive rocks, such as granite and gneiss; then the transition rocks, such as slates and some limestones; and finally, recent rocks, such as sandstone and other types of limestone. As the fluid lost its chemical content, it gradually became the salty oceans that we know today.

In Werner's theory, which came to be called Neptunism, volcanoes were local phenomena caused by burning coal seams. He did not postulate an inner heat.

The rock basalt became a particular issue between Werner's many followers and other scientists who had identified basalt as essentially cooled lava. These scientists came to be called Vulcanists, after the god who lived in a volcano. Werner explained such basalt structures as the Giant's Causeway in Ireland by saying that a local volcano had melted preexisting basalt, which had then cooled in this formation.

Although Werner's ideas had many attractive features, especially in explaining the sequence of rock strata, they suffered several fatal flaws. Werner never offered an explanation of where the rest of the fluid had gone. Even more alarming was the implication from his theory that all the rock that ever would exist had formed already. It was clear that the continents were gradually being washed into the oceans. If no new rock were formed, the continents would eventually disappear.

James Hutton announced his alternative to Werner's ideas in 1785. He specifically addressed the problem of the continents' wearing away—the denudation problem. Hutton gave a great role to the interior heat of Earth, which was well known from mines. As a result, his theory came to be called Plutonism (after the god of the underworld), to contrast with Werner's Neptunism (after the god of the sea). Hutton thought that Earth's heat could cause sediments in the ocean to rise up and become new continents. This was just as wrong as any of Werner's ideas, but Hutton thought it was testable. In 1787 he identified some rock formations that looked to him like sedimentary rock that had been changed by heat (what we call metamorphic rocks today) and felt that his theory was confirmed.

Since Hutton, too, was wrong, how did he come to be the father of modern geology? He was the first to recognize that changes taking place in the present could account, over a long period of time, for observed features of Earth's crust. Werner's Earth was created and traveled along a foreordained path toward a universal desert. Hutton's Earth always existed and would continue forever to exist very much as it was in 1785, with only the details changed. Of course, neither concept was correct, but Charles Lyell was able to take Hutton's idea of small changes taking place over long periods of time to develop specific theories that are held today.

William Herschel's *On the construction of the heavens* gives the first reasonably correct description of the shape of the Milky Way galaxy, although he underestimates its size by three orders of magnitude and places the sun near the center instead of where we now believe it to be, fairly near the edge

Pierre-Simon Laplace's *Théorie des attractions des sphéroïdes et de la figure des planètes* (Theory on the

Lazzaro Spallanzani performs artificial insemination on a dog

Claude-Louis Berthollet (b Talloire, France, Dec 9, 1748) introduces *eau de Javel*, that is, chlorine, for chemical bleaching

Cavendish determines the composition of the atmosphere; he finds that when all the oxygen and nitrogen have been removed there is a small residue of an unknown gas; however, this discovery is not followed up for over a hundred years, when the unknown gas is found to be

				Inventor Henry Cort (b Lancaster, England, 1740) of Fontley Forge, Hampshire, develops the puddling method of turning coke-smelted iron into satisfactory wrought iron, making production of useful iron completely independent of the forest for the first time; he also perfects the rolling mill with grooved rollers	**1784** cont
				Benjamin Franklin introduces bifocal eyeglasses	
				Vincent Lunardi makes the first English balloon ascent, using a balloon filled with hydrogen	
				Andrew Meikle invents the threshing machine	
				William Murdock (b Auchinleck, Scotland, Aug 21, 1754), an employee of steam-engine manufacturer Boulton & Watt, builds a working model of a steam-powered carriage	
				Henry Shrapnel (b England, Jun 3, 1761) invents the shrapnel shell	
				James Watt uses steam pipes to heat his office, the first use of steam heat	
Theory of the Earth by James Hutton (b Edinburgh, Scotland, Jun 3, 1726) explains the principle of uniformitarianism: all geological features can be explained by changes now observable taking place over very long periods of time; in fact, Hutton thinks there is no sign of Earth's beginning or prospect for its end	Legendre's *Recherches d'analyse indéterminée* (Researches in indeterminant analysis) contains the first statement of the law of quadratic reciprocity, an important theorem concerning prime numbers, but his proof of the theorem is flawed Marie-Jean Condorcet's *Essai sur l'applications de l'analyse à la probabilité* (Essay	William Withering's *Account of the foxglove* reports on his discovery of the use of digitalis in the treatment of heart disease	Coulomb makes precise measurements of the forces of attraction and repulsion between charged bodies and between magnetic poles, using a torsion balance, demonstrating conclusively that electric charge and magnetism obey inverse-square laws like that of gravity; he also discovers that electrically charged bodies discharge	Jean-Pierre Blanchard (b Les Andelys, France, Jul 4, 1753) and Dr. J. Jeffries make the first balloon crossing of the English Channel Edmund Cartwright (b Nottingham, England, Apr 24, 1743) invents a power loom and uses the machine in his own factory in 1787; the machine, however, is imperfect and power	**1785**

1785 cont

(ASTRONOMY) attraction of spheroids and the shape of planets) contains the partial differential equation named after him; this equation describes gravitational, electromagnetic, and other potentials

(CHEMISTRY) a mixture of argon, neon, krypton, and xenon; he also produces nitric acid by passing an electric spark through air

1786

(GENERAL) Michel-Gabriel Paccard and Jacques Balmat are the first to climb Mont Blanc, on August 8, opening the tops of mountains to human exploration

(ASTRONOMY) William Herschel publishes his *Catalogue of nebulae* which eventually will be expanded by Herschel, his son, and J.L.E. Dreyer into the *New general catalogue* or NGC, which is still used by astronomers today

Thomas Wright d Byers Green, England, Feb 25

(CHEMISTRY) Chemist Karl Wilhelm Scheele d Köping, Sweden, May 21

1787

(ASTRONOMY) Pierre-Simon Laplace's "*Mémoire sur les inégalités séculaires des planètes et des satellites*" (Note on secular irregularities of planets and satellites) is an early work leading to his mammoth *Mécanique céleste*

(CHEMISTRY) Berthollet discovers the composition of ammonia, prussic acid, and hydrogen sulfide

The first of four volumes of *Méthode d'une nomenclature chimique* (Method of chemical nomenclature), written by Guyton de Morveau, with Lavoisier, Berthollet, and Fourcroy, presents a system for naming chemicals based on scientific principles

1788

(BIOLOGY) Naturalist Georges-Louis Leclerc, comte de Buffon, d Paris, Apr 16

(CHEMISTRY) Physician Sir Charles Blagden (b Gloucestershire, England, Apr 17, 1748) discovers that the depression of the freezing point of a solution is proportional to the concentration of solute

EARTH SCIENCE	MATHEMATICS	MEDICINE	PHYSICS	TECHNOLOGY	
	on the applications of analysis to probability) is an important early work in the science of probability

Legendre's extensive work on elliptic integrals and elliptic functions begins; it will continue until 1830, shortly before his death, culminating in the third volume of *Traité des fonctions elliptiques* (Theory of elliptic functions) | | spontaneously; in the twentiethth century, it is found that cosmic radiation is responsible for this discharge | looms will not be used widely until the nineteenth century, after they have been improved

Pilâtre de Rozier and Romain are killed trying to cross the English Channel by balloon; they are the first casualties of flight | 1785 cont |
| Geologist Jean-Etienne Guettard d Paris, Jan 6

German-Swiss geologist Johann von Charpentier b Freiberg (Germany), Dec 7; he works out the details of the European ice age, first proposed by Ignatz Venetez, and uses them to convince Louis Agassiz, who eventually convinces everyone of the existence of the ice age | | Benjamin Rush's *Observations on the cause and cure of the tetanus* suggests that some illnesses may be psychosomatic | Inventor and physicist Abraham Bennet (b England, 1750) invents the gold-leaf electroscope | The first experiments with gas lighting are conducted by the English and Germans

Ezekiel Reed invents a machine for making nails | 1786 |
| | Mathematician, physicist, and astronomer Ruggiero Giuseppe Boscovich d Milan (Italy), Dec 13 | Physician Sir William Watson d London, May 10 | Jacques-Alexandre Charles shows that different gases expand by the same amount for a given rise in temperature (Charles's law)

German physicist Ernst Florens Friedrich Chladni (b Wittenberg, Nov 30, 1756) experiments with sound patterns on vibrating plates | Friedrich Krupp b; he establishes a steel plant at Essen (Germany) in 1810 that will be the foundation of the Krupp enterprises

John Wilkinson builds the first iron barge

John Fitch (b Windsor, CT, Jan 21, 1743) successfully tests his steamboat on the Delaware River; by 1790 one of Fitch's steamboats is in regular service for several weeks during the summer, but it is a commercial failure | 1787 |
| | | Surgeon Percivall Pott d

John Brown, Scottish physician, d London, Oct 17 | Joseph-Louis Lagrange's *Mécanique analytique* (Analytical mechanics) is a study of mechanics carried out at its highest level of generality, using | | 1788 |

1788
cont

1789

Thomas Day, one of the two literary members of the Birmingham Lunar Society, d 1789

The Bastille prison in Paris is stormed by a French mob on Jul 14; this event is generally considered the beginning of the French Revolution

William Herschel completes his great reflecting telescope, with a mirror 124.5 cm (49 in.) in diameter and a focal length of 12.2 m (40 ft); although a technological marvel for the age, it is too cumbersome to be used frequently

Carpology: or treatise on the fruits and seeds of plants, by German biologist Joseph Gaertner (b 1732), treats over a thousand species of plants and illustrates 180

Johann Wolfgang Goethe's *Versuch, die Metamorphose der Pflanzen zu erklären* (Attempt to explain the metamorphosis of plants) claims incorrectly that all plant structures are modified leaves, but clearly espouses evolution

Antoine-Laurent de Jussieu's *Genera plantarum* (Genera of plants) classifies plants into families on bases that are still accepted today (for example, grasses, lilies, and palms)

Gilbert White's *The natural history and antiquities of Selborne*, a series of notes about animal and plant life in Selborne, England, is the first recognizable work of ecology; it goes through about 200 different editions and becomes the fourth best-selling book in English

German chemist Martin Heinrich Klaproth (b Wernigerode, Dec 1, 1743) discovers uranium and zirconium

Lavoisier publishes a "Table of thirty-one chemical elements," to which he adds light and heat, which he believes to be material, but without mass

Lavoisier publishes the textbook *Traité elémentaire de chimie* (Elementary treatise on chemistry), which firmly establishes the oxygen theory and the new chemical nomenclature; it also states the law of conservation of mass

English chemist James Marsh b Sep 2; he develops the sensitive tests for arsenic and for antimony that are named after him

1790

Archaeologist John Frere (b Westhorpe, Suffolk, England, Aug 10, 1740) discovers flint tools that he believes were made by prehistoric humans, but few agree with him; it is not until 1846 that Jacques Boucher de Crèvecoeur de Perthes (b Rethel, France, Sep 10, 1788) is able to convince other scientists that humans once used flint tools

William Herschel discovers planetary nebulas

EARTH SCIENCE	MATHEMATICS	MEDICINE	PHYSICS	TECHNOLOGY	
			only algebraic and analytic methods; the book does not contain a single illustration; it also includes his version of the principle of least action		1788 cont
			Alessandro Volta introduces the concepts of electrical tension and capacity		
		Abraham Bennet invents a simple electric induction machine		The United States introduces its first patent law	1789
		Ernst Chladni invents the euphonium, a musical instrument in the tuba family		The first cotton factory driven by steam is opened in Manchester, England	
				Oliver Evans (b Newport, DE, Sep 13, 1755) takes out the first US patent for a steam-propelled land vehicle	
				James Watt invents the governor, a centrifugal device that controls the speed of a steam engine by a feedback mechanism	
				Sir William Fairbairn b Scotland, Feb 19; he introduces the riveting machine to the manufacture of steam boilers and designs and builds about a thousand steel bridges	
	Physiologist Marshall Hall b Basford, England, Feb 18; he is the first to identify and study reflexes; in 1830 he denounces bloodletting as a treatment for disease		Friedrich Gren founds the *Journal der Physik*, today known as the *Annalen der Physik* William Cullen d	The first rolling mill driven by steam is opened in England	1790
				George Washington's dentist, John Greenwood, invents the dental drill	
				James Watt builds a pressure gauge	
				Scottish inventor John Campbell d London, Dec 16	

1790 cont

Johann Friederich Blumenbach's *Collectionis suae craniorum diversarum* gives descriptions of 60 human crania

British explorer William Parry b; he makes three trips in an effort to find the Northwest Passage and another aiming at the North Pole; all are unsuccessful, but he is knighted for his tries

Statesman and scientist Benjamin Franklin d Philadelphia, PA, Apr 17

1791

The metric system of measurement is proposed in France, Mar 30

1792

The metric system and beyond

As early as 1670, when Jean Picard was establishing a modern value for the length of a meridian of Earth, it was suggested that measurement should be based on the meridian. This suggestion appealed to the scientific bent of the revolutionaries, who established the unit of length at one ten millionth of the meridian from the pole to the equator. They called this unit the meter. They also established a basic unit of mass, called the gram, that is based on a volume of pure water at a given temperature.

Metric units are all related to each other by powers of 10. The same prefixes—milli- (0.001), centi- (0.01), deci- (0.1), deka- (10), hecto- (100), and kilo- (1,000)—are used in all parts of the system. Thus, a millimeter is a thousandth of a meter and a milligram is a thousandth of a gram. The liter, used for capacity, was originally 1000 cubic centimeters. A kiloliter is 1000 liters and a kilometer is 1000 meters. Later this system was extended to smaller and larger prefixes: micro- (0.000001), nano- (0.000000001), pico- (0.000000000001), femto- (0.000000000000001), atto- (0.000000000000000001), mega- (1,000,000), giga- (1,000,000,000), and tera- (1,000,000,000,000). Thus, a micrometer is a millionth of a meter, an attogram is a quintillionth of a gram, and a gigaliter is a billion liters.

Nations were slow to abandon their customary measures for the metric system despite its advantages. Even in France the change did not come easily. By 1875, however, the metric system was sufficiently established to make it worthwhile to set up an International Bureau of Weights and Measures to oversee it. This was established on May 20 of that year in Sèvres, France.

The new international bureau decided that it was too inaccurate for scientific purposes to define length in terms of the size of Earth and mass in terms of pure water. Instead, the bureau replaced these measurements with a standard length on a platinum-iridium bar and a standard mass of platinum-iridium. In the process, the bureau was unable to make the standard liter come out as originally intended. The result was that a liter became 1000.028 cubic centimeters instead of 1000.

Jean-Baptiste Delambre (b Amiens, France, Sep 19, 1749) and Pierre Mechain begin their measurement of the arc of the meridian from Dunkirk to Barcelona, leading to the establishment of a uniform system of measures

Giuseppe Piazzi (b Ponte de Valtellina, Switzerland, Jul 16, 1746) observes the unusually rapid proper motion of the star 61 Cygni

William Gregor (b Trewarthenick, England, Dec 25, 1761) discovers titanium

Richard Kirwan (b Cloughbnallmore, Ireland, Aug 1, 1733), one of the last defenders of the phlogiston theory of fire, gives up in the face of overwhelming evidence that it is wrong

German chemist Jeremias Richter (b Hirschberg, Mar 10, 1762) establishes the principle of fixed chemical reactions called stoichiometry; for example, acids and bases in neutralizing each other always do so in the same proportion

John Mercer b England, Feb 21; he is the inventor of the process for treating cotton cloth known as *mercerizing*

The metric system and beyond (continued)

As scientists found ways to measure the frequency of light with great accuracy, the meter was redefined on the basis of the frequency of the light emitted by a particular isotope of krypton. This standard persisted for about 25 years.

In 1967 the availability of atomic clocks of unprecedented accuracy led scientists to abandon the rotation of Earth as a measure of time. Instead, a second was defined as the time it takes microwaves emitted by hot cesium to oscillate 9,192,631,770 times. Since Earth's rotation has slowed somewhat since then, it has been necessary occasionally to add a leap second to the year.

In 1983 the International Bureau of Weights and Measures settled on a new basis for a standard for length as well—the speed of light in a vacuum. This speed is one of the fundamental constants of the universe, as has been known since 1905 when Einstein developed the special theory of relativity. By making a meter exactly the distance light travels in a vacuum in 1/299,792,458th second, the current standards were reached. Another way to say this is that the speed of light is *exactly* 299,792.458 kilometers per second, since it is true by definition. Before this new definition, the speed of light was given as approximately 299,792.5 kilometers per second (186,292 miles per second). The new definition, therefore, is not a very big correction in the length of the meter. Note also that this definition depends on the 1967 definition of the second. Therefore, both measures are completely divorced from Earth.

The new definition of length fixes the definition of area and volume. Capacity—measured in liters—also has been regularized. A liter is once more 1000 cubic centimeters. But a gram is still defined in terms of the platinum-iridium kilogram in Sèvres.

This definition also fixes length, area, and volume in the customary US system. Long ago, Congress set the US inch at exactly 2.54 centimeters. Since most other length, area, and volume measures can be determined from the inch, these measures are now exact in terms of the speed of light in both systems.

1791

Traité medico-philosophique sur l'aliénation mentale (Medical and philosophic treatise on mental illness) by Phillippe Pinel (b Tarn, France, Apr 20, 1745) advocates a more humane treatment of the insane

Luigi Galvani announces that electricity applied to severed frogs' legs causes them to twitch and that frogs' legs twitch in the presence of two different metals with no electric current present; the latter discovery eventually leads to Alessandro Volta's developing the electric battery

Pierre Prévost (b Geneva, Switzerland, Mar 3, 1751) develops his theory of exchanges of radiation of heat; he correctly shows that cold is merely the absence of heat and that all bodies continually radiate heat; if they seem not to radiate heat, it means that they are in heat equilibrium with their environment

Abraham Darby III, who built the first cast-iron bridge, d

1792

Claude Chappe (b Brûlon, France, Dec 25, 1763) invents the semaphore, an optical system for the transmission of messages

Thomas Henry Maudslay b England; he is the creator of the first machine tools

Inventor William Murdock invents coal-gas lighting and uses it to light his home

	GENERAL	ASTRONOMY	BIOLOGY	CHEMISTRY
1792 cont		Guillaume Le Gentil d Paris, Oct 22		
1793	In August, France abolishes the *Académie des Sciences* and its universities		German botanist Christian Konrad Sprengel (b Spandau, Sep 22, 1750) publishes his findings on plant fertilization by insects and by wind Naturalist Charles Bonnet d Genthod, Switzerland, May 20	The world's first chemical society is founded in Philadelphia, PA Carl Friedrich Wenzel d Feb 27
1794	The *Conservatoire National des Arts et Métiers* is founded in Paris; it is a museum of science and technology in which a model of every new invention is deposited Marie-Jean de Condorcet's *History of the progress of the human spirit* expresses his belief in the perfectability of humanity	Ernst Chladni shows that meteors are extraterrestrial	John Dalton's *Extraordinary facts relating to the vision of colors* gives the earliest account of color blindness, which he refers to as Daltonism, since he is afflicted with the condition *Zoonomia* by Erasmus Darwin (b Elston Hall, Nottinghamshire, England, Dec 12, 1731) contains his ideas about evolution, which are Lamarckian in that they assume the environment has a direct influence on organisms, causing permanent changes in the germ line (a belief now known to be incorrect) Karl Friedrich Phillip von Martius b; he becomes an expert on the botany of Brazil Swedish botanist Elias Magnus Fries b Aug 15; his *Systemia mycologicum* becomes a standard work on fungi	Johan Gadolin (b Åbo, Finland, Jun 5, 1760) discovers yttrium in a Swedish mineral deposit; it is not completely separated from other rare earths and other elements until the 1840s, when Carl Gustav Mosander breaks apart many of the rare earths Chemist Antoine-Laurent Lavoisier d Paris, May 8, executed by the radical faction of the French Revolution
1795	The *Institut National des Sciences et des Arts* in Paris replaces abolished academies in France; among its 48 scholars are Lagrange and Laplace, but not Le-		Georges Léopold Chrétien Frédéric Dagobert Cuvier, b Montbéliard (now France), Aug 23, 1769, develops a method of classifying mammals	Adair Crawford, Scottish chemist, d Lymington, England, Jul

EARTH SCIENCE	MATHEMATICS	MEDICINE	PHYSICS	TECHNOLOGY	
				Inventor Sir Richard Arkwright d Cromford, England, Aug 3	**1792** cont
				Inventor and industrialist John Smeaton d Oct 28	
Meteorological observations and essays by John Dalton (b Eaglesfield, England, about Sep 6, 1766) reflects Dalton's weather observations since 1787, a practice he continues for the rest of his life (until 1844) Jean-Baptiste Lamarck argues that fossils are the remains of organisms that once were living Geologist John Michell d Thornhill, England, Apr 21		An epidemic of yellow fever in Philadelphia kills about 10 percent of the population John Hunter is the first to ligate the femoral artery for treatment of aneurysm		Jean-Pierre Blanchard makes the first successful parachute jump from a balloon In Apr Eli Whitney (b Westboro, MA, Dec 8, 1765) invents the cotton gin	**1793**
William Smith (b Churchill, England, Mar 23, 1769) publishes the first large-scale geological map of England	Legendre's *Eléments de géométrie* (Elements of geometry) is a simplification and rearrangement of Euclid's work that becomes the basic geometry textbook in France and the United States for most of the nineteenth century Marie-Jean-Antoine-Nicolas, Marquis de Condorcet, d Bourg-la-Reine, France, Apr 8	German physiologist Kaspar Wolff d St. Petersburg (Leningrad, USSR) Feb 22 Jean-Pierre-Marie Flourens b France, Apr 13; he studies the nervous system, locates the center of respiration, and demonstrates that the cerebellum controls muscular movements Scottish physician James Lind d Hampshire, England, Jul 13	Daniel Rutherford designs the first maximum-minimum thermometer Alessandro Volta demonstrates that the electric force observed by Galvani is not connected with living creatures, but can be obtained whenever two different metals are placed in a conducting fluid	Inventor John Roebuck d Jul 17 The Ecole Polytechnique opens in Paris in Dec; it will produce the scientific elite of France during the early years of the nineteenth century	**1794**
Georges Cuvier identifies bones found between 1766 and 1780 in a quarry near the Meuse River in the Netherlands as be-		*Morbid anatomy of some of the most important parts of the human body* by anatomist Matthew Baillie (b Shotts,		Joseph Bramah invents an hydraulic press	**1795**

1795 cont

gendre, although Legendre is quickly elected to membership by his colleagues

France adopts the metric system as its official system of measurement

Naturalist Christian Ehrenberg b Delitzsch (Germany), Apr 19; he makes pioneering studies of invertebrates, especially protists, but his work is handicapped by his belief that protists have organ systems like those of vertebrates and other multicellular organisms and by his refusal to accept evolution

1796

Pierre-Simon Laplace's *Exposition du système du monde* (Explanation of the system of the world) states his "nebular hypothesis:" the solar system evolved from a condensing cloud of gas; in a modern form, this idea is still accepted by most astronomers

Johann Titius d Wittenberg (Germany), Dec 16

Botanist John Torrey b US; he becomes the state botanist of New York before a career as a teacher and researcher at West Point, Princeton, and Columbia

J.T. Lowitz prepares pure ethyl alcohol

longing to a giant, prehistoric reptile, now known as the *Mosasaur*, a 14-m-long (46-ft-long) sea reptile

Lanarkshire, Scotland, Oct 27, 1761) is the first work to treat pathology as a subject in itself

Physician Sir Gilbert Blane (b Blanefield, Scotland, Aug 29, 1749) makes the use of lime juice to prevent scurvy mandatory in the British navy as proposed by James Lind in 1753; this is the origin of the nickname "limey" for the British sailor

Surgeon James Braid b Rylawhouse, Scotland; he renames mesmerism *hypnotism* and gives the practice some medical respectability by correctly explaining why it works

French surgeon Pierre-Joseph Desault d; he is famous for teaching improved surgical techniques and for improvements in instruments used in surgery

Napoleon offers a prize for a practical method of preserving foods, which is eventually won by François Appert (b Châlons-sur-Marne, France, Oct 23, 1752); he introduces a sterilization process for food by bottling or canning, heating, and sealing

Inventor Josiah Wedgwood d Jan 3

1795 cont

On Mar 30 German mathematician Karl Friederich Gauss (b Braunschweig, Apr 30, 1777) discovers how to construct a regular polygon of 17 sides using just a compass and a straightedge; this discovery convinces Gauss to become a mathematician instead of a linguist

German mathematician Jakob Steiner b 1796; considered the greatest geometer of modern times, he showed, among many other discoveries, that all constructions possible with a straightedge and compass are equally possible with just a straightedge and a single given circle

C. W. Hufeland publishes *Macrobiotics, or the art to prolong one's life*

Physician Edward Jenner (b Berkeley, England, May 17, 1749) performs the first inoculation against smallpox by infecting a boy with cowpox (vaccinia virus)

The Rumford Medals of the Royal Society and the American Association for the Advancement of Science are instituted

Johan Carl Wilcke, German physicist, d Stockholm, Sweden, Apr 18

Johann Christian Poggendorf, founder and editor of *Annalen der Physick und Chemie* (Annals of Physics and Chemistry) b Dec 29

1796

1796
cont

1797

Caroline Herschel, b Hanover (Germany), Mar 16, 1750, discovers her eighth comet in eleven years

Treatise concerning the easiest and most convenient method of determining the orbit of a comet by Heinrich Wilhelm Mathäus Olbers, b Arbergern (Germany), Oct 11, 1758, introduces new methods for calculating parabolic orbits

Wood engraver Thomas Bewick's *History of British birds* is published

Marie-François-Xavier Bichat (b Thoirette, France, Nov 16, 1771) identifies 21 different types of tissues in animals, including bone, muscle, and blood

Adrien-Laurient-Henri de Jussieu b; his textbook on elementary botany, *Botanique,* is printed in nine editions in French and in many editions in other languages

William Nicholson (b London, 1753) begins publication of *Journal of Natural Philosophy, Chemistry and the Arts*

Louis-Nicolas Vauquelin (b St. André, Calvados, France, May 16, 1763) discovers the element chromium

After attacking a celebration of the second anniversary of the fall of the Bastille in Birmingham, a crowd of angry English people set Joseph Priestley's church and house on fire; Priestley and his family escape, although his laboratory and scientific papers are destroyed; Priestley later emigrates to the United States

1798

Essay on the principle of population as it affects the future improvement of society, published anonymously by Thomas Malthus (b near Guilford, Surrey, England, Feb 13, 1766), projects an expanding population that can only be controlled by famine, disease, and war

The first volume of James Bradley's catalog of stellar positions is published posthumously

H.W. Brandes and J.F. Benzenberg use triangulation to measure the height of meteors as they burn in the atmosphere

Pierre-Simon Laplace predicts the existence of black holes

Vauquelin discovers beryllium in the gems beryl and emerald, but does not isolate the element; beryllium is finally isolated by Friedrich Wöhler in 1828

EARTH SCIENCE	MATHEMATICS	MEDICINE	PHYSICS	TECHNOLOGY	
	Gauss succeeds in proving the law of quadratic reciprocity after a year-long effort; he had discovered this theorem dealing with prime numbers independently of Adrien-Marie Legendre, who first stated it in 1785, but failed in two published attempts to prove it				**1796** cont
In a series of experiments, Sir James Hall (b Haddington, Scotland, Jan 17, 1761) proves that igneous rock cools to form crystalline rock Geologist James Hutton d Edinburgh, Scotland, Mar 26	Joseph-Louis Lagrange's *Théorie des fonctions analytiques* (Theory of analytic functions) attempts to avoid infinitesimals by use of the infinite Taylor series derived from functions (hence the name "derivative"); it introduces the notation $f'(x)$, $f''(x)$, etc., for the first, second, etc., derivatives Lorenzo Mascheroni's *Geometria del compasso* (Geometry of compasses) contains his surprising result that all geometric constructions possible with a straightedge and compass are possible with a compass alone (anticipated 125 years previously by Georg Mohr of Denmark)	The Royal Society rejects Jenner's inoculation technique for smallpox Physician Richard Brocklesby d London, Dec 11	Physician Jean Poiseuille b Paris, Apr 22; in the process of studying blood pressure, for which he invents new measuring tools, he works out the laws by which fluids travel through narrow tubes	Some English roadways in Shropshire are converted to iron rails along which wagons are drawn by horses Engineer Henry Maudslay (b Woolwich, England, Aug 22, 1771) perfects the slide rest for lathes, called one of the great inventions of history; it permits the lathe operator to operate the lathe without holding the metal-cutting tools in his hands Jedediah Strutt d	**1797**
	Caspar Wessel, a Norwegian surveyer, publishes in the *Transactions of the Danish Academy* the method of representing complex numbers in the plane that is generally attributed to Gauss or Argand, who developed the same ideas later	Anatomist Luigi Galvani d Bologna (Italy), Dec 4	Henry Cavendish determines the mass of Earth by measuring the gravity between two small masses and two large masses; this gives the gravitational constant G, which was the only unknown in Newton's equations; solving for G enables Cavendish to establish that Earth is about 5.5 times as dense as water *Enquiry concerning the source of heat which is excited by friction* by Count Rumford (Benjamin	Robert Fulton (b Little Britain, PA, Nov 14, 1765) builds a four-person submarine William Murdock demonstrates the use of coal gas for lighting Boulton & Watt's plant in Soho, England Aloys Senefelder invents lithography John Fitch, American inventor, d Bardstown, KY, Jul 2	**1798**

**1798
cont**

1799

Soldiers in Napoleon's army, digging near the Rosetta branch of the Nile, uncover a stone engraved in three different scripts; the Rosetta Stone turns out to be the key to unlocking Egyptian hieroglyphics

Baron Alexander von Humboldt, b Berlin (Germany) Sep 14, 1769, starts his five-year exploration of the Spanish colonies of the New World

The Great Leonid meteor shower is observed by Wilhelm von Humboldt

Pierre-Simon Laplace's first of five volumes, *Traité de mécanique céleste* (Celestial mechanics), begins his summary of the advances in mathematical astronomy made during the eighteenth century; the last volume will not be issued until 1825

A perfectly preserved mammoth is found frozen in Siberia

In *The naturalists miscellany* George Shaw gives the first scientific account of the duck-billed platypus, based on a skin and a sketch provided by a former governor of New South Wales, Australia

Biologist Lazzaro Spallanzani d Pavia (Italy), Feb 11

Antoine-François, comte de Fourcroy (b Paris, Jun 15, 1755) isolates urea

Joseph-Louis Proust (b Angers, France, Sep 26, 1754) finds that the elements in a compound always combine in definite proportions by mass (Proust's law)

Charles Tennant invents bleaching powder

Joseph Black d Edinburgh, Scotland, Dec 6

1800

The "Celestial Police," presided over by Johann Schröter, are founded at Lilienthal to search for a "missing planet" between Mars and Jupiter

Louis-Jean-Marie Daubenton d Paris, Jan 1

William Nicholson and Anthony Carlisle, using the newly developed Voltaic battery, perform the first electrolysis of water, reversing earlier experiments that showed that hydrogen and oxygen combine to form water

EARTH SCIENCE	MATHEMATICS	MEDICINE	PHYSICS	TECHNOLOGY	
			Thompson, b Woburn, MA, Mar 26, 1753) describes his experiments with boring cannons that show that the caloric theory of heat cannot be true, and that heat should be considered a kind of motion		**1798** cont
Alexander von Humboldt identifies the Jurassic Period in Earth's history	Gauss's doctoral dissertation, ''New proof of the theorem that every integral rational function of one variable can be decomposed into real factors of the first or second degree'' is the first successful proof of the fundamental theorem of algebra—that every polynomial equation has a solution Gaspard Monge's *Géométrie descriptive* (Descriptive geometry) consists of his 1794 to 1795 lectures at the Ecole Normale in Paris and provides the first printed account of descriptive geometry (geometry in three dimensions, the basis for mechanical drawing) *Teoria generale delle equazioni* (General theory of equations) by Italian mathematician Paolo Ruffini (b Valentano, Sep 22, 1765) contains the first proof that the general equation of the fifth degree cannot be solved by algebraic methods based on radicals, although the proof is flawed in some details	Dutch physician and plant physiologist Jan Ingenhousz d Wiltshire, England, Sep 7 Physician William Withering d Birmingham, England, Oct 6	Abraham Bennet, British physicist and inventor, d Horace-Bénédict de Saussure d Geneva, Switzerland, Jan 22	Phillipe Lebon (b Brachay, France, May 29, 1767) receives a patent for his ''Thermo-lampe,'' which uses gas for lighting	**1799**
		Marie-François-Xavier Bichat studies postmortem changes in human organs Chlorine is used to purify water by William Cruikshank in England	William Herschel's ''An investigation of the powers of prismatic colours to heat and illuminate objects'' tells of his discovery of infrared radiation; while investigating the power of different parts of	Jute is cultivated in India British engineer Richard Trevithick (b Illogan, England, Apr 13, 1771) builds the first steam engine to use high-pressure steam	**1800**

1800 cont

CHEMISTRY

William Cruikshank determines composition of carbon monoxide

1801

ASTRONOMY

Johann Elert Bode's *Uranographia* is an atlas of 17,240 stars and nebulas

Bibliographie astronomique by Joseph-Jérôme Le Français de Lalande (b Bourg-en-Bresse, France, Jul 11, 1732) lists the positions of 47,000 stars

Giuseppe Piazzi discovers the first asteroid, Ceres, at Palermo (Italy), Jan 1; from data obtained in a brief 41 days of observation, Karl Friederich Gauss calculates the orbit of the new planet, a feat he will repeat for the next three asteroids to be discovered, Pallas, Juno, and Vesta

BIOLOGY

Jean-Baptiste Lamarck's *Système des animaux sans vertèbres* (System for animals without vertebras) includes a classification system for invertebrates and a preliminary view of his idea of evolution

Zoologist Félix Dujardin b Tours, Indre-et-Loire, France, Apr 5; he is the first to realize that protists do not have organ systems as vertebrates and other multicellular organisms do

CHEMISTRY

John Dalton formulates his law of partial pressures for gases, which states that each component of a mixture of gases in a given region produces the same pressure as if it occupied the region by itself

Andrès del Rio (b Madrid, Spain, Nov 10, 1764) discovers the element vanadium, although he fails to convince other chemists at the time

Charles Hatchett (b London, Jan 2, 1765) discovers niobium, which he names columbium; later William Hyde Wollaston disputes the identity of the metal, claiming that Hatchett has really rediscovered tantalum; Hatchett is vindicated, but his name is lost and the metal is named for Niobe, a daughter of Tantalus

Jean d'Arcet d Paris, Feb 12

1802

GENERAL

On a bet with some drinking companions, Georg Friedrich Grotefend becomes the first to translate a cuneiform text, a Persian inscription from Persepolis; credit for such translations is usually given to Henry Creswicke Rawlinson, whose translation of old Persian cuneiform is published in 1846

ASTRONOMY

William Herschel publishes his third list of nebulas; he also establishes the existence of binary star systems

Wilhelm Olbers discovers Pallas, the second asteroid

BIOLOGY

Jean-Baptiste Lamarck and Trevirons simultaneously coin the term *biology*

Lamarck starts publishing *Mémoire sur les fossiles des environs de Paris* (On the fossils in the surroundings of Paris), completed in 1806

CHEMISTRY

Joseph-Louis Gay-Lussac (b Haute Vienne, France, Dec 6, 1778) shows that all gases at a given pressure increase by the same percentage in volume when subjected to the same increase in temperature

Anders Gustaf Ekeberg (b Stockholm, Sweden, Jan 15, 1767) discovers the element tantalum

1800 cont

Benjamin Waterhouse (b Newport, RI, Mar 4, 1754) is the first US physician to use smallpox vaccine (on his son)

Humphry Davy (b Penzance, England, Dec 17, 1778) discovers nitrous oxide ("laughing gas") on Apr 9 and suggests its use as an anesthetic

the spectrum to heat a thermometer, he finds that invisible light beyond the red produces the most heat

Alessandro Volta announces his invention, made in 1799, of the electric battery, also known as the Voltaic pile; it consists of a stack of alternating zinc and silver disks separated by felt soaked in brine; it is the first source of a steady electric current

1801

Geologist Dieudonné de Gratet de Dolomieu d Château-neuf, France, Nov 28

Gauss proves that every natural number is the sum of at most three triangular numbers, a conjecture first made by Pierre de Fermat in 1636

Gauss's *Disquisitiones arithmeticae* (Arithmetical researches) greatly extends number theory and introduces his concept of congruences, numbers that have the same remainder when divided by a particular number—for example, 17 and 23 are congruent modulo (with respect to) 3

Marie-François-Xavier Bichat's *Anatomie générale appliquée à la physiologie et à la médecine* (General anatomy applied to physiology and medicine) investigates and names the tissues of the body and stresses the study of the different tissues making up organs

Thomas Young (b Milverton, England, Jun 13, 1773) discovers the cause of astigmatism

German physicist Johann Ritter (b Samitz, Dec 16, 1776), working with silver chloride, finds that radiation beyond the spectrum of visible light—ultraviolet radiation—exists

Thomas Young rediscovers interference phenomena of light: light passing through two narrow slits produces an interference pattern, proving light is a wave phenomenon (previously discovered in the seventeenth century by Francesco Grimaldi, but not taken seriously at the time)

James Finney of Pennsylvania builds the first modern suspension bridge

Robert Hare (b Philadelphia, PA, Jan 17, 1781) invents the hydrogen-oxygen blowpipe, the ancestor of modern-day welding torches

Joseph-Marie Jacquard (b France, Jul 7, 1752) develops an early version of his loom in which each decision about the design still must be carried out by a human operator

Richard Trevithick builds a full-scale steam-powered carriage, completed on Dec 24; it runs well for four days, but burns up when he carelessly allows all the water in the boiler to evaporate

1802

Franz Ulrich Theodosius Aepinus, German mathematician and physicist, d Dorpat, Estonia (USSR), Aug 10

German physician Franz Joseph Gall (b Tiefenbrunn, Mar 9, 1758) is required to stop lecturing about his research in phrenology by Austrian emperor Francis I

Thomas Young's *On the theory of light and colors* is the first of three pivotal papers describing his wave theory of light

The chemist Archard founds the first factory for manufacturing beet sugar

Nathaniel Bowditch's *New American practical navigator* sets the standards for navigation of sailing ships

	GENERAL	ASTRONOMY	BIOLOGY	CHEMISTRY
1802 cont		William Hyde Wollaston (b East Dereham, England, Aug 6, 1766) observes dark lines in the solar spectrum, but he fails to follow through on their significance; Joseph von Fraunhofer, who observes them in 1814, is the first to realize they are important	Physician Erasmus Darwin d Breadsall Priory, near Derby, England, Apr 18	*System of chemistry* (first edition) by Thomas Thomson (b Scotland, 1773) introduces a symbolic representation of minerals Gerardus Johannes Mulder b Utrecht, Holland, Dec 27; he studies proteins and coins the word, but fails to appreciate their complexity
1803	Thomas Malthus, in his second edition of *Essay on the principles of population*, suggests that "moral restraint" might prevent famine from overpopulation	Jean-Baptiste Biot (b Paris, Apr 21, 1774) makes a study of meteorites found at L'Aigle, France, and determines that they did not originate on Earth	John C. Otto conducts the first banding studies on wild American birds	Claude-Louis Berthollet's *Essai de statique chimique* (Essay on static chemistry) shows that reaction rates not only depend on affinities but also on the amounts of reacting substances; it is the first systematic attempt to deal with the physics of chemistry Jöns Jakob Berzelius (b Sörgård, Sweden, Aug 20, 1779), Wilhelm Hisinger (b Skinnskatteberg, Västmanland, Sweden, Dec 23, 1766), and (separately) Martin Heinrich Klaproth discover the element cerium English chemist John Dalton puts forward his atomic theory of matter, stating that since chemicals combine only in integral proportions, atoms must exist William Henry (b Manchester, England, Dec 12, 1775) formulates the law named after him that states that the mass of gas dissolved in a liquid is directly proportional to the pressure Smithson Tennant (b Selby, England, Nov 30, 1761) discovers the elements iridium and osmium English chemist William Hyde Wollaston discovers the elements palladium and rhodium; palladium is named after the asteroid Pallas

Physician Marie-
François-Xavier Bichat
d Lyon, France, Jul 22

John C. Stevens builds
a propeller-driven
steamboat

**1802
cont**

Lazare Carnot's
Géométrie de position
contains generaliza-
tions of well-known
theorems of geometry
as well as the
development of
various coordinate
systems, including in-
trinsic coordinates—
systems that are not
bound to any particular
choice of axes

Robert Fulton's steam-
boat makes a suc-
cessful trip on the
Seine in France

William Murdock
lights his main factory
with coal gas, the first
building to be routine-
ly lighted in this way;
Murdock had been
the first to realize (in
1792) that coal gas
might be used for
light

Inventor Aimé Argand
d Geneva, Oct 24

1803

1804

GENERAL

Napoleon is crowned emperor of France

Philosopher Immanuel Kant d Königsberg (Kalingrad, USSR), Feb 12

ASTRONOMY

German astronomer Friedrich Wilhelm Bessel (b Minden, Jul 22, 1784) calculates the orbit of Halley's comet from 200-year-old observations by Thomas Harriot

Karl Ludwig Harding discovers the third asteroid, Juno

Wilhelm Olbers discovers the fourth asteroid, Vesta

BIOLOGY

Nicholas de Saussure's *Recherches chimiques sur la végétation* (Chemical researches on vegetation) shows that plants require carbon dioxide from the air and nitrogen from the soil; earlier researchers had assumed that plants got carbon from the soil instead of from carbon dioxide in the air

A.D. Thaer's *Grundsätze der rationellen landwirtschaft* (Foundations of rational agriculture) introduces the concept of rotation of crops

Botanist Antonio José Cavanilles d Madrid, Spain, May 4

CHEMISTRY

Hans Christiaan Oersted (b Rudkøbing, Denmark, Aug 14, 1777) proposes a theory of chemical forces based on combustion and compares these forces to electrical phenomena

English chemist Joseph Priestley d Northumberland, PA, Feb 6

Antoine Baumé d Paris, Oct 15

1805

GENERAL

Napoleon approves the foundation of the *Société d'Arcueil;* members include Claude Berthollet, Pierre-Simon Laplace, François Arago, Jean-Baptiste Biot, Pierre-Louis Dulong, Louis-Joseph Gay-Lussac, and Alexander von Humboldt

German anthropologist Johann Friedrich Blumenbach (b Gotha, May 11, 1752) founds the science of anthropology with his *Handbuch der vergleichenden Anatomie*

BIOLOGY

Georges Cuvier's *Leçons d'anatomie comparée* (Lessons on comparative anatomy) founds the science of comparative anatomy

Pierre André Latreille publishes his *Comprehensive natural history of crustaceans and insects*

German chemist Friedrich Sertürner (b Neuhaus, Jun 19, 1783) isolates morphine from opium

CHEMISTRY

Gay-Lussac proves that water consists of two parts hydrogen and one part oxygen by volume, and that other gases also combine to form compounds in ratios that are also integral

Theodor von Grotthuss announces his theory of electrolysis in which the negative pole attracts hydrogen atoms by an electric field

EARTH SCIENCE	MATHEMATICS	MEDICINE	PHYSICS	TECHNOLOGY	
Jean-Baptiste Biot and Joseph Gay-Lussac ascend in a balloon to study terrestrial magnetism and Earth's atmosphere	Mathematician Carl Gustav Jacob Jacobi b Potsdam (Germany), Dec 10; he does significant work with elliptic functions	Austrian pathologist Karl Rokitansky b Bohemia (Czechoslovakia), Feb 19; dissecting more than 30,000 cadavers during his career, he is the first to recognize many pathological changes caused by disease and the first to discover bacteria as the cause of malignant endocarditis	John Leslie's *An experimental inquiry into the nature and propagation of heat* establishes that the transmission of heat through radiation has the same properties as the propagation of light	Nicolas Appert invents food canning as a means of preparation and opens the world's first canning factory; he also invents the bouillon cube Frederick Winsor (b Brunswick, England, 1763) patents an oven for the manufacture of coal gas Richard Trevithick develops a steam locomotive that runs on iron rails and successfully hauls 10 tons of iron 16 km (10 mi) Engineer Nicholas-Joseph Cugnot d Paris, Oct 2 Pioneer of aerial navigation George Cayley (b Scarborough, Yorkshire, England, Dec 27, 1773) develops his first instrument to measure air resistance on Dec 1; during this year he also begins to build a series of gliders that will determine the basic principles of aerodynamics Phillipe Lebon d Dec 2	1804
Meteorologist Robert Fitzroy b Ampton Hall, Suffolk, England, Jul 5; he is best known as the captain of the *Beagle*, which carried Charles Darwin as its chief scientist on its scientific voyage around the world		Franz Joseph Gall's *Über die verrichtungen des gehirns* (On the activities of the brain), while correct in showing that different parts of the brain have different functions, incorrectly states that the brain can be studied by examining the shape of a person's skull—through phrenology, as it came to be known	Pierre-Simon Laplace measures molecular forces in liquids and announces his theory of capillary forces	Joseph-Marie Jacquard develops a method of controlling the operation of a loom based on punched cards, an idea that will later be used in early computers Inventor Claude Chappe d Paris, Jan 23	1805

1806

Nicolas Clément (b France, 1779) and Charles Désormes show how saltpeter acts as a catalyst in producing sulfuric acid by facilitating the transfer of oxygen from the air to burning sulfur

Louis-Nicolas Vauquelin and Robiquet isolate asparagine, the first amino acid to be discovered; Vauquelin and Valentin Rose isolate quinic acid

Chemist Nicolas Leblanc d by his own hand at Saint-Denis (near Paris), Jan 16; the French revolutionary government had forced him to make his Leblanc chemical process public property and he was reduced to poverty

1807

Archaeologist John Frere d East Dereham, Norfolk, England, Jul 12

Nathaniel Bowditch determines that the Connecticut meteor of 1807 once weighed 6 million tons

Astronomer Joseph-Jérôme Le Français de Lalande d Paris, Apr 4

Lorenz Oken, German naturalist, (b Bohlsbach, Baden, Aug 1, 1779) elaborates on Johann Wolfgang Goethe's theory that the skull in vertebrates evolved from enlargement and fusion of vertebrae (proved incorrect in 1858 by Thomas Huxley)

Jöns Jakob Berzelius classifies chemicals as either organic or inorganic

Chemistry applied to the arts by Jean-Antoine-Claude Chaptal, comte de Chanteloup (b Nogaret, France, Jun 4, 1756) is the first book on industrial chemistry

Thomas Thomson's *System of chemistry* (third edition) contains Dalton's theory of atoms explained for the first time

Jeremias Benjamin Richter d Berlin (Germany), Apr 4

Humphry Davy discovers the element potassium by electrolysis of potash on October 6 and the element sodium by electrolysis of soda on October 13

				1806
	Essay on a method of representing imaginary quantities in geometric constructions by Jean-Robert Argand (b Geneva, Switzerland, Jul 18, 1768) uses complex numbers to show that all algebraic equations have roots, but the proof is not rigorous; Argand also introduces the Argand diagram representation of complex numbers as points in the coordinate plane		Charles-Augustin Coulomb d Paris, Aug 23	Joseph Bramah invents a numerical printing machine for bank notes

Adrien-Marie Legendre's *Nouvelles méthodes pour la détermination des orbites des comètes* (New method for finding comets' orbits) contains the first printed description of the method of least squares, a technique for finding the best approximation to observed data, discovered about ten years earlier by Gauss, but not published; Legendre does not offer a proof that the method works, but one is supplied by Gauss in 1809

				1807
The Geological Society of London becomes the first scientific society devoted to the newly named science of geology			Hans Christian Oersted announces that he is looking for a connection between electricity and magnetism	London streets begin to be illuminated by the coal-gas lighting invented by William Murdock

Swiss-American geographer Arnold Henry Guyot b Boudevilliers, Switzerland, Sep 28; he is known primarily because flat-topped undersea mountains (guyots) were named after him in 1946

Thomas Young introduces the concept of and is the first to use the word *energy*

Ferdinand Berthoud, Swiss inventor, d Jun 20

Robert Fulton's *North River Steam Boat* is tested in the East River off New York City on Aug 9; later renamed the *Clermont*, it makes its maiden voyage from Manhattan to Clermont, NY on Aug 17; although not the first steamboat, it is the first practical and economical one

1808

Siméon-Denis Poisson (b Pithiviers, France, Jun 21, 1781) publishes his work on perturbations of planetary orbits

Humphry Davy discovers the elements barium, strontium, and calcium and infers the existence of magnesium, although pure magnesium will not be isolated until 1829; he is also a codiscoverer of boron with Joseph-Louis Gay-Lussac and Louis-Jacques Thénard

Joseph-Louis Proust identifies three sugars in plant juices: glucose, fructose, and sucrose

Gay-Lussac announces on December 31 that gases combine chemically in definite proportions of volumes; previously it was only known that elements combine in definite proportions of masses (weights)

1809

Jean-Baptiste Lamarck's *Philosophie zoologique* states his view that animals evolved from simpler forms; the mechanisms he proposes for evolution are vague and, where specific, later disproved on the basis of experiment; for example, he supports the discredited inheritance of acquired characteristics theory

Chemist Antoine-François, comte de Fourcroy, d Paris, Dec 16

1810

The University of Berlin is founded; it is the first university in which the goal of research is more important than education; most teaching is done in large lectures or in small seminars of advanced students; Berlin and universities modeled on it will put Germany ahead in science during the nineteenth century

William Hyde Wollaston investigates Francesco Grimaldi's seventeenth-century discovery of sounds made by muscles and concludes that human muscles have a frequency of about 23 beats per second; his investigation involves running a carriage over London streets until the pitch of its rumble is the same as that of the muscles

Augustin Fresnel (b Broglie, France, May 10, 1788) develops a method of making soda using limestone and common salt

Physicist and chemist Henry Cavendish d London, Feb 24

EARTH SCIENCE	MATHEMATICS	MEDICINE	PHYSICS	TECHNOLOGY	
		Franz Joseph Gall, founder of phrenology, presents his theory that bumps on the head indicate character	Etienne-Louis Malus (b Paris, Jun 23, 1775) discovers that reflected light is polarized and introduces the term *polarization*	Humphry Davy develops the first electric-powered lamp, the arc light; while it produces a great deal of smoke and heat, it also produces a brilliant bright light so long as the two carbon rods can be kept the proper distance apart Richard Trevithick builds a circular passenger railway in London Inventor John Wilkinson d Jul 14	**1808**
William Maclure publishes the first geological survey of the eastern United States	Gauss's *Theoria motus corporum coelestium in sectionibus conicus solem ambietium* (Theory of the motion of the heavenly bodies about the sun in conic sections) includes Gauss's method of least squares for minimizing errors in calculations	Physician Leopold Auenbrugger d Vienna, Austria, May 18	Dominique-François Arago (b Estagel, France, Feb 26, 1786) rediscovers that the blue sky is polarized; he finds the neutral point (where polarization is absent), which is named after him Ernst Chladni introduces sand figures (Chladni figures), obtained by sprinkling sand on vibrating metal sheets, to demonstrate the sound vibrations of solid objects	The first Western suspension bridge capable of carrying vehicles, with a span of 74 m (244 ft), is built across the Merrimac River in Massachusetts Aeronaut Jean-Pierre-François Blanchard d Paris, Mar 7 Industrialist Matthew Boulton d Birmingham, England, Aug 18	**1809**
	Jean-Baptiste Joseph Fourier (b Auxerre, France, Mar 21, 1768) presents his results on representing functions by infinite trigonometric series to the French Academy Joseph-Diaz Gergonne starts the *Annales de Mathématiques Pures et Appliquées,* the first privately established mathematics journal	Samuel Friedrich Hahnemann's *Organon of rational healing* introduces homeopathy	Johann Wolfgang von Goethe's *Zur Farbenlehre* (On the theory of color) rejects Newton's theory that white light is a mixture of colors and contains a theory of color that is psychologically oriented and is a revival of Aristotle's view Johann Wilhelm Ritter d Munich, Bavaria (Germany), Jan 23	Nicolas Appert's *Le livre de tous les ménages, ou l'art de conserver pendant plusieurs années toutes les substances animales et végétales* (Art of saving animal and vegetable substances for several years) explains heat sterilization of food Inventor Joseph-Michel Montgolfier d Balaruc-les-Bains, France, Jun 26	**1810**

1811

The University of Breslau (Wroclaw, Poland) is founded on the model of the University of Berlin

Navigator Louis Antoine de Bougainville d Paris, Aug 31

William Herschel develops his theory of development of stars from nebulas, occasioned by his Nov 13, 1790, observation of a planetary nebula, NGC 1514, with a star embedded in it; he believes that clouds of gas collapse into cluster of stars, which themselves collapse, creating an explosion that results in clouds of gas

Astronomer Nevil Maskelyne d Greenwich, England, Feb 9

Amedeo Avogadro, comte de Quarenga, proposes that equal volumes of any gas at any given temperature and pressure always contain equal numbers of molecules (Avogadro's law)

Jöns Jakob Berzelius introduces a system of chemical symbols that forms the basis of the system used today

Bernard Courtois (b Dijon, France, Feb 8, 1777) discovers the element iodine in seaweed, but does not recognize it as a new element; it is so recognized by Humphry Davy in 1814

German-Russian chemist Gottlieb Sigismund Constantin Kirchhoff, b Teterow (Germany), Feb 19, 1764, prepares glucose by heating starch with sulfuric acid

Louis-Nicolas Vauquelin discovers uric acid in the excrement of birds

1812

Jöns Jakob Berzelius's *Versuch über die Theorie der chemischen Proportionen und über die chemische Wirkung der Elektrizität* (Theory of chemical proportions and the chemical action of electricity) assumes that electrical and chemical forces are identical and atoms have electrical charges

Humphry Davy's *Elements of chemical philosophy* is published

Richard Kirwan d Dublin, Ireland, Jun 22

Understanding fossils (part two)

Little progress in understanding fossils took place during the eighteenth century, although comte de Buffon recognized that different groups of fossilized organisms appeared at different times in the past. Thomas Jefferson, who collected fossils, firmly believed that they were the remains of living organisms, but he thought that any organisms that he did not recognize were alive somewhere on Earth. He did not believe the Creator would permit extinction.

In the nineteenth century, geologists began to understand that Earth had lasted long enough for many forms of life to have arisen and become extinct. Georges Cuvier in 1811 showed that fossil bones previously ascribed to human victims of Noah's flood were actually the remains of a giant salamander. Cuvier also recognized various other fossils, including marine reptiles and flying reptiles. In 1825 he put forth three hypotheses about the origins of such fossils, each of which had merit, although he considered them to be mutually exclusive.

The first was the same as that of Thomas Jefferson: these creatures are still alive in some unexplored part of Earth. As it happens, every so often such a "living fossil" is found. An example is the coelacanth, a fossil fish believed to have been extinct for 65 million years when it was found alive in 1938. The second hypothesis was that organisms change shape over the course of time. This is a version of the theory of evolution. Finally, he suggested that some organisms may be extinct, which is the actual case for most organisms found as fossils. Cuvier rejected his first two hypotheses in favor of the third. He believed that from time to time great catastrophes caused all life to become extinct. Following each catastrophe, he alleged, the Creator put a new set of organisms on Earth.

At the same time as Cuvier's book on the subject was published, Gideon Mantell recognized some fossil teeth as belonging to a giant reptile. Other fossils with similar characteristics were found. In 1842 Richard Owen coined the word *dinosaur* to describe these common reptiles. In 1854, Owen prepared an exhibition of dinosaurs for display at the Crystal Palace. Dinosaurs excited the public enormously.

Darwin's theory of evolution, which became known a few years later, finally gave the philosophical basis for fossils that we still use today.

A series of earth-quakes centered around New Madrid, MO, continuing into 1812, devastates much of the poorly settled American Midwest; estimated at about 8 on the Richter scale, these earthquakes are considered the most powerful in US history because of the wide region affected

The *Transactions of the Geological Society of London* start publication

Twelve-year-old Mary Anning discovers the 10-m-long (33-ft-long) fossil of an ichthyo-saur, the first ichthyosaur fossil known

Georges Cuvier and Alexandre Brogniart publish a geological map of the region around Paris

Charles Bell's *New anatomy of the brain* discusses the difference between sensory and motor nerves and contains the law named after him: the anterior spinal roots are motor and the posterior roots are sensory

Samuel Hahnemann publishes a catalog of homeopathic drugs

François Arago discovers the optical activity of quartz

Siméon-Denis Poisson develops a mathematical theory of heat based on the work of Jean-Baptiste Fourier

Georges Cuvier completes *Recherches sur les ossements fossiles des quadrupèdes* (Researches on the fossil bones of quadrupeds) in four volumes; he explains his theory of the extinctions of animal groups in catastrophes and founds comparative vertebrate paleontology

Cuvier identifies from a drawing of the fossil the first known ptero-saur, the pterodactyl

Pierre-Simon Laplace's *Théorie analytique des probabilités* (Analytic probability theory) contains a complete account of probability theory, but little further work on the subject is done for three-quarters of a century; one factor for this neglect is that Laplace applies the theory in inappropriate contexts

Benjamin Rush's *Medical inquiries and observations upon the diseases of the mind* contains one of the first modern attempts to explain mental disorders

François Arago builds a polarization filter from a pile of glass sheets

Sir David Brewster (b Jedburgh, Scotland, Dec 11, 1781) discovers the law named after him; it gives the relationship between the index of refraction and the angle of incidence at which light reflected becomes completely polarized

William Hyde Wollaston invents the camera lucida, a device for projecting an image onto a flat surface, such as drawing paper, on which the object can then be traced

Etienne-Louis Malus d Paris, Feb 24

The London Gas Light & Coke Company is formed, leading to the lighting of the city of London by coal gas over the next quarter century

	GENERAL	ASTRONOMY	BIOLOGY	CHEMISTRY
1813		Italian-French astronomer and mathematician Joseph-Louis, comte de Lagrange d Paris, Apr 10	Swiss-French botanist Augustin de Candolle (b Geneva, Switzerland, Feb 4, 1778) introduces the word *taxonomy* in his lifelong project of a 21-volume plant encyclopedia; seven volumes are published during his lifetime, while the remainder are seen through the press by his son	Humphry Davy's *Elements of agricultural chemistry* is published Anders Gustaf Ekeberg d Uppsala, Sweden, Feb 11 Chemist Jean Servais Stas b Louvain, Belgium, Aug 21; he devotes most of his professional career to establishing atomic masses, using 16 for oxygen as a reference point; this practice continues well into the twentieth century, when chemists return to basing atomic masses on carbon-12 Physical chemist Thomas Andrews b Belfast, Ireland, Dec 19; he identifies ozone as a form of oxygen and, independent of Mendeléev, discovers the critical point for liquefaction of gases
1814		German astronomer Joseph von Fraunhofer (b Straubling, Mar 6, 1787) makes his first detailed map of the solar spectrum, showing many lines		André-Marie Ampère suggests, as did Avogadro in 1811, that equal volumes of different gases contain equal numbers of particles Edmond Frémy b Versailles, France, Feb 28; he studies many of the compounds of fluorine, but is unable to isolate the element
1815	Napoleon loses the battle of Waterloo to the Duke of Wellington, Jun 16	Fraunhofer makes a more detailed map of the solar spectrum; it contains 324 lines	Jean-Baptiste Lamarck's *Histoire naturelle des animaux sans vertèbres* (Natural history of invertebrates) contains a number of new elements in animal classification; the work will be completed in 1822 Naturalist Edward Forbes b Douglas, Isle of Man, England, Feb 12; he establishes that living organisms exist deep in the sea, well below the distance that light will penetrate	English chemist Smithson Tennant d Boulogne, France, Feb 22 William Nicholson d London, May 21 Michel Chevreul (b Angers, France, Aug 31, 1786) shows that the sugar in diabetics' urine is glucose, an important step toward understanding the disease

1813

EARTH SCIENCE

Robert Bakewell's *An introduction to geology* is one of the first and most influential textbooks on the subject

Mineralogist Peter Jacob Hjelm d Stockholm, Sweden, Oct 7

MEDICINE

Physician Benjamin Rush d Philadelphia, PA, Apr 19

TECHNOLOGY

Inventor Frederick Scott Archer b Bishop's Stortford, Hertfordshire, England; he develops the idea of using a negative and making prints in photography

1814

MATHEMATICS

Laplace's *"Essai philosophique sur les probabilités,"* which is the introduction to the second edition of *Théorie analytique des probabilités* contains his calculation of the mathematical expectation of sunrise based on the assumption that the sun has risen each day for the past 5000 years

PHYSICS

American-British physicist Benjamin Thompson, Count Rumford, d Auteuil (near Paris), Aug 21

TECHNOLOGY

The Times in London is printed by a steam-driven cylinder press

British engineer George Stephenson (b Wylam, England, Jun 9, 1781) introduces his first steam locomotive, capable of hauling 30 tons at speeds faster than possible with a horse-drawn system

Inventor Joseph Bramah d London, Dec 9

1815

EARTH SCIENCE

William Smith's *The geological map of England* is the first book to identify rock strata on the basis of fossils; it enables geologists far distant from each other to know that they are working with the same time period

Mount Tambora on Sumbawa (Indonesia) explodes on April 7, killing 12,000 people and sending enough dust in the air to lower temperatures for a year or so; it

MATHEMATICS

The Analytical Society, devoted to reforming British mathematics along Continental lines, begins meeting at Cambridge

MEDICINE

Physician Franz Anton Mesmer d Meersburg, Germany, Mar 5

Physician Carl Reinhold Wunderlich b Sulz, Germany, Aug 4; he is the first to realize that taking accurate temperatures of patients with a thermometer is useful

PHYSICS

Jean-Baptiste Biot discovers the circular polarization of light; he also discovers the optical activity (rotation of polarization plane) in fluids, such as turpentine, and the strong dichroism in tourmaline (different colors seen from two different axes of the crystal)

Augustin-Louis, Baron Cauchy (b Paris, Aug 21, 1789) presents his formulas for describing turbulence to the French Academy

TECHNOLOGY

The first warship powered by steam is built in the United States

Humphry Davy invents the safety lamp to be used in coal mines

Scottish engineer John Loudon McAdam (b Ayr, Scotland, Sep 21, 1756) originates paving roads with crushed rock; although the name macadam is in his honor, he does not use tar or asphalt as

1815 cont

Chemistry: William Prout (b Gloucestershire, England, Jan 15, 1785) claims that specific gravities of atomic species are integral multiples of the value for hydrogen, suggesting that the known chemical elements were formed from hydrogen atoms

1816

Astronomy: Lewis Morris Rutherfurd b New York, NY, Nov 25; he develops telescopes that are dedicated to taking photographs and rules the best diffraction gratings of the mid-nineteenth century

Biology: Mystical German naturalist Lorenz Oken founds the influential journal *Isis*

Botanist Christian Konrad Sprengel d Berlin, Apr 7

Chemistry: Humphry Davy discovers the catalytic action of platinum and other metals on reactions of organic gases with oxygen

Andrew Ure (b Scotland, May 18, 1778) invents the alkalimeter

Chemist Baron Louis Bernard Guyton de Morveau d Paris, Jan 2

Thomas Henry d Manchester, England, Jun 18

Charles Frédéric Gerhardt b Aug 21; he revives the theory of acid radicals and contributes to the developing concept of atomic masses

1817

Astronomy: The first of Jean-Baptiste Delambre's six-volume *Histoire de l'astronomie* (History of astronomy) begins his full-scale technical history of astronomy

Charles Messier d Paris, Apr 11

Biology: Georges Cuvier's *Le règne animal distribué d'après son organisation* (The animal kingdom, distributed according to its organization) gives an account of the whole animal kingdom, dividing it into four groups

Russian zoologist Christian Pander (b Riga, Jul 24, 1794) discovers the three different layers that form in the early development of the chick embryo

Pierre Pelletier (b Paris, Mar 22, 1788) and Joseph Bienaimé Caventou (b Saint-Omer, France, Jun 30, 1795) isolate chlorophyll

Chemistry: Johan August Arfwedson (b Sweden, 1792) discovers lithium, but fails to isolate it

German chemist Leopold Gmelin (b Göttingen, Aug 2, 1788) publishes the first edition of the three-volume *Handbook of chemistry*, the first systematic treatment of chemistry since Lavoisier's

German chemist Friederich Strohmeyer (b Göttingen, Aug 2, 1776) discovers the element cadmium

German chemist Martin Heinrich Klaproth d Berlin, Jan 1

produces the "year
without a summer" in
New England in 1816

Geologist Nicolas
Desmarest d Paris,
Sep 28

in modern macada-
mized roads

Robert Fulton,
American inventor, d
New York, NY, Feb 24

1815
cont

Parker Cleaveland's
*An elementary
treatise on mineralogy
and geology* is an ear-
ly and influential text-
book in the United
States

William Smith's *Strata
identified by orga-
nized fossils contain-
ing prints on colored
paper of the most
characteristic speci-
mens in each strata*
provides a list of
characteristic fossils
that can be used to
identify rock strata

Lord Byron's poem
"Darkness," written in
Jun, describes his
feelings in the "year
without a summer"
that follows the erup-
tion of Tambora in
1815

Théophile René
Laënnec (b Finistère,
France, Feb 17, 1781)
invents the
stethoscope

Physiologist François
Magendie (b
Bordeaux, France, Oct
6, 1783) publishes
Précis de physiologie

David Brewster in-
vents the
kaleidoscope

Augustin Fresnel
demonstrates with his
mirror experiment the
wave nature of light;
he also gives an ex-
planation of
polarization

Engineer Sir Marc
Isambard Brunel (b
Hacqueville, France,
Apr 25, 1769) invents
the round stocking
frame

Johann Maelzel
patents the
metronome

1816

Geologist Abraham
Gottlob Werner d
Dresden (Germany),
Jun 30

Mineralogist William
Gregor d Creed,
England, Jul 11

Bernhard Bolzano's
*Rein analytischer
Beweis* (Pure
analytical proof) con-
tains his proof of
what we now call the
Bolzano-Weierstrass
theorem, the essential
step in making
analysis provable by
turning it into
arithmetic; Bolzano
also defines a con-
tinuous function
without using
infinitesimals

A pandemic, or
epidemic that spreads
over a wide region, of
cholera begins in In-
dia, spreading to East
Africa and most of
Asia, including Japan
and the Philippines

*An essay on the shak-
ing palsy* by James
Parkinson (b London,
Apr 11, 1755) gives a
clinical description of
the disease that now
bears his name

Thomas Young and
Augustin Fresnel
demonstrate that light
waves must be
transverse vibrations

1817

1818

The University of Bonn (Germany) is founded on the model of the University of Berlin

Chemist Benjamin Silliman (b North Stratford, CT, Aug 8, 1779) founds the *American Journal of Science and Arts*

Friedrich Wilhelm Bessel's *Fundamenta astronomiae* contains a catalog of over 3000 stars

Jean-Louis Pons (b Peyres, France, Dec 24, 1761) discovers Encke's comet, the comet with the shortest period on record; it is named for Johann Encke, who calculated its orbit

Studying a comet observed this year by Jean-Louis Pons, German astronomer Johann Encke (b Hamburg, Sep 23, 1791) shows that it returns every 3.29 years, which it continues to do

Jöns Jakob Berzelius completes the discovery of lithium by separating out the pure element

Pierre-Louis Dulong (b Rouen, France, Feb 12, 1785) and Alexis Thérèse Petit (b Vesoul, France, Oct 2, 1791) formulate the law named after them: in solid elements the product of the specific heat and the atomic mass is a constant; the Dulong-Petit law can be used to compute atomic masses

Louis-Jacques Thénard (b La Louptière, France, May 4, 1777) accidentally discovers hydrogen peroxide

Louis-Nicolas Vauquelin isolates cyanic acid

August Wilhelm von Hofmann b Giessen, Germany, Apr 8; his dramatic lectures in England set William Perkin on the road to discovering mauve dye and other chemicals; after returning to Germany, Hofmann discovers the dyes called Hofmann violets and becomes a leader in the German dye industry

1819

François Arago discovers that light from comet tails is polarized

Naturalist Henri Braconnot (b Commercy, France, May 29, 1781) obtains glucose from sawdust, linen, and bark, showing that simple plant materials contain some substance built from subunits of glucose, previously obtained from starch; we now know that cellulose was the precursor of glucose

Chemist and physician John Kidd (b London, Sep 10, 1775) derives naphthalene from coal tar, the first of the many important substances to be found in or made from coal tar

German chemist Eilhardt Mitscherlich (b Oldenburg, Jan 7, 1794) formulates his law of isomorphism, which states that chemical composition is related to crystalline form

Daniel Rutherford d Edinburgh, Scotland, Nov 15

1818

Mineralogist Johann Gottlieb Gahn d Falun, Kopparburg, Sweden, Dec 8

Gaspard Monge d Paris, Jul 28

Jean-Baptiste Dumas (b Alais, France, Jul 14, 1800) treats goiter with iodine

Jean-Baptiste Biot discovers biaxial crystals

David Brewster discovers "Brewster's brush" in pleochroic crystals, those that show different colors in different forms of light

Augustin Fresnel's *Mémoire sur la diffraction de la lumière* (Memoir of the diffraction of light) reports his demonstration that light is a transverse wave phenomenon; Fresnel's wave theory of light is at first strongly opposed, especially by Jean-Baptiste Biot, but in due time it is accepted by the scientific community

In England, the Institution of Civil Engineers is founded

Canvass White, American engineer, discovers rock in Madison County, New York, that produces natural hydraulic cement

1819

John Farrar of Harvard University makes the first translation of Adrien-Marie Legendre's *Eléments de géométrie* into English

W.G. Horner (b 1786) develops Horner's rule for approximating the solution to equations; the same rule was known to the Chinese five centuries earlier

John Playfair d Edinburgh, Scotland, Jul 20

Traité de l'auscultation médiate (Treatise on diagnosis by listening to sounds) by physician Théophile René Laënnec treats the use of the stethoscope for investigating the lungs, heart, and liver

William Lawrence's *Lectures on physiology, zoology and the natural history of man* argues that the human species is unique, but consists of several races

Pierre-Louis Dulong and Alexis Petit develop a list of 12 atomic weights

Hans Christian Oersted accidentally discovers that a magnetized needle is deflected by a nearby electric current; he publishes his information in 1820

The paddle steamer *Savannah* becomes the first steamship to cross the Atlantic (in 27 days 100 hours), although it is propelled by its sails 87 percent of the trip

Practical essay on roads by engineer John McAdam describes his invention of road paving with crushed rock

Inventor Oliver Evans d Apr 15

Scottish engineer James Watt d Heathfield, near Birmingham, England, Aug 19

OVERVIEW

Nineteenth-century science

Scientists in the eighteenth century had become increasingly fascinated with electricity, especially as better means of generating and storing it became available. Their work culminated in the Voltaic pile, or battery, in 1800. Before this invention, electricity was only available for the brief moment of a discharge or as small amounts of static electricity. After 1800, scientists could work with an electric current. Even so, there was little progress toward understanding the nature of electricity until Hans Christian Oersted's accidental discovery, published in 1820, that electricity and magnetism are linked. As soon as word of this discovery spread, André-Marie Ampère, François Arago, and Michael Faraday began both to explain electricity and to try to put it to practical use. Indeed, Faraday had built the first prototype of the electric motor by 1821 and the first form of electric generator ten years later. It was not until late in the century, however, that practical applications became possible.

Much nineteenth-century science started with the discovery of electromagnetism, including Maxwell's mathematical laws that describe it (possibly the high point of nineteenth-century physical science), and from the relationship of electrolysis to chemistry. As the twentieth century dawned, the more developed nations were utilizing the technological results of this scientific effort: electric motors, electric lights, the telegraph and telephone, the radio, and many other milestones.

At the end of the century, work with cathode-ray tubes led directly to the discovery of X rays and the electron and indirectly to the discovery of natural radioactivity. The passage from electricity to electrons—or from Oersted and Volta in 1820 to Wilhelm Konrad Roentgen, J.J. Thomson, and Antoine Becquerel from 1895 to 1897—marks the true beginning of twentieth-century science. The new era begins with the surprises of 1895.

The nature of science

The nineteenth century was a period in which science and the teaching of science underwent a number of changes, giving it much of the form it has today. Science expanded enormously during the period and many of the new fields of science, such as anthropology, archaeology, cell biology, psychology, and organic chemistry originated during the first half of the century. Many other sciences, such as geology and chemistry, matured during the period.

During the eighteenth century, scientists who were not wealthy had to depend largely on patrons. There were no institutions that supported scientists, although there were a few teaching jobs and a little money could be made from books. This situation changed during the nineteenth century as the occupation of scientist became a paid profession. This development first took place in Germany, where in a few decades the universities developed into centers where science flourished. Justus von Liebig set up his research laboratory for chemistry at the University of Giessen, initiating a trend that was soon followed at the other universities. The teaching of science became linked to scientific research, a practice that later was followed by most of the universities in the world. The universities and scientific societies also started publishing scientific information, a development that was imitated in other countries. Scientific papers became important. In Germany, with Lorenz Oken taking the lead, scientists started meeting at national scientific congresses from 1822 on. First viewed with suspicion, these congresses later became instruments for increasing national unity in Germany.

Science continued to be truly international in this century. During the Napoleonic wars at the beginning of the century, scientists could freely travel between France and Britain. During the second half of the century, scientists started to travel to international conferences. The exchange of scientific information between nations increased considerably.

France was the leading scientific country during the first decades of the nineteenth century, mainly through the impetus of Napoleon and the influence of the Ecole Polytechnique, the breeding ground for much scientific talent. In Germany during the eighteenth century, science was much less developed than in France or England and was still strongly influenced by mystical concepts from the Middle Ages and the Renaissance. Many of the great German scientists had to leave Germany for their education: Friedrich Wöhler went to Sweden to study with Jöns Jakob Berzelius, and Liebig went to Paris to study with Joseph-Louis Gay-Lussac.

Wilhelm von Humboldt, director of the Prussian Department of Education and brother of Alexander von Humboldt, started the reorganization of higher education in Germany and founded the Berlin University in 1809. This university soon became the model of the other universities throughout Prussia. While the French universities were closed during the revolution and their replacements began to suffer from the French bureaucracy, teaching at German universities became more liberalized. Although the German universities were state controlled, the concept of academic freedom, the freedom of professors and students to choose their topics of study and research, became important and contributed to the strong development of science in Germany. The German term for science, *Wissenschaft*, is a broad term that includes all systematic knowledge, such as philosophy, philology, and history. This broad view encouraged an exchange among different scientific disciplines and also the development of entirely new fields, such as psychology.

National differences in style of research

During the nineteenth century differences in style of research between countries became fully apparent. In Germany, where science was strongly organized, the university system favored pure science; in England scientists traditionally dealt with practical problems.

The research style in England was similar to that of the eighteenth century and the number of scientists stayed relatively small. Science was the domain of a few talented individuals. Many scientists had no academic positions. Training in science was generally lacking in England, although a large number of amateur associations of scientists existed. Several technical schools ("mechanic's institutes") were founded around the middle of the century. In some of them science teaching was more advanced than at the universities of Oxford or Cambridge, hampered by

their connection to the Church of England, which required (in theory at least) that instructors be clergymen and that dissections not be practiced.

In England, the Analytical Society began a trend toward reviving mathematics that culminated in the start of abstract algebra before the end of the century. Started by undergraduates at Cambridge, the society was committed to bringing Continental mathematics to England. They recognized that almost no one in England could read or understand Continental mathematics, in part because the symbolism used was not the same.

In Scotland the situation was somewhat different. The University of Edinburgh was not connected to the church; consequently, it was free to develop a medical school that became central to the scientific enterprise throughout Great Britain (see "The Lunar Society," p 216). Aspiring to a national identity, the Scottish universities developed quickly and established a tradition for learning that is still strong today. The Scottish universities attracted large numbers of students studying chemistry, but as professors' income were based on the number of students attending courses, the transition to practical classes and scientific research was never made.

Interest in science grew in England throughout the century. By the end of the century, there were about a hundred times as many people enrolled in the various English scientific societies as there had been in the single Royal Society of the preceding century.

In the United States, relatively little attention was given to science. The population in the United States grew steadily because of immigration, but the interests of the immigrants were largely oriented toward practical ventures rather than the pursuit of theoretical knowledge. The country excelled, however, in technology. Labor-saving devices were especially valued because of the comparatively low population density and the resulting high salaries for manual workers. Therefore the role of technological inventors and entrepreneurs in industrial development was large: many major US industries were founded by inventors such as Alexander Graham Bell, George Westinghouse, Thomas Edison, and George Eastman. In 1848 the

Although the ancient Greeks and probably earlier peoples knew about magnetism and static electricity, not much was accomplished with these interesting phenomena until the Chinese began to use the magnetic compass for navigation around 1000 AD. Some of the basic laws of magnets were written down in 1269 and William Gilbert studied both static electricity and magnetism at the end of the sixteenth century. Then the subject was ignored by most scientists for a long time.

In the eighteenth century, Stephen Grey and Charles François Du Fay revived the study of static electricity, but it was not until Pieter van Musschenbroek received the first known electric shock (apart from those caused by lightning) in 1745 that the subject gained attention. Van Musschenbroek was one of the inventors of the Leiden jar, a device for storing static electricity. A Leiden jar releases all of its electricity at once, causing a shock if it passes through a grounded person. Van Musschenbroek reported, "In a word, I thought it was all up with me" after his shocking experience. The discovery that electricity could cause shocks prompted Benjamin Franklin's famous kite experiment, in which he showed that lightning is electricity. It could have been "all up with" Franklin during that experiment too. Other scientists who repeated the experiment were killed.

Charles-Augustin Coulomb made careful studies of the forces exerted by both static electricity and magnetism in the 1780s. As early as 1733, Du Fay had discovered that there are two types of charge and that like charges repel while unlike charges attract. Coulomb showed that both magnets and electric charges obey the same rule, an inverse-square law. This implied a connection between electricity and magnetism. In 1807 Hans Christian Oersted announced that he would search for that connection.

Somewhat before this, there had been a major development in electricity that would facilitate Oersted's search. In 1791, while studying the reactions of muscle to electricity, Luigi Galvani accidentally discovered that the muscles behaved as if a current was passing through them when they were placed in contact with brass and iron at the same time. Although Galvani thought the muscles were the sources of the current, when Alessandro Volta learned of this phenomenon, he found that a simple chemical solution could be substituted for the muscles. With this knowledge, he cleverly constructed the first electric battery at the end of the eighteenth century. The battery was the first source of current electricity, electricity that moves steadily through a conductor as opposed to being quickly released from a Leiden jar.

The history of science is replete with stories of someone finding something by luck or accident. What is often overlooked in these tales is that the scientist in question had already been looking for something like the phenomenon he or she "accidentally" discovered. If Galvani had not been studying electric actions on muscle, he would not have been prepared to notice the reaction of the muscle with the metals. Similarly, Oersted discovered the connection he sought by accident while he was performing a classroom demonstration. He placed a compass over a wire carrying a current. As he did, the needle suddenly moved to become perpendicular to the current. Electricity and magnetism really were connected.

As soon as Oersted's discovery became known in 1820, there was a burst of creative activity. In 1820 alone, André-Marie Ampére and Dominique Arago discovered that wires carrying currents attract or repel each other and that an electric current in a wire will attract iron, just as a magnet does. Three years later the thermoelectric effect was discovered by Thomas Seebeck. Also at that time, William Sturgeon built the first electromagnet.

A notable series of investigations of the relationship between electricity and magnetism was conducted almost in parallel in England by Michael Faraday and in America by Joseph Henry. Both Faraday and Henry discovered the principle of the dynamo in 1830–1831, for example. Although they independently discovered many of the same connections and devices, Faraday's work was to have the greater theoretical impact while Henry's had more immediately practical application.

Faraday had observed how iron filings form patterns of lines under the influence of a magnet and similar phenomena. He concluded that space is filled with these invisible lines, forming what we now call a field. He used his idea to explain in 1845 why some substances are diamagnetic (developing a magnetic field opposite to the one surrounding them) and others paramagnetic (developing a magnetic field parallel to the one surrounding them). When Faraday discovered that a magnetic field can affect the polarization of light, he proposed in 1846 that light may be waves in the lines of force of electromagnetism.

There were other clues to a relationship between light and electromagnetism. In 1857, for example, Gustav Robert Kirchhoff calculated the relationship between the forces of static electricity and magnetism, finding that the speed of light in a vacuum is a constant in his formula. In 1864, James Clerk Maxwell followed up Faraday's ideas with mathematical formulas that described light as waves of electromagnetism and that implied other forms of electromagnetic waves. Maxwell's work was experimentally verified in 1888 when Heinrich Hertz, following a suggestion from George Francis Fitzgerald, discovered radio waves by directly applying Maxwell's formulas.

In America, Henry developed the first practical electric motor and powerful electromagnets in 1831. In the 1830s he also developed the electrical relay that made the first practical telegraph possible. Samuel Finley Breese Morse, usually given credit for the telegraph, worked directly with Henry and incorporated his ideas into the new device.

In 1879 Thomas Alva Edison and Joseph Swan independently invented practical lights that use electricity. With this invention, electricity moved into the home, where the many discoveries of the nineteenth century soon found application.

American Association for the Advancement of Science was founded; it played an important role in the development of American science. But theoretical science remained in a secondary role in American education until well into the twentieth century.

The philosophical basis of nineteenth-century science

During the beginning of the nineteenth century, especially in Germany, there was a romantic reaction against the mechanistic and materialistic philosophy behind scientific development. The influence of Hegel, whose philosophy of nature was not based on experimentation but on *a priori* concepts, remained strong throughout the first half of the century in Germany. Scientists such as the great German poet Johann Wolfgang von Goethe tried to find scientific explanations by using general philosophical principles. For example, Goethe fought Isaac Newton's idea of white light as a mixture of colors, claiming that in principle white light should be simpler and purer in composition than colored light.

The idea that science would ultimately explain all phenomena in nature became stronger. The *Naturphilosophie* eventually lost influence and was replaced by materialism: *Kraft und Stoff* (force and matter) became the tenet of the new philosophical outlook. Matter and force, and later, matter and motion, were viewed as the ultimate explanations of reality.

The French philosopher Auguste Comte became the strongest spokesman of the philosophical school termed "positivism." He viewed science as the most advanced stage of knowledge. In positivism the use of scientific principles to explain the laws governing all phenomena supersedes the two earlier stages of knowledge: a theological or imaginative stage in which phenomena are explained by divine powers, and a metaphysical or abstract stage in which phenomena are explained by general philosophical ideas.

Science and the public

Scientific thought was not only much better known, but for the first time it was opposed by segments of the public, especially those who held certain religious beliefs. This reaction was different from the official condemnation of Galileo by the church, which was essentially an institutional response, or Bishop George Berkeley's criticism of Newton and Edmund Halley. It also had little in common with the popular reaction against labor-saving technology of the eighteenth century. The nineteenth century saw general public criticism of some scientific ideas, along with newspaper cartoons and well-attended public meetings.

The first issue to cause this kind of problem was the age of the Earth (*see* "The age of the Earth," p 404). Geologists offered good evidence that Earth is far older than most Christians, Jews, and Moslems believed (at the time, about 6000 years, because 6000 years easily encompasses all the events of the Bible). Although Buddhists and Hindus believed Earth to be far older, they were not influential in science at the time. Opposition in this case was not so much the organized opposition of the church, as was the case when the Copernican theory was championed by Galileo, but rather the refusal at the level of the ordinary minister or churchgoer to accept the concept of an older Earth.

Opposition to this idea, however, was muted compared to the great reaction against Charles Darwin and Alfred Wallace's theory of evolution by natural selection. People did not want to admit that humans were descendants, however long ago, of animals. The popular conception—never advocated by scientists—that human ancestors were monkeys was particularly the subject for ridicule. This controversy has continued to agitate some religious groups in the United States throughout the twentieth century, although in the rest of the world it was settled in favor of evolution by the end of the nineteenth century.

Science and technology

At the end of the eighteenth century, discoveries by Antoine Lavoisier and Nicolas Leblanc in France had propelled a small chemical industry. But it was in Germany, which became the leading country in theoretical chemistry, that chemical research had the biggest impact on industry. By the end of the nineteenth century, the country had developed into the largest manufacturer of such chemicals as dyes, fertilizers, and acids used in industrial processes.

The relationship between scientific education and technological progress became fully understood during the nineteenth century. Following the example of the Ecole Polytechnique in France, Germany, and later the United States, also founded technical schools with the idea of applying science to technology. At the end of the century, these technical universities played an essential role in the rapid expansion of Germany's industry. They developed the various kinds of engineers who used science to solve technological problems rather than to advance knowledge.

The scientific method

In Cambridge in 1883, William Whewell suggested to the British Association for the Advancement of Science that their members be called scientists, a term analogous to the word artist for those who practice the arts. Today we would call few of those members scientists, since most were amateurs or supporters of science. The word gradually caught on, however, and began to displace "natural philosopher," although many of the scientists of the nineteenth century resented Whewell's new terminology.

Gradually, during the course of the century, the concept of a scientist became codified, at least in the public mind. The scientist observes, forms a hypothesis, conducts an experiment to test it, and announces a theory. In practice, most science has never been conducted according to this strict "scientific method." Some scientists never conduct experiments, while others determine many facts by experiment and never produce a theory.

Part of the development of the concept of a scientific method resulted from a change during the century in how students learned science. Laboratories were set up at the University of Edinburgh and, later, at other universities. William Thomson (later Lord Kelvin) was the first to have his physics students at the University of Glasgow conduct experiments.

Major advances

Anthropology and archaeology. Interest in archaeology, spurred in the second half of the eighteenth century by the excavation of Pompeii, increased after Napoleon's expedition to Egypt in 1798. Napoleon took scholars with him to study the antiquities known to be there. One of their discoveries, the Rosetta stone, proved to be a seminal object. Jean Champollion in 1822 recognized that the stone contained the same inscription in two Greek scripts and in Egyptian hieroglyphics and was able to translate the previously untranslatable Egyptian inscriptions. Soon after, Georg Friedrich Grotefend and Henry Rawlinson translated the first texts in the cuneiform system used in Mesopotamia. Around the same time (1839), John Lloyd Stephens and Frederick Catherwood discovered the Maya civilization, the beginning of serious archaeology in the New World.

The most notable excavation of the century was the unearthing of many previously unknown tombs and temples in Egypt. Also notable was the discovery that the mounds of Iraq were actually such fabled cities of Mesopotamia as Nineveh and Ur. The most surprising find was Hermann Schliemann's excavation of Troy, which he followed with locating the ruins of the Mycenaean civilization.

In the New World, E. George Squier moved from surveying the mounds of the Mississippi valley in the 1840s, to Central America in the 1850s, and finally to Peru in the 1870s, discovering previously unknown civilizations all

---

Non-Euclidean geometry

Even in Hellenistic times, Euclid's fifth postulate was viewed suspiciously. Euclid himself arranged his *Elements* so that the fifth postulate was not used in the first 25 propositions, although the 16th assumes something equivalent to it. The suspect postulate was much more complex and less self-intuitive than the other postulates, which are in the nature of, "through two distinct points one and only one straight line can be drawn." The fifth postulate states, "If a straight line falling on two straight lines makes the interior angles on the same side less than two right angles, the two straight lines if produced indefinitely meet on that side on which are the angles less than the two right angles." One objection to the fifth postulate was that the meeting point could be as far as one liked from the original line; in essence, the point could be infinitely far away. By the time of Proclus, 750 years after Euclid, geometers were bent on getting rid of this objectionable postulate—but no one could figure out how.

There were essentially two strategies used: replacing the fifth postulate with a different equivalent postulate or establishing it as a mere theorem, a result proved from the other postulates. For the first strategy, the most common version of the postulate used today is known as Playfair's postulate after the eighteenth-century version of British mathematician John Playfair. It states, "Through a given point, not on a given line, only one parallel can be drawn to the given line." Although Playfair's *Elements of geometry* popularized this postulate, Proclus had also used it as an alternative to the fifth postulate in the fifth century AD. Playfair's postulate is a little easier to understand than Euclid's fifth postulate, but because the two postulates are equivalent, it is just as suspect. Two postulates are equivalent when each one implies the other.

The second strategy also involves equivalent postulates. For each time that someone tried to prove the fifth postulate as a theorem, it was found that he had to invoke a result that was equivalent to the fifth postulate. These equivalent postulates include "A line that intersects one of two parallel lines intersects the other"; "The sum of the angles of a triangle is 180°"; "For any triangle there exists a similar triangle not congruent to the first triangle"; and "There exists a circle passing through any three noncollinear points." Invoking any of these in a proof of the fifth postulate amounts to circular reasoning since each is equivalent to the fifth postulate.

In the eighteenth century, Girolamo Saccheri suggested a third strategy, indirect proof of the fifth postulate by contradiction. He assumed that one of the equivalents to the fifth postulate was *not* true. There were two ways in which this equivalent postulate could be not true, so Saccheri needed to deal with two assumptions. Saccheri developed a number of geometric theorems from each of these assumptions combined with Euclid's other postulates, looking for a contradiction between two of these new theorems. Finding such a contradiction from one assumption would mean that either the fifth postulate or the other assumption had to be true. Saccheri convinced himself that he had found such contradictions for both assumptions, leaving the truth of the fifth postulate the only remaining possibility. His report of his efforts was called *Euclid vindicated of every blemish.*

Other mathematicians agreed that one of the assumptions easily produces contradictory theorems, provided one also assumes that lines are infinite. But to contradict the other assumption, Saccheri struggled. In the process of trying to find the contradictions, he developed a large amount of what we now recognize as a non-Euclidean geometry. Saccheri finally convinced himself that he had found a contradiction using the other assumption, but it rested on some very shaky geometry that other mathematicians rejected.

Although Saccheri is best known for this approach today, his work was not well known in his own time. Other mathematicians, notably Johann Heinrich Lambert and and Adrien-Marie Legendre, also tried the same strategy, with approximately the same results.

When he was 15, Karl Friedrich Gauss began to work on the problem of the fifth postulate. Gauss recapitulated the approaches of others with the same results, except that he was better than they had been at seeing that his attempted proofs did not work. After 25 years of working on the problem, he evidently reached the conclusion that the fifth postulate is independent of the others. This means that a contradiction to the fifth postulate can be used to develop a consistent geometry. Gauss proceeded to do this to his own satisfaction, but did not publish his work, although he told a few friends about his conclusion.

Shortly after (around 1830), two other gifted mathematicians also reached the conclusion that the fifth postulate is independent of the others. Both Nikolai Ivanovich Lobachevski and Janos Bolyai independently discovered and published their non-Euclidean geometries. In all three versions (including that of Gauss), the mathematicians assumed that more than one line passing through the same point could be parallel to another line, which is equivalent to the assumption that gave Saccheri such doubtful results. Although both Lobachevski and Bolyai campaigned for their views, the concept of non-Euclidean geometry had little impact for the next quarter of a century.

In 1854, however, Bernhard Riemann discussed the foundations of geometry in a general way, giving very little mathematical detail, but making non-Euclidean geometry more accessible. Riemann also suggested that there were several non-Euclidean geometries. For example, one could avoid the contradictions of Saccheri's first assumption by changing Euclid's first and second postulates along with the fifth. The result appeared to be consistent—that is, no contradictory theorems can arise.

In short order, Eugenio Beltrami was able to prove that the original non-Euclidean geometry of Lobachevski and Bolyai is consistent if Euclidean geometry is itself consistent. Felix Klein soon showed that two different versions of Riemann's geometry were also as consistent as Euclid's. The method employed in each case was to find a model of the non-Euclidean geometry that was within Euclid's geometry. The easiest such model to understand takes the surface of a sphere as the plane, which is undefined in the modern treatment of Euclidean geometry. Then if great circles are taken to be lines, they obey such postulates as "Two distinct points determine at least one line" (replaces Euclid's first postulate), "A straight line is finite in length" (replaces Euclid's second postulate), and "No two lines are parallel" (replaces Euclid's fifth postulate). The new postulates not only describe one form of non-Euclidean geometry, they also accurately describe geometry on a sphere (with the understanding that a line is really a great circle); but geometry on a sphere can also be described by Euclid's original postulates when great circles are understood as circles, not lines.

Despite the mathematical validity of non-Euclidean geometries, they were not deemed of much practical value until 1915, when Albert Einstein showed that gravity could be explained by treating space as a four-dimensional Riemann-type geometry. In other words, space itself is non-Euclidean, despite our impression of it. This is because we view it on a small scale. On a small scale, Earth looks flat. On a larger scale, with a different definition of line, it is non-Euclidean.

---

along the way. In the last quarter of the nineteenth century, archaeologists became aware of the civilizations of the American Southwest.

Just as archaeology was being born, another closely related science came into being. Unlike archaeology, which at least had some ancient texts, such as Herodotus, Homer, and the Bible, to guide it, the other new science, anthropology, was starting from scratch. It had occurred to only a few people that human beings might have existed long before any of the known civilizations. When the first Neanderthal remains appeared in 1856, there was no context in which to put them. They were initially interpreted as recent. The peculiarities, or differences from a modern skeleton, were attributed to disease. Then, in 1859, Charles Darwin's theory of evolution provided a theoretical context. Furthermore, Charles Lyell had already offered convincing arguments that the age of Earth was greater than

most people thought. There was time for various precursor cultures of human beings to exist. The important 1879 discovery of cave paintings at Altamira in Spain was too spectacular for most people to accept. It was acceptable for early people to make primitive stone tools (after all, American Indians had used stone tools when they were first encountered), it was unacceptable for them to be great artists.

As time passed and Darwin's ideas gained influence, the idea of an ancestor to human beings took hold. For one thing, more Neanderthal remains were discovered, making Neanderthals a likely candidate as a different species of human (although today most scientists consider them a subspecies of the modern human). Furthermore, geological evidence suggested that some clearly human remains might be as old as 35,000 years. Eugène Dubois decided he would look for the "missing link" between people and the great apes. He believed that Java would be a good place to look, so he managed to get a job in Java. Astonishingly, he found the remains of a new species, closely related to modern humans, on the island.

At the end of the nineteenth century, despite these discoveries, physical anthropology was still not much of a science. It would gradually develop in the twentieth century, joined by the even newer science of cultural anthropology.

Astronomy. Progress in the nineteenth century was marked by two technical discoveries that each did something unexpected. One was the discovery that light can be recorded by chemicals (in photographs) and the other was that light from different hot elements is itself different—so that, for example, one can tell the difference between hot hydrogen and hot sodium merely by examining the light each gives off. The first discovery meant that astronomers no longer had to depend on looking through a telescope and recording what they saw in words or drawings. They could, instead, photograph what they saw. The second discovery meant that something that everyone thought to be completely impossible was in fact quite possible: a scientist could determine which elements are in a star. The two important new tools of photography and spectrography were eventually combined, and photography of stellar spectrums was routine by the end of the century.

Ancient astronomers had realized that the stars must be very far away, since they showed no measurable parallax (apparent movement as seen from the opposite ends of Earth's orbit). Indeed, the absence of measurable parallax was used as an argument that Earth does not move. The discovery by Frederich Bessel and others that stars are so far away that light takes years to reach Earth considerably expanded our understanding of the size of the universe.

Other discoveries were closer to home. Mathematics was the key in locating the eighth planet, Neptune, while careful observation at a propitious time helped Asaph Hall find the moons of Mars. An excellent telescope at an excellent location helped Edward Barnard locate the fifth known moon of Jupiter. Since Barnard's discovery in 1892, all subsequent discoveries of planetary satellites have been made using photography or by space probes.

A key change in the way observatories were sited also affected astronomy at the end of the nineteenth century. Before 1887, telescopes were placed wherever it was convenient, usually in the vicinity of major cities. In 1887, the Lick Observatory was built on top of California's Mount Hamilton, an inconvenient place that offered good viewing. Almost every optical observatory built since has been on a mountain top.

Biology. One of the great events in the history of science occurred when Darwin and Wallace presented the theory of evolution to the world in 1858 (*see* "The theory of evolu-

tion," p 275). This was the watershed event in biology in the nineteenth century, but there were many other significant discoveries.

One was the development of the cell theory and the understanding of the importance of the chemistry of cells. Schleiden and Schwann proposed the cell theory for plants and animals in 1838-1839. The technique of staining cells allowed biologists to begin to understand how the parts of a cell work.

Louis Pasteur made important advances in chemistry, biology, and medicine, many of them connected to each other. He showed that fermentation is not a purely chemical process, but one that is caused by yeast, and he convinced most biologists that spontaneous generation does not occur.

The borderline between chemistry and life was discovered late in the century as biologists realized that some agents of disease could pass through their finest filters. These unseen and unknown agents came to be known first as filterable viruses and later simply as viruses. Their exact nature would not be known until the twentieth century.

Important advances in understanding heredity occurred during the century, but were not realized at the time. The most significant, Mendel's laws, were published but not noticed. They were rediscovered in 1900 (*see* "A remarkable coincidence," p 396). Progress was also made toward discovering the mechanisms of heredity, but most of this progress can only be recognized in hindsight (*see* "Discovering DNA," p 492).

Chemistry. In some ways, chemistry is the archetypal nineteenth-century science. Chemists of the eighteenth century had paved the way for modern chemistry by ridding themselves of phlogiston, discovering elements, and accepting the atom. In the twentieth century, chemistry would be rethought in terms of physics. In the nineteenth century, however, chemistry held sway as the leading science.

But nineteenth-century chemistry was still emerging, with little understanding of why certain rules worked. For example, the concept of valence was introduced in 1852 and the periodic table in 1869, but it was not until the discovery of Wolfgang Pauli's exclusion principle in 1925 that either of these could be understood from first principles. The same holds true for another major advance, spectroscopy. With no understanding of how electrons behave—indeed, without any suspicion of electrons—the wonder of spectroscopy, identifying elements from the light they give off when heated, emerged without a theoretical basis. Chemistry is not alone in running ahead of its theories. A similar situation existed with gravity, for example; it was an enormously successful concept long before Albert Einstein explained its mechanism in the twentieth century.

For a hit-or-miss science, however, chemistry had man

Charles Darwin may be safely called the greatest biologist of the nineteenth century, but his ideas about evolution were neither entirely new nor complete. Indeed, it is one of the famous stories of science that Darwin's theory of evolution was made public in 1858 only because Alfred Wallace had reached the same conclusions as Darwin and mailed these to Darwin himself. Ever generous, Darwin arranged to present his work and Wallace's in the same session, so that neither would have priority, although Darwin had formulated the basic ideas many years earlier. It is unlikely, however, that evolution would have had the immediate impact that it did if the theory had been put forth by Wallace alone. Darwin was already well known for his account of the voyage of the *Beagle* and for his scientific work on that voyage. Furthermore, Darwin's *On the origin of species by means of natural selection or the preservation of favoured races in the struggle for life*—better known simply as *The origin of species*—published November 24, 1859, is one long, persuasive argument for the theory, which convinced many biologists of its truth through many examples.

But before there was *The origin of species* or Wallace's conclusions, biology had to progress to a point where the existence of evolution seemed reasonable and therefore deserved a scientific explanation. For one thing, there was no point to explaining the origin of species until a species was defined. The modern concept of species began with John Ray's 1686 definition, based on common descent. It was put into the form we still teach today by Comte de Buffon in 1749: a species is a group of interbreeding individuals who cannot breed successfully outside the group.

This concept of species assumed that species are immutable. Even though dogs or cabbages could be bred to many different shapes, for example, neither could be bred to the point that members of the species could not interbreed. In other words, one cannot develop a breed of dog that is of the cat species. Ancient writers who had a less clear concept of species believed that wheat seed could sometimes sprout as millet.

Most scientists of the seventeenth and eighteenth centuries also believed in a doctrine called *preformation*. This existed in two much argued versions, namely that the adult was preformed in the egg and that the adult was preformed in the sperm. In either case, preformation would seem to preclude one species giving birth to a member of another species. As the science of embryology became more precise, however, the doctrine of preformation—a potential barrier to a theory of evolution—was replaced by the doctrine of *epigenesis*, the development of the embryo from undifferentiated tissues.

A second factor in creating the climate for evolution was the emerging understanding of fossils as the remains of living creatures and the realization that many fossils were of species that no longer existed .

Late in the eighteenth century, a third precondition emerged—an understanding that the time required for evolution to occur was present in Earth's history. Both Abraham Gottlob Werner and James Hutton proposed processes for the development of geological features that required great stretches of time for Earth to have existed. Large amounts of time meant that evolution could take place so slowly that it would not be observed among living creatures. This was a necessary precondition, since evolution at the species level is not observable during a lifetime.

With that background, many scientists and near scientists of the first half of the nineteenth century believed in some form of evolution. Lamarck, one of the first evolutionists, thought, correctly, that the environment causes species to evolve. But he also thought that acquired characteristics of the parent could be inherited by the offspring; this was greeted with much disbelief in Lamarck's day and is still discounted today. Charles Darwin's own grandfather, Erasmus Darwin (one of the near scientists), offered a general theory of evolution by generation; it included not only living organisms but also Earth itself.

In England the most influential evolutionist before Charles Darwin was Robert Chambers. His 1844 *Vestiges of creation* influenced both Darwin and Wallace. Indeed, Wallace recognized that Chambers had the fact of evolution but did not have an explanation for it, setting Wallace on his own search for an explanation. Chambers was not a careful scientist, however, and his many errors put off many professional scientists.

After publication of *Origin of species* many—although not all—scientists were converted both to evolution and to the Darwin-Wallace theory of natural selection. The public was less convinced at first, especially those who took Darwinism to be atheistic (which Darwin did not consider it or himself to be). Many people also rejected Darwin's explicitly stated idea that humans and great apes share an ancestor. Two notable opponents of natural selection were Samuel Butler, the author of *The way of all flesh,* and George Bernard Shaw. Butler wanted to return to the ideas of Erasmus Darwin, and Shaw believed in a mystical life force. In the end, however, the theory of evolution and natural selection won all except those who continued to think that it contradicts the biblical account of Creation.

The theory of evolution did not arise fully formed in 1859, however. Modern evolutionists continue to improve on Darwin and Wallace, and a few have even rejected their ideas. New ideas have emerged, such as the concept of mutation and the laws of heredity. These new ideas combined with natural selection to form the basis for neo-Darwinism, the prevalent theory of evolution during most of the twentieth century. Since the 1940s, a theory of evolution has arisen that calls for rapid bursts of evolutionary change between long periods of species stability. This theory, often called "punctuated evolution," has most recently been articulated by Stephen Jay Gould and Niles Eldredge. The discovery of how heredity works has more exactly explained the mechanisms of the theory of evolution as well.

Chemists often heat substances to see what will happen. Some solids melt; other solids and some liquids vaporize. If the liquid or vapor is trapped and the heat lowered, the same material comes back. But some substances burn, char, or coagulate. After that happens, cooling will not bring back the original substance. For the most part, this second class of materials can be recognized as products of living creatures, while the class that only melts includes metals and other items produced from the earth. In 1807, Jöns Jakob Berzelius named the first class—those that melt—*inorganic* and the second—those that burn—*organic*.

Although it was known that new chemicals could be produced from organic sources—Pierre-Eugène-Marcelin Berthelot made the first of these *synthetics* as early as 1778—it was widely believed at the beginning of the nineteenth century that organic chemicals could not be synthesized from inorganic ones. Consequently, Friedrich Wöhler was astonished when in 1828 he heated some ammonium cyanate, classed as inorganic, and got urea, an organic chemical. We now know that ammonium cyanate is not, strictly speaking, inorganic, but Wöhler generally gets the credit for the first synthesis from inorganic material. People were astounded in his time. The popular idea that some mysterious *vital* principle was in organic chemicals had been dealt a severe blow.

Even if the vitalists were wrong, it was clear that organic chemicals were very different in some ways from inorganic chemicals. For one thing, chemicals that seemed to be exactly the same behaved differently. Jean-Baptiste Biot observed in 1815 that tartaric acid produced by grapes polarized light, while seemingly the same acid produced in the laboratory did not polarize light—but both acids had the same chemicals in the same proportions, or the same chemical formula. Justus von Liebig and Wöhler found other similar situations in the 1820s. When they analyzed various different organic compounds, they found that apparently different substances had exactly the same chemical formulas. In 1830, Berzelius named such pairs of compounds *isomers.*

Louis Pasteur's first major project as a young chemist was to try to unravel the mystery of why the two varieties of tartaric acid behaved differently. He observed a tiny difference among the crystals of the type of tartaric acid that did not affect light. After painstakingly separating the crystals into two groups, he discovered that one group polarized light just the way tartaric acid from grapes did. The other group also polarized light, but in the opposite direction. Pasteur correctly realized in 1844 that the two types of polarization cancelled each other out in the laboratory-made substance. He also understood that two different organic chemicals might have different properties and the same formula because the shape of the molecule might be different between the isomers.

In 1845 Adolph Wilhelm Hermann Kolbe became the first chemist to synthesize an organic compound (acetic acid) directly from chemical elements. Shortly after, the concepts of valence and bonding were introduced into chemistry.

Friedrich August Kekulé began to use diagrams based on bonding in organic chemistry in 1861. Kekulé's diagrams showed that Pasteur was correct. The shape of an organic molecule determines its properties.

Even before a basic understanding of organic molecules developed, chemists were beginning to synthesize new organic compounds with important properties not available in inorganics. The first such synthetic, nitrocellulose, was found by accident by Christian Schönbein in 1846. Also known as guncotton, it was very explosive. Nitrocellulose was discovered when Schönbein's wife's apron, which he had used to wipe up a spilled mixture of acids, exploded and vanished in a puff of smoke. When others tried to manufacture guncotton in quantity, many were killed by premature explosions. Another synthetic discovered the same year, nitroglycerine, was only marginally safer. Eventually, however, both substances were tamed into cordite and dynamite. The modern age of high explosives was at hand.

Ten years later a young Englishman accidentally started another industry. William Perkin was trying to synthesize quinine when he produced the first synthetic dye, which we know as mauve. Perkin got rich on mauve and then went on to synthesize other chemicals. In 1875 he started his second industry by creating the first synthetic perfume ingredient, coumarin.

Although Perkin was English, he was something of an anomaly, for most of the organic chemists of the second half of the nineteenth century were German. In fact, Perkin's chemistry teacher, August Wilhelm von Hofmann, was a German chemist teaching in England. Hofmann synthesized his first dye, magenta, in 1858. After he returned to Germany, Hofmann continued to work on dyes and developed a number of violets. Other chemists in Germany worked on producing natural dyes from easily available chemicals, obtaining a red called alzarin in 1869 and indigo in 1880. All of these dyes became the basis of an immense German chemical industry. They also had an impact on biology, for biologists discovered that coloring bacteria or cells with dyes made previously invisible structures apparent.

Another group of organic chemicals got their start in England and the United States. In 1865 English chemist Alexander Parkes found a way to convert nitrocellulose to a nonexplosive (but still quite flammable) substance that we know as celluloid, the first plastic. This was improved on by American inventor John Wesley Hyatt, who was looking for a replacement for ivory billiard balls. In the twentieth century, English and American chemists continued to dominate the plastics industry, creating rayon, Bakelite, nylon, Teflon, Lucite, and polyester, among other synthetics.

In the late nineteenth century and the early twentieth, the raw materials for most of these synthetics were coal, water, and air. Later in the twentieth century, petroleum replaced coal, for there are generally fewer steps in a chemical process that starts with petroleum.

more hits than misses. The atomic theory, first proposed at the beginning of the nineteenth century, gradually became accepted over the course of the century, although there were a few skeptics even into the 1900s. As early as 1824, chemists discovered isomers—compounds with the same chemical formulas but different molecular structures. By 1848 Pasteur had worked out the mechanism by which two otherwise identical isomers behave differently in living organisms. Organic chemistry became a part of the broader field of chemistry in 1848 when Wöhler synthesized the first organic compound from inorganic components. Within organic chemistry itself, the first work with dyes gradually led to an enormous chemical industry. Chemists also began to understand the structure of organic molecules.

Earth science.

Geology moved from its tentative beginnings at the end of the eighteenth century to become a major science in the nineteenth century. In particular, Charles Lyell convinced many of the truth of James Hutton's view of the ancient history of Earth. In turn, Lyell influenced Darwin in the development of the theory of evolution. One of the major developments that started with Lyell's work was the acceptance of a standard geologic time scale based largely on the presence of characteristic fossils in rock strata.

Major excitement came from discoveries of fossils, which were just beginning to be understood as the remains of species that no longer existed on Earth. The dinosaurs, first found and interpreted during this period, were especially exciting and prompted the rivalries of the bonehunters of the American West. As an explanation of the extinctions that clearly had taken place, Georges Cuvier proposed that catastrophes had destroyed these early species. Darwin's view that small changes had removed some species superseded Cuvier's by the end of the period.

This is not to say that the catastrophe theory was rejected completely. Louis Agassiz promoted the idea that Earth had suffered an Ice Age, a concept that was gradually accepted as the century wore on and new evidence accumulated.

In fact, the nineteenth century had its own small catastrophe—the eruption and explosion of Krakatoa in what is now Indonesia, the major volcanic event of the period. It raised waves and colored sunsets around the world and was widely heralded as the loudest noise ever on Earth.

Fossils had another important use during the period. Geologists beginning with William Smith realized that different assemblages of fossils could be used to date rock strata that were many kilometers apart. If the same groups of fossils were present, the rocks had to be about the same age. This tool was used to work out the geologic history of Earth and to identify such ages as the Devonian and Cambrian.

The seismometer, which records the various types of waves produced by an earthquake, was developed during the period. It would contribute to a much greater understanding of the interior of Earth in the twentieth century.

Mathematics.

The reform movement in analysis was an attempt to put the calculus of Newton and Leibniz and their followers on a logical basis. It was clear that the calculus (and the larger field of analysis, which includes calculus) worked, but it was not clear why. In the 1820s Augustin-Louis Cauchy redefined the calculus without using infinitesimals or appeals to intuition. Later in the century, Karl Weierstrass, Richard Dedekind, Georg Cantor, and others worked to make analysis completely dependent on the arithmetic of natural numbers and their ratios, returning in a sense to the goals of the Pythagoreans (see "Mathematics and mysticism," p 36). This effort is considered to have been successful by most mathematicians.

Notable achievements at the beginning of the period were the proof that quintics (polynomial equations of the fifth degree) cannot be solved by algebraic methods and the development of elliptic functions, both by Niels Abel who died at 26 of tuberculosis. Elliptic functions are functions that have two periods, similar to trigonometric functions that have one period.

Before the beginning of this period, Karl Friedrich Gauss had revived number theory and added considerably to it. His work was the basis for number theory during the rest of the nineteenth century. Gauss also developed new uses for differential equations, which became a major topic of study during the century.

Early in the period, projective geometry emerged as a new variety of generalized geometry. It is concerned with the properties of figures that do not change when the figure is projected from one plane to another; it is, therefore, concerned with questions about the intersections of figures, but not with questions of measurement or parallelism, since measures are not preserved under projection.

Abstract systems, such as group theory (see "The legend of Galois," p 296), developed into whole branches of mathematics in their own right. At first mathematicians thought there might be some simple set of universal rules for numbers and numberlike entities. This view became untenable, however, when Sir William Rowan Hamilton realized in 1843 that there were numberlike entities of interest for which a times b does not necessarily equal b times a. Thereafter, algebra began increasingly to be concerned with sets of rules that describe specific structures such as rings and fields as well as groups. The study of such structures is termed abstract algebra.

A development related to abstract algebra and sometimes considered a part of it is Boolean algebra, a system developed by George Boole in 1854 for applying an algebraic system to the laws of thought. To some degree, however, Boolean algebra has been subsumed by two other major developments of the century, the rise of formal systems of symbolic logic and the creation of set theory.

Although Boolean algebra is a formal symbolic logic, it was not used when Gottlob Frege and Giuseppe Peano began their attempts to find a logical basis for arithmetic. They created new symbolic logics for this purpose. Symbolic logic in the sense of Frege and Peano was to become vital in the development of the foundations of mathematics in the twentieth century.

Set theory is almost purely the creation of Georg Cantor, although later mathematicians have axiomatized and expanded it. Set theory brought the actual infinite into mathematics for the first time. Since the time of the ancient Greeks, the infinite was viewed in terms of potential—a line could be extended as far as one liked, a larger number could always be found, and so forth. Cantor found ways to consider such sets as the complete set of natural numbers or the complete set of points on a plane, sets that are actually infinite. His work was not well accepted by his contemporaries, but it has been extraordinarily influential in the twentieth century. Basic operations with sets are similar to operations in Boolean algebra, although the symbolism is different.

Medicine. While the nineteenth century was a period of steady progress in understanding how the human body works, the most significant advances of the period were the discovery of anesthetics and the articulation of the germ theory of disease with the related development of vaccines for diseases other than smallpox.

Doctors had known from ancient times that alcohol ingestion could have a pain-deadening effect, but alcohol is not suitable as a way to kill pain during surgical procedures. It is too slow, for one thing. For another, the amount needed to produce unconsciousness is close to the amount that can cause death. In the 1840s, doctors and dentists found three substances that worked much better than alcohol: ether, nitrous oxide, and chloroform. Any of these anesthetics produces complete unconsciousness with no arousal by pain. Although all are dangerous, few patients were lost to the new anesthetics. With immobilized patients who felt no pain during the procedure, surgery could proceed toward developing the efficiency and delicacy that make it one of the most important branches of medicine today.

Previously, scientists believed that disease was caused by imbalances (the humoral theory of disease), bad air (the miasmal theory), or other conditions. The germ theory of disease replaced these ideas and formulated the concept that water supplies and air could transport disease, that doctors themselves could spread disease, and that certain small specific organisms could be found in conjunction with specific diseases. Louis Pasteur and Robert Koch not only were able to explain and extend these ideas, they were also able to develop methods of immunization against some diseases (see "The germ theory of disease," p 356).

The theory became so entrenched by the end of the period that it was assumed even in cases where no disease-causing agent could be found, leading to the concept of filterable viruses.

Pasteur's success in producing immunity by vaccination is even more remarkable when one considers that the concept of an immune system was just being developed. In the 1880s Ilya Ilich Mechnikov discovered one aspect of the immune system. He found that white blood cells ingest and destroy foreign particles in the blood. Today we know that there are a host of different types of white blood cells, each with its own set of duties, forming the backbone of the immune system. Paul Ehrlich also studied white blood cells and later developed one of the first theories of antibodies near the end of the century.

Other important body systems just beginning to be understood in the nineteenth century were the two systems of hormones, the endocrine and the exocrine. In the 1820s and 1830s William Beaumont became the first person to study the digestive process in a living human when his patient Alexis St. Martin incurred a stomach wound that failed to close completely. Beaumont was able to perform a number of experiments on St. Martin that demonstrated that powerful chemicals in the stomach turn food into simpler substances. These chemicals are hormones of the exocrine system. Much later, in 1889, came the beginnings of knowledge of the endocrine system when it was observed that removal of a dog's pancreas produced symptoms of diabetes. Similarly, in 1894 it was found that ground-up adrenal glands produce physiological changes. Early in the twentieth century, these and similar observations were the bases of the development of the hormone theory.

Physics. As noted above, a major influence on both the science and technology of the nineteenth century was the discovery of the relationship between electricity and magnetism (see "Electricity and magnetism," p 270, and "From action at a distance to fields," p 334). Another thread that runs through the century is the science of heat, or thermodynamics. Thermodynamics was introduced as a concept as early as 1824 and its main laws worked out by 1865.

Often an important concept must be defined before progress can be made. Just as the concept of force was unclear before Newton's time, the concept of energy was unclear before the nineteenth century. Today children in elementary school are taught that a battery changes chemical energy to electrical energy or that friction changes the energy of motion (kinetic energy) to heat energy. These insights were made possible from 1842 through 1847 when Julius Mayer, James Prescott Joule, and Hermann von Helmholtz independently studied energy transformation and concluded that energy is not gained or lost as it is transformed from one type to another (the law of conservation of energy). By 1853 it was possible to relate these ideas to potential energy, such as the energy of position. The concept of energy was an essential prerequisite to Einstein's 1905 proof that energy and mass (matter) are two forms of the same thing.

The study of waves, especially light waves, also developed during the century, beginning with the discovery in 1842 of the Doppler effect: waves produced by a moving source are raised in frequency (higher in pitch for sound waves) when the source is moving toward an observer and lowered when the source is moving away. In 1848, Hippolyte Fizeau showed that the same effect occurs with light, implying that light from a source moving away from an observer will appear shifted to the red. Fizeau and Léon Foucault were able to measure the velocity of light using other methods, but in 1881 Albert Michelson developed the interferometer, which measures slight changes in a wave by comparing two versions of the same wave (see "Measuring with waves," p 520). Not only could this new device measure the speed of light with greater accuracy than previous measurements, it could also use a form of the Doppler effect to determine the absolute motion of Earth through space—assuming that there is a stationary substance called ether pervading space (see "Does the ether exist?," p 366). The negative conclusion of Michelson's measurement with Edward Morley in 1887 was not fully explained until Einstein's special relativity theory in 1905 (see "Relativity," p 384). In the meantime, James Clerk Maxwell's explanation of electromagnetism in terms of waves, discussed earlier, led to Heinrich Hertz's discovery of radio waves.

Toward the end of the century, many physicists began to experiment with the first vacuum tubes, based on the work of Heinrich Geissler, who developed the means of producing good vacuums starting in 1855. Using such a tube, the mathematician Julius Plücker discovered that rays could be produced in a vacuum tube. Named cathode rays by Eugen Goldstein in 1876, the rays were investigated extensively by William Crookes, who developed even better vacuums than Geissler had. Although Crookes came close to making fundamental discoveries with his Crookes tube, as it came to be known, others were to start the so-called second scientific revolution in 1895 based on discoveries made with the Crookes tube (see "Discovering new rays," p 388).

Technology. The impact of electromagnetism on the technology of communications and daily life was especially great. At the beginning of the nineteenth century, messages across a continent or across an ocean might take weeks to reach their recipients; well before the century was out, such messages were carried virtually instantaneously by telegraph, even across oceans. Even more revolutionary, the telephone connected people in a brand new way. After 1879 the electric light not only turned night into day, it also brought electric power into people's homes, where it would soon run small motors or heaters of all sorts.

Transportation was revolutionized as well. At the beginning of the century, the canal was still the best way to transport freight (the Erie Canal opened in 1824) and the steamboat was just beginning to replace sailing ships. But a year after the Erie Canal opened, the first practical locomotive service began. By 1869, railroads crossed North America and were ubiquitous in Europe. Ten years earlier, however, another element entered the transportation picture when Jean Lenoir developed the first internal-combustion engine. By 1885, Karl Benz was driving the first automobile, the destined major form of transportation of the twentieth century.

The third big technological change came from new construction materials and techniques, which included Portland cement, cheap steel, suspension bridges, and skyscrapers. The suspension bridge was pioneered as early as 1825, but it achieved dominance after the innovative techniques used to produce the Brooklyn Bridge, opened in 1883. As an alternative to a bridge, one could build a tunnel, with the first one under the Thames completed in 1843. Joseph Paxton's Crystal Palace in London in 1851 presaged techniques that would be used to build the first skyscrapers in Chicago, Illinois, starting in 1885.

A fourth change came from the advent of cheap steel with the Bessemer process. Closely connected to this was the use of steel in the new engines and machines that were developed and in the new construction techniques.

As noted in the discussion of chemistry above, the chemical industry arose during this century, especially in Germany.

New agricultural machines contributed to the ability of fewer farmers to feed and clothe a growing population.

Certain technological developments had a direct impact on other sciences. Various forms of photography, starting in 1822, were especially important to astronomy. Mechanical calculators, which became truly practical and available near the end of the century, would greatly simplify any scientific study that required computation—even though such pioneering calculators as the Burroughs adding machine and the Comptometer were intended for business use.

One attempted technology of the century did not get very far; however, it can be viewed from our vantage point as a singular effort. Charles Babbage designed a machine he called the Difference Engine in 1822 and even built a model of it. It was intended to calculate various functions automatically. But Babbage abandoned his Difference Engine in 1832 when he conceived the Analytical Engine, which we recognize today as a general-purpose mechanical computer. A working Analytical Engine was never achieved, but in 1890 an ingenious Herman Hollerith found a way to simplify handling the information from the US Census based on punched cards. Hollerith later founded the company that was to become IBM.

1820

Russian explorer Fabian Gottlieb von Bellingshausen (b Arensburg on the island of Oesel, Aug 30, 1779), American Nathaniel B. Palmer, and Englishman Edward Bransfield, independently become the first humans known to sight the continent of Antarctica

The Royal Astronomical Society is founded by, among others, John Herschel (b Slough, England, Mar 7, 1792) and Charles Babbage

Botanist Sir Joseph Banks d Isleworth, England, Jun 19

James Keir d West Bromwich, England, Oct 11

Science and its former view of women

Many great men of science had a low opinion of women. Aristotle, who used "pure logic" to infer that women have fewer teeth than men, but never bothered to count them, said that women are passive and men active. He considered women as "mutilated men." Darwin and Freud also believed in the innate inferiority of women, and during the nineteenth century many scientists were convinced that women had to be less intelligent because of differences in brain structure.

In 1861, the famous French physical anthropologist, Paul Broca, weighed the brains of deceased men and women in four Parisian hospitals and concluded that the brains of women, on the average, were 200 grams (7 ounces) lighter than the brains of men. Gustave Le Bon, a student of Broca, found that the brain of a woman was in size "closer to the brain of the gorilla than the brain of men." And according to Le Bon, "all psychologists who have studied the intelligence of women recognize that they represent an inferior form of evolution and that they are closer to the child and the savage."

Some scientists from the nineteenth century were convinced that intelligence was located in the frontal lobe of the brain, and therefore believed that women should have smaller frontal lobes. The year 1884 saw the publication of a scientific book in which woman is called *Homo parietalis* and man is called *homo frontalis*. It was soon found, however, that the frontal lobes in women were generally larger than those of men, and therefore male scientists concluded that not the frontal lobe but the parietal lobe of the brain should be the seat of intelligence.

In the early twentieth century scientists found that the size of the brain had nothing to do with intelligence, disproving all those learned theories from the nineteenth century as pure nonsense. Women are generally smaller than men, so they have smaller brains, but not smaller intelligence.

1821

Anthropologist Louis-Laurent Mortillet b Meylan, Isère, France, Aug 29; he will be the first to separate the paleolithic (Old Stone Age) into such periods as the Acheulian and

The Catholic church lifts its ban on teaching the Copernican system

Austrian chemist Johann Joseph Loschmidt b Putschirn, Bohemia, Mar 15; he introduces the practice of using single lines for single bonds, double lines

1820

In his opening lecture at Oxford University, William Buckland (b England, 1784) proposes that geology be directed toward confirming Noah's flood and other biblical accounts

John Frederic Daniell (b London, Mar 12, 1790) invents a dew point hygrometer

The science of electrodynamics is born with the announcement of Hans Christian Oersted's discovery of electromagnetism

Dominique-François Arago discovers the magnetic effect of electricity passing though a copper wire, demonstrating that iron is not necessary for magnetism

André-Marie Ampère formulates one of the basic laws of electromagnetism, the right-hand rule for the influence of an electric current on a magnet and demonstrates that two wires that are each carrying an electric current will attract or repel each other, depending on whether the currents are in the same or opposite directions

Johann Salomo Christoph Schweigger (b Erlangen, Germany, Apr 8, 1779), on learning of Oersted's discovery of electromagnetism, uses the effect to build the first galvanometer, a tool for measuring the intensity and direction of an electric current

Oersted invents an instrument for measuring electric currents, the amperemeter

Augustin-Jean Fresnel invents the so-called Fresnel lens, a lens used in lighthouses

Alexis-Thérèse Petit d Paris, Jun 21

Natural fertilizer (guano) is used in Europe

Agriculturist Arthur Young d Apr 20

1821

Mary Anning, who discovered the first ichthyosaur at age 12, now discovers the first known fossil of a plesiosaur at age 21

Augustin Cauchy's *Cours d'analyse de l'Ecole Polytechnique* (Analysis course for the Polytechnique School) is the first of three textbooks that puts elementary cal-

Charles Bell (b Edinburgh, Scotland, Nov, 1744) gives the first description of the facial paralysis named for him (Bell's palsy)

Russian-German physicist Thomas Johann Seebeck, b Revel (USSR), Apr 9, 1770, discovers thermoelectricity, the conversion of heat into electricity that occurs when a

Michael Faraday (b Newington, England, Sep 22, 1791) reports his discovery of electromagnetic rotation in the paper "On some new electromagnetical motions,

| 1821 cont | | Mousterian, based on the different tool kits used in the different periods | Friedrich Wilhelm Bessel starts his program of determining the positions of 50,000 stars, completed in 1833

Alexis Bouvard (b Contamines, France, Jun 27, 1767) uses various observations of Uranus made both before and after Herschel's 1781 discovery of the planet to plot its orbit; he finds that earlier positions do not agree with later ones, an observation that leads to the discovery of Neptune | | for double bonds, and so forth, into organic chemistry; he also determines that aromatic compounds have a benzene ring as part of their structure |
| 1822 | | English antiquarian William Bullock explores Teotihuacán and other ruins in the Valley of Mexico; he brings back relics and casts of Aztec works of art and monuments for display in England; the exhibit, in 1824, creates new interest among the English in New World civilizations

Jean-François Champollion translates the hieroglyphic panel of the Rosetta stone, the first translation of Egyptian hieroglyphics | Jean-Baptiste Delambre d Paris, Aug 19

The Roman Catholic Church removes Galileo's *Dialogue concerning the two chief world systems* from the *Index of Prohibited Books* 190 years after it was first published

German-English astronomer Sir William Herschel d Slough, Buckinghamshire, England, Aug 25

The first calculated return of a short-period comet comes with the observation of Encke's comet by Carl Ludwig Christian Rümker

Heinrich Christian Schumacher founds the *Astronomische Nachrichten* | Jean Lamarck's *Histoire naturelle des animaux sans vertèbres* (Natural history of invertebrates) is the first work to distinguish between vertebrates and invertebrates | Jöns Jakob Berzelius begins publication of his *Jahres-Bericht*, an annual review reporting the research of himself and others

Anselme Payen (b Paris, Jan 6, 1795) discovers that charcoal can be used to remove impurities from sugar

Chemist Claude-Louis Berthollet d Arcueil, near Paris, Nov 6 |

Ignatz Venetz (b Visperterminen, Switzerland, Mar 21, 1788) is the first to propose that glaciers once covered much of Europe

Gas bubbles are observed in a well in Fredonia; soon afterwards, William Aaron Hart drills the first natural gas well

culus into substantially the form that it is taught today, although it does not contain the modern definition of a limit

junction of certain metals is heated

and on the theory of magnetism" and creates the first two motors that are powered by electricity

Engineer and architect John Rennie d Oct 4

1821 cont

William Daniel Conybeare and William Phillips identify the Carboniferous Period in Earth's history

Jean-Baptiste-Julien Omalius d'Halloy identifies the Cretaceous Period in Earth's history

Mary Ann Mantell discovers the first fossil to be recognized as a dinosaur, named iguanodon by her husband Gideon Algernon Mantell (b Lewes, England, Dec 3, 1790); however, recent research indicates that Gideon Mantell may have been the actual discoverer

Grundriss der Mineralogie (Outline of mineralogy) by German mineralogist Friedrich Mohs (b Gernode, Jan 29, 1773) introduces his system of classifying minerals and his scale of mineral hardness (Mohs' scale)

Mineralogist René Just Haüy d Paris, June 1

Karl Wilhelm Feuerbach publishes what is now known as Feuerbach's theorem (discovered the year before by Jean-Victor Poncelet and Charles Julien Brianchon): the nine-point circle of a triangle is tangent internally to the inscribed circle of the triangle and externally to the three escribed circles

Jean-Baptiste Joseph Fourier's *La théorie analytique de la chaleur* (Analytic theory of heat) proves that any continuous function can be found as the sum of sine and cosine curves; this is the foundation of what is now known as Fourier analysis

Scottish essayist and historian Thomas Carlyle makes the second translation of Adrien-Marie Legendre's *Eléments de géométrie* (Elements of geometry) into English; it becomes the most popular geometry textbook in the United States during the nineteenth century

William Beaumont (b Lebanon, CT, Nov 21, 1785) starts his experimental study of digestion in the exposed stomach of a wounded man

Charles Babbage (b Teignmouth, England, Dec 26, 1792) develops the Difference Engine, a machine for calculating values of logarithms and trigonometric functions; it does not work well owing to an inability to make parts that live up to Babbage's design; soon he will abandon the Difference Engine for a general-purpose computer

Joseph Nicéphore Niepce (b Chalôns, France, Mar 7, 1765), using silver chloride, produces the first fixed positive image that could be called a photograph

1822

1822
cont

1823

		Astronomy	Biology	Chemistry
		Joseph von Fraunhofer observes spectra of fixed stars and detects dark lines in them that are different from those observed in his study of dark lines in the spectrum of the sun	Zoologist Spencer Fullerton Baird b Feb 3; Baird, a systematic person, as secretary of the Smithsonian Institution amasses astonishingly large collections of preserved organisms, especially birds	Berzelius discovers silicon

Botanist Nathanael Pringsheim b Nov 30; he makes the first scientific studies of algae

Michael Faraday's *On fluid chlorine* describes the liquefaction of chlorine by cooling, and properties of the fluid

Charles Macintosh (b Glasgow, Scotland, Dec 29, 1766) dissolves rubber in low-boiling naphtha and invents the waterproof fabric that bears his name

---

1824

The yard is defined by government decree in England on the basis of the length of a pendulum with a period of one second placed in Greenwich

The first school of science and engineering opens in the United States; later it will be called the Rensselaer Polytechnic Institute

Marie-Charles-Théodore Damoiseau creates the *Tables lunaires* (Lunar tables) published by the Bureau des Longitudes in France from 1824 to 1828

Joseph von Fraunhofer builds the first telescope to be mounted equatorially with a clock drive, the Dorpat refractor; equatorial mounts and clock drives for astronomical instruments were both introduced by the Chinese, but they did not have telescopes with lenses or mirrors

Berzelius isolates zirconium, which was discovered in the form of zirconium oxide in 1789 by Martin Klaproth

German chemists Justus von Liebig (b Darmstadt, May 1, 1803) and Friedrich Wöhler (b Eschersheim, Jul 1, 1800) each find the formula of a chemical compound; Joseph-Louis Gay-Lussac notes that although the compounds are chemically different, the two formulas are the same; it is the first discovery of chemical isomers, chemicals that differ only in molecular structure

					1822 cont
	Traité des propriétés projectives des figures (Treatise on the projective properties of figures) by Jean-Victor Poncelet (b Metz, France, Jul 1, 1788) is the first systematic development of projective geometry				

EARTH SCIENCE	MATHEMATICS	MEDICINE	PHYSICS	TECHNOLOGY	1823
Reliquiae diluvianae: Observations on the organic remains contained in caves, fissures, and diluvial gravel, and on other geological phenomena, attesting the action of a universal deluge by William Buckland (b England, 1800) sets the time of Noah's flood at about 5000 to 6000 years before the present John Frederic Daniell's *Meteorological essays and observations* presents a study of the atmosphere and trade winds Denison Olmsted conducts the first state geological survey in North Carolina		Physician Edward Jenner d Berkeley, England Jan 26 Physician Matthew Baillie d Sep 23	André-Marie Ampère develops a theory relating electricity to magnetism; he proposes that magnetism is caused by the movement of very small electrical charges in bodies; although something like this is now thought to be true, his contemporaries were not impressed William Sturgeon (b Whittington, England, May 22, 1783) makes the first electromagnet Jacques-Alexandre-César Charles d Paris, Apr 7	Inventor Edmund Cartwright d Hastings, England, Oct 30	

EARTH SCIENCE	MATHEMATICS	MEDICINE	PHYSICS	TECHNOLOGY	1824
Mineralogist Hieronymus Theodor Richter b Dresden, Saxony, Germany, Nov 21; he plays a role in discovering indium by spotting the indigo line in the spectrum for the color-blind Ferdinand Reich, whose assistant he is	Niels Henrik Abel (b Finnoy Island, near Stavanger, Norway, Aug 5, 1802) proves that it is impossible to find roots for an equation of the fifth degree or greater; an overlooked proof in 1799 by Paolo Ruffini was unknown to Abel and is considered to be less convincing than Abel's proof Karl Friedrich Gauss reveals in a letter to Franz Taurinus that he has discovered and accepted non-Euclidean geometry, but wishes to keep it secret	Charles Bell's *Injuries of the spine and thigh bone* is published Henry Hickman uses carbon dioxide on an animal as a general anesthetic Physician James Parkinson d London, Dec 21	*Réflexions sur la puissance motrice du feu* (On the motive power of fire) by Nicholas L.S. (Sadi) Carnot (b Paris, Jun 1, 1796) shows that work is done as heat passes from a high temperature to a low temperature; defines work; hints at the second law of thermodynamics; and suggests internal combustion engines	The Erie Canal from Albany, NY, to Buffalo, NY, is completed; it had been started in 1817 English mason and building contractor Joseph Aspdin patents Portland cement	

1825

		John Herschel describes the actinometer, a device for measuring the intensity of solar radiation	Naturalist Henry Walter Bates b Leicester, England, Feb 8; his theory of insect mimicry, developed in an 11-year stay in the jungles of South America, contributes to the acceptance of Charles Darwin's and Alfred Wallace's theory of evolution	Berzelius isolates the chemical element titanium, discovered as an oxide in 1795 by Martin Klaproth

Michel-Eugène Chevreul and Joseph-Louis Gay-Lussac patent a method of making candles from fatty acids; these candles are a great improvement on tallow candles commonly in use, and the two chemists are hailed as benefactors of humankind

Michael Faraday isolates benzene by fractionally distilling whale oil

Carl Löwig discovers bromine, not announced until the rediscovery by Antoine Jérôme Balard the following year

Hans Christian Oersted discovers aluminum by using electric currents and chlorine to produce anhydrous aluminum chloride from alumina; he then dissolves this compound in mercury and distills metallic aluminum from the solution

Richard August Erlenmeyer b Wehen, Germany, Jun 28: he synthesizes many important organic compounds and popularizes the commonly used way to indicate single, double, and triple bonds; he is best known for the invention of the conical, flat-bottomed Erlenmeyer flask

The French *Corps Royal des Mines* starts field work on the *Carte géologique de la France* (Geological map of France), which becomes the first national geological survey

Georges Cuvier announces his catastrophe theory: great catastrophes have caused the extinction of large groups of animal species and altered Earth

Mineralogist Franz Joseph Müller d Vienna, Austria Oct 12

Elliptic functions, functions with a double periodicity, are discovered by Niels Henrik Abel and published in 1827; they are independently discovered by Carl Jacobi in 1829 and probably also by Gauss as early as 1808, although never published by Gauss

Peter Gustav (P.G.) Lejeune Dirichlet (b Düren, Germany, 1805) presents a flawed proof that Fermat's "last theorem" is true for $n = 5$; the flaws are soon corrected by Adrien-Marie Legendre; previously the theorem had been shown true for $n = 3$ and $n = 4$ (the case $n = 4$ having been used by Fermat to illustrate his method of infinite descent)

Sophie Germain (b Paris, Apr 1, 1776), nicknamed the Hypatia of the nineteenth century, proves that Fermat's "last theorem" is true for odd-prime exponents less than 100 that do not divide the product xyz, where p is the prime and $x^p + y^p = z^p$

Pierre Bretonneau (b St. Georges-sur-Cher, France, Apr 3, 1778) successfully performs the first tracheotomy to restore breathing to a child suffering from diphtheria

Anatomist Max Johann Sigismund Schultze b Freiburg, Germany, Mar 25; he studies cytoplasm, then known as protoplasm, and shows that it is approximately the same for all forms of life

John George Appolt invents the chamber gas-producing retort

Thomas Drummond (b England, Oct 10, 1797) and Goldsworth Gurney (b England, 1798) invent limelight, an intense beam of light focused by a parabolic mirror and produced by burning lime in an alcohol flame enriched by additional oxygen

A suspension bridge over Menai Straits in Wales, built by Thomas Telford (b Scotland, Aug 9, 1757), with a single span of 176 m (579 ft) inaugurates the age of modern bridge construction

Inventor Eli Whitney d New Haven, CT, Jan 8

George Stephenson's "Locomotion No. 1" makes its initial trip, the first steam locomotive to carry regularly both passengers and freight

1826

The University of Munich (Germany) is founded, modeled on the University of Berlin

Statesman and scholar Thomas Jefferson d Monticello, VA, Jul 4

Wilhelm von Biela (b Rossla, Austria, Mar 19, 1782) calculates the orbit of a short-period comet that by 1846 has split in two and by 1866 has disappeared altogether, showing for the first time that some comets have short lifetimes

Johann Elert Bode d Berlin (Germany), Nov 23

Heinrich Olbers formulates the paradox named after him: If stars are distributed evenly through infinite space, why then is the night sky dark? (twentieth-century cosmology and the understanding that the universe has a limited age has brought a solution to this paradox)

Giuseppe Piazzi d Naples (Italy), Jul 22

Heinrich Samuel Schwabe, b Dessau (Germany), Oct 25, 1789, starts observing sunspots, which leads him to the discovery of the sunspot cycle

German physiologist Ernst Heinrich Weber (b Wittenberg, Jun 24, 1795) starts his experiments with the two-point threshold of sensation: when two points of simulation on the skin are close, they are perceived as single-point sensation

Antoine-Jérôme Balard (b Montpellier, France, Sep 30, 1802), studying seaweed, rediscovers the chemical element bromine

Sir Humphry Davy's last lecture, *On the relation of electrical and chemical changes*, results in his receiving the Royal Medal

Jean-Baptiste-André Dumas develops a method to measure vapor densities of substances that are liquid or solid at normal temperatures

Henri Dutrochet studies osmosis and discovers the laws governing it

Leopoldo Nobili (b 1784) gives a description of the rings obtained electrolytically with lead and salt solutions that are now named after him

Otto Unverdorben (b Oct 13, 1806) discovers aniline by distilling indigo

Joseph-Louis Proust d Angers, France, Jul 5

German geologist Christian Leopold von Buch (b Stolpe, Apr 25, 1774) makes the first geologic map of Germany

Niels Henrik Abel gives an example of an integral equation (an equation that contains an unknown under the integration sign): find the curve described by a mass when that mass slides down the curve from a position at rest toward its lowest point, given that the time for reaching that point is known

German mathematician August Leopold Crelle (b 1780) launches *Journal für die reine und angewandte Mathematik* (Journal for Pure and Applied Mathematics), commonly known as *Crelle's Journal*, the first periodical devoted exclusively to research in pure mathematics (the "Applied" in the title was regularly ignored)

Jean-Victor Poncelet and Joseph-Diaz Gergonne discover the principle of duality; that is, in theorems of projective geometry one can exchange the words *point* and *line* to obtain a new, correct theorem; this theorem is proved by Julius Plücker in 1829

On Feb 23 Nikolai Ivanovich Lobachevski (b near Nizhni Novgorod, Russia, Dec 2, 1793) delivers a paper to the mathematics and physics staff at Kazan University that is believed to have contained an outline of this ideas about non-Euclidean geometry

Pierre Bretonneau describes the symptoms of and names diphtheria

Théophile-René-Hyacinthe Laënnec's *De l'auscultation médiate et des maladies des poumons et du coeur* (On using sound to diagnose maladies of the lungs and heart) presents an extension of his work of 1819

Physician Théophile-René-Hyacinthe Laënnec d Kerbouarnec, Brittany, France, Aug 13

Anatomist Karl Gegenbaur b Würzburg, Germany, Aug 21; he shows that all vertebrate cells arise from divisions of the egg and sperm

Physician Philippe Pinel d Paris, Oct 26

Joseph von Fraunhofer d Munich (Germany), Jun 7

1827

Jacques Babinet (b Lusignon, France, Mar 5, 1794) suggests using the wavelength of light as a basis for standard measures; his suggestion is adopted between 1960 and 1983, when the meter is based on the wavelength of light from the gas krypton

Baron Friedrich Wilhelm von Humboldt starts his popular lectures on astronomy in Berlin

Astronomer and mathematician Pierre-Simon, marquis de Laplace d Paris, Mar 5

Félix Savary calculates the orbit of the binary star zeta Ursae Majoris and shows that the orbit is governed by Newton's law of gravitation

John James Audubon, b Les Cayes, Santo Domingo (Haiti), Apr 26, 1785, starts the publication of *Birds of America*, a collection of lifelike drawings that are published in Europe

De ovi mammalium et hominis genesi (On the origin of the mammalian and human ovum) by Russian biologist Karl Ernst von Baer (b Piep, Estonia, Feb 28, 1792) reports his discovery of eggs in mammals and humans, and that mammals develop from eggs

Michael Faraday's *Chemical manipulation* is a manual of chemical processing dealing with distilling and similar topics

Friedrich Wöhler develops a new method to prepare aluminum in pure form; because of the complexities of the chemistry required, aluminum remains the most expensive metal on Earth; some aluminum jewelry is made during the nineteenth century

Pierre-Eugène-Marcelin Berthelot b Paris, Oct 27; he succeeds in synthesizing many organic compounds, including the first that do not occur naturally

Predicting the planets

The ancients knew the planets Mercury, Venus, Mars, Jupiter, and Saturn (and Earth, although they did not think of it as a planet). Some of them offered reasons why there were these and no more. Pythagoreans, for example, argued on the basis of number properties. Much later, Johannes Kepler claimed that the number of planets is fixed at six by the number of the five regular solids. The world was astounded, then, when William Herschel discovered a seventh planet, Uranus, in 1781.

In fact, Uranus had been seen by other astronomers for years. It can be seen with the naked eye and had even been included on star maps. Since no one expected a new planet, however, it had gone unidentified as one. Herschel was the first observer whose telescope was good enough to reveal the "star" as a disk. (All stars except the sun are so far away that they appear as points in even the largest telescopes.)

Herschel's discovery was the result of his systematic searching of the sky, his good telescope and good vision, and a certain amount of luck. The other two planets, Neptune and Pluto, that have been discovered since were both predicted. Nevertheless, the same qualities, especially the luck, were vital in finding each.

In 1821 Alexis Bouvard used all the known sightings of Uranus to calculate its orbit. The calculations from earlier sightings could not be made to agree with later ones, however. Something was wrong.

Something was wrong with Mercury's orbit, also. Like that of Uranus, the orbit of Mercury could not be made to fit with Newton's laws. Either Newton was not quite correct or there must be other bodies, most likely planets, near enough to Mercury and Uranus to affect their orbits. Most astronomers felt strongly that Newton must be correct and that other planets would be found.

Among the first to calculate the position of such a planet was Urbain Leverrier in 1845. He predicted a planet inside the orbit of Mercury. He even named the planet—Vulcan. A few astronomers had been searching for such a planet already. Even with the help of Leverrier's calculations, no planet was found.

In 1843 John Couch Adams was calculating the position of a planet outside the orbit of Uranus. But when he offered his calculations to the astronomer royal, George Airy, Airy paid no attention. Airy was one of the few astronomers who thought the irregularities were caused by defects in Newton's theory.

Jean-Baptiste Fourier suggests that human activities have an effect on Earth's climate

Karl Friedrich Gauss's *Untersuchungen zur Differentialgeometrie* introduces the subject of differential geometry, by means of which the small features of curves and surfaces are described using the techniques of analysis

August Möbius's *Der barycentrische Calcul* (Barycentric calculus) is probably the first work to use homogeneous coordinates, although K.W. Feuerbach, Julius Plücker, and Etienne Bobillier also invent the same idea around the same time

Richard Bright (b Bristol, England, Sep 28, 1789) gives a description of the symptoms of Bright's disease, a kidney disorder

Charles-Eduard-Ernest Delezenne develops his technique of "just noticeable differences" to study hearing

André-Marie Ampère's *Mémoire sur la théorie mathématique des phénomènes électrodynamiques uniquement déduite de l'expérience* (Memoir on the mathematical theory of electrodynamic phenomena deduced solely from experiment) contains the inverse square law for magnets

Botanist Robert Brown (b Montrose, Scotland, Dec 21, 1773), using a microscope, discovers that tiny particles suspended in a liquid are constantly in motion; this motion is eventually known as Brownian motion; nearly 100 years later it is the first concrete indication that molecules really exist

Theory of systems of rays by William Rowan Hamilton (b Dublin, Ireland, Aug 4, 1805) is a unification of the study of optics through the principle of "varying action"; it contains his correct prediction of conical refraction; when his prediction is verified, he becomes well known and is knighted

Die galvanische kette, mathematisch bearbeitet (The galvanic circuit investigated mathematically) by German physicist Georg Simon Ohm (b Erlangen, Mar 16, 1789) contains the first statement of what is eventually called Ohm's law—that the current of electricity is equal to the ratio of the voltage to the resistance

Inventor Samuel Crompton d Jun 26

John Walker invents the "match;" its tip was coated with antimony sulfide and potassium chlorate

---

Predicting the planets (continued)

In 1846, although disappointed in the search for Vulcan, Leverrier also turned to the planet beyond Uranus. He produced essentially the same result as Adams had. Unlike Adams, however, Leverrier took his calculations to someone who was willing to conduct the search. Johann Gottfried Galle immediately found a planet where Adams and Leverrier had said it would be, although at a different distance. Both calculators had relied upon Bode's law to locate the planet, but the "law" is apparently the result of chance and does not apply to the new planet, which Leverrier named Neptune.

When the orbit of Neptune was calculated from these observations, it became clear that Leverrier and Adams were also lucky. Neptune could not account for all of the deviations in the orbit of Uranus after all. Furthermore, Neptune's orbit also deviated.

Scientists set out to find the next planet, which they believed would resolve these problems. The most dedicated was Percival Lowell, but he was unable to find a planet either by calculation or observation. Lowell even built his own observatory.

After Lowell's death in 1916, the search continued at Lowell Observatory. Finally, Clyde Tombaugh located a "star" on a pair of photographs that was in a slightly different place from one day to the next. He had found Pluto.

As usual, a lot of luck as well as systematic work went into the discovery. However, finding Pluto did not solve the problems with the orbits of Uranus and Neptune. For one thing, Pluto is too small, actually the smallest planet. Some astronomers predict that there is another planet beyond Pluto, which they call Planet X.

As an aside, in 1915 Albert Einstein developed a new theory of gravity, one that superseded Newton. In Einstein's theory, Mercury's orbit was perfectly all right. There was no need for a planet to perturb it.

1827
cont

1828

The observed position of Uranus differs so much from its calculated orbit that astronomers start looking for another planet that may disturb its orbit; this leads to the discovery of Neptune in 1846

Karl Ernst von Baer's *Über Entwicklungsgeschichte der Tiere* (The developmental history of animals) discusses his "germ layer theory," according to which eggs develop to form four layers of tissue

Botanist Karl Thunberg d Uppsala, Sweden, Aug 8

Jöns Jakob Berzelius publishes a revised version of his table of atomic weights, with many elements having atomic weights close to presently accepted values

Friedrich Wöhler prepares the organic compound urea from inorganic compounds, proving wrong the idea that organic compounds can be produced only by living organisms

Wöhler and Antoine-Alexandre-Brutus Bussy prepare pure beryllium

Chemist Alexander Mikhailovich Butlerov b Chistopol' (USSR), Sep 6; he is among the first to investigate the structure of organic compounds

Chemist and physician William Hyde Wollaston d London, Dec 22

1829

The Royal Observatory in Cape Town (South Africa) is completed

Nathaniel Bowditch begins publication of the English translation of *Mécanique céleste* by Pierre-Simon Laplace

Naturalist Jean-Baptiste Pierre Antoine de Monet, chevalier de Lamarck, d Paris, Dec 28

Berzelius discovers the chemical element thorium, named after the Scandinavian god Thor

Wolfgang Döbereiner develops a system of periodically returning characteristics of elements, a precursor of the periodic table

EARTH SCIENCE	MATHEMATICS	MEDICINE	PHYSICS	TECHNOLOGY	
			Alessandro Giuseppe Antonio Anastasio, Count Volta, d Como (Italy), Mar 5 German physicist Ernst Florens Chladni d Breslau (Wroclaw, Poland), Apr 3 Augustin-Jean Fresnel d Ville-d'Avray, France, July 14		**1827** cont
Paul Erman (b Feb 29, 1764) measures the magnetic field of Earth; his measurements become the basis for Gauss's theory of Earth's magnetic field	George Green's *Essay on the application of mathematical analysis to the theories of electricity and magnetism* introduces the theorem named after him, which reduces certain volume integrals to surface integrals, making them easier to handle; it is rediscovered by several others before it enters the main stream of mathematics	German physician Franz Joseph Gall d Montrouge, near Paris, Aug 22			**1828**
Alexander von Humboldt travels to Siberia and explores Altai and Dsungarei	Nikolai Lobachevski's version of non-Euclidean geometry, *On the foundations of geometry*, is the first to be published, although he was anticipated by Jano Bolyai (1823) and Karl Friedrich Gauss (1816)	William E. Horner's *A treatise on pathological anatomy* is the first text on the subject published in the United States James Mill's *Analysis of the phenomena of the human mind* tries to show that the mind is nothing more than a machine without any creative function	Gustave-Gaspard Coriolis (b Paris, May 21, 1792) coins the term *kinetic energy* in *On the calculation of mechanical action* Joseph Henry (b Albany, NY, Dec 17, 1777) shows that wire wrapped into coils produces a greater magnetic field when an electric current goes through it than a straight wire does,	The first electro-magnetically driven clock is made Jacog Bigelow coins the term *technology* in *The elements of technology* William Austin Burt invents an early typewriter	**1829**

1829
cont

The work of Thomas Graham (b Glasgow, Scotland, Dec 21, 1805) with the diffusion of gases leads to the law named after him; the law states that the rate of diffusion of a gas is inversely proportional to the square root of the gas's density

English chemist Sir Humphry Davy d Geneva, Switzerland, May 29

Louis-Nicolas Vauquelin d St.-André-d'Hébertot, France, Nov 14

1830

Cours de philosophie positive by Auguste-Marie-François-Xavier Comte (b Jan 19, Montpellier, France, 1798) establishes positivism as a philosophical school; the six-volume work, completed in 1842, contains a hierarchical division of the sciences

The July Revolution in France puts Louis-Phillipe of the House of Orleans on the throne

Mary Fairfax Somerville's *The mechanisms of the heavens* is a popularization in English of Laplace's *Mécanique céleste*; although Somerville is self-educated, Laplace declares that she is the only woman to have understood his work

Berzelius coins the term *isomerism* to describe compounds of identical chemical composition but different structure and properties, first discovered accidentally by Justus von Liebig and Friedrich Wöhler in 1824

J.B.A. Dulong determines the percentage of hydrogen and carbon in organic compounds by his perfected incineration method

German mathematician Julius Plücker (b Eberfeld, Jun 16, 1801) establishes the principle of duality for projective geometry: for any theorem you can exchange the words *point* and *line* and obtain a different theorem that is also true

Niels Henrik Abel d Froland, Norway, Apr 6

In May, Evariste Galois (b Bourg-la-Reine, France, Oct 25, 1811) presents his paper introducing group theory to the French Academy of Sciences, but Augustin-Louis Cauchy, who is to referee it, mysteriously does nothing about it; according to some sources, Cauchy lost the manuscript

Johann Schönlein describes hemophilia, a blood disease

and that an insulated wire wrapped around an iron core can produce a powerful electromagnet

Physicist and physician Thomas Young d London, May 10

1830

The first volume of *The principles of geology* by Charles Lyell (b Kinnordy, Scotland, Nov 14, 1797) begins a massive study that shows that Earth must be several hundred million years old; volume three is published in 1833; the whole trilogy is republished many times over, with a dozen revisions during the author's lifetime alone

About this time William Nicol (b Scotland, 1768) begins work on what will eventually become the petrological microscope, a device for viewing thin sections of rock; Nicol's first experiments use thin sections of wood, however

Treatise on algebra by George Peacock (b Durham, England, Apr 9, 1791) is one of the first attempts to formulate the fundamental laws of numbers

In February, Evariste Galois presents a second version of his theory of groups to the French Academy of Sciences, but the referee, Joseph Fourier, dies before looking at it; the paper is never found among Fourier's papers

Jean-Baptiste Joseph, Baron Fourier, d Paris, May 16

Charles Bell's *The nervous system of the human body* distinguishes different types of nerves

Joseph Henry discovers the principle of the dynamo shortly before Michael Faraday does, but fails to publish; he publishes his results after learning of Faraday's

Joseph Jackson Lister (b London, Jan 11, 1786) succeeds in developing an achromatic lens for the microscope about 70 years after the first achromatic telescope lenses were made

Charles Sauria and J.F. Kammerer discover how to make matches that light when struck, using yellow phosphorus, sulfur, and potassium chlorate

1831

The Royal Society starts publishing *Proceedings of the Royal Society*, abstracts of papers from its members; the first issue contains papers published since 1800

French astronomer Jean-Louis Pons d Florence (Italy), Oct 14

Robert Brown discovers the nucleus of the cell

Charles Robert Darwin (b Shrewsbury, England, Feb 12, 1809) joins the crew of the H.M.S. *Beagle* as the ship's naturalist; the *Beagle* plans to take a two-year voyage to map the coast of South America; this turns out to be a five-year voyage

Robiquet and Colin discover the red dye alizarin

Nils Sefström (b Hälsingland, Sweden, Jun 2, 1787) rediscovers vanadium, which had been discovered in 1801 by Andrès Del Rio, but Del Rio was persuaded by other chemists that he had merely rediscovered chromium; eventually Friedrich Wöhler sorts all this out

Eugène Soubeiran, Justus von Liebig, and Samuel Guthrie independently discover chloroform; Guthrie develops a process for producing it

The legend of Galois

Popular legend has it that 20-year-old Evariste Galois, thinking he might be killed in a duel over a woman's honor, invented group theory—one of the most basic and important concepts of modern mathematics—on the night of May 29, 1832. Galois used his new concept to prove that equations of the fifth degree—quintics—and higher could never be solved. He was indeed killed the following day, but because his ideas were so radical it took 150 years for mathematicians to work out their implications.

It is a dramatic legend, but almost entirely untrue. Paolo Ruffini proved that the general quintic could not be solved by algebraic means before Galois was born, and independently Niels Abel gave even a better proof in 1824. Although Galois contributed to group theory, he cannot lay claim to being its inventor. The notes he wrote out that night were not new mathematics; over the previous five years Galois had submitted three versions of the same material to the French Academy, but two great mathematicians had managed to ignore or misplace the first two versions, and the third was turned back for lack of clarity. It is even unlikely that the duel was over a woman's honor.

Group theory can be traced back to origins in geometry, number theory, and permutations, as well as the algebraic studies of Ruffini, Abel, and Galois, although it was Galois who first used the term *group*. Later, a group was defined as a set of objects that obeys four simple rules for a given operation (for the mathematically inclined, groups are closed and associative and possess an identity element and inverses with respect to it). Possibly the earliest use of group theoretic ideas was by Leonhard Euler in 1761 in a paper on number theory, although he did not explicitly label the concepts as anything special.

Galois's accomplishment, while misrepresented in popular legend, was quite real. Nearly 20 years after his death, the concepts of group theory had advanced enough so that his obscurely written paper could be clearly understood. He had developed a general method for determining the solvability of equations, going considerably beyond Ruffini and Abel. Along the way, he had independently created large chunks of group theory.

Group theory came to pervade all of mathematics. In 1872, Felix Klein suggested that different types of geometry be defined in terms of groups in his famous Erlangen Program. This was largely accomplished in the late nineteenth and early twentieth centuries. In fact, a purely geometric concept, such as topology (the branch of geometry concerned with connections rather than shapes), evolved to produce a purely algebraic theory of topology

James Clark Ross (b London, Apr 15, 1800) reaches the magnetic North Pole on Jun 1

Augustin-Louis Cauchy announces that an analytic function of a complex variable can be expanded about a point in a power series that is convergent for all values of the variable in a circle about the point that passes through the nearest singularity of the function

Karl Friedrich Gauss warns that the actual infinite (as opposed to the potential infinite) must be kept out of mathematics, which would then be able to stay free of contradictions

Julius Plücker extends the principle of duality to three dimensions, where the words *point* and *plane* can be interchanged to produce new theorems, while the word *line* remains unchanged

Lambert Adolphe Jacques Quetelet (b Gent, Belgium, Feb 22, 1796) studies how such factors as sex, age, education, climate, and season affect the crime rate in France

In January, Evariste Galois presents a third version of his paper on group theory, now titled *"Un Mémoire sur les conditions de résolubilité des équations par radicaux"* to the French Academy of Sciences, where it is rejected by referee Simeon-Denis Poisson as not sufficiently clear

Sophie Germain d Paris, Jun 27

A cholera epidemic starts in Europe and lasts until 1832

Chemist and physician Samuel Guthrie (b Brimfield, MA, 1782) discovers chloroform

David Brewster's "Treatise on optics" is published in *Dr. Lardner's cabinet encyclopaedia* and is reprinted until 1854

Independently, Michael Faraday and Joseph Henry discover that electricity can be induced by changes in a magnetic field (electromagnetic induction), a discovery leading to the first electric generators

Russian-German physicist Thomas Johann Seebeck d Berlin, Dec 10

Joseph Henry describes a practical electric motor

Charles Wheatstone (b Gloucester, England, Feb 6, 1802) and William Fothergill create the first telegraph, a machine with an arrow that points to letters of the alphabet

Cyrus Hall McCormick (b Walnut Grove, VA, Feb 15, 1809) demonstrates his first version of the reaper in Jul

---

The legend of Galois (continued)

based on groups. Modern algebra books introduce groups in the early chapters and use the concept thereafter. Even branches of mathematics with no explicit use of group theory are structured in ways that emerged from the theory. One textbook that addresses all of classical mathematics states in its introduction, "If a person is not ready and anxious to explore the concept of a group further, we recommend that he reconsider his relationship to mathematics."

Despite its pervasive influence in mathematics, group theory was not seen as particularly useful in applications until recently. Although it was suggested as a tool in chemistry, nothing much emerged. Crystallographers were the main beneficiaries outside of mathematics. Then, in 1926, Eugene Paul Wigner introduced groups to particle physics. After World War II, a number of physicists began to think about how group theory could be used. In the early 1960s Sheldon Glashow, Murray Gell-Mann, and others discovered that group theory enabled them to predict the properties of particles before the particles were observed and to make sense out of the many new particles that were being discovered. By the 1980s a "standard model" of particle physics had emerged, based on group theory, and physicists were using groups as a tool in trying to develop Grand Unified Theories (GUTs) that combine all the known forces except gravity. The even more ambitious superstring theory, which includes gravity, is also based on groups.

Meanwhile, the very mathematicians who specialized in one part of pure group theory worked themselves out of a job. It had become apparent that there were only a finite number of simple groups that have a finite number of members. Group theorists doggedly worked out each possible group in succession. Finally, in 1980, Robert L. Griess, Jr., worked out the last possible finite group, nicknamed the "monster" because it is so large.

	GENERAL	ANTHROPOLOGY/ ARCHAEOLOGY	ASTRONOMY	BIOLOGY	CHEMISTRY
1832	The Reform Bill in England gives the lower classes more representation in Parliament Philosopher Jeremy Bentham d London, Jun 6 Poet and scientist Johann Wolfgang von Goethe d Weimar, Thuringia, Germany, Mar 22			Anatomist Georges Léopold Chrétien Frédéric Dagobert, Baron Cuvier d Paris, May 13	Jean-Baptiste-André Dumas and Augustin Laurent discover anthracine in coal tar Eilhardt Mitscherlich prepares nitrobenzene in the laboratory French chemist Charles Friedel b Strasbourg, Bas-Rhin, Mar 12; he attempts unsuccessfully to make artificial diamonds and investigates organic compounds that contain silicon Jean-Antoine-Claude Chaptal, comte de Chanteloup, d Paris, Jul 30
1833	William Whewell (b Lancaster, England, May 24, 1794) coins the term *scientist* at a meeting of the British Association for the Advancement of Science; he popularizes his new word further in *The philosophy of the inductive sciences* in 1840		The first magnetic observatory is founded in Göttingen (Germany) A great meteor shower hits the United States, causing astronomers to inquire into the causes of such showers	The first volume of the five-volume *Recherches sur les poissons fossiles* (Researches on fossil fishes) by Jean-Louis-Rudolphe Agassiz (b Môtier, Switzerland, May 28, 1807) is published	Jean-Baptiste-André Dumas develops a method for determining the nitrogen content of organic compounds Michael Faraday produces aluminum by electrolysis Anselme Payen discovers the first enzyme, diastase, an extract of malt that speeds the conversion of starch to sugar German chemist Karl Reichenbach (b 1788) discovers creosote German-Russian chemist Gottlieb Sigismund Kirchhof d St. Petersburg, Feb 14

1832

Chemist and geologist Sir James Hall d Edinburgh, Scotland, Jun 23

The appendix by Janos Bolyai (b Kolozsvár, Hungary, Dec 15, 1802) to his father's *Tentamen*, the *Appendix scientiam spatii absolute veram exhibens* (Appendix explaining the absolute science of space), is the second published account of non-Euclidean geometry

Mary Fairfax Somerville's *A preliminary dissertation on the mechanisms of the heavens* provides the mathematical background necessary to understand her popularization of Laplace's *Mécanique céleste*

Evariste Galois d Paris, May 31, after being shot in a duel over a personal matter on May 30; he leaves behind a summary of his paper on group theory addressed to a friend

The Warburton Anatomy Act legalizes the sale of bodies for dissection in England, ending the practice of body snatching and sometimes murder to provide bodies

Thomas Hodgkin gives a description of Hodgkin's disease, a cancer of the lymph nodes

Anatomist Antonio Scarpa d Bonasco (Italy), Oct 31

Joseph Henry discovers self-induction or inductance, the process of a second current in a coil through which one current is passing being induced by the first current

Nicolas Léonard Sadi Carnot d Paris, Aug 24

Charles Babbage conceives of the first computer, the Analytical Engine; it is a mechanical calculating machine driven by an external set of instructions or program; although strikingly modern in concept, the computer is never built in workable form

Hippolyte Pixii demonstrates a hand-driven "magneto-electric machine," a generator in which a horseshoe magnet rotates in front of two coils

Charles Wheatstone invents the stereoscope

1833

Charles Lyell identifies the Recent, Pliocene, Miocene, and Eocene periods in Earth's history

Mineralogist Don Fausto D'Elhuyar d Madrid, Spain, Jan 6

Adrien-Marie Legendre's *Réflexions sur différentes manières de démontrer la théorie de parallèles* (Reflexions on different ways to prove the theory of parallels) summarizes a dozen faulty proofs of Euclid's fifth, or parallel, postulate

Adrien-Marie Legendre d Paris, Jan 9

Handbuch der Physiologie by German physiologist Johannes Peter Müller (b Koblenz, Jul 14, 1801) summarizes physiological research of the period; it contains the theory that each nerve has its own specific energy

Michael Faraday introduces the law of electrolysis; that is, the amount of substance decomposed by an electric current is proportional to the length in time and magnitude of the electric current producing electrolysis

Heinrich Friedrich Emil Lenz, b Dorpat, Estonia (Tartu, USSR), Feb 24, 1804, discovers that the resistance of a metallic conductor increases as temperature is raised and decreases as temperature is lowered

Siméon-Denis Poisson's *Traité de mécanique* (Treatise on mechanics) becomes a standard text

Karl Friedrich Gauss and Wilhelm Weber build an electric telegraph operating over a distance of 2 km (1.25 mi)

Inventor Richard Trevithick d Dartford, Kent, England, Apr 22

Inventor Joseph Nicéphore Niepce d Saint-Loup-de-Varennes, France, Jul 5

1833
cont

1834	Economist Thomas Robert Malthus d Haileybury, Somersetshire, England, Dec 23	Christian Jorgensen Thomsen (b Copenhagen, Denmark, Dec 29, 1788) divides the time people have existed into a Stone Age, a Bronze Age, and an Iron Age	The British Admiralty publishes the first "modern" *Nautical Almanac* Marie-Charles-Théodore Damoiseau publishes *Tables écliptiques des satellites de Jupiter* (Ecliptical tables of Jupiter's satellites) The first exhaustive survey of the southern stars is begun by John Frederick Herschel at Cape Colony, South Africa Ernst Heinrich Weber discovers the law governing the sensibility of the eye and the ear; Gustav Theodor Fechner applies it to stellar magnitudes in astronomy and popularizes the rule, now known as the Weber-Fechner law: the percentage difference in intensity is more important in perception than absolute differences	German-Swiss physiologist Gabriel Gustav Valentin, b Breslau (Wroclaw, Poland), Jul 8, 1810, and Jan Purkinje, b Libochovice (Czechoslovakia), Dec 17, 1787, discover that cilia in the oviduct move the ovum from place to place	The British Association for the Advancement of Science recommends the adoption of the system of chemical symbols devised by Jöns Jakob Berzelius German chemist Robert Wilhelm Bunsen (b Göttingen, Mar 31, 1811) discovers an antidote against arsenic poison Jean-Baptiste-André Dumas formulates his law of substitution: halogens can replace hydrogens in organic compounds Anselme Payen extracts the cellulose from wood and names it Friedlieb Ferdinand Runge (b Germany, Feb 8, 1795) discovers phenol (carbolic acid) in coal tar German chemist Heinrich Caro b Posen (now Poland) Feb 13; after learning about synthesis of dyes from William Perkin in England, Caro returns to Germany to lead the development of the German dye industry
1835	Entrepreneur and philanthropist Andrew Carnegie b Dumfernline, Scotland, Nov 25; his many donations to libraries in the United States will greatly further the advancement of learning in the nineteenth century and beyond		Halley's comet makes its second predicted return; it is first spotted by Dumouchel in Rome Auguste Comte declares that knowledge of the chemical composition of stars will forever be denied to man	Charles Darwin, while scientific officer on the voyage of the *Beagle*, visits the Galápagos Islands; he observes that closely related finches (known today as Darwin's finches) seem to have developed from a common ancestor found on the	Jean-Baptiste-André Dumas and Eugène-Melchior Péligot prepare methyl alcohol Friedrich Strohmeyer d Göttingen, Germany, Aug 18

EARTH SCIENCE	MATHEMATICS	MEDICINE	PHYSICS	TECHNOLOGY	
			In correspondence Michael Faraday and William Whewell introduce the terms *electrode, anode, ion, cathode, anion, cation, electrolyte,* and *electrolysis*		**1833** cont
Friedrich August von Alberti identifies the Triassic Period in Earth's history	Sir William Rowan Hamilton transforms the Lagrange formulas called Lagrangians into the form of "canonical equations," now called Hamiltonians

John Venn b Hull, Yorkshire, England, Aug 4; he is best known for his development of diagrams that can be used to indicate the union or intersection of sets or to show logical consequences | Amalgam (mercury alloy) is used as filling material for teeth

Physician Sir Gilbert Blane d London, Jun 26 | Benoit-Pierre Clapeyron (b Paris, Feb 26, 1799) develops the first version of the second law of thermodynamics (often expressed as entropy tends to increase in a closed system) on the basis of his study of the thermodynamics of steam engines; the law is later generalized by Rudolf Clausius

Heinrich Friedrich Emil Lenz discovers that a current produced by electromagnetic forces always creates an effect that opposes those forces; this is now known as the Lenz law

Jean-Charles-Athanase Peltier discovers the Peltier effect, that a current flowing across a junction of dissimilar metals causes heat to be absorbed or freed, depending on the direction in which the current is flowing | E.M. Clarke produces a commercial electromagnetic generator

Joseph-Marie Jacquard, French inventor, d Aug 7

Thomas Telford, Scottish engineer, d Sep 2 | **1834** |
| Roderick Impey Murchison (b Tarradale, Scotland, Feb 19, 1792) identifies the Silurian Period in Earth's history

Adam Sedgwick (b Yorkshire, England, Mar 22, 1785) identifies the Cambrian Period in Earth's history | | | Gustave-Gaspard Coriolis's *Mémoire sur les équations du mouvement relatif des sytèmes de corps* (Memoir on the equations of relative movement of systems of bodies) describes the Coriolis effect: the deflection of a moving body caused by Earth's rotation; the | Samuel Colt (b Hartford, CT, Jul 19, 1814) patents his revolver

William Gossage (b England, 1799) invents a tower that absorbs hydrogen chloride, an important step in the development of the chemical industry | **1835** |

1835 cont

BIOLOGY

mainland of South America, the nearest continent

Jan Purkinje notes that animal tissues are, like plant tissues, made from cells

Lambert Adolphe Jacques Quetelet's *Anthropométrie* uses measurements to determine that physical sizes of body parts fall into what we now recognize as the normal curve

1836

ASTRONOMY

Francis Baily (b Newbury, England, Apr 28, 1774) describes Baily's beads, the bright spots along the moon's edge seen during a total eclipse

Wilhelm Beer, b Berlin (Germany), Jan 4, 1797, and Johann Heinrich Mädler produce a map of the visible lunar surface; it remains the best for several decades

BIOLOGY

The first living lungfish species, now viewed as an important link between fish and amphibians, is discovered

Elements of botany by Asa Gray (b Sauquoit, NY, Nov 18, 1810) is published

Botanist Antoine-Laurent de Jussieu d Paris, Sept. 17

German biologist Theodor Schwann (b Neuss, Dec 7, 1810) discovers the enzyme pepsin, the first known animal enzyme; he reports the discovery in *Über das Wesen des Verdauungsprozesses* (On the essence of digestion)

CHEMISTRY

Edmund Davy discovers acetylene gas, which he obtains from potassium carbide

Auguste Laurent (b St. Maurice, France, Nov 14, 1807) demonstrates that a chlorine atom can be substituted for a hydrogen atom in a chemical with little change in properties; as this refutes the prevailing concept, his work is rejected by the chemical community at first; gradually Laurent's ideas are accepted

1837

ANTHROPOLOGY/ARCHAEOLOGY

Henry Creswicke Rawlinson (b Chadlington, England, Apr 11, 1810), after heroic labors to read the "Mountain of the Gods" inscription of Darius the Great— hundreds of feet above the ground— manages to translate two paragraphs of Persian cuneiform text, 35 years after the obscure Grotefend had accomplished a similar feat without the physical daring

ASTRONOMY

Johann Franz Encke discovers the small gap in Saturn's outer ring that is named after him

Mensura micrometricae (Micrometric measurement of double stars) by Friedrich Georg Wilhelm von Struve, b Altona (Germany), Apr 15, 1793, is the first good catalog of double stars

BIOLOGY

Henri Dutrochet shows that only those parts of plants that contain chlorophyll absorb carbon dioxide and that they do so only in the presence of light

EARTH SCIENCE	MATHEMATICS	MEDICINE	PHYSICS	TECHNOLOGY	
			Coriolis effect is important in the study of wind	Joseph Henry develops the basic principles of the telegraph, put into more practical form 11 years later by Samuel F.B. Morse Henry invents the electrical relay, enabling a current to travel long distances from its origin	**1835** cont
	Joseph Liouville (b St. Omer, France, Mar, 24, 1809) founds *Journal de Mathématiques Pures et Appliquées*; it becomes one of the most important mathematical journals in France Mikhail Ostrogradsky rediscovers Green's theorem on the conversion of multiple integrals; it continues to be largely overlooked until it is discovered for the fourth time—Gauss had also discovered it—by Lord Kelvin in 1846 Mathematician and physicist André-Marie Ampère d Marseille, France, Jun 10	Anatomist Heinrich Wilhelm Gottfried von Waldeyer-Hartz b Braunschweig, Germany, Oct 6; he is the first to note that the nervous system is built from separate cells and that the nerve cells do not actually touch each other; he also coins the word *chromosome*	John Frederic Daniell invents the Daniell cell, the first reliable source of electric current, based on the interactions of copper and zinc	The combine harvester is introduced in United States Engineer John Loudon McAdam d Moffat, Dumfriesshire, Scotland, Nov 26	**1836**
Louis Agassiz begins to use the term *die Eiszeit* (Ice Age) to describe his view that the whole of Europe had been covered with glaciers in the past	Joseph Liouville discusses integral equations, but little attention is paid to his work Pierre Wantzel proves that it is impossible to trisect an arbitrary angle with compass and straightedge, resolving one of the three classic problems of antiquity			Samuel Finley Breese Morse (b Charlestown, MA, Apr 27, 1791) patents his version of the telegraph, a machine that sends letters in codes made up of dots and dashes	**1837**

1838

GENERAL

Publication of *Proceedings* is started by the American Philosophical Society

ASTRONOMY

Friedrich Bessel is the first to determine how far away a star (other than the sun) is by measuring the parallax of 61 Cygni

BIOLOGY

German biologist Matthias Jakob Schleiden (b Hamburg, Apr 5, 1804) recognizes that cells are the fundamental components of plants

Theodor Schwann shows that yeast is made of small living organisms, but he is not believed by biologists or chemists, who are finally convinced by the work of Louis Pasteur in 1856 and in subsequent years

CHEMISTRY

Pierre Louis Dulong d Paris, Jul 18

Bernard Courtois d Paris, Sep 27

1839

GENERAL

Russian explorer Nikolay Mikhaylovich Przhevalsky b Kimbarovo, Smolensk region, Apr 12; he explores central Asia, bringing back many previously unknown species; his most famous discovery will be Przhevalsky's horse, the last truly wild horse (other wild horses have domesticated ancestors)

ANTHROPOLOGY/ARCHAEOLOGY

John Lloyd Stephens and artist Frederick Catherwood embark on an expedition to Central America with the goal of locating the ruins of Copán and investigating them; after buying Copán they discover that it is the remains of a civilization largely unknown to Europeans, the Maya

ASTRONOMY

Harvard College Observatory is founded, the first official observatory in the United States, with William Cranch Bond (b Portland, ME, Sep 9, 1789) as its first director; a 38-cm (15-in.) refractor is installed there in 1847; it and one other are the largest in the world at the time

Thomas Henderson (b Dundee, Scotland, Dec 28, 1798) measures parallax of alpha-Centauri, the second star other than the sun whose distance from Earth is found; Henderson had the measurements as early as 1832, but failed to interpret them correctly until after Bessel's determination of the distance to a star in 1838

Pulkovo Observatory in Russia is founded; Friedrich von Struve becomes the first director

BIOLOGY

Charles Darwin's *Journal of researches into the geology and natural history of the various countries visited by HMS Beagle 1832-36* is an account of Darwin's work as a scientist, principally in South America, where he finds fossils, collects plants and animals, and studies the geology of the continent

Jan Purkinje coins the word *protoplasm* to describe the contents of a cell

Theodor Schwann's *Mikroskopische Untersuchungen über die Übe berinstimmung in dem Struktur und dem Wachstum der Tiere und Pflanzen* (Microscopical researches on the similarity in the structure and growth of animals and plants) discusses the existence of cells in animals and lays the foundations of cell biology

CHEMISTRY

Carl Mosander (b Kalmar, Sweden, Sep 10, 1797) discovers the chemical element lanthanum

1838

Sir Roderick Impey Murchison's *The silurian system* details the geological history of the strata underlying the Old Red Sandstone

In an encyclopedia article on mathematical induction, English mathematician Augustus De Morgan (b Madras, India, Jun 27, 1806) makes the first known use of that term for the method of proof, despite the fact that De Morgan proposes in the article that the method be called successive induction

William Beaumont's *Experiments and observations on the gastric juice and the physiology of digestion* relates his study started in 1822 of digestion *in vivo* and *in vitro* with a wounded man whose stomach remained partially open through a healed hole in the abdomen

Michael Faraday is the first to discover phosphorescent glow produced by electric discharges in gases at low pressure; he also discovers the so-called Faraday dark space, a dark area before the cathode

---

1839

Charles Lyell identifies the Pleistocene Period in Earth's history

A treatise on crystallography by William Hallowes Miller (b Llandovery, Wales, England, Apr 6, 1801) introduces the Millerian indices for crystals, a coordinate system that uses three numbers to describe each type of crystal face; the system is still in use

Sir Roderick Impey Murchison and Adam Sedgwick's *On the physical structure of Devonshire* is an important study of the rocks of the Devonian Era

Geologist William Smith d Northampton, England, Aug 28

German geologist Friedrich Mohs d Agardo (Italy), Sep 29

Karl Friedrich Gauss lays the foundations of potential theory, which becomes an independent mathematical theory

Gabriel Lamé proves Fermat's last theorem is true for $n = 7$, bringing the known proved cases to $n = 3, 4, 5,$ and 7

Jan Evangelista Purkinje becomes the director of the first institute for physiology in Breslau (Wroclaw, Poland)

Pierre Prévost d Geneva, Switzerland, Apr 8

Isaac Babbit invents an antifriction alloy (called Babbit metal) that becomes extensively used in bearings

Louis Jacques Daguerre (b Seine-et-Oise, France, Nov 18, 1789) announces his process for making photographs (a silver image on a copper plate), known as daguerrotypes

Charles Goodyear (b New Haven, CT, Dec 29, 1800) discovers vulcanization of rubber (the addition of sulfur and processing to rubber to make it stable during both heat and cold) by accident, although he was seeking such a process

William Robert Grove (b Swansea, Wales, Jul 11, 1811) develops the first fuel cell, a device that produces electrical energy by combining hydrogen and oxygen; although theoretically an excellent way to produce electricity, fuel cells have failed to become practical for most applications

1839
cont

1840

Anthropologist Johann Blumenbach d Göttingen, Saxony (Germany), Jan 22

Before this date John William Draper (b Saint Helens, England, May 5, 1811) is the first to take daguerrotypes of the moon

Heinrich Wilhelm Matthäus Olbers d Bremen (Germany), Mar 2

Friedrich von Struve measures the parallax of Vega (Alpha Lyrae), becoming the third astronomer to measure the distance of a star other than the sun

Jean-Baptiste-Joseph-Dieudonné Boussingault (b Paris, Feb 2, 1802) shows that plants obtain nitrogen from nitrates in the soil

Charles Darwin's *Zoology of the voyage of the* Beagle describes the animals collected during his work on a scientific expedition

Russian embryologist Alexander Onufriyevich Kovalevski b Dünaburg, Latvia, Nov 19; his investigations of the borderline between vertebrates and invertebrates lead to the discovery that an invertebrate embryo begins with the same three layers as a vertebrate embryo and to identification of the notochord

Jean-Baptiste-André Dumas's theory of structural types advances the idea that the properties of organic compounds are due to the structure of their molecules more than to the elements that make up the molecules

Swiss-Russian chemist Germain Henri Hess (b Geneva, Switzerland, Aug 8, 1802) announces the law that will be named after him; the law states that the amount of heat involved in changing one substance to another depends on the substances and not on the reactions involved

Justus von Liebig's *Die organische Chemie in ihre Anwendung auf Agrikultur und Physiologie* (Organic chemistry and its application to agriculture and physiology) explains his theory of the exchanges of carbon and nitrogen in plants and animals

German chemist Christian Schönbein (b Württemberg, Oct 18, 1799) discovers ozone

---

The cell theory

Before the nineteenth century, when people thought of the structure of living organisms, they recognized tissues and organs but assumed that the tissues were simple substances similar to nonliving materials. They did not expect a level of organization *between* simple substances and recognizable tissues. Even when Robert Hooke observed and named cells (because they looked like the cells in a monastery) in cork in 1665, he did not realize that he was making a fundamental discovery about life.

Plant cells are easier to observe than animal cells because plants have cell walls and animals do not. As microscopes continued to improve, scientists observed cells in various plant tissues. Even so, as late as 1831, when Robert Brown first noted that plant cells all contain a small dark body that he named the nucleus ("little nut"), there was no recognition of the importance of cells. In 1835, Jan Evangelista Purkinje noted that animal tissues are also made of cells, but this observation attracted little attention.

This all changed suddenly in 1838 when Matthias Jakob Schleiden proposed that all plant tissues are made from cells. The following year, Theodor Schwann proposed that eggs are cells and that all animal tissues are also made from cells. Schwann also suggested that all life starts as a single cell. Thus the cell theory is generally attributed to Schleiden and Schwann, since each contributed vital parts to the theory.

EARTH SCIENCE	MATHEMATICS	MEDICINE	PHYSICS	TECHNOLOGY	
				William Talbot (b Dorsetshire, England, Feb 11, 1800) invents photographic paper for making negatives	**1839** cont
				Inventor William Murdock d Birmingham, Warwickshire, England, Nov 15	
Louis Agassiz's *Etudes sur les glaciers* (Study of glaciers) describes motions and deposits of glaciers, confirming his theories on the ice ages	Siméon-Denis Poisson d Paris, Apr 25	*Pathologischen Untersuchungen* (Pathological investigations) by German pathologist and anatomist Friedrich Gustav Jakob Henle (b Fürth, Jul 19, 1809) expresses his conviction that diseases are transmitted by living organisms, although he offers no hard evidence	Alexandre-Edmond Becquerel (b Paris, Mar 24, 1820) shows that light can initiate chemical reactions that produce an electric current	Italian physicist Giovanni Battista Amici (b Modena, Mar 23, 1786) invents the oil-immersion microscope, one of his several innovations in microscope building that result in instruments with an enlarging power of 6000 times	**1840**
Paleontologist Edward Drinker Cope b Philadelphia, PA, Jul 28; he directs many searches for fossils, especially of dinosaurs, in the American West in competition with Othniel Charles Marsh of Yale University		Johannes Peter Müller's *Handbuch der Physiologie des Menschen* (Handbook of human physiology; volume 1 published in 1834) is among the first to give a mechanistic explanation of human thinking		John William Draper takes what is today the oldest surviving photograph of a person	
				Mémoire sur l'artillerie des anciens et sur celle du Moyen Age (Ancient and medieval artillery) by Guillaume-Henri Dufour is the most important work by the Swiss general, cartographer, and military writer	
				Thomas Drummond, Scottish engineer, d Apr 15	

The cell theory (continued)

Schleiden and Schwann were right about the basic idea of the cell theory, but other workers had to correct the details and extend the theory. For example, Schleiden thought that new cells are formed as buds on the surfaces of existing cells. Karl Wilhelm von Nägeli showed that this is not true, although it then took many years before cell division was fairly well understood. In 1845 Karl Theodor Ernst von Siebold extended the cell theory to protists, single-celled creatures. He thought, however, that multicellular organisms are made from protists. Rudolf Albert von Kölliker demonstrated in the 1840s that sperm are cells and that nerve fibers are parts of cells.

The cell theory was promoted by Rudolph Carl Virchow in the 1850s. His credo was "all cells arise from cells." Virchow also suggested that disease is caused when cells revolt against the organism of which they are a part. While this might be one way to view cancer, Virchow regarded all forms of disease this way, which caused him to resist Pasteur's germ theory. Despite such occasional setbacks, the cell theory became one of the main foundations of modern biology.

1841

GENERAL

ANTHROPOLOGY/ARCHAEOLOGY

John Lloyd Stephens's *Incidents of travel in Central America, Chiapas, and Yucatan,* illustrated by Frederick Catherwood, reports on his discoveries of the Maya civilization at Copán, Palenque, and Uxmal; it is immensely popular and introduces the Maya to the United States and Europe

ASTRONOMY

Friedrich Bessel determines dimensions of Earth with geodetic degree measurements

Bessel interprets measurements dating from 1834 to show that the star Sirius has an unseen companion; later this companion, Sirius B, will be the first white dwarf to be identified

Popular astronomy by German astronomer Johann Heinrich Mädler (b Berlin, May 29, 1794) is an account of astronomy for the lay person that goes through six editions in the author's lifetime

BIOLOGY

The Berlin zoo opens

Jean-Baptiste Boussingault shows that the amount of carbon, hydrogen, oxygen, and nitrogen in plants is larger than the amounts supplied by manure

Swiss-French botanist Augustin-Pyrame de Candolle d Geneva, Sep 9

CHEMISTRY

Jöns Jakob Berzelius observes chemical allotropy (two different forms of the same element) by converting charcoal into graphite

German physical chemist Hermann Franz Moritz Kopp (b Hanau, Oct 30, 1817) starts his four-volume history of chemistry, completed in 1847

Friedrich Wilhelm Sertürner d Hamelm, Germany, Feb 20

1842

ASTRONOMY

An important total solar eclipse is observed by astronomers; it is inferred from it that the corona and prominences are solar rather than lunar; Majocci makes the first attempt to photograph totality, but misses

BIOLOGY

Hermann Ludwig von Helmholtz (b Potsdam, Prussia, Germany, Aug 31, 1821) studies the relation between nerve cells and nerve fibers

Samuel Dana describes how phosphates play a role in manure as a fertilizer

Charles Darwin writes a 35-page abstract of his theory of the evolution of species

Justus von Liebig's *Organic chemistry in relation to physiology and pathology* is published

Karl Wilhelm von Nägeli's *Zur Entwicklungsgeschichte des Pollens* is published

CHEMISTRY

Sir John Bennett Lawes (b Rothamsted, England, Dec 28, 1814) develops the artificial fertilizer superphosphate

Eugène-Melchior Péligot (b Feb 24, 1811) discovers the element uranium

Pierre-Joseph Pelletier d Paris, Jul 19

1841

Arnold Escher von der Linth describes structures in the Alps where tens of kilometers of rock layers have been folded and laid over other layers "like a napkin"; geologists come to call such giant folds *nappes*

Sir Roderick Impey Murchison identifies the Permian Period in Earth's history

Geologist Clarence Edward Dutton b Wallingford, CT, May 15; his study of volcanoes and earthquakes leads him to the correct conclusion that the continents are made from lighter rock than the oceans; he proposes the name isostasy, still used, to describe how a mountain floats in denser rock

Lambert Adolphe Jacques Quetelet establishes a central statistical bureau in Belgium that is widely imitated around the world

Charles Thomas Jackson (b Plymouth, MA, Jun 21, 1805) discovers that ether is an anesthetic

Swiss surgeon Emil Theodor Kocher b Aug 25; he develops surgical removal of the thyroid as a cure for goiter

US inventor James Buchanan Eads (b Lawrenceburg, IN, May 23, 1820) patents a new type of diving bell about this time and uses it for extensive salvage operations in the Mississippi River

Joseph Whitworth (b England, Dec 21, 1803) introduces the standard screw thread

Inventor Nicolas Appert d Massy, near Paris, Jun 3

1842

Charles Darwin's *The structure and distribution of coral reefs, being the first part of the geology of the voyage of the* Beagle classifies coral reefs into three types and presents Darwin's theory of the formation of atolls by subsidence of islands

Richard Owen (b Lancaster, England, Jul 20, 1804) coins the name *dinosaur* to describe what we now recognize as two unusual groups of reptiles that were the dominant animals for about 175 million years

The first use of ether in surgery is by Crawford Williamson Long (b Danielsville, GA, Nov 1, 1815) on Mar 30, but lack of publication allows credit for the discovery to go to William Morton in 1846; Long publishes his own results in 1849

Christian Johann Doppler (b Salzburg, Austria, Nov 29, 1803) discovers that the frequency of waves emitted by a moving source changes when the source moves relative to the observer; this is called the Doppler effect

German physician and physicist Julius Robert Mayer (b Heilbronn, Nov 25, 1814) is the first to state the law of conservation of energy, noting specifically that heat and mechanical energy are two aspects of the same thing

William Thomson, later Lord Kelvin, publishes *On the uniform motion of heat in homogeneous solid bodies*

Alexander Bain proposes a picture transmitter and receiver

Werner von Siemens (b Dec 13, 1816) invents an electroplating process

Henry Shrapnel, English soldier and inventor of the shrapnel shell, d Mar 31

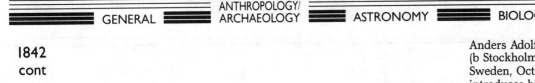

	GENERAL	ANTHROPOLOGY/ ARCHAEOLOGY	ASTRONOMY	BIOLOGY	CHEMISTRY
1842 cont				Anders Adolf Retzius (b Stockholm, Sweden, Oct 13, 1796) introduces his theory on head forms, including the cranial index to differentiate races (such as the Nordic, Mediterranean, and Alpine "races") Matthias Jakob Schleiden's *Principles of scientific botany* is published	
1843	William Hallowes Miller develops new standard measures for England, replacing the ones destroyed in the fire of the Houses of Parliament in 1834	Paul Botta excavates Assyrian sculpture from a mound near Khorsabad (Iraq), the site of an ancient Assyrian city	In Oct John Couch Adams (b Laneast, England, Jun 5, 1819) calculates the position of the then unknown planet Neptune from irregularities in the orbit of Uranus Alexis Bouvard d Paris, Jun 7 John William Draper takes the first daguerrotype of the solar spectrum Heinrich Samuel Schwabe discovers a 10-year (later corrected to 11-year) period in the number and activity of sunspots		
1844			Francis Baily d London, Aug 30 Thomas Henderson d Edinburgh, Scotland, Nov 23	Robert Chambers's *Vestiges of the natural history of creation*, published anonymously, influences both Alfred Wallace's and Charles Darwin's thinking on evolution; although it contains many errors and does not suggest a mechanism for evolution, *Vestiges* clearly promotes the evolution concept	Russian chemist Carl Claus (b Dorpat, Estonia, Jan 23, 1796) discovers the chemical element ruthenium Justus von Liebig's *Chemische Briefe* (Chemical Letters) is published John Dalton d Manchester, England, Jul 27

1843

Geologist Robert Bakewell d Hampstead, England, Aug 15

Geologist Thomas Chrowder Chamberlin b Mattoon, IL, Sep 25; he is among the first to realize that there were several ice ages, not just one

While walking with his wife along the Royal Canal, Sir William Rowan Hamilton suddenly realizes that he can create a mathematical system that is not commutative (one in which *ab* does not necessarily equal *ba*); this insight leads to the development of quaternions, a mathematical system similar to that of vectors

Joseph Liouville announces to the French Academy of Sciences on Jul 4 that he has found a remarkable proof among the papers of Evariste Galois, who had died 11 years earlier in a duel; it is the foundation of group theory, published in heavily edited form in 1846

Emil Heinrich du Bois-Reymond (b Berlin, Germany, Nov 7, 1818) demonstrates that electricity is used by the nervous system to communicate between different parts of the body

Oliver Wendell Holmes (b Cambridge, MA, Aug 29, 1809) advises doctors to prevent spreading puerperal, or childbed, fever (a common disease of mothers after childbirth at the time) by washing their hands and wearing clean clothes

Neurologist Sir David Ferrier b Aberdeen, Scotland, Jan 13; he uses the brains of living primates and other animals to locate motor and sensory regions and to map them

James Prescott Joule (b Salford, England, Dec 24, 1818) determines the mechanical equivalent of heat by measuring the rise in temperature produced in water by stirring it

Charles Wheatstone develops his Wheatstone bridge, a sensitive device for measuring electrical resistance, and the shunt, a device that enables a galvanometer to measure large electrical currents

Gustave-Gaspard Coriolis d Paris, Sep 19

Joseph Fowle develops the first tunnel drill to used compressed air for power

Isambard Kingdom Brunel builds the first tunnel under the Thames at London, which opens on Mar 25

Charles Macintosh, Scottish inventor, d Jul 25

1844

Charles Darwin's *Geological observations on volcanic islands visited during the voyage of H.M.S. Beagle, being the second part of the geology of the Beagle* confirms that Lyell's theories apply to places that Lyell never visited

Ausdehnungslehre (The study of extension) by Hermann Günther Grassman, b Stettin, Pomerania (Poland), Apr 15, 1809, deals with vectors in more than 3 dimensions, but is poorly understood by his contemporaries

The Commission for Enquiring into the State of Large Towns establishes a connection between dirt and epidemic disease in England

Charles Thomas Jackson suggests using ether to deaden pain to dentist William Thomas Green Morton

Gustave-Gaspard Coriolis's *Treatise on the mechanics of solid bodies* is published posthumously

Jean-Bernard-Léon Foucault (b Paris, Sep 18, 1819) and Armand Hippolyte Fizeau (b Paris, Sep 23, 1819) formulate the theory of allotropy, explaining how crystals can exist in two or more different forms

William Siemens develops a mechanical copying method

Samuel F.B. Morse uses his telegraph system to send a famous message from Washington to Baltimore: "What hath God wrought?"

1844 cont			Rudolf Albert von Kölliker (b Zurich, Switzerland, Jul 6, 1817) shows that the egg is a cell and that all cells in an organism originate by divisions from the egg cell Gabriel Gustav Valentin discovers that pancreatic juice breaks down food in digestion	
1845	Alfred Beach founds *Scientific American*, a US science magazine for the general reader Edward Frankland (b Churchtown, England, Jan 18, 1825) estimates that about 20 people are trained in chemistry per year in England and none in physics	Austen Layard begins to excavate the ruins of Nineveh, the capital of Assyria that fell to Babylon and the Medes in 612 BC	Hippolyte Fizeau and Léon Foucault take good quality daguerrotypes of sunspots Karl Hencke discovers the fifth asteroid, Astrea, and the sixth, Hebe Urbain Leverrier (b St. Lô, France, Mar 11, 1811), independently of John Couch Adams, uses small irregularities in the orbit of Uranus to postulate the existence and position of an eighth planet, found the next year by Johann Galle and named Neptune William Parsons, the third earl of Rosse, (b York, England, Jun 17, 1800) completes a 183-cm (72-in.) reflecting telescope at Birr Castle, with which he discovers the spiral forms of galaxies (spiral nebulas) Edgar Allan Poe is one of the first to explain why the night sky is dark by assuming that the universe has a finite age; Olbers' paradox notes that an infinite universe would imply a night sky filled equally with light at all points, but we observe individual stars	German physician Robert Remak (b Posen, Poland, Jul 30, 1815) corrects Karl Ernest Baer's theories of the development of the embryo by showing that there are only three layers present in early development, which he names the ectoderm, mesoderm, and endoderm	The Royal College of Chemistry is founded in England Adolph Wilhelm Hermann Kolbe (b Elliehausen, Germany, Sep 27, 1818) synthesizes acetic acid from nonorganic compounds Christian Schönbein discovers guncotton (nitrocellulose) when he accidentally wipes up a spilled acid solution with his wife's cotton apron; the apron vanishes in a puff as it dries Peter Spence (b Feb 19, 1806) obtains a patent for manufacturing alum by treating burnt shale and iron pyrites with sulfuric acid John Frederic Daniell d London, Mar 13 Nils Gabriel Sefström d Stockholm, Sweden, Nov 30

EARTH SCIENCE	MATHEMATICS	MEDICINE	PHYSICS	TECHNOLOGY	
Charles Wilkes publishes a report of the US Exploring Expedition (1838 to 1842) to Antarctica and the Pacific Northwest		Horace Wells (b Jan 21, 1815) is the first to use nitrous oxide as an anesthetic in dentistry Physician Sir Patrick Manson b Old Meldrum, Scotland, Oct 3; a specialist in tropical medicine, he studies elephantiasis and is the first to suggest that the mosquito may be the vector in malaria	Ludwig Edward Boltzmann b Vienna, Austria, Feb 20; he is the founder of statistical mechanics, having worked out the kinetic theory of gases independently of James Clerk Maxwell; he also develops the statistical interpretation of the second law of thermodynamics		**1844** cont
		German pathologist Rudolph Carl Virchow, b Schivelbein, Pomerania (Poland), Oct 13, 1821, is the first physician to describe leukemia	Michael Faraday discovers diamagnetism and paramagnetism, explaining each in terms of his concept of a magnetic field Faraday relates magnetism to light after finding that a magnetic field affects polarization of light in crystals; he proposes that light may be waves of electromagnetism Jean-Charles-Athanase Peltier, French scientist, d Oct 27 Polish physicist Zygmunt von Wroblewski b Grodno, Oct 28; he works with liquefying gases, becoming the first to liquefy hydrogen	Erastus Bigelow invents a power loom for weaving carpets and tapestries Robert William Thomson (b Stonehaven, Scotland, 1822) invents the rubber tire	**1845**

1846

GENERAL

The Smithsonian Institution is founded in Washington, DC, with a £100,000 bequest from English chemist and mineralogist James Smithson (b France, 1765)

Joseph Henry is elected first secretary of the Smithsonian Institution

ANTHROPOLOGY/ARCHAEOLOGY

E. George Squier and Edward Hamilton Davis explore the mounds built by American Indians in the Mississippi River valley, locating among other great mounds the Serpent Mound in Highland County, OH; the mound is about 6 m (20 ft) wide, 1.5 m (5 ft) tall, and 382 m (1254 ft) long in the shape of a snake eating an egg

ASTRONOMY

German astronomer Friedrich Bessel d Königsberg, Prussia (Kalingrad, USSR), May 17

The comet named after Wilhelm von Biela splits in two

On September 23, Johann Galle, b Pabsthaus (Germany), Jun 9, 1812, discovers the planet Neptune using the predictions of its position by Urbain Leverrier and John Couch Adams along with Bode's "law"

William Lassell (b Bolton, England, Jun 18, 1799) discovers Triton, Neptune's large satellite

Macedonio Melloni, b Parma (Italy), Apr 11, 1798, measures the heating effect of the sun's light reflected from the moon that reaches Earth during the night

BIOLOGY

German biologist Hugo von Mohl (b Stuttgart, Apr 8, 1805) identifies protoplasm (cytoplasm) as the principal living substance of a cell

Karl Wilhelm von Nägeli (b Kilchberg, Switzerland, Mar 27, 1817) shows that plant cells are not formed as buds from the surfaces of existing cells, as was proposed by Theodor Schwann

Lehrbuch der vergleichenden anatomie by German biologist Karl Theodor Ernst von Siebold (b Würzburg, Feb 16, 1804) claims that protists are made from a single cell; it also claims, incorrectly, that they form the basis of other organisms

CHEMISTRY

Ira Remsen b New York, NY, Feb 10; he introduces German methods of laboratory instruction in the United States at Johns Hopkins University and later becomes president of the university

English scientist James Marsh d Jun 21

1847

GENERAL

William Redfield (b Middletown, CT, Mar 26, 1789) becomes president of the newly founded American Association for the Advancement of Science (AAAS)

ASTRONOMY

John Herschel's *Results of observations at the Cape of Good Hope* completes the survey of the southern sky started by Edmund Halley; his work is the first to measure the brightness of stars with true precision

Maria Mitchell (b Nantucket, MA, Aug 1, 1818) discovers a comet on Oct 1; this calls attention to the role of American women in science

BIOLOGY

René-Joachim-Henri Dutrochet, French chemist and biologist, d Feb 4

Asa Gray's *Manual of botany* lists all the known plants in the northern United States

Wilhelm Friedrich Benedikt Hofmeister (b Leipzig, Germany, May 18, 1824) describes in detail how a fertilized plant ovule develops into an embryo

CHEMISTRY

Ascanio Sobrero, b Casale (Italy), Oct 12, 1812, discovers nitroglycerine

Charles Hatchett d Chelsea, England, Mar 10

1846

William Thomson, later Lord Kelvin (b Belfast, Ireland, Jun 26, 1824) uses the temperature of Earth to determine (incorrectly) that Earth is about 100 million years old; he does not take into account heat from radioactivity, unknown in his time, which makes his estimate very short of the true age

Evariste Galois's paper on group theory is finally published in Oct, more than 14 years after his death

William Thomas Morton (b Charlton City, MA, Aug 9, 1819) uses ether as an anesthetic during operations as advised by Charles Jackson, who discovered that ether is an anesthetic in 1841

A measles epidemic occurs in the Faeroe Islands; Peter Panum's study of how the disease progresses from person to person and island to island is a classic in epidemiology

Sir James Simpson (b Bathgate, Scotland, Jun 7, 1811) discovers that chloroform is a better anesthetic than ether or nitrous oxide; he starts the use of chloroform in childbirth; in 1847 his *Account of a new anesthetic agent* describes his discovery

Psychologist Granville Hall b Ashfield, MA, Feb 1; he establishes the first experimental psychology laboratory in the United States at Johns Hopkins University

Physician Benjamin Waterhouse d Cambridge, MA, Oct 2

James Prescott Joule discovers that the length of an iron bar changes slightly when the bar is magnetized

German physicist Wilhelm Eduard Weber (b Wittenberg, Oct 24, 1804) works out a system of inferring the magnetic force on a single charged particle caused by a current and devises a logical system of fundamental units for electricity

Alexander Bain develops a method of sending telegraph messages using punched paper tape, greatly improving the speed of transmission

Elias Howe (b Spencer, MA, Jul 9, 1819) patents the lock-stitch sewing machine

---

1847

George Boole (b Lincoln, England, Nov 2, 1815) publishes his first ideas on symbolic logic

Karl Christian von Staudt's *Geometrie der Lage* develops projective geometry without regard to lengths or numbers

Ernst Kummer introduces the concept of an ideal into number theory, a generalization of prime numbers that

The American Medical Association is founded

Karl Friedrich Wilhelm Ludwig (b Witzenhausen, Germany, Dec 29, 1816) develops a device that continuously records blood pressure which he uses to show that the circulation of the blood is purely mechanical; no mysterious vital processes outside of ordinary physics need to be invoked

John William Draper shows that all substances begin to glow a dull red at about 525° C (977° F), and that they progressively change color as the temperature is raised, eventually reaching a bright white

"Über die Erhaltung der Kraft" (On the conservation of energy) by Hermann Ludwig von Helmholtz, is one of the three basic papers

The Institution of Mechanical Engineers is founded in England

Charles Ellet (b Jan 1, 1810) builds a suspension bridge over the Ohio, one of the first major suspension bridges in the United States

Richard March Hoe (b Sep 12, 1812) invents both the rotary and web printing presses

1847 cont

Thomas Savage and Jeffries Wyman give the first detailed description of the gorilla

Henry Sorby publishes information on the role of sulfur and phosphorus in crops

---

1848

The revolution in France restores the republic

Karl Marx and Friedrich Engels write the *Communist manifesto*

Maria Mitchell is the first woman elected to join the American Academy of Science

Lieutenant J.H. Simpson observes the cliff houses of Mesa Verde in Colorado from a distance, but does not investigate

George Phillips Bond (b Dorchester, MA, May 20, 1925) discovers the eighth moon of Saturn, Hyperion

Joseph Henry projects the sun's image onto a screen and establishes that sunspots are cooler than the surrounding surface

German-English astronomer Caroline Herschel d Hanover (Germany), Jan 9

Julius Mayer calculates that the sun would cool down in 5000 years in the absence of an energy source and proposes that the solar surface is continuously struck by meteorites that heat it

Rudolf Albert von Kölliker is the first to isolate cells of smooth muscle tissue

Richard Owen's *On the archetype and homologies of the vertebrate skeleton* incorrectly proposes that the vertebra of different animals are modified to form skulls and other parts of the body

Louis Pasteur (b Dole, France, Dec 27, 1822) uses a slight difference in the shape of the crystals to separate the two forms of tartaric acid into a form that bends light to the right and a form that bends light to the left, founding the field of stereochemistry

Jöns Jakob Berzelius d Stockholm, Sweden, Aug 7

Chemist and physician Samuel Guthrie d Sackets Harbor, NY, Oct 19

makes the fundamental theorem of arithmetic (that all integers can be factored into primes in only one way) applicable to complex numbers

In May, Kummer writes Joseph Liouville to describe his invention of ideal complex numbers, an important step in modern number theory and also a step toward the development of modern algebra with its emphasis on structures, such as rings

In Oct, Kummer uses his concept of ideal complex numbers to show that Fermat's "last theorem" holds true for a large—posssibly infinite—class of prime-number exponents, the regular primes

Hungarian physician Ignaz Philipp Semmelweiss (b Budapest, Jul 1, 1818) discovers that puerperal fever (childbed fever) is contagious; he has doctors working for him wash their hands in the hopes of reducing the number of cases of the disease in his hospital

developing the concept of conservation of energy (along with papers by Julius Mayer and James Prescott Joule)

James Prescott Joule (independently of Julius Mayer) discovers the law of conservation of energy, which he refers to as the transformation of mechanical energy into heat and vice versa

---

Rudolph Carl Virchow coins the terms *thrombosis* for the formation of blood clots and *embolus* for a blood clot that may detach and block blood vessels

Dentist Horace Wells d Jan 28

Surgeon Sir William Macewen b Jun 22

Hippolyte Fizeau suggests that light from a source moving away from the observer will be shifted toward the red end of the spectrum, a phenomenon known as redshift; this is closely related to, but not exactly the same as, the Doppler effect

The Illinois-Michigan Canal is completed, linking the Great Lakes to the Mississippi

Inventor George Stephenson d Derbyshire, England, Aug 12

1849

ASTRONOMY

Léon Foucault observes that the sodium, or D, line in the solar spectrum is absorbed by the D line from an electric arc, the first clue as to the nature of spectral lines

Benjamin Apthorp Gould founds the first American astronomical periodical, the *Astronomical Journal*

Edouard-Albert Roche (b Monpellier, France, Oct 17, 1820) develops the concept of Roche's limit— within two and one half times the radius of a planet a satellite would tend to break up into small particles; this concept suggests that Saturn's rings could not form into a moon, since they are within Roche's limit

BIOLOGY

Berthold demonstrates the effect of hormones by showing that the implanting of testes on castrated cocks prevents the appearance of signs of castration

Naturalist Luther Burbank b Lancaster, MA, Mar 7; he develops the Burbank potato, new varieties of plums and berries, the Shasta daisy, and many other plant varieties, despite his belief in the inheritance of acquired characteristics, proved wrong many times over

Rudolf Albert von Kölliker shows that nerve fibers are extensions of nerve cells

CHEMISTRY

Edward Frankland isolates amyl alcohol

Johann Wolfgang Döbereiner d Thuringia, Germany, Mar 24

1850

ANTHROPOLOGY/ ARCHAEOLOGY

Auguste-Ferdinand-François Mariotte arrives in Egypt and begins a program of surveying and excavation which leads him to locate the burial places of the sacred Apis bulls in Saqqara, the necropolis of Memphis, an early capital of Egypt

E. George Squier reports on his discovery of mysterious statues on islands in Lake Nicaragua in Central America that do not resemble the art of the nearby Maya; later archaeologists relate them to art from the northern part of South America

ASTRONOMY

William Cranch Bond and George Bond discover a ring of Saturn inside the B ring, called the Crêpe ring or C ring

William Cranch Bond takes the first clear daguerrotype of the moon

BIOLOGY

German botanist Ferdinand Julius Cohn, b Breslau (Wroclaw, Poland, Jan 24, 1828, shows that plant and animal cytoplasm are essentially the same substance

German physicist Gustav Theodor Fechner (b Gross-Särchen, Apr 19, 1801) discovers that an increase in the intensity of a stimulus does not produce a one-to-one increase in the intensity of sensation; the sensation increases as the logarithm of the excitation

CHEMISTRY

Thomas Graham distinguishes between crystalloids and colloids, founding the science of collodial chemistry; as part of this work he also discovers dialysis

1849

Handbook of terrestrial magnetism by Scottish-German astronomer Johann von Lamont (b Braemar, Scotland, Dec 13, 1805) is a pioneering study in the field

Spanish-Mexican mineralogist Andrès Manuel Del Rio d Mexico City, Mar 23

Thomas Addison (b Longbenton, England, Apr, 1793) describes pernicious anemia

Aloys-Antoine Pollender discovers the anthrax bacillus

Hippolyte Fizeau measures the velocity of light in air by measuring the time it takes for a beam of light to pass between the teeth of a rotating gear; the light is reflected by a mirror and stopped by the next tooth of the gear; the result, 315,000 km/sec (196,000 mi/sec), is within 5 percent of today's accepted value

James Thomson (b Feb 16, 1822) predicts that applying pressure to water lowers its freezing point, using Sadi Carnot's theory of heat

William Thomson, later Lord Kelvin, coins the term *thermodynamics* in an account of Sadi Carnot's theory of heat

The first bombing raid from the air occurs when a pilotless Montgolfier balloon is used to drop bombs on Venice (Italy)

George Henry Corliss (b Jun 2, 1817) patents a more efficient steam engine that uses four valves instead of one

Sir Marc Isambard Brunel, British engineer, d Dec 12

1850

Johann von Lamont discovers periodic variations in the magnetic field of Earth and relates them to the occurrence of sunspots

Matthew Fontaine Maury (b near Fredericksburg, VA, 1806) prepares a chart of the Atlantic Ocean and notes for the first time that it is shallower in the center than it is nearer the edges, the first indication of the mid-Atlantic Ridge

Paradoxien des Unendlichen (Paradoxes of the infinite) by Bernhard Bolzano contains the first use of the word *set* in mathematics and deals with infinite sets in a manner made popular later by Georg Cantor and Richard Dedekind; however, Bolzano's work is overlooked at the time

Francis Guthrie discovers that it seems unnecessary to use more than four colors in coloring any map to prevent adjacent regions of the same color

Über die bewegende Kraft der Wärme by German physicist Rudolf J.E. Clausius, b Köslin, Pomerania (Koszalin, Poland), Jan 2, 1822, is the first statement of the second law of thermodynamics, restated by Clausius in 1865 as "entropy always increases in a closed system," or energy in a closed system will change toward heat and disorder

William Armstrong develops an hydraulic accumulator, making hydraulic machinery independent of a water source

Francis Galton (b Birmingham, England, Feb 16, 1822) invents a teletype printer

1851

Baron von Humboldt's *Kosmos* gives currency to Heinrich Samuel Schwabe's 1843 discovery of the 11-year solar sunspot cycle

French-American ornithologist John James Audubon d New York, NY, Jan 27

Bacteriologist Charles Chamberland b Chilly-le-Vignoble, France, Mar 12; his specialties include improving means for sterilizing medical equipment and production of special filters for bacteria; his filters lead to the discovery of viruses

Wilhelm Friedrich Hofmeister discovers the alternation of generations in such nonflowering plants as mosses and ferns

Hugo von Mohl's *Grundzüge der anatomie und physiologie der vegetabilischen zelle* (Principles of the anatomy and physiology of the vegetable cell) describes his view of protoplasm (now called cytoplasm) and generally advances cell theory for plants, noting the fibrous structure of cell walls

German naturalist Lorenz Oken d Zurich, Switzerland, Aug 11

Chemist and physician John Kidd d Oxford, England, Sep 17

The nature of heat

In Aristotle's physics, heat is one of the active qualities of a body. Hotness, as well as dryness, and their opposites, coldness and moistness, were the qualities that defined an element. Later, heat and temperature came to be viewed as lesser fundamental properties of matter, but opinions about the nature of heat varied sharply. Francis Bacon defined heat as local motion, and René Descartes and Robert Hooke described heat as the ceaseless motion of particles. Others, such as the Dutch physician Hermann Boerhaave, described heat as a subtle fluid, later called *caloric,* that could pass from one object to another.

During the beginning of the nineteenth century, the theory that heat was a form of motion won over the liquid, or caloric, theory. Even earlier, at the end of the eighteenth century, Count Rumford (Benjamin Thomson) had observed the large amount of heat produced during the boring of cannons and reached the conclusion that heat was added to the cannon by motion.

The German physician Julius Robert Mayer was one of the first to relate heat directly to mechanical work, writing of his discovery in 1842. That heat and mechanical work were both forms of energy was a very important discovery. It became the basis of a new science that gained quickly in importance during the nineteenth century: thermodynamics.

Sadi Carnot, intrigued by why British steam engines were more efficient than French ones, started investigating the physical processes that take place in a steam engine. He was the first to apply physics to a technical problem. He published the results in 1824 as *"Reflexions sur la puissance motrice du feu"* (Reflections on the motive power of fire). However, this paper remained little known for 10 years, until Clapeyron discussed it in the *Journal de l'Ecole Polytechnique.*

Carnot had found that the efficiency of a mahcine depended on the difference in temperature between the heat source that supplies the work (hot reservoir) to the machine and the temperature reservoir into which the machine discharges excess heat (cold reservoir). The mechanical equivalent of heat is the amount of heat that can produce a given amount of motion. Although Carnot had found an approximate value for the mechanical equivalent of heat, he died too early, at the age of 36, to publish this result.

Contrary to Carnot, James Prescott Joule believed that heat was a quantity that was destroyed in a machine to produce mechanical work. In a series of famous experiments, he measured the increase of temperature of water stirred by a determined amount of mechanical work. As a result, he also determined the mechanical equivalent of heat, five years after the similar work of Julius Mayer. His work was more quickly recognized than Mayer's, however, since Joule was a physicist and Mayer a physician.

Paul Erman, German physicist, d Oct 11

Joseph Liouville establishes that transcendental numbers exist; transcendental numbers are numbers such as π and e that are not solutions of algebraic equations; Liouville's transcendental numbers are typified by 0.10100100010000 . . . in which the number of zeros between ones increases regularly

Georg Friedrich Bernhard Riemann (b Hanover, Germany, Sep 17, 1826) relates the theory of functions to the theory of n-dimensional surfaces in a complex domain, now called Riemann surfaces

Carl Jacobi d Berlin, Germany, Feb 18

Claude Bernard (b Saint-Julien, France, Jul 12, 1813) discovers that nerves controlling the dilation of the blood vessels control the body's temperature in humans

Hermann Ludwig von Helmholtz reinvents the ophthalmoscope independently of Charles Babbage's version of 1847

Hippolyte Fizeau shows that the velocity of light is higher in water flowing in the direction of the beam than that of light propagating in the direction opposite the direction of flow

Léon Foucault uses a pendulum suspended in a church to demonstrate Earth's rotation

George Stokes (b Skreen, Ireland, Aug 13, 1819) develops the formula that describes how a small body, such as a droplet of oil or water, falls through a fluid, such as air

William Thomson, later Lord Kelvin, proposes the concept of absolute zero, a temperature that is the least possible because, according to Charles' law, the volume of a gas would be zero at a temperature of −273 degrees C (−460 degrees F); Thomson suggests that the energy of molecules instead of the volume would be zero

Thomson deduces a form of the second law of thermodynamics from the work of Sadi Carnot—that is, that energy in a closed system tends to become unusable as it gradually becomes uniform heat

Hans Christian Oersted d Copenhagen, Denmark, Mar 9

Oliver Joseph Lodge (b Staffordshire, England, Jun 12, 1851) suggests correctly that the sun may emit radio waves, although these are not detected until the 1940s

William Nicol d Edinburgh, Scotland, Sep 2

The Great International Exhibition is held in London; it features the latest technical innovations in industry and promotes the application of science to technology

William Channing and Moses Farmer develop an electric fire alarm system

Isaac Merrit Singer (b Oct 27, 1811) patents a continuous-stitch sewing machine

The Crystal Palace, designed and built by Joseph Paxton, is opened by Queen Victoria; a remarkable building of glass and iron, it is one of the highlights of the Great Exhibition in London

Artist and inventor Louis Daguerre d Petit-Brie-sur-Marne, France, Jul 12

	GENERAL	ANTHROPOLOGY/ ARCHAEOLOGY	ASTRONOMY	BIOLOGY	CHEMISTRY
1852	Fabian von Bellingshausen, Russian explorer, d Kronstadt, Jan 25 President Louis-Napoleon of France declares that he is the emperor Napoleon III			Potassium sulfate, mined at Srassfurt, Germany, is used as a fertilizer James Gana formulates his theory of cephalization: the more highly evolved an animal, the more its head region is developed Hermann Ludwig von Helmholtz becomes the first to discover how fast a message travels along a nerve, measuring the speed of transmission in a frog's nerve cell Rudolf Albert von Kölliker discovers the cellular origin of spermatozoa, which were believed to originate from a fermentation process	Edward Frankland introduces the concept of valence of elements into chemistry; that is, that each atom has a definite capacity to combine with other atoms Abraham Gesner discovers kerosene and gives it its name Henri-Victor Regnault (b Aix-la-Chapelle, France, Jul 21, 1810) shows that the behavior of gases deviates slightly from Boyle's law and determines absolute zero to be $-273°$ C ($-460°$ F), although he does not have a clear idea of what the behavior of a gas at that temperature might be Johan Gadolin d Wirmo, Finland, Aug 15
1853				Bacteriologist Pierre Roux b Charente, France, Dec 17; he discovers that diphtheria is not caused directly by a bacterium but by a toxin produced by the bacterium	Hans Peter Thomsen (b Copenhagen, Denmark, Feb 16, 1826) develops a method of producing sodium carbonate from cryolite Leopold Gmelin d Heidelberg, Germany, Apr 13 Auguste Laurent d Paris, Apr 23

1852

Sir Edward Sabine (b Dublin, Ireland, Oct 14, 1788) shows that changes in sunspots cause parallel changes in Earth's magnetic field

Mineralogist Wilhelm Hisinger d Skinnskatteberg, Sweden, Jun 28

Geologist Gideon Mantell d London, Nov 10

Michel Chasles's *Traité de géométrie supérieure* is influential in establishing the use of directed line segments in geometry

Karl Vierodt makes the first accurate count of red blood cells; such counts later become an important tool in diagnosing anemia

James Prescott Joule and William Thomson, later Lord Kelvin, establish that an expanding gas becomes cooler; this is now known as the Joule-Thomson effect

The American Society of Civil Engineers is founded in New York

Henri Giffard builds and flies the first steam-powered dirigible

William Kelly (b Aug 21, 1811), working in a small town near Eddyville, KY, and aided by four Chinese steel-making experts, invents a new process for making steel that anticipates the Bessemer process that will be developed only four years later

Elisha Graves Otis (b Halifax, VT, Aug 3, 1811) invents the first passenger elevator that has a safety guard that keeps the elevator from falling even when the cable is completely cut

Samuel Wetherhill develops an economic method for removing zinc oxide from ore

1853

James Coffin describes three distinct wind zones that occur in the Northern Hemisphere

Geologist Christian von Buch d Berlin, Germany, Mar 4

William Shanks (b Corsenside, England, Jan 25, 1812) completes his calculation of the value of pi to 707 places; in 1944 it will be discovered, however, that his mistake in the 528th place caused all subsequent places to be incorrect

Surgeon William Beaumont d St. Louis, MO, Apr 25

Léon Foucault demonstrates that the velocity of light is less in water than in air, tending to confirm the wave theory of light

Johann Wilhelm Hittorf (b Bonn, Rhenish Prussia, Mar 27, 1824) suggests that different ions in a solution impelled with an electric current travel at different rates; this would account for more ions of one type appearing at one of the two electrodes

William John Macquorn Rankine (b Edinburgh, Scotland, Jul 5, 1820) introduces the concept of potential energy, or energy of position

Austrian physicist Christian Johann Doppler d Venice (Italy), Mar 17

Dominique-François Arago d Paris, Oct 2

Sir George Cayley's helicopter "toys" modeled on Chinese toy helicopters are shown to go 27 m (90 ft) into the air

1854

Astronomy

Dominique-François Arago's *Astronomie populaire* (Popular astronomy) is published posthumously

Hermann Ludwig von Helmholtz proposes that the sun is heated by gravitational contraction and calculates that if the sun contracted by 60 m (200 ft) per year, it would keep radiating for 15 million years; his calculations show that the sun has existed for only 25 million years, far too short a time for Earth's geological record

Helmholtz predicts the "heat death" of the universe: the universe will ultimately reach a state of a universally uniform temperature

Biology

Naturalist Edward Forbes d Wardic, Scotland, Nov 18

Naturalist Henri Braconnot d Nancy, France, Jan 13

Chemistry

Experiments by David Alter (b US Dec 3, 1807) lead him to the conclusion that each element can be identified from its spectrum

Alexander William Williamson (b London, May 1, 1824) is the first to explain how a catalyst works

Physical chemist Hendrik Roozeboom b Alkmaar, Netherlands, Oct 24; he establishes experimentally the mathematical phase rule of Josiah Gibbs, an equation connecting the solid, liquid, and gaseous phases of a substance that also accounts for many other rules that had been experimentally discovered

1855

General

Friedrich Karl Christian Ludwig Büchner's *Kraft und Stoff* (Force and matter) revives the idea that force and matter are the ultimate realities in nature

Astronomy

William Parsons, the third earl of Rosse observes the spiral structure of galaxies

Chemistry

About this time Robert Bunsen begins using the gas burner that is named for him; he is not the inventor, however, for the burners Bunsen uses are developed by his technician, C. Desaga;

Sir George Biddel Airy (b Alnwick, England, Jul 27, 1801) measures the mass of Earth from the top and bottom of a coal mine

Heinrich Ernst Beyrich identifies the Oligocene Period in Earth's history

George Boole's *An investigation of the laws of thought, on which are founded the mathematical theories of logic and probabilities* develops the first form of symbolic logic, known today as Boolean algebra

Geologist W.K. Loftus finds two clay tablets in Babylonian that show that Mesopotamian peoples used a place-value system of numeration based on 60

Bernhard Riemann proposes a definition of the integral that does not require the function to be integrated to be continuous

Karl Weierstrass (b Ostenfelde, Westphalia, Germany, Oct 31, 1815) publishes *"Zur Theorie der Abelschen Functionen"* (On the theory of Abel's functions), which earns him instant recognition in the mathematical community

Riemann's *Über die Hypothesen welche der Geometrie zu Grunde liegen* (On the hypotheses that underlie the foundation of geometry), delivered as a lecture Jun 10, but not published until 1868, generalizes the concept of non-Euclidean geometry to show that various non-Euclidean geometries are possible

John Snow (b York, England, Mar 15, 1813) shows that removing the pump handle of a well contaminated by sewage reduces the incidence of cholera in the vicinity of the well

Georg Simon Ohm d Munich, Germany, Jul 7

Macedonio Melloni d Portici (Italy), Aug 11

Johannes Rydberg b Halmstad, Sweden, Nov 8; he develops an empirical equation to describe the position of lines in the spectrum that is a generalized version of the better known Balmer equation; Rydberg creates his equation without knowing of the Balmer equation

An electric telegraph is installed between Paris and London

Inventor William Hampson b Cheshire, England; he develops methods of mass-producing liquid air

Elisha Otis installs the first safe and workable elevator at New York's Crystal Palace

1854

Matthew Maury's *Physical geography of the sea* is the first textbook of oceanography

Luigi Palmieri, b Faicchio (Italy), Apr 22, 1807, invents the first crude seismometer

Johann Karl Friedrich Gauss d Göttingen, Germany, Feb 23

P.G. Lejune Dirichlet succeeds Gauss as professor of mathematics at Göttingen University

The third known pandemic of bubonic plague (the others started in 541 AD and 1346) begins in China

Thomas Addison describes the hormone deficiency disease that results

Robert Bunsen invents the absorptiometer

German inventor Heinrich Geissler (b Igelshieb, May 26, 1814) develops his mercury pump, which he uses to produce

The International Exhibition is held in Paris

Henri-Etienne Sainte-Claire Denville (b St. Thomas, Virgin Islands, Mar 11, 1818) develops a new method of producing

1855

	GENERAL	ANTHROPOLOGY/ ARCHAEOLOGY	ASTRONOMY	BIOLOGY	CHEMISTRY
1855 cont					furthermore, essentially the same type of burner was developed earlier by Michael Faraday

Lord Dunaldson suggests the use of poison gas (sulfur dioxide) in the Crimean War

Charles Adolphe Wurtz (b Wolfisheim, Alsace, France, Nov 26, 1817) develops the method of synthesizing long-chain hydrocarbons using hydrocarbon iodides and sodium, a method still known as the Wurtz reaction |
| **1856** | | A skeleton found in a cave in the Neander valley near Düsseldorf, although derided by many as recent, is actually the first known remains of what we now call the Neanderthals, a unique subspecies of *Homo sapiens* that flourished in the ice ages | Austrian astronomer Wilhelm von Biela d Venice (Italy), Feb 18

George Phillips Bond discovers that photographs of stars reveal their magnitudes | John William Draper publishes a work on physiology that contains some of the earliest known photomicrographs

Louis Pasteur's *Recherches sur la putréfaction* (Researches on fermentation) reports his discovery that fermentation is caused by microorganisms (yeast), not by chemical means, as previously supposed

Zoologist Edmund Wilson b Geneva, IL, Oct 19; he studies how fertilized eggs develop into embryos and is the first to note the X and Y chromosomes of mammals | Sir William Henry Perkin (b London, Mar 12, 1838) synthesizes the first artificial (aniline) dye, mauve, and starts a fashion craze; the next few years in England are known as the Mauve Age

Charles-Frédéric Gerhardt d Aug 19 |
| **1857** | | | George Phillips Bond is the first to photograph a double star, Misar, and obtain images of both components

Léon Foucault starts the production of silver-coated glass telescope mirrors; | Hermann Ludwig von Helmholtz proposes the resonance theory of hearing; he argues that the transverse fibers of the basilar membrane in the cochlea of the inner ear act as tuned resonators | Louis Thénard d Paris, Jun 21 |

EARTH SCIENCE	MATHEMATICS	MEDICINE	PHYSICS	TECHNOLOGY	
German-Swiss geologist Johann von Charpentier d Bex, Switzerland, Sep 12		from deterioration of the adrenal gland; the disease is named after him Claude Bernard's *Leçons de physiologie expérimentale* (Lessons in experimental psychology) advances his theories of homeostasis in the body, the concept that the body maintains a constant environment even though external conditions change Physiologist François Magendie d Seine-et-Oise, France, Oct 7	the first good vacuum tubes; these tubes, as modified by Sir William Crookes, become the first to produce cathode rays, leading to the discovery of the electron German physicist Heinrich Daniel Rühmkorff (b Hanover, Jan 15, 1803) invents the version of the induction coil that bears his name William Thomson, later Lord Kelvin, develops a theory of transmission of electrical signals through submarine cables that is then applied to the first submarine telegraph cables	aluminum and casts a 7-kg (15-lb) ingot; over the next four years the price of aluminum is reduced by a factor of 100, although it is still more expensive than steel	**1855 cont**
William Thomas Blanford observes that the Talchir conglomerates of India were caused by glaciation, an idea leading to the realization that ice ages occurred many times in the past and that both hemispheres were involved	Nikolai Ivanovich Lobachevski d Kazan (USSR), Feb 24	Claude Bernard discovers glycogen, used by the body to store glucose in the liver; when glucose is needed for energy, the liver converts the glycogen back to glucose Karl Friedrich Wilhelm Ludwig is the first to keep animal organs alive outside the body, which he does by pumping blood through them	Amedeo Avogadro, Count of Quarenga, d Turin (Italy), Jul 9 Robert Bunsen and Henry Roscoe invent the actinometer	English inventor Henry Bessemer (b Charlton, England, Jan 19, 1813) develops the Bessemer process for producing inexpensive steel Sir William Siemens (b Lenthe, Germany, Apr 4, 1823) develops the regenerative furnace; it burns previously unburnt gases for greater efficiency and is the forerunner of the open-hearth steel process Engineer Frederick Winslow Taylor b Philadelphia, PA, Mar 20; he is among the first to make time-and-motion studies of workers with the idea of eliminating unneeded steps	**1856**
Meteorologist William Redfield d New York, NY, Feb 12	Augustin-Louis, Baron Cauchy, d Sceaux, France, May 23	Physiologist Marshall Hall d Brighton, England, Aug 11	*"Über die Art der Bewegung, welche wir Wärme nennen"* (On the type of motion termed heat) by Rudolf J.E. Clausius establishes his kinetic theory of heat on a mathematical basis; it also explains how evaporation occurs	Inventor Frederick Archer d London, May 2	**1857**

1857 cont

ASTRONOMY

these will replace metal mirrors used in previous reflecting telescopes

Peter Andreas Hansen's *Tables de la lune* (Lunar tables) are printed by the British government and used thereafter in nautical almanacs

James Clerk Maxwell (b Edinburgh, Scotland, Nov 13, 1831) proves that Saturn's rings must be composed of discrete particles

BIOLOGY

Gregor Johann Mendel, b Heinzendorf, Silesia (Poland), Jul 22, 1822, starts the experiments with peas in his garden that will lead to his working out the laws of heredity

1858

ASTRONOMY

Usherwood takes the first photograph of a comet (Donati's)

Warren De la Rue (b Guernsey, Channel Islands, England, Jan 15, 1815) develops the photoheliograph, which makes taking photographs of the sun routine

BIOLOGY

Sir Jagadischandra Bose b Mymensingh, India, Nov 30; he uses his training in physics to develop instruments for studying plant growth and other extremely small motions of plants

The secretary of the Linnean Society reads communications from Charles Darwin and Alfred Russel Wallace, an unpublished manuscript of Darwin's from 1844, a letter from Darwin to Asa Gray written in 1857, and a letter from Wallace to Darwin from Feb, 1858; all three describe the theory of evolution by natural selection

George Bentham (b England, Sep 22, 1800) publishes the *Handbook of British flora*

Botanist Robert Brown d London, Jun 10

Rudolph Carl Virchow propounds the theory that all cells arise from cells, which is correct, and that disease is caused when cells refuse to cooperate with other cells, which is true of some diseases, such as cancer and lupus, but not true of infectious diseases

CHEMISTRY

Italian chemist Stanislao Cannizzaro (b Palermo, Sicily, Jul 13, 1826) learns of the hypothesis of Avogadro that all gases at a given temperature contain the same number of particles; this idea leads Cannizzaro to clarify the concept of a molecule

Archibald Scott Couper (b Kirkintillach, Scotland, Mar 3, 1831) introduces the concept of bonds into chemistry; he also assumes that carbon atoms form the backbone of organic compounds

August Wilhelm von Hofmann makes the artificial dye magenta from coal tar; it is named the following year by French dyers after the French defeat the Italians at the battle of Magenta

German chemist Friedrich August Kekulé (b Darmstadt, Hesse, Sep 7, 1829) shows that the four bonds of carbon can account for the different isomers of organic compounds

Robert Hare d Philadelphia, PA, May 15

Carl Mosander d Angsholm, Sweden, Oct 15

Gustav Robert Kirchhoff, b Königsberg (Kalingrad, USSR), Mar 12, 1824, discovers that static electric forces and magnetic forces are related to each other by a constant that turns out to be the speed of light in a vacuum, a clue that electromagnetism has something to do with light

Johann Salomo Schweigger d Halle, Germany, Sep 6

1857 cont

Arthur Cayley (b Richmond, England, Aug 16, 1821) founds the algebra of matrices in the context of the theory of invariant transformations

Florence Nightingale's *Notes on matters affecting the health, efficiency and hospital administration of the British army* contains numerous innovative statistical diagrams, including polar-area charts, which Nightingale invents

A. Henry Rhind purchases the Egyptian papyrus subsequently known as the Rhind papyrus; it proves to be one of the two main keys to understanding Egyptian mathematics

George Peacock d Ely, England, Nov 8

Franciscus Cornelis Donders (b Tilburg, Netherlands, May 27, 1818) discovers that farsightedness can be caused by too shallow eyeballs

German psychologist Wilhelm Wundt (b Neckray, Aug 16, 1832) begins publication of *Beiträge der Sinneswahrnehmungen* (Contribution to the theory of sensory perception), which is completed in 1862; it introduces the concept of experimental psychology and becomes one of the founding works of psychology as a science

Physiologist Johannes Peter Müller d Berlin (Germany), Apr 28

Physician John Snow d London, Jun 16

Physician Richard Bright d London, Dec 16

Julius Plücker shows that cathode rays bend under the influence of a magnet, suggesting that they are connected with charge in some way; this is an early step along the path that will lead in 1897 to the discovery that cathode rays are composed of electrons

The Foreland lighthouse in Kent, England, is the first to be equipped with electrically powered arc lights

The first aerial photograph is taken from the balloon *Nadir* as it flies over Paris

The first Atlantic telegraphic cable is laid

1858

1859		Pierre-Paul Broca (b Sainte-Foy-la-Grande, France, Jun 28, 1824) starts the *Société d'Anthropologie*	German astronomer Friedrich Wilhelm August Argelander, b Memel (Klaipeda, USSR), Mar 22, 1799, begins publication of the *Bonner Durchmusterung* (Bonn survey); when complete in four volumes in 1862, it will list the positions of 457,848 stars	Charles Darwin's *On the origin of species by means of natural selection or the preservation of favoured races in the struggle for life*, usually known as *The origin of species*, explains in detail his principle of natural selection and its influence on the evolution of different species	Adolph Wilhelm Hermann Kolbe develops the reaction named after him; it leads to the large-scale synthesis of acetylsalicylic acid—aspirin
			William Cranch Bond d Cambridge, MA, Jan 29	Naturalist Alexander, Baron von Humboldt d Berlin, Germany, May 6	
			Richard Christopher Carrington (b London, May 26, 1826) discovers that the sun does not rotate as a single unit; instead the equator rotates in 27.5 days, while regions near the pole rotate in 25 days	Biochemist Thomas Osborne b New Haven, CT, Aug 5; his work demonstrates that proteins come in a vast number of varieties, some of which contain amino acids essential for life; he also discovers vitamin A independently of Elmer McCollum	
			Gustav Kirchhoff and Robert Bunsen introduce the spectroscope for chemical analyses of metals placed in a flame; they also use it to study the chemical structure of the sun		
1860		Jacob Burkhardt's *The civilization of the Renaissance in Italy* popularizes the concept of a Renaissance period in history	Edouard-Armand-Isidore Hippolyte Lartet (b Saint Guiraud, France, Apr 15, 1801) about this time discovers a mammoth tooth engraved with a sketch of a mammoth, convincing evidence that whoever engraved the tooth had actually seen a live mammoth	Warren De la Rue shows that the prominences visible in a solar eclipse come from the sun and not from the moon	Louis Agassiz attacks Darwin's *The origin of species*, rejecting the idea of evolution of the species and arguing that each species was created separately	Friedrich Kekulé organizes the First International Chemical Congress in Karlsruhe, Germany, the first international meeting of scientists ever held
				Canadian-American astronomer Simon Newcomb (b Wallace, Nova Scotia, Mar 12, 1835) demonstrates that it is unlikely that the asteroids orbiting between Mars and Jupiter are the survivors of a planet that once occupied an orbit in that region but which somehow broke into pieces	Zoologist Félix Dujardin d Rennes, France, Apr 8	Stanislao Cannizzaro revives Avogadro's hypothesis that all gases at a given temperature contain the same number of molecules at the First International Chemical Conference, making the hypothesis popular for the first time; he also emphasizes atomic weights
					Thomas Henry Huxley (b Ealing, Middlesex, England, May 4, 1825) clashes with Bishop "Soapy Sam" Wilberforce about evolution at the annual meeting of the British Association for the Advancement of Science	Gustav Kirchhoff and Robert Bunsen announce on May 10 the discovery of

1859

Swiss geologist Ignatz Venetz d Saxon-les-Bains, Valais, Apr 20

Geophysicist Harry Reid b Baltimore, MD, May 18; he develops the elastic rebound theory of the cause of earthquakes, in which one fault moving against another causes the quakes; previous theories assumed that the quakes caused faults rather than the other way around

Edwin Laurentine Drake (b Greenville, NY, Mar 29, 1819) drills the world's first oil well in Titusville, PA, striking oil on Aug 28, starting a process that will eventually fund much large-scale geological research in the search for more oil

Gustav Kirchhoff finds that one consequence of the second law of thermodynamics is that, at a uniform temperature, it is impossible to distinguish between objects by the light they radiate, even if they are glowing hot; an outside light source must be provided

On Oct 27 Kirchhoff and Robert Bunsen announce the first measurements of frequencies of spectral lines, a powerful new way to identify elements from the light they emit or absorb

Gustav Kirchhoff recognizes that sodium is found on the sun and discovers his black body radiation law

Jean-Joseph-Etienne Lenoir (b Mussy-la-Ville, Belgium, Jan 12, 1822) develops the first internal combustion engine that works, using coal gas as a fuel; it is, however, very inefficient

Gaston Planté (b Orthez, France, Apr 22, 1834) invents the storage battery, which can be recharged again and again

William John Macquorn Rankine's *Manual of the steam engine* introduces engineers to thermodynamics and coins most of the modern terms used in the field

Ferdinand de Lesseps (b France, Nov 19, 1805) turns over the first shovelful of earth to start construction of the Suez Canal, which is completed ten years later on Aug 15, 1869

Isambard Kingdom Brunel, British engineer, d Sep 15

1860

Herman von Meyer discovers a rare fossil of soft tissue; in this case, it is a feather, the first sign of what Meyer names *Archaeopteryx*, recognized for many years as the earliest bird; later in the same year, an almost complete fossil of the same early bird is found

Janos Bolyai d Marosvásárhely, Hungary, Jan 17

Gustav Theodor Fechner's *Elemente der Psychophysik* presents psychology as an exact science

Scottish surgeon James Braid d Manchester, England, Mar 25

Anders Retzius, Swedish anatomist, d Stockholm, Apr 18

Physician Thomas Addison d Bristol, England, Jun 29

James Clerk Maxwell and, separately, Ludwig Edward Boltzmann develop the study of the statistical behavior of molecules in a gas; this is known today as the Maxwell-Boltzmann statistics

Michael Faraday describes his experiment with lowering the freezing point of water by applying pressure in a paper "Pressure melting effect" at a meeting of the Royal Society

Jean-Joseph-Etienne Lenoir builds the first "horseless carriage" that uses an internal combustion engine for power (previous versions used steam engines)

Elisha Otis patents a steam-driven elevator

Inventor Charles Goodyear d New York, NY, Jul 1

	GENERAL	ANTHROPOLOGY/ ARCHAEOLOGY	ASTRONOMY	BIOLOGY	CHEMISTRY
1860 cont				About this time Gregor Mendel discovers the laws of heredity, which he communicates to Karl Wilhelm von Nägeli, who is not interested and does not pursue the matter	cesium, the first element found as a result of spectroscopy; the name *cesium* means "sky blue," referring to the color of its spectrum
1861	Italy is unified under the rule of Victor Emmanuel of Sardinia		Gustav Kirchhoff's map of the solar spectrum shows that elements such as sodium can be detected in the atmosphere of the sun by means of the dark lines they cause in the spectrum Zöllner invents the astronomical photometer for measuring stellar magnitudes	Rudolf Albert von Kölliker is among the first to interpret the development of the embryo in terms of cell theory	William Crookes (b London, Jun 17, 1832) discovers thallium as a bright green line in the spectrum of a selenium ore he is investigating; *thalliu* means "green twig" Gustav Kirchhoff and Robert Bunsen discover a bright red line in a spectrum that represents a new element, which they name rubidium for "red" Ernest Solvay (b Rebecq, Belgium Apr 16, 1838) discovers a process for making sodium bicarbonate from salt water, ammonia, and carbon dioxide, which is much more economical than preceding processes for developing sodium bicarbonate
1862	Sir James Clark Ross, Scottish explorer, d Aylesbury, England, Apr 3	Edwin Smith, an American Egyptologist, purchases a papyrus that, when translated in 1930, turns out to be a summary of the surgical practices of ancient Egypt based on a text that was probably originally written around 2500 BC (although the Edwin Smith Surgical Papyrus itself dates from about 1550 BC)	The first great refracting telescopes are constructed, including the Newall 64-cm (25-in.) made by Cooke While testing a new telescope lens, Alvan Clark (b Fall River, MA, Jul 10, 1832) and his son Alvan Graham Clark observe Sirius B, the dark companion (now known to be a white dwarf) of the bright star Sirius Heinrich Ludwig d'Arrest (b Berlin, Germany, Aug 13, 1822) discovers the asteroid Freia Léon Foucault measures the distance from Earth to the sun	Charles Darwin's *The various contrivances by which orchids are fertilized by insects* shows how various parts of orchids' flowers have evolved to increase pollination by insects Biochemist Ernst Felix Hoppe-Seyler (b Freiburg-an-der-Unstrut, Germany, Dec 26, 1825) prepares a crystalline form of hemoglobin Karl Georg Friedrich Rudolf Leuckart (b Helmstedt, Braunschweig, Germany, Oct 7, 1822) publishes the first of his two volumes describing the	Anders Jonas Ångström (b Medelpad, Sweden, Aug 13, 1814) discovers from the solar spectrum that the element hydrogen is found on the sun Alexandre-Emile Beguyer de Chancourtois (b Paris, Jan 20, 1820) develops an early version of the periodic table of the elements, but his work is poorly understood until after Mendeléev's table becomes known Swiss-American chemist John Ulric Nef b Herisau (Switzerland), Jun 14; he establishes that

EARTH SCIENCE	MATHEMATICS	MEDICINE	PHYSICS	TECHNOLOGY	
	Bernhard Riemann's *Über eine Frage der Wärmeleitung* (On a question in the conduction of heat) develops the mathematics of quadratic differential forms, which will become important in general relativity	Pierre-Paul Broca is the first to demonstrate that a particular region in the brain (Broca's area) is connected to a particular faculty, in this case the faculty of speech, by discovering a lesion in the brain during an autopsy on a man who could not speak intelligibly Physiologist Karl von Voit (b Amberg, Germany, Oct 31, 1831) demonstrates that different foods do not provide energy for different body functions; for example, proteins break down at the same speed whether or not work is being done		New York and San Francisco are connected by a telegraph line Inventor Elisha Graves Otis d Yonkers, NY, Apr 8 James Clerk Maxwell makes the first color photograph, a form of projection slide	1861
		French physician Pierre Bretonneau d Passy, Feb 18	Physicist Jean-Baptiste Biot d Paris, Feb 3 Physician Allvar Gullstrand b Landskrona, Sweden, Jun 5; he specializes in the physics of vision and develops eyeglasses to correct astigmatism and for use after lenses have been removed in cataract operations	In the American Civil War, the ironclad *Monitor*, designed by Swedish-American inventor John Ericsson (b Långbanshyttan, Sweden, Jul 31, 1803), defeats but does not sink the ironclad *Merrimac* in the first naval engagement between ironclad ships Alexander Parkes (b Birmingham, England, Dec 29, 1813) displays at an international exhibition in London many items made from what came to be known as celluloid, a plastic on which he had been working throughout the 1850s Samuel Colt, American inventor, d Jan 10	1862

1862 cont

BIOLOGY

parasites of human beings; volume two is published in 1876

CHEMISTRY

carbon can have a valence of two as well as its more common valence of four

1863

GENERAL

The National Academy of Sciences is founded in the United States

Thomas Henry Huxley's *Man's place in nature* is published

ASTRONOMY

Friedrich Argelander founds the first international organization of astronomers, the *Astronomische Gesellschaft*

Astronomer Annie Jump Cannon b Dover, DE, Dec 11; her work will form the basis of the *Henry Draper catalogue* of 225,300 stars

Sir William Huggins (b London, Feb 7, 1824) uses the spectra of stars to show that the same elements exist in the stars as on Earth, disproving the Greek idea that the stars are composed of something different from elements on Earth

BIOLOGY

Louis Pasteur discovers the microorganism responsible for the souring of wine

CHEMISTRY

Johann Friedrich Wilhelm Adolf von Baeyer (b Berlin, Germany, Oct 31, 1835) develops the first barbiturate, which he supposedly names for his girlfriend Barbara

John Alexander Newlands on Feb 7 announces what he calls the "law of octaves," essentially an early version of the periodic table; his work is met with derision

German chemist Ferdinand Reich (b Bernburg, Feb 19, 1799) discovers what he thinks is a new element; since he is colorblind, he asks his assistant, Hieronymus Theodor Richter, to examine its spectrum; indeed, it is a new element, which is named indium for the indigo color of its spectrum

Johannes Wislicenus (b Klein-Eichsted, Germany, Jun 24, 1835) discovers two isomers of lactic acid that differ only in the way they behave with regard to polarized light; he uses this difference a few years later to fight for an organic chemistry based on three-dimensional concepts of molecules

German chemist Eilhardt Mitscherlich d Schönberg, Feb 28

Geochemist Vladimir Vernadsky b St. Petersburg (Leningrand, USSR), Mar 12; he specializes in the overall chemistry of Earth's crust

---

From action at a distance to fields

Early theories of motion, such as Aristotle's, assumed that some physical object must push or pull on another to cause motion. The ancient Greeks and others were aware of exceptions to this hypothesis, since they knew of magnets and static electricity. In both cases, objects move without any apparent contact—action at a distance.

In the first scientific revolution of the sixteenth and seventeenth centuries, the problem of action at a distance was "solved" by postulating various subtle media through which the action is transmitted. Generically, these are called aethers or ethers. The spelling *aether* is preferred for the whole class, while *ether* is generally used for the aether of electromagnetism. Different scientists postulated different aethers as needed, so there was an aether for heat, another for gravity, another for the magnetic force, and so forth—although Newton's theory of gravity really involved action at a distance, since Newton made no hypotheses as to how gravitational force was transmitted.

Michael Faraday was not satisfied with the theory of the aethers. He had observed the indications of lines of force that can be made visible by the effect of a magnet on iron filings. Similar lines can also be detected in three dimensions by carefully moving a compass near a magnet.

Charles Ellet, American engineer, d Jun 21

1862 cont

In Nov Richard Jordan Gatling (b Maney's Neck, NC, Sep 12, 1818) develops the first machine gun

1863

Meteorographica by Francis Galton introduces the term *anticyclone* and founds the modern method of mapping the weather

Sir Charles Lyell's *Geological evidence of the antiquity of man* argues for the existence of early humans on the evidence of stone tools found in various ancient deposits

Physiologist Etienne-Jules Marey (b Beaune, France, Mar 5, 1830) invents the sphygmograph, the predecessor of the sphygmometer used to measure blood pressure today

Wilhelm Wundt's *Vorlesungen über die Menschen und Tierseele* (Lectures on the minds of men and animals) becomes a popular text reprinted until the 1920s

Heat as a mode of motion by John Tyndall (b Leighinbridge, Ireland, Aug 2, 1820) popularizes James Clerk Maxwell's theory of heat as the motion of molecules

Geophysicist Augustus Love b Weston-super-Mare, England, Apr 17; he discovers the type of earthquake wave now called a Love wave, which enables scientists to measure the thickness of Earth's crust

---

From action at a distance to fields (continued)

Faraday pictured these lines of force as thin tubes reaching from one pole of a magnet to another. By 1846, Faraday had developed evidence that suggested that the same concept could be used to explain the actions caused by static and current electricity as well as light. He used his lines of force as a device to explain various magnetic phenomena that he discovered.

Mathematically, Faraday's idea corresponds to assigning a number and a direction to every point in space. When you move a freely suspended compass near a magnet, the needle will tend to point in a particular direction and the amount of force on the compass will be a definite positive or negative number (depending on which pole you are approaching). Such an array of numbers and directions pervading space is called a *field*.

In the 1860s James Clerk Maxwell used the mathematical representation of Faraday's lines of force to explain light in terms of electric and magnetic fields. The two fields are bound together in this way: When a magnet moves, it introduces an electric current; when an electric current flows, it induces magnetism. Maxwell's equations show that the interaction of electricity and magnetism produces waves that propagate through space. He identified these waves as light waves.

Maxwell's equations implied that waves longer and shorter than those of light could be produced by moving electric currents; this was confirmed in 1888 when Heinrich Hertz used this method to produce radio waves. Infrared radiation was also explained as electromagnetic waves and accounted for the transmission of heat through space. Thus, one result of Maxwell's field theory was the unification of the various aethers into a single ether that was believed to transmit electromagnetic waves.

Maxwell next turned to gravity. He thought he could develop a field theory of gravity, since each point in space can have a number and a direction assigned to it to represent the total effect of gravitational forces on it. The number is always positive, however, which makes it very difficult to see how such a gravitational field would resemble an electric or a magnetic field. Maxwell abandoned the effort. Einstein solved the problem in 1915, although the mathematical difficulties continue to be immense.

Einstein and others thought that there must be a single field theory that would include both electromagnetism and gravity. Einstein tried to develop such a unified field theory for many years, but was ultimately unsuccessful.

1864

ASTRONOMY

Astronomers agree on 147 million km (91 million mi) as the sun's distance from Earth; the presently accepted value for the mean distance is 149.6 million km (92.96 million mi)

Italian astronomer Giovanni Battista Donati (b Pisa, Dec 16, 1826) makes the first spectroscopic examination of a comet, Tempel's comet, demonstrating that near the sun a comet begins to radiate light while farther from the sun it only reflects light

John Herschel publishes a catalog of nebulas and star clusters that contains more than 5000 items, based on his and his father William's observations

Sir William Huggins demonstrates that bright nebulas, such as the nebula in Orion, consist of gases

Friedrich Georg Wilhelm von Struve, German-Russian, d Pulkovo (USSR), Nov 23

BIOLOGY

Hermann Ludwig von Helmholtz's *Lehre von dem Tonempfindungen* (Theory of the perception of sound) advances his theory that pitch is detected by a series of resonators of different sizes in the cochlea and that overtones and beats based on different wavelengths determine the quality of perceived sound

CHEMISTRY

Agricultural chemist George Washington Carver b near Diamond Grove, MO; as head of the Department of Agricultural Research at Tuskegee Institute in Alabama, Carver works to develop techniques for regenerating land by growing sweet potatoes and peanuts; he also develops hundreds of new uses for both plants

Cato Maximilian Guldberg (b Christiana, now Oslo, Norway, Aug 11, 1836) and Peter Waage (b Flekkefjord, Norway, Jun 29, 1833) publish on Mar 11 the law of mass action: a reaction is dependent both on the mass and the volume of the constituents

Russian chemist Carl Claus d Dorpat, Estonia, Mar 24

Chemist Benjamin Silliman d New Haven, CT, Nov 24

1865

GENERAL

Early history of mankind by Edward Burnett Tylor (b England, 1832) is published

Archaeologist Christian Jorgenson Thomsen d Copenhagen, Denmark, May 21

ASTRONOMY

Jules Verne's novel *From the Earth to the moon* depicts three scientists and a journalist being shot to the moon from a great cannon situated at Cape Canaveral, FL

George Phillips Bond, American, d Cambridge, MA, Feb 17

Astronomer Johann Encke d Spandau, Germany, Aug 26

BIOLOGY

Claude Bernard's *Introduction à l'étude de la médecine expérimentale* argues that all phenomena in living systems follow the same rigorous laws as those in the inanimate world, and denies the existence of a "vital force."

Gregor Mendel publishes his theories of genetics in the obscure *Transactions of the Brünn Natural History Society*

Hybridization experiments by Charles Naudin (b Aug 14, 1815) show certain

CHEMISTRY

Friedrich August Kekulé finds the structural formula for benzene in a dream, establishing that organic molecules can form rings or other complex forms

Jean Servais Stas (b Louvain, Belgium, Aug 21, 1813) produces the first modern table of atomic weights using oxygen as a standard (set at 16); he shows conclusively that atomic weights are not always integral

| | Benjamin Pierce's "Linear associative algebra" is read to the American Association for the Advancement of Science; it is best known for its opening sentence: "Mathematics is the science which draws necessary conclusions"

Karl Weierstrass is appointed a full professor at Berlin University, a position from which he makes his ideas known more in lectures than in publications

Mathematician George Boole d Ballintemple, Ireland, Dec 8 | Franciscus Cornelis Donders discovers that astigmatism is caused by an uneven curvature of the lens or cornea of the eye | James Clerk Maxwell's *A dynamical theory of the electromagnetic field* is the first of his publications to use Michael Faraday's concept of a field as the basis of a mathematical treatment of electricty and magnetism; it introduces Maxwell's equations to describe electromagnetism | Engineer Benoit Clapeyron d Paris, Jan 28 | **1864** |

| Meteorologist Robert Fitzroy d London, Apr 30 | The London Mathematical Society is founded and begins to issue its *Proceedings*

German mathematician August Ferdinand Möbius (b Schulpforte, Nov 17, 1790) presents his discovery of a figure that has only one side and one edge, now called the Möbius strip

About this time Karl Weierstrass proves the Bolzano-Weierstrass theorem: every bounded infinite set | Joseph Lister (b Upton, England, Apr 5, 1827) introduces phenol as a disinfectant in surgery, reducing the surgical death rate from 45 percent to 15 percent

Jean-Antoine Villeman shows that tuberculosis is a contagious disease

Hungarian physician Ignaz Semmelweiss d Vienna, Austria, Aug 13, ironically of childbed fever, which he had tried so hard to prevent | Rudolf J.E. Clausius invents the term *entropy* to describe the degradation of energy in a closed system

Russian physicist Heinrich Lenz d Rome, Italy, Feb 10 | German-American electrical engineer Charles Proteus (Karl August) Steinmetz b Breslau (Wroclaw, Poland), Apr 9; his work on the theory of AC current is the foundation of today's electrical industry | **1865** |

1865 cont

The periodic table

In 1799, Joseph-Louis Proust established that the compound copper carbonate always has the same proportions of copper to carbon to oxygen by mass, no matter how the compound is prepared. He proceeded to carry out similar analyses of other compounds over the next ten years, showing that in every case the proportions of the elements by mass stayed constant for a given compound. John Dalton was the first to explain why Proust's law of definite proportions held true. Dalton postulated in 1800 that the chemical elements are made from atoms that differ largely by mass. If the mass of an atom of hydrogen, the lightest element, is taken as 1, then the other elements have atoms whose masses are multiples of hydrogen. Dalton tried to calculate these masses, but got in trouble because he was not sure of the exact atomic composition of compounds, such as water. In 1828, however, Jöns Jakob Berzelius published a list of atomic masses (called atomic weights at that time) that was quite accurate. No one paid very much attention to this idea at the time.

In 1860, Friedrich Kekulé reached the conclusion that chemistry was in chaos because different formulas were being used for the same compounds. He organized the first international scientific congress, the First International Chemical Congress, to straighten things out. The highlight of the meeting was an address by Stanislao Cannizzaro that stressed the importance of atomic masses.

A few scientists began to list the known elements in the order of their atomic masses. When they did, they found that roughly every eighth element was similar. The first to publish this infor-

---

BIOLOGY

regularities in the inheritance of characteristics in plants

Zoologist Christian Pander d St. Petersburg (Leningrad, USSR), Sep 22

German botanist Julius von Sachs, b Breslau (Wroclaw, Poland), Oct 2, 1832, discovers that chlorophyll in plants is found only in small bodies (later termed chloroplasts) and that chlorophyll is the key compound that turns carbon dioxide and water into starch while releasing oxygen

Karl von Voit shows that the pathways by which food is converted to energy are complicated; food is not simply burned to produce energy, but many intermediate reactions take place instead

---

1866

GENERAL

Scholar William Whewell d Cambridge, England, Mar 6

ANTHROPOLOGY/ARCHAEOLOGY

Scottish-South African paleontologist Robert Broom b Paisley, Scotland, Nov 30; his work in finding the fossils of various australopithecines is essential in the development of understanding of human prehistory

British anthropologist Sir Arthur Keith b Feb 5; it is likely that the Piltdown hoax was intended to make Keith look foolish

ASTRONOMY

Great Leonid meteor showers occur

Daniel Kirkwood (b Harford County, MD, Sep 27, 1814) shows that the influence of Jupiter's gravity causes gaps, now known as Kirkwood gaps, in the distribution of the orbits of asteroids

Giovanni Virginio Schiaparelli establishes that there is an association between some comets and meteor showers

BIOLOGY

Generelle Morpologie der Tiere by Ernst Heinrich Philipp August Haeckel (b Potsdam, Germany, Feb 16, 1834) popularizes the incorrect notion that ontology recapitulates phylogeny; that is, that the embryo goes through all the stages of evolution from the beginning of life to the present species

Haeckel is the first to use the word *ecology* (spelled *Oecologie* in German)

Hermann Ludwig von Helmholtz's *Physiological optics* is completed in 1866 and is so influential that it is translated into English 60 years later

Karl von Voit develops the basal metabolism test between this year and 1873

CHEMISTRY

John Mercer, English chemist, d Nov 30

of points in a Euclidean space has at least one point for which every neighborhood of that point contains a point in the set (that is, there is an accumulation point); this was proved earlier by Bernhard Bolzano

Mathematician Sir William Hamilton d near Dublin, Ireland, Sep 2

Robert Remak d Kissingen, Germany, Aug 29

The periodic table (continued)

mation was Alexandre-Emile Beguyer de Chancourtis in 1862, but the article failed to reproduce his diagram. Without a diagram, the periodicity of the elements was far from clear. The next year John Alexander Newlands announced his version, which he called the law of octaves. People were unimpressed and claimed that just as much periodicity could be seen if the elements were listed in alphabetical order.

Finally, in 1869 and 1870, two scientists, Dmitri Mendeléev and Julius Lothar Meyer, published clear versions of the idea. Not only was Mendeléev the first to publish, but he also announced in 1871 that the gaps in his periodic table would be filled as new elements were discovered. He specified three gaps, all of which were filled by discoveries between 1875 and 1885. As a consequence, Mendeléev gets most of the credit for the periodic table.

No one knew, however, why the properties of elements were periodic. After electrons and protons were discovered, Henry Moseley showed in 1914 that each element had a definite number of protons, that normally corresponded to the same definite number of electrons. This atomic number, not the atomic mass at all, was the basis of the periodic table. With Moseley's work, it was clear that gaps existed between whole numbers of protons.

Gabriel-August Daubrée (b Metz, France, Jun 25, 1814) suggests that the center of Earth is a core of iron and nickel

Othniel Charles Marsh (b Lockport, NY, Oct 29, 1831) is appointed professor of vertebrate paleontology at Yale University, the first such post at a US university; from this position, Marsh directs many searches for fossils, especially those of dinosaurs, throughout the western United States

Niccolò Paganini, a 16-year-old schoolboy, startles mathematicians with his discovery of the amicable pair 1184 and 1210; although many pairs of amicable numbers had been found in the seventeenth and eighteenth centuries, the pairs were all much larger, usually in the millions or billions

Georg Friedrich Bernhard Riemann, German mathematician, d Selasca, Italy, Jul 20

Physician Sir Thomas Clifford Allbutt (b Dewsbury, England, Jul 20, 1836) develops the clinical thermometer; previously thermometers used in medicine were very long and took about 20 minutes to determine the temperature

William Budd (b North Tawton, England, Sep 14, 1811) demonstrates in Bristol, England, that limiting the contamination of a town's water supply can stop a cholera epidemic

Physician Jesse Lazear b Baltimore County, MD, May 2; he becomes a part of Walter Reed's team fighting yellow fever in Cuba at the turn of the century

August Adolph Eduard Eberhard Kundt (b Schwerin, Germany, Nov 18, 1839) devises a method for measuring the speed of sound in different gases by analyzing patterns sound waves cause in a fine dust scattered inside a tube filled with the gas being investigated

Cyrus West Field (b Stockbridge, MA, Nov 30, 1819) succeeds in laying a telegraph cable across the Atlantic Ocean

Mahlon Loomis (b Oppenheim, NY, Jul 21, 1826) sends telegraph messages over radio waves between two mountains in West Virginia using aerials held in the air by kites

Engineer Robert Whitehead (b Jan 3, 1823) invents the torpedo

	GENERAL	ANTHROPOLOGY/ ARCHAEOLOGY	ASTRONOMY	BIOLOGY	CHEMISTRY
1867	Karl Marx's *Das Kapital* (Capital) develops the theory of evolution of society as a result of Hegelian synthesis		Astronomer Andrew Douglass b Windsor, VT, Jul 5; he carefully matches rings on ancient pieces of wood to develop dating from tree rings, a procedure known as dendrochronology William Parsons, third earl of Rosse, d Monkstown, Cork, Ireland, Oct 31 Italian astronomer Pietro Angelo Secchi (b Reggio, Emila, Jun 18, 1818) classifies the spectra of stars into four classes		
1868	Explorer Robert Scott b Devonport, England, Jun 6; he becomes the second person to reach the South Pole	Workmen building a road in France discover five skeletons in a cave named Cro-Magnon; the skeletons appear to be about 35,000 years old, giving rise to the name Cro-Magnon for early *Homo sapiens* Archaeologist Jacques Boucher de Crève-coeur de Perthes d Abbeville, France, Aug 5	Anders Jonas Ångström's *Recherches sur le spectre solaire* (Researches on the solar spectrum) is a detailed map of the solar spectrum Sir William Huggins, using the spectrogram of the star, finds from the Doppler effect that Sirius is moving away from us Pierre-Jules-César Janssen (b Paris, Feb 22, 1824) and Joseph Norman Lockyer (b Rugby, England, May 17, 1836) independently describe a method for observing the solar prominences at times of noneclipse Pietro Angelo Secchi completes the first spectroscopic survey of the stars, cataloging the spectrograms of about 4000 stars	Charles Darwin's *The variation of animals and plants under domestication* is a detailed expansion of the subject treated in the first chapter of *The origin of species* Zoologist Sir Charles Wyville Thomson (b Bonsyde, Scotland, Mar 5, 1830) starts deep-sea dredging operations that reveal that there are many different forms of life at the bottom of the oceans	Karl James Peter Graebe (b Frankfurt-am-Main, Germany, Feb 24, 1841) discovers the structure of the molecule of alizarin, an important natural dye; this is the key to synthesizing the dye from coal tar, which is accomplished the following year; William Perkin independently finds the structure While observing an eclipse of the sun, Pierre-Jules-César Janssen observes a line in the sun's spectrum that does not belong to any known element; he sends a copy of the spectrum to Sir Joseph Norman Lockyer, who confirms that it is a new element; Lockyer names it helium, which means "the sun" William H. Perkin synthesizes coumarin, a scent and flavoring used in foods until 1954, when it is discovered to cause liver poisoning German-Swiss chemist Christian Schönbein d Sauersberg, Germany, Aug 29

EARTH SCIENCE	MATHEMATICS	MEDICINE	PHYSICS	TECHNOLOGY	
	Mathematician Jean-Victor Poncelet d Paris, Dec 23	Wilhelm Wundt teaches a course on physiological psychology at Heidelberg, the first course of its kind Jean-Pierre-Marie Flourens, French physiologist, d Dec 8	William Thomson, later Lord Kelvin, proposes a vortex model of the atom to Hermann Helmholtz in a letter dated Jan 22; this model remains popular with many scientists for about 20 years, and then is abandoned Physicist Michael Faraday d Hampton Court, England, Aug 25	Belgian-French inventor Zénobe Théophile Gramme (b Jehay-Bodegnée, Belgium, Apr 4, 1826) builds the first commercially practical generator for producing alternating current American engineer George Westinghouse (b Central Bridge, NY, Oct 6, 1846) solves a major problem of the railways by inventing the air brake Inventor Elias Howe d Brooklyn, NY, Oct 3	**1867**
A violent earthquake accompanies repeated eruptions of Mauna Loa and Kilauea, killing more than 100 and resulting in a fundamental change in the type of lava produced by Mauna Loa	*Saggio del interpretazione della geometria non-Euclidea* (Essay on the interpretation of non-Euclidean geometry) by Eugenio Beltrami (b 1835) shows that non-Euclidean geometry is consistent if Euclidean geometry is consistent; his method is to redefine "line" and to develop a Euclidean model for a non-Euclidean geometry Julius Plücker, German mathematician and physicist, d Bonn, Rhenish Prussia, May 22 August Ferdinand Möbius d Leipzig, Germany, Sep 26	Sir Francis Galton shows that mental abilities of human beings form a normal distribution, lying along the familiar bell-shaped curve William Thomas Morton, American dentist, d New York, NY, Jul 15	Sir David Brewster, Scottish physicist, d Allerly, Roxburghshire, Feb 10 Jean-Bernard-Léon Foucault, French physicist, d Paris, Feb 11 Giovanni Battista Amici, Italian physicist, d Florence, Apr 10 Wallace Clement Ware Sabine, American physicist, b Richwood, OH, Jun 13; he becomes the first acoustical engineer and uses acoustic principles to design Boston's Symphony Hall	Thomas Alva Edison (b Milan, OH, Feb 11, 1847) patents his first invention, a device to record votes in Congress developed with the intent of speeding up proceedings; Congress informs Edison that slow votes have advantages over fast ones and does not buy his invention Georges Leclanché invents the zinc-carbon battery, a precursor to the dry cell and the familiar "flashlight battery" John Fillmore Hayford, American engineer, b Rouses Point, NY, May 19; he develops modern methods, called geodesy, to determine the shape of Earth	**1868**

	GENERAL	ANTHROPOLOGY/ ARCHAEOLOGY	ASTRONOMY	BIOLOGY	CHEMISTRY
1869	The Suez Canal opens Publication of the scientific journal *Nature* begins		The International Earth Measurement is started Charles Augustus Young (b Hanover, NH, Dec 15), in studying the solar eclipse of this year and the one of 1870, notes that the dark lines of the sun's spectrum brighten just before totality	Père Armand David is the first scientist to observe and describe the giant panda Johann Frederick Miescher (b Basel, Switzerland, Aug 13, 1844) discovers nuclein, the portion of a cell that contains nucleic acids, but is discouraged from publishing his findings until 1871 because they are so unexpected	Richard Wilhelm Heinrich Abegg, German chemist, b Danzig (Gdańsk, Poland), Jan 9; he is among the first to see that the outer electron shell governs the chemical properties of an atom Dmitri Ivanovich Mendeléev (b Tobol'sk (USSR), Feb 7, 1834) publishes his first version of the periodic table of elements on Mar 6; Lothar Meyer (b Varel, Germany, Aug 19, 1830) also develops a version of the periodic table this year, but publishes in 1870 Thomas Graham, Scottish physical chemist, d London, Sep 16
1870	Otto von Bismarck unifies Germany			Gustave T. Fritsch (b 1838) and Julius Eduard Hitzig (b Berlin, Feb 6, 1838) discover that the cerebral cortex has different compartments for different functions; they also introduce the study of the brain by electrical stimulation	
1871		Edouard-Armand-Isidore Hippolyte Lartet, French paleontologist, d Seissan, France, Jan 28	*Società degli Spettroscopisti Italiana* (The Italian spectroscopic society) is founded Sir John Frederick Herschel, English astronomer, d Collingwood, Hawkhurst, Kent, May 11	Charles Darwin's *The descent of man and selection in relation to sex* discusses evidence of evolution of human beings from lower forms of life, comparing people to animals; he introduces the concepts of sexual selec-	Organic chemist Viktor Meyer (b Berlin, Germany, Sep 8, 1848) discovers that molecules of iodine and bromine, each of which contain two atoms, break up into single-atom molecules when heated

EARTH SCIENCE	MATHEMATICS	MEDICINE	PHYSICS	TECHNOLOGY	
Meteorologist Cleveland Abbe (b New York, NY, Dec 3, 1838) on Sep 1 begins to send out weather bulletins from the Cincinnati observatory where he is director		Physician Paul Langerhans (b Berlin, Germany, Jul 25, 1847) makes a careful dissection of the pancreas and discovers the small groups of cells now called the islets of Langerhans; later it is discovered that these groups of cells are the source of insulin Jan Evangelista Purkinje, Czech physiologist, d Prague, Jul 28 Jean-Léonard-Marie Poiseuille, French physician, d Paris, Dec 26	John Tyndall discovers the effect named after him, namely that a beam of light passing through a colloidal solution can be observed from the side	Thomas Alva Edison invents a stock ticker that he sells to a Wall Street firm for $40,000, a fortune at the time Belgian-French inventor Zénobe Théophile Gramme builds the first commercially practical generator for producing direct current John Wesley Hyatt (b Starkey, NY, Nov 28, 1837) develops celluloid, the first artificial plastic, independently of Alexander Parkes Ferdinand de Lesseps completes construction of the Suez Canal, linking the Mediterranean and Red seas J.F. Tretz is the first to use a chain drive to propel bicycles The golden spike is hammered at Promontory Point, UT, completing the first railway line from the Atlantic to the Pacific Joseph Jackson Lister, English optician, d West Ham, Essex, Oct 24	**1869**
	Camille Jordan deals with the theory of substitutions and algebraic equations, which contains a theory of permutation groups	Sir James Young Simpson, Scottish obstetrician, d London, May 6		Germain Sommeiller's Mont Cenis Tunnel through the Alps is the first major railroad tunnel	**1870**
Cleveland Abbe becomes the scientific assistant in the first US weather bureau, which is a part of the US Army Sir Roderick Impey Murchison, Scottish geologist, d London, Oct 22	Christian Felix Klein (b Düsseldorf, Germany, Apr 25, 1849) publishes the first of two monographs entitled *Über die Sogenannte nich-Euclidische Geometrie* (On the so-called non-Euclidean	Walter Bradford Cannon, American physiologist, b Prairie du Chien, WI, Oct 19; he devises the use of bismuth compounds to make soft internal organs visible on X rays early in the twentieth century;	James Clerk Maxwell explains how his statistical theory of heat works by inventing Maxwell's demon, a mythical creature that can see and handle individual molecules; by opening a gate between two	**1871**

1871 cont

Primitive culture by Edward Burnett Tylor discusses his analysis of animism, a religion in which spirits are supposed to inhabit both animate organisms and inanimate objects

tion to explain hairlessness and group selection to explain humans' lack of natural weapons

Ernst Felix Hoppe-Seyler discovers invertase, an enzyme that speeds conversion of sucrose into glucose and fructose

Fritz Richard Schaudinn, German zoologist, b Röseningken, East Prussia, Sep 19

Dmitri Mendeléev announces on Jan 7 that the gaps in his periodic table represent undiscovered elements; the gaps are filled by discoveries of new elements in 1875, 1879, and 1885, making Mendeléev famous

Anselme Payen, French chemist, d Paris, May 12

François-Auguste Grignard, French chemist, b Cherbourg, Manche, May 6

Max Bodenstein, German chemist, b Magdeburg, Jul 15; he develops the concept of chain reactions, where one change in a molecule causes the next change and so on

1872

Cracker Jacks are invented

Roald Engelbregt Amundsen, Norwegian explorer who becomes the first to reach the South Pole, b Borge, Ostfold, Jul 16

Henry Draper (b Prince Edward County, VA, Mar 7, 1837) is the first astronomer to photograph the spectrum of a star, Vega

Forest Ray Moulton, American astronomer, b Le Roy, MI, Apr 29; along with Thomas Chrowder Chamberlin, Moulton espouses an early version of the planetesimal formation of the solar system; he assumes that the planetesimals arose from an encounter with a passing star, not from condensation

E. Weiss predicts on the basis of the orbit of a comet that vanished six years earlier that there will be a meteor shower on Nov 28; the predicted shower actually occurs on Nov 27

Ferdinand Cohn begins publication of a three-volume work that is the first to treat bacteria systematically by dividing them into genera and species

Charles Darwin's *Expressions of the emotions in man and animals* characterizes emotion in evolutionary terms and argues that emotions are the result of the inheritance of behavior from animals

Hugo von Mohl, German botanist, d Tübingen, Baden-Württemberg, Apr 1

geometry) that develop Euclidean models for various types of non-Euclidean geometries; the second is published in 1873

Augustus De Morgan, English mathematician, d London, Mar 18

Charles Babbage, English mathematician, d London, Oct 18

after World War I he develops the theory of homeostasis, the idea that the body tries to maintain a constant internal state

vessels containing a gas only when a fast molecule is passing into one, the demon would make heat flow from cold to hot

George Johnstone Stoney (b Oakley Park, Ireland, Feb 15, 1826) notes that the wavelengths of three lines in the hydrogen spectrum are found to have simple ratios, an anticipation of Balmer's formula, an important step toward understanding the structure of the atom

---

Charles Thomson starts his tour as chief scientist on the four-year voyage of the *Challenger*, the first major oceanographic expedition

German mathematician Georg Cantor (b St. Petersburg, Russia, Mar 3, 1845) establishes that a function can be represented uniquely by a trigonometric series even if it fails to converge for an infinite set of points in an interval; this result leads Cantor to the study of infinite sets

Stetigkeit und irrationale Zahlen (Continuity and irrational numbers) by Richard Dedekind (b Braunschweig, Germany, Oct 6, 1831) introduces the "Dedekind cut" as a definition for a real number

H.Eduard Heine's *Elemente*, influenced by Karl Weierstrass's lectures, gives the modern "epsilon-delta" definition of a limit

Felix Klein presents his "Erlanger Programm," in which he calls for geometry to be based on groups of transformations

Claude Bernard's *Physiologie générale* (General psychology) is published

Physician Jean Martin Charcot (b Paris, Nov 29, 1825) uses hypnosis as part of his treatment for therapy; in 1885, Sigmund Freud is a student of Charcot's and learns this use of hypnotism from him

Austrian physicist Ernst Mach (b Chirlitz-Turas, Moravia, Feb 18, 1838) develops his philosophy that all knowledge is simply sensation

James Clerk Maxwell remarks that atoms remain in the precise condition in which they first began to exist

Jacques Babinet, French physicist, d Paris, Oct 21

In Germany, George M. Pullman introduces the "sleeper car" to railroad travel

Samuel Finley Breese Morse, American artist and inventor, d New York, NY, Apr 2

William John Macquorn Rankine, Scottish engineer, d Glasgow, Dec 24

The first long-distance pipeline carrying natural gas is completed in Rochester, NY; it is made of wood and is 25 mi (40 km) long

1872 cont

1873	Publication of *Comptes Rendus* starts in France	Georg M. Ebers, a German Egyptologist, purchases a 20 m (65 ft) scroll that, it is claimed, had been buried with a mummy at Thebes; it turns out to be a 1600 BC copy of a summary of Egyptian medicine that had originally been written much earlier Heinrich Schliemann (b Neu Buckow, Mecklenburg-Schwerin, Germany, Jan 6, 1822) announces on June 15 that he has found Troy at a site in Turkey; his finding is confirmed by later archaeologists as to site, although they believe that he has misidentified the level at that site that was the Troy of Priam	Giovanni Battista Donati, Italian astronomer, d Florence, Sep 20 Richard Anthony Proctor (b London, Mar 23, 1837) suggests that the craters of the moon were formed by the impact of meteorites; previously, astronomers assumed that the craters were formed by volcanoes	Jean-Louis-Rodolphe Agassiz, Swiss-American naturalist, d Cambridge, MA, Dec 12 Italian histologist Camillo Golgi (b Corteno, Jul 7, 1843) discovers a new method of staining cells using silver salts that enables staining of a few neurons or neuronal systems in a tissue, making it possible to study a few out of the many thousands of systems that touch each other in the brain	Justus von Liebig, German chemist, d Munich, Bavaria, Apr 18 Nevil Vincent Sidgwick, English chemist, b Oxford, May 8
1874	Francis Galton publishes *English men of science*	William Henry Jackson is the first non-Indian to enter Mesa Verde, a collection of cliff houses in southwestern Colorado near the Four Corners region where four US states meet	The Astrophysical Observatory of Potsdam (Germany) is founded Johann Heinrich Mädler, German astronomer, d Hanover, Mar 14	Karl Wilhelm von Nägeli develops equations that describe competition among plants; these are ignored by biologists until the 1920s, when they are rediscovered by Alfred Lotka and Vito Volterra	Jacobus Henricus van't Hoff (b Rotterdam, Netherlands, Aug 30, 1852) shows that the four bonding directions of the carbon atom point to the four vertices of a regular tetrahedron; Joseph Achille Le Bel (b Merkwiller-Péchelbronn, Jan 24,

EARTH SCIENCE	MATHEMATICS	MEDICINE	PHYSICS	TECHNOLOGY	
	Charles Méray's *Nouveau précis d'analyse infinitési-male* (New summary of infinitesimal analysis) develops a method of describing limits without using irrational numbers Karl Weierstrass announces to the Berlin Academy of Sciences his discovery of a function that is continuous but nowhere differentiable; this "pathological curve" was probably discovered as early as 1861, but, typically, Weierstrass did not publish it and used it only in his lectures				**1872** cont
Adam Sedgwick, English geologist, d Cambridge, Jan 27 Matthew Fontaine Maury, American oceanographer, d Lexington, VA Feb 1	*Sur la fonction exponentielle* (On the exponential function) by Charles Hermite (b Dieuz, France, Dec 24, 1822) shows that *e* is not the solution to any algebraic equation; it is what mathematicians call a transcendental number	Wilhelm Wundt begins publication of what is considered the most important book in the history of psychology: *Grundzüge der physiologischen Psychologie* (Principles of physiological psychology)	James Clerk Maxwell's *Electricity and magnetism* contains the basic laws of electromagnetism and predicts such phenomena as radio waves and pressure caused by light rays in great detail Johannes Diderik Van der Waals (b Leiden, Netherlands, Nov 23, 1837) develops a version of the gas laws that is closer to actual behavior of gases than the idealized Boyle's law and Charles' law William Weber Coblentz, American physicist, b North Lima, OH, Nov 20; he shows that compounds can be analyzed by the way they absorb infrared radiation	Robert William Thomson, Scottish engineer, d Edinburgh, Mar 8	**1873**
Wilhelm Philipp Schimper identifies the Paleocene Period in Earth's history	Georg Cantor's *Über eine Eigenshaft des Inbegriffes aller reelen algebraischen Zahlen* (On a property of the system of all the real algebraic numbers) is his first paper on set theory	Franz Brentano's *Psychologie vom empirischen Standpunkt* (Psychology from an empirical standpoint) dissents strongly with Wilhelm Wundt's views by being empirical instead of experimental in his approach to psychology	The Cavendish Laboratory at Cambridge (England) is completed; although widely believed to have been named after the eighteenth-century physicist Henry Cavendish, it is, in fact, named after the entire Cavendish family	Gail Borden, American inventor and food technologist, d Borden, TX Jan 11 Sir William Fairbairn, Scottish engineer, d Aug 18	**1874**

1874 cont		Lambert Adolphe Jacques Quetelet, Belgian astronomer and statistician, d Brussels, Feb 17		1847) independently makes the same discovery

Othman Zeidler prepares DDT, but does not discover its insecticidal properties |
| **1875** | The official kilogram at the International Bureau of Weights and Measures at the Pavillion de Breteuil near Paris is installed; this bar of platinum-iridium alloy continues as the official kilogram to this day (unlike the official meter, which has been redefined several times since 1875) | Friedrich Wilhelm August Argelander, German astronomer, d Bonn, Feb 17

Heinrich Ludwig d'Arrest, German astronomer, d Copenhagen, Denmark, Jun 14

Richard Christopher Carrington, English astronomer, d Churt, Surrey, Nov 27

Heinrich Samuel Schwabe, German astronomer, d Dessau, Apr 11 | Ernst Felix Hoppe-Seyler develops a method for classifying proteins that is still used today

Biochemist Henry Clapp Sherman b Ash Grove, VA, Oct 16; he shows that the ratio of calcium to phosphorus in a diet is as important as the total amount of either mineral; he also investigates the amounts needed for various vitamins in the diets of animals

Eduard Adolf Strasburger (b Warsaw, Poland, Feb 1, 1844) and Walther Flemming (b Sachsenberg, Germany, Apr 21, 1843) study the mechanism of cellular division and the associated motions of the chromosomes | Paul-Emile Lecoq de Boisbaudran (b Cognac, France, Apr 18, 1838) discovers the element gallium, the first discovery of an element predicted by Dmitri Mendeléev on the basis of his periodic table |
| **1876** | Johns Hopkins University is founded as a graduate school; it provides a new model for education in the United States and causes older schools to begin to modernize their curriculum and methods | Sir William Huggins uses dry plates to make pictures of Vega | Louis Pasteur's *Etudes sur la bière, ses maladies, et causes qui les provoquent* (Studies of beer, its diseases, and the causes that provoke them), tells of Pasteur's discovery of anerobic organisms, organisms that live in | Antoine-Jérôme Balard, French chemist, d Paris, Mar 30

Germain Henri Hess, Swiss-Russian chemist, d St. Petersburg (Leningrad, USSR), Dec 13 |

					1874 cont
	Sonya Kovalevsky (b Moscow, Russia, Jan 15, 1850) becomes the first woman to receive a German University doctorate (awarded without examinations and in absentia for outstanding work in partial differential equations, Abelian integrals, and mathematical astronomy)	Max Johann Sigismund Schultze, German anatomist, d Bonn, Jan 16

Schack August Steenberg Krogh, Danish physiologist, b Gronå, Jutland, Nov 15; he discovers that capillaries in muscles are partially closed when a muscle is resting and that the capillaries control blood flow in various parts of the body | because the nineteenth-century steel-making descendant of Henry, William Cavendish, financed the laboratory

George J. Stoney estimates the charge of the then unknown electron to be about 10^{-20} coulomb, close to the modern value of $1.6021892 \times 10^{-19}$ coulomb; he also introduces the term *electron*

Anders Jonas Ångström, Swedish physicist, d Uppsala, Jun 21 | | |

| Sir Charles Lyell, Scottish geologist, d London, Feb 22 | | | William Crookes invents the radiometer, a device consisting of vanes blackened on one side while the other side is shiny; the vanes are kept in a partly evacuated glass bulb; light falling on the radiometer causes the vanes to turn, pushed by molecules of air that rebound more from the dark side than from the shiny one

James Clerk Maxwell notes that atoms have a structure far more complex than that of a rigid body

Sir Charles Wheatstone, English physicist, d Paris, Oct 19 | Sidney Thomas invents a method for removing phosphorus from iron

Isaac Merrit Singer, American inventor, d Jul 23 | 1875 |

| The Mineralogical Society of Great Britain and Ireland is founded | Edouard A. Lucas determines the first ironclad rule since the sieve of Eratosthenes in Hellenistic times for establishing whether a given natural number is a prime | Robert Koch (b Klausthal-Zellerfeld, Hanover, Germany, Dec 11, 1843) discovers that the microorganism responsible for cattle anthrax can be grown in culture | Eugen Goldstein shows that the radiation in a vacuum tube produced when an electric current is forced through the tube starts at the cathode; Goldstein introduces the term *cathode ray* to describe the light emitted | Alexander Graham Bell (b Edinburgh, Scotland, Mar 3, 1847) patents the telephone

Paul Jablochkoff develops the "electric candle," an improved version of the arc light that will burn for two hours with no mechanical adjustment | 1876 |

| 1876 cont | | | | the absence of free oxygen (and, indeed, perish in its presence)

Christian Gottfried Ehrenberg, German naturalist, d Berlin, Jun 27 |

| 1877 | E. George Squier's *Peru illustrated; incidents of travel and exploration in the land of the Incas* tells of his discovery that there were South American civilizations that preceded that of the Inca, notably the Chimú and Moche | Asaph Hall (b Goshen, CT, Oct 15, 1829) discovers the two satellites of Mars, which had been foretold (accidentally) by Jonathan Swift in 1726; Hall names them Phobos (fear) and Deimos (terror) after the two sons of Ares, the Greek equivalent of the god Mars

Urbain Leverrier, French astronomer, d Paris, Sep 23

Giovanni Schiaparelli thinks that he discovers "canals," or channels, on Mars; eventually this observation fails to be confirmed, but it is considered a possibility for the next 100 years | Charles Darwin's *The different form of flowers on plants of the same species* contains his concept of the evolutionary purpose of some species of plants having two different, incompatible types of flowers

Wilhelm Friedrich Benedikt Hofmeister, German botanist, d Leipzig, Jan 12

Botanist Wilhelm Pfeffer (b Grebenstein, Hesse, Germany, Mar 9, 1845) begins his pioneering work with semipermeable membranes, using them to measure the molecular weight of proteins | Louis-Paul Cailletet (b Châtillon-sur-Seine, France, Sep 21, 1832) and Raoul Pictet (b Geneva, Switzerland, Apr 4, 1846) liquefy oxygen

James Mason Crafts (b Boston, MA, Mar 8) and Charles Friedel discover that aluminum chloride is an important catalyst for reactions that bind two rings of carbon atoms together (the Friedel-Crafts reaction)

Joseph Bienaimé Caventou, French chemist, d Paris, May 5 |

| 1878 | Heinrich Schliemann claims to have found the tomb of Agamemnon in Mycenae in Greece, recovering various valuable artifacts from the site; later investigators believe that these treasures date some 300 years before Agamemnon | Simon Newcomb improves the tables that show the motions of the moon

Johann F.J. Schmidt produces the last lunar map based on visual observations

Pietro Angelo Secchi, Italian astronomer, d Rome, Feb 26 | Warrington shows that microorganisms convert ammonium compounds in fertilizers into nitrites and nitrates, which can be absorbed by plants | Louis-Marie-Hilaire Bernigaud, comte de Chardonnet (b Besançon, Doubs, France, May 1, 1839) develops rayon, which he patents in France one year after Joseph Swan patents the same process in England |

1876 cont

MATHEMATICS

(none)

MEDICINE

Physiologist Wilhelm (Willy) Friedrich Kühne (b Hamburg, Germany, Mar 28, 1837) discovers the enzyme trypsin in pancreatic juice and invents the term *enzyme* to distinguish those compounds that work equally well outside the cell as in it; he reserves the older name *ferment* for vital processes in cells

Karl Ernst von Baer, German-Russian embryologist, d Dorpat, Estonia (USSR), Nov 28

PHYSICS

Willem Hendrik Keesom, Dutch physicist, b Texel, Netherlands, Jun 21; he is the first to solidify helium and to recognize that there are two different forms of liquid helium

TECHNOLOGY

Karl Paul Gottfried von Linde (b Berndorf, Germany, Jun 11, 1842) builds the first practical refrigerator, using liquid ammonia as a coolant

William Thomson, later Lord Kelvin, shows that machines can be programed for all sorts of mathematical problems

---

1877

MATHEMATICS

Georg Cantor succeeds in proving that the number of points on a line segment is the same as the number of points in the interior of a square; publication is held up for a year because mathematicians refuse to believe the result

The distinguished British mathematician James Joseph Sylvester (b England, Sep 3, 1814) is appointed the first professor of mathematics at Johns Hopkins, raising the level of mathematics throughout the United States

Hermann Günther Grassman, German mathematician, d Stettin (Poland), Sep 26

MEDICINE

Charles Darwin's *Biographical sketch of an infant*, the diary of the development of his son, is the first source of child psychology

Robert Koch develops a way of obtaining pure cultures of bacteria

Louis Pasteur notes that some bacteria die when cultured with certain other bacteria, indicating that one bacterium gives off substances that kill the other; it is not until 1939 that this observation is put to use, when René Jules Dubos discovers the first antibiotics produced by a bacterium

Carl Reinhold Wunderlich, German physician, d Leipzig, Saxony, Sep 25

PHYSICS

Johann Christian Poggendorf, German physicist and chemist, d Jan 24

Heinrich Daniel Ruhmkorff, German physicist and inventor, d Dec 20

TECHNOLOGY

Otto Lilienthal, German aeronautical engineer (b Auklam, Prussia, May 23, 1848) develops the first glider to use arched wings like those of a bird

Nikolaus August Otto (b Holzhausen, Hesse-Nassau, Germany, Jun 10, 1832) develops the four-cycle internal combustion engine, the basis of the most common type of engine today

William Henry Fox Talbot, English inventor and archaeologist, d Lacock Abbey, Wiltshire, Sep 17

The Bell Telephone Company is founded on July 9, 1877

1878

EARTH SCIENCE

The International Union of Geological Sciences is founded and holds its first meeting in Paris

Swedish geologist Nils Adolf Nordenskiöld (b Helsinki, Finland, Nov 18, 1832) starts on a voyage in which he will become the

MATHEMATICS

James Joseph Sylvester founds the *American Journal of Mathematics*, the first US research journal in mathematics

MEDICINE

Physiologist Paul Bert (b Auxerre, France, Oct 17, 1833) announces that dissolved nitrogen in the blood of people working under pressurized air causes the disease commonly known as the "bends" or caisson disease; he proposes that if air

PHYSICS

Julius Robert Mayer d Heilbronn, Germany, Mar 20

Joseph Henry, American physicist, d Washington, DC, May 13

TECHNOLOGY

The first commercial telephone exchange is started in New Haven, CT

1878
cont

| | | | | Henri-Victor Regnault, French chemist and physicist, d Auteuil, Jan 19 |

1879

Pope Leo XIII reinstalls Thomas Aquinas as a ruling authority for Roman Catholic thought

Principles of sociology by Herbert Spencer (b Derby, England, Apr 27, 1820) applies Darwin's theory of evolution to social life, a philosophy that becomes known as Spencerism

Maria Sautuola, while exploring Altamira cave in Spain with her father, is the first person known to discover Cro-Magnon cave paintings; smaller than her father, she could stand upright in places where he had to crawl; as a result, she is the first person in 10,000 years to see the bisons painted on the cave's ceiling

George H. Darwin (b Down House, Kent, England, Jul 9, 1845) proposes that the moon was formed from material thrown from Earth's crust when Earth was rotating very fast; subsequently, tidal friction slowed Earth's rotation and caused the moon to move farther away; this theory remained plausible until the 1960s

Sir William Huggins obtains the ultraviolet spectrum of white stars

Johann von Lamont, Scottish-German astronomer, d Munich, Germany, Aug 6

Biochemist Albrecht Kossel (b Mecklenburg, Germany, Sep 16, 1853) begins his studies of nuclein, a substance found in cells, which leads to his discovery of nucleic acids

Per Teodor Cleve (b Stockholm, Sweden, Feb 10, 1840) discovers the elements thulium and holmium

Discovering a sweet taste on his hands after working with coal tar derivatives, Constantin Fahlberg is led to the discovery of saccharin

Paul-Emile Lecoq de Boisbaudran discovers the element samarium

Russian chemist Vladimir Markovnikov (b Knyaginino, Dec 25, 1837) shows that carbon rings with four atoms can be formed; before this feat, it was believed that all carbon-based molecules could have rings of six atoms only

Lars Fredrik Nilson (b Östergötland, Sweden, May 27, 1840) discovers the element scandium, which he names after Scandinavia; Per Teodor Cleve shows that scandium is one of the elements predicted by Dmitri Mendeléev on the basis of his periodic table

first person to pass
through the Northeast
passage and circum-
navigate Asia

pressure is lowered
by stages the disease
will be prevented,
which is correct

Ernst Heinrich Weber,
German physiologist,
d Leipzig, Saxony, Jan
26

Claude Bernard,
French physiologist, d
Paris, Feb 10

Crawford Williamson
Long, American
physician, d Athens,
GA, Jun 16

Karl Rokitansky,
Austrian pathologist,
d Jul 23

Charles Lapworth set-
tles a dispute between
followers of Roderick
Impey Murchison and
those of Adam
Sedgwick by putting
Murchison's lower
Silurian and
Sedgwick's upper
Cambrian into a new
period called the
Ordovician

Richard Dedekind is
the first to expressly
define a field in
mathematics: a set
that obeys certain
simple rules for addi-
tion and multiplica-
tion, such as the
distributive rule that
$a(b + c) = ab + ac$;
examples include the
rational, real, and
complex number
systems

Begriffsschrift (Con-
ceptual notation) by
Gottlob Frege (b
Wismar, Germany,
Nov 8, 1848) is the
first serious attempt
to reduce mathe-
matics to statements
in formal logic

Karl Weierstrass ex-
pands the theory of
exponential series of
analytic functions to
analytic functions
with several complex
variables

Louis Pasteur
discovers by accident
that weakened
cholera bacteria fail to
cause disease in
chickens and that
chickens previously
infected with the
weakened virus are
immune to the nor-
mal form of the virus,
thus paving the way
for the development
of vaccines against
many diseases, not
just smallpox

Wilhelm Wundt
founds the first
laboratory for
psychology in Leipzig
(Germany)

German physician
Albert Neisser (b
Schweidnitz, Jan 22,
1855) discovers the
bacterium that causes
gonorrhea

Edwin H. Hall
discovers the effect
named for him; an
electric current in a
conductor that is
perpendicular to a
magnetic field causes
electric charges to
build up on one side
of the conductor

Josef Stefan (b St.
Peter, Carinthia,
Austria, Mar 24, 1835)
discovers that the
total radiation of a
body is proportional
to the fourth power of
its absolute
temperature; this is
now known as
Stefan's fourth-power
law or as the Stefan-
Boltzmann law, since
Ludwig Boltzmann
proves it from first
principles in 1884

Max Theodor Felix
von Laue, German
physicist, b Pfaffen-
dorf, Prussia, Oct 9

James Clerk Maxwell,
Scottish mathemati-
cian and physicist, d
Cambridge, England,
Nov 5

An electric railway is
demonstrated at a
Berlin exhibition

In America Thomas
Alva Edison and in
England Joseph
Swan (b Sunderland,
England, Oct 31,
1828) each produce
carbon-thread electric
lamps that can burn
for practical lengths
of time

Heinrich Geissler,
German inventor, d
Bonn, Rhenish
Prussia, Jan 24

1880

Claude-Joseph-Désiré Charnay finds wheeled toys at Tenenepango, high on Mount Popacatepetl, the first evidence of wheels of any kind in the pre-Columbian New World	William Lassell, English astronomer, d Maidenhead, Berkshire, Oct 5	Francis Maitland Balfour (b Edinburgh, Scotland, Nov 10, 1851) suggests that animals with backbones or notochords be classified as the phylum Chordata, which has been accepted by taxonomists; animals with true backbones are classified as Vertebrata, a subphylum	Friedrich Konrad Beilstein, b St. Petersburg (Leningrad, USSR), Feb 17, 1838, starts publishing his *Handbook of organic chemistry*, with the second volume in 1882; later editions run to more volumes, and the German Chemical Society takes over the project in 1900; the handbook continues to be published
		Robert Koch starts using solid cultures for growing microbes (gelatin or agar)	Gerardus Johannes Mulder, Dutch chemist, d Bennekom, Apr 18
			Charles Thomas Jackson, American chemist, d Somerville, MA, Aug 28

Lights and lighting

The production of light is probably the second oldest technology of humankind, coming about a million years after the development of stone tools. Prehistoric people used fire for heating and cooking, but also for the production of light. The simplest lighting device was the torch.

Lighting devices became more practical with the introduction of the oil lamp. In its simplest form, the oil lamp consists of a small receptacle with a nozzle in which burns a wick. Lamps of this design have been found in caves dating from the end of the Old Stone Age.

The Romans developed a form of lamp that has no container—the candle. Candles, made from tallow, beeswax, sperm oil, and bayberry wax, continued to be a major source of illumination through the eighteenth century.

Oil lamps became more sophisticated, but the basic principle of a wick and an inflammable liquid continued in use for thousands of years. Kerosene lamps were the main source of light in rural parts of the United States until rural electrification in the 1930s and 1940s. The better kerosene lamps used a mantle instead of a wick; the mantle is a cloth net that absorbs gas and glows with a bright, white light when lit. Campers today often use lamps that combine a volatile fuel with a mantle.

At the end of the eighteenth century, gas light was introduced in cities. Natural gas, or more commonly, gas made from wood or coal, was piped to street lamps and into houses. The Gaslight Era did not last long, however, for the electric lamp displaced gas quickly after it was invented.

However, the first electric lamps had great disadvantages. In 1808, Humphry Davy developed the arc light. Although the electric arc between two carbon electrodes produced a brilliant light, it generated large amounts of smoke and heat. Furthermore, the carbon was gradually consumed, so the distance between the electrodes needed to be shortened frequently to maintain the proper amount of gap. Despite these problems, arc lamps were used in Paris, London, and in lighthouses during the nineteenth century.

The greatest breakthrough in lighting came in 1879 with the invention of the filament lamp by Thomas Alva Edison in the United States and Joseph Swan in England. It had been known since 1838 that an electric current through a thin filament of carbon in a vacuum produces light. The achievement of Edison and Swan was to develop this method into a practical incandescent lamp that could burn for 40 or more hours.

The fluorescent lamp was developed in the 1930s. It consists of a tube that is coated on the inside with a fluorescent material and filled with mercury vapor. An electric discharge in the vapor produces ultraviolet light, which in turn causes the coating to fluoresce. By adjusting the coating, fluorescent lamps can be produced to give light with wavelengths that are closer to natural sunlight than can be achieved with the incandescent bulb. The lamps also operate on less electric power.

1880

The Seismological Society of Japan is founded

John Milne invents the modern seismograph for measuring earthquake waves

William Hallowes Miller, English mineralogist, d Cambridge, May 20

A. Amthor makes the first serious progress in solving Archimedes' problem of the "cattle of the sun," which Archimedes first proposed to Eratosthenes around 300 BC

Physician Josef Breuer (b Vienna, Austria, Jan 15, 1842) treats a patient suffering from psychological disabilities by having her relate her fantasies, sometimes using hypnosis; this relieves her difficulties; Breuer tells Austrian psychiatrist Sigmund Freud, b Freiberg (Pribor, Czechoslovakia), May 6, 1856, of his experience, and Freud soon begins similar treatments for his patients

Charles-Louis-Alphonse Laveran (b Paris, Jun 18, 1845) discovers a parasite in the blood of a soldier suffering from malaria; it is the first known case of a disease caused by a protist

Louis Pasteur's "On the extension of the germ theory to the etiology of certain common diseases" develops the germ theory of disease; he also demonstrates his findings on vaccination to the Academy of Medicine

William Budd, English physician, d Clevedon, Somerset, Jan 9

Arnold Lucius Gesell, American psychologist, b Alma, WI, Jun 21; he leads a group at Yale University that studies the stages of mental development of children, employing motion pictures to record the behavior of more than 12,000 children

Pierre-Paul Broca, French surgeon and anthropologist, d Paris, Jul 9

Pierre Curie (b Paris, May 15, 1859) discovers the piezoelectric effect; certain substances produce an electric charge as a result of pressure on them; also, an electric current can change the dimensions of these substances slightly; the effect has many applications

Jacques-Arsène d'Arsonval invents an improved form of the galvanometer, now known as the d'Arsonval galvanometer

Graham G. Bell patents the photophone; a phone circuit sets a mirror in vibration and reflected sun rays are detected by a selenium detector, thus allowing the transmission of sound by light

Thomas Alva Edison's first electric generation station, designed mainly for lighting, is opened in London

Edwin Laurentine Drake, American petroleum engineer, d Bethlehem, PA, Nov 8

1881

Emile Brugsch becomes the first archaeologist to enter the place where 36 members of the Egyptian royal family, including many of the most famous, had been reburied; the site, near Luxor, had been located about ten years earlier by Arabs, who then removed small pieces from it and sold them

The first photographic discovery of a comet is made by Edward Emerson Barnard (b Nashville, TN, Dec 16, 1857); a philanthropist was giving $200 to each American who discovered a comet; Barnard went on to earn $3200 in this way, discovering 20 comets in his career

N.I. Kibaltchich designs a rocket; he also makes the bomb used to kill the czar of Russia and is executed

Paul Ehrlich (b Strehlen, Germany, Mar 14, 1854) introduces staining of bacteria with methylene blue

Matthias Jakob Schleiden, German botanist, d Frankfurt, Jun 23

Kutcherov discovers acetaldehyde by passing acetylene gas through sulfuric acid with mercury as a catalyst

Henri-Etienne Sainte-Claire Deville, French chemist, d Boulogne-sur-Seine, Jul 1

The germ theory of disease

In the middle of the nineteenth century, some surgeons realized that doctors could spread disease from one person to another. Ignaz Semmelweiss in 1847 reduced the number of cases of childbed fever by having surgeons wash their hands between deliveries. Joseph Lister in 1865 went further and proposed using carbolic acid (also called phenol) on patients' wounds during surgery. In this he was inspired by the work of Louis Pasteur, who had shown that antiseptics could stop the spread of a disease of silkworms.

Doctors were not the only agents spreading disease, of course. John Snow, a doctor in London, noticed that cholera cases were clustered where people used water from particular sources. He persuaded the authorities to remove the pump handle from a well at the center of the largest outbreak area; the number cases in that area dropped immediately. William Budd learned of Snow's success and repeated it, stopping a cholera epidemic in Bristol in 1866 by managing the town's water supply.

Pasteur moved from studying disease in silkworms to urging French army surgeons to sterilize their instruments and their patients during the Franco-Prussian War of 1870. After the war, Pasteur began an extensive study of the causes of disease. He started with anthrax—a disease that affects both farm animals and humans—because another proponent of the newly developing germ theory of disease, Robert Koch, had already found the bacterium that causes anthrax in 1876. Pasteur first helped stop the spread of anthrax by proposing sterilization and burial of the bodies of animals that had died from it. Later he was to find something even more effective.

In 1879 Pasteur discovered by accident that bacteria could be weakened. In their weakened state, they failed to cause disease but still provoked immunity. He soon learned that the anthrax bacterium could be weakened in this way and by 1881 was able to demonstrate that an effective immunization against anthrax resulted from injection with the weakened bacteria. Pasteur called the process of producing immunization in this way vaccination, in recognition of Edward Jenner's use of cowpox (vaccinia) to prevent smallpox. Pasteur, Koch and others then developed vaccines of varying degrees of effectiveness against many of the other major communicable diseases of the time—cholera, tuberculosis, tetanus, diphtheria, rabies.

Originally the germ theory applied only to bacteria, but about the same time as Pasteur was writing his major exposition of the theory, Charles Laveran discovered that the organism causing malaria is a protist (sometimes called a protozoan), not a bacterium. The agents of some diseases, especially skin diseases, may be funguses. In other communicable diseases, such as measles, neither a bacterium, protist, nor fungus could be found in the late nineteenth or early twentieth centuries. Because of the success of the germ theory, organisms too small to be seen or trapped in filters—filterable viruses—were postulated to explain these diseases. With the invention of the electron microscope in the twentieth century, such viruses were finally observed directly.

Today the germ theory of disease is so entrenched that even diseases that are not communicable—such as cancer, diabetes, or multiple sclerosis—are often suspected of being caused by some "germ."

Elements of vector analysis by Josiah Willard Gibbs (b New Haven, CT, Feb 11, 1839) develops a system of vectors in three dimensions; it is primarily a pamphlet Gibbs produces for the use of his students at Yale

Leopold Kronecker (b Liegnitz, Silesia, Poland, Dec 7, 1823) gives various examples of systems that form mathematical fields

Ferdinand Cohn's *Bacteria, the smallest of living organisms* relates his pioneering work in bacteriology

Carlos Finlay (b Dec 3, 1833) suggests that the mosquito is the carrier of yellow fever

Edwin Klebs detects the typhoid bacillus

Louis Pasteur develops the first artificially produced vaccine, against anthrax, a deadly disease that affects both animals and humans

On May 5 Louis Pasteur successfully demonstrates that vaccination of sheep and cattle against anthrax prevents their falling ill with the disease after injection with live bacteria; unvaccinated animals die when given the same amount of live bacteria

Philosophische Studien, the official organ reporting research at Wilhelm Wundt's laboratory of psychology and the first journal of psychology, is founded

Walter Rudolf Hess, Swiss physiologist, b Frauenfeld, Thurgau, Mar 17; he develops the technique of using small electrodes to stimulate specific regions of the brain; working with dogs and cats, he identifies various regions in the brain

Hermann Ludwig von Helmholtz shows that the electrical charges in atoms are divided into definite integral portions, suggesting the idea that there is a smallest unit of electricity

Samuel Pierpont Langley (b Roxbury, MA, Aug 22, 1834) invents the bolometer, an ultrasensitive device for measuring temperature; he uses it to measure the radiation from the sun in both the visible and infrared regions

German-American physicist Albert Michelson (b Strelno, Prussia, Dec 19, 1852) invents the interferometer, a device for measuring distances by observing interference patterns of a light beam that has been split in two; his *Relative motion of the Earth and the luminiferous ether* suggests that no such ether may exist

Joseph John Thomson (b Manchester, England, Dec 18, 1856) introduces electromagnetic mass: he deduces from Maxwell's equations that the mass of an object changes when electrically charged

Theodore von Kármán, Hungarian-American physicist, b Budapest, May 11; he and his students develop the science of aerodynamics and apply it to aircraft design

Heinrich Barkhausen, German physicist, b Bremen, Dec 2; he determines why the magnetization of an iron bar proceeds in small jerks accompanied by tiny clicks:

The first practical electric generator and electric distribution system are built

The first electric streetcar is introduced in Berlin

Etienne-Jules Marey develops a camera that will take many pictures of the same scene in a short period of time; it is a predecessor of the motion-picture camera; Marey uses it to learn about animal locomotion

Hiram Maxim (b Sangerville, ME, Feb 5, 1840) builds a self-regulating electrical generator

Ernst Mach proposes the use of electric discharges to produce photographs with extremely short exposure

1881

1881
cont

1882

		James Challis, English astronomer, d Cambridge, Dec 3	*Cell substance, nucleus, and cell division* by Walther Flemming, German biologist, reports his discovery of chromosomes and mitosis (cell division)	Viktor Meyer discovers thiophene when a common test for benzene fails to work on a sample of benzene; careful work shows that all previous samples of benzene known had been contaminated with thiophene, which is what the test actually detects

Henry Draper, American astronomer, d New York, NY, Nov 20

David Gill (b Aberdeen, Scotland, Jun 12, 1843) makes a classic photograph of the Great Comet of 1882 that shows so many stars that the idea of stellar cataloging by photography is born; Gill proceeds to map most of the stars visible in the Southern Hemisphere using his new method

Edward Charles Pickering (b Boston, MA, Jul 19, 1846) develops the method of photographing the spectra of several stars at a time by placing a large prism in front of the photographic plate

Friedrich Löffler (b Frankfurt an der Oder, Germany, Jun 24, 1852) discovers the bacterium causing glanders

Eduard Adolf Strasburger coins the terms *cytoplasm* for the part of a cell within the membrane but outside the nucleus, and *nucleoplasm* for the material within the nucleus

Theodor Schwann, German physiologist, d Cologne, Rhenish Prussia, Jan 11

Sir Charles Wyville Thomson, Scottish zoologist, d Bonsyde, Mar 10

Charles Robert Darwin, English naturalist, d Down, Kent, Apr 19

Francis Maitland Balfour, Scottish biologist, d Mont Blanc, Switzerland, Jul 19

John William Draper, English-American chemist, d Hastings-on-Hudson, NY, Jan 4

Friedrich Wöhler, German chemist, d Göttingen, Sep 23

1883

Accepting an 1879 report from meteorologist and astronomer Cleveland Abbe, the United States sets up four time zones based on zones previously developed by the railroad industry

The pyramids and temples of Gizeh by Flinders Petrie (b Charlton, England, Jun 3, 1853) reveals for the first time the precision with which the Great Pyramid is aligned, only a little more than a twelfth of a degree away from a perfect north-south alignment

Edouard-Albert Roche, French astronomer, d Montpellier, Hérault, Apr 18

Hermann Carl Vogel lists the spectra of 4051 stars

The last known quagga, a close relative of or subspecies of the plains zebra of Africa, dies in the Amsterdam zoo

Camillo Golgi discovers "Golgi cells," a type of cell in the nervous system

Barlow predicts the structure of sodium chloride from studying the shape of crystals

Johann Kjeldahl (b Jagerpris, Denmark, Aug 16) develops a simple way to analyze the nitrogen content of organic compounds

EARTH SCIENCE	MATHEMATICS	MEDICINE	PHYSICS	TECHNOLOGY	
			as small magnetic domains become aligned, one at a time, they rub against their neighbors, causing the clicks		**1881** cont
Balfour Stewart (b Edinburgh, Scotland, Nov 1, 1828) suggests the existence of the ionosphere to account for small daily changes in Earth's magnetic field Ferdinand Reich, German mineralogist, d Freiberg, Apr 27	*Über die Zahl π* (On the number π) by Ferdinand Lindemann (b Hanover, Germany, Apr 12, 1852) proves that π is a transcendental number, implying that the circle cannot be squared with straightedge and compass *Vorlesungen über Neuere Geometrie* (Lectures on more recent geometry) by Moritz Pasch (b 1843) is the first attempt to reform Euclid's geometry by changing the undefined terms and postulates William Shanks, English mathematician, d Houghton-le-Spring, Durham Joseph Liouville, French mathematician, d Paris, Sep 8	Paul Ehrlich introduces his diazo reaction for diagnosing typhoid fever Robert Koch discovers the bacterium that causes tuberculosis, the first definite association of a germ with a specific human disease	Women begin to be admitted to the Cavendish Laboratory in Cambridge, England, on the same terms as men Henry Augustus Rowland (b Honesdale, PA, Nov 27, 1848) constructs a machine that will make precision diffraction gratings for use in advanced spectroscopy John William Strutt, Lord Rayleigh (b Terling Place, England, Nov 12, 1842) discovers that the ratio of the atomic mass of oxygen to that of hydrogen is not 16 exactly, as had been assumed, but 15.882	The first attempt to build a tunnel beneath the English Channel is halted for political reasons, not technical ones The first hydroelectric power plant goes into operation in Appleton, WI Thomas Alva Edison patents a three-wire system for transporting electrical power that is still in use today Hiram Maxim (b Sangerville, ME, Feb 5, 1840) develops the Maxim machine gun The Pearl Street power station in New York brings electric lighting to the United States	**1882**
Krakatoa, a volcanic island between Java and Sumatra, explodes on Aug 27, sending waves around the world and killing nearly 40,000 people	Georg Cantor's *Foundations of a general manifold theory* claims that every set can be made into a well-ordered set, the proof of which, supplied by Ernst Zermelo in 1904, requires the controversial axiom of choice	Antipyrene, a powder used to reduce fever and relieve pain, is synthesized Francis Galton's *Enquiries into human faculty* introduces the term *eugenics* and suggests that human beings can be improved by selective breeding		Inventor Gottlieb Wilhelm Daimler (b Schorndorf, Germany, Mar 17, 1834) develops the first of his high-speed internal combustion engines and uses it on a boat, the first true motorboat	**1883**

George John Romanes's *Animal intelligence* is the first book on comparative psychology

The value of π

It is surprising, but many college students, when asked which is greater, 3.14 or 22/7, will say that the two numbers are the same. That is because they think that both numbers are the same as the number π. Neither number is actually π. Since 22/7 starts off 3.1428571 . . ., it is the greater number.

There is some evidence that the ancient Hebrews and Babylonians were even less accurate than today's college students. In the Bible (First Kings 7:23) we learn of the model of a sea made by Hiram of Tyre for King Solomon: "it was round, ten cubits from brim to brim . . . and a line of thirty cubits measured its circumference." The implication is that π is 3, since π is the ratio of the circumference of a circle to its diameter. There is some evidence that Babylonian mathematicians used a better value for π, namely 3.125. It was clear to the ancient Greeks and Chinese that one could get a good approximation of π by comparing a circle to a straight-sided figure that was approximately a circle. It is relatively easy to find the length of the perimeter of such a regular polygon if you know the distance from the polygon's center to one of its sides or to one of its vertices. Using this method, Archimedes calculated that π is between 3-10/71 and 3-10/70 (22/7), while Chinese scholars around 500 AD showed that π is between 3.1415926 and 3.14152927. In 1596, Ludolph of Cologne used this method to calculate π to 32 places. His result was engraved on his tombstone and to this day Germans call π the Ludolphine number.

Although everyone knew that these values for π were not exact (since they were based on perimeters of polygons, not the circumference of a circle), it was not clear whether an exact value could be found. Around the time of Ludolph, Vieta developed the first simple numerical expression for π. It was not expressed as a decimal numeral or as a fraction, however. It was an infinite product. Later mathematicians also found other infinite products and infinite series (sums) for π. Two with especially easy patterns are actually for π/2 and π/4. In the seventeenth century, John Wallis discovered $\pi/2 = 2/1 \times 2/3 \times 4/3 \times 4/5 \times 6/5 \times 6/7 \times$ . . ., in which the numerators are the even numbers from 2 given twice, while the denominators are a similar pattern of odd numbers. James Gregory and Wilhelm Gottfried Leibniz discovered an even simpler pattern for an infinite sum: $\pi/4 = 1/1 - 1/3 + 1/5 - 1/7 + 1/9 - 1/11 + . . .$. This pattern is known as the Leibniz series, although Gregory was the first to find it. Note that these patterns carried to infinity yield exact values for π, but they still do not tell whether π can be expressed as a finite decimal. Many infinite products and series converge to finite decimals.

These infinite products and series, and others like them, however, provided an easier way to compute approximations to π than using polygons. At the end of the seventeenth century, Abraham Sharp found 71 decimal places. In the nineteenth century, π was gradually extended, reaching 707 places in the calculation of William Shanks in 1853 that took him 15 years to complete. When computers were invented, however, it was found that Shanks had made a mistake in the 528th place, causing every place afterward to be wrong.

In the meantime, in the eighteenth century, Johann Lambert finally solved one of the problems connected with π. He showed that π is irrational; in other words, it cannot be expressed as a finite decimal, nor can it have a simple repeating pattern as a decimal.

A related problem was still unsolved. Since the time of Anaxagoras at least, in the fifth century BC, people had been trying to use a straightedge and compass to construct a square the same area as a given circle. By 1775 the ranks of people trying to solve this famous problem were so great that the Academy of Paris passed a resolution that it could no longer examine purported successes.

This problem was effectively solved in 1882, when the mathematician Ferdinand Lindemann showed that π is a member of a large class of numbers of which only a few are commonly known. These numbers are called *transcendental*. There are more of them than of any of the more familiar numbers. Their defining characteristic is that they are not the solutions to algebraic equations with integer coefficients. Constructing a line with a straightedge and compass implies that its length is the solution to such an equation. Since π is transcendental, it cannot be that kind of solution. Squaring the circle is impossible.

This did not stop people from calculating the value of π to more and more decimal places. When electronic computers became available in the 1940s and 1950s, some people used the calculation of π as a kind of demonstration of how powerful these computers were. By 1949, in 70 hours of computer time (as opposed to Shanks's 15 years of paper-and-pencil time), π was extended to 2037 places. By 1988 Japanese computer scientist Yasumasa Kanada had reached 201,326,000 decimal places, and he was planning to go further. The 1988 computation only took six hours of supercomputer time.

1883
cont

Victor Horsley studies growth disorders and shows that they are caused by lack of thyroid secretion

Edwin Klebs and Friedrich Löffler identify the diphtheria bacillus

Robert Koch discovers *Cholera vibrio*, the bacterium that causes cholera, and shows that cholera can be transmitted by food and drinking water

Sydney Ringer (b Norwich, England, 1835) discovers that an isolated frog heart kept in a saline solution will beat longer if calcium and potassium are added to the solution, the combination known today as Ringer's solution; Ringer also finds that other activities of cells require calcium

Carl Stumpf's *Tonpsychologie* (Psychology of tone) becomes his most influential book; it contains some of the tenets of the philosophical school of phenomenology (Edmund Husserl, the founder of phenomenology, was Stumpf's student)

Sir Cyril Lodowic Burt, English psychologist, b Stratford-on-Avon, Warwickshire, Mar 3; the founder of Mensa, a society for the intellectually gifted, Burt firmly believed that intelligence is inherited, so firmly that he made up data that would demonstrate that he was correct

Gabriel Gustav Valentin, German-Swiss physiologist, d Bern, Switzerland, May 24

George Francis Fitzgerald (b Dublin, Ireland, Aug 3, 1851) points out that Maxwell's theory of electromagnetic waves indicates that such waves can be generated by periodically varying an electric current; later Heinrich Hertz demonstrates that this is true; radio waves are still generated this way today

Sir Edward Sabine, British physicist, d Richmond, Surrey, Jun 26

Metallurgist Sir Robert Abbott Hadfield (b Sheffield, England, Nov 28, 1858) patents manganese steel, a superhard alloy that is the first of the specialty alloy steels

Albert and Caston Tissandier design the first dirigible that is capable of being steered along a specific course

The Brooklyn Bridge, which introduces a revolutionary method of cable spinning, is dedicated on May 24

Sir William Siemens, German-British inventor, d London, Nov 19

1884

GENERAL

Herbert Spencer suggests that the principle of survival of the fittest implies that people who are burdens on society should be allowed to die rather than be helped by society

ASTRONOMY

An international meeting in Washington, DC, sets the prime meridian through Greenwich, England

Samuel P. Langley's *The new astronomy* introduces the general public to the first results of astrophysics

BIOLOGY

Bacteriologist Hans Christian Joachim Gram (b Copenhagen, Denmark, Sep 13, 1853) develops the dye that is used to classify bacteria into two groups, the grampositive that retain the dye and the gramnegative that do not retain it; in the 1940s it is discovered that the two types react differently to antibiotics

Friedrich Löffler discovers the bacillus that causes diphtheria and that some animals are naturally immune to the disease

Gregor Johann Mendel, Austrian botanist, d Brünn, Bohemia (Brno, Czechoslovakia), Jan 6

George Bentham, English botanical taxonomist, d Sep 10

Otto Fritz Meyerhof, German-American biochemist, b Hanover, Germany, Apr 12; he recognizes that exertion causes glycogen stored in muscles to break down into lactic acid and that lactic acid combines with oxygen during a muscle's resting phase to restore the glycogen level

Arthur Nicolaier discovers the bacterium that causes tetanus

CHEMISTRY

Svante August Arrhenius (b Vik, Sweden, Feb 19, 1859) proposes his dissociation theory: electrolytes dissociate into positively and negatively charged ions when in solution

Jacobus Henricus van't Hoff's *Etudes de dynamique chimique* (Studies in chemical dynamics), published in English in 1896, contains his results in the research on the theory of solutions, especially reaction kinetics and chemical equilibrium

German organic chemist Otto Wallach, b Königsberg (Kaliningrad, USSR), Mar 27, 1847, starts a long project of isolating terpenes from various essential oils, such as menthol and camphor; his work forms the basis of much of the perfume industry

Jean-Baptiste-André Dumas, French chemist, d Cannes, Alpes-Maritimes, Apr 10

Charles Adolphe Wurtz, French chemist, d Paris, May 12

Adolph Wilhelm Hermann Kolbe, German chemist, d Leipzig, Nov 25

1885

ASTRONOMY

A supernova appears in M31, the Andromeda galaxy; this is the only recorded extragalactic supernova to reach the fringe of naked-eye visibility until the supernova of 1987

BIOLOGY

Laurent Chabry starts his embryology experiments using a micromanipulator he devised

Charles Darwin (son of the author of *The origin of species*) designs the microtome, an apparatus to

CHEMISTRY

Karl Auer, later Baron von Welsbach (b Vienna, Austria, Sep 1, 1858) discovers that one supposed element is actually two, which he names neodymium ("new twin") and praseodymium ("green twin" for the color of its spectrum)

1884

Vladimir Köppen produces a world map of temperature zones

Arnold Henry Guyot, Swiss-American geographer, d Princeton, NJ, Feb 8

Gottlob Frege's *Die Grundlagen der Arithmetik* (The foundations of arithmetic) bases all of natural-number arithmetic on his definition of a natural number (that is, 1, 2, 3, . . .) as the class of all classes that can be put into one-to-one correspondence with a given class

Czech-American surgeon Carl Koller (b Schüttenhofen, Bohemia, Dec 3, 1857) uses cocaine as a local anesthetic

Russian-French bacteriologist Ilya Illich Mechnikov, b Ivanovka, Ukraine, May 16, 1845, discovers phagocytes in the body, mobile white blood cells that attack and devour invading organisms

Physiologist Max Rubner (b Munich, Germany, Jun 2, 1854) discovers that the body gets energy from carbohydrates, fats, and proteins after stripping away nitrogen for other uses

Endocrinologist Philip Edward Smith b De Smet, SD, Jan 1; a specialist in the pituitary gland, he showed that its removal causes other endocrine glands to cease functioning

Alexander M. Ljapunow publishes an equilibrium theory for rotating fluids

The ammeter is introduced in electrical engineering

The Washington Monument is completed

Telephone wires connect Boston to New York

Louis-Marie-Hilaire Bernigaud begins to produce fibers made from cellulose, which comes to be known as rayon

Ottmar Mergenthaler (b May 11, 1854) invents the Linotype typesetting machine

Croatian-American engineer Nikola Tesla, b Smiljan (Yugoslavia), Jul 10, 1856, invents the electric alternator

Paul Nipkow invents the scanning disk named after him, a precursor of television

Charles Algernon Parsons (b London, Jun 12, 1854) designs and installs the first steam turbine generator for electric power

Frank Julian Sprague develops a direct-current motor for electric locomotives

W.H. Walker invents the roll film

Lewis E. Waterman patents a fountain pen

Cyrus Hall McCormick, American inventor, d May 13

1885

Lord Rayleigh identifies the type of earthquake waves now called Rayleigh waves

The first volume of *Das Antlitz der Erde* (The face of the Earth) by Austrian geologist Eduard

Austrian psychiatrist Sigmund Freud studies hypnotism with Jean-Martin Charcot, the beginning of his path toward the development of psychoanalysis

Johann Jakob Balmer (b Lausen, Switzerland, May 1, 1825) discovers the formula for the hydrogen spectrum that will later inspire Niels Bohr to develop his model of the atom

Karl Benz (b Nov 25, 1844) builds a three-wheel automobile

Gottlieb Daimler installs one of his internal-combustion engines on a bicycle, creating the world's first motorbike

1885 cont		*Uranometria nova oxoniensis* (The new Oxonian Uranometria) by Charles Pritchard (b Feb 29, 1808) catalogs stars using his wedge photometer to measure relative brightness	make thin slices of specimen to be observed by the microscope Hermann Ebbinghaus's *Über das Gedächtnis* (On memory) contains his "curve of forgetting," which reflects how much is retained in human memory with time Paul Ehrlich discovers the blood-brain barrier, which prevents many substances dissolved in blood from reaching the brain Camillo Golgi and others elucidate the asexual cycle of the malaria parasite Karl Theodor Ernst von Siebold, German zoologist, d Munich, Bavaria, Apr 7	*Lehrbuch der allgemeinen Chemie* by Friedrich Wilhelm Ostwald, b Riga (USSR), Sep 2, 1853, is considered to mark the founding of physical chemistry Jacobus Henricus van't Hoff develops his formula for osmotic pressure Clemens Alexander Winkler (b Freiberg, Saxony, Germany, Dec 26, 1838) discovers germanium, the third element predicted by Mendeléev on the basis of his periodic table Thomas Andrews, Irish physical chemist, d Belfast, Nov 26
1886	The second discovery of skeletons of the Neanderthal type occurs	Henry Augustus Rowland uses diffraction gratings prepared by his improved method to map the solar spectrum, giving the precise wavelength of about 14,000 lines in the spectrum	William Rutherford proposes the "telephone theory" of hearing, in which it is assumed that the whole cochlear partition is put into movement by sound; this theory is superseded by the discovery that tiny hairs in the cochlea are put into motion and thus convey sound	Hermann Hellriegel (b Mausitz, Saxony, Germany, Oct 21, 1831) announces his discovery that legumes remove nitrogen from the air, unlike most other plants, which must use nitrogen fixed in compounds in the soil; later it is shown that specialized bacteria on the roots of legumes actually fix the nitrogen Paul-Emile Lecoq de Boisbaudran discovers the element dysprosium Jean-Charles Marignac discovers gadolinium Alfred Bernhard Nobel (b Stockholm, Sweden, Oct 21, 1833) discovers ballistite, a nitroglycerine explosive that does not produce smoke

Suess (b London, Aug 20, 1831) is published; this five-volume work, completed in 1909, attempts to explain such geological features as mountain ranges in terms of Earth's contraction as it cools

Pierre Marie describes acromegaly

Louis Pasteur develops a vaccine against hydrophobia (rabies) and uses it to save the life of a young boy, Joseph Meister, bitten by a rabid dog

Friedrich Gustav Jakob Henle, German pathologist and anatomist, d Göttingen, May 13

James Dewar (b Kincardine, Scotland, Sep 20, 1842) invents a thermos bottle in which heat is prevented from leaking by a vacuum between two glass walls; the bottle becomes known as the Dewar flask

George Eastman (b Waterville, NY, Jul 12, 1854) patents a machine for producing continuous photographic film

Dorr Felt develops the Comptometer, a key-driven adding and subtracting machine

William Stanley invents the electric transformer

Charles S. Tainter develops the Dictaphone, a machine for recording dictation

The first transcontinental rail link across Canada is opened

**1885
cont**

A major earthquake strikes Charleston, SC; although measuring only 6.5 to 7.0 on the Richter scale, its effects are felt strongly as far away as Chicago because the rock structure of the eastern part of the United States transmits earthquake waves more efficiently than that of the western part

Alexandre-Emile Beguyer de Chancourtois, French geologist, d Paris, Nov 14

Antifibrin is synthesized

Psychopathia sexualis (Sexual psychopathy) by neurologist Baron Richard von Krafft-Ebing (b Mannheim, Germany, Aug 14, 1840) is the first account of abnormal sexual practices in humans and introduces such terms as *paranoia, sadism,* and *masochism*

Paul Bert, French physiologist, d Hanoi (Vietnam), Nov 11

William Crookes proposes that atomic weights measured by chemists are averages of the weights of different kinds of atoms of the same element (although it will not be until 1910 that Frederick Soddy identifies these different kinds of atoms as isotopes)

Eugen Goldstein observes that a cathode-ray tube produces, in addition to the cathode ray, radiation that travels in the opposite direction—away from the anode; these rays are called canal rays because of holes (canals) bored in the cathode; later these will be found to be ions that have had electrons stripped in producing the cathode ray

Alexander Graham Bell uses wax disks for recording sound with a modified version of Edison's phonograph

Hamilton Y. Castner invents the sodium process of aluminum production

Charles M. Hall (b Thompson, OH, Dec 6, 1863) and Paul-Louis-Toussaint Héroult (b Thury-Harcourt, Calvados, France, Apr 10, 1863) independently discover an economical way to make aluminum from abundant alumina and electric power

English-American inventor Elihu Thomson (b Manchester England, Mar 29, 1853) patents a welding method using heating by electrical resistance

1886

1886 cont

Does the ether exist?

In 1887 Albert A. Michelson and Edward Morley tried to measure the absolute velocity of Earth with respect to the rest of the universe. The result of their measurement showed that Earth is at rest in space, but no one believed that that could be true. The idea of a motionless Earth had been given up 300 years earlier. Something had to be wrong with the experiment.

The problem at first glance seemed to be with light. Ever since Thomas Young had demonstrated in 1803 that light produces interference patterns, it was assumed that light must be waves, a view previously held by Christiaan Huygens and others. All those who espoused the wave theory—the alternative theory was that light consisted of particles—assumed that light must be waves in something. When Augustin-Jean Fresnel began to work on the wave theory of light, he named the something *ether,* after Aristotle's fifth element. Fresnel was able to show that light waves are more like three-dimensional water waves than like sound waves.

Michelson had taken a suggestion of James Clerk Maxwell, made in 1875, to develop a sensitive measuring tool based on the interference patterns of waves of light. The Michelson-Morley experiment was designed to use the interferometer, Michelson's new tool, to measure how Earth moves through the ether, which was assumed to be stationary. Michelson and Morley measured the speed of light in two perpendicular directions. It was found to be the same, implying that Earth stands still.

Others had different explanations. Both George Francis Fitzgerald and Hendrick Antoon Lorentz suggested that the problem could be resolved by assuming that objects contract when they

---

François-Marie Raoult (b Fournes-en-Weppes Nord, France, May 10, 1830) introduces the law named for him; the law states that the lowering of the vapor pressure of a solvent is proportional to the concentration of the substance dissolved, unless the solvent is an electrolyte

After many unsuccessful and dangerous attempts by other chemists, Ferdinand-Frédéric-Henri Moissan (b Paris, Sep 28, 1852) succeeds on Jun 26 in isolating the element fluorine; fluorine's poisonous nature is believed to contribute to Moissan's early death at age 54

Alexander Mikhailovich Butlerov, Russian chemist, d Butlerovka Kazanskaya, Aug 17

1887

Ernst Mach revives the idea that science only gives information that is supplied by the senses, and that the ultimate nature of reality is beyond the reach of our intelligence

Flinders Petrie begins his excavation of the pyramid of Senurset II

The International Congress for Astrophotography is held in Paris

The Lick 91-cm (36-in.) refracting telescope is completed and installed on Mount Hamilton near San Francisco, CA, on Dec 31; it is the world's first mountaintop telescope

William Abney (b Derby, England, Jul 24, 1843) develops methods to detect infrared radiation with photographs and uses them to observe the spectrum of the sun

Joseph Lockyer's *Chemistry of the sun* describes the results of his spectroscopic analysis of the sun which had resulted in the discovery of helium in 1868

Spencer Fullerton Baird, American zoologist, d Aug 19

Edouard-Joseph-Louis-Marie van Beneden (b Louvain, Belgium, Mar 5, 1846) discovers that each species has a fixed number of chromosomes; he also discovers the formation of haploid cells (cells with half the original number of chromosomes) during cell division of the sperm and ova (meiosis)

Emil Hermann Fischer (b Euskirchen, Rhenish Prussia, Germany, Oct 9, 1852) analyzes the structure of sugars

German-American chemist Herman Frasch (b Württemberg, Germany, Dec 25, 1851) develops a method for removing sulfur compounds from petroleum

Jean-Baptiste-Joseph-Dieudonné Boussingault, French agricultural chemist, d Paris, May 12

Does the ether exist? (continued)

move. By assigning the correct value to the contraction, the speed of light could be made to appear the same in two different directions, since the apparatus used to measure it would contract in the direction in which Earth is moving.

In 1905, Albert Einstein resolved the issue in a different way. Apparently, although he knew of the Michelson-Morley experiment, it was not a major influence on his thinking. Einstein perceived that the same physical laws must be true in all frames of reference in steady motion with respect to each other. But this requirement also means that light must have the same speed no matter how a body is moving. There can be no fixed frame of reference against which Earth moves, so Michelson and Morley got the correct result. Furthermore, in a separate paper, Einstein showed that light behaves like a particle as well as like a wave. Although this wave-particle duality remains difficult to picture, it has become quite clear since 1905 that light has both wave and particle properties. A particle does not require a medium the way that a wave does.

Does the ether exist? Einstein's theory of special relativity suggests that the question is irrelevant. Certainly no one thinks today that a solid medium pervades space. On the other hand, it is widely believed that there is no real vacuum. Arno Penzias and Robert Wilson detected in 1964 the cosmic background radiation, a leftover of the Big Bang. This radiation is pervasive and basically the same in all directions. It is possible to measure Earth's movement against this background radiation. As it turns out, Earth moves.

Richard March Hoe, American inventor, d Jun 7

Mahlon Loomis, American dentist and pioneer in radiotelegraphy, d Oct 13

1886 cont

Vito Volterra founds functional analysis

A German glassblower develops the first form of contact lens, which covers the whites of the eye as well as the cornea; the corneal contact lens is developed by American optician Kevin Tuohy in 1948, who accidentally breaks off the corneal part of the lens as he is making the older version.

Granville Stanley Hall founds the *American Journal of Psychology*, the first American journal in the field

Wolfgang Köhler, Russian-German-American psychologist, b Revel, Estonia (USSR) Jan 21; one of the founders of the Gestalt school of psychology, he is best known for his experiments with chim-

Robert Bunsen invents the vapor calorimeter

Ernst Mach notes that airflow becomes disturbed at the speed of sound

Albert Michelson and Edward Morley (b Newark, NJ, Jan 29, 1838) measure the velocity of light in two directions, attempting to detect the proper motion of Earth through the ether; the Michelson-Morley experiment reveals no evidence of motion

Gustav Robert Kirchhoff, German physicist, d Berlin, Oct 17

Gustav Theodor Fechner, German physicist, d Leipzig, Saxony, Nov 18

Gottlieb Daimler uses his internal-combustion engine to power a four-wheeled vehicle, one of the first automobiles

Hannibal W. Goodwin invents the celluloid photographic film and develops a production process for this film

Robert Lanston patents the Monotype typesetting machine

James Buchanan Eads, American engineer, d Mar 8

Alfred Krupp, German metallurgist, d Jul 14

1887

1887
cont

1888

The Pasteur Institute is founded

Clinton M. Merriam founds the National Geographical Society and starts publishing *National Geographic Magazine*

Explorer Fridtjof Nansen (b Store-Froen, Norway, Oct 10, 1861) and his team of explorers become the first people to cross Greenland by land

Nikolay Mikhaylovich Przhevalsky, Russian explorer, d Karakol, Kirgiz, Nov 1

Richard Wetherill discovers Cliff Palace, the largest of the cliff dwellings at Mesa Verde; it contains about 200 ruins and 23 kivas (round buildings used for religious ceremonies)

Johann L.E. Dreyer's *A new general catalogue of nebulas and clusters of stars* is published; the designation NGC for an astronomical object refers to this catalog, which contains 7840 nebulas and star clusters

Richard Anthony Proctor, English astronomer, d New York, NY, Sep 12

Hermann Carl Vogel makes the first spectrographic measurements of the radial velocities of stars, using Hippolyte Fizeau's principle of the red shift (also known as the Doppler effect)

Theodor Boveri (b Bamberg, Germany, Oct 12, 1862) discovers and names the centrosome, the small body that appears during cell division and that seems to control the process of division

Asa Gray, American botanist, d Cambridge, MA, Jan 30

Emile Hansen introduces new methods for obtaining yeast; they are adopted by breweries

Friedrich Wilhelm Ostwald discovers that catalysts affect only reaction speed and not reaction equilibrium

F. Reinitzer discovers the liquid crystal characteristics of cholesteryl benzoate; O. Lehmann discovers that one liquid phase is optically anisotropic

Ascanio Sobrero, Italian chemist, d Turin, May 26

1889

Warren De la Rue, British astronomer, d London, Apr 19

William Harkness measures the mass of the planets Mercury, Venus, and Earth

Russian physiologist Ivan Petrovich Pavlov (b Ryazan', Sep 26, 1849) demonstrates that the stimulus that causes secretions from the stomach is mediated by the nervous system

Frederick Augustus Abel (b Woolwich, England, Jul 17, 1827) and James Dewar invent cordite, smokeless gunpowder

1887 cont

MEDICINE

panzees that demonstrate their problem-solving ability

Bernardo Alberto Houssay, Argentinian physiologist, b Buenos Aires, Apr 10; he demonstrates that the pituitary gland produces a hormone that has the opposite effect of insulin, showing the complexity of the endocrine system

PHYSICS

Balfour Stewart, Scottish physicist, d near Drogheda, Ireland, Dec 19

1888

EARTH SCIENCE

The first seismograph in the United States is installed at the Lick Observatory in California

Victor Moritz Goldschmidt, Swiss-Norwegian geochemist, b Zurich, Switzerland, Jan 27; as a pioneer geochemist, he is the first to be able to predict the formation of specific minerals from specific combinations of elements and geological conditions

MATHEMATICS

The New York Mathematical Society is founded; in 1894 it will become the American Mathematical Society

Richard Dedekind's *Was sind und was sollen die Zahlen* (The nature and meaning of numbers) introduces the definition of an infinite set as a set that can be put into one-to-one correspondence with a proper subset of itself

Francis Galton introduces the concept of correlation; it is a measure of the interdependence of two sets of variables

Sonya Kovalevsky receives the Prix Bourdin from the French Academy for her anonymously submitted paper *On the rotation of a solid body about a fixed point*

Sophus Lie (b Norway, Dec 17, 1842) develops the theory of monotonic transformation groups

MEDICINE

Ernst von Bergmann's *Die chirurgische Behandlung bei Hirnkrankheiten* (The surgical treatment of brain disorders) becomes an important guide to brain surgery

James McKeen Cottell is appointed professor of psychology at the University of Pennsylvania, the first professorship of the kind in the world

Paul Langerhans, German physician, d Funchal, Madeira, Portugal, Jul 20

PHYSICS

Heinrich Hertz produces and detects radio waves for the first time; radio waves will be called Hertzian waves until renamed by Marconi, who calls them radiotelegraphy waves

Zygmunt Florenty von Wroblewski, Polish physicist, d Cracow, Apr 19

Rudolf Julius Emmanuel Clausius, German physicist, d Bonn, Aug 24

TECHNOLOGY

William S. Burroughs patents an adding machine

John Boyd Dunlop introduces air-filled rubber tires in England

George Eastman introduces the first commercial roll-film camera

Oliver Schallenberger invents an electric meter for alternating current

Nikola Tesla invents an alternating current induction motor

William Kelly, American inventor, d Feb 11

George Henry Corliss, American inventor, d Feb 21

1889

MATHEMATICS

Francis Galton's *Natural inheritance*, reporting on his research in genetics, includes the concept of the correlation coefficient and the formula for the standard error

MEDICINE

Charles-Edouard Brown-Séquard (b Apr 8, 1817) injects himself during old age with hormones from testicles in the hope of experiencing a return of vigor

PHYSICS

Gaston Planté, French physicist, d Paris, May 21

James Prescott Joule, English physicist, d Sale, Cheshire, Oct 11

TECHNOLOGY

On the Willamette River at Oregon City, OR, the first dam is built to produce power to drive a hydroelectric plant

1889 cont

Maria Mitchell, American astronomer, d Lynn, MA, Jun 28

Hermann Vogel discovers spectroscopic binaries, stars for which the light is shifted alternately to the red and the violet, indicating that there are two stars revolving about each other; a similar discovery is made at Harvard Observatory about this same time by Edward C. Pickering

Vladimir Markovnikov constructs molecules that have carbon rings with seven atoms

Michel-Eugène Chevreul, French chemist, d Paris, Apr 9

1890

James George Frazer's *The golden bough* attempts to unravel the connections between the world's myths

Heinrich Schliemann, German archaeologist, d Naples, Italy, Dec 26

The Russian Astronomical Society is founded

Thomas Alva Edison makes unsuccessful attempts to detect radio waves from the sun

Sir Harold Spencer Jones, English astronomer, b London, Mar 29

In *The Henry Draper catalogue* of stellar spectra Edward C. Pickering and Williamina ("Mina") Fleming (b Dundee, Scotland, May 15, 1857; d Boston, MA, 1911) introduce a stellar classification based on an alphabetic system (known as the Harvard Classification)

Alexander Parkes, English chemist, d London, Jun 29

1889 cont

MATHEMATICS

I principii di geometria, logicamente expositi (The principles of geometry, explained logically) by Giuseppe Peano, b Piedmont (Italy), Aug 27, 1858, translates the reformed Euclidean geometry of Moritz Pasch into symbolic logic, making it virtually impossible to understand

Arithmetices principia, nova methedo exposita (Principles of arithmetic, an exposition of new methods) by Giuseppe Peano includes his axioms for the establishment of the system of natural numbers and introduces much of the notation used for set theory and logic, including the common signs for contains, union, and intersection

MEDICINE

Baron Shibasabura Kitasato (b Kumamoto, Japan, Dec 20, 1856) isolates the tetanus bacillus

Oskar Minkowski and Baron Joseph von Mering attempt to find out if the pancreas has a function essential to life by removing the gland from dogs; noticing that flies are attracted to the urine of their experimental dogs, they discover that the pancreas supplies a hormone essential to glucose metabolism (insulin)

Franciscus Cornelia Donders, Dutch physiologist, d Utrecht, Mar 24

PHYSICS

George Francis Fitzgerald formulates the principle that objects shrink slightly in the direction they are traveling, now known as the Fitzgerald-Lorentz contraction, since Hendrik Antoon Lorentz reaches the same conclusion a few year laters

TECHNOLOGY

Gustave Eiffel's tower in Paris, named after him, at 303 m (993 ft), is the tallest free-standing structure built to this time

Dorr Felt's Comptometer is equipped with a built-in printer

John Ericsson, Swedish-American inventor, d New York, NY, Mar 8

1890

EARTH SCIENCE

Arthur Holmes, English geologist, b Hebburn on Tyne, Durham, Jan 14; he uses radioactivity to date various rock formations, establishing that Earth is about 4.6 billion years old

MATHEMATICS

Oliver Heaviside invents the operational calculus

Giuseppe Peano discovers a curve that, while one-dimensional and continuous, passes through every point in the interior of a square

MEDICINE

Emil von Behring (b Deutsch-Eylau, Germany, Mar 3, 1854) develops a vaccine against tetanus and diphtheria and introduces the concepts of passive immunization and antitoxins

Paul Ehrlich standardizes the diphtheria antitoxin, establishing the field of immunology

Surgeon William Halsted (b New York, NY, Sep 23, 1852) introduces the practice of wearing sterilized rubber gloves during surgery

The principles of psychology by William James (b New York, NY, Jan 11, 1842) becomes an enormous success; it describes psychology as a natural science

PHYSICS

Roland, Baron von Eötvös (b Budapest, Hungary, Jul 27, 1848) conducts experiments in relating the gravitational and inertial masses of various substances; he shows that the two ways of calculating mass give the same results

J. Alfred Ewing discovers the phenomenon of hysteresis in magnetic materials

Hendrik Antoon Lorentz (b Arnhem, Netherlands, Jul 18, 1853) proposes that atoms may consist of charged particles that produce visible light by oscillating, essentially a correct idea

TECHNOLOGY

Herman Hollerith develops an electrically driven system based on punched cards that is used in counting the census

William Kemmer becomes the first person to be executed in the electric chair, a device that uses alternating current; Thomas Alva Edison had arranged for prisons to have alternating current for the electric chair as an element in his fight to have direct current adopted for household use

Clement Adler's *Eole* is the first full-size aircraft to leave the ground under its own power

	GENERAL	ANTHROPOLOGY/ ARCHAEOLOGY	ASTRONOMY	BIOLOGY	CHEMISTRY
1890 cont					
1891			Seth Carlo Chandler discovers the Chandler period, an oscillation of Earth's polar axis with a period of 14 months George Ellery Hale (b Chicago, IL, Jun 29, 1868) invents the spectroheliograph, an instrument that allows a photograph of the sun to be made in the light of a single spectral line Norman Robert Pogson, English astronomer, d Jun Maximilian Wolf (b Heidelberg, Germany, Jun 21, 1863) makes the first discovery of an asteroid from photographs, finding Brucia; he goes on to use the photographic method to discover about 500 more asteroids, about a third of all asteroids known	Karl Wilhelm von Nägeli, Swiss botanist, d Munich, Germany, May 10 Karl von Voit shows that various ingested sugars are converted to glycogen by the body for storage until needed	Edward Goodrich Acheson (b Washington, PA, Mar 9, 1856) discovers a process for making carborundum (silicon carbide), a material almost as hard as diamond German-Swiss chemist Alfred Werner (b Mulhouse, France, Dec 12, 1866) in his sleep develops the first inkling of his theory of secondary valence, a new way of explaining chemical structures beyond ionic and covalent bonds Jean Servais Stas, Belgian chemist, d Brussels, Dec 13
1892	The University of Chicago is founded		International mapping of stars begins John Couch Adams, English astronomer, d Cambridge, Jan 21 Sir George Biddell Airy, English astronomer and mathematician, d Greenwich, Jan 2 Edward Emerson Barnard observes that a nova emits a cloud of gas when it brightens, the first clear evidence that novas are exploding stars	Mosaic disease of tobacco is correctly believed to be caused by a "virus," an organism that is too small to be seen in the microscope and that passes through filters, the first known instance of a virus Henry Walter Bates, English naturalist, d London, Feb 16 Sir Francis Galton's *Finger prints* is a practical study of the differences in fingerprints and the use of fingerprints in identification	Charles Frederick Cross (b Brentford, England, Dec 11, 1855) develops viscose rayon, an improvement over the first types of rayon Hermann Franz Moritz Kopp, German physical chemist, d Heidelberg, Feb 20 Archibald Scott Couper, Scottish chemist, d Kirkintilloch, Mar 11 August Wilhelm von Hofmann, German chemist, d Berlin, May 2

EARTH SCIENCE	MATHEMATICS	MEDICINE	PHYSICS	TECHNOLOGY	
		Robert Koch announces the discovery of tuberculin as a cure for tuberculosis James McKeen Cattell coins the term *mental test* and develops a test that measures bodily or sensory-motor responses	Arthur Schuster calculates the ratio of charge to mass of the particles making up cathode rays (now known as electrons) by measuring the magnetic deflection of cathode rays		**1890 cont**
The US Weather Bureau leaves the military to become a separate department with Cleveland Abbe as the meteorologist in charge	Sonya Kovalevsky, Russian mathematician, d Stockholm, Sweden, Feb 10 Leopold Kronecker, German mathematician, d Berlin, Dec 29	An antitoxin for diphtheria is tested for the first time on humans Paul Ehrlich uses methylene blue for treating malaria	Samuel Langley's *Experiments in aerodynamics* is published Alexandre-Edmond Becquerel, French physicist, d Paris, May 11 Wilhelm Eduard Weber, German physicist, d Göttingen, Jun 23	In a public lecture about space flight, eccentric German inventor Hermann Ganswindt proposes a spaceship that would be propelled by firing a cannon in the direction opposite to the one in which one wishes to travel Nikola Tesla invents the Tesla coil, which produces high voltage at high frequency Nikolaus August Otto, German inventor, d Cologne, Jan 26	**1891**
	Georg Cantor describes his "diagonal method," which he uses to prove that the infinity of real numbers (all numbers that can be represented by infinite decimals) is different from and greater than the infinity of natural numbers (1, 2, 3, . . .)	The American Psychological Association is founded Paul Ehrlich distinguishes between active and passive immunity Camillo Golgi shows that in intermittent malaria, the malaria parasites develop in the blood while in pernicious malaria, the parasites develop in the organs and brain	The fine structure in the spectral lines of hydrogen is detected by Albert Michelson George Francis Fitzgerald and Hendrik Lorentz formulate the Lorentz-Fitzgerald contraction to explain the negative result of the Michelson-Morley experiment: objects that move at velocities close to that of light undergo a contraction	William S. Burroughs introduces and adding-subtracting machine with printer Henri Moisson introduces an improved electrical arc furnace for metallurgy Cyrus West Field, American businessman, d New York, NY, Jul 12 Ernst Werner von Siemens, German electrical engineer and inventor, d Dec 6	**1892**

1892 cont

ASTRONOMY

On Sep 9 Edward Emerson Barnard discovers a fifth moon of Jupiter, the first satellite of Jupiter found since Galileo; Amalthea, as the moon is named, is the last planetary satellite to be discovered without the aid of photography or spaceprobes

Lewis Morris Rutherfurd, American astronomer, d Tranquility, NJ, May 30

Thomas Jefferson Jackson See proposes his theory of how binary star systems originate

BIOLOGY

Russian-British bacteriologist Waldemar Haffkine (b Odessa, Mar 15, 1860) develops an attenuated strain of the cholera bacteria, which he tests on himself; the next year he uses it on 45,000 people in India, where it reduces the death rate by 70 percent among those inoculated

Sir Richard Owen, English zoologist, d Richmond Park, London, Dec 18

Das Keimplasma (Germ plasm) by German biologist August Friedrich Leopold Weismann (b Frankfurt-am-Main, Jan 17, 1834) states that the germ plasm is unchanged from generation to generation and that changes in the body do not affect it; Weismann theorizes that the germ plasm is in the chromosomes

CHEMISTRY

Sir William Ramsay (b Glasgow, Scotland, Oct 2, 1852) announces on Aug 13 the discovery of argon, based on a suggestion from Baron John William Strutt Rayleigh, who in 1882 discovered that nitrogen from air seems to have a different atomic weight from nitrogen prepared in other ways

1893

ANTHROPOLOGY/ARCHAEOLOGY

In southern Utah Richard Wetherill discovers mummies of the Basket Makers, an American Indian culture that preceded the cliff dwellers who built Mesa Verde

ASTRONOMY

The 71-cm (28-in.) Greenwich refracting telescope is completed

Edward Walter Maunder (b London, Apr 12, 1851) discovers in the historical records of sunspots the Maunder minimum, a period from 1645 to 1715 when almost no sunspots are observed; tree rings and other evidence suggest that this period is the middle of the "Little Ice Age," when global temperatures are low

BIOLOGY

Santiago Ramón y Cajal (b Petilla de Aragón, Spain, May 1, 1852) proposes that learning occurs as a result of increased connections between neurons

CHEMISTRY

Henri Moisson announces that he has made artificial diamonds from carbon; later researchers learn that Moisson did not have enough heat or pressure to produce diamonds; it is believed that someone perpetrated a hoax on Moisson

Russian biologist Dmitri Ivanovsky (b Gdov, Nov 9, 1864) shows that a virus exists

Robert Koch introduces filtration of water for controlling a cholera epidemic in Hamburg, Germany

Ilya Ilich Mechnikov's *The comparative pathology of inflammation* is published

Max Joseph von Pettenkofer (b Lichtenheim, Germany, Dec 3, 1818) miraculously fails to get sick after swallowing cholera-causing bacteria in an effort to disprove the germ theory of disease

Pathologist Theobald Smith (b Albany, NY, Jul 31, 1859) discovers that Texas cattle fever is spread by ticks; this is the first known arthropod vector for disease and paves the way for the discovery of how such diseases as malaria, typhus, and Lyme disease are spread

Jean-Antoine Villemin d Oct 6

Heinrich Hertz, who has concluded (incorrectly) that cathode rays must be some form of wave, shows that the rays can penetrate thin foils of metal, which he takes to support the wave hypothesis

Hungarian-German physicist Philipp von Lenard, b Pozsony (Bratislava, Czechoslovakia), Jun 7, 1862, develops a cathode-ray tube with a thin aluminum window that permits the rays to escape, allowing the rays to be studied in open air

Hendrik Antoon Lorentz independently discovers what we now call the Fitzgerald-Lorentz contraction: bodies contract slightly as they move, with the contraction becoming noticeable near the speed of light

---

The first volume of Gottlob Frege's *Grundgesetze der Arithmetik* (Fundamental laws of arithmetic) continues his efforts to derive arithmetic from logic, but his complex notation results in the book attracting little attention; volume two is published in 1903

Jean-Martin Charcot's papers on the use of hypnosis in medicine are published

The principles and practice of medicine by English-American-Canadian physician William Osler (b Canada, 1849) becomes a standard textbook in the United States

American surgeon Daniel Williams performs the first open-heart surgery on a patient injured by a knife wound

Wilhelm Wien (b Gaffken, Germany, Jan 13, 1864) shows mathematically that the intensity of the maximum wavelength emitted by a hot body varies inversely with its absolute temperature; Wien's law becomes useful in establishing the temperature of stars; the problems he has with deriving an equation to describe black body radiation lead to Planck's introduction of the quantum in 1900

The "Millionaire," the first efficient four-function calculator, appears on the market

Belgian-American chemist Leo H. Baekeland (b Ghent, Belgium, Nov 14, 1863) develops a photographic paper that is sensitive enough for printing by artificial light (Velox)

Rudolf Diesel (b Paris, Mar 18, 1858) describes the engine that will be named after him

1893 cont

Ernst Julius Öpik, Soviet astronomer, b Port Kunda, Estonia (now USSR), Oct 23; he works out the theory of meteor ablation as meteors enter the atmosphere

Charles Pritchard, British astronomer, d May 28

1894

Marie Eugène Dubois (b Eijsden, Netherlands, Jan 28, 1858) publishes his discovery of fossil remains in Java of a precursor of *Homo sapiens*; called *Pithecanthropus erectus* by Dubois, the species in now known as *Homo erectus*

Percival Lowell (b Boston, MA, Mar 13, 1855) founds his observatory at Flagstaff, AZ, and starts searching for a hypothetical ninth planet

Max Rubner demonstrates that the amount of energy produced by food used by the body is the same as the energy produced if the food is burned after urea, which the body does not consume in respiration, is removed

Eduard Adolf Strasburger shows that spore-bearing generations in nonflowering plants, such as ferns and mosses, are diploid (with pairs of chromosomes), while the sexual generation has haploid egg and sperm cells (with only one chromosome of each type)

Botanist Nathanael Pringsheim d Berlin, Germany, Oct 6

Edmond Frémy, French chemist, d Paris, Feb 2

Jean-Charles Galissard de Marignac, Swiss chemist, d Geneva, Apr 15

Herman Frasch begins to develop the method of removing sulfur from deep deposits by melting it with superheated water

Jean-Martin Charcot, French physician, d Lake Settons, Nievre, Aug 16

Physician Niels Finsen (b Faeroe Islands, Denmark, Dec 15, 1860) claims that red light reduces the symptoms of smallpox; although this idea was later abandoned, Finsen goes on to establish that ultraviolet light kills bacteria and also to cure the skin disease *lupus vulgaris* with ultraviolet light

Sigmund Freud's and Josef Breuer's collaboration in studying the psychic mechanism of hysterical phenomena becomes the foundation of psychoanalysis

The first issue of the *Physical Review* is published in Jul

Sir Franz Eugen Francis Simon, German-British physicist, b Berlin, Jul 2; he uses the paramagnetism of molecules and their nuclear spins to lower temperatures to very near absolute zero

John Tyndall, Irish physicist, d Hindhead, England, Dec 4

Josef Stefan, Austrian physicist, d Vienna, Jan 7

The New York Mathematical Society becomes the American Mathematical Society

Emile Borel introduces the concept of "metrics of groups" (Borel measure)

Elie Cartan (b Dolomieu, France, Apr 9, 1869) in his doctoral thesis classifies all known finite groups

Baron Shibasaburo Kitasato and Alexandre Yersin discover separately the bubonic plague bacterium

Charles-Edouard Brown-Séquard, French physician and physiologist, d Apr 1

Oliver Wendell Holmes, American author and physician, d Boston, Oct 7

J.J. Thomson announces that he has found that the velocity of cathode rays is much lower than that of light

Heinrich Rudolf Hertz, German physicist, d Bonn, Jan 1

August Adolph Eduard Kundt, German physicist, d Israelsdorf, May 21

Hermann Ludwig Ferdinand von Helmholtz, German physiologist and physicist, d Charlottenburg, Sep 8

The Manchester Ship Canal is completed, linking industrial Manchester to the Atlantic Ocean

B.F.S. Baden-Powell uses kites to lift human beings into the air

Italian electrical engineer Marchese Guglielmo Marconi (b Bologna, Apr 25, 1874) builds his first radio equipment, a device that will ring a bell from 10 m (30 ft) away

Vicomte Ferdinand-Marie de Lesseps, French engineer, d Dec 7

OVERVIEW

Science in the twentieth century through World War II

Beginning just before the start of the twentieth century, a series of related developments in physics—the discovery of X rays, radioactivity, subatomic particles, relativity, and quantum theory (see "Discovering new rays," p 388)—led to a profound revolution in how scientists view matter and energy. In turn, these developments affected to various degrees chemistry, astronomy, geology, biology, medicine, technology, and ultimately the fate of Earth, since they culminated with the first nuclear weapon, the atomic bomb, in 1945.

The growth of twentieth-century science

Science during the nineteenth century was still the occupation of only a few persons. During the twentieth century, however, the number of scientists became so large that it has almost become a cliché that more scientists have lived in the twentieth century than in all previous eras together. The nature of scientific research had profoundly changed as well; science became much more of a communal effort. Its progress was not only determined by the great discoveries of a talented few, such as Einstein, Bohr, and Rutherford, but also by the numerous small steps made by specialized researchers who did not have a famous formula or law named after them. Many scientific advances were also made by teams of researchers, each working on a small piece of the puzzle.

Many of the observations and discoveries made during the nineteenth century, such as the periodic table, Mendel's laws, and the negative result of Michelson's and Morley's experiment to measure the velocity of Earth against the ether (See "Does the ether exist?," p 366), were explained by new scientific theories that emerged in the twentieth century.

Not only the size of the scientific enterprise changed drastically, but also its influence on society at large. Science during the Renaissance and the Enlightenment had strongly influenced the philosophical outlook but had had little effect on society itself. During this century, scientific research became not only firmly entrenched in the universities but also in industry. After the example of the first industrial chemical laboratories in Germany from the nineteenth century, several industries founded their own research laboratories, seeing that not only applied research but also fundamental research was extremely important to technological progress. Among the most important of these was Bell Labs in the United States, founded by American Telephone and Telegraph.

Much of the methodology worked out by the great scientists during the nineteenth century started to bear fruit in the twentieth century. For example, the microscopic staining techniques developed during the late nineteenth century led to the discovery of many new organisms that cause disease. The synthesis of new chemical compounds became a daily occurrence.

During the nineteenth century society was transformed through technology; many of the great inventions, from the cotton gin to the electric light, were produced by people with little interest in or knowledge of basic science. But in the twentieth century, science itself started having an effect on society directly. For example, the timespan between a discovery and its technical application became much shorter. The discovery of the electron resulted in the construction of electron tubes in less than 20 years, with the ensuing revolution in communications, including broadcasting and long-distance telephoning.

Around 1900 Germany played the leading role in science and technology. Since the middle of the nineteenth century, Germany had been more successful in applying chemistry to industry; it became the leader in the production of dyestuffs and other chemical products. The development of the German pharmaceutical industry was a consequence of this industrial leadership. Germany lost its leading role in the sciences when Hitler came into power. Largely because of the influx of European scientists, the United States became the leading country of science. Between 1932 and 1938, eight of the 28 Nobel laureates in science were American, but these were native-born and not emigrés. After 1938, although the United States continued to be heavily represented among the Nobel laureates, many of the US winners had lived in Germany longer than they had lived in America. A country that failed to receive the influx of German emigrés, the Soviet Union, despite the native ability of its scientists, managed only two Nobel prizes in science from 1932 until today.

World War I had showed that science could play an important role in the outcome of a war. In Germany, more than 100 laboratories were involved in scientific research for the military. During the years after World War I, governments in the East and West started actively funding science, thus largely contributing to the enormous growth of science during the twentieth century. Except in Germany, where many scientists had fled the Nazi regime, scientists played a major role during World War II, their work culminating in the development of nuclear weapons.

New philosophies

Science in the first half of the twentieth century became highly successful in explaining the nature of matter, mechanisms of chemical reactions, fundamental processes of life, and the general structure of the universe. These successes of science started to exert a profound philosophical influence on the outlook of human beings, a phenomenon somewhat similar to the philosophical swing that occurred during the Enlightenment. The American philosophers C.S. Peirce and William James, founders of the philosophical school of pragmatism, held similar ideas to those of Ernst Mach: they believed that reality could be understood by experience alone. At the same time it became clear that science, especially mathematics, was also based on logical reasoning. The philosophical school termed "New Realism" combined the strictly empiricist views of James and Mach with Hegel's idea that knowledge can be attained only from *a priori* concepts by logic (although Mach believed that science is not only based on empirical facts, but also on underlying principles). The scientific view became the only acceptable view of reality and in many cases the model for philosophical thinking.

Darwin's theory of evolution also had a major impact on philosophers, especially Herbert Spencer, Henri Bergson, and John Dewey. Natural selection, which is the basis of Darwin's theory of evolution (see "The theory of evolu-

tion," p 275), became a principle accepted in the social sciences and psychology. Spencer's ideas had become almost a philosophical system: evolution and natural selection were not only the fundamental laws of the living world, but also of society, an idea that resulted in the laissez-faire political philosophies of the first decades of the twentieth century.

Changes in mathematics and the development of psychology also had important philosophical implications.

Quantum reality

Another prominent scientific concept that changed the way philosophers think about the universe is quantum theory. Around the end of the nineteenth century, Newton's mechanics proved to fail in several areas. It became clear that one cannot trace all physical phenomena back to classical mechanics. To solve these problems physicists developed quantum mechanics and abandoned several of the tenets of classical physics. Wolfgang Pauli and Werner Heisenberg introduced theories in which the visualization of phenomena (as was possible in classical physics) was eliminated. Erwin Schrödinger developed an equivalent theory that was later interpreted in terms of waves of probability. In this theory, the electron could not be viewed as a point mass orbiting the nucleus, but as a wave train showing where the electron might be. The idea that each particle could be associated with a wave came from Louis de Broglie, and experiments had shown that electrons could behave as waves when they pass through crystals and are diffracted.

In 1927 Heisenberg introduced one of the most fundamental principles of quantum mechanics: It is not possible to simultaneously observe the position and the speed of a particle (such as the electron) with absolute accuracy. This principle is known as the principle of uncertainty, and Bohr incorporated it into his concept of complementarity: If you observe a system, you interact with it, disturbing it. Bohr elevated the concept of complementarity to a fundamental principle of the natural sciences; it includes the complementarity between the wave and the particle theory of light. Light can be viewed as a wave, for example, when it is diffracted passing through a narrow slit, or as a particle, when ejecting an electron from a metal surface in the photoelectric effect.

These changes affected the way scientists viewed reality in fundamental ways.

Major advances

Anthropology and archaeology. In 1895 Eugène Dubois brought to Europe his Javanese fossils of what we now call *Homo erectus* (the "Java ape-man"), the first hominid that most anthropologists considered a different species from *H. sapiens.* Dubois' discovery met considerable resistance and he eventually took his specimens and hid them under floorboards of his house. In the early twentieth century, however, more fossils of *H. erectus* were found in China and Africa, confirming the new species. Sadly, one of the greatest collections of *H. erectus* fossils, the "Peking man" specimens, were lost during World War II and have never been found.

The other great discovery in physical anthropology during the period started when Raymond Dart was given the fossilized head of a child to examine. He correctly recognized that the "Taung child" represented a new genus, closely related to *Homo,* which he named *Australopithecus.* As with the Dubois discovery, Dart's ideas were rejected at first, but a search for other fossils of australopithecines, as the members of the new genus are called, produced many new fossils and was eventually convincing. Interpretation of the meaning of the new species is still going on.

Our view of our more recent ancestors was changed considerably by the discovery of a young girl in 1879, when the Altamira cave paintings were found. However, this discovery was rejected as a fake until the beginning of our period. Discovery of the La Mouthe cave paintings by four boys in 1895 reduced the cries of forgery to whispers. With other examples found in 1896 and 1901, the reality of paleolithic art was beyond question. When four other young boys found the Lascaux cave paintings in 1940, the greatness of some of the paleolithic artists was also realized.

This period also includes virtually all of the development of cultural anthropology. Although the earlier nineteenth century included some studies of American Indians, the history of cultural anthropology is usually taken to begin in 1896 with Franz Boas's department of anthropology at Columbia University, the first such department anywhere. Another major factor was the expeditions of the American Museum of Natural History (in New York), starting in 1900, to study the peoples of eastern Siberia to see whether they are the ancestors of American Indians. The major

---

The quantum

One of the most important tools of science had its start in 1665 or 1666 when young Isaac Newton observed that a prism demonstrates that white light is a mixture of the colors of the rainbow. Although scientists realized earlier that the rainbow is formed when light is broken into different colors (the spectrum), Newton was the first to make a thorough study of the subject. Later, other scientists investigated this effect for light produced by heating different elements: from this investigation we learned of the composition of the stars. It is well known that study of the spectrum reveals the composition of a heated body or gas. Much less well known is that this study also reveals much about the atom.

The spectrum not only appears in the rainbow, it is hidden in another effect. An observation that must have been made in antiquity is that as iron is heated to forge it, it first becomes dull red, then brighter red, and gradually white. Other solid materials that do not burn behave much the same way. After formulation of the electromagnetic theory of light, theorists tried to explain this phenomenon from first principles. It was apparent that longer wavelengths appear at moderate temperatures. As the temperature rises, shorter and shorter wavelengths begin to appear. When the material becomes white hot, all the wavelengths are represented. Studies of even hotter bodies—stars—showed that in the next stage the longer wavelengths drop out, so that the color gradually moves toward the blue part of the spectrum.

Efforts to make theoretical sense of the way the spectrum gradually appears and disappears were at first unsuccessful. For one thing, theory suggested that a perfectly black body—one that absorbs every wavelength of electromagnetic radiation equally well—would, upon heating, radiate every wavelength equally well. Experiments with simulated black bodies, however, showed that they behave in the same way that iron does when it is heated. In the 1890s Wilhelm Wien and Lord Rayleigh each tried to find a formula to explain these phenomena, but each failed in a different way. Wien's formula worked near the blue end of the spectrum and above, but failed for long wavelengths. Rayleigh's formula was just the opposite, good for long wavelengths and not for short.

In 1900 Max Planck found an explanation that worked for all wavelengths, but little attention was paid to it. Planck made the assumption—which seemed quite odd at the time, even to Planck—that electromagnetic radiation could only be emitted in packets of a definite size, which he called quanta. People took notice of Planck's quantum only when Albert Einstein, in 1905, used the idea to explain the photoelectric effect, to reconcile theory and experiment for heat, and to account for the propagation of light without relying on an "ether." It appeared that Planck's quanta went beyond theory and had a physical reality.

By 1911 Ernest Rutherford had established that the atom has a positive nucleus surrounded by orbiting electrons. Like the black-body problem, however, the theory of the atom did not match experiment. Electrons orbiting a nucleus should give off radiation constantly, resulting in the electron falling into the nucleus. But atoms do not give off that kind of energy and they are usually quite stable.

Niels Bohr turned to Planck's quantum to salvage the theory. The size of the quantum, based on a pure number called Planck's constant, could be calculated. Starting in 1913, Bohr calculated the quantum of the simplest case, hydrogen, in which a single electron orbits a proton. He showed that experiment and theory could be reconciled by saying that the quantum restricts the electron to particular orbits. For each counting number (1, 2, 3, . . .) there was one permissible orbit. For a given electron, the orbit it was in could be assigned that number, called its quantum number. Bohr based his calculations on the lines that form the spectrum of hydrogen gas (when a pure gas is heated, the spectrum consists of discontinuous lines, not a full rainbow). Bohr explained the lines by saying the light is emitted when the electron changes from a higher quantum number to a lower one. There was not a continuous spectrum because the electron moved from orbit to orbit in "quantum jumps." More complex atoms were beyond direct calculations, but approximations indicated that the same approach was correct.

But there were minor complications. Bohr's model explained the large lines in the spectrum, but these lines are broken into smaller lines, called the fine structure of the spectrum. In 1915, Arnold Sommerfeld introduced a second quantum number to explain the fine structure. This was based on the idea that orbits allowed to electrons are ellipses, not circles. Next, it was observed that since the spectrum is affected by a magnet—the Zeeman effect—so there needs to be a third quantum number to account for the magnetic state of the electron. Finally, in 1925 George Uhlenbeck and Samuel Goudsmit found that electrons spin, necessitating yet another quantum number. Each number is an integer that describes the specific state of the electron. If you know that the numbers are, say, 3, 1, 1, 2, then you have a precise description of the electron it its orbit.

The discovery of spin was a major breakthrough, resulting in the rapid development of what is now known as the quantum theory. In 1925 Wolfgang Pauli determined that four quantum numbers is just right. Everything known about an electron in an atom can be reduced to the four numbers. Furthermore, no two electrons in an atom can have the same numbers. This Pauli exclusion principle, as it became known, accounts for how electrons are arranged in all atoms and tells why sulfur has different properties than tin.

About the same time, Werner Heisenberg found that arrays of quantum numbers could be used to calculate lines in the spectrum. This is called the Heisenberg matrix mechanics.

Earlier, Louis de Broglie had proposed that every particle has a wave associated with it. Erwin Schrödinger used de Broglie's idea to calculate the spectral lines. Later it was shown that the Heisenberg matrix mechanics and the Schrödinger wave equation were equivalent.

In 1927 Heisenberg put forward the idea that it was theoretically impossible to determine the position and the momentum of an electron at the same time. The greater the degree of accuracy about one quantity, the less the accuracy of the other. This uncertainty principle, as it is known, was later extended to other particles and other quantities. Max Born suggested that the Schrödinger equation could be interpreted as giving the probability that an electron is located in a particular orbit. This interpretation is still the most common in quantum theory.

Although Schrödinger's wave equation gives good results, they are not perfect. The wave equation does not take spin or the theory of relativity into effect. In 1928, Paul Adrien Maurice Dirac revised the equation to include spin and relativity. Dirac's theory was important because it revealed for the first time the existence of antimatter, but the mathematics is so complicated that physicists still use the Schrödinger wave equation. Dirac's theory essentially completed classical quantum theory.

After World War II, physicists developed quantum electrodynamics, a method of calculating the behavior of electrons and other particles that is even more precise than classical quantum theory.

---

works of Bronislaw Malinowski during World War I and immediately after, Margaret Mead during the 1920s, and Ruth Benedict in the 1930s seemed to define cultural anthropology.

Certainly the most astonishing achievement of archaeology during this period was the discovery of the Minoan civilization on Crete by Arthur Evans beginning in 1900. Like Schliemann before him, Evans was guided in part by legend and literature. The palaces and plumbing of Knossos, the principal city of the Minoan civilization, dazzled the world, although inability to read the inscriptions frustrated everyone. (This was partially resolved in 1953 when Michael Ventris deciphered one of the two forms of Minoan script.)

Almost as astonishing was the discovery in 1911 of Machu Picchu in the Peruvian Andes by Hiram Bingham. Bingham was looking for the retreat to which the last Inca (the ruler of the people also called Inca) had fled during the Spanish conquest, a retreat never seen by Europeans. On top of a high mountain, he found a vast city that dated from the *first* Inca, about 1000 AD. Later, he realized that it was the retreat of the last Inca as well.

Finally, we cannot leave out the most famous discovery of all, Howard Carter's finding of the tomb of Tutankhamen in 1922. Because of the treasure and legends associated with the find, "King Tut" came to represent the type of discovery that was the goal of archaeologists in the early twentieth century.

At the same time, however, archaeology was turning away from searches for spectacular finds and turning toward a more scientific study of the past, a trend that would accelerate after World War II.

Astronomy. At the beginning of this century astronomers held that the sun was one of the many stars comprising a huge system called the Milky Way that filled the entire universe.

Many of the fuzzy objects that had been observed for over 200 years, such as the Andromeda nebula, were believed to be objects within the Milky Way galaxy. An important breakthrough—the discovery that one could determine the distance of a certain type of variable star, the Cepheids, by observing their variation period—allowed astronomers to measure the distance of some of the closer "nebulas". By observing Cepheid variables in the Andromeda nebula, Edwin Powell Hubble discovered that it was at an enormous distance from our galaxy, and that in fact it was a galactic system comparable in size and shape to our own.

From the spectra of galaxies Hubble could derive their velocity relative to us. From this, he made a second surprising discovery. During the 1920s he found that the greater the distance of a galaxy, the faster it is moving away from us (see "The size of the universe," p 440). This discovery formed the observational basis of the model of the expanding universe and the Big Bang.

In solar system astronomy, the most notable achievement was the expansion of the solar system to nine planets, with the discovery of Pluto in 1930. Today, however, Pluto still seems to be an unlikely addition to the system, and it is not clear what its significance is.

A major achievement of astronomy during the period was the development of understanding of the life cycle of a typical star. As stars were classified into gas giants, white dwarfs, and so forth it became apparent that stars of different ages tended to fall into certain categories. The Hertzsprung-Russell diagram showed these relationships graphically.

Biology. The foundations of experimental biology were laid during the nineteenth century. Fundamental research in heredity started with the rediscovery of Mendel's work by de Vries, Correns, and Tschermak in 1900. The concept of a "gene," a unit of inherited characteristics, such as the color of a flower, or in humans the color of the eyes, made it possible to understand how these characteristics were transmitted through generations. The role of genes in the production of enzymes in the cells of organisms also became understood and led to the "one gene—one enzyme" theory. What the compound was that contained the genetic information became clear only during the early 1940s, when biologists discovered that DNA is the substance that transmits genetic information (see "Discovering DNA," p 492).

The role of mutations, sudden changes in the transfer of inherited characteristics, became clear in the mechanism of evolution. The fruit fly, in which the four chromosomes are clearly visible, and mutants are clearly distinguishable (for example, by the shape of wings), became the most important subject for the study of mutations.

Biology gained enormously from the advances made in chemistry, especially from the study of organic compounds. The application of chemistry to biology gave rise to a new important discipline, biochemistry.

The role of certain substances, such as enzymes and hormones produced by living organisms, became understood, and the importance of hormones to many disorders, such as diabetes, was recognized.

Chemistry. With the understanding of the structure of the atom, chemists could explain many of the chemical properties of elements and compounds.

Scientists recognized that the periodicity of the chemical characteristics in Mendeléev's table reflected a periodicity in the configuration of the electrons in the outer layer of an atom. It is possible to explain the chemical properties of elements uniquely by the configuration of these outer electrons of the atom. The development of the theory of chemical bonding by Linus Pauling in the 1930s explained the role of electrons in the formation of molecules: atoms could be bound together either by electrostatic forces (ionic bonding) or by sharing electrons (covalent bonding). The newly developed quantum mechanics formed the theoretical underpinnings of the study of the chemical bonding and the interpretation of atomic and molecular spectra, making spectroscopy a powerful analytical tool. Spectroscopy was used not only in visible light but also in infrared and microwave regions of the spectrum.

Reaction mechanisms and velocities also were the objects of a large number of studies. The study of polymerization reactions led to the development of many compounds made up of macromolecules, such as artificial fibers and the first plastic materials. The chemistry of silicone, which can form complex compounds similarly to carbon, led to the development of the synthesis of silicones, which became important industrial compounds after World War II.

Earth science. From our perspective today, it is easy to see Alfred Wegener's theory of continental drift as the main event in earth science in the first part of the twentieth century; but geologists during this period generally rejected Wegener's ideas. They were not to be generally accepted (in a somewhat modified form) until the 1960s.

Most of the progress in geology that was recognized at the time came from the use of earthquake waves to determine the internal structure of Earth. It was during this period that geologists discovered that Earth has a crust, a mantle, an outer core, and an inner core. The knowledge of radioactivity quickly led to the idea that the age of a rock could be determined from the ratio of a radioactive element to its stable decay product (*see* "The age of the Earth," p 404). This period also marks the beginnings of the systematic study of volcanoes.

Oceanography benefited from the German *Meteor* expedition, which used sonar to discover the mid-Atlantic ridge. William Beebe also began the first efforts to explore the deeper parts of the ocean. Oceanography would not fully become a science, however, until after World War II.

Another part of earth science that saw progress was meteorology. The role of air masses was identified and groundwork was laid for numerical methods of weather prediction.

Mathematics. During the end of the nineteenth century, mathematicians had engaged in a massive effort to develop a purely logical basis to mathematics. One of the first attempts to find an axiomatic basis for mathematics was undertaken by David Hilbert. He proposed that a system of mathematical foundations should satisfy three requirements: it should be consistent, complete, and decidable. The existence of infinite sets, however, led to paradoxes. In 1931 Kurt Gödel proved that Hilbert's ideas could not be realized: mathematics could not be both consistent and complete (*see* "The limits of mathematics," p 462).

Alfred North Whitehead, Bertrand Russell, and Giuseppe Peano extended algebra from symbols for numbers to symbols for concepts, creating symbolic logic. In France during the 1930s, a group of mathematicians working under the pseudonym of N. Bourbaki started the task of giving mathematics an axiomatic basis by searching for those structures that form the basis of the different mathematical theories (*see* "The mathematics of N. Bourbaki," p 471). Another group, led by Emmy Noether at Göttingen in Germany, developed abstract algebras that could represent any type of system.

The concept of an integral had been generalized by T.J. Stieltjes in the nineteenth century, but the definitive expansion of the definition came in 1902 in connection with the measure theory of Henri Lebesgue. In turn, Lebesgue's work combined with set theory and topology, also nineteenth-century ideas, merged into the new discipline of functional analysis, the mainstream of analysis in the twentieth century.

Other developments were so numerous that they can only be mentioned here: the axiomatization and extension of probability, algebraic geometry, optimization theory, analytic number theory, the theory of integral equations, and stochastic processes, among others. Mathematics in the twentieth century reached the point that science had in the nineteenth. No longer could any one mathematician be competent in all branches of the subject.

Medicine. The identification of organisms causing disease, started in the nineteenth century by a number of scientists such as Koch and Pasteur, continued throughout the twentieth century. Also, several of the toxins secreted by microorganisms, often the compounds causing the disease, became identified. Several antimicrobial agents, such as those based on sulfamides, were developed; the most important of these was penicillin, a fungus with antibiotic properties discovered by accident by Alexander Fleming. The introduction of penicillin and other antibiotics made many formerly fatal diseases, such as tuberculosis, curable.

By the 1930s several infectious diseases were known whose agent, called a virus, was so small that it remained invisible in the ordinary microscope. The first virus to become isolated was the tobacco mosaic virus, which could be crystallized, and of which it is now known that the main constituents are nucleic acid and proteins. With the introduction of the electron microscope during the 1940s, it became possible to photograph viruses directly.

The role of hormones and certain important agents in food, called vitamins, became clear in health care. During the first decade it was discovered that the absence of thiamine is the cause of beriberi and the absence of insulin production is the cause of one type of diabetes. Diabetes became treatable by the administration of insulin extracted from animals.

Psychology made further progress and was often based on experiments with animals. The importance of childhood experiences and sex in the development of affective disorders was emphasized in the work of Sigmund Freud, who influenced society at large, as well as science. Several schools of thought developed, each with different forms of treatments for psychic disorders.

Physics. Physics underwent a revolution during the first decades of this century. Ideas about space and time, continuity, and cause and effect, which were the underpinnings of Newtonian mechanics, changed fundamentally because of two important developments: the introduction of relativity theory by Einstein and the advent of quantum mechanics. The first development, however, was the understanding of the structure of the atom, brought about by a series of important discoveries.

Until the end of the nineteenth century, physicists assumed that all phenomena encountered in physics could be explained by the motion of particles following Newton's laws of motion. One notable example was that the classical theory could not explain the repartition of energy in the molecules of a gas and the energy distribution of radiation emitted by hot bodies. These problems led Max Planck to announce in 1900 a revolutionary postulate in physics: energy can only be given off by matter in small packets, called quanta (*see* "The quantum," p 380). What exactly these quanta were became clearer in 1905 when Einstein introduced the concept of the photon: light travels in small

Until the end of the nineteenth century, physicists believed that all physical phenomena, ranging from the motion of atoms to that of celestial bodies, were governed by one set of laws: the laws of motion formulated by Newton in the *Principia*. Newton's theory implied that these laws were also valid for systems that move at constant speed relative to each other. That laws stay the same for such systems is known as the principle of relativity. Consequently, it is impossible to find out whether a system is uniformly moving or not by performing mechanical experiments. A fundamental concept of Newtonian physics is the existence of absolute space. During the nineteenth century, when Thomas Young showed that light is a wave phenomenon, an invisible substance called "ether," linked to absolute space, was believed to be the medium that carried these waves.

During the 1880s Albert A. Michelson and Edward Williams Morley attempted to measure the velocity of Earth relative to the ether by measuring the velocity of light. Their experiments showed that the velocity of light is exactly the same in every direction and thus does not depend on the proper motion of Earth. Some physicists argued that this result showed that the principle of relativity did not apply to electromagnetic radiation. The Dutch physicist Hendrik Antoon Lorentz and and the Irish physicist George FitzGerald tried to explain the result that the velocity of light seemed independent from the motion of Earth by assuming that everything contracts in the direction in which one is moving. They argued that the instrument that Michelson and Morley had used contracted imperceptibly in the direction of Earth's motion, thus falsifying the measurement of the velocity of light. Perhaps the most important aspect of their theory is that they held on to the idea of an ether.

In 1905 Albert Einstein published a theory based on the notion that it is impossible to determine the absolute motion of a moving object. Einstein's concern, however, was not the failure of Michelson and Morley to measure the motion of Earth relative to the ether, but the validity of James Clerk Maxwell's electromagnetic theory in systems that move at speeds close to the velocity of light. Einstein's theory did not require the presence of an ether and was based on the following assumptions: (1) absolute speed cannot be measured, only speed relative to some other object; (2) the measured value of the speed of light in a vacuum is always the same no matter how fast the observer or light source is moving; and (3) the maximum velocity that can be attained in the universe is that of light.

Einstein's theory was called the "special" theory of relativity because it applied the principle of relativity only to systems in uniform motion relative to each other.

Because of the principle of relativity, passengers who are traveling smoothly in a train cannot tell whether they are moving or not unless they look out of a window. The situation becomes different if the train is uniformly accelerated. The passengers will feel a slight push in the direction opposite to that in which the train is moving. This is termed *acceleration force:* because of this extra force it appears that the laws of physics would be different for bodies accelerated with respect to each other.

In 1916 Einstein published his general theory of relativity, which he based on the assumption that the laws of physics would also be the same in systems that are accelerated relative to each other. To formulate this theory, he introduced the principle of equivalence: acceleration forces and gravitational forces are not distinguishable from each other. Einstein argued that if one were in a closed elevator that is uniformly accelerated upward, one would perceive a force that is indistinguishable from gravitation. The principle of equivalence can also be expressed by saying that the inertia of an object (its reluctance to be set in motion) is proportional to its mass. The principle of equivalence was already known to Galileo, who had deduced it from his experiments with wooden balls rolling down sloping planes. In 1891 the Hungarian scientist Roland Eötvös made precise measurements and established that inertial and gravitational mass are equivalent to an extremely high degree. Because acceleration forces and gravitational forces are equivalent, they should not be distinguishable, but viewed as a property of space.

In a formulation of the special theory, the mathematician Hermann Minkowski had introduced a four-dimensional space in which the fourth dimension is time. In this space-time continuum as adapted to the general theory, gravitation corresponds to the amount of curvature in a non-Euclidean, four-dimensional space. Near a large mass, space becomes more curved, and objects moving near that mass will follow the curvature of space.

One of the interesting consequences of the equivalence of gravitation and acceleration force is the bending of light rays by the presence of large masses, such as a star or planet. For example, if light enters through a small hole on one side of an spaceship that is accelerated, the light ray will reach the other side after the spaceship has moved. The same effect would exist if the spaceship were to come close to a massive planet or star. To an observer in the spaceship, the light ray will appear curved in either case. But the observer cannot tell whether the spaceship is being accelerated or is near a planet or star.

In 1919, during a solar eclipse, Arthur Eddington showed that stars whose light passes close to the sun appear to be displaced by a minute amount that corresponds to the value calculated by Einstein. This was the first experimental proof of the general theory of relativity. Several other experiments more recently have had results that eliminate most doubts about either theory of relativity among physicists.

packets called photons; this was reminiscent of Newton's idea that light consists of vibrating particles. According to Einstein, light is only emitted in small packets, but it also can only be absorbed in small packets. Physicists had observed that certain metals eject electrons when placed in a strong light and that the speed of these ejected electrons does not depend on the intensity of light but on its color. This is called the photoelectric effect. Einstein explained the photoelectric effect by assuming that an electron is only ejected when directly hit by a photon, and that the energy of the photon does not depend on the intensity of the light but on its wavelength (color).

One of the major problems that physics solved at the end of the nineteenth century was the nature of cathode rays. The discovery that these rays consist of minute particles with negative electrical charge, called electrons, and the discoveries of X rays and radioactivity opened the way to the understanding of the structure of the atom during the first decades of the twentieth century (see "The electron and the atom," p 392). Rutherford succeeded in identifying the particles that were emitted by radioactive substances and the changes in the atoms that emitted these particles. By bombarding atoms with alpha particles, Rutherford found that some alpha particles were deflected from their path almost directly back along the same path. From this observation he concluded that an atom must consist of a very dense nucleus of positive charge around which revolve electrons like planets in a miniature planetary system.

The model of the atom described by Ernest Rutherford in 1911, a nucleus around which pointlike electrons orbited like planets, had an important flaw. The moving electrons should emit electromagnetic waves, gradually lose energy, and ultimately fall onto the nucleus. Niels Bohr solved this problem by introducing a model of the atom that incorporated a principle similar to Planck's hypothesis of quanta: electrons occupy fixed energy levels in the atom and can only absorb or emit energy by jumping from one energy level to another. Bohr's model of the atom also explained the spectrum of the hydrogen atom.

Bohr's theory, however, was still fraught with theoretical problems. For example, his theory could not explain the spectra of atoms more complex than hydrogen. Also, certain results of Maxwell's electrodynamic theory were inconsistent with other existing theories. In 1905 Albert Einstein published his theory of special relativity, a theory of mechanics consistent with electrodynamics. In 1915 Einstein published his general relativity theory, which solved problems with gravity that were not explained by the special theory (see "Relativity," p 384). Special relativity theory also explained the negative result of the Michelson-Morley experiment, while general relativity explained the minute rotation of Mercury's perihelion, which had been observed earlier. Einstein's general theory also predicted that light would be bent by massive objects. The bending of light by mass was experimentally confirmed by an ex-pedition led by Arthur Eddington that measured the displacement of a star close to the limb of the sun during an eclipse in 1919.

The first artificial nuclear reactions were achieved shortly before the Second World War by Otto Hahn and Fritz Strassmann. This led to the development of the atomic bomb during World War II (see "Scientists and World War II," p 480).

Technology. In the twentieth century the extent of the dependence of technology on basic science became clear. The discovery of the electron gave rise to an entirely new technology, electronics. This was comparable to the development of the chemical industry in Germany during the nineteenth century based on discoveries in chemistry.

The technological device that transformed society most profoundly during this period was the electronic vacuum tube, an extremely versatile device that became the heart of the development of electronics during the first half of the twentieth century. The most important aspect of the vacuum tube was that it could amplify electric audio signals (for example, in telephone lines) and that it could generate, amplify, and detect high-frequency signals (radio waves). The first applications of vacuum tubes were in the amplification of telephone signals, making long-distance telephone connections possible. The vacuum tube was at the heart of the enormous development of radio broadcasting during the 1920s and 1930s and of television in the 1940s.

Transportation also underwent a revolution. Automobiles, fragile and impractical at the beginning of the century, became a widely popular and dependable mode of transportation. Both world wars accelerated technological development enormously. Airplanes, developed shortly before World War I, received a strong impetus during that war; the modern jet plane is the consequence of developments during World War II. Electronics also had a strong impetus during both wars: voice radio communication developed during World War I and radar during World War II.

Some of the first working electronic computers were also the result of military needs. The mathematician Norbert Wiener developed an electronic gun-pointing device based on the feedback mechanism, and Alan Turing developed an electronic computer that could break successfully the almost unbreakable "Enigma" code used by German forces. Others, however, had preceded them. The first electronic digital computer was developed in the late 1930s and early 1940s by John V. Atanasoff and Clifford E. Berry for the purpose of solving systems of equations. In the case of the Atanasoff-Berry Computer, or ABC, the war prevented complete development, as both inventors were drafted for other wartime duties. Nevertheless, ideas based on the ABC were used in creating ENIAC, the first general-purpose electronic digital computer, operational by 1945.

1895

GENERAL

Grace Chisholm Young's doctoral degree in mathematics from the University of Göttingen is the first doctoral degree in any subject awarded by a German university through the regular examination process to a woman

ANTHROPOLOGY/ARCHAEOLOGY

Eugène Dubois brings the fossils he had found of *Homo erectus*, which he calls *Pithecanthropus* and the public calls "the ape man of Java," to Europe, where they are met with considerable skepticism

Sir Henry Creswicke Rawlinson, English archaeologist, d London, Mar 5

ASTRONOMY

James Keeler (b La Salle, IL, Sep 10, 1857) observes Saturn's rings and recognizes that they do not rotate as a unit, suggesting that they are formed of discrete particles, as James Clerk Maxwell had predicted

Daniel Kirkwood, American astronomer, d Riverside, CA, Jun 11

Simon Newcomb's *Astronomical constants* contains calculations of the constants of precession, nutation, yearly aberration, and solar parallax

Konstantin E. Tsiolkovsky publishes his first scientific papers about space flight; because of his priority, Russians to refer to him as the father of space flight

BIOLOGY

Ernst Felix Immanuel Hoppe-Seyler, German biochemist, d Lake Constance, Bavaria, Aug 10

Thomas Henry Huxley, English biologist, d Eastbourne, Sussex, Jun 29

Johann Friedrich Miescher, Swiss biochemist, d Davos, Aug 26

Rabl asserts that chromosomes keep their identity during cell division; this is an important point leading to the assumption that chromosomes are carriers of heredity

CHEMISTRY

Hermann Hellriegel, German chemist, d Bernburg, Anhalt-Bernburg, Sep 24

Johann Joseph Loschmidt, Austrian chemist, d Vienna, Jul 8

Julius Lothar Meyer, German chemist, d Tübingen, Württemberg, Apr 11

Louis Pasteur, French chemist, d St.-Cloud, Sep 28

William Ramsay discovers helium on Earth; it had been found in 1868 as an element in the sun from observations of the sun's spectrum; the element is also discovered independently by Per Teodor Cleve

Russian-German chemist Paul Walden (b Rosenbeck, Latvia, Jul 26, 1863) discovers a method to change the polarization of malic acid from clockwise to counterclockwise by a series of chemical reactions, now called the Walden inversion

Rev Jeanette Piccard, American scientist who launches the first balloon into the stratosphere, b Jan 5

Francis W. Reichelderfer, American meteorologist, b Aug 6

About this time Georg Cantor discovers the contradiction that comes to be known as the paradox of Burali-Forti, since Cesare Burali-Forti rediscovers it in 1897 and publishes it; the paradox concerns infinite ordinal numbers; the ordinal number of the series of ordinal numbers is not itself found in the series

Georg Cantor's *Beiträge zur Begründung der Tranfiniten Mengenlehre* (Contributions to the founding of the theory of transfinite numbers) is a survey of his work in set theory

Arthur Cayley, English mathematician, d Cambridge, Jan 26

Giuseppe Peano's *Formulaire de mathématiques* uses the special symbolism for symbolic logic developed by Peano to express many of the fundamental definitions and theorems of mathematics; by its fifth and last edition in 1908 (as *Formulario mathematico*) it contains about 4200 theorems

Analysis situs by Jules-Henri Poincaré (b Nancy, France, Apr 29, 1854) founds topology as a branch of mathematics, although a few theorems of topology had been proved previously

Karl Friedrich Wilhelm Ludwig, German physiologist, d Leipzig, Saxony, Apr 27

Pierre Curie shows that as the temperature of a magnet is increased, there is a level at which the magnetism is disrupted and ceases to exist; this temperature is still called the Curie point

Hendrik Antoon Lorentz shows that a moving charged particle will have a force exerted on it perpendicular to the direction of its motion by electric and magnetic fields (the Lorentz force)

Louis Carl Heinrich Paschen (b Schwerin, Germany, Jan 22, 1863) confirms that the helium found on Earth is the same as that found in the sun

Jean-Baptiste Perrin (b Lille, France, Sep 30, 1870) shows that cathode rays deposit a negative electric charge where they impact, refuting Heinrich Hertz's concept of cathode rays as waves and showing that they are particles

Wilhelm Konrad Roentgen (b Lennep, Germany, Mar 27, 1845) discovers X rays on Nov 8

Charles Thomson Rees Wilson (b Glencorse, Scotland, Feb 14, 1869) develops the cloud chamber, a box containing a gas that is saturated; when a charged particle passes through the gas, small droplets are formed that make the track of the particle visible; the cloud chamber becomes a powerful tool in particle physics

Robert Buckminster Fuller, American inventor, b Jul 12

Otto Lilienthal and his brother Gustav design and fly the first glider that can soar above the height of takeoff

Auguste and Louis Lumière invent the cinematograph

David Schwartz builds an airship with a rigid aluminum frame of 36,800-cu m (130,000-cu ft) capacity

Russian physicist Konstantin Tsiolkovsky (b Izhevsk, Sep 17, 1857) proposes that liquid-fueled rockets can be used to propel vehicles in space

1895

1896

Berlin University is reorganized and continues to be a model for German universities

The first lunar photographic atlas is published by Lick Observatory, Mount Hamilton, CA

The Meudon 83-cm (33-in.) refracting telescope is completed in France

John Martin Schaeberle (b Württemberg, Germany, Jan 10, 1853) observes the dark companion of the star Procyon for the first time, finding the second white dwarf known (after Sirius B); this leads to the correct idea that such dwarf stars are common

German chemist Eugen Baumann (b Württemberg, Dec 12, 1845) discovers iodothyrin, an active agent secreted by the thyroid gland that contains essentially all of the iodine found in the human body; this leads to the use of iodine to treat and prevent thyroid disorders, such as goiter

Eugen Baumann, German chemist, d Freiburg, Nov 3

Eugène-Anatole Demarçay (b Paris, Jan 1, 1852) begins the line of research that leads him to the discovery of the element europium

Friedrich August Kekulé von Stradonitz, German chemist, d Bonn, Prussia, Jul 13

Robert Sanderson Mulliken, American chemist and physicist, b Newburyport, MA, Jun 7; he helps develop the theory of electron orbitals

Ida Eva Tacke Noddack, German chemist, b Wesel, Rhenish Prussia, Feb 25

Discovering new rays

The perception that there is invisible radiation pervading the universe is relatively new. In 1800 William Herschel discovered that invisible rays, now called infrared radiation, heat a thermometer better than visible sunlight. The following year Johann Ritter found the other side of the story, that invisible ultraviolet radiation affects silver chloride more than light does.

Like most discoveries of invisible rays, the discoveries of Herschel and Ritter were accidental. The next discovery, however, was predicted. In 1873 James Clerk Maxwell incorporated infrared and ultraviolet radiation along with light into his theory of electromagnetic fields. The same theory forecast that there should be radiation with wavelengths longer than infrared and shorter than ultraviolet. In 1888, Heinrich Hertz was able to use Maxwell's theory to produce the long waves, which we now know as radio waves. But no one knew how to make the very short waves.

There was, however, another form of invisible radiation that was known. Scientists trying to find whether electricity would pass through a vacuum had discovered that a ray could be detected by the fluorescence it produced. This was particularly pronounced in the vacuum tubes made by William Crookes, which were so good that similar tubes are still called Crookes tubes. Crookes did careful work during the 1870s that offered convincing evidence that the rays produced in his tube, called *cathode rays*, were streams

of invisible charged particles. In 1897, J.J. Thomson was to firmly establish that Crookes was right. Thomson is generally given the credit for the discovery of the electron, which is the charged particle that makes up the cathode ray.

Cathode rays were to lead to the next two accidents that uncovered new forms of radiation, which in turn started what some have called the second scientific revolution. In 1895, Wilhelm Roentgen was investigating materials that fluoresce when exposed to cathode rays. He had wrapped his tube in black paper and was going to view a sheet coated with a fluorescent material in the dark. But he noticed that the sheet started to fluoresce even before he removed the black paper from his Crookes tube. Since Crookes had established that cathode rays do not penetrate such a shield, something else must be causing the fluorescence. Roentgen took his fluorescent material into a closet, where it continued to glow when the Crookes tube was turned on. A series of experiments showed that the invisible radiation could pass through metals as well as through layers of paper. Roentgen had discovered X rays, the very short waves predicted by Maxwell's theory, although he failed to recognize them as short electromagnetic radiation. Within days of his announcement of these rays, doctors began to use them to aid in seeing inside the human body.

Svante Arrhenius discovers that the amount of carbon dioxide in the atmosphere determines the global temperature and theorizes that the ice ages occurred because some process had reduced the level of carbon dioxide

Gabriel-Auguste Daubrée, French geologist, d Paris, May 29

Jacques Hadamard proves his prime number theorem: When a is very large, the number of primes less than a is about $a/\log a$

Emil Heinrich Du Bois-Reymond, German physiologist, d Berlin, Dec 26

Michael I. Pupin of Columbia University takes the first diagnostic X-ray photograph in the United States

Antoine-Henri Becquerel (b Paris, Dec 15, 1852) discovers rays produced by uranium on Mar 1, the first observation of natural radioactivity

Armand-Hippolyte Fizeau, French physicist, d Venteuil, Seine-et-Marne, Sep 18

Sir William Robert Grove, British physicist, d London, Aug 1

French physicist Charles Edouard Guillaume (b Fleurier, Switzerland, Feb 15, 1861) discovers invar, an alloy of nickel and iron that resists change in volume when temperature changes

Less than three months after the discovery of X rays by Roentgen, Eddie McCarthy of Dartmouth, MA, becomes the first person to have a broken arm set with their help

Luigi Palmieri, Italian physicist, d Naples, Sep 9

Pieter P. Zeeman (b Sonnemaire, Netherlands, May 25, 1865) discovers that spectral lines of gases placed in a magnetic field are split, a phenomenon called the Zeeman effect; Hendrik Antoon Lorentz explains this effect by assuming that light is produced by the motion of charged particles in the atom

Lorentz uses Zeeman's observations of the behavior of light in a magnetic field to calculate the mass/charge ratio of the electron in an

Selective weed killers are used in France

Herman Hollerith, after successfully applying his punched-card technique to the US census, founds Tabulating Machine Company, which later changes its name to International Business Machines, and still later becomes familiar as IBM

Samuel Pierpont Langley tests his steam-driven flying machine on the Potomac, flying for 1.2 km (0.75 mi) before crashing

Otto Lilienthal, German aeronautical engineer, d of injuries sustained in a glider crash near Rhinow, Germany, Aug 10

Alfred Bernhard Nobel, Swedish inventor, d San Remo, Italy, Dec 10

Discovering new rays (continued)

Within three months of Roentgen's announcement of X rays, another form of invisible radiation was found. Henri Becquerel thought that there might be X rays produced by fluorescent materials. He was working with a uranium compound that became fluorescent when exposed to sunlight. When he exposed a photographic plate that was wrapped in black paper to the fluorescing uranium compound, he got a faint image on the plate. Luck brought Becquerel a series of cloudy days, so he could not expose the compound to sunlight. Becquerel left the sample and the wrapped plate in drawer. When the cloudy weather persisted, he decided to develop the photograph anyway, hoping for a slight exposure. Instead, he found that the plate was heavily fogged. Radiation had been emitted by the uranium compound continuously, without the stimulation of sunlight.

Becquerel and other physicists soon found that the radiation given off by the uranium compound—which in 1898 Marie Curie named radioactivity—consisted of three parts, alpha, beta, and gamma rays. Becquerel himself was the first to show that beta rays are the same as cathode rays; that is, streams of electrons. In 1900, Ernest Rutherford was able to show that gamma rays are basically the same as X rays, but with even shorter wavelengths. A few years later, Rutherford and Hans Geiger found that alpha rays are helium atoms that have been stripped of their electrons. In a sense, then, Becquerel's radioactivity was not a new type of radiation. Instead, its significance was that it was radiation spontaneously emitted by atoms. From that discovery, many others have flowed.

1896

cont

1897

		ASTRONOMY	BIOLOGY	CHEMISTRY
		Alvan Graham Clark, American astronomer, d Cambridge, MA, Jun 9	Eduard Buchner (b Munich, Germany, May 20, 1860) accidentally discovers that a cell-free extract of yeast, which he names zymase, can convert sugar into alcohol; this is the beginning of biochemistry; previously, chemists believed that vital processes could only take place inside living cells	Viktor Meyer, German organic chemist, d Heidelberg, Aug 8

George Ellery Hale sets up the Yerkes Observatory in Williams Bay, WI; the Yerkes telescope at 1 m (40 in.) is still the largest refracting telescope on Earth

Antonia Caetana Maury refines the spectral classification of stars using the sharpness of lines

Henry Rowland photographs the solar spectrum using a concave diffraction grating, producing a 20-m (66 ft-) long photograph of the spectrum

Otto Struve, Russian-American astronomer, b Kharkov (USSR), Aug 12; he discovers thin clouds of gas and dust between the stars, and suggests mechanisms for planet formation around stars

Julius von Sachs, German botanist, d Würzburg, Bavaria, May 29

Paul Sabatier (b Carcassonne, France, Nov 5, 1854) uses catalysts, such as nickel or cobalt, to combine hydrogen with unsaturated compounds

1898

GENERAL	ANTHROPOLOGY/ ARCHAEOLOGY	ASTRONOMY	BIOLOGY	CHEMISTRY
Andrew Lang's *The making of religion* is published	Louis-Laurent-Gabriel de Mortillet, French anthropologist, d St. Germain-en-Laye, Yvelines, Sep 25	Simon Newcomb finds a more accurate value for precession	Martinus Willem Beijerinck declares that tobacco mosaic disease is caused by an infective agent that he names a filterable virus; this is the first identification of a virus	Marie Sklodowska Curie (b Warsaw, Poland, Nov 7, 1867) and Pierre Curie discover polonium on Jul 18 and radium (with Gustave Bémont) on Dec 26

Jules-Jean-Baptiste Vincent Bordet (b Soignes, Belgium, Jun 12, 1870) discovers

Johann (Hans) Goldschmidt (b Berlin, Germany, Jan 18, 1861) develops thermite, a mixture of

EARTH SCIENCE	MATHEMATICS	MEDICINE	PHYSICS	TECHNOLOGY	
			atom, a year before electrons are discovered and 15 years before it is known that electrons are constituents of atoms		**1896** cont
Jacob Aall Bonnevie Bjerknes, Norwegian-American meteorologist, b Stockholm, Sweden, Nov 2; working at first with his father, Vilhelm, Bjerknes helps develop the mathematical theory of weather forecasting and identifies the jet stream after pilots in World War II report that they are being blown about

Edward Drinker Cope, American paleontologist, d Philadelphia, PA, Apr 12

R.D. Oldham discovers that seismic waves consist of two components: compressional and distortional waves, as was predicted by Poisson in 1829 | Cesare Burali-Forti publishes the paradox known by his name (although discovered by Georg Cantor two years earlier): the ordinal number of the series of ordinal numbers does not appear in the series

German mathematician David Hilbert, b Königsberg (Kalingrad, Poland), Jan 23, 1862, introduces his theory of algebraic number fields

James Joseph Sylvester d Mar 15

Karl Weierstrass, German mathematician, d Berlin, Feb 19 | Paul Ehrlich publishes an influential paper on the dosage of diphtheria antitoxin

Ehrlich develops his "side-chain" theory of immunity, in which the protein molecule is viewed having unstable side chains

Christiaan Eijkman (b Nijkerk, Netherlands, Aug 11, 1858) shows the relationship between the occurrence of beriberi and the consumption of polished rice; however, he does not attribute the disease to the absence of a vitamin in polished rice

English physician Ronald Ross (b Almora, India, May 13, 1857) locates the malaria parasite in the Anopheles mosquito and determines that the mosquito transmits the parasite from one human to another

Walter B. Cannon discovers that a bismuth compound can be used to make intestines visible on X-ray photographs | Walter Kaufmann determines the ratio of the mass to charge for cathode rays in Apr, about the same time that J.J. Thompson does, but Kaufmann fails to consider that the rays might be subatomic particles

Russian physicist Alexander Popov (b Bogoslavsky, Mar 16, 1859) uses an antenna to transmit radio waves over a distance of 5 km (3 mi)

The first statement that there may exist particles about 2000 to 4000 times lighter than the hydrogen atom is made by Emil Wiechert (b Tilsit, Germany, Dec 26, 1861) on Jan 7

Joseph John Thomson discovers the electron, the first known particle that is smaller than an atom, in part because he has better vacuum pumps than were previously available; he, and independently, Emil Wiechert, determine the ratio of mass to charge of the particles by deflecting them by electric and magnetic fields | Charles Algernon Parson's *Turbina*, the first steamship to use turbines for power, demonstrates its superiority over conventional steamships by unexpectedly scooting through a formal naval review in front of Queen Victoria; a conventional steam launch sent to catch *Turbina* cannot even come close

Adolf Spittler discovers, probably by accident, casein plastics | **1897** |
| Hieronymus Theodor Richter, German mineralogist, d Freiberg, Sep 25 | Johann Jakob Balmer, Swiss mathematician and physicist, d Basel, Mar 12

Jules Drach develops new methods for the solution of differential equations

A treatise on universal algebra by Alfred North Whitehead (b Ramsgate, England, | On Jun 2 Paul-Louis Simond, fighting the bubonic plague pandemic in Bombay, India, realizes that fleas on rats transmit the disease to humans | Marie and Pierre Curie discover that thorium gives off "uranium rays," which Marie renames *radioactivity*

James Dewar liquefies hydrogen | Sir Henry Bessemer, English metallurgist, d London, Mar 15

Inventor Hamilton Young Castner d Oct 11 | **1898** |

1898
cont

The electron and the atom

Although the first notion that matter might be composed of atoms goes back to Leucippus of Miletus in the fifth century BC (*see* "Early atomists," p 32), it was not until the seventeenth century that evidence began to accumulate that atoms really exist. Robert Boyle, for example, believed that gases must be made of small particles to behave the way they do when compressed.

Little progress was made in atomic theory during most of the eighteenth century, but in 1797 Joseph Proust stated that the ratios of one element to another are always the same in a given chemical compound. This law of definite proportions suggested to John Dalton that solids and liquids, as well as gases, must also be made of atoms. In 1803, Dalton showed that atomic theory could explain the law of definite proportions. An "atom" of a compound (what we now call a molecule) always contains a definite number of atoms of each element of which the compound is composed. In 1816, William Prout went further by explaining that atoms of elements, like "atoms" of compounds (molecules), can be made of smaller parts. Noting that the atomic masses of the known elements are close to integral multiples of hydrogen, the lightest element, Prout hypothesized that an atom of, say, oxygen, with an atomic mass of about 16, is composed of 16 hydrogen atoms. Unfortunately, as new elements were discovered, many of them had atomic masses about halfway between integral multiples of the mass of hydrogen. Still, chemists noted that about half of all elements have atomic masses that are very close to integral, a ratio that seemed too great to be merely a coincidence.

In 1854 a series of apparently unrelated discoveries eventually unraveled the truth behind Prout's hypothesis. Heinrich Geissler found a way to make a fairly good vacuum in a glass tube. Experiments quickly revealed that an electric current applied to the vacuum produced a glow in the glass at the other end of the tube. Apparently, invisible rays were traveling through the near vacuum. It was shown that the rays cast sharp shadows when masks were inserted into the tube. These rays were named cathode rays in 1869, since they emerged from the electric terminal called the cathode. There was great interest in cathode rays in the latter part of the nineteenth century. The interest was well placed, for it led to a number of important discoveries (*see* "Discovering new rays," p 388).

complement, a component of blood needed for antibodies to react with bacteria

Ferdinand Julius Cohn, German botanist, d Breslau (Wroclaw, Poland), Jun 25

Camillo Golgi describes the Golgi apparatus, a reticular formation in the cytoplasm of cells

Karl Georg Friedrich Rudolf Leuckart, German zoologist, d Leipzig, Saxony, Feb 6

Friedrich Löffler and Paul Frosch demonstrate that foot-and-mouth disease is caused by a filterable virus, the first known instance of a viral disease in animals (viral diseases were known but it was not yet known that these were caused by viruses)

Trofim Denisovich Lysenko, Soviet biologist, b Karlovka, Poltava Oblast, Ukraine, Sep 29; with Joseph Stalin as his chief supporter, Lysenko gains control of Soviet biological research in the period between 1928 and 1965, imposing his incorrect view that acquired characteristics can be inherited

aluminum powder and iron or chromium that burns at high temperatures, leaving a residue of pure iron or chromium; its most common use is in welding

John Alexander Newlands, English chemist, d London, Jul 29

The elements krypton, xenon, and neon are discovered by Alexander Ramsay and Morris William Travers (b London, Jan 24, 1872)

1899

Ernst Heinrich Haeckel's *Die Weltträtsel* (The riddle of the universe) expresses his view that mind, a product of creation depends on the body and does not survive it

The *Astronomischer Jahresbericht* (Bibliography of world astronomical literature) is founded in Germany

The International Latitude Service is founded

S.I. Bailey discovers short-period cluster variables

German-American physiologist Jacques Loeb (b Mayen, Germany, Apr 7, 1859) demonstrates parthenogenesis by raising unfertilized sea urchin eggs to maturity after changing their environment

Charles Naudin d Apr 19

Robert Wilhelm Bunsen, German chemist, d Heidelberg, Aug 16

André-Louis Debierne (b Paris, 1874) discovers actinium

Sir Edward Frankland, English chemist, d Golaa, Norway, Aug 9

Feb 15, 1861) concerns the symbolic structure of algebra

The electron and the atom (continued)

Physicists were divided on what cathode rays might be. In Germany, most physicists thought that the rays might be waves, similar to electromagnetic waves. In England, where particle theories had always been popular (Newton and light, Dalton and atoms), most physicists thought that they were particles. In 1881 English physicist George Stoney even named the particle: the electron.

Finally, in 1897, J.J. Thomson settled the particle-wave controversy for the next few years. He was able to measure the deflection of the rays in an electric field, indicating that they were composed of particles. Furthermore, by measuring the amount of deflection, he was able to work out the mass of the electrons, which turned out to be about 2000 times smaller than that of a hydrogen atom. Electrons were subatomic.

Since the electrons came from materials, it was quickly assumed that electrons must be parts of atoms. In 1898, Thomson proposed the "raisin-pudding" model of an atom. A sphere of positive charge is embedded with electrons, rather like raisins in a pudding. Others had different ideas. Philipp Lenard found experimental evidence that suggested that atoms are mostly empty space. As a result, he proposed in 1903 that atoms are electrons paired with similarly small positive charges in configurations that are mostly empty. In Japan, Hantaro Nagaoka stated the following year that atoms look like the planet Saturn, with rings of electrons circling a positive core.

By 1911, Ernest Rutherford and his associates had good experimental evidence that Nagaoka had come closest, but the Nagaoka model called for thousands of electrons in each atom. Other evidence showed that there could be only a few electrons. Therefore, the Rutherford model was more like the solar system than like Saturn: a central positive core, or nucleus, with a few electrons orbiting it in circles. This is the image of the atom that most people have today.

Unfortunately, nature turned out to be a lot more complicated. The Rutherford atom was not stable, and various adjustments to the model have been made over time. Furthermore, it was established that electrons are waves at the same time that they are particles. The Germans had a point after all! Today the most common picture of an atom involves cloudy regions in which electron "wavicles" may be found according to the laws of probability and quantum mechanics (see "The quantum," p 380).

William Thomson (Lord Kelvin) in "The age of the Earth as an abode fitted for life" argues that life could have evolved quickly, so that there is no necessity for geological time to exceed the 100 million years that since 1862 he had calculated repeatedly for the age of Earth

Georg Cantor develops a paradox (similar to Russell's great paradox of 1902) that suggests that there are problems with Cantor's new theory of sets

David Hilbert's *Grundlagen der Geometrie* (Foundations of geometry) develops the basic concepts of geometry

The American Physical Society is founded

The laws of gases by Emile Hilaire Amagat (b Saint-Satur, France, Jan 2, 1841) reports his experiments with gases subjected to very high pressures

Fritz Geisel, Antoine-Henri Becquerel, and Marie Curie prove that beta rays consist

Inventor Ottmar Mergenthaler d Oct 28

1899

cont

ASTRONOMY

William Wallace Campbell discovers with a spectroscope that Polaris is a system of three stars

William Henry Pickering (b Boston, MA, Feb 15, 1858) discovers Phoebe, the ninth satellite of Saturn

CHEMISTRY

Charles Friedel, French chemist, d Montauban, Tarn-et-Garonne, Apr 20

Lars Fredrik Nilson, Swedish chemist, d Stockholm, May 14

William Jackson Pope (b London, Oct 31, 1870) discovers the first compound that polarizes light, or is optically active, that does not contain carbon

---

1900

ANTHROPOLOGY/ARCHAEOLOGY

Arthur Evans discovers the palace at Knossos, the central site of the Minoan civilization on Crete

Luigi Pernier discovers a second great palace of the Minoan civilization at Phaistos

ASTRONOMY

James Edward Keeler photographs a large number of nebulas and discovers that some have a spiral structure

James Edward Keeler, American astronomer, d San Francisco, CA, Aug 12

BIOLOGY

During this year, Hugo Marie De Vries (b Haarlem, Netherlands, Feb 16, 1848), Karl Franz Joseph Correns (b Munich, Germany, Sep 19, 1864), and Erich Tschermak von Seysenegg (b Vienna, Austria, Nov 12, 1871) independently rediscover Gregor Mendel's work on genetics, ignored for 40 years

CHEMISTRY

German physicist Friedrich Ernst Dorn, b Guttstadt (Dobre Miasto, Poland), Jul 27, 1848, discovers radon

Russian-American chemist Moses Gomberg (b Elizavetgrad, Ukraine, Feb 8, 1866) develops a carbon compound that has one valence location open, the first known free radical (although free radicals form often, they are usually destroyed quickly by interacting

Thomas Chrowder Chamberlin attacks the basic framework of William Thomson's argument that Earth is only about 100 million years old; among other flaws he notes that the ice age, which Thomson takes as evidence of steady cooling, was several ice ages, broken up by warmer weather

Othniel Charles Marsh, American paleontologist, d New Haven, CT, Mar 18

from undefined concepts of point, line, and plane, overcoming many logical difficulties that had surfaced in Euclid's *Elements*

Marius Sophus Lie d Feb 18

of high-speed electrons

British physicist Ernest Rutherford (b Brightwater, New Zealand, Aug 30, 1871) observes that thorium produces a gas, which he calls thorium emanation (we now know it as radon); the same discovery is also made independently by Friedrich Ernst Dorn

Rutherford discovers that radioactivity from uranium has at least two different forms, which he calls alpha and beta rays

Joseph John (J.J.) Thomson, using Charles Wilson's condensation chamber, proves that cathode particles carry the same amount of charge as hydrogen ions in electrolysis

J.J. Thomson measures the charge of the electron and thus completes his discovery of the electron; he also recognizes ionization to be a splitting of atoms and that particles emitted by the photoelectric effect have the same mass/charge ratio as cathode rays

Sir John Bennett Lawes, English agricultural scientist, d Rothamsted, Aug 31

Emil Wiechert invents the inverted pendulum seismograph, essentially the type used today

The American Mathematical Society begins publishing *The Transactions of the American Mathematical Society*, a journal of research

At the International Congress of Mathematicians in Paris, David Hilbert suggests 23 problems that he hopes will be solved in the twentieth century; many mathematicians take Hilbert's list as an agenda, with the result that most of the

Walter Reed (b Belroi, VA, Sep 13, 1851) directs James Carroll, Jesse William Lazear, and Aristides Agramonte in the study of the yellow fever epidemic in Havana and establishes that it is caused by the mosquito of the genus *Aedes*; Carlos Finlay had proposed this a few years earlier

On Aug 27, James Carroll (b Woolwich, England, Jun 5, 1854) deliberately allows

Becquerel discovers that part of the radiation produced by uranium (and identified as beta rays by Rutherford) is identical to the electrons identified in cathode-ray experiments

Paul Karl Ludwig Drude shows that moving electrons conduct electricity in metals

Max Planck (b Kiel, Germany, Apr 23, 1858) announces the first step toward

The first offshore oil wells are drilled

The first Browning revolvers are manufactured

Gottlieb Wilhelm Daimler, German inventor, d Kannstatt, Württemberg, Mar 6

Thomas Alva Edison invents the nickel-alkaline accumulator

Paul Kollsman, American aeronautical engineer, b Feb 22

1900
cont

Frederick Hopkins (b Eastbourne, England, Jun 30, 1861) discovers tryptophan, an amino acid; he demonstrates that it is essential to rats, the first known essential amino acid

with other compounds; Gomberg's compound is stable)

Russian-American chemist Vladimir Ipatieff (b Moscow, Nov 21, 1867) discovers that catalysts can affect organic chemicals during explosions

Johann Gustav Kjeldahl, Danish chemist, d Tisvildeleje, Jul 18

Peter Waage, Norwegian chemist, d Oslo, Jan 13

A remarkable coincidence

The monk Gregor Mendel performed his famous experiments with peas that led to his discovery of the laws of heredity at about the same time that Charles Darwin was explaining evolution. Darwin's ideas were soon known around the world. Mendel, however, had trouble getting published. He first sent his work to a prominent biologist, who, as it happened, did not like mathematics. The biologist sent the paper back to Mendel with negative comments. In 1865 and 1869 Mendel's work was published—by the local natural history society. After this Mendel was promoted to abbot, which kept him busy at the same time that it allowed him to grow fat. He gave up both gardening and science.

Darwin never got the chance to learn of Mendel's work, which is unfortunate, since Mendel's laws neatly fill a major gap in Darwin's theory. Darwin knew that variation occurred, but he did not know how it was inherited. Mendel's laws described the mechanism by which many traits pass from generation to generation.

In 1900, however, an astonishing coincidence put Mendel's work into the scientific mainstream. Three different biologists working in three different countries—Hugo de Vries in the Netherlands, Karl Correns in Germany, and Erich Tschermak von Seysenegg in Austria—worked out Mendel's laws for themselves. Each searched the scientific literature for prior discoveries of these laws and each somehow found the obscure papers from over 30 years before. When they published their work, they each unselfishly credited Mendel.

problems are either solved or proved unsolvable in the century

David Hilbert introduces the "direct method" in the calculus of variations

mosquitoes that have fed on yellow fever victims to bite him in his effort to prove that the disease is carried by mosquitoes; he develops the fever, but survives, only to die a few years later from yellow fever-induced heart disease

Jesse William Lazear is killed by yellow fever in a successful attempt to show that it is transmitted by mosquitoes

Paul Ehrlich and Julius Morgenroth make a study of hemolysis (the destruction of red blood cells) and introduce the terms complement and amboceptor

Sigmund Freud's *Die Traumdeutung* (The interpretation of dreams) accounts for dreams in terms of hidden symbols that reveal the unconscious mind

Wilhelm Kühne, German physiologist, d Heidelberg, Jun 10

Austrian doctor Karl Landsteiner (b Vienna, Jun 14, 1868) shows that there are at least three different types of human blood (A, B, and O) some of which are incompatible; the serum of a person can agglutinate red globules of a donor from an incompatible group

William Leishman (b Glasgow, Scotland, Nov 6, 1865) discovers the cause of a tropical disease then known as kala-azar, but now known as leishmaniasis; it is caused by a protist that is spread by sandflies

quantum theory on Dec 14; he states substances can emit light only at certain energies, which implies that some physical processes are not continuous, but occur only in specified amounts called quantums

Paul Ulrich Villard (b Lyon, France, Sep 28, 1860) is the first to observe a radiation that is more penetrating than X rays, now called gamma rays

Jean-Joseph-Etienne Lenoir, Belgian-French inventor, d Varenne-St. Hilaire, France, Aug 4

Hyman George Rickover, Polish-American naval officer, b Makov, Russia (Poland), Jan 27; he is the leader in the development of nuclear submarines in the 1950s

Count Ferdinand von Zeppelin (b Konstanz, Germany, Jul 8, 1838) constructs his first dirigible, which flies successfully on Jul 2

1901

ASTRONOMY

Annie Jump Cannon completes the Harvard Classification of stars by introducing spectral subclasses of stars

Peter Van de Kamp, Dutch-American astronomer, b Kampen, Netherlands, Dec 26; he attempts to prove that several nearby stars have planetary systems, but his work is largely refuted by other astronomers

BIOLOGY

The okapi, a relative of the giraffe, is discovered in Africa

Jules Bordet discovers that complement is used up when an antibody reacts with an antigen

Hugo De Vries' *Die Mutationstheorie* (The mutation theory) introduces the idea that changes in species occur in jumps, which he calls mutations; volume 1 is published this year and volume 2 in 1903

Russian biologist Ilya Ivanovich Ivanov (b Shirgry, Aug 1, 1870) founds a center for artificial insemination, introducing the first practical application of the technique developed by Lazzaro Spallanzini in 1785

Alexander Onufriyevich Kovalevski, Russian embryologist, d St. Petersburg (Leningrad, USSR), Nov 22

Jokichi Takamine (b Takaoka, Japan, Nov 3, 1854) discovers adrenaline and synthesizes it; Thomas Bell independently makes the same discovery

Edward Bradford Titchener's *Experimental psychology* is considered the most erudite work in psychology in the English language in his time

CHEMISTRY

The first synthetic vat dye, indanthrene blue, is manufactured

Eugène Demarçay discovers the chemical element europium

Victor Grignard introduces organic magnesium halides (Grignard reagents) to prepare organic compounds

Max Joseph von Pettenkofer, German chemist, d near Munich, Feb 10

François-Marie Raoult, French physical chemist, d Grenoble, Isère, Apr 1

Jacobus van't Hoff of the Netherlands wins the Nobel Prize for Chemistry for his discovery of the law of chemical dynamics of weak solutions

EARTH SCIENCE

Johann Phillip Ludwig Elster (b Bad Blankenburg, Germany, Dec 24, 1854) and Hans Geitel demonstrate radioactivity in rocks, springs, and air

Nils Adolf Erik Nordenskiöld, Swedish geologist, d Dalbyö, Sweden, Aug 12

MATHEMATICS

Charles Hermite, French mathematician, d Paris, Jan 14

MEDICINE

The journal *Biometrika*, in which psychologists support the idea of eugenics, is founded

Emil von Behring of Germany wins the Nobel Prize for Physiology or Medicine for his discovery of diphtheria antitoxin

Havelock Ellis starts his *Studies in the psychology of sex*; the sixth and last volume is published in 1910

Sigmund Freud's *Zur Psychopathologie des Alltags* (The psychopathology of everyday life) introduces the famous concept of the "Freudian slip"

Gerrit Grijns shows that beriberi is caused by the removal of a nutrient from rice during polishing

Ilya Ilich Mechnikov's *Immunity in infectious diseases* is published

Eugene L. Opie discovers the relationship of the islets of Langerhans to diabetes mellitus

Walter Reed establishes that a virus causes yellow fever; he had shown that the disease was carried by mosquitoes the previous year

Adolf Windaus (b Berlin, Dec 25, 1876) shows that the molecule of vitamin D can be affected by sunlight

PHYSICS

The US National Bureau of Standards is established

George Francis Fitzgerald, Irish physicist, d Dublin, Feb 21

Pyotr Nicolaievich Lebedev (b Moscow, Mar 8, 1866) measures the radiation pressure of light, confirming Maxwell's theory of electromagnetism; this is independently measured by E.F. Nichols and G.F. Hull

Wilhelm K. Roentgen of Germany wins the Nobel Prize for Physics for his discovery of X rays

Henry Augustus Rowland, American physicist, d Baltimore, MD, Apr 16

Ernest Rutherford and Frederick Soddy (b Eastbourne, England, Sep 2, 1877) discover that thorium left to itself changes into another form (another element, in fact); however, they do not realize at the time that thorium is changing into an isotope of radium

TECHNOLOGY

The first electric typewriter is produced, the Blickensderfer Electric

Motor-driven bicycles are introduced

The first vacuum cleaner is invented by Hubert Booth

Ferdinand Braun uses a crystal detector for the detection of radio waves

King Camp Gillette and William Nickerson patent the first safety razor

Zénobe Théophile Gramme, Belgian-French inventor, d Bois-Colombes, Hauts-de-Seine, France, Jan 20

Peter C. Hewitt invents the mercury vapor arc lamp

Guglielmo Marconi receives the letter S in St. Johns, Newfoundland, that has been transmitted from England, the first transatlantic telegraphic radio transmission

G. Whitehead performs the first flight on a motor-driven airplane

The Wright brothers, Wilbur (b Millville, IN, Apr 16, 1867) and Orville (b Dayton, OH, Aug 19, 1871) fly their first glider

1901

	GENERAL	ANTHROPOLOGY/ARCHAEOLOGY	ASTRONOMY	BIOLOGY	CHEMISTRY
1902	Animals crackers go on sale; the box is intended as a Christmas ornament	Pierre Boule reconstructs the skeleton of a Neanderthal man

A French expedition at Susa, the ancient capital of Elam, discovers tablets engraved with the code of Hammurabi, the first known set of laws | | William M. Bayliss (b Wolverhampton, England, May 2, 1860) and Ernest Henry Starling (b London, Apr 17, 1866) discover secretin, a hormone released by the walls of the small intestines that controls the pancreas, and establish the role of hormones

Hans Ernst Angass Buchner, German bacteriologist, d Munich, Apr 8

Ivan Petrovich Pavlov formulates his law of reinforcement, or learning by conditioning; he demonstrates that a dog trained by giving it food at the same time a bell is rung will soon learn to salivate at the sound of the bell with no food present

Walter Stanborough Sutton (b Utica, NY, Apr 5, 1877) states that chromosomes are paired and may be the carriers of heredity | Sir Frederick Augustus Abel, English chemist, d London, Sep 6

Emil Fischer of Germany wins the Nobel Prize for Chemistry for sugar and purine synthesis

Herman Frasch fully develops his method of removing sulfur from deep deposits by melting it with superheated water

Cato Maximilian Guldberg, Norwegian chemist and mathematician, d Christiania (Oslo), Jan 14

William Pope develops optically active compounds based on sulfur, selenium, and tin

Fritz Strassman, German chemist, b Boppard, Rhine, Feb 22; in the late 1930s he replaces Lise Meitner as Otto Hahn's assistant in investigating nuclear fission after Meitner flees the Nazi regime in Germany

Johannes Wislicenus, German chemist, d Leipzig, Saxony, Dec 5 |

Mount Pelée on Martinique erupts, killing almost all of the 38,000 people in Saint Pierre; legend has it that one prisoner lived through the deadly cloud of hot gases that flowed from the volcano, although later research suggests that two people survived

Oliver Heaviside and A.E. Kennelly predict the existence of an electrified layer in the atmosphere that reflects radio waves; this layer later becomes known as the ionosphere

Léon Teisserenc de Bort (b Paris, Nov 5, 1855) is the first to discover that Earth's atmosphere has at least two different layers, which he names the troposphere and the stratosphere

Henri Lebesgue introduces a new method of integration that allows the integration of functions over domains not possible by the Riemann method

Beppo Levi makes the first explicit statement of the axiom of choice: one can choose exactly one element from any given collection of disjointed nonempty sets; this axiom eventually separates mathematics into two types, one that accepts it and one that rejects it

Bertrand Russell (b Trelleck, England, May 18, 1872) discovers his "great paradox" concerning the set of all sets: it either contains itself or it does not, but if it does, then it does not and vice versa; this paradox undermines the foundations of Gottlob Frege's and Russell's attempts to base mathematics on logic

Alfred Adler (b Penzing, Austria, Feb 7, 1870) joins Freud and others to form the first psychoanalytic society

American surgeon Harvey William Cushing (b Apr 8, 1869) investigates the pituitary gland

Hermann Ebbinghaus's *Die Grundzüge der Psychologie* (The principles of psychology) becomes a highly successful general textbook in psychology

Baron Richard von Krafft-Ebing, German neurologist, d Mariagrun, Austria, Dec 22

Karl Landsteiner discovers a fourth blood group, AB

Walter Reed, American military surgeon, d Washington, DC, Nov 23

Sir Ronald Ross of England wins the Nobel Prize for Physiology or Medicine for his work on malaria infections

Charles W. Stiles names the American variety of the hookworm, which was discovered by Allan J. Smith

Rudolf Carl Virchow, German pathologist, d Berlin, Sep 5

Josiah Gibbs's *Elementary principles in statistical mechanics* is published

British-American electrical engineer Arthur Edwin Kennelly (b Bombay, India, Dec 17, 1861) discovers that there is a layer of electrically charged particles in the upper atmosphere that reflects radio waves, this is independently discovered by Oliver Heaviside a few months later

Philipp von Lenard discovers that the energy of transmitted electrons in the photoelectric effect is a function of the wavelength of the light and not of the intensity, which will be explained by Einstein in 1905

Hendrik Antoon Lorentz and Pieter Zeeman of the Netherlands win the Nobel Prize for Physics for their discovery of the effect of magnetism on electromagnetic radiation

Owen Willans Richardson (b Dewsbury, England, Apr 26, 1879) tries to refract X rays with the edge of a Gillette safety-razor blade; his failure to observe refraction tends to confirm the (incorrect) theory that X rays are a beam of neutral particles; had he used a crystal, he would have discovered X-ray diffraction

Rutherford and Soddy publish the paper "The cause and nature of radioactivity"; it contains the atomic disintegration theory of radioactivity, saying that atomic nuclei split to form other elements

The first practical airship, *Le Jaune*, is launched in France by the Lebaudy brothers

The radio magnetic detector is introduced

Light bulbs with osmium filaments are produced commercially

Robert Bosch invents the spark plug

Willis H. Carrier invents the first air conditioner, although the name is first used in 1906 to describe a different device

G. Honold develops high-voltage ignition for internal combustion engines based on electromagnetic induction

Millar Hutchinson invents the first electrical hearing aid in New York

Valdemar Poulsen (b Copenhagen, Denmark, Nov 23, 1869) invents the arc generator for the production of radio waves

Louis Renault develops the drum brake

Richard Adolf Zsigmondy (b Vienna, Austria, Apr 1, 1865) invents the ultramicroscope, a device for seeing the small particles in a colloidal solution

1902

1903

Herbert Spencer, English sociologist, d Brighton, Sussex, Dec 8

Walter Sutton's *The chromosomes theory of heredity* contains detailed studies that support his proposal of the previous year that genes are carried by chromosomes; he argues that each egg or sperm cell contains only one of each chromosome pair, which accounts for the random factor in heredity

Svante Arrhenius of Sweden wins the Nobel Prize for Chemistry for his theory of ionic dissociation

K. Birkeland and K. Eyde develop a method for producing nitric acid from atmospheric nitrogen

Lars Onsager, Norwegian-American chemist, b Oslo, Norway, Nov 27; during World War II he works out the details of the gaseous diffusion method of separating uranium-235 from uranium-238

W.H. Stearn and F. Topham develop a method for producing artificial silk (viscose)

Ivar Fredholm makes a systematic study of integral equations

Gottlob Frege's *Grundgesetze der Arithmetik* (Basic laws of arithmetic) is the first attempt to derive the laws of mathematics from formal logic, but it is flawed by its failure to account for Bertrand Russell's "great paradox" of 1902, sent to Frege just as he was finishing the second volume of his book

Bertrand Russell's *The principles of mathematics* is an outline of a program to find an axiomatic foundation to mathematics by basing mathematics on pure logic

Sir George Gabriel Stokes, British mathematician and physicist, d Cambridge, Feb 1

Dutch physiologist Willem Einthoven (b Semarang, Java, May 22, 1860) develops in the Netherlands the string galvanometer, the forerunner of the electrocardiograph, used to measure tiny electrical currents produced by the heart

Niels Ryberg Finsen of Denmark wins the Nobel Prize for Physiology or Medicine for his treatment of skin disease with light

Karl Gegenbaur, German anatomist, d Heidelberg, Jun 14

German surgeon George Perthes discovers that X rays inhibit the growth of tumors and proposes X-ray treatment for cancer

Gregory Pincus, American biologist, b Woodbine, NJ, Apr 9; he becomes one of the pioneers of the birth-control pill

Antoine-Henri Becquerel, Pierre Curie, and Marie Curie of France share the Nobel Prize for Physics, Becquerel for his discovery of natural radioactivity and the Curies for their later study of radioactivity

Josiah Willard Gibbs, American physicist, d New Haven, CT, Apr 28

Ernst Mach's *Geschichte der Mechanik* (History of mechanics) criticizes the concepts of absolute space and absolute time in Newtonian mechanics and profoundly influences Einstein

Henri Poincaré recognizes that very small inaccuracies in initial conditions can lead to vast differences in a short order, the fundamental idea of chaos (which was not to be explored much further until the 1970s and 1980s)

William Ramsay and Soddy discover that helium is formed by the radioactive decay of radium; specifically, alpha particles are the nuclei of helium atoms, although Ramsay and Soddy do not know this at the time

Rutherford shows that a strong magnetic field can deflect alpha particles, meaning that they are charged

Rutherford and Soddy state that radioactivity is caused by one element changing into another; Rutherford names the third kind of radioactivity, *gamma rays*

J.J. Thomson's *Conduction of electricity through gases* is published

Richard Jordan Gatling, American inventor, d New York, Feb 26

W. Siemens develops an electric locomotive

Konstantin Tsiolkovsky proposes that liquid oxygen be used for space travel

The first successful airplane is launched at Kitty Hawk by Wilbur and Orville Wright on Dec 17; its best flight of the day lasts 59 seconds

1903

1904

Charles Guillaume discovers that a kilogram of water has a volume of 1000.028 cu cm at 4°C, not 1000 cu cm as planned by the designers of the metric system; since this is also the definition of a liter, the liter is redefined as being 1000.028 cu cm

George Ellery Hale sets up the Mount Wilson Observatory in California

Johannes Franz Hartmann (b Erfurt, Germany, Jan 11, 1865) reports the discovery of the stationary calcium lines in the spectrum of the binary Delta Orionis caused by the presence of an interstellar cloud of atoms; this is the first discovery of interstellar matter

Jacobus Cornelius Kapteyn (b Barneveld, Netherlands, Jan 19, 1851) discovers two distinct star streams in the galaxy

American-Argentine astronomer **Charles Dillon Perrine** (b Steubenville, OH, Jul 28, 1867) discovers the sixth moon of Jupiter

Arthur Harden (b Manchester, England, Oct 12, 1865) discovers the first coenzyme, a compound that is not a protein but which is required for the protein to act as an enzyme

Nuttall's *Blood immunity and blood relationship* shows that phyla, classes, orders, and genera of organisms can be distinguished by serological methods

Ivan Pavlov of Russia wins the Nobel Prize for Physiology or Medicine for his study of the physiology of digestion

Santiago Ramón y Cajal summarizes a lifetime of studies of brain cells, establishing the theory that the nervous system is composed only of nerve cells and their processes

Friedrich Giesel discovers "actinium X," an isotope of radium

Frederick Stanley Kipping (b Manchester, England, Aug 16, 1863) discovers silicones

Vladimir Markovnikov, Russian chemist, d Moscow, Feb 11

Sir **William Ramsay** of England wins the Nobel Prize for Chemistry for his discovery of inert gas elements and their placement in the periodic table

Alexander William Williamson, English chemist, d Hindhead, Surrey, May 6

Clemens Alexander Winkler, German chemist, d Dresden, Oct 8

The age of the Earth

While some people, including Aristotle, have assumed that Earth always existed, most people do not believe that. In the Judeo-Christian heritage, Earth is created at a specific time. Curious people could use the evidence of the Bible to work out what that time was. The most famous timetable was developed by Archbishop James Ussher in 1650 and further refined by John Lightfoot in 1654. This chronology reported that Earth was created at 9:00 AM on October 26, 4004 BC. The chronology was for the next 200 years or so printed in the margins of the Authorized, or King James, Version of the Bible. Many people in England and other English-speaking countries thought it had always been a part of the Bible. Charles Darwin, for one, was surprised to learn that the biblical chronology was a recent work of humans.

Mikhail Lomonosov in the eighteenth century was one of the first to see things differently. He thought from geological evidence that Earth must be hundreds of thousands of years old. Perhaps the first attempt to determine the age of Earth on scientific grounds was made by comte de Buffon, who assumed that Earth was originally hot and was slowly cooling. By measuring the rate at which iron balls cooled, Buffon decided in 1779 that Earth is about 75,000 years old.

James Hutton in 1785 had suggested that Earth had "no vestige of a beginning." His followers, Charles Darwin and Charles Lyell, did not try to find Earth's beginning, but they made some informed guesses as to how long ago certain fossils formed. Darwin dated the end of the Cretaceous Period at 300 million years ago (a bit too far back; the currently accepted time is 65 million years ago). Lyell erred on the other side in 1867, placing the start of the Ordovician Period at 240 million years ago; we now think it was about 500 million years ago.

William Thomson (later Lord Kelvin), disagreed with these dates. Using an improved version of the Buffon idea about Earth's cooling, he proposed in 1862 that the crust formed between 20 million and 98 million years ago. He also calculated how long the sun could have existed, assuming that it got its energy as a result of gravitational contraction. This figure was 100 million years. Geologists did not accept his work, although they had a hard time refuting it.

Vilhelm Bjerknes's *Weather forecasting as a problem in mechanics and physics* is one of the first scientific studies of weather forecasting

Johan Vogt's *The molten silicate solution* is an important study of igneous rocks

David Hilbert develops a model for Euclidean geometry within the system of arithmetic, demonstrating that if there are no contradictions in arithmetic, then there are also none in geometry

Ernst Zermelo uses the axiom of choice to prove Cantor's well-ordering theorem; the axiom of choice remains suspect until 1963, when Paul Cohen shows that it can either be accepted or rejected without losing consistency in mathematics

Niels Ryberg Finsen, Danish physician, d Copenhagen, Sep 24

Army surgeon William Gorgas (b Mobile, AL, Oct 3, 1854) is put in charge of disease control during the building of the Panama Canal; he develops such effective measures for controlling mosquitoes that both malaria and yellow fever are wiped out in the region around the canal

Stanley Hall's *Adolescence: Its psychology, and its relations to physiology, anthropology, sociology, sex, crime, religion, and education* argues that the child, in its development, recapitulates the life history of the race

Etienne-Jules Marey, French physiologist, d Paris, May 15

Ronald Ross's *Researches on malaria* tells of his 1897 discovery that the Anopheles mosquito transmits malaria from person to person

Charles Glover Barkla (b Widnes, England, Jun 27, 1877) discovers that X rays are transverse waves, like those of light, and not longitudinal waves, as had been previously thought

Bertram Borden Boltwood (b Amherst, MA, Jul 27, 1870), Herbert McCoy, and Robert Strutt independently discover that radium is a decay product of uranium

The dynamical theory of gases by James Hopwood Jeans (b Ormskirk, England, Sep 11, 1877) is published

Hendrik Antoon Lorentz anticipates relativity in a paper by introducing the notion of "corresponding states," although he maintains the notion that there is an ether

"Kinetics of a system of particles illustrating the line and band spectrum and the phenomena of radioactivity" by Hantaro Nagaoka (b Nagasaki, Japan, Aug 15, 1865) includes his "Saturnian model" of the atom, in which a positive nucleus is surrounded by a ring of thousands of electrons

Ludwig Prandtl discovers a flowing liquid in a tube has a thin boundary layer adjacent to the walls of the tube that does not flow as fast as the rest of the liquid

John Strutt, Lord Rayleigh, of England wins the Nobel Prize for Physics for his discovery of argon

Rutherford's *Radioactivity* is published

Ultraviolet lamps are introduced

Work begins on the Panama Canal

German-American inventor Emile Berliner (b Hanover, Germany, May 20, 1851) invents the flat disk form of the phonograph, an improvement over Thomas Alva Edison's wax cylinder system that is quickly adopted by the record industry

Johann Phillip Ludwig Elster devises the first practical photoelectric cell

John Ambrose Fleming (b Lancaster, England, Nov 29, 1849) files a patent for the first vacuum tube; it is a diode that acts as a rectifier, a device that makes current flow in a single direction instead of alternating back and forth; hence, it changes alternating current (AC) to direct current (DC)

French scientist Leon Guillet develops the first stainless steels, but unaccountably fails to note that they resist corrosion

Arthur Korn telegraphs photographs from Munich to Nuremberg

W. Rubel invents offset printing

Engineer and inventor Friedrich Siemens d May 24

The age of the Earth (continued)

In 1896, however, radioactivity was discovered (*see* "Discovering new rays," p 388). Radioactive elements produce heat, making Thomson's calculations meaningless.

Radioactivity also provided the solution to finding out how old rocks really are. In 1904, Bertram Boltwood proved that one element changes into another during radioactive decay. By 1907 he realized that he could calculate the age of a rock by measuring the ratio of uranium, a fairly common element, to lead, its final decay product. He applied his method to a mineral from Glastonbury, Connecticut, and got an age of 410 million years. (More recent calculations, based on an improved knowledge of isotopes, gives the Glastonbury mineral as only 265 million years old.)

Other radioactive elements can be used to date rocks that contain no uranium. Using various adaptations of Boltwood's method, scientists have dated rocks from all periods of Earth's past. In 1983, zircon crystals embedded in younger rock were found to be about 4.2 billion years old, the current recordholder. The current best estimate for the age of Earth is 4.6 billion years.

1904

cont

1905

Sigmund Freud's *Jokes and their relation to the unconscious* and *Three essays on the theory of sexuality* are published

Ejnar Hertzsprung (b Frederiksberg, Denmark, Oct 8, 1873) is the first to note that there is a relationship between color and luminosity of stars, but his work is not well known; 10 years later Henry Norris Russell makes the same observation and gives it currency

Percival Lowell predicts the existence of a ninth planet with an orbit beyond Neptune

Charles Dillon Perrine discovers the seventh moon of Jupiter

International rules of zoological nomenclature is the first worldwide attempt at bringing order to zoological classification

William Bateson (b Whitby, England, Aug 8, 1861) demonstrates that some characteristics are not independently inherited, a finding that will be confirmed and explained by Thomas Hunt Morgan

Austrian-American biochemist Erwin Chargaff b Czernowitz, Austria, Aug 11; his work in the late 1940s demonstrates that the number of adenine bases in DNA is equal to the number of thymine bases and the number of cytosine bases is equal to the number of guanine bases; this is an important clue to DNA's structure

Arthur Harden discovers that intermediate compounds often form in metabolism; these compounds are required for the reaction to take place but are not involved as the original compounds or as the final ones

Clarence McClung finds that female mammals have two X chromosomes and males have an X paired with a Y

Adolf von Baeyer of Germany wins the Nobel Prize for Chemistry for his work on organic dyes

J. Edwin Brandenburger invents cellophane

Per Teodor Cleve, Swedish chemist and geologist, d Uppsala, Jun 18

Richard Willstätter (b Karlsruhe, Germany, Aug 13, 1872) discovers the structure of chlorophyll

EARTH SCIENCE	MATHEMATICS	MEDICINE	PHYSICS	TECHNOLOGY	
			J.J. Thomson's "On the structure of the atom" proposes the "plum-pudding model" of the atom in which the electrons are embedded in a sphere of positive electricity		**1904** cont
Daniel Barringer (b Raleigh, NC, May 25, 1860) proposes that the large crater in Arizona was caused by a meteor, not by a volcano; it is now known as the Great Barringer Meteor Crater	L. E. Dickson defines a new class of algebras termed *cyclic algebras* Emanuel Lasker, better known as a chess champion than as a mathematician, establishes that every polynomial ideal is a finite intersection of primary ideals; in 1921 Emmy Noether will generalize this result as part of the foundation of abstract algebra	Alfred Binet (b Nice, France, Jul 8, 1857), V. Henri, and T. Simon develop the intelligence test Alexis Carrel (b Lyon, France, Jun 28, 1873), working at the Rockefeller Institute in New York City, develops techniques for rejoining severed blood vessels, paving the way for organ transplantation George Washington Crile performs the first direct blood transfusion Albert Einhorn introduces a local anesthetic, novocain Walther Flemming, German anatomist, d Kiel, Schleswig, Aug 4 Robert Koch of Germany wins the Nobel Prize for Physiology or Medicine for his tuberculosis research Rudolf Albert von Kölliker, Swiss anatomist and physiologist, d Würzburg, Bavaria, Nov 2 J.B. Murphy develops the first artificial joints for use in the hip of an arthritic patient Fritz Schaudinn and Eric Hoffmann discover *Spirocheta pallida*, the cause of syphilis	An international conference on electrical units is held in Berlin Albert Einstein (b Ulm, Germany, Mar 14, 1879) postulates the light-quantum in his Mar 17 paper explaining the photoelectric effect; light behaves like particles and thus can liberate by impact electrons from a metal surface; later the light particle will come to be known as the photon Einstein calculates how the movement of molecules in a liquid can cause the Brownian motion of small particles suspended in the liquid; this is viewed by many as the first proof that atoms actually exist Einstein submits his first paper on the special theory of relativity, "On the electrodynamics of moving bodies," to the *Annalen der Physik* on Jun 30; the theory calls the speed of light constant for all conditions and states that time passes at different rates for objects in constant relative motion Einstein's second paper on special relativity, "Does the inertia of a body depend on its energy content?," submitted on Sep 27, states the famous relationship between mass and energy: $E = mc^2$	Safety glass is patented The first German U-boat submarine is launched Frederick G. Cottrell invents the electrical precipitator for the removal of dust that is named after him Guglielmo Marconi invents the directional radio antenna American undertaker Almon Brown Strowger invents the dial telephone Gabriel Voisin, Ernest Archdeacon, and Louis Blériot start the first airplane factory in Billancourt, near Paris Engineer Robert Whitehead d Nov 19	**1905**

1905

cont

Ernest Starling coins the term *hormone* in his paper "On the chemical correlation of the functions of the body"

1906

A mysterious explosion occurs near Tunguska in Siberia, flattening a huge region and knocking down millions of trees; no meteorite remains are found in the area and the cause is never determined; later speculation about possible causes ranges from an impact with a comet to an impact with a small black hole

Burnham's double star catalog contains data on 13,665 double star systems

Carl Wilhelm Ludvig Charlier proposes a universe consisting of a system of systems of systems in order to find a correct Newtonian treatment of the universe; applying Newtonian equations, the universe should collapse

Jacobus Cornelius Kapteyn uses statistical methods to determine the distribution of stars, analyzing their proper motion in 252 selected regions of the sky

Samuel Pierpont Langley, American astronomer and aviation pioneer, d Aiken, SC, Feb 22

Astronomer William Wilson Morgan b Bethesda, TN, Jan 3; he is the first astronomer to

Camillo Golgi of Italy and Santiago Ramón y Cajal of Spain win the Nobel Prize for Physiology or Medicine for their study of the structure of the nervous system and of nerve tissue

Frederick Hopkins suggests that food contains ingredients essential to life that are not proteins or carbohydrates; as such substances are found in succeeding years they come to be called vitamins

Fritz Richard Schaudinn, German zoologist, d Hamburg, Jun 22

Friedrich Konrad Beilstein, Russian chemist, d St. Petersburg (Leningrad, USSR), Oct 18

Pierre Curie, French chemist, d Paris, Apr 19

Henri Moissan of France wins the Nobel Prize for Chemistry for the isolation of fluorine

Russian botanist Mikhail Tsvett (b Asti, Italy, May 14, 1872) develops paper chromatography as a means of separating dyes; its main importance will come when it is applied to organic molecules starting in the 1940s

Otto Hahn (b Frankfurt-am-Main, Germany, Mar 8, 1879) discovers radiothorium, found in 1907 to be an isotope of thorium

Philipp von Lenard of Germany wins the Nobel Prize for Physics for his work on cathode rays

---

| | | | | | 1906 |

The great San Francisco earthquake on the morning of Apr 18 kills about 700 people and destroys buildings as far away as 600 km (400 mi) from the epicenter; it is estimated to have a strength of about 8.3 on the Richter scale

Bernard Brunhes discovers that sediments that are baked into rock by lava contain the same residual magnetism from Earth's magnetic field as the lava does; this shows that the residual magnetism is not altered by the direction of flow of the lava, but is a direct result of Earth's field when the lava cooled

Clarence Dutton suggests that pockets of radioactive minerals might account for volcanoes, an idea that is wrong, but which calls attention to the role of radioactive heating in the processes of Earth

R.D. Oldham shows that Earth has a core from studying earthquake waves; one kind of waves, called primary waves or P waves, will travel through both liquids and solids; another kind, the secondary or S waves, will not travel through liquids; differences in P and S waves were the key to finding the core

Maurice Fréchet develops the functional calculus

Jules Bordet discovers *Bacillus pertussis*, the bacterium causing whooping cough

The integrative action of the nervous system by Charles Sherrington (b London, Nov 27, 1857) divides the study of the nervous system into three regions—the mechanical level, the level at which thought occurs, and the mind-body level; this influential work has its fifth edition in 1947

German bacteriologist August von Wasserman (b Bamberg, Feb 21, 1866) develops his famous test for syphilis

Charles Barkla discovers that each element has a characteristic X ray, produced by scattering of X-ray beams; this is the key discovery that eventually leads to the concept of atomic number

Ludwig Edward Boltzmann, Austrian physicist, d Duino, near Trieste, Italy, Sep 5

Marie Curie becomes the first woman professor at the Sorbonne, taking her late husband's position

Harvey Fletcher claims in a posthumous note that he suggested to Robert Andrews Millikan the experiments measuring the charge of the electron with oil drops that contributed to Millikan's winning the Nobel Prize in 1923

Otto Hahn and Lise Meitner begin to investigate the energy levels of beta rays, or the beta spectra, of radioactivity

James Hopwood Jeans's *Theoretical mechanics* is published

Alexander Stepanovich Popov, Russian physicist, d St. Petersburg (Leningrad, USSR), Jan 13

The tungsten-filament light bulb is introduced

A. d'Arsonval and F. Bordas develop freeze-drying in Paris

Thomas Alva Edison invents the "cameraphone" for the synchronization of a phonograph and a projector

Canadian-American physicist Reginald Aubrey Fessenden (b Milton, Quebec, Oct 6, 1866) invents AM radio and transmits music and voice via radio waves

1906 cont

ASTRONOMY

demonstrate that the Milky Way galaxy has a spiral structure like that of M31

Karl Schwarzschild (b Frankfurt-am-Main, Germany, Oct 9, 1873) develops the theory that the transmission of heat in stellar atmospheres occurs principally by radiation

Max Wolf discovers the asteroid Achilles

1907

GENERAL

William James's *Pragmatism* is his major book on philosophy: he supports an idea of reality based only on experience

ASTRONOMY

Asaph Hall, American astronomer, d Goshen, CT, Nov 22

Pierre-Jules-César Janssen, French astronomer, d Meudon, Dec 23

Hermann Carl Vogel, German astronomer, d Potsdam, Aug 13

BIOLOGY

Thomas Hunt Morgan (b Lexington, KY, Sep 25, 1866) begins his work with fruit flies (*Drosophila melanogaster*) that will prove that chromosomes have a definite function in heredity, establish mutation theory, and lead to a fundamental understanding of the mechanisms of heredity

Charles L.A. Laveran of France wins the Nobel Prize for Physiology or Medicine for his discovery of the role of protozoa in disease generation

Hans Hugo Selye, Austrian endocrinologist, b Jan 26; his work shows that stress can affect the physical functioning of the body

CHEMISTRY

Pierre-Eugène Berthelot, French chemist, d Paris, Mar 18

Eduard Buchner of Germany wins the Nobel Prize for Chemistry for the discovery of noncellular fermentation

Emil Hermann Fischer synthesizes a polypeptide, or small protein, from amino acids; containing just 18 amino acids, it is shown to be broken up by digestive juices just as natural proteins are

Dmitri Ivanovich Mendeléev, Russian chemist, d St. Petersburg (Leningrad, USSR), Feb 2
Ferdinand-Frédéric-Henri Moissan d Paris, Feb 20

Sir William Henry Perkin, English chemist, d Sudbury, Middlesex, Jul 14

EARTH SCIENCE	MATHEMATICS	MEDICINE	PHYSICS	TECHNOLOGY	
			Ernest Rutherford discovers alpha-particle scattering		**1906** **cont**
			Rutherford improves previous measurements of the ratio of mass to charge in alpha particles, which leads him to think (correctly) that alpha particles are the nuclei of helium atoms		
			Joseph John Thomson of England wins the Nobel Prize for Physics for his discovery of the electron		
			J.J. Thomson demonstrates that a hydrogen atom has only a single electron; some previous theories allowed as many as a thousand electrons per hydrogen atom		
Bertram B. Boltwood discovers how to use uranium to obtain the age of rocks by determining the ratio of uranium to its final decay product, lead; since the rate of decay is known and is unaffected by any chemical or mechanical processes, the age can be determined by working in reverse from the ratio	Andrei Andreyevich Markov develops a theory of linked probabilities (Markov chains) in which the probability of one event affects the probability of the following event Paul Montel starts developing his theory of normal families of functions	Vladimir Bektherev's *Objective psychology* argues for a completely objective approach to psychological phenomena James Carroll, English-American physician, d Washington, DC, Sep 16 John Scott Haldane (b England, May 3, 1860) develops a method for deep-sea divers to rise to the surface safely R.G. Harrison demonstrates the *in vitro* growth of living animal tissue Julius Eduard Hitzig, German physiologist, d Luisenheim zu St. Blasien, Aug 20 William Osler and Thomas McCrae's *Modern medicine, its theory and practice* (seven volumes) is published	Potassium and rubidium are found to be radioactive Albert Abraham Michelson of the United States wins the Nobel Prize for Physics for his spectroscopic studies of and measurements of light Pierre Weiss (b Mulhouse, France, Mar 25, 1865) develops the domain theory of ferromagnetism, explaining that iron and other ferromagnetic materials form small domains of a given polarity; when the poles of the domains are aligned, a strong magnetic force occurs; this explanation is still accepted William Thomson, Baron Kelvin, Scottish mathematician and physicist, d near Largs, Ayr, Dec 17	The paint spray gun is invented French bicycle dealer Paul Cornu builds the first helicopter that can take off vertically carrying a human; after a 20-second flight at 309 cm (1 ft), it breaks up on landing Lee De Forest (b Council Bluffs, IA, Aug 26, 1873) and R. von Lieben invent the amplifier vacuum tube (triode) Reginald A. Fessenden and E.F.W. Alexanderson invent a high-frequency electric generator Auguste and Louis Lumière develop color photography in a form that can be used by amateur photographers	**1907**

1907 cont			Zoologist Nikolaas Tinbergen b The Hague, The Netherlands, Apr 15; one of the founders of the new science of ethology, he studies herring gulls and discovers the submissive posture	Hendrik Willem Bakhuis Roozeboom, Dutch physical chemist, d Amsterdam, Netherlands, Feb 8 Georges Urbain (b Paris, Apr 12, 1872) discovers the element lutetium, named after the Roman village that preceded Paris, and ytterbium, named for a village in Sweden Adolf Windhaus synthesizes histamine
1908	Black cowboy George McJunkin discovers the first known Folsom points, associated with fossil bison bones in New Mexico	George Ellery Hale discovers magnetic fields in sunspots by observing Zeeman splitting of spectral lines; further work by various scientists eventually shows that sunspots are largely caused by magnetism Hale installs a 1.5-m (60-in.) reflecting telescope at Mount Wilson Observatory Giant and dwarf stellar divisions are described by Ejnar Hertzsprung; he is the first to suggest identifying the absolute magnitude of stars by considering the magnitude as if the star were at a standard distance of 3.25 light-years Charles Augustus Young, American astronomer, d Hanover, NH, Jan 3	William M. Bayliss's *The nature of enzyme action* describes the action of hormones Paul Ehrlich of Germany and Ilya Ilich Mechnikov of Russia win the Nobel Prize for Physiology or Medicine for their pioneering research in the mechanics of immunology Upon a request from R.C. Punnet, G.H. Hardy works out the mathematical law governing the frequency of a dominant trait in successive generations of a population obeying Mendelian laws of heredity, a problem already solved by W. Weinberg; the result is known as the Hardy-Weinberg law: random mating within one generation will produce an approximately stationary genotype distribution with an unchanged gene frequency Robley Cook Williams, American biophysicist, b Santa Rosa, CA, Oct 13; he develops the technique of spraying tiny objects with an opaque metal before taking electron-microscope photographs of them	Fritz Haber, b Breslau (Wroclaw, Poland), Dec 9, 1868, begins work in Germany on the Haber process for extracting nitrogen from the air and making ammonia for use as a cheap fertilizer Sir Ernest Rutherford of England wins the Nobel Prize for Chemistry for his studies of radioactivity, alpha particles, and the atom

L.E.J. Brouwer's *On the unreliability of logical principles* challenges the application of traditional Aristotelian logic to infinite sets; specifically, he rejects the law of the excluded middle (that either a statement or its negation is true) for infinite sets

W.S. Gosset's paper "The probable error of a mean," published under the pseudonym "Student," discusses the so-called "Student's test," a test for establishing the significance of individual data in a series of observations

Henri Poincaré expresses his famous view of set theory: "Later mathematicians will regard set theory as a disease from which one has recovered"

German mathematician P. Wolfskehl leaves 100,000 marks to the Göttingen Academy of Science as a prize for the first person to publish a complete proof of Fermat's "last theorem;" this leads to a spate of over a thousand incorrect proofs before inflation makes the prize virtually worthless

The chemical sulfanilamide is discovered, but its bacteria-killing property is not known until after 1936, when it is discovered to be the active part of Prontosil, the first "wonder drug"

Hermann Ebbinghaus's *Abriss der Psychologie* (A summary of psychology) becomes his most popular textbook

William McDougall's *Introduction to social psychology* contains a theory of behavior based on instinct

Karl von Voit, German physiologist, d Munich, Jan 31

The international ampere is adopted by the International Conference on Electrical Units and Standards; the current balance, based on a design by W.E. Ayrton and J. Viriamu Jones, is used at the National Physical Laboratory in England to measure the absolute ampere; the Weston cell is adopted as a standard voltage source

Antoine-Henri Becquerel, French physicist, d Le Croisic, Loire Inférieur, Aug 25

James Hopwood Jeans's *The mathematical theory of electricity and magnetism* is published

Heike Kamerlingh Onnes (b Groningen, Netherlands, Sep 21, 1853) succeeds in liquefying helium

French physicist Gabriel Jones Lippmann (b Holerich, Luxembourg, Aug 16, 1845) wins the Nobel Prize for Physics for the invention of a method of color photography

Space and time by Hermann Minkowski (b Alexotas, Russia, Jun 22, 1864) provides

The Holt Company of California develops the first tractor to use moving treads

The Hughes Tool Company develops the steel-toothed rock-drilling bit, revolutionizing the oil industry by enabling drilling through hard rock for the first time

H. Anschütz-Kaempfe invents the gyroscopic compass

George Harold Brown, American engineer, b Milwaukee, WI, Oct 14

Charles Frederick Cross invents cellophane

Hans Geiger (b Neustadt-an-der-Haardt, Germany, Sep 30, 1882) and Ernest Rutherford develop the radiation detector that comes to be known as the Geiger counter

Nikolay Alekseyevich Pilyugin, Soviet electrical engineer, b May 18

M. Wilm invents duraluminium

Orville Wright makes the first airplane flight that lasts an hour

1908

1908

cont

1909	Robert Peary (b Cresson, PA, May 6, 1856) and Matthew Hensen become the first people, on Apr 6, to reach the North Pole	Karl Bohlin suggests that the sun is at an off-center position in the Milky Way galaxy	Wilhelm Johannsen (b Copenhagen, Denmark, Feb 3, 1857) coins the terms *gene* to describe the carrier of heredity; *genotype* to describe the genetic constitution of an organism; and *phenotype* to describe the actual organism, which results from a combination of the genotype and various environmental factors	Karl Bosch develops the process invented by Fritz Haber for obtaining nitrogen compounds from air into an industrial operation

Karl Bohlin suggests that the sun is at an off-center position in the Milky Way galaxy

Jesse Leonard Greenstein, American astronomer, b New York, NY, Oct 15; he studies the different elements and isotopes that make up individual stars, showing that one star is not just like another

Simon Newcomb, Canadian-American astronomer, d Washington, DC, July 11

Wilhelm Johannsen (b Copenhagen, Denmark, Feb 3, 1857) coins the terms *gene* to describe the carrier of heredity; *genotype* to describe the genetic constitution of an organism; and *phenotype* to describe the actual organism, which results from a combination of the genotype and various environmental factors

William Bateson's *Mendel's principles of heredity: A defense* contains the first application of Mendel's laws to animals

Archibald Garrod discovers that some genes function by blocking steps in life processes that would otherwise take place

Emil Theodor Kocher of Switzerland wins the Nobel Prize for Physiology or Medicine for his work on the thyroid gland

August Friedrich Weismann's *Die Selektionstheorie* is published

Karl Bosch develops the process invented by Fritz Haber for obtaining nitrogen compounds from air into an industrial operation

Richard August Erlenmeyer, German chemist, d Aschaffenburg, Bavaria, Jan 22

Russian-American chemist Phoebus Aaron Theodor Levene, b Sager (USSR), Feb 25, 1869, discovers that the sugar ribose is found in some nucleic acids, those that we now call ribonucleic acids (RNA)

Wilhelm Ostwald of Germany wins the Nobel Prize for Chemistry for his work on catalysis

Austrian chemist Fritz Pregl, b Laibach (Yugoslavia), Sep 3, 1869, develops the first techniques for analyzing tiny amounts of organic chemicals

Ernst Zermelo develops an axiomatic treatment of set theory that will become the basis of modern mathematics; it uses just two undefined terms, set and membership, and contains only seven axioms

an addition to relativity theory; Minkowski shows that relativity theory makes more sense if the universe has four dimensions, with time as the fourth dimension

Louis Paschen discovers the series of lines in the hydrogen spectrum that is named after him

Jean-Baptiste Perrin uses Einstein's formula for Brownian motion to calculate the size of a molecule of water

Ernest Rutherford and T.D. Royds collect enough alpha particles to show that they are helium nuclei

Andrija Mohorovičić, b Volosko (Yugoslavia), Jan 23, 1857) discovers the change in earthquake waves that occurs about 30 km (18 mi) below Earth's surface that is now known as the Mohorovicic discontinuity, or "Moho;" it is the boundary between Earth's crust and mantle

R.D. Carmichael establishes the properties of composite numbers that are pseudoprimes for every given base (they pass a particular test for primeness, but are not actually prime); these numbers come to be known as Carmichael numbers

Maurice Fréchet introduces the concept of abstract space

Hermann Minkowski, Russian-German mathematician, d Göttingen, Germany, Jan 12

Charles-Jules-Henri Nicolle (b Rouen, France, Sep 21, 1884) discovers that typhus fever is transmitted by the body louse

Edward Bradford Titchener's A text book of psychology is based on structuralism and introspection

Lev Andreevich Artsimovich, Soviet physicist, b Moscow, Feb 25; he will be the main physicist involved in developing the Tokamak device for nuclear fusion, still considered among the most promising designs, although it has not achieved fusion

Guglielmo Marconi of Italy and Karl Ferdinand Braun of Germany win the Nobel Prize for Physics for wireless telegraphy

Ernest Marsden, under the direction of Hans Geiger and Ernest Rutherford, determines that some alpha particles bounce back from a thin gold foil, an experiment that leads to Rutherford's correct view of the atomic nucleus surrounded by electrons (announced in 1911)

General Electric begins to market the world's first electric toaster

Leo Baekeland patents Bakelite, the first plastic that solidifies on heating and the first to be widely used to replace more traditional materials such as wood, ivory, and hard rubber

On Jul 25 Louis Blériot of France becomes the first human to fly across the English Channel; the trip from Calais to Dover takes 37 minutes

English aviator Henri Farman completes the first airplane flight of 161 km (100 mi)

Enrico Forlanini develops the hydrofoil ship, which uses winglike foils to lift the main hull out of the water, reducing drag

1909

cont

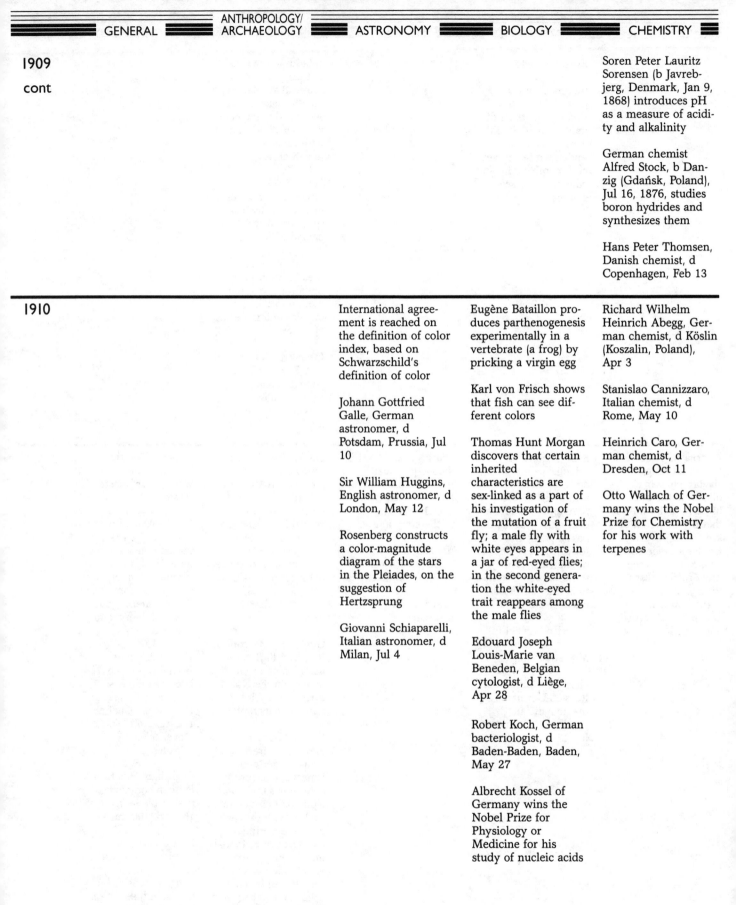

Soren Peter Lauritz Sorensen (b Javrebjerg, Denmark, Jan 9, 1868) introduces pH as a measure of acidity and alkalinity

German chemist Alfred Stock, b Danzig (Gdańsk, Poland), Jul 16, 1876, studies boron hydrides and synthesizes them

Hans Peter Thomsen, Danish chemist, d Copenhagen, Feb 13

1910

International agreement is reached on the definition of color index, based on Schwarzschild's definition of color

Johann Gottfried Galle, German astronomer, d Potsdam, Prussia, Jul 10

Sir William Huggins, English astronomer, d London, May 12

Rosenberg constructs a color-magnitude diagram of the stars in the Pleiades, on the suggestion of Hertzsprung

Giovanni Schiaparelli, Italian astronomer, d Milan, Jul 4

Eugène Bataillon produces parthenogenesis experimentally in a vertebrate (a frog) by pricking a virgin egg

Karl von Frisch shows that fish can see different colors

Thomas Hunt Morgan discovers that certain inherited characteristics are sex-linked as a part of his investigation of the mutation of a fruit fly; a male fly with white eyes appears in a jar of red-eyed flies; in the second generation the white-eyed trait reappears among the male flies

Edouard Joseph Louis-Marie van Beneden, Belgian cytologist, d Liège, Apr 28

Robert Koch, German bacteriologist, d Baden-Baden, Baden, May 27

Albrecht Kossel of Germany wins the Nobel Prize for Physiology or Medicine for his study of nucleic acids

Richard Wilhelm Heinrich Abegg, German chemist, d Köslin (Koszalin, Poland), Apr 3

Stanislao Cannizzaro, Italian chemist, d Rome, May 10

Heinrich Caro, German chemist, d Dresden, Oct 11

Otto Wallach of Germany wins the Nobel Prize for Chemistry for his work with terpenes

Frank B. Taylor proposes the continents move from place to place on Earth's surface and that Africa and South America were once joined; these ideas are essentially the same as those of Alfred Wegener in 1912, but Wegener gathered much evidence to support them

Bertrand Russell and Alfred North Whitehead start *Principia mathematica*, a three-volume work that will be completed in 1913; it attempts to derive all of mathematics from pure logic, an attempt that ultimately fails

Ernst Steinitz writes an early important paper on modern algebra; it is a study of abstract mathematical entities called fields, which are defined by a set of axioms

James Angell, a functionalist psychologist, proposes that the term *consciousness* should disappear from psychology, much as the term *soul* already had disappeared

Paul Ehrlich introduces "salvarsan" (arsphenamine), also known as 606, as a "magic bullet," or cure, for syphilis, the beginning of modern chemotherapy

William James, American psychologist, d Chocorua, NH, Aug 26

Howard Taylor Ricketts (b Findley, OH, Feb 9, 1871) demonstrates that Mexican typhus is transmitted by the body louse, as Charles Nicolle had shown for European typhus the year before

Howard Taylor Ricketts, American pathologist, d Mexico City, May 3, from typhus he catches during research on the transmission of the disease

Sydney Ringer, English physician, d Lastingham, Yorkshire, Oct 14

William Henry Bragg (b Cumberland, England, Jul 2, 1862) discovers how X rays and gamma rays cause rarified gases to conduct electricity as a result of the rays knocking electrons out of the gas molecules; he also notes that the X-ray and gamma-ray waves behave like particles in this process

Marie Curie's *Traité de radioactivité* (Treatise on radioactivity) is published

Amos Emerson Dolbear, American physicist, d Medford, MA Feb 23

J.J. Thomson uses positive cathode rays to measure atomic masses of several substances and discovers that neon has two isotopes, neon-20 and neon-22, the first confirmation that isotopes, predicted by Frederick Soddy, are possible

Johannes Diderik van der Waals of the Netherlands wins the Nobel Prize for Physics for the equation of state for gases

The Holt Company introduces the first gas-propelled combine harvester

William Grey Walter, American-British neurologist, b Kansas City, MO, Feb 19; he develops the "turtle," a precursor of modern robots

Electric washing machines are introduced

Rayon stockings for women are manufactured in Germany

Georges Claude (b Paris, Sep 24, 1870) introduces the neon light in Paris

Eugene Ely becomes the first person to take off in an airplane from the deck of a ship, showing that aircraft carriers are possible

Charles Proteus Steinmetz warns in *Future of electricity* about air pollution from burning coal and water pollution from uncontrolled sewage disposal into rivers

1910

1910 cont

1911

GENERAL

The National Biscuit Company (later Nabisco) introduces Oreos, which will become the world's most popular cookie

Roald Amundsen reaches the South Pole

ANTHROPOLOGY/ARCHAEOLOGY

The first fragments of the "Piltdown man" hoax are discovered in England

Franz Boas's *The mind of primitive man* is published

Sir Francis Galton, English anthropologist, d Haslemere, Surrey, Jan 17

ASTRONOMY

A meteorite the size of a basketball kills a dog in Nakhla, Egypt, the only recorded case of a meteorite killing a mammal; 75 years later, scientists decide that the meteorite originated on Mars

Halm discovers the relationship between the absolute magnitude and the masses of the stars

Ejnar Hertzsprung discovers that Polaris is a Cepheid variable

BIOLOGY

Alvar Gullstrand of Sweden wins the Nobel Prize for Physiology or Medicine for his work on refraction of light in the eye

Thomas Hunt Morgan explains the separation of certain inherited characteristics that are usually linked as caused by the breaking of chromosomes during the process of cell division

Thomas Hunt Morgan begins to map the positions of genes on chromosomes of the fruit fly

R.C. Punnet's *Mendelism* contains Punnet's famous diagram for showing heredity

Biochemist David Shemin b New York, NY, Mar 18; he uses carbon-14 as a tracer to determine chemical reactions in the body and to develop a method for synthesizing heme, part of the hemoglobin molecule

Alfred Henry Sturtevant (b Jacksonville, IL, Nov 21, 1891) produces the first chromosome map, showing five sex-linked genes

CHEMISTRY

William Burton introduces thermal cracking for refining petroleum

Marie Curie of France wins the Nobel Prize for Chemistry for her discovery of radium and polonium

Jacobus Henricus van't Hoff, Dutch physical chemist, d Berlin, Germany, Mar 1

Russian-British-Israeli chemist Chaim Weizmann, b Motol' (USSR), Nov 27, 1874, discovers how to obtain acetone from bacteria involved in fermenting grain, providing an essential ingredient to Britain for cordite during World War I

Francis Peyton Rous (b Baltimore, MD, Oct 5, 1879) discovers that some animal cancers are caused by viruses

Major Frank Woodbury of the US Army Medical Corps introduces the use of tincture of iodine as a disinfectant for wounds

---

L.E.J. Brouwer establishes that a mapping from one dimension to another cannot occur if the mapping is both one-to-one and continuous; Georg Cantor's famous mapping of the line to the square is not continuous, for example

Alfred Binet, French psychologist, d Paris, Oct 18

Harvey Cushing's *The pituitary body and its disorders* is published

London doctor William Hill develops the first gastroscope, a tube that can be swallowed by a patient so that the doctor may look at the inside of the patient's stomach through the tube

The first Solvay conference is held, bringing together many of the physicists involved in the new study of the atom

Ladislaw Natanson states that Planck's law is a result of the indistinguishability of light-quanta

Niels Henrik Bohr (b Copenhagen, Denmark, Oct 7, 1885) makes his first attempt to link atomic structure to Planck's constant

Dutch physicist Heike Kamerlingh Onnes discovers superconductivity in mercury cooled close to absolute zero

Owen Richardson develops the theory of the Edison effect: heated metals emit electrons, the basis of such vacuum tubes as diodes and triodes

Ernest Rutherford presents his theory of the atom, consisting of a positively charged nucleus surrounded by negative electrons, to the Manchester Literary and Philosophical Society on May 7

Frederick Soddy observes that whenever an atom emits an alpha particle, it changes to an element that is two places down in the

The first escalators are introduced at the Earl's Court underground station in London; a man with a wooden leg is hired to ride up and down regularly to demonstrate their safety

US inventor Charles Franklin Kettering (b Loudonville, OH, Aug 29, 1876) develops the first practical self-starter for automobiles

German scientist P. Monnartz becomes the first to realize that stainless steels resist corrosion

Elmer Ambrose Sperry (b Cortland, NY, Oct 12, 1860) invents the gyrocompass

1911
cont

1912

GENERAL	ANTHROPOLOGY/ ARCHAEOLOGY	ASTRONOMY	BIOLOGY	CHEMISTRY
Cracker Jacks adds a prize to each package The *Titanic* sinks on its maiden voyage and 1500 people perish Robert Falcon Scott, English explorer, d Antarctica, about Mar 29, after reaching the South Pole on Jan 17 and finding that Amundsen had beaten him by a month	Charles Dawson finds further fossils of "Piltdown Man;" the skull resembles that of modern man while the lower jaw has characteristics of an ape; although this accords with prejudices of the time, later discoveries make this combination seem unlikely; in 1953 it will be established that the fossil is a hoax	The age-old tradition of Chinese court astrologers observing the sky and recording events is abolished Sir George Howard Darwin, English astronomer, d Cambridge, Dec 7 Austrian-American Victor Franz Hess (b Schloss Waldstein, Austria, Jun 24, 1883) discovers that ionization of air increases with height, indicating the existence of what we now call cosmic rays, during a balloon ascent Studies of short-period variable stars by Henrietta Swan Leavitt (b Lancaster, MA, Jul 4, 1868) in the Small Magellanic Cloud lead to the period-luminosity law of Cepheid variables; that is, the luminosity of a Cepheid variable star can be determined from its period; this turns out to be the key to unlocking the distances to galaxies outside the Milky Way	Biochemist Sidney Walter Fox, b Los Angeles, CA, Mar 24; he attempts to show that life originated when intense heat caused amino acids to form polymers that become tiny spheres when dissolved in water Polish-American biochemist Casimir Funk (b Warsaw, Feb 23, 1884) coins the term *vitamin* for a class of substances that Frederick Hopkins had found to be important to health and which had previously been called accessory food factors Jacques Loeb's *The mechanistic conception of life* uses physics and chemistry to explain the origin of life Lloyd Morgan's *Instinct and experience* is published Eduard Adolf Strasburger, German botanist, d Poppelsdorf, May 19	Victor Grignard and Paul Sabatier of France win the Nobel Prize for Chemistry, Grignard for his discovery of Grignard reagents and Sabatier for his catalytic hydrogenation compounds Paul Lecoq de Boisbaudran, French chemist, d Paris, May 28

list of atomic masses (an important precursor of the concept of atomic number)

George Johnstone Stoney, Irish physicist, d London, Jul 5

John Archibald Wheeler, American physicist, b Jacksonville, FL, Jul 9; he is the inventor of the term *black hole* to describe an object so massive that nothing, not even light, can escape its gravitational attraction

Wilhelm Wien of Germany wins the Nobel Prize for Physics for determining the laws of radiation of black bodies

1911 cont

Earth Science

Clarence Edward Dutton, American geologist, d Englewood, NJ, Jan 4

Thomas A. Jagger is the first scientist to work at the newly opened Hawaiian Volcano Observatory, situated on the rim of the Kilauea caldera

Alfred Lothar Wegener (b Berlin, Germany, Nov 1, 1880) first proposes his theory of continental drift and the supercontinent of Pangaea

Mathematics

L.E.J. Brouwer's *Intuitionism and formalism* details the principles of the intuitionist movement in mathematics, a movement to base mathematics on the natural numbers, which are understood intuitively, and to reject mathematical entities that cannot be actually constructed

Arnaud Denjoy develops the "concept of totalization" and the Denjoy integral, a generalization of the Lebesgue integral

Jules-Henri Poincaré, French mathematician, d Paris, Jul 17

Medicine

Alexis Carrel of France wins the Nobel Prize for Physiology or Medicine for vascular grafting of blood vessels

Joseph, Baron Lister, English surgeon, d Walmer, Kent, Feb 10

Ernest Henry Starling's *Principles of human physiology* becomes the standard textbook on the subject

Physics

Max von Laue demonstrates that X rays are a form of electromagnetic radiation by creating diffraction patterns with crystals; his work also shows that crystals do have the periodic structure of molecules that had been predicted

William Henry Bragg, working with his son William Lawrence (b Adelaide, Australia, Mar 31, 1890), measures the wavelength of X rays using the structure of crystals to scatter the X rays; they are inspired by the work of Max von Laue who uses the same technique to study crystals earlier in the year

Nils Gustaf Dalen of Sweden wins the Nobel Prize for Physics for inventing automatic gas regulators for lighthouses and sea buoys

Technology

American doctor Sidney Russell invents the heating pad, which later grows (in 1930) to become the electric blanket

Wilbur Wright, American inventor, d Dayton, OH, May 30

1912

1912 cont

ASTRONOMY

Vesto Melvin Slipher (b Mulberry, IN, Nov 11, 1862) is the first to obtain the spectrum of a distant galaxy, the Andromeda galaxy

BIOLOGY

Heinrich Otto Wieland (b Pforzheim, Germany, Jun 4, 1877) begins a study of bile acids that reveals that they are steroids based on cholesterol

1913

ASTRONOMY

The Harvard Classification of stars is adopted by the International Solar Union

Guthnick, Rosenberg, Meyer, Kunz, and Stibbins introduce photoelectric photometry in astronomy

Enjar Hertzsprung is the first to use Cepheid variable stars to estimate distances to the stars

Henry Norris Russell (b Oyster Bay, NY, Oct 25, 1877) announces his theory of stellar evolution; independently discovered by Hertzsprung in 1905, this concept becomes the famous Hertzsprung-Russell diagram of how stars change with time

BIOLOGY

Biophysicist Britton Chance, b Wilkes-Barre, PA, Jul 24; his work with the enzyme peroxidase in the 1940s results in confirmation of the theory that enzymes work by binding briefly to their substrate (the material that the enzyme catalyzes)

Archibald Vivian Hill (b Bristol, England, Sep 26, 1886) discovers that muscle cells respire, or use oxygen, after a contraction is finished, not while the contraction is taking place

Elmer Verner McCollum (b near Fort Scott, KS, Mar 3, 1879) and Marguerite Davis identify a fat-soluble vitamin later called vitamin A to distinguish it from the water-soluble vitamin discovered by Christiaan Eijkman, termed vitamin B

High-school teacher Johann Regan ingeniously uses the newly available telephone to determine whether it is true that a male cricket's call is a mating signal to a female cricket; when a male cricket chirps over the telephone, the female immediately heads for the ear-

CHEMISTRY

German chemist Friedrich Karl Bergius, b Goldschmieden (Poland), Oct 11, 1884, introduces hydrogenation of coal at high pressure for producing gasoline

Kasimir Fajans discovers the chemical element protactinium

György Hevesy (b Budapest, Hungary, Aug 1, 1885) and Friedrich A. Paneth (b Vienna, Austria, Aug 31, 1887) use radioactive lead to trace the solubility of lead salts, demonstrating that even a small amount of the chemical can be used as a radioactive tracer

German-American chemist Leonor Michaelis (b Berlin, Jan 16, 1875) and his assistant work out the Michaelis-Menten equation that describes the rate at which enzymes catalyze reactions

Fritz Pregl demonstrates that he can chemically analyze samples of organic materials as small as 3 g (0.1 oz)

Theodore William Richards (b Germantown, PA, Jan 31, 1868), who has

| | | | Pyotr Nicolaievich Lebedev, Russian physicist, d Moscow Apr 1 | **1912** |
| | | | | **cont** |

Dayton Miller (b Strongsville, OH, Mar 13, 1866) invents the photodeik, a device that makes sound visible as patterns of light

Frederick Soddy's *Matter and energy* is published

1913

EARTH SCIENCE

Physicist Charles Fabry (b Marseilles, France, Jun 11, 1867) discovers that there is an ozone layer in the upper atmosphere

John Milne, English geologist, d Shide, Isle of Wight, Jul 30

Léon-Philippe Teisserenc de Bort, French meteorologist, d Cannes, Alpes-Maritimes, Jan 2

MATHEMATICS

Srinivasa Ramanujan (b Tanjore district, Madras, India, Dec 22, 1887) writes on Jan 16 to English mathematician G.H. Hardy and encloses a number of remarkable results in infinite series that he has developed; Hardy soon invites the self-taught Ramanujan to England, where he keeps up his flow of new ideas

MEDICINE

John Jacob Abel (b Cleveland, OH, May 19, 1857) develops the first artificial kidney

Alfred Adler develops a theory of human personality based on power, as opposed to Sigmund Freud's theory, based on sex

Johannes Andreas Grib Fibiger (b Apr 23, 1867) induces cancerous growths in rats

Frank Mallory discovers the bacterium causing whooping cough

Charles Robert Richet of France wins the Nobel Prize for Physiology or Medicine for his work on anaphylaxis allergy

German surgeon A. Salomen develops mammography for diagnosing breast cancer

Bela Schick introduces the Schick test for diphtheria

John Broadus Watson (b Greenville, SC, Jan 9, 1878) publishes his first paper on behaviorism in *Psychological Review*; he argues for a psychology based only on the observation of behavior

PHYSICS

Antonius Johannes van den Broek notes that atomic mass and the number of electrons in an atom are independent

Niels Bohr's trilogy on the constitution of atoms and molecules is written; Bohr assumes that electrons orbit the nucleus in fixed orbits and give off or absorb fixed amounts of energy (quanta) by jumping from one orbit to another

Bohr proposes that beta-decay is a nuclear process

Louis-Paul Cailletet, French physicist, d Paris, Jan 5

Hans Geiger and Ernst Marsden establish that Rutherford's picture of an atomic nucleus surrounded by electrons is correct; they direct a beam of alpha particles on a gold foil and find that a small number are reflected backward, proving the existence of a dense nucleus in the atom

Heike Kamerlingh Onnes of the Netherlands wins the Nobel Prize for Physics for his work

TECHNOLOGY

Vacuum triodes are used in repeaters in telephone lines for the amplification of weak signals

Fritz Haber develops nitrogen fixation from air in Germany

The cascade-tuning radio receiver and the heterodyne radio receiver are introduced

The first home refrigerator goes on sale in Chicago

William David Coolidge (b Hudson, MA, Oct 23, 1873) invents a hot-cathode X-ray tube

Rudolf Diesel, German inventor, d English Channel, Sep 30

Henry Ford (b Greenfield Village, MI, Jul 30, 1863) introduces the first true assembly line, where cars are carried along a conveyer belt at a speed slow enough for workers to assemble them, but fast enough to reduce the assembly time from 12.5 to 1.5 hours

Hans Geiger invents a radiation detector for alpha particles

1913

cont

piece, confirming the hypothesis

Alfred Russel Wallace, English naturalist, d Broadstone, Dorset, Nov 7

specialized all his life in determining atomic masses with great accuracy, finds that the element lead from different minerals can have different atomic masses; this result supports the theory of isotopes announced by Frederick Soddy earlier in the year

Alfred Werner of Switzerland wins the Nobel Prize for Chemistry for his study of the bonding of atoms

in low-temperature physics and liquefaction of helium

Robert Andrews Millikan (b Morrison, IL, Mar 22, 1868) completes a seven-year study of the value of the charge of the electron (the famous "oil drop" experiment), which contributes to the Nobel Prize he wins in 1923

Henry Gwyn-Jeffreys Moseley (b Weymouth, England, Nov 23, 1887) deduces the place of elements in the periodic table from their X-ray spectra and formulates his law of numbers of electrons: an element has the same number of electrons as its atomic number

Frederick Soddy, K. Fajans, and A.S. Russell note that when an atom emits a beta particle, it changes into an atom one place up in a list of atoms by atomic mass

Soddy coins the term *isotope* on Feb 18 to describe atoms of the same element that have different atomic masses

Johannes Stark (b Schickenhof, Germany, Apr 15, 1874) discovers that spectral lines of a light source split when the light source is placed in an electric field (the Stark effect)

Karl Fritiof Sundmann finds a theoretical solution to the three-body problem, destroying the generally held belief that such a solution is impossible

American engineer Frederick Kolster (b Geneva, Switzerland, Jan 13, 1883) develops a radio-compass system with transmitters on the New Jersey coast

Irving Langmuir (b Brooklyn, NY, Jan 31, 1881) develops an improvement in the tungsten lamp—filling the bulb with an inert gas so that atoms of tungsten will evaporate more slowly from the filament

A. Meissner invents a radio transmitter with vacuum tubes

Igor Sikorsky builds and flies a multi-engine airplane

1913

cont

	GENERAL	ANTHROPOLOGY/ARCHAEOLOGY	ASTRONOMY	BIOLOGY	CHEMISTRY
1914	The assassination of Austrian archduke Francis Ferdinand and his wife on Jun 28 in Sarajevo (Yugoslavia) precipitates World War I		American astronomer Walter Sydney Adams (b Kessab, Turkey, Dec 20, 1876) introduces the misnamed method of spectroscopic parallax; it is a method of deducing a star's real magnitude from its spectrum, which can then be compared with its apparent magnitude to obtain an approximation of its distance Arthur Stanley Eddington (b Kendal, England, Dec 28, 1882) suggests in *Stellar movements and the structure of the universe* that spiral nebulas are galaxies Sir David Gill, Scottish astronomer, d London, Jan 24 Harlow Shapley's "On the nature and cause of Cepheid variation" in the *Astrophysical Journal* explains that Cepheid stars vary in brightness because of pulsations; Shapley also correlates absolute magnitude with period in Cepheid variables Vesto M. Slipher discovers evidence of interstellar dust clouds by studying the spectrum of the nebula NGC 7023	Henry Hallet Dale (b London, Jun 9, 1875) proposes that acetylcholine is a compound involved in the transmission of nerve impulses; he isolates it in 1929 Edwin Calvin Kendall (b South Norwalk, CT, Mar 8, 1886) identifies thyroxin, a hormone secreted by the thyroid gland Fritz Albert Lipmann (b Kalingrad, USSR, Nov 12, 1899) elucidates the role of adenine triphosphate (ATP): it forms the link between energy generation and energy utilization in the cell John Broadus Watson's *Behavior: An introduction to comparative psychology* proposes the use of animal experiments in psychology August Friedrich Weismann, German biologist, d Freiburg-im-Breisgau, Baden, Nov 5	P. Duden and J. Hess synthesize acetic acid Herman Frasch, German-American chemist, d Paris, May 1 Charles Martin Hall, American chemist, d Daytona Beach, FL, Dec 27 Johann Wilhelm Hittorf, German chemist and physicist, d Münster, Rhenish Prussia, Nov 28 Theodore Richards of the United States wins the Nobel Prize for Chemistry for his determination of atomic weights
1915	The Panama Canal is formally opened on Jul 12, although ships had been using it since Aug 15, 1914	Joseph Hatzidhakis begins the excavation of the Cretan palace at Mallia, the third great palace known from the Minoan civilization	Walter Sydney Adams measures the temperature of the surface of Sirius B; it is 2000°C hotter than the sun, although Sirius B is nearly invisible from Earth; combined with evidence of its mass, this implies that Sirius B is a tiny, extremely dense, hot star—what we now know as a white dwarf	Theodor Boveri, German cytologist, d Würzburg, Bavaria, Oct 15 Friedrich August Löffler, German bacteriologist, d Berlin, Apr 9 Thomas Hunt Morgan, Calvin Bridges, Alfred Sturtevant, and Hermann Muller's *The mechanism of Mendelian heredity* is a classic in genetics	John Ulric Nef, Swiss-American chemist, d Carmel, CA, Aug Richard Willstätter of Germany wins the Nobel Prize for Chemistry for his research on chlorophyll in plants

1914

German-American geologist Beno Gutenberg (b Darmstadt, Germany, Jun 4, 1889) discovers a discontinuity in the behavior of earthquake waves at a depth of about 3000 km (1860 mi); this *Gutenberg discontinuity* marks the boundary between Earth's mantle and the outer core

Sir John Murray, British oceanographer, d Mar 16

Eduard Suess, Austrian geologist, d Marz, Burgenland, Apr 26

Felix Hausdorff's *Gründzuge der Mengenlehre* (Basic features of set theory) introduces the concept of topological spaces

Roger Adams synthesizes Adamsite (phenarsazine chloride), a substance that causes sneezing

Robert Bárány of Austria (b Vienna, Apr 22, 1876) wins the Nobel Prize for Physiology or Medicine for his study of inner ear function and pathology

Alexis Carrel performs the first successful heart surgery on a dog

James Chadwick (b Manchester, England, Oct 20, 1891) detects the continuous beta-spectrum

James Franck (b Hamburg, Germany, Aug 26, 1882) and Gustav Hertz (b Hamburg, Germany, Jul 22, 1887) confirm Niels Bohr's model of the atom by bombarding mercury vapor with electrons and measuring the frequencies of the emitted radiation, which correspond to the transitions between different energy levels

Max von Laue of Germany wins the Nobel Prize for Physics for the crystal diffraction of X rays

Ernest Rutherford discovers the proton

Sir Joseph Wilson Swan, English physicist and chemist, d Warlingham, Surrey, May 27

Bernard Vonnegut, American physicist, b Indianapolis, IN, Aug 29; he is most noted for his discovery that silver iodide can be used to produce precipitation when seeded into clouds

The first modern sewage plant, designed to treat sewage with bacteria, opens in Manchester, England

Red and green traffic lights are introduced for the first time, in Cleveland, OH

The brassiere is patented

Radio transmitter triode modulation is introduced

Edwin Howard Armstrong (b New York, NY, Dec 18, 1890) patents a radio receiver circuit with regeneration (positive feedback)

Engineer Robert Hutchings Goddard (b Worcester, MA, Oct 5, 1882) starts developing experimental rockets

Paul-Louis-Toussaint Héroult, French metallurgist, d off the coast of Antibes, May 9

Edward Kleinschmidt invents the teletypewriter

George Westinghouse, American engineer, d New York, NY, Mar 12

1915

Alfred Wegener's *Die Entstehung der Kontinente und Ozeane* (The origin of continents and oceans) gives several lines of evidence for his 1912 proposal of drifting continents and the supercontinent of Pangaea

Carlos J. Finlay d Aug 20

Austrian-American physician Joseph Goldberger, b Girált (Czechoslovakia), Jul 16, 1874, establishes that a vitamin deficiency causes pellagra

Arnold Sommerfeld introduces the fine structure constant

Emile-Hilaire Amagat, French physicist, d Saint-Satur, Feb 15

Sir William Henry Bragg and Sir William Lawrence Bragg of England win the Nobel Prize for Physics for their study of crystals with X rays

The radio tube oscillator is introduced

The Corning Glass Works develops Pyrex, a heat-resistant glass

The first transatlantic radiotelephone conversation takes place between Arlington, VA, and the Eiffel Tower in Paris

1915

cont

Robert Innes (b Edinburgh, Scotland, Nov 10, 1861) discovers a faint companion to the double star system Alpha Centauri; the companion is currently the closest known star to Earth except the sun; therefore, it is often called Proxima (nearest) Centauri

R.C. Punnet's *Mimicry in butterflies* contains the first mathematical analyses of the effects of selection on populations that obey Mendelian laws of heredity

D'Arcy Thompson's *On growth and form* is the first important study of how differential growth rates or different sizes can affect organisms

Frederick William Twort (b Camberley, England, Oct 22, 1877) discovers viruses that prey on bacteria

1916

Edward Emerson Barnard discovers "a small star with larger proper-motion," meaning that it appears to move across the sky much faster than normal; later astronomers call the the small star Barnard's runaway star, or simply Barnard's star; it has the largest proper motion of any star ever found

Percival Lowell, American astronomer, d Flagstaff, AZ, Nov 12

Félix-Hubert D'Hérelle (b Montreal, Canada, Apr 25, 1873) discovers viruses that prey on bacteria independently of Twort's discovery; D'Hérelle names them *bacteriophages*, or "bacteria eaters;" today they are often called phages

Ilya Ilich Mechnikov, Russian-French bacteriologist, d Paris, Jul 15

Kotaro Honda (b Aichi-jen Prefecture, Japan, Feb, 1870) discovers that adding cobalt to tungsten steel produces increased strength when the alloy is magnetized, leading to the development of various powerful magnetic alloys, culminating in alnico

Gilbert Newton Lewis (b West Newton, MA, Oct 25, 1875) explains chemical bonding and valence of chemical elements by his theory of shared electrons; in his *The atom and the molecule* he

EARTH SCIENCE	MATHEMATICS	MEDICINE	PHYSICS	TECHNOLOGY	
			William and Lawrence Bragg's *X rays and crystal structure* describes their use of X rays to determine the structure of crystals, an important technique that continues to be useful today	Fokker aircraft become the first airplanes to be equipped with a machine gun that can fire between the blades of a moving propeller	**1915** cont
			Albert Einstein completes his theory of gravitation, known as the general theory of relativity, on Nov 25; he had been working on the mathematical problems presented by the theory since 1911; the theory is submitted in final form to *Annalen der Physik* on Mar 20, 1916	The first North American transcontinental telephone call is between Alexander Graham Bell in New York and Thomas A. Watson in San Francisco	
				American physicist Manson Benedicks discovers that a germanium crystal can convert AC current to DC, a discovery that leads to the development of the "chip"	
			Irving Langmuir invents a mercury-vapor condensation pump that creates a nearly perfect vacuum	Detroit blacksmith August Fruehauf invents the tractor trailer	
			Henry Gwyn-Jeffreys Moseley, English physicist, d in action at Gallipoli, Turkey, Aug 10	French scientist Paul Langevin (b Paris, Jan 23, 1872) invents sonar, mainly as a method for ships to detect icebergs	
				Frederick Winslow Taylor, American engineer, d Philadelphia, Mar 21	
Cleveland Abbe, American meteorologist, d Chevy Chase, MD, Oct 28	Julius Wilhelm Richard Dedekind, German mathematician, d Braunschweig, Feb 12	Sigmund Freud's *Vorlesungen zur Einführung in die Psychoanalyse* is published	Peter J.W. Debye (b Maastricht, Netherlands, Mar 24, 1884) uses powdered crystals to study their structure by observing the X-ray diffraction patterns (powder crystallography)	Windshield wipers for automobiles are introduced in the United States	**1916**
		Albert Ludwig Neisser, German physician, d Breslau (Wroclaw, Poland), Jul 30	Friedrich Ernst Dorn, German physicist, d Halle, Jun 13	American engineer John Fisher develops the first modern washing machine	
			Ernst Mach, Austrian physicist, d Vaterstetten, Germany, Feb 19	Sir Hiram Stevens Maxim, American-English inventor, d London, Nov 24	
			Robert Andrews Millikan confirms Planck's constant (one of the fundamental constants in physics)		

1916
cont

CHEMISTRY (1916 cont)

introduces the even-number rule: the number of electrons in compounds is almost always even

Sir William Ramsay, Scottish chemist, d High Wycombe, Buckinghamshire, Jul 23

1917

GENERAL

The National Research Council is established by the US National Academy of Sciences

The Russian Revolution occurs in Oct as communists assume control of the government

ASTRONOMY

Heber Doust Curtis (b Muskegon, MI, Jun 27, 1872) determines the distance to the Andromeda galaxy

Albert Einstein's "Cosmological considerations on the general theory of relativity" introduces a model of a static universe and his "cosmological constant" that keeps it static; he later calls the cosmological constant his worst mistake, as evidence suggests that the universe is expanding

George Ellery Hale installs a 2.5-m (100-in.) refracting telescope at Mount Wilson Observatory; it will be the world's largest telescope until 1948, but will be retired from duty in 1987 because of light and air pollution around the observatory

James Hopwood Jeans develops the theory that a passing star might have caused the creation of the planets and other bodies of the solar system as a result of its gravitational attraction (now thought unlikely)

Karl Schwarzschild develops the equations that predict the

BIOLOGY

Emil Adolf von Behring, German bacteriologist, d Marburg, Rhenish Prussia, Mar 31

Wilhelm Johannsen's *Erblichkeit* (Heredity) is published

Plough demonstrates the rearrangement of chromosomes known as "crossing over"

CHEMISTRY

Johann Friedrich Wilhelm Adolf von Baeyer, German chemist, d Starnberg, Bavaria, Aug 20

Eduard Buchner, German chemist, d Focsani, Rumania, Aug 13

James Mason Crafts, American chemist, d Ridgefield, CT, Jun 20

EARTH SCIENCE	MATHEMATICS	MEDICINE	PHYSICS	TECHNOLOGY	
			using the photoelectric effect; he also confirms Einstein's theoretical work of 1905 that explains the photoelectric effect Arnold Sommerfeld, b Königsberg (Kalingrad, USSR), Dec 5, 1868, develops the second quantum number by modifying Niels Bohr's model of the atom, in which electrons travel in circular orbits, to a model in which the orbits are elliptical		**1916** **cont**
	G.H. Hardy and Srinivasa Ramanujan write on theory of numbers Tullio Levi-Civita (b Padua, Italy, Mar 29, 1873) introduces the concept of parallel displacement to the absolute differential calculus	The natural anti-coagulant heparin is discovered Surgeon Emil Theodor Kocher d Jul 27 Ralph Parker develops a vaccine against Rocky Mountain spotted fever Margaret Sanger and M.C. Stopes write on birth control *Psychology of the unconscious* by Carl Gustav Jung (b Kesswil, Switzerland, Jul 26, 1875) is published	Charles Barkla of England wins the Nobel Prize for Physics for work on X-ray scattering by elements	Clarence Birdseye develops freezing as a way of preserving foods Ferdinand Adolf August Heinrich, Count von Zeppelin, German inventor, d Charlottenburg, Prussia, Mar 8	**1917**

1917 cont

existence of black holes from Einstein's equations for the general theory of relativity

Willem de Sitter (b Sneek, Netherlands, May 6, 1872) demonstrates that Einstein's theory of general relativity shows that the universe must be expanding (provided Einstein's "correction factor" is left out); this calculation is later modified by Alexander Friedmann and Arthur Eddington

1918

World War I ends on Nov 11 when the Germans surrender

Leonard Woolley (b London, April 17, 1880), later Sir Charles Leonard Woolley, starts excavations that lead to the uncovering of Ur

Geschichte und Literatur des Lichtwechsels der bis Ende 1915 als sicher veränderlich anerkannte Sterne (History and literature of light variation of stars known by the end of 1915 to be definitely variable) is published

Robert G. Aitken's *The binary stars* is published

Arthur Stanley Eddington gives a theoretical basis to the pulsation theory for Cepheids, first formulated by Harlow Shapley in 1914

Eddington's *Report on the relativity theory of gravitation* is a general discussion of Einstein's general theory of relativity; it forms the basis for his 1923 book *Mathematical theory of relativity*

Harlow Shapley makes the first accurate estimate of the size of the Milky Way galaxy; his paper "Remarks on the arrangement of the sidereal universe" places the solar system near the outer edge of the galaxy

Jesse M. Coulter's *Plant genetics* is published

Herbert M. Evans finds (incorrectly) that human cells contain 48 chromosomes

Georg Theodor August Gaffky, German bacteriologist, d Hanover, Oct 23

Fritz Haber of Germany wins the Nobel Prize for Chemistry for his synthesis of ammonia from atmospheric nitrogen

Jaroslav Heyrovský (b Prague, Czechoslovakia, Dec 20, 1890) begins to develop polarography, a delicate system for measuring the concentration of ions in a solution

Georg Cantor, German mathematician, d Halle, Jan 6

Karl Ferdinand Braun, German physicist, d New York, NY, Apr 20

English physicist Francis William Aston (b Harborne, England, Sep 1, 1877) builds the first mass spectrograph, with which he discovers that there are different forms of the same element, now called isotopes; different isotopes have the same chemical properties, but the masses of individual atoms are not the same

Emmy Noether proves that every symmetry in physics implies a conservation law and vice versa; eventually this becomes a powerful tool in developing new physical theories

Max Planck of Germany wins the Nobel Prize for Physics for his quantum theory of light

The first radio link between England and Australia is opened

The radio crystal oscillator is introduced

The electric beater for mixing foods is introduced

Edwin H. Armstrong develops the superheterodyne radio receiver

Bell invents a high-speed hydrofoil boat

1918

	GENERAL	ANTHROPOLOGY/ ARCHAEOLOGY	ASTRONOMY	BIOLOGY	CHEMISTRY
1919		F. Wood Jones argues in "The origin of man" that humans diverged from the great apes at a stage similar to that of the modern tarsier; in this view no common ancestor of humans could be like a monkey or a great ape	The International Astronomical Union is founded Edward Emerson Barnard's catalog of dark nebulas is published Ernest W. Brown's improved lunar tables are based on 30 years of work and his theory of lunar motion Arthur Stanley Eddington and co-workers observe the solar eclipse of May 29 in an effort to determine whether Einstein's general theory of relativity or Newton's theory of gravity best predicts how light from a star will be affected by the gravitational attraction of the sun; Einstein predicts twice as much bending as Newton On Nov 6 at a joint meeting of the Royal Society and the Royal Astronomical Society in London, the astronomer royal, Sir Frank Watson Dyson (b Jan 8, 1868), announces that the photographs of the eclipse of the sun taken by the Eddington expedition confirm Einstein's theory of gravitation, not Newton's Edward Charles Pickering, American astronomer, d Cambridge, MA, Feb 3	Jules Bordet of Belgium wins the Nobel Prize for Physiology or Medicine for his studies in immunology Karl von Frisch discovers that bees communicate through body movements Ernst Heinrich Haeckel, German naturalist, d Jena, Thuringia, Aug 8 Thomas Hunt Morgan's *The physical bases of heredity* is published Mikhail Semenovich Tsvett, Russian botanist, d Voronezh, Jun 26	Frederick G. Cottrell invents a technique for separating helium from natural gas Emil Hermann Fischer, German chemist, d Berlin, Jul 15 Irving Langmuir develops his theory of covalent bonding of atoms Alfred Werner, German-Swiss chemist, d Zurich, Switzerland, Nov 15
1920	Robert Edwin Peary, American explorer, d Washington, DC, Feb 20		Sir William de Wiveleslie Abney, English astronomer, d Folkestone, Kent, Dec 2 Eddington's *Space, time, and gravitation* is published	Dmitri Iosifovich Ivanovsky, Russian botanist, d USSR, Jun 20 Shack August Krogh of Denmark wins the Nobel Prize for	Walther Nernst of Germany wins the Nobel Prize for Chemistry for his study of thermodynamics

1919

Joseph Larmor develops a theory explaining the magnetic fields of Earth and the sun by assuming circular motion in those bodies functioning as a self-exciting dynamo

Bertrand Russell's *Introduction to the philosophy of mathematics* is written while Russell is in prison during World War I for his pacifist ideas

John Broadus Watson's *Psychology from the standpoint of a behaviorist* advocates behavioral conditioning for research

Sir William Crookes, English physicist, d London, Apr 4

Roland, Baron von Eötvös, Hungarian physicist, d Budapest, Apr 8

John William Strutt, 3rd Baron Rayleigh, English physicist, d Witham, Essex, Jun 30

Ernest Rutherford reports on his discovery of two years earlier that alpha particles striking nitrogen can knock out protons, the first form of artificial atomic fission; the experiment also proves that protons are constituents of the nucleus of the atom

Johannes Robert Rydberg, Swedish physicist, d Lund, Malmöhus, Dec 28

Wallace Clement Ware Sabine, American physicist, d Cambridge, MA, Jan 10

Johannes Stark of Germany wins the Nobel Prize for Physics for his study of the spectral lines in electrical fields

Shortwave radio is developed

A method of reaching extreme altitudes by Robert Hutching Goddard suggests sending a small vehicle to the moon using rockets; adverse press comments make Goddard disinclined to expose himself to further ridicule

W.H. Eccles and F.W. Jordan publish a paper on flip-flop circuits

1920

Geologist Charles Lapworth d Mar 13

Srinivasa Ramanujan d Apr 26

William Crawford Gorgas, American army surgeon, d London, Jul 3

Wilhelm Max Wundt, German psychologist, d Grossbathen, Saxony, Aug 31

Francis William Aston discovers that all atomic masses, when isotopes are taken into account, are integral multiples of the same number (a number now taken to be one-twelfth the mass of carbon-12)

The first regular licensed radio broadcast takes place

John Wesley Hyatt, American inventor, d Short Hills, NJ, May 10

1920 cont

ASTRONOMY

Sir Joseph Norman Lockyer, English astronomer, d Salcombe Regis, Devonshire, Aug 16

Albert Michelson measures the diameter of the star Betelgeuse using a stellar interferometer; it is the first measurement of the diameter of any star other than the sun

After years of study, Vesto M. Slipher announces the presence of red shifts in the spectra of galaxies

German-American astronomer Walter Baade (b Schrötting-hausen, Germany, Mar 24, 1893) discovers the asteroid Hidalgo

BIOLOGY

Physiology or Medicine for his discovery of the motor mechanism of blood capillaries

Wilhelm Pfeffer, German botanist, d Kassel, Hesse, Jan 31

1921

ASTRONOMY

Henrietta Swan Leavitt, American astronomer, d Cambridge, MA, Dec 12

Edward Arthur Milne (b Hull, Yorkshire, England, Feb 14, 1896) develops his theory of solar atmosphere, determining the sun's temperature in various layers and predicting the solar wind

Meghnad Saha gives the relationship between temperature and degree of ionization (obtained from the spectrum) in stellar atmospheres

BIOLOGY

German-American physiologist Otto Loewi (b Frankfurt-am-Main, Germany, Jun 3, 1873) discovers that chemicals can stimulate nerves to fire and develops his theory of chemical transmission of nervous impulses

Frederick Hopkins discovers glutathione, a compound of three amino acids required for the utilization of oxygen by the cell

CHEMISTRY

Thomas Midgley, Jr. (b Beaver Falls, PA, May 18, 1889) discovers that tetraethyl lead prevents knock in gasoline engines

Frederick Soddy of England wins the Nobel Prize for Chemistry for his discovery of isotopes of radioactive elements

EARTH SCIENCE	MATHEMATICS	MEDICINE	PHYSICS	TECHNOLOGY	
			Nicolaas Bloembergen, Dutch-American physicist, b Dordrecht, Netherlands, Mar 11	Retired US Army officer John T. Thompson patents his submachine gun, later famous as the "tommy gun"	**1920** **cont**
			Niels Bohr establishes the Copenhagen Institute of Theoretical Physics		
			Johann Philipp Ludwig Elster, German physicist, d Bad Harzburg, Saxony, Apr 6		
			Charles Guillaume of Switzerland wins the Nobel Prize for Physics for his discovery of invar, a special nickel-steel alloy that resists change in volume with change of temperature, useful in many applications		
			William Draper Harkins (b Titusville, PA, Dec, 28, 1873) proposes the existence of the neutron, an uncharged particle that is discovered in 1932 by James Chadwick		
			Augusto Righi, Italian physicist, d Bologna, Jun 8		
Vilhelm Bjerknes's *On the dynamics of the circular vortex with applications to the atmosphere and to the atmospheric vortex and wave motion* shows that the atmosphere is made from sharply differentiated air masses	John Maynard Keynes's *Treatise on probability* is published Emmy Noether's "The theory of ideals in rings" develops the axiomatic approach to algebra	Frederick Grant Banting (b Alliston, Ontario, Canada, Nov 14, 1891), Charles Best (b West Pembroke, ME, Feb 27, 1899), John McLeod, and James Collip extract insulin from the human pancreas and start experiments on dogs in an effort to develop a treatment for diabetes Alexander Fleming (b Lochfield, Scotland, Aug 6, 1881) discovers the antibacterial substance lysozyme in saliva, mucus, and tears; as with his later discovery of penicillin, a fortuitous accident was involved; he had a cold and some of his mucus	Charles Drummond Ellis shows that Einstein's explanation of the photoelectric effect can explain some forms of beta radiation (the production of electrons by radioactive atoms) Arthur Holly Compton (b Wooster, OH, Sep 10, 1892) conjectures that ferromagnetism is due to the rotation of an electron around its own axis Albert Einstein of Germany wins the Nobel Prize for Physics for his discovery of the law of photoelectric effect	Cultured pearls are introduced Playwright Karl Câpek coins the term *robot* to describe the mechanical people in his play *RUR* Albert W. Hull invents the magnetron, an electron tube that produces microwaves Canadian-American medical student John Augustus Larson (b Shelbourne, Nova Scotia, Dec 11, 1892) invents the polygraph (lie detector)	**1921**

1921

cont

1922

GENERAL
Benito Mussolini takes power in Italy

ASTRONOMY
Alexander A. Friedmann, b St. Petersburg (Leningrad, USSR), Jun 29, 1888, on the basis of Albert Einstein's general theory of relativity, predicts that the universe must be expanding, improving on a similar prediction in 1919 by Willem de Sitter and paving the way for a further revision in 1930 by Arthur Stanley Eddington

Jacobus Cornelius Kapteyn d Amsterdam, Netherlands, Jun 18

John Stanley Plaskett (b Hickson, Ontario, Nov 17, 1865) discovers that a star previously thought to be single is actually a pair of massive stars revolving about each other; for the next 50 years these are the most massive stars known

BIOLOGY
Herbert McLean Evans and K.J. Scott discover vitamin E

Archibald V. Hill of England and Otto Meyerhof of Germany share the Nobel Prize for Physiology or Medicine, Hill for his discovery of muscle heat production and Meyerhof for his discovery of oxygen-lactic acid metabolism

Elmer McCollum discovers vitamin D in cod liver oil and uses it for treating rickets

CHEMISTRY
Francis William Aston of England wins the Nobel Prize for Chemistry for his discovery of isotopes with mass spectrograph and related work with atomic masses

Ernest Solvay, Belgian chemist, d Brussels, May 26

Jokichi Takamine, Japanese-American chemist, d New York, NY, Jul 22

1921 cont

dripped onto a culture plate, where it dissolved the bacteria present

Ernst Kretschmer's *Physique and character* suggests that the mental state of a person is related to his body build

Hermann Rorschach (b Zurich, Switzerland, Nov, 8, 1884) introduces his inkblot test for the study of personality in *Psychodiagnostics*

Heinrich Wilhelm Gottfried von Waldeyer-Hartz, German anatomist, d Berlin, Jan 23

Alfred Landé introduces half-integer quantum numbers; the two previous quantum numbers known were assumed to be always integral (whole numbers)

Gabriel Jonas Lippmann, French physicist, d aboard ship on the Atlantic Ocean, Jul 13

1922

The German *Meteor* oceanographic expedition, designed to recover gold from the oceans to pay Germany's war debts (an unsuccessful enterprise), is equipped with sonar that can measure the depths of the oceans

Lewis Fry Richardson's *Weather prediction by numerical process* gives equations for the prediction of rainfall and transfer of heat and moisture; these will become practical after the invention of the computer

Abraham A. Fraenkel improves the axioms of set theory developed by Ernst Zermelo by making it clearer how a set can be defined by a definite property

Frederick Banting and Charles Best's *Internal secretions of the pancreas* is published

Stanley Hall's *Senescence* is the first large-scale geriatric survey of a psychological nature

William Stewart Halsted, American surgeon, d Baltimore, MD, Sep 7

Charles-Louis-Alphonse Laveran, French physician, d Paris, May 18

Sir Patrick Manson, Scottish physician, d London, Apr 9

Hermann Rorschach, Swiss psychiatrist, d Herisau, Apr 2

Francis William Aston's *Isotopes* is published

Niels Bohr of Denmark wins the Nobel Prize for Physics for his work on atomic structure and radiation

Alexander Graham Bell, Scottish-American inventor, d Beinn Bhreagh, Nova Scotia, Aug 2

1923

Edward Emerson Barnard, American astronomer, d Williams Bay, WI, Feb 6

George Ellery Hale invents the spectro-helioscope

German biochemist Otto Heinrich Warburg (b Freiburg-im-Breisgau, Oct 8, 1883) develops a method for studying respiration in thin slices of tissue

Johannes Nocolaus Bronstead (b Varde, Denmark, Feb 22, 1879) suggests that acids give up a hydrogen ion in solution and that bases accept a hydrogen ion in solution, virtually the modern concept of acids and bases

Sir James Dewar, Scottish chemist and physicist, d London, Mar 27

Johann (Hans) Wilhelm Goldschmidt, German chemist, d Baden-Baden, May 20

György Hevesy and Dirk Coster (b Amsterdam, Netherlands, Oct 5, 1889) discover the element hafnium, named for the Latin version of Copenhagen, where the discovery is made

Thermodynamics and the free energy of chemical substances by Gilbert N. Lewis and Merle Randall introduces the concepts of thermodynamics in chemical settings

Edward Williams Morley, American chemist, d West Hartford, CT, Feb 24

Fritz Pregl of Austria wins the Nobel Prize for Chemistry for microanalysis of organic substances

Theodor H.E. Svedberg (b Fleräng, Sweden, Aug 30, 1884) develops the ultracentrifuge, a device that spins so rapidly that it removes particles from colloidal suspensions; from the speed of removal it is possible to determine the molecular weight of the particles, essential in studying large organic molecules

The size of the universe

During the first decades of the twentieth century, astronomers realized that many fuzzy patches that had been observed since the introduction of better telescopes at the end of the eighteenth century were huge systems of stars like ours. This idea was not entirely new. The German philosopher Immanuel Kant wrote in 1755 in his book *General natural history and theory of the heavens* that our solar system belongs to a flat, rotating system that is only one of the many other "island universes" that exist in the universe.

In 1912, the American astronomer Henrietta Leavitt studied Cepheid variables in the Large and Small Magellanic clouds. Cepheids are a type of variable star in which the absolute brightness is periodic over time. By measuring the variation period one can calculate the brightness of these stars and thus determine their distance. Leavitt found that the Magellanic clouds were well outside our own galaxy, at approximately a distance corresponding to twice the diameter of our galaxy. In 1923 the American astronomer Edwin Powell Hubble succeeded in observing Cepheid variables in the spiral nebula in Andromeda and other spiral nebulas. From the apparent magnitudes of these variables, Hubble was able to determine the distance of these nebulas; he found that they were much further from our galaxy than even the Magellanic clouds.

Some years before this discovery, the American astronomer Vesto M. Slipher had been studying the spectra of spiral nebulas. The spectrum of a star or nebula not only discloses its chemical composition, but from the Doppler shift (red shift) one can deduce its velocity relative to Earth. When Hubble compared the distances of a number of galaxies with their red shifts, he made an astonishing discovery: the farther a galaxy, the faster it is moving away from Earth. He also found that the recession velocity is proportional to the distance. This observation became the basis of the theory that the universe was created by a huge explosion and that the galaxies are the debris flying in all directions.

Vening Meinesz makes gravity measurements at sea using a pendulum aboard a submarine

Stefan Banach generalizes the concept of a vector space to what come to be known as abstract Banach spaces

John Venn, English mathematician and logician, d Cambridge, Apr 4

Sir Frederick Banting of Canada and John J.R. Macleod of England win the Nobel Prize for Physiology or Medicine for the discovery of insulin

Albert Calmette and Camille Guérin develop the tuberculosis vaccine BCG (Bacillus Calmette-Guérin)

Georges and Gladys Dick, independently of A. Dochez, find that scarlet fever is caused by streptococci; the Dicks develop an antitoxin for the disease

Sigmund Freud's *The ego and the id* is published

Arthur Holly Compton discovers the "Compton effect": the wavelength of X rays and gamma rays increases when these photons collide with electrons; this also demonstrates the particulate nature of X rays

Louis-Victor de Broglie (b Dieppe, France, Aug 15, 1892) introduces particle-wave duality for matter: an electron or any other subatomic particle can behave either as a particle or a wave; his theory is proven in 1927 by the electron diffraction experiment of Clinton Davisson

Robert Andrews Millikan of the United States wins the Nobel Prize for Physics for his discovery of elementary electrical charge and for his work with the photoelectric effect

Wilhelm Konrad Roentgen, German physicist, d Munich, Bavaria, Feb 10

Johannes Diderik Van der Waals, Dutch physicist, d Amsterdam, Mar 9

The first photoelectric cell (or "electric eye") is produced

Spanish inventor Juan de la Cierva develops the basic idea of the autogiro

Charles Proteus Steinmetz, German-American electrical engineer, d Schenectady, NY, Oct 26

John B. Tytus develops continuous hot-strip rolling of steel

The size of the universe *(continued)*

One of the first to propose that the universe was created by such an explosion was the Belgian priest Georges Lemaître. In 1927 he published a paper in which he showed that the concept of an expanding universe corresponds with the cosmological models based on relativity that had been introduced by the Dutch astronomer Willem de Sitter and the Russian mathematician Alexander Friedmann. Lemaître proposed that the universe had started by the explosion of a "primeval atom,"—the concentration of all the mass of the universe in an extremely small space. Subsequently this explosion became known as the Big Bang.

The strongest confirmation of the Big Bang theory was the discovery in 1965 of the cosmic background radiation by two physicists from Bell Telephone, Arno H. Penzias and Robert W. Wilson. While experimenting with a very sensitive microwave antenna, they discovered a hiss at a wavelength of 7.35 cm (2.89 in.) with an intensity that seemed independent of the orientation of the antenna. The hiss corresponds to radiation emitted by a body at a temperature of 3°K (–454°F), the temperature to which the universe is believed to have cooled as a result of its expansion since the Big Bang.

1924

GENERAL

(none)

ANTHROPOLOGY/ARCHAEOLOGY

Australian-South African anthropologist and surgeon Raymond Arthur Dart (b Brisbane, Australia, Feb 4, 1893) is given the skull of the Taung child, the first recognized fossil of an australopithecine; Dart concludes that the Taung child is a hominid that is not a human and not an ape; he names the species *Australopithecus africanus* and announces his discovery in 1925

ASTRONOMY

The Henry Draper catalogue, compiled by Annie Jump Cannon and listing the spectra of 225,300 stars, is published by Harvard Observatory

Arthur Stanley Eddington formulates the mass-luminosity law for stars, relating the luminosity of a star to its mass

Eddington suggests that white dwarf stars are made up of degenerate matter in which the electrons have collapsed from their orbits

Edwin Powell Hubble (b Marshfield, MO, Nov 20, 1889) demonstrates that the galaxies are true independent systems rather than parts of our Milky Way system by resolving the spiral arms of the Andromeda nebula into stars

John Martin Schaeberle, German-American astronomer, d Ann Arbor, MI, Sep 17

BIOLOGY

Russian-English biochemist David Keilin (b Moscow, Mar 21, 1887) discovers cytochrome, an important cell respiratory enzyme

Hans Spemann and H. Mangold show that if an amphibian embryo is grafted to another embryo, the graft is formed in part from tissue of the host, demonstrating "organizing action" by the second embryo; this is one of a series of experiments by Spemann that demonstrates that cells affect other cells near them

Harry Steenbock discovers that the ultraviolet component of sunlight increases the amount of vitamin D in food

CHEMISTRY

Louis-Marie-Hilaire Bernigaud, comte de Chardonnet, French chemist, d Paris, Mar 12

1925

GENERAL

Arthur S. Heineman opens the world's first motel, the Milestone Motel, in Monterey, CA, on Dec 12

ASTRONOMY

Nicolas-Camille Flammarion, French astronomer, d Juvisy-sur-Orge, Jun 3

Vesto M. Slipher determines the radial velocities of 41 galaxies from their red shifts

BIOLOGY

Statistical methods for research workers by Ronald Aymler Fisher (b London, Feb 17, 1890) becomes a standard work on statistical methods for genetics researchers

A. J. Lotka's *Elements of physical biology* contains simple mathematical models for competition between species and

CHEMISTRY

German chemist Karl Bosch invents a process for manufacturing hydrogen

Walter Karl Friedrich Noddack (b Berlin, Germany, Aug 17, 1893), Ida Tacke (later Ida Tacke Noddack), and Otto Berg discover the element rhenium

1924

Acetylene is used as an anesthetic

Sir William Maddock Bayliss, English physiologist, d London, Aug 27

Willem Einthoven of the Netherlands wins the Nobel Prize for Physiology or Medicine for his invention of the electro-cardiograph

Granville Stanley Hall, American psychologist, d Worcester, MA, Apr 24

Jacques Loeb, German-American physiologist, d Hamilton, Bermuda, Feb 11

Surgeon Sir William Macewen d Mar 22

Bell Laboratories for research in physics is founded jointly by American Telephone and Telegraph and General Electric

Satyendranath N. Bose (b Calcutta, India, Jan 1, 1894) introduces new statistics for light-quanta (photons); Albert Einstein applies the Bose statistics to matter and deduces that matter should exhibit wave properties

Karl Manne Georg Siegbahn (b Örebro, Sweden, Dec 3, 1886), whose previous work with X rays included developing methods to create X rays of varying wavelengths, becomes the first to refract X rays with a prism

Karl Siegbahn of Sweden wins the Nobel Prize for Physics for X-ray spectroscopy

German-American physicist Otto Stern, b Sohrau (Poland), Feb 17, 1888, and Gerlach demonstrate the effect of an inhomogeneous magnetic field on neutral atoms and molecules; they establish that the atoms and molecules behave like tiny magnets as a result of the charges on the proton and electron

The spiral-bound notebook is first produced

Kimberley Clark introduces the first version of Kleenex, known then as "Celluwipes"

The self-winding watch is patented

Insecticides are used for the first time

Photographs are transmitted by radio from New York to London

The rocket into interplanetary space by Hermann Oberth is the first truly scientific account of space-research techniques; it is the first book to contain the notion of escape velocity

Russian-American physicist Vladimir Kosma Zworykin (b Mourom, Jul 30, 1889) develops the iconoscope, an early type of television system

---

1925

The German *Meteor* expedition, using sonar, discovers the Mid-Atlantic Ridge

Edward Victor Appleton (b Bradford, England, Sep 6, 1892) discovers a layer in the ionosphere that reflects radio waves; the layer is named after him

Alexander Alexandrovich Friedmann, Russian mathematician, d Leningrad, Sep 16

Friedrich Ludwig Gottlob Frege, German mathematician, d Bad Kleinen, Mecklenburg, Jul 26

Sir Thomas Clifford Allbutt, English physician, d Cambridge, Feb 22

Josef Breuer, Austrian physician, d Vienna, Jun 20

Harvey A. Carr's *Psychology* defines the subject matter of psychology as mental activity, and processes

Patrick M.S. Blackett (b London, Nov 18, 1897) starts his experiments with colliding atoms in a Wilson condensation chamber in Apr, making the first photographs of nuclear reactions

Vannevar Bush and coworkers develop the first analog computer, a machine designed to solve differential equations

John Fillmore Hayford, American engineer, d Evanston, IL, Mar 10

1925

cont

predator-prey interactions, making it a serious attempt at mathematical biology

The Scopes "monkey trial" is held in Dayton, TN; John T. Scopes, a high-school teacher, is prosecuted for teaching evolution

August von Wasserman, German bacteriologist, d Berlin, Mar 16

George Hoyt Whipple (b Ashland, NH, Aug 28, 1878) finds that iron is an important constituent of red blood cells

Robert Robinson (b Bufford, England, Sep 13, 1886) locates the positions of all but one of the atoms in morphine

Richard Zsigmondy of Germany wins the Nobel Prize for Chemistry for his study of colloidal solutions

EARTH SCIENCE

Christian Felix Klein, German mathematician, d Göttingen, Jun 22

MEDICINE

such as memory, perception, feelings, imagination, judgment, and will

John Broadus Watson's *Behaviorism* is a semipopular book containing his program for the improvement of society

Albert P. Weiss reduces psychology to physical science in *A theoretical basis of human behavior*

PHYSICS

Wolfgang Pauli (b Vienna, Apr 25, 1900) discovers the exclusion principle in Jan: two electrons with the same quantum numbers cannot occupy the same atom; application of this principle accounts for the chemical properties of the elements

Clinton Joseph Davisson (b Bloomington, IL, Oct 22, 1881) accidentally creates large nickel crystals that exhibit X-ray diffraction, confirming Louis de Broglie's theories that particles always have waves associated with them

James Franck and Gustav Hertz of Germany win the Nobel Prize for Physics for their discovery of the laws of electron impact on atoms

Samuel Goudsmit (b The Hague, Netherlands, Jul 11, 1902) and George Uhlenbeck (b Batavia, Indonesia, Dec 6, 1900) formulate the hypothesis of the electron spin in Oct

Oliver Heaviside, English physicist and electrical engineer, d Paignton, Devonshire, Feb 3

Werner Heisenberg's first paper on quantum mechanics is published in Jul

Robert Andrews Millikan names cosmic rays and investigates how they are absorbed by the atmosphere and water in deep lakes

Werner Heisenberg, and Pascual Jordan in Nov give a comprehensive treatment of matrix mechanics and recognize the concept of second quantization

TECHNOLOGY

Vladimir Zworykin files for a patent on a color television system; the patent is granted in 1928

1925

cont

1925

cont

1926

Richard Evelyn Byrd (b Winchester, VA, Oct 25, 1888) is the first person to fly over the North Pole

Roy Chapman Andrews (b Beloit, WI, Jan 26, 1884) writes *On the trail of ancient man* to describe his experiences in the Gobi Desert, where he finds no traces of ancient humans, but does discover the first fossilized dinosaur eggs known to science, as well as other important fossils

The Bruce Proper Motion Survey starts, where plates showing the same region of the night sky a quarter century apart are compared with the blinking microscope to see which stars have moved relative to the others

Arthur Stanley Eddington's *The internal constitution of the stars* is the first major work on stellar structure; Eddington uses the concept of radiation pressure from the interior of the star as one of the main factors involved in the luminosity

Edwin P. Hubble's *A general study of diffuse galactic nebulae* deals with galactic and extragalactic nebulas; it also introduces a classification scheme for extragalactic nebulas

Bertil Lindblad (b Örebro, Sweden, Nov 26, 1895) proposes a dynamical model of the Milky Way in which the center core rotates as a unit and the outer portions rotate more slowly; later studies of other galaxies suggest that the visible portion of a galaxy rotates with the same degree of angular velocity

William Bateson, English biologist, d Merton, Surrey, Feb 8

Luther Burbank, American naturalist, d Santa Rosa, CA, Apr 11

Camillo Golgi, Italian histologist, d Pavia, Lombardy, Jan 21

Thomas Hunt Morgan's *The theory of the gene* is the culmination of work on the physical basis for Mendelian genetics based on breeding studies and optical microscopy; later advances will come from X-ray studies and molecular biology

Hermann Joseph Muller (b New York, NY, Dec 21, 1890) discovers that X rays induce genetic mutations; he increases the mutation rate of the fruit fly by a factor of 150 by irradiating them with X rays

James Batcheller Sumner (b Canton, MA, Nov 19, 1887) crystallizes urease, the first enzyme to be crystallized, and establishes that it is a protein

Hermann Staudinger (b Worms, Germany, Mar 23, 1881) begins work that leads to the realization that plastics are all formed from long chains of monomers that have been polymerized

Theodor Svedberg of Sweden wins the Nobel Prize for Chemistry for his development of and work with the ultracentrifuge

		Dutch-American physicist George Uhlenbeck and Samuel Goudsmit introduce half-integer quantum numbers for the hydrogen atom in Aug	**1925** cont

James B. Collip discovers parathormone, a hormone secreted by the parathyroid gland

Paul de Kruif's *The microbe hunters* becomes a popular book about bacteriology

Johannes Fibiger of Denmark wins the Nobel Prize for Physiology or Medicine for his discovery of Spiroptera carcinoma

Sir William Boog Leishman, Scottish physician, d London, Jun 2

Otto Fritz Meyerhof's *The chemical dynamics of life phenomena* is published

George Richards Minot (b Boston, MA, Dec 2, 1885) and William Parry Murphy (b Stoughton, WI, Feb 6, 1892) establish the use of liver in the diet as a successful treatment for anemia, based on earlier work with dogs by George Hoyt Whipple

The measurement of intelligence by Edward Lee Thorndike (b Williamsburg, MA, 1874) describes how to use tests to develop numerical measures of intelligence

Eugene Paul Wigner (b Budapest, Hungary, Nov 17, 1902) introduces group theory into quantum mechanics in Nov

John Desmond Bernal develops the "Bernal chart," an important tool for deducing the structure of a crystal from photographs of X-ray diffraction patterns

German-British physicist Max Born (b Breslau-Wroclaw, Poland, Dec 11, 1882) writes his first paper on the probability interpretation of quantum mechanics in Jun

Enrico Fermi (b Rome, Italy, Sep 29, 1901) introduces Fermi-Dirac statistics in Feb, in which gas particles obey the exclusion principle of Wolfgang Pauli

Paul Adrien Maurice Dirac (b Bristol, England, Aug 8, 1902) relates symmetric wave functions to Bose-Einstein statistics and antisymmetric wave functions to Fermi-Dirac statistics; from this he derives in Aug Planck's law from first principles

Heike Kamerlingh Onnes, Dutch physicist, d Leiden, Feb 21

Gilbert N. Lewis introduces the name *photon* for the light-quantum in Oct

The movie *The jazz singer*, starring Al Jolsen, introduces the era of talking motion pictures

The pop-up toaster is introduced in the United States

John Logie Baird (b Helensburgh, Scotland, Aug 13, 1888) produces television images of moving objects using the Nipkow disk

Robert Hutchings Goddard launches the first liquid-fuel propelled rocket, which reaches a height of 56 m (184 ft) and a speed of 97 km (60 mi) per hour

William Hampson, English inventor, d London, Jan 1

1926

1926

cont

1927

Davidson Black (b Toronto, Canada, Jul 25, 1884) discovers a single hominid tooth that he names *Sinanthropus pekinensis*; this earliest discovery of "Peking man" has been viewed as the discovery of a race of *Homo erectus*

Georges F. Lemaître (b Charleroi, Belgium, Jul 17, 1894) proposes that the universe was created by the explosion of a concentration of matter and energy which he called the "cosmic egg" or "primeval atom;" this is the first version of the currently accepted Big Bang theory of the origin of the universe

Bernard Ferdinand Lyot (b Paris, Feb 27, 1897) measures the polarization of light from planets

Dutch astronomer Jan Hendrik Oort (b Franeker, Apr 28, 1900) gives a dynamical proof that our galaxy has a spiral structure by measuring the velocity of stars

The basis of sensation by Edgar Douglas Adrian (b London, Nov 30, 1889) is published

Ronald Aymler Fisher's paper on mimicry and genetics is published

Gustav Theodor Fritsch d Jun 12

Animal biology by John Burdon Sanderson Haldane (b Oxford, England, Nov 5, 1892) is published

Frank A. Hartman isolates "cortin" (cortisone) from the adrenal glands and suggests that Addison's disease may be caused by the absence of it

Wilhelm Ludwig Johannsen, Danish botanist, d Copenhagen, Nov 11

Svante August Arrhenius, Swedish chemist, d Stockholm, Oct 2

Bertram Borden Boltwood, American chemist and physicist, d Hancock Point, ME, Aug 15

Karl James Peter Graebe, German chemist, d Frankfurt-am-Main, Jan 19

Walter Heitler and Fritz London, b Breslau (Wroclaw, Poland), Mar 7, 1900, demonstrate that a substantial fraction of the binding energy of the hydrogen molecule can be derived from the rules of quantum mechanics, providing a theoretical bases for explaining the covalent bond

EARTH SCIENCE	MATHEMATICS	MEDICINE	PHYSICS	TECHNOLOGY	
			Jean-Baptiste Perrin of France wins the Nobel Prize for Physics for discoveries of the discontinuous structure of matter and equilibrium of sedimentation The first paper on wave mechanics, written by Erwin Schrödinger (b Vienna, Austria, Aug 12, 1887) in Jan, replaces the electron in Niels Bohr's model of the atom by wave trains, applying Louis de Broglie's theory that electrons behave as waves; the theory is embodied by the Schrödinger equation Llewellyn Thomas completes the development of the concept of the spin of an electron in Feb by introducing a factor of two that had been missing in previous studies		**1926** cont
Rudolf Geiger's *The climate near the ground* founds the study of microclimatology	Emmy Noether's "Abstract construction of ideal theory in the domain of algebraic number fields" becomes a cornerstone of modern abstract algebra and completes her earlier axiomatic treatment of abstract rings	Philip Drinker and Louis Shaw develop the "iron lung," a device for mechanical artificial respiration Willem Einthoven, Dutch physiologist, d Leiden, Sep 29 Julius Wagner von Jauregg (b Wels, Austria, Mar 7, 1857) wins the Nobel Prize for Physiology or Medicine for the treatment of some forms of paralysis using malaria inoculation to induce the fever Ernest Henry Starling, English physiologist, d at sea near Kingston, Jamaica, May 2	Niels Bohr states the notion of complementarity in Sep Percy Williams Bridgman (b Cambridge, MA, Apr 21, 1882) argues in *The logic of modern physics* that all physical concepts must be defined in precise and rigid terms, and that all concepts lacking physical references must be discarded Louis de Broglie's hypothesis that subatomic particles behave like waves is confirmed by experiment; Clinton Davisson and, independently, George Paget Thomson (b Cambridge, England, May 3, 1892) show that electrons can be diffracted as light waves by crystals	The pentode, a vacuum tube with five electrodes, is introduced The Holland Tunnel opens, linking New York and New Jersey Harold Stephen Block introduces negative feedback in audio amplifiers, thus reducing distortion Charles A. Lindbergh (b Detroit, MI, Feb 4, 1902) makes the first nonstop solo flight across the Atlantic Ocean, in 33.5 hours Using ideas about rockets and space from a 1924 book by Hermann Oberth, *Verein für Raumschiffart* (The Society for Space Travel) is founded in Germany; among its early members are Wernher	**1927**

1927 cont

BIOLOGY

Karl Martin Leonhard Albrecht Kossel, German biochemist, d Heidelberg, Jul 5

Thomas Hunt Morgan's *Experimental embryology* is published

Ivan P. Pavlov's *Conditioned reflexes* relates his experimental work with reflexes in dogs

CHEMISTRY

Ira Remsen, American chemist, d Carmel, CA, Mar 4

Heinrich O. Wieland of Germany wins the Nobel Prize for Chemistry for his development of steroid chemistry

1928

GENERAL

Roald Engelbregt Amundsen, Norwegian explorer, d near Spitsbergen, Jun 16, crashing while on a flight to search for shipwreck victims

Richard Evelyn Byrd establishes Little America on the ice shelf of Antarctica; from this base he flies over the South Pole, beginning an extensive program of exploration of the continent that will cover five expeditions, the last in 1955

Arthur Stanley Eddington's *The nature of the physical world* argues that the mind can manipulate behavior of matter and so enforce its own free will

ANTHROPOLOGY/ARCHAEOLOGY

Franz Boas's *Anthropology and modern life* is an attack on theories of racial superiority; he defends the need for strong leaders and for equality between the sexes, but he questions monogamy and Marxian economic determinism

Margaret Mead's *Coming of age in Samoa* describes the passage from childhood to adulthood in a primitive society

ASTRONOMY

Edwin P. Hubble estimates that the Crab nebula may be 900 years old

Edward Walter Maunder, English astronomer, d London, Mar 21

Henry Norris Russell determines the abundance of elements in the solar atmosphere by studying the spectrum of the sun

BIOLOGY

Fred Griffith observes that certain strains of bacteria acquire new characteristics that remain present in their offspring when grown in an extract from another strain possessing these characteristics, showing that genetic information is transmitted by a chemical compound

Hans Fischer (b Höchst-am-Main, Germany, Jul 27, 1881) determines all of the atoms in heme, the part of hemoglobin that is not made from amino acids

J. Gray's *Experimental cytology* is the founding work in biophysics

Hungarian-American biochemist Albert Szent-Györgyi (b Albert Szent-Györgyi von Nagyrapolt, Budapest, Hungary, Sep 16, 1883) discovers vitamin C independently of Charles Glen King, who has priority by two weeks

CHEMISTRY

German chemists Otto Paul Hermann Diels (b Hamburg, Jan 23, 1876) and Kurt Alder (b Königshütta, Silesia, Jul 10, 1902) develop diene synthesis, more commonly known as the Diels-Alder reaction, a technique for combining atoms into molecules that is useful in forming many compounds, especially synthetic rubber and plastics

Sir Cyril Norman Hinshelwood (b London, Jun 19, 1897), independently of Nikolay Nikolaevich Semenov, shows that below a certain temperature hydrogen and oxygen gases do not explode because of reactions with the walls of the vessels that contain the gases

Theodore William Richards, American chemist, d Cambridge, MA, Apr 2

Adolf Windaus of Germany wins the Nobel Prize for Chemistry for his study of cholesterol

1927 cont

PHYSICS

Arthur Holly Compton of the United States and Charles Wilson of England share the Nobel Prize for Physics, Compton for studies of the effect of wavelength change in X rays and Wilson for his cloud chamber

Werner Heisenberg postulates his uncertainty principle: it is impossible to determine accurately two variables, such as position and momentum, of an electron simultaneously

George Paget Thomson discovers diffraction of electrons independently of Max von Laue

TECHNOLOGY

von Braun, b Wirsitz (Poland), Mar 23, 1912, who will later develop the first rockets to travel in space, and Willy Ley (b Berlin, Germany, Oct 2, 1906), who will write books that make rocket concepts easily understood by the average US citizen

The Golden Gate Bridge is completed in San Francisco

1928

EARTH SCIENCE

Thomas Chrowder Chamberlin, American geologist, d Chicago, IL, Nov 15

Emil Wiechert, German seismologist, d Göttingen, Mar 19

MATHEMATICS

Richard von Mises develops a philosophical approach to probability theory in *Probability, statistics and truth*

John von Neumann (b Budapest, Hungary, Dec 3, 1903) develops his minimax theory, a game theory

MEDICINE

Sir David Ferrier, Scottish neurologist, d London, Mar 19

Johannes Andreas Grib Fibiger, pathologist and physician, d Jan 30

Alexander Fleming discovers penicillin in molds; its clinical use in therapy starts only in the 1940s, when Howard Florey and Ernst Chain further develop it and it is learned how to manufacture it in quantity

Charles Nicolle of France wins the Nobel Prize for Physiology or Medicine for his research on typhus

Greek-American George Papanicolau develops the Pap test for diagnosing uterine cancers

John Broadus Watson's *Psychological care of the infant and child* proposes a regulatory, instead of permissive, system of child rearing

PHYSICS

Swiss-American physicist Felix Bloch (b Zurich, Oct 23, 1905) proves that some electrons can pass through a crystalline array without being scattered

Paul Adrien Maurice Dirac develops an equation that combines the ideas of quantum mechanics with special relativity (the relativistically invariant equation of the electron)

C.D. Ellis and W.A. Wooster show that beta radiation apparently does not obey the law of conservation of energy; later it will be shown that an unnoticed particle, the neutrino, is involved in beta radiation and that conservation of energy is not violated

George Gamov (b Odessa, Russia, Mar 4, 1904) develops a theory of radioactive decay by emission of an alpha particle that

TECHNOLOGY

The radio beacon is introduced

The Minnesota Mining and Manufacturing Corporation introduces Scotch Tape

American inventors J. W. Horton and W. A. Marrison develop the first quartz crystal clock

Joseph Schick invents the electric razor

Penicillin and other antibiotics

From antiquity people have tried to find substances that would cure disease or heal wounds. Most of the recipes called for exotic ingredients, such as eye of newt. Folk wisdom and early doctors occasionally knew of something that seemed to work some of the time—garlic, molds, herbs, even soil—but most of the time the substances were ineffective. This did not stop people from looking for what we now think of as "wonder drugs."

In the sixteenth century, Paracelsus advocated strong chemicals to cure disease, including mercury compounds and other poisons. But little progress was made in this direction. In the early twentieth century, Paul Ehrlich came close in his search for chemicals based on dyes that would kill germs without killing people. In 1907 he found the dye trypan red would do this if handled properly for the protists that cause African sleeping sickness. Two years later he found an arsenic compound, which he called compound 606 or Salvarsan, that would kill the bacteria that cause syphilis, although it too had to be handled very carefully to work properly. Finally, in 1932, Gerhard Domagk discovered the first sulfa drug in a different red dye. It killed streptococci without killing mice. When his daughter developed a dangerous streptococcal infection, Domagk desperately tried his discovery on her, effecting a cure. Other scientists carried on investigation of the chemical, leading to four different sulfa drugs by the end of World War II.

Before that time, however, scientists had learned that looking for inorganic compounds was not the best way to solve the problem. They knew as early as 1877, for example, that some bacteria release substances that kill other bacteria. They also knew that soil, which is filled with many harmless bacteria and fungi, does not harbor most disease-causing bacteria. Substances in soil, probably produced by common soil bacteria or fungi, seem to kill other bacteria. Searching for these substances might be more productive than searching for inorganic compounds that would do the same thing. After all, it was known that the soil substances existed. However, there are so many different substances in soil that no one tried this approach for many years.

In the meantime, Alexander Fleming was seeking such a substance in a different known source. Tears contain something that kills bacteria and causes no physical damage to the body. In 1922 he isolated the substance lysozyme from tears. It had no medical application, but Fleming continued to work on the problem of killing bacteria. Quite by accident he found another substance in 1928. This substance was produced by a mold that had somehow landed on one of his cultures of staphylococcus. It was evident in a ring of dead bacteria around each speck of mold. John Tyndall had observed the same effect while investigating bacteria and molds on airborne dust in the nineteenth century, but had not pursued it.

Fleming investigated and found that the substance produced by the mold, which he named penicillin, killed some bacteria, but not others. In addition, the substance did not damage white blood cells. He published a report on penicillin and went back to other work. Howard Florey and Ernst Chain, who were both experimenting with lysozyme, turned their attention to penicillin when World War II started, hoping to isolate the substance for use against war wounds. They succeeded, and scientists at the US Northern Research Laboratory in Peoria, Illinois, found strains of mold that could be grown in large tanks to produce the new drug in quantity. Penicillin saved thousands of lives in World War II and continues to be, in one form or another, one of the most powerful and useful antibiotics.

Shortly before Florey and Chain began to work with penicillin, the first scientist to search for antibiotics in soil had achieved success. René Jules Dubos discovered two antibiotics, gramicydin and tyrocidin, in a substance produced by a soil bacterium. But it was the success of penicillin that started many teams of scientists searching soil samples. In 1943 Selman A. Waksman found streptomycin, an effective antibiotic for some bacteria although somewhat toxic to humans. Benjamin Duggar and coworkers found Aureomycin, the first tetracycline, in soil samples the following year. Since then, a variety of antibiotics have been found in soil samples and produced by various fungi. Antibiotics—although very mild—have even been found in garlic. By analyzing the structures of antibiotics and how they work, scientists have created many synthetic antibiotics.

While antibiotics have greatly reduced the effects of infectious disease, their very success has led to the problem of resistant bacteria. Bacteria can easily exchange genetic material though structures called plasmids. If a few bacteria are resistant to a particular antibiotic, they soon pass this trait on to others. Frequent use of antibiotics and sulfa drugs has created resistant populations of many disease-causing bacteria. The effectiveness of sulfa drugs against some diseases dropped by 50 percent over a 20-year period. Consequently, the search for new antibiotics continues, since bacteria resistant to one antibiotic may not be resistant to another that works by a different mechanism.

contains the first ideas of electron tunneling

R.H. Fowler and Lothar W. Nordheim explain why electrons are emitted from solids in high-intensity external electric fields; the electrons are able to tunnel through the potential barrier at the surface of the solid; tunneling is a quantum effect equivalent to walking through a brick wall

Edwin Herbert Land (b Bridgeport, CT, May 7, 1909) develops a polarizing filter

Hendrik Antoon Lorentz, Dutch physicist, d Haarlem, Feb 4

Chandrasekhara Venkata Raman (b Tiruchirappalli, India, Nov 7, 1888) discovers the scattering effect named after him; light, because of its particle nature, is scattered by molecules of matter; this scattering can be used to determine what the molecules are

Sir Owen Richardson of England wins the Nobel Prize for Physics for his work on the effect of heat on electron emission (the Edison effect)

Arnold Sommerfeld uses quantum mechanics to show that electrons in a conductor behave as a degenerate gas; most electrons cannot participate in conduction; only a few high-energy electrons are involved in conducting electricity; today the border between the two types of electrons is called the Fermi level

1928

cont

1929

Teeth and part of a skull are found near Peking (Beijing), China

Edwin P. Hubble determines the distance of the Andromeda nebula (930,000 light-years) using Cepheid variables

Hubble establishes that the more distant a galaxy is, the faster it is receding from Earth (Hubble's law), confirming that the universe is expanding

Swiss astronomer Fritz Zwicky (b Varna, Bulgaria, Feb 14, 1898) proposes that the red shift of distant galaxies is caused by the "aging" of photons: the photons, during their long trip through the universe, lose energy and their frequency shifts toward the red; this theory is not generally accepted

Fritz Albert Lipmann isolates adenosine triphosphate (ATP) from muscle tissue

Thomas Burr Osborne d New Haven, CT, Jan 29

W. Vogt publishes the first "fate map" of a vertebrate embryo

J.H. Woodger's *Biological principles* is the first full-scale analysis of theoretical biology

Karl, Baron von Welsbach Auer, Austrian chemist, d Welsbach Castle, Carinthia, Aug 8

Adolf Friedrich Butenandt (b Bremerhaven-Lehe, Germany, Mar 24, 1903) isolates estrone, a female sex hormone

Sir Arthur Harden and Hans von Euler-Chelpin of Sweden (b Augsburg, Germany, Feb 15, 1873) share the Nobel Prize for Chemistry for their research into sugar fermentation and the role of enzymes in it

Phoebus Levene discovers that a previously unknown sugar, deoxyribose, is found in nucleic acids that do not contain ribose; such nucleic acids are now known as deoxyribonucleic acids, or DNA

Raoul Pierre Pictet, Swiss chemist, d Paris, Jul 27

Richard Adolf Zsigmondy, Austrian-German chemist, d Göttingen, Germany, Sep 23

			Wilhelm Wien, German physicist, d Munich, Bavaria, Aug 30	**1928** cont

Eugene Paul Wigner introduces the concept of parity for atomic states in Feb; essentially he shows that there are two states between which atoms can fluctuate, and that the atom must be in either one or the other

---

1929

EARTH SCIENCE

M. Matuyama shows that rocks of different strata have their magnetic fields reversed in some instances; he concludes that Earth's magnetic field reverses direction from time to time

Daniel Moreau Barringer, American mining engineer and geologist, d

MEDICINE

German psychiatrist Hans Berger (b Neuses, Bavaria, May 21, 1873) develops the electroencephalogram (EEG)

Christiaan Eijkman of the Netherlands and Sir Frederick G. Hopkins of England win the Nobel Prize for Physiology or Medicine for their work with vitamins

Joseph Goldberger, Austrian-American physician, d Washington, DC, Jan 17

Manfred J. Sakel introduces insulin shock for the treatment of schizophrenia

PHYSICS

Walther (Walter) Wilhelm Georg Bothe (b Oranienburg, Germany, Jan 8, 1891) develops the idea of using two Geiger counters at the same time to detect the direction of cosmic rays, an important technique in working with subatomic particles

Prince Louis de Broglie of France wins the Nobel Prize for Physics for his discovery of the wave character of electrons

Paul Adrien Maurice Dirac introduces the notion of hole theory, incorrectly identifying a hole of negative energy in a sea of electrons with a proton; he later corrects this idea and predicts the existence of antimatter

George Gamov introduces his "liquid drop" model of the atomic nucleus of heavy elements

William Francis Giauque (b Niagara Falls, Ontario, May 12, 1895) discovers that oxygen is a mixture of three isotopes, O-16, O-17, and O-18, eventually resulting in abandonment of the oxygen-16 standard for atomic mass in 1961 and its replacement by the carbon-12 standard

TECHNOLOGY

FM radio is introduced

Construction is started on the Empire State Building in New York; it will be finished in 1931

The Dunlop Rubber Company produces the first foam rubber

Engineer Karl Benz d Apr 4

Emile Berliner, German-American inventor, d Washington, DC, Aug 3

Georges Claude develops the first power plant to use the difference in temperature between the upper and lower layers of the ocean to generate electricity

Robert Goddard launches the first instrumented rocket; it carries a barometer, a thermometer, and a small camera

W. A. Morrison introduces the quartz-crystal clock

Hermann Oberth's *Way to space travel* is published

German engineer Felix Wankel patents a rotary engine, although it will not be until 1951 that the Wankel engine becomes practical

1929

cont

1930				
Louis Bamberger and his sister Caroline found the Princeton Institute for Advanced Study	Margaret Mead's *Growing up in New Guinea* tells of her research among primitive tribes in New Guinea Andrew Ellicott Douglas determines the date of an American Indian site by using tree rings observed in artifacts from the site, thus establishing the science of dendrochronology	Bernard Lyot builds the coronagraph, a special telescope in which a baffle stops the light from the solar disk and allows the observation of the corona Marcel Gilles Minnaert and Gerard Mulders determine the abundance of elements in the solar atmosphere by studying the solar spectrum Edison Pettit and Seth Nicholson measure the temperature of the moon by measuring the increase in temperature of a thermocouple placed in the focus of a telescope Bruno Benedetto Rossi, Italian-American physicist (b Venice, Apr 13, 1905), points out that cosmic rays should be deflected by Earth's	Waldemar Mordecai Wolfe Haffkine, Russian-British bacteriologist, d Lausanne, Switzerland, Oct 25 Karl Landsteiner of the United States wins the Nobel Prize for Physiology or Medicine for his definition of four human blood groups	I G. Farberindustrie in Germany develops polystyrene W.L. Semon of the B.F. Goodrich Company invents polyvinyl chloride William Thomas Astbury, English physical biochemist (b Stoke-on-Trent, Feb 25, 1898) discovers that wool has a different X-ray diffraction pattern when it is stretched; this leads him to use X-ray diffraction techniques to study the three-dimensional structure of proteins Hans Fischer of Germany wins the Nobel Prize for Chemistry for his analysis of blood heme Joseph Achille Le Bel d Paris, Aug 6

1929
cont

Werner Heisenberg and Wolfgang Pauli give their formulation of quantum field theory based on Lagrangians

Walter Heitler and Gerhard Herzberg demonstrate that there must be an even number of protons in nitrogen, implying that a neutral particle must exist in the nucleus of most atoms; this demonstration is not taken seriously at the time, although Heitler and Herzberg were correct in their reasoning

Dmitri Skobeltzyn notes that cosmic rays often occur in groups called showers

Hermann Weyl formulates gauge invariance and discovers its relation to charge conservation

---

1930

William Beebe (b Brooklyn, NY, Jul 29, 1877) and Otis Barton dive to 417 m (1368 ft) in their new submersible, the bathysphere

Fridtjof Nansen, Norwegian explorer, d Lysaker, May 13

Alfred Lothar Wegener, German geologist and meteorologist, d Greenland, Nov

L. G. Schnirelman proves some number k exists such that every sufficiently large number is the sum of at most k prime numbers, but the proof yields no clues as to the value of k; this is a generalization of Goldbach's conjecture that every even number greater than two is the sum of two primes

B.L. van der Waerden's *Moderne algebra* (Modern algebra) is a two-volume treatment of abstract algebra, much of it based on the ideas of Emmy Noether, that becomes the standard approach to the subject

Hermann Weyl succeeds David Hilbert at Göttingen University

Christiaan Eijkman, Dutch physician, d Utrecht, Nov 5

Allvar Gullstrand, Swedish physician, d Uppsala, Jul 21

Hans Zinsser develops an immunization against typhus

Paul Adrien Maurice Dirac's *Principles of quantum mechanics* formulates a general mathematical theory in which matrix mechanics and wave mechanics are special cases

Eugen Goldstein d Berlin, Dec 25

The lambda point of helium is discovered: at 2.2°K (−456°F) helium becomes superfluid, having a viscosity about one thousandth that of the least viscous gas at room temperature; the term superfluidity is coined in 1941 by Peter Kapitza to describe this

Wolfgang Pauli proposes that an unknown particle (christened the neutrino by Enrico Fermi in 1932)

The Postum Company begins marketing frozen foods for the first time

Sliced bread is introduced

The tape recorder using magnetized plastic tape is developed in Germany

Vannevar Bush completes the differential analyzer, an analog computer

Elmer Ambrose Sperry, American inventor, d Brooklyn, NY, Jun 16

British engineer Frank Whittle patents the jet engine

1930 cont

ASTRONOMY

magnetic field—to the east if positive and to the west if negative; studies show the deflection to the east, showing that most cosmic rays are positive

Russian-German optician Bernhard Voldemar Schmidt (b Neissaar Island, Estonia, Mar 30, 1879) invents the corrector for the Schmidt camera and telescope, a device that allows very wide-angle views with little distortion; the Schmidt telescope comes to dominate astronomy because it is free from coma, an aberration

Clyde William Tombaugh (b Streator, IL, Feb 4, 1906) discovers the planet Pluto on Feb 18

Swiss-American astronomer Robert Julius Trumpler (b Zurich, Oct 2, 1886) shows the existence of diffuse interstellar dust by studying its dimming effect on star clusters; as a result, he shows the Milky Way galaxy is about three-fifths the size of previous estimates, which assumed dimming was caused by distance

CHEMISTRY

Thomas Midgley, Jr., develops Freon as a gas for use in refrigerators

John Howard Northrop (b Yonkers, NY, Jul 5, 1891) crystallizes the enzyme pepsin, an important step toward understanding the chemical nature of enzymes

Fritz Pregl, Austrian chemist, d Graz, Dec 13

Arne Wilhelm Tiselius (b Stockholm, Sweden, Aug 10, 1902) introduces electrophoresis, a method for separating proteins in suspension using electric currents

1931

ASTRONOMY

Subrahmanyan Chandrasekhar (b Lahore, India, Oct 19, 1910) predicts the existence of white dwarf stars and determines that a white dwarf can exist only if its mass is less than 1.4 times that of the sun; this comes to be known as Chandrasekhar's limit

Experiments by Karl Jansky (b Norman, OK, Oct 22, 1905) with an improvised aerial, trying to locate

BIOLOGY

Martinus Willem Beijerinck, Dutch botanist, d Gorssel, Jan 1

Adolf Butenandt isolates androsterone, a male sex hormone

Eloise B. Cram suggests that some parasites may change their host's behavior; specifically she notes that a nematode may cause a grasshopper it has infected to be an easier prey for chickens

CHEMISTRY

The first Freons are produced by the Kinetic Chemical Corporation; much later they will be implicated as possible causes of destruction of the ozone layer that protects Earth from ultraviolet radiation

Karl Bosch and Friedrich Bergius of Germany win the Nobel Prize for Chemistry for their invention of high-pressure methods of chemical production

1930
cont

accounts for the apparent violation of the law of conservation of energy in beta decay

Chandrasekhara Raman of India wins the Nobel Prize for Physics for his laws of light diffusion

---

1931

George David Birkhoff proves the general ergodic theorem, building on the mean ergodic theorem proved by John von Neumann earlier in the year

Austrian-American mathematician Kurt Gödel, b Boünn (Brno, Czechoslovakia), Apr 28, 1906, proves that any formal system strong enough to include the laws of arithmetic is either incomplete or inconsistent

Ernest Goodpasture grows viruses in eggs, making the production of such vaccines as polio vaccine possible for viral diseases

Walther Bothe and Wilhelm Becker bombard beryllium with alpha particles and discover a highly penetrating radiation that is not deflected by magnetic forces; James Chadwick will show in 1932 that this radiation consists of neutrons; Irène Joliot-Curie and Frédéric Curie make a similar discovery

The George Washington Bridge over the Hudson River, with double the span (1066 m) of the previous recordholder, opens

Edward Goodrich Acheson, American inventor, d New York, NY, July 6

Thomas Alva Edison, American inventor, d West Orange, NJ, Oct 18

1931 cont

ASTRONOMY

sources of inter-ference in radio telephone links, lead to the founding of radio astronomy; Jansky publishes his first results in 1932 and in 1933 finds that the radio emission comes from the Milky Way

Harold Spencer Jones determines solar parallax and finds it to be 499.15 arc seconds

Bruno Rossi demonstrates that cosmic rays are powerful enough to penetrate a meter (yard) or more into solid lead

J.P. Schafer and W.M. Goodal discover reflections of radio waves by meteors

BIOLOGY

Swiss chemist Paul Karrer (b, Moscow, USSR, Apr 21, 1889) establishes the structure and function of vitamin A

Baron Shibasaburo Kitasato, Japanese bacteriologist, d Nakanojo, Gumma, Jun 13

Otto H. Warburg of Germany wins the Nobel Prize for Physiology or Medicine for his discovery of respiratory enzymes

CHEMISTRY

Belgian-American chemist Julius Arthur Nieuwland (b Hansbeke, Belgium, Feb 14, 1878) develops neoprene, a synthetic rubber

Linus Carl Pauling (b Portland, OR, Feb 28, 1901) proposes that the phenomenon of resonance causes the stability of the benzene ring

Otto Wallach, German organic chemist, d Göttingen, Feb 26

1932

ANTHROPOLOGY/ARCHAEOLOGY

Edward Lewis finds in India part of the jaw of a primitive primate that he names *Ramapithecus*; for the next 50 years most anthropologists will believe that *Ramapithecus* is the earliest known ancestor of human beings; in the 1980s evidence accumulates that *Ramapithecus* is the ancestor of the orangutan

ASTRONOMY

Astronomers decide during an astrometric conference at Pulkovo to use distant galaxies as an inertial reference frame

Robert Grant Aitken publishes a catalog with 17,180 visible binaries

T. Dunham discovers carbon dioxide in the atmosphere of Venus

Maximilian Franz Joseph Wolf, German astronomer, d Heidelberg, Oct 3

BIOLOGY

Edgar Adrian's *The mechanism of nervous action* discusses electrical changes in the brain and nerves

Julian S. Huxley's *Problems of relative growth* contains the important rule in embryology stating that the specific growth rates of the organs of the body stand in constant ratios to each other

Ilya Ivanovich Ivanov, Russian biologist, d Alma-Ata, USSR, Mar 20

Charles Glen King (b Entiat, WA, Oct 22, 1896) isolates vitamin C from lemon juice

German-British biologist Hans Adolf Krebs (b Hildesheim,

CHEMISTRY

American engineer Edwin Land invents a synthetic substance that will polarize light

Irving Langmuir of the United States wins the Nobel Prize for Chemistry for his study of monomolecular films

John Northrop crystallizes the enzyme trypsin

Friedrich Wilhelm Ostwald, Russian-German physical chemist, d Leipzig, Saxony, Apr 4

Richard von Mises's *Wahrscheinlichkeits-rechnung* (Mathematical theory of probability and statistics) introduces the concept of sample space into probability theory

Paul Adrien Maurice Dirac proposes the positron in May, a particle almost exactly like the electron but with a positive charge

Ernest Orlando Lawrence (b Canton, SD, Aug 8, 1901) develops the first workable particle smasher, the cyclotron, in Apr

Albert Abraham Michelson, German-American physicist, d Pasadena, CA, May 9

Ernst August Friedrich Ruska (b Heidelberg, Germany, Dec 26, 1906) discovers that a magnetic coil can focus a beam of electrons; he creates the first crude electron microscope and achieves a magnification of 400 power; the electron microscope is independently invented by Rheinhold Ruedenberg about the same time

Sir Charles Algernon Parsons, British engineer, d on a ship off Kingston, Jamaica, Feb 11

Winkler makes the first European firing of a liquid-propelled rocket

1931 cont

Carl-Gustav Arvid Rossby introduces the so-called Rossby diagram for plotting air mass properties; it becomes widely used

The Fields Medal of the International Mathematical Congress is introduced, a prize equivalent in prestige to the Nobel Prize, which is not offered to mathematicians (Alfred Nobel disliked the most prominent Norwegian mathematician of the day and so did not provide funds for a mathematics prize)

Abraham Adrian Albert proves the existence of noncyclic division algebras of degree four

John von Neumann gives a proof of the ergodic theorem

Giuseppe Peano, Italian mathematician and linguist, d Turin, Apr 20

German chemist Gerhard Domagk, b Lagow (Poland), Oct 30, 1895, discovers the first sulfa drug, Prontosil

John Eccles (b Melbourne, Australia, Jan 27, 1903) publishes *Reflex activity of the spinal cord*

Armand Quick introduces the Quick test to measure the clotting ability of blood

Sir Ronald Ross, English physician, d London, Sep 16

Max Rubner, German physiologist, d Berlin, Apr 27

Carl David Anderson (b New York, NY, Sep 3, 1905) discovers the positron (a positively charged electron and the first form of antimatter to be discovered) in cosmic-ray tracks, fulfilling a prediction by P.A.M. Dirac

Louis Winslow Austin, American physicist, d Washington, DC, Jun 27

James Chadwick discovers the neutron, which he announces first at the Kapitza Club at the Cavendish Laboratory and then more formally in a paper in the Feb 27 *Nature*

RCA demonstrates a television receiver with a cathode-ray picture tube

George Eastman, American inventor, d Rochester, NY, Mar 14

Auguste Piccard (b Basel, Switzerland, Jan 28, 1884) becomes the first human to enter the stratosphere; his balloon climbs to 16,201 m (53,153 ft)

1932

1932
cont

The limits of mathematics

In the nineteenth century, mathematics both purged itself of the murkiness of many of the basic concepts of the calculus and liberated itself with the discovery of non-Euclidean geometry. It seemed possible by the end of the century that the process could reach completion by showing that mathematics was consistent within itself and complete in the sense that everything provable could be proved. In a famous speech in 1900, the mathematician David Hilbert proposed 23 unsolved problems for twentieth-century mathematics to handle. Proving the consistency of arithmetic with integers was problem number two in his list.

Georg Cantor, the founder of set theory, had discovered 17 years earlier some unsettling problems that suggested that there were roadblocks in the way of Hilbert's goal, but Hilbert did not think they would be insurmountable. For one thing, Cantor's discoveries concerned contradictions—generally called paradoxes by mathematicians—in his own set theory. One of these contradictions came from Cantor's proof that for every set, there is a set with more members. The paradox occurs when one considers the set of all sets. How can there be another set with more members than that?

A year after Hilbert's speech, Bertrand Russell discovered an even more devastating paradox: Is the set of all sets that are not elements of themselves an element of itself? Russell's paradox was so close to the heart of set theory that he, and others, made major revisions and restrictions in the theory in an effort to avoid it.

In the 1920s, Hilbert launched a major effort in collaboration with other mathematicians to prove that mathematics is consistent. His group, the formalists, seemingly achieved large parts of their goals, but the price was too high. They had to assume ideas that were far more complicated than the ones they were trying to prove.

In 1931, Kurt Gödel ended many of the arguments and effectively halted the efforts of the formalists. That year, he proved what is now known as Gödel's incompleteness theorem: for any consistent axiomatic theory that is strong enough to produce the properties of natural numbers (1, 2, 3, . . .), there exists a statement whose truth or falsity cannot be proved (or else the system is inconsistent). In other words, mathematics is either inconsistent or incomplete.

Since Gödel's incompleteness theorem, other mathematicians have used methods similar to his to show specific examples of statements that cannot be proved from widely accepted sets of axioms that are supposed to underlie all of mathematics. Despite incompleteness, however, mathematics continues to progress. Some mathematicians think mathematics is even more interesting now that they know that it is not unlimited in scope.

Hanover, Germany, Aug 25, 1900) discovers the urea cycle, which transforms ammonia to urea in mammals

Thomas Hunt Morgan's *The scientific basis of evolution* is published

Sir Charles Sherrington of England and Edgar D. Adrian of the United States win the Nobel Prize for Physiology or Medicine for multiple discoveries in the function of neurons

Swedish biochemist Axel Hugo Teodor Theorell (b Linköpin, Östergötland, Jul 6, 1903) succeeds in isolating myoglobin crystals

Edward Chace Tolman's *Purposive behavior in animals and men* contains both behavioristic and Gestalt concepts

John Cockcroft (b Todmorden, England, May 27, 1897) and Ernest Walton (b Bungarvan, Ireland, Oct 6, 1903) become the first to use a particle accelerator; they accelerate protons to bombard lithium, thereby producing two alpha particles; their achievement is commonly known as splitting the atom; the results are published on Apr 30

Reginald Aubrey Fessenden, Canadian-American physicist, d Hamilton, Bermuda, Jul 22

Werner Heisenberg proposes in Jun a model of the atomic nucleus in which protons and the newly discovered neutrons are held together by exchanging electrons, the model will give rise to Yukawa's more successful picture of the nucleus; he introduces certain matrices as formal mathematical devices; these matrices eventually lead to the concept of isospin

Werner Heisenberg of Germany wins the Nobel Prize for Physics for his discovery of quantum mechanics

Harold Clayton Urey (b Walkerton, IN, Apr 29, 1893) discovers deuterium, or heavy hydrogen; in addition to a proton, the nucleus of deuterium contains a neutron

Eugene Paul Wigner introduces invariance under time reversal into quantum mechanics in Nov

1933

GENERAL

Adolf Hitler is appointed chancellor of Germany on Jan 30

ASTRONOMY

Arthur Stanley Eddington's *The expanding universe* is one of the first books popularizing modern cosmology

Robert Thorburn Ayton Innes, Scottish astronomer, d Surbiton, England, Mar 13

Donald Menzel (b Florence, CO, Apr 11, 1901) discovers oxygen in the solar corona

BIOLOGY

The last known Tasmanian wolf dies in a zoo, although unverified reports of sighting of this marsupial in the Tasmanian jungle continue

Karl Franz Joseph Correns, German botanist, d Berlin, Feb 14

John Burdon Sanderson Haldane's *The causes of evolution* is published

Thomas Hunt Morgan of the United States wins the Nobel Prize for Physiology or Medicine for his discovery of the relation of chromosomes to heredity

Tadeusz Reichstein (b Wloclawek, Poland, Jul 20, 1897) synthesizes ascorbic acid (vitamin C), which Walter Haworth also synthesizes independently in 1934

Pierre-Paul Roux, French bacteriologist, d Paris, Nov 3

CHEMISTRY

Hubert James and Albert Sprague Coolidge calculate the strength of the covalent bond from the basic principles of quantum mechanics to a high degree of accuracy, demonstrating a close agreement between theory and experiment

Gilbert N. Lewis obtains deuterium oxide (heavy water) a year after Harold Urey discovers deuterium, or hydrogen-2, the isotope of hydrogen that has a nucleus of one proton and one neutron

				1933
Hermann Weyl resigns at Göttingen	Grantley Dick-Read's *Natural childbirth* advocates exercises and procedures for childbirth without drugs	Harold DeForest Arnold, American physicist, d Summit, NJ, Jul 10	High-intensity mercury vapor lamps are introduced	

Hermann Weyl resigns at Göttingen

Grantley Dick-Read's *Natural childbirth* advocates exercises and procedures for childbirth without drugs

Harold DeForest Arnold, American physicist, d Summit, NJ, Jul 10

John Desmond Bernal and R.H. Fowler study the structure of water and ionic solutions, using X-ray diffraction patterns; their findings explain the increase of density of water when its temperature is increased from 0 to 4 degrees Celsius

Paul Adrien Maurice Dirac of England and Erwin Schrödinger of Germany win the Nobel Prize for Physics for their discoveries in wave mechanics

Enrico Fermi proposes a theory of beta decay in Dec; it is the first suggestion of the existence of the weak interaction and it includes the neutrino, a particle first proposed by Pauli

W. Meissner discovers that the magnetic field is excluded from the interior of a superconductor; this effect is known as the Meissner effect

Ernst Ruska builds the first electron microscope that is more powerful than a conventional light microscope; he obtains a magnification of 12,000 power

Otto Stern, using molecular beams, measures the magnetic moment of a proton and finds it to be 2.5 times larger than predicted by Paul Adrien Maurice Dirac

Robert Jemison Van de Graaff (b Tuscaloosa, AL, Dec 26, 1901) develops a static

High-intensity mercury vapor lamps are introduced

American engineer Edwin H. Armstrong perfects FM (frequency modulation) radio

1933

cont

1934

Ruth Benedict's *Patterns of culture* is a classic of comparative anthropology that is based on her work with the Zuni and Hopi in the American Southwest

Davidson Black, Canadian anthropologist, d Peking, China, Mar 15

Walter Baade and Fritz Zwicky discover that novas found in any galaxy are comparable in brightness, and that they differ from supernovas; they also propose that neutron stars are formed during supernova eruptions; this is confirmed in 1968 when a neutron star is located in the debris of the Jul 4, 1054, supernova

Dutch-American astronomer Bart Jan Bok (b Hoorn, Netherlands, Apr 28, 1906) shows that stellar associations are young systems

Arthur Stanley Eddington's *New pathways of science* is published

W. Grotrian discovers zodiacal dust by studying the spectrum of the solar corona

Willem de Sitter, Dutch astronomer, d Leiden, Nov 20

Fritz Zwicky and Walter Baade develop the theory that stars larger than 1.4 times the mass of the sun (Chandrasekhar's limit) will collapse into neutrons, becoming neutron stars

John Desmond Bernal takes the first X-ray diffraction photograph of a protein crystal, pepsin

Adolf Butenandt obtains the first preparation of progesterone, a female sex hormone, in pure form

Russian ecologist Gause is the first to state the principle that two similar species cannot occupy similar ecological niches for long periods of time (known as Gause's principle)

György Hevesy uses the radioactive phosphorus isotope to study phosphorus metabolism in plants

Santiago Ramón y Cajal, Spanish histologist, d Madrid, Oct 18

American chemist Robert Runnels Williams (b Nellore, India, Feb 15, 1886) extracts about a third of an ounce of thiamin from rice polishings

Arnold O. Beckman develops the pH meter, the first instrument that accurately measures the amount of acidity or alkalinity in a solution using electronic means

Marie Sklodowska Curie, Polish-French chemist, d Haute Savoie, France, Jul 4

Fritz Haber, German chemist, d Basel, Switzerland, Jan 29

Walter Norman Haworth (b Chorley, England, Mar 19, 1883) is among the first to synthesize vitamin C, which he names ascorbic acid

American biochemist J.P. Lent discovers an anticoagulant in spoiled clover, now known as coumarin

Karl Paul Gottfried von Linde, German chemist, d Munich, Bavaria, Nov 16

Harold C. Urey of the United States wins the Nobel Prize for Chemistry for the discovery of heavy hydrogen

EARTH SCIENCE	MATHEMATICS	MEDICINE	PHYSICS	TECHNOLOGY	
			electricity generator that can produce 7 million volts		**1933** cont
			Clarence Zener explains the electrical breakdown of insulators in terms of tunneling effects; an electron tunnels through a potential barrier by a quantum effect similar to disappearing on one side of a barrier and reappearing on the other; it "tunnels through," rather than "climbs over," the barrier		
William Beebe and Otis Barton set a depth record by diving to 1001 m (3038 ft) below the ocean's surface in a tethered sphere called a bathysphere	Aleksander Osipovich Gelfand and T. Schneider show that an irrational power of an algebraic number (other than zero or one) is transcendental, a feat that David Hilbert had predicted would come long after the proof of Fermat's last theorem; Hilbert was, for once, wrong	George R. Minot, William P. Murphy, and George H. Whipple of the United States win the Nobel Prize for Physiology or Medicine for the discovery and development of liver treatment for anemia Theobald Smith, American pathologist, d New York, NY, Dec 10	Jesse W. Beams develops an ultracentrifuge that works in a vacuum James Chadwick and Austrian-American physicist Maurice Goldhaber (b Lemberg, Austria, Apr 18, 1911) measure the mass of a neutron and find it to be greater than the mass of a proton plus an electron, suggesting that neutrons are truly elementary particles Pavel Cherenkov (b Voronezh, USSR, Jul 15, 1904) discovers that when a particle traveling close to the speed of light passes through a liquid or transparent solid, it emits light; this phenomenon occurs when the velocity of the particle is greater than that of light in the medium; it is known as the Cherenkov effect Louis de Broglie introduces the general term *antiparticle* in Jan Enrico Fermi suggests that slow, or thermal, neutrons can be used to split atoms	The first streamlined car, the Chrysler Airflow, is introduced German engineer Wernher von Braun develops a liquid-fuel rocket that achieves a height of 2.4 km (1.5 mi) Charles Francis Jenkins, American TV pioneer, d Washington, DC, Jun 5 George Owen Squier, American radio technician, d Washington, DC, Mar 24	**1934**

1934

cont

1935

German paleoanthropologist Ralph von Koenigswald finds the first evidence—fossil teeth in a Hong Kong apothecary shop—of *Gigantopithecus*, the largest primate known

Henry Norris Russell shows that conservation laws are not obeyed in the theory of Jeffrey and Jeans concerning the origin of the universe

Bernhard Voldemar Schmidt, Russian-German optician, d Hamburg, Germany, Dec 1

Bengt Strömgren formulates his theory about the internal structure of stars

Rupert Wildt (b Munich, Germany, Jun 25, 1905) discovers methane and ammonia on large planets

Riboflavin (vitamin B$_2$) is synthesized

Henrik Dam (b Copenhagen, Denmark, Feb 21, 1895) discovers vitamin K

Edward Calvin Kendall studies the cortical hormones from the adrenal cortex and isolates one that later becomes known as cortisone

Konrad Lorenz (b Vienna, Nov 7, 1903) publishes a general study of social behavior of animals

Charles Frederick Cross, English chemist, d Hove, Sussex, Apr 15

François-Auguste-Victor Grignard, French chemist, d Lyons, Rhône, Dec 13

Frédéric and Irène Joliot-Curie of France win the Nobel Prize for Chemistry for their synthesis of new radioactive elements

John Northrop crystallizes the enzyme chymotrypsin

French physicists Frédéric (b Paris, Mar 19, 1900) and Irène (b Paris, Sep 12, 1897) Joliot-Curie develop the first artificial radioactive element, a radioactive form of phosphorus

Marcus Laurence Oliphant, Australian physicist (b Adelaide, Oct 8, 1901), discovers tritium, the radioactive isotope of hydrogen, by bombarding heavy water (deuterium) with deutrons (particles composed of a single proton and a single neutron); tritium contains two neutrons and one proton

Wolfgang Pauli and Victor F. Weisskopf demonstrate that certain subatomic particles must have antiparticles; not many particles were known at this time, and the demonstration was largely overlooked

Paul Ulrich Villard, French physicist, d Bayonne, Basses-Pyrénées, Jan 13

1935

Sydney Chapman determines the lunar air tide (the effect of the moon's gravitation on Earth's atmosphere)

William Maurice Ewing (b Lockney, TX, May 12, 1913) starts the seismic study of the sea bottom using refraction of waves caused by explosions

Charles Francis Richter develops a scale for measuring the strength of earthquakes based on seismograms; it is still known as the Richter scale

Alan Mathison Turing (b London, Jun 23, 1912) develops the hypothetical machine named after him as a method of determining what kinds of mathematical results can be proved

Gerhard Domagk uses the first sulfa drug, Prontosil, on his youngest daughter to prevent her death from a streptococcal infection, the first use on a human being; its success in this and other instances make Prontosil famous worldwide as the first "wonder drug"

Antonio Caetano de Abreu Freire Egas Moniz (b Avanca, Portugal, Nov 29, 1874) develops prefrontal lobotomy as a treatment for mental illness

The first commercial electron microscope becomes available in England

Jesse W. Beams obtains the first separation of isotopes by centrifuging

Sir James Chadwick of England wins the Nobel Prize for Physics for his discovery of the neutron

Arthur Jeffrey Dempster (b Toronto, Canada, Aug 14, 1896) discovers the U-235 isotope of uranium

William Francis Giauque uses a magnetic field to cool

The beer can is introduced in New Jersey

Wallace Hume Carothers (b Burlington, IA, Apr 27, 1896) patents nylon, which he invented in 1934

John Gibbon and his wife develop the first prototype of the heart-lung machine

British scientists, led by Robert Alexander Watson-Watt (b Brechin, Scotland, Apr 13, 1892), develop the first radar

1935

cont

William Cumming Rose (b Greenville, SC, Apr 4, 1887) discovers threonine, the last of the essential amino acids to be found

Croatian-Swiss chemist Leopold Ružička , b Vukovar (Yugoslavia), Sep 12, 1887, finds the structure of testosterone, a male sex hormone

German biochemist Rudolf Shoenheimer (b Berlin, May 10, 1898) is the first to use isotopes of elements that naturally occur in the body as a tracer; specifically, he uses deuterium (heavy hydrogen) to trace fat storage and metabolism

Hans Spemann of Germany wins the Nobel Prize for Physiology or Medicine for his discovery of the "organizer effect" in embryonic development

Wendell Meredith Stanley (b Ridgeville, IN, Aug 16, 1904) purifies and crystallizes tobacco mosaic virus

Albert Szent-Györgyi locates four acids involved in the respiration of muscle cells, the beginning of the discovery of the Krebs cycle

Axel Theorell shows that the coenzyme for an enzyme involved in respiration has a structure like that of vitamin B_2

Hugo Marie De Vries, Dutch botanist and geneticist, d Lunteren, May 21

John Peters is the first to suggest that there must be some mechanism in or near the heart that acts as a blood-pressure sensor and that can adjust the blood volume; it will be more than 40 years before the mechanism is discovered

Charles Robert Richet, French physiologist, d Paris, Dec 3

Sir Edward Albert Sharpey-Schäfer, English physiologist, d North Berwick, Scotland, Mar 29

helium to a temperature of 0.1 degree from absolute zero

Soviet physicist Lev Davidovich Landau (b Baku, Azerbaijan, Jan 22, 1908) publishes a mathematical explanation of the behavior of atomic magnets, clarifying ferromagnetism

Michael Idvorsky Pupin, Yugoslavian-American physicist, d New York, Mar 12

Konstantin Tsiolkovsky, Russian physicist, d Kaluga, Sep 19

Hideki Yukawa (b Hideki Ogawa, Kyoto, Japan, Jan 23, 1907) proposes in Nov that a new particle, called a meson, causes the attraction between particles in the atomic nucleus; this "exchange particle" is believed to have a mass of 200 electron masses

The mathematics of Nicolas Bourbaki

In 1939 the first volume of *Elements de mathématique* (Elements of mathematics) by Nicolas Bourbaki appeared. This influential work, which eventually ran to more than 20 volumes, sought to recast all of mathematics into a consistent whole with the same symbolism used throughout. The author was identified in a 1949 biographical note as having been formerly with the Royal Poldavian Academy. Somewhat later he moved to the Mathematics Institute of the University of Nancago.

If these institutions seem a bit unfamiliar, do not be surprised. Like the redoubtable M. Bourbaki, they do not exist. Instead, Bourbaki was a pseudonym used by a group of mainly French mathematicians, including Jean Dieudonné, Henri Cartan, and Samuel Eilenberg, with other mathematicians, such as André Weil, loosely affiliated. Each volume is written by a committee and critiqued at a *congrès* featuring fine wine and gourmet food, paid for by Bourbaki's royalties.

Although the true identity of Nicolas Bourbaki has been known to mathematicians for many years, the members of the group have continued to pretend. When the mathematician Ralph P. Boas wrote an article about the group in a popular publication, the editors received a letter from Nicolas Bourbaki himself, protesting the article and claiming that there was no real Ralph P. Boas, but only a group of mathematicians claiming to be Boas.

	GENERAL	ANTHROPOLOGY/ARCHAEOLOGY	ASTRONOMY	BIOLOGY	CHEMISTRY
1936	The British philosopher Alfred Jules Ayer publishes *Language, Truth and Logic*; the book is the first presentation in English of Logical Positivism and the ideas of the "Vienna Circle"		Johannes Franz Hartmann, German astronomer, d Göttingen, Sep 13	Andrei Nikolaevitch Belozersky isolates DNA in the pure state for the first time	Catalytic cracking is developed for refining petroleum

Further Biology entries:

John Joseph Bittner (b Meadville, PA, Feb 25, 1904) shows that a mouse's mother's milk can transmit cancer, the first significant evidence since 1911 the some cancers might be caused by viruses

Sir Henry Dale of England and Otto Loewi of Austria win the Nobel Prize for Physiology or Medicine for their work on the chemical transmission of nerve impulses

Russian biochemist Alexander Ivanovich Oparin (b near Moscow, Mar 3, 1894) addresses the origin of life in his book *The origin of life on Earth*; he puts forward the idea that life evolved in random chemical processes in the ocean, which became a biochemical soup conducive to early life forms, still the most popular theory

Ivan Petrovich Pavlov, Russian physiologist, d Leningrad, Feb 27

Tadeusz Reichstein isolates cortisone and finds its structure

Wendell Meredith Stanley isolates nucleic acids from the tobacco mosaic virus, which later will be found to cause the viral activity

Robert R. Williams synthesizes vitamin B_1 (thiamine), an essential ingredient in fighting beriberi

Etienne Wolff performs his studies of experimental anomalies in embryological development

Further Chemistry entries:

Peter J.W. Debye of the Netherlands wins the Nobel Prize for Chemistry for his study of dipolar moments

Henri-Louis Le Châtelier, French chemist, d Miribel-les-Echelles, Isère, Sep 17

Julius Arthur Nieuwland, Belgian-American chemist, d Washington, DC, Jun 11

Lehmann proves the existence of the inner core of Earth by observing diffracted P waves

Andrija Mohorovičić, Croatian geologist, d Zagreb (Yugoslavia), Dec 18

Alan Turing solves Hilbert's twenty-third problem by showing that it is not possible to devise a method for determining the truth or falsity of all statements that can be made in a part of formal logic called the predicate calculus; therefore, there is no single way to prove or disprove all logical statements

Alonzo Church, independently of Alan Turing, shows that there is no single method for determining whether a statement in mathematics is provable or even true

English mathematician Karl Pearson d Coldharbour, Surrey, Apr 27

Robert Bárány, Austrian physician, d Uppsala, Sweden, Apr 8

Daniele Bovet (b Neuchâtel, Switzerland, Mar 23, 1907) and coworkers discover that the wonder drug Prontosil breaks down in the body and that the part that kills streptococci is a known chemical, sulfanilamide; they show that sulfanilamide is as effective as Prontosil

Alexis Carrel, working with Charles A. Lindbergh, develops a form of artificial heart that is used during cardiac surgery

Charles-Jules-Henri Nicolle, French physician, d Tunis, Tunisia, Feb 28

Felix Bloch suggests a method for polarizing neutrons by passing them through magnetized iron

Victor F. Hess of Austria and Carl D. Anderson of the United States win the Nobel Prize for Physics, Hess for his discovery of cosmic radiation and Anderson for his discovery of the positron

German-American physicist Erwin Wilhelm Mueller (b Berlin, Jun 13, 1911) develops the field-emission microscope, a device for observing a needle tip by using the electrons it emits as it is heated to make an image on a fluorescent screen; with improvements, it is the first device to show individual atoms

Merle A. Tuve (b United States, Jun 27, 1901), N. Neydenberg, and L.R. Hafstad show that the force between two protons is about the same as the force between a proton and a neutron, suggesting that neutrons cannot be composed of protons joined to electrons

Fluorescent lighting is introduced

Penguin Books introduces the first paperbacks

Germany builds a secret place for construction of experimental liquid-fueled rockets

Regular public television transmission starts in Great Britain

Boulder Dam, later christened Hoover Dam for a while, is completed on the Colorado River, creating Lake Mead, the world's largest reservoir

George Harold Brown invents the turnstile antenna for television broadcasting

German engineer Heinrich Focke develops the first practical helicopter

Konrad Zuse builds a primitive form of digital computer using electromagnetic relays instead of tubes (or transistors)

1936

| 1937 | Japan invades China | Franz Weidenreich discovers complete skulls of Peking man | *Dritter fundamental Katalog* (FK3), a catalog of 1587 stars with high-precision coordinates, is published | English plant pathologist Frederick Charles Bawden (b North Tawton, Aug 18, 1908) and coworkers discover that tobacco mosaic virus contains ribonucleic acid (RNA) | William Astbury is the first to use X-ray diffraction to study a nucleic acid |

Benjamin Boss publishes *General catalogue* with 33,342 stars and their positions

N.U. Mayall discovers a supernova remnant by Doppler-split spectral lines

The first intentional radiotelescope is built by Grote Reber (b Wheaton, IL, Dec 22, 1911), a "dish" 9.4 m (31 ft) in diameter

Fritz Zwicky discovers three supernovas in other galaxies using a Schmidt camera

The asteroid Hermes, probably a half-mile in diameter, passes a half-million miles from Earth—the closest approach of a large heavenly body (other than the moon)

Albert Francis Blakeslee (b Geneseo, NY, Nov 9, 1874) discovers that the alkaloid colchicine, derived from the autumn crocus, produces mutations in plants by allowing chromosomes to double without the cell dividing

Genetics and the Origin of Species is published by Theodosius Dobzhansky (b Nemirov, Ukraine, Jan 25, 1900)

Conrad Arnold Elvehjem (b McFarland, WI, May 27, 1901) discovers vitamin A

Hans Adolf Krebs discovers the Krebs cycle, the cycle of oxidation and energy production of all food in living cells

William Cumming Rose establishes that only 10 of the approximately 20 amino acids found in proteins are essential for rats; later he finds that only eight of these are essential for humans

Albert Szent-Györgyi of Hungary wins the Nobel Prize for Physiology or Medicine for his study of vitamin C and respiration

J.H. Woodger's *The axiomatic method in biology* is an axiomatic approach to theoretical biology

Wallace Hume Carothers, inventor of nylon, d Philadelphia, PA, Apr 29

Sir Walter N. Haworth of England and Paul Karrer of Switzerland win the Nobel Prize for Chemistry, Haworth for his work on carbohydrates and vitamin C and Karrer for his work on carotenoids, flavins, and vitamins

Italian-American physicist Emilio Segrè (b Tivoli, Italy, Feb 1, 1905) produces technetium, the first artificial element

A.V. Kazakov explains the origin and distribution of beds of phosphate rock as a result of ocean currents, with the rock being deposited where cold currents upwell; such places encourage vast populations of living things whose remains and fecal pellets are the source of the phosphorus in the rock

Alan Turing's *On computable numbers* describes the "Turing machine," an imaginary machine developed in 1935 that can solve all problems that are computable; in fact, the definition of "computable" is that the problem can be solved by a Turing machine

Ivan M. Vinogradov (b Sep 14, 1891) proves that every sufficiently large even integer is the sum of four primes; this is the closest result of its time to Goldbach's conjecture that every even integer greater than two is the sum of two primes

Based on the model of the successful drug sulfanilamide, scientists create sulfapyridine, the second sulfa drug

Alfred Adler, Austrian psychiatrist, d Aberdeen, Scotland, May 28

Pharmacologist Daniele Bovet, working at the Pasteur Institute in France, develops the first antihistamine

Italian doctors Ugo Cerlutti and Lucio Bini develop the first form of electroconvulsive therapy (ECT), often known as "shock treatment," for treating schizophrenia

South African-American microbiologist Max Theiler (b Pretoria, South Africa, Jan 30, 1899) introduces a vaccine against yellow fever

The Nobel Institute of Physics is founded in Stockholm, Sweden

Carl David Anderson discovers the mu meson (muon) in cosmic radiation; it is a subatomic particle with mass in between that of the proton and electron (later measured at 207 times the mass of the electron); this discovery leads physicists to think (incorrectly) that it is the Yukawa meson, a particle that mediates the strong force; instead it behaves as a "fat electron"

Sir Jagadischandra Bose, Indian physicist, d Giridh, Nov 23

Clinton Davisson of the United States and George P. Thomson of England win the Nobel Prize for Physics for their discovery of electron diffraction by crystals

Canadian-American physicist James Hillier (b Brantford, Ontario, Aug 22, 1915) and Albert F. Prebus improve the electron microscope, developing the basic design that is used in most later versions

Russian physicist Peter Leonidovich Kapitza (b Kronshtadt, Jul 8, 1894) explains superfluidity in helium II (helium cooled below 2.2°K)

A. Kramers introduces the charge conjugation operation in Nov and declares that particle interactions are invariant under it; in the charge conjugation operation, positive charges are replaced by negative ones and vice versa

The British start building a chain of 20 radar stations reaching from the Solent to the Firth of Tay

John V. Atanasoff starts work on the first electronic computer, a machine for solving systems of linear equations; the first operational prototype is completed in Oct, 1939, and an operational version known as the ABC, which fails frequently because of problems with the punched-card input, is working by 1942

The first rocket tests are performed at the Baltic research station at Peenemünde; one of the leaders of the team is Wernher von Braun

American law student Chester Carlson invents xerography, the first method of photocopying

Walter R. Dornberger (b Germany, Sep 6, 1895) organizes the construction of the V-2 rocket

Marchese Guglielmo Marconi, Italian electrical engineer, d Rome, Jul 20

Elihu Thomson, English-American inventor, d Swampscott, MA, Mar 13

Frank Whittle builds the first working jet engine

1937

cont

BIOLOGY

John Zachary Young studies propagation speed in nerves

1938

ASTRONOMY

German physicist Hans Bethe, b Strasbourg (France), Jul 2, 1906, and, independently, Carl von Weizsäcker (b Kiel, Germany, Jun 28, 1912) propose a new theory of the cause of the energy produced by stars—nuclear fusion of hydrogen into helium; this basic concept, with minor modifications, continues to be accepted today

Paul Adrien Maurice Dirac links the Hubble constant, which describes the age and size of the universe, to physical constants that describe subatomic particles

George Ellery Hale, American astronomer, d Pasadena, CA, Feb 21

Dutch-American astronomer Gerard Peter Kuiper (b Harenkarspel, Netherlands, Dec 7,

BIOLOGY

John Jacob Abel, American biochemist, d Baltimore, MD, May 26

Calvin Bridges produces a map of 1024 genes of the X chromosome of the fruit fly (*Drosophila*)

Captain Hendrik Goosen catches a living coelacanth in the Indian Ocean off the African coast; previously the species had been believed to have been extinct for 60 million years

Hans Christian Joachim Gram, Danish bacteriologist, d Copenhagen, Nov 14

Corneille Heymans of Belgium wins the Nobel Prize for Physiology or Medicine for his discoveries in respiratory regulation

CHEMISTRY

Austrian-German chemist Richard Kuhn (b Vienna, Austria, Dec 3, 1900) isolates vitamin B_6 from skim milk

Richard Kuhn of Germany wins the Nobel Prize for Chemistry for his carotenoid and vitamin research

Du Pont chemist Roy Plunkett on Apr 6 accidentally discovers the first form of Teflon in the residue of refrigeration gases

Arthur Stoll and Albert Hofman discover lysergic acid diethylamide (LSD)

Georges Urbain, French chemist, d Paris, Nov 5

			Isidor Rabi and Polykarp Kusch use the magnetic resonance method for studying the atomic nucleus		**1937** cont

Ernest Rutherford, 1st Baron Rutherford of Nelson, British physicist, d London, Oct 19

Russian physicist Igor Yevgenyevich Tamm (b Vladivostok, Jul 8, 1895) develops a theoretical explanation of Cherenkov radiation, the electromagnetic radiation caused when electrons accelerate

---

EARTH SCIENCE	MATHEMATICS	MEDICINE	PHYSICS	TECHNOLOGY	
G.S. Callendar determines that human activities are causing an increase in the amount of carbon dioxide in Earth's atmosphere	Kurt Gödel proves that if the Zermelo-Fraenkel axioms for set theory are consistent with each other, they will continue to be consistent if the controversial axiom of choice is added to them	Burrhus Frederick Skinner's *The behavior of organisms* is published	Stückelberg formulates the law of baryon conservation; that is, the number of heavy particles, such as protons and neutrons, stays the same in any closed system	The radio altimeter is introduced	**1938**

Medicine (cont.):

Tom D. Spies shows that pellagra is a deficiency disease that can be treated with niacin

English surgeon Philip Wiles develops the first total artificial hip replacement, using stainless steel

Physics (cont.):

Albert Einstein and Leopold Infeld's *The evolution of physics* describes the development of physical science for the layman; it is an enormous success

Enrico Fermi of Italy wins the Nobel Prize for Physics for his work with thermal neutrons

Charles-Edouard Guillaume, Swiss-French physicist, d Sèvres, near Paris, Jun 13

German physicist Otto Hahn is the first to split the atom of uranium, opening up the possibility of a chain reaction and

Technology (cont.):

Hungarian Lazlo Biró patents the first ball-point pen

Germany's liquid-fueled rocket experiments, under the direction of Wernher von Braun, succeed in producing a rocket that can travel 18 km (11 mi)

George Harold Brown develops the vestigial sideband filter for use in television transmitters, doubling the horizontal resolution of television pictures at any given bandwidth

Nylon, invented in 1934 by Wallace Hume Carothers of Du Pont, goes on sale in the United States

German engineer Ferdinand Porsche introduces the prototype of the Volkswagen beetle

1938
cont

1905) finds the relationship between spectral type and biometric correction for a star, making it possible to determine the luminosity of a star over all wavelengths

J. Robert Oppenheimer (b New York, NY, Apr 22, 1904) and George Volkoff predict the existence of rapidly rotating neutron stars; these are discovered in 1967 by Jocelyn Bell and named pulsars

William Henry Pickering, American astronomer, d Mandeville, Jamaica, Jan 16

1939

On Sep 1, German and Soviet forces begin occupying Poland, precipitating World War II

Albert Einstein writes the letter to President Roosevelt, dated Aug 2, that will lead to the US effort to develop an atomic bomb

By comparing photographs taken in 1909, 1921, and 1938, John C. Duncan discovers that the Crab nebula expands, proving that the nebula is some 800 years old

Sir Frank Watson Dyson d May 25

J. Robert Oppenheimer calculates that if the mass of a star is more than 3.2 times the mass of the sun, a collapse of the star caused by lack of internal radiation would lead to the star's mass being concentrated at a point, creating what would come to be known as a black hole

Andrei Nikolaevitch Belozersky starts his experimental work showing that both DNA and RNA are always present in bacteria

Edward Adelbert Doisy (b Humer, IL, Nov 13, 1893) isolates vitamin K

Edmund Beecher Wilson, American zoologist, d New York, NY, Mar 3

Du Pont begins to market nylon

The British firm ICI begins to manufacture polythene

Adolf Butenandt of Germany and Leopold Ružička of Switzerland share the Nobel Prize for Chemistry, Butenandt for his study of sexual hormones and Ružička for his work with atomic ring structures and terpenes

Paul Müller (b Olten, Switzerland, Jan 12, 1899) discovers that DDT is a potent insecticide

Linus Pauling's *The nature of the chemical bond, and structure of molecules*

atomic bombs; however, when he publishes his results in Jan, 1939, he is too cautious to call it nuclear fission

In Mar, Lise Meitner (b Vienna, Austria, Nov 7, 1878) flees Hitler-controlled Austria for Sweden, taking the problem of uranium atom-splitting with her; in Sweden she determines that she and Otto Hahn had indeed been splitting uranium; her paper on the subject in Jan, 1939, starts the drive to develop an atomic bomb

Nevill Mott and R.W. Gurney formulate a theory explaining image formation in photographic emulsions

Fritz Zernike (b Amsterdam, Netherlands, Jul 16, 1888) invents the phase contrast microscope

US engineer T. Ross develops the first machine that can learn from experience (to find its way through mazes)

P. Schlack develops "perlon," a synthetic fiber

A symbolic analysis of relay and switching circuits by Claude Elwood Shannon (b Gaylord, MI, Apr 30, 1916) is a founding document of the mathematical theory of information

Konrad Zuse completes the Z_1, a binary calculating machine

1938

cont

Hannes Alfvén (b Norrköping, Sweden, May 30, 1908) publishes a theory relating auroras to magnetic storms

Edward Victor Appleton measures the height of the E layer in the ionosphere

Walter Maurice Elsasser (b Mannheim, Baden, Germany, Mar 20, 1904) suggests that the liquid iron core of Earth has eddy currents that set up Earth's magnetic field

The first volume of mathematics attributed to Nicolas Bourbaki appears; this attempt to synthesize all known mathematics, which continues to appear in numerous volumes, is actually written by a group of mostly French mathematicians who use Bourbaki as a pseudonym

Carl Louis Ferdinand von Lindemann, German mathematician, d Munich, Mar 6

Scientists create sulfathiazole, the third sulfa drug

Surgeon and author Harvey William Cushing d Oct 7

Gerhard Domagk of Germany wins the Nobel Prize for Physiology or Medicine for discovery of the first sulfa drug, Prontosil

René-Jules Dubos (b Saint-Brice, France, Feb 2, 1901) searches for and finds two compounds produced by a soil bacterium that kill other bacteria—antibiotics; these are the first antibiotics to have been deliberately sought for this property

Sigmund Freud, Austrian psychiatrist, d London, Sep 23

Lise Meitner and Austrian-British physicist Otto Robert Frisch (b Vienna, Oct 1, 1904) advance the theory that uranium, when bombarded by neutrons, breaks into smaller atoms; the term *fission* is their word for this process

Philip Hauge Abelson (b Tacoma, WA, Apr 27, 1919) identifies the products of uranium fission: three radioactive isotopes of antimony, six of tellurium, and four of iodine

Homi Bhabha suggests the name *meson* for the middle-weight particle predicted by Yukawa

Felix Bloch measures the magnetic properties of the neutron

The handheld electric slicing knife is introduced in the United States

The complex number calculator is built at Bell Labs

Pan American institutes the first regular commercial flights across the Atlantic Ocean

The first precooked frozen foods are marketed under the Birds Eye label

Arthur Edwin Kennelly, British-American electrical engineer, d Boston, MA, Jun 18

1939

1939

cont

Scientists and World War II

Several developments in science and technology, such as radar, early computers, large liquid-fueled rockets, penicillin, and DDT, were the result of the war effort during World War II. The harnessing of nuclear energy and the construction of the atomic bomb is probably the most important scientific and technical result of the scientific effort during the war.

Because of fear that Nazi Germany might develop an atomic bomb—later proven unfounded—Leo Szilard persuaded Albert Einstein to write a letter to US President Franklin D. Roosevelt, in which Einstein suggested that the fission of uranium could be used to produce an atomic explosion. This letter resulted in the establishment of the largest single enterprise in the history of science. The project, later known as the Manhattan project, comprised 37 installations in 19 states and Canada, and employed 43,000 people. It ran on a budget of $2.2 billion. A large number of physicists, among them the best scientists who had fled the Nazis, joined the project.

New cities emerged as a direct result of the Manhattan project. Oak Ridge in Tennessee, where the gaseous diffusion plant for the separation of uranium-235 from uranium-238, was located, reached a population of 79,000 in less than two years. Hanford in the state of Washington housed the nuclear reactors for the transmutation of uranium-238 to plutonium-239 and reached a population of 60,000. Los Alamos, New Mexico, was the site where the atomic bombs were designed and built under the directorship of the physicist J. Robert Oppenheimer.

Notwithstanding the enormous size of the organization, the American government succeeded in keeping the whole operation a secret. Most of the employees, and even some scientists, did not know the exact aim of the organization. Secrecy was strictly enforced. Scientists traveled under assumed names. For example, Enrico Fermi was disguised as Henry Farmer, and Eugene Wigner as Eugene Wagner. All telephone conversations at Los Alamos were under surveillance and were interrupted if the military authorities judged them a threat to secrecy. Sensitive matters were referred to by code words; plutonium was called 49 (94 is its atomic number) and uranium "tube alloy."

The massive effort resulted in the first experimental explosion of a nuclear device in July 1945, quickly followed by the dropping of atomic bombs on the Japanese cities Hiroshima and Nagasaki on August 6 and 9, respectively, 1945.

and crystals becomes a classic work of chemistry, still used today

Marguerite Perey discovers the chemical element francium

Sir William Jackson Pope, English chemist, d Cambridge, Oct 17

Soren Peter Lauritz Sorensen, Danish chemist, d Copenhagen, Feb 12

Niels Bohr proposes a liquid-drop model of the atomic nucleus

John Ray Dunning (b Shelby, NE, Sep 24, 1907) is the first to confirm Meitner's theory of uranium fission experimentally

W. Conyer Herring develops a method of calculating properties of material objects from quantum principles; he succeeds in explaining why beryllium behaves as a metal

French scientists Frédéric and Irène Joliot-Curie demonstrate that the fission of the uranium atom can lead to a chain reaction

Ernest Lawrence of the United States wins the Nobel Prize for Physics for his invention of the cyclotron

American biophysicist Richard Brooke Roberts (b Titusville, PA, Dec 7, 1910) discovers that a fissioning uranium atom does not release all its neutrons at the same time; the "delayed neutrons" are an important element in controlling nuclear reactors

Hungarian-American physicist Leo Szilard (b Budapest, Feb 11, 1897) and Walter Zinn confirm that fission reactions can be self-sustaining because of chain reactions, first observed by Otto Hahn and Fritz Strassman in Germany in 1938

Vladimir Zworykin improves the electron microscope to the point that it enlarges 50 times as much as the best light microscope

German engineer Pabst von Ohain's jet engine becomes the first such engine actually to fly an airplane, the He 178

Igor Sikorsky constructs the first helicopter designed for mass production; it flies for the first time on Sep 14

1939

cont

	GENERAL	ANTHROPOLOGY/ ARCHAEOLOGY	ASTRONOMY	BIOLOGY	CHEMISTRY
1940		Four young men discover cave paintings, 17,000 years old, in Lascaux cave in France in Sep Marie Eugène François Thomas Dubois, Dutch paleontologist, d Halen, Belgium, Dec 16	Marcel Minnaert, Gerard Mulders, and Jaap Houtgast's *Photometric atlas of the solar system* is the graphic recording of the Fraunhofer lines in the solar spectrum; the atlas becomes a widely used research tool Rupert Wildt suggests that absorption in the solar atmosphere is mainly caused by negative hydrogen ions	Sir Arthur Harden, English biochemist, d Bourne End, Buckinghamshire, Jun 17 Russian botanist Nikolay Vavilov (b Moscow, Nov 25, 1887) is arrested on Aug 6 for opposing the views of Soviet geneticist Trofim Lysenko that acquired characteristics can be inherited; he is sentenced to death but his sentence is later changed to imprisonment Woods isolates para-aminobenzoic acid, which proves to counteract growth inhibitory effects of sulfanilamide on bacteria	Philip Hauge Abelson and Edwin Mattison McMillan (b Redondo Beach, CA, Sep 18, 1907) create the first element with an atomic number higher than that of uranium; element number 93 is christened neptunium Karl Bosch, German chemist, d Heidelberg, Apr 26 Vincent Du Vigneaud, American biochemist (b Chicago, May 18, 1901) identifies biotin, the compound previously known as vitamin H Sir Robert Abbott Hadfield, British metallurgist, d London, Sep 30 Martin David Kamen, Canadian-American biochemist (b Toronto, Ontario, Aug 27, 1913) discovers carbon-14, the long-lived radioactive isotope of carbon that has become the most useful of all the radioactive tracers, with a special use in dating ancient sites Phoebus Aaron Theodor Levene, Russian-American chemist, d New York, NY, Sep 6
1941	The Japanese attack United States forces at Pearl Harbor on Dec 7 On the day before the Pearl Harbor attack by the Japanese that plunges the United States into World War II, US President Franklin D. Roosevelt signs the order that leads to the development of the nuclear-fission bomb; this will end the war with Japan in the major destruction of the Japanese cities of Hiroshima and Nagasaki		The Maxutov telescope, in which aberration is corrected by a spherical meniscus, is invented in Moscow Annie Jump Cannon, American astronomer, d Cambridge, MA, Apr 13 Bengt Edlen shows that the "coronium" lines in the solar corona are caused by strongly ionized iron, calcium, and nickel lines; from the ionization the temperature	George Wells Beadle (b Wahoo, NE, Oct 22, 1903) and Edward Lawrie Tatum demonstrate that genes control chemical reactions in cells; their theory becomes known as the one-gene, one-enzyme hypothesis Rudolf Schoenheimer, German-American biochemist, d New York, NY, Sep 11 Hans Spemann, German zoologist, d Freiburg-im-Breisgau, Baden, Sep 12	The German I.G. Farberindustrie begins to produce polyurethanes, plastics first developed between 1937 and 1939 by Otto Bayer Arnold O. Beckman invents a spectrophotometer that measures the chemical composition of a sample based on the wavelengths of light reflected by the sample American scientists L.D. Goodhue and W.N. Sullivan develop

EARTH SCIENCE	MATHEMATICS	MEDICINE	PHYSICS	TECHNOLOGY	
Harold Jeffreys (b Fatfield, England, Apr 22, 1891) and K.E. Bullen publish the "J-B tables" for the travel times of P and S seismic waves through Earth Augustus Edward Hough Love, English geophysicist, d Oxford, Jun 5	Kurt Gödel shows that Cantor's continuum hypothesis (that there is no number of an infinite set between the set of natural numbers and the set of real numbers) is consistent with the axioms of set theory	Herbert M. Evans uses radioactive iodine to prove that iodine is used by the thyroid gland Howard Florey (b Adelaide, Australia, Sep 24, 1898) and Ernst Chain (b Berlin, Germany, Jun 19, 1906) develop penicillin as an antibiotic in England Karl Landsteiner and Alexander S. Wiener discover a relationship between human and rhesus blood cells and discover the rhesus factor, more commonly called the Rh factor Julius Wagner von Jauregg, Austrian psychiatrist, d Vienna, Sep 27	John Ray Dunning shows that Niels Bohr's theory that uranium-235 is more fissionable than uranium-238 is correct; during World War II, Dunning develops the gas-diffusion method of separating uranium-235 from uranium-238, making the first nuclear bomb Maurice Goldhaber discovers that beryllium slows down fast neutrons and makes them more likely to fission uranium Donald William Kerst (b Galena, IL, Nov 1, 1911) builds the first usable betatron Sir Oliver Joseph Lodge, English physicist, d Lake, Wiltshire, Aug 22 Sir Joseph John Thomson, English physicist, d Cambridge, Aug 30 Pierre Weiss, French physicist, d Lyons, Oct 24	The collapse of the four-month-old Tacoma Narrows Bridge leads engineers to consider aerodynamic stability in bridges and buildings for the first time Freeze drying, developed earlier for medicines, is used for food preservation for the first time in the United States Paul Nipkow, German TV pioneer, d Berlin, Aug 24 The first color television broadcast takes place using a system developed by Peter Carl Goldmark	1940
	Abraham Adrian Albert starts his work on nonassociative algebras	Scientists create sulfadiazine, the fourth sulfa drug Sir Frederick Grant Banting, Canadian physiologist, d Near Musgrave Harbour, Newfoundland, Feb 21 Hans Berger, German psychiatrist, d Jena, Germany, Jun 1 Canadian-American surgeon Charles Branton Huggins (b Halifax, Nova Scotia, Sep 22, 1901) shows	Soviet physicist Georgii Nikolaevich Flerov (b Mar 2, 1913) discovers spontaneous fission of uranium Dayton Clarence Miller, American physicist, d Cleveland, OH, Feb 22	Frank Conrad, American broadcaster, d Miami, FL, Dec 10 In Britain, John Rex Whinfield (b Feb 16, 1901) and J.T. Dickson invent the artificial fiber terylene, better known in the United States as dacron German scientist Konrad Zuse's Z_2 computer is the first to use electromagnetic relays and a punched tape for data entry	1941

| 1941 cont | | of the corona is shown to be 1 million degrees Celsius

Bernard Lyot invents the Lyot filter (monochromator)

John Stanley Plaskett, Canadian astronomer, d Esquimalt, British Columbia, Oct 17 | | the "bug bomb," or aerosol spray for insecticides

Hermann Walther Nernst, German physical chemist, d Bad Muskau, near Berlin, Nov 18

Paul Sabatier, French chemist, d Toulouse, Haute-Garonne, Aug 14

Glenn Theodore Seaborg (b Ishpeming, MI, Apr 19, 1912) and Edwin Mattison McMillan create element 94, plutonium |
| 1942 | Sir William Matthew Flinders Petrie, English archaeologist, d Jerusalem, Palestine, Jul 28 | Heber Doust Curtis, American astronomer, d Ann Arbor, MI, Jan 9

Solar radio emission is detected by M.H. Hey and his colleagues on Feb 27-28; the emission had previously been attributed to intentional jamming by the Germans

Grote Reber makes the first radio maps of the universe, locating individual radio sources, including the radio galaxy Cygnus A, about 700 million light-years away | Salvador Edward Luria, Italian-American microbiologist (b Turin, Italy, Aug 13, 1912) obtains the first good electron photomicrographs of a bacteriophage | Max Bodenstein, German chemist, d Berlin, Sep 3

Vincent Du Vigneaud deduces the complex two-ring structure of biotin; the following year, chemists following the description of Du Vigneaud succeed in synthesizing biotin

American chemist Louis Fieser develops napalm

American chemist Frank Harold Spedding (b Hamilton, Ontario, Canada, Oct 22, 1902), a specialist in purifying chemicals, produces in Nov two tons of extremely pure uranium for use in developing the nuclear bomb

Richard Willstätter, German chemist, d Locarno, Switzerland, Aug 3 |

that administration of female sex hormones can be used to control prostate cancer

Russian-American microbiologist Selman Abraham Waksman (b Priluki, Jul 22, 1888) coins the term *antibiotic* to describe substances that kill bacteria without injuring other forms of life

---

After the yellow fever vaccine used to inoculate thousands of US Army personnel is shown to have caused hepatitis B in many of them, a new vaccine is developed that does not require the use of human blood serum

Sir William Henry Bragg, English physicist, d London, Mar 12

A team headed by Italian physicist Enrico Fermi creates the first controlled chain reaction in a pile of uranium and graphite at the University of Chicago on Dec 2

Physicist Stephen William Hawking b Oxford, UK, Jan 8; despite a severe physical handicap, Hawking become one of the leaders in studying black holes and various aspects of cosmology

Jean-Baptiste Perrin, French physicist, d New York, NY, Apr 17

Shoichi Sakata and T. Inoue present their two-meson theory to account for the differences between the as yet unobserved Yukawa meson (now known as the pion) and the cosmic-ray meson (now known as the muon); when first discovered, the cosmic-ray meson was thought to be the particle predicted by Yukawa because of similarities in mass

US Army Engineers build the Alcan Highway, the 2450.5-km (1519.3-mi) link between the United States and Alaska

LORAN (Long Range Air Navigation) begins operation for the first time along the Atlantic seaboard of North America

John V. Atanasoff and Clifford Berry complete the ABC or Atanasoff-Berry Computer, considered the prototype of all later electronic computer designs; however, the ABC is not fully operational when Atanasoff and Berry are called to other duties connected with the war effort

Valdemar Poulsen, Danish inventor, d Copenhagen, Jul

1943

ASTRONOMY

Albrecht Unsöld determines the abundance of elements in tau Scorpii

BIOLOGY

Henrik Dam of Denmark and Edward A. Doisy of the United States share the Nobel Prize for Physiology or Medicine, Dam for the discovery of and Doisy for determining the composition of vitamin K

Nikolay Ivanovich Vavilov, Russian botanist, d Saratov, Jan 26, as a result of maltreatment by Soviet prison guards; he had been imprisoned for opposing the views of Trofim Lysenko, who believed that acquired traits could be inherited

CHEMISTRY

The Dow Corning Corporation in the United States begins to manufacture the first silicones

George Washington Carver, American agricultural chemist, d Tuskegee, AL, Jan 5

György Hevesy of Hungary wins the Nobel Prize for Chemistry for his use of isotopes as tracers

Swiss chemist Albert Hofmann discovers that LSD is hallucinogenic

1944

ASTRONOMY

Walter Baade discovers that there are two populations of stars: population I consists of younger stars found in the arms of galaxies; population II stars are older and found in the nuclei of spiral galaxies

Sir Arthur Stanley Eddington, English astronomer and physicist, d Cambridge, UK, Nov 22

Hendrik Christoffel van de Hulst (b Utrecht, Netherlands, Nov 19, 1918) suggests that interstellar hydrogen must emit radio waves at a wavelength of 21.1 cm (8.3 in.), the key to mapping the spiral arms of our galaxy

Gerard Peter Kuiper discovers the atmosphere of Titan

BIOLOGY

Oswald Theodore Avery (b Halifax, Canada, Oct 21, 1877), Colin MacLeod, and Maclyn McCarthy determine that deoxyribonucleic acid (DNA) is the hereditary material for almost all living organisms

R.B. Cowles and C.M. Bogert find that desert reptiles regulate their body temperature by specific behavior patterns

Joseph Erlanger (b San Francisco, CA, Jan 5, 1874) and Herbert Spencer Gasser (b Platteville, WI, Jul 5, 1888) win the Nobel Prize for Physiology or Medicine for work conducted in the 1920s on such functions of a single nerve fiber as speed of transmission of a message in different fibers

CHEMISTRY

Leo Hendrik Baekeland, Belgian-American chemist, d Beacon, NY, Feb 23

Otto Hahn of Germany wins the Nobel Prize for Chemistry for his discovery of atomic fission

British biochemists Archer John Porter Martin (b London, Mar 1, 1910) and Richard Lawrence Millington Synge (b Liverpool, Oct 28, 1914) develop paper chromatography, an important new tool in identifying the chemistry of organic compounds

Thomas Midgley, Jr., American chemist, d Worthington, OH, Nov 2

In Mexico farmers recognize on Feb 20 that a volcano, later known as Mount Parícutan, is growing in a cornfield about 320 km (200 mi) west of Mexico City	David Hilbert, German mathematician, d Göttingen, Feb 14	Dutch doctor Wilhelm Kolff develops the first kidney dialysis machine Karl Landsteiner, Austrian-American physician, d New York, NY, Jun 26 Selman A. Waksman discovers the antibiotic streptomycin, produced by a mold that grows in soil; previous antibiotics worked only against Gram-positive bacteria, but streptomycin is effective against Gram-negative bacteria	The world's first operational nuclear reactor is activated at Oak Ridge, TN Marcus Oliphant proposes the proton synchrotron, a powerful type of particle accelerator Otto Stern of the United States wins the Nobel Prize for Physics for his work on the magnetic momentum of molecular beams Shin'ichiro Tomonaga (b Kyoto, Japan, Mar 31, 1906) develops a renormalizable quantum electrodynamics (QED) four years before a similar theory is created by Richard Feynman and Julian Schwinger Pieter Zeeman, Dutch physicist, d Amsterdam, Oct 9	Jacques-Yves Cousteau (b St. André-de-Cubzac, France, Jun 11, 1910) and Gagnan invent the Aqualung, commonly known as scuba gear (for self-contained underwater breathing apparatus) Continuous casting of steel is developed by German engineer S. Junghans Nikola Tesla, Croatian-American electrical engineer, d New York, NY, Jan 7 A team headed by Alan Turing develops Colossus, the first all-electronic calculating device (it uses vacuum tubes); unlike a general-purpose computer, however, Colossus is dedicated to cracking German codes—and is very good at it, possibly influencing the course of World War II	**1943**
Harry Fielding Reid, American geophysicist, d Baltimore, MD, Jun 18	Oskar Morgenstern, German-American economist, b Görlitz, Silesia (Poland), Jan 24, 1902, and John von Neumann develop the mathematical theory of games in their *Theory of games and economic behavior*	Alfred Blalock performs the first "blue baby" operation, correcting blood supply to the lungs of a female infant Alexis Carrel, French-American surgeon, d Paris, Nov 5 Benjamin Minge Duggar (b Gallion, AL, Sep 1, 1872) and coworkers discover the antibiotic Aureomycin, the first of the tetracyclines, as a result of checking many soil samples for antibacterial action Carl Koller, Austrian-American physician, d New York, NY, Mar 21	Charles Glover Barkla, English physicist, d Edinburgh, Scotland, Oct 23 Jacob Bigeleisen and Maria Goeppert Mayer develop a theory explaining differences in chemical behavior of isotopes in which they consider the vibrations of atoms in molecules Isidor Isaac Rabi of the United States, b Rymanów, Austria (now Poland), Jul 29, 1898, wins the Nobel Prize for Physics for his studies of the magnetic properties of molecular beams	Early in the year, the German armed forces begin to use the V-1 flying bomb, propelled by a jet engine and controlled by an autopilot, against the United Kingdom; in Sep the V-2, a liquid-fueled rocket-propelled bomb, also goes into operation The Greenwich Royal Observatory installs its first quartz-crystal clock, providing ten times the accuracy of the previous pendulum system The Germans introduce a rocket-powered airplane, the Me 163B-1 Komet, into World War II, but its habit of exploding spontaneously makes it a poor weapon	**1944**

1944

cont

ASTRONOMY

Carl Friedrich von Weizsäcker revises the nebular hypothesis of the origin of the solar system in a way that causes other astronomers to accept it; in von Weizsäcker's theory, small bodies called planetesimals are attracted to each other to form the planets

CHEMISTRY

Glenn T. Seaborg and his coworkers create elements 95 and 96, which they name americium and curium

1945

GENERAL

Germany surrenders to the Allies on May 7, ending the European portion of World War II

The Japanese city of Hiroshima is bombed on Aug 6 with a nuclear fission bomb based on uranium-235; this provides the first knowledge to the world at large of the existence of the "atomic bomb"

A plutonium-based fission bomb is exploded over Nagasaki, Japan on Aug 9; this exhausts the US supply of "atomic bombs," although this is kept secret

Japan surrenders to the Allies on Aug 14, ending the Pacific part of World War II

ASTRONOMY

The first radar signals are reflected from the moon by Z. Bay of Hungary and independently by the US Army Signal Corps

Science-fiction writer Arthur C. Clarke proposes the idea of communications satellites that will be stationary above a particular part of Earth; called synchronous satellites or satellites in a geosynchronous orbit, they become the principal means of intercontinental communication, starting in 1965

BIOLOGY

Melvin Calvin (b St. Paul, MN, Apr 8, 1911) starts using the Carbon-14 isotope for investigating photosynthesis

Albert Claude, Belgian-American cytologist (b Longlier, Belgium, Aug 24, 1898) completes electron-microscope studies of the cell; these show for the first time such structures as the endoplasmic reticulum and the details of mitochondria

Karl Landsteiner's *The specificity of serological reactions* is the first large study of chemical specificity in immunology

Salvador Luria shows that the same spontaneous mutations occur in bacteriophages and in the bacteria on which the phages prey, suggesting that the genetic material of the phage somehow gets mixed into the genetic material of the bacterium

Thomas Hunt Morgan, American geneticist, d Pasadena, CA, Dec 4

CHEMISTRY

The herbicide 2, 4-D is introduced

Francis William Aston, English chemist and physicist, d Cambridge, Nov 20

Hans Fischer, German chemist, d Munich, Bavaria, Mar 31

Artturi Virtanen (b Helsinki, Finland, Jan 15, 1895) wins the Nobel Prize for Chemistry for his invention of fodder preservation

EARTH SCIENCE	MATHEMATICS	MEDICINE	PHYSICS	TECHNOLOGY	
				The second electronic digital computer, the Automatic Sequence Controlled Calculator, or Mark I, is completed by Howard Aiken and a team of engineers from IBM; it uses punched paper tape for programming and vacuum tubes to calculate problems, but breaks down frequently from problems with the vacuum tubes Stuart Ballantine, American physicist, d Morristown, NJ, May 7	**1944** cont
Sir William Napier Shaw, British meteorologist, d Mar 23 Vladimir Ivanovich Vernadsky, Russian geochemist, d Moscow, Jan 6		Fluoridation of a water supply to prevent dental decay is introduced in the United States Walter Bradford Cannon, American physiologist, d Franklin, NH, Oct 1 Sir Alexander Fleming, Sir Howard W. Florey, and Ernst Boris Chain of England win the Nobel Prize for Physiology or Medicine for the discovery of penicillin and research into its value as a weapon against infectious disease	M. Conversi, E. Pancini, and O. Piccioni demonstrate that the known meson (now known as the muon) interacts too weakly with protons and neutrons to be the Yukawa particle predicted in 1935 Charles Fabry, French physicist, d Paris, Dec 11 Hans Wilhelm Geiger, German physicist, d Potsdam (E Germany), Sep 24 Wolfgang Pauli of Austria wins the Nobel Prize for Physics for the Pauli exclusion principle for subatomic particles Soviet physicist Vladimir Iosifovich Veksler (b Zhitomir, Ukraine, Mar 4, 1907) suggests the design of an improved particle accelerator, the synchrocyclotron	The White Sands proving ground for US rocket research is established in New Mexico John Presper Eckert (b Philadelphia, PA, Apr 9, 1919) and John W. Mauchly develop ENIAC, The Electronic Numerical Integrator and Computer, now generally conceded to be the first all-purpose stored-program electronic computer, although preceded by two specialized electronic computers Sir John Ambrose Fleming, English electrical engineer, d Sidmouth, Devonshire, Apr 18 Robert Hutchings Goddard, American rocket pioneer, d Baltimore, MD, Aug 10 John von Neumann publishes his *First draft of report on the EDVAC*, placing the concept of the EDVAC (Electronic Discrete Variable Computer) in the public domain	**1945**

OVERVIEW

Science after World War II

During World War II scientific development was vastly accelerated as a result of the war effort. Among the discoveries and inventions that reached practical application as a result of the war (although they all had roots in prewar research) were synthetic rubber, radar, DDT, penicillin, nuclear fission, jet-powered aircraft, helicopters, ballistic missiles, and the electronic digital computer. After the war, this technology quickly reached the public in developed countries. The cumulative impact of the advances changed the environment in fundamental ways. There had been no period of comparably swift technological change since the Industrial Revolution or perhaps not even since the adoption of farming some 10,000 years earlier.

In some cases, inventions or discoveries not directly involved with the war effort actually slowed rather than accelerated. One such example is that of television. But after the war, when the growth of television resumed, it quickly became ubiquitous in advanced countries.

Although the changes that occurred were phenomenal, the scientists of the time were not very good at predicting which changes would occur in the postwar period. Some said that helicopters would replace automobiles. No one foresaw that DDT would be banned in the United States after more than 20 years of use. The real significance of the electronic computer was misunderstood in the immediate postwar period (one prediction was that by the 1980s, a few large companies would own their own computers), while people expected nuclear power to solve all of the world's energy problems.

In pure science as well, the unexpected continued to occur. Few scientists in the brief optimistic period between the end of World War II and the start of the Cold War would have predicted that before the end of the century, our evolutionary history would be revealed in the test tube instead of in fossils; people would walk on the moon; the beginnings of the universe would be explained; the secrets of heredity would be unraveled; the discarded continental drift theory would resurface as the most vital part of earth science; many famous conjectures in mathematics would be resolved; the elementary particles would almost

make sense; and solid-state devices would replace vacuum tubes in most applications.

Big science

The way science was conducted also changed after (in some instances during) World War II. Before that time, almost all advances could be ascribed to an individual working on his or her own or with a single partner or mentor. After World War II, it was the exception rather than the rule that a concept or device would be developed by a single scientist—except for studies of whole organisms and their behavior and for mathematics, which remained the preserve of the individual. In other disciplines, teams did it all.

There were good reasons for this shift. After 1945, it was not possible to have a major program in particle physics without expensive particle accelerators and detectors. To a lesser degree, similar equipment needs were felt in other "hard" sciences. Expensive equipment must be shared to be affordable. Even in the life sciences, the cost of science soared. By the mid-1980s the average grant from the US National Institutes of Health had reached $137,000 a year, paying for ever more complicated studies. In addition, a phenomenon often needed to be viewed from many angles before it could be understood. Interdisciplinary teams fulfilled this requirement.

Science became big in another way, also. There are far more scientists today, both in actual numbers and as a percent of the world's population, than there ever existed in the past. The reasons for this growth are complex. For one thing, the world's population is larger, and much less of it is needed to produce food and manufactured goods, freeing more of the population for other pursuits. More institutions perceived the need for scientists on their staffs. In the 1980s, half of all the scientists in America were employed in business or industry. Some large corporations, such as AT&T and IBM, even support their own basic research institutions. Government funding of science has made it possible for universities to have larger science departments and for individual scientists to make more money.

Furthermore, scientists at research institutions are writing more. Before the 1970s, top scientists usually wrote about two dozen articles in their scientific lifetimes. As competition for jobs increased among the scientists, each person felt the need to write more to make himself or herself more attractive to employers. In many cases, the results of research were broken into small bits with each published separately in one of the thousands of journals available by the 1980s. In a few cases, the same pressures resulted in fraud, as scientists published results without taking the time and effort to gather or analyze data.

Most scientists, however, were producing a lot of knowledge, so much that no one could keep up with the details and only a few generalists could even follow the major trends in all areas of science. This has resulted in another trend, specialization.

Specialization and changing categories

Over the history of science since the Renaissance, as the totality of scientific knowledge has grown, scientists have specialized more and more. Newton investigated all aspects of the physics of his time and also contributed greatly to mathematics. Einstein limited his work to theoretical physics, but he contributed to virtually all parts of that field. A theoretical physicist today must specialize in particles, materials, nuclear physics, astrophysics, biophysics, or perhaps just gravity. Only the very best, such as Enrico Fermi, can bridge the gaps between even two of these specialties. The same situation is true in other parts of science as well. A "whole-organism biologist," who studies such whole organisms as birds or plants and looks at behavior or anatomy or taxonomy, has little in common with a molecular biologist, who studies how events happen within cells. This kind of dichotomy was not so significant in prewar science.

At the same time as there has been specialization, there has also been a merging of disciplines. Astrophysics and biophysics are only two examples. Chemistry has been especially partitioned into new disciplines. On one side, chemistry has increasingly become a part of physics, as physicists have redeveloped chemistry from the proton, neutron, and electron on up to substances. In another direction, biologists have gradually moved into chemistry in their search for how living things work at the molecular level. One result is that much of what used to be chemistry is materials science (physics) and biochemistry (a part of biology). The leftover part is largely chemical engineering, for the theoretical part of chemistry has been taken over by the materials scientists.

In a few cases, new sciences have emerged that were not perceived as separate disciplines to any significant degree before World War II. Ecology existed, for example, but there were no professional ecologists as such; they were biologists who specialized in ecology. Ethology, the scientific study of animal behavior, and limnology, the study of lakes and freshwater systems, are similar cases. In physics, the separation of nuclear and particle physicists did not take place until after the war. Some astronomers (prewar) became cosmologists (postwar), and so did some physicists. Space science is still emerging. There are even sciences with nothing known at present to study, such as exobiology, the study of extraterrestrial living things.

New concerns

Although the growth of science would seem to suggest that most people regard science in a positive fashion, it is far from clear that that is the case. The development of nuclear weapons led to a perception by some of scientists as amoral seekers after knowledge. Among scientists themselves there was considerable horror at what they had accomplished. It was apparent that nuclear weapons could destroy life on Earth, especially after the development of fusion bombs. Some scientists formed the Union of Concerned Scientists to work against the misuse of nuclear energy. Others refused to work on defense-related projects. The public perception of nuclear energy as dangerous was further enhanced by a number of accidents in nuclear power plants; the most spectacular, in Chernobyl in the

Soviet Union, killed a few dozen people outright and may have caused cancer in hundreds more. Around the same time, interdisciplinary studies of the effects of nuclear war began to predict periods of total darkness, or rain with the strength of battery acid, in addition to the problems of radioactive fallout.

But it was not only nuclear energy that was perceived as dangerous. Probably as a result of the concerns about nuclear energy, other parts of the scientific endeavor were also viewed with suspicion. The unsophisticated worried that space travel was affecting the weather, while the somewhat more sophisticated worried that genetic engineering would accidentally loose plagues upon Earth. Before 1946, nearly everyone seemed to think that science was ultimately beneficial, although a few (such as Aldous Huxley, who wrote *Brave New World* in 1932) were concerned. After 1945, many people were no longer sure that science would ultimately benefit humankind. A growing number worried that it would destroy it.

Major advances

Anthropology and archaeology. Physical anthropologists discovered two previously unknown species of hominid, one definitely on the line to modern humans (*Homo habilis*) and the other possibly ancestral to that line (*Australopithecus afarensis*). By the mid-1980s, nearly complete skeletons of both species were known. Skeletal remains of modern humans dating back almost 100,000 years were also found. But much of the excitement in physical anthropology was about molecules, not bones. Starting from a suggestion of Linus Pauling, a host of investigators used proteins and DNA to analyze the relationship of humans to the great apes. The results of this molecular approach—mainly that humans and chimpanzees, especially pygmy chimpanzees, are more closely related than chimpanzees or humans to gorillas, orangutans, or gibbons, and that the split between the human and ape lines took place only a few million years ago—were

---

Discovering DNA

Today it is common knowledge that DNA, a nucleic acid, directs the development of cells. Scientists gradually learned about DNA in a curiously twisted fashion that is commonplace in science. For one thing, the discovery of DNA required progress on three separate fronts: cytology (the study of cells through a microscope), genetics, and chemistry.

After Gregor Mendel's laws of heredity were rediscovered in 1900, considerable interest developed in what causes heredity. The fundamental structures involved—the chromosomes—had been discovered and studied by Walther Flemming in the 1880s, but no one knew that they were connected to heredity. They were just long thin structures that appeared when cells were stained during cell division. Also, Friederich Miescher had discovered nucleic acids in cell nuclei as early as 1869, but they were not connected either to heredity or to chromosomes—although Miescher's later discovery that salmon sperm are almost entirely nucleic acid plus a simple protein should have been a clue to the connection with heredity.

In 1907 Thomas Hunt Morgan, who was somewhat skeptical about genetics, began to use fruit flies in breeding experiments. Within a short time he found that Mendel's laws worked, but also that some inherited characteristics appeared to be linked together. These linkages behaved as if the units of heredity, the genes, lined up in long rows. A suitable long thin part of the cell that could physically contain the genes was the chromosome, as had earlier been suggested on other grounds by August Weismann. By 1911 Morgan was able to show that genes strung along the chromosomes are the agents of heredity.

While this development was occurring on the genetic front, there was also some progress being made in chemistry. In 1909, Phoebus Aaron Theodor Levene was the first to determine that nucleic acids contain a sugar, ribose. Twenty years later, he found that other nucleic acids contain a different sugar, deoxyribose. Hence, there are two types of nucleic

acid: ribonucleic acid (RNA) and deoxyribonucleic acid (DNA). Levene also worked out the chemical nature of the other compounds that were in RNA and DNA. This chemistry was then explored in detail in the 1930s by Alexander Todd.

Chromosomes, like other cell structures, contain proteins. They also contain DNA. Proteins were known to be complex molecules that are biologically very active, so everyone thought that genes must be proteins—until 1944 when Oswald Avery and coworkers showed that hereditary characteristics could be induced by pure DNA, without a protein involved. He showed that genes must consist of DNA in some way.

By the early 1950s a few scientists from the different fronts were tackling the problem of understanding DNA. Among these was Linus Pauling who was at the time probably the most accomplished chemist. In 1951, Pauling, working with B.B. Corey, determined that the structure of a class of proteins is a helix, which is a three-dimensional spiral. This was the first determination of the physical structure of a large biological molecule. At about this time, Pauling turned to the study of DNA, hoping to discover its structure as well.

In England, there were several scientists interested in the structure of DNA. Maurice Wilkins and Rosalind Franklin were doing X-ray diffraction studies of DNA in hopes of elucidating its structure. Diffraction studies had proved successful in analyzing crystal structures, and DNA could be crystallized. The complexity of DNA was a frustration, even as Franklin produced better and better photographs showing the diffractions.

Another English scientist interested in the subject was Francis Crick, a 35-year-old graduate student. With an undergraduate degree in physics, he too would have liked to do X-ray diffraction studies; but English custom kept him from competing with Wilkins and Franklin. Instead, he just thought and talked incessantly about DNA.

A fourth interested scientist was James Watson, an American. Watson was working as a postgraduate student, trying to learn about genetics from studying organisms. But he realized that the true solution to the problem was more likely on the chemical front, so he abandoned what he was doing and applied for work in X-ray diffraction. He was lucky to be taken on at the same Cambridge laboratory where Francis Crick was pursuing his degree, not far from London, where Wilkins and Franklin were working.

News of Pauling's discovery of a helical structure in proteins set all of the English group (except Franklin) thinking that DNA might be a helix as well. Wilkins thought it might be several helices twisted together.

Watson and Crick decided to try using the method by which Pauling had found the helix in proteins. He had stuck together models of the subunits of the molecule, rather as one puts a tinker toy set together. The models need to be constructed so that they fit together according to Pauling's theory of the chemical bond. Watson and Crick acquired a copy of Pauling's 1939 book on the chemical bond and came up with a model for DNA of three helices twisted together. But when they showed it to Wilkins and Franklin, Franklin pointed out that it disagreed with her diffraction data and had other deficiencies as well.

Watson gradually learned to do X-ray diffractions and established to his and Crick's satisfaction that DNA does have a helical structure. Crick figured out that the bases in DNA are always paired in the same way. Franklin insisted on the correct location of the sugars, which Watson and Crick ignored.

Meanwhile, Pauling produced two versions of his model of DNA. It contained three twisted helices and was clearly wrong. One of the best chemists of the century had made a mistake in his chemistry.

After another false step, Watson finally built a model that incorporated two helices, paired bases, and the sugar structure recommended by Franklin. Crick did calculations that showed that this model was feasible. Wilkens and Franklin produced X-ray diffraction calculations that confirmed the structure. On a visit to Cambridge, Pauling agreed.

The true nature of DNA had finally been discovered. Watson, Crick, and Wilkins won the Nobel Prize a few years later (Franklin died before the prize was awarded, and was thus disqualified). By then, Pauling already had a Nobel Prize for Chemistry for his work on chemical bonds; that same year he won a Nobel Peace Prize for his opposition to atmospheric nuclear weapons testing.

---

gradually accepted by the more traditional physical anthropologists. Before the molecular results were available, most anthropologists believed *Ramapithecus*, an ape that lived as long as 30 million years ago, to be our most recent common ancestor.

Other theories of physical anthropologists also changed greatly during the period. Before World War II it was generally assumed that the main characteristic of hominids was the large brain relative to the body. In part this was based on the Piltdown hoax, which combined a modern human brain with an orangutan jaw, and on human vanity. When Piltdown man was discredited in 1953 and evidence accumulated that the small-brained australopithecines walked erect, the definition of hominid shifted from the brain to the legs and pelvis. Most physical anthropologists of the 1960s believed that the australopithecines gradually evolved into the earliest humans; but the discovery in the 1970s that the earliest humans and australopithecines occupied the same territory at the same time changed that idea. The known australopithecines and humans must have a common ancestor (now believed by some anthropologists to be *A. afarensis*). In the 1980s the idea that early humans were hunters began to give way to the idea that they were scavengers.

Archaeologists made several striking finds, including an ice-age religious sanctuary, boats designed to carry an ancient Egyptian pharaoh's soul to heaven, and armies of statues guarding the tomb of the first emperor of China. One of the main trends in archaeology emerged from the invention of better diving equipment by Jacques-Yves Cousteau. Scuba gear made it possible for archaeologists to expand their work into the sea, where many essential artifacts and even whole villages have been preserved. Increasingly, archaeologists took care to see that their work was scientifically valuable and not just a search for trophies and treasures. For example, parts of sites were deliberately left untouched so that future generations of archaeologists, who might have better tools, could investigate the undisturbed portions.

Astronomy and space. Although all phases of astronomy have advanced continuously since World War II, it is convenient to relate the advances by decade. In the 1940s, the Hale telescope was put into operation at Mount Palomar in California. As of this writing, it is still the best optical telescope in the world, although that will change in the 1990s. In the 1950s radio astronomy began to probe the universe, which was found, by other means, to be twice the size scientists had thought it was. It was not clear in those days whether the universe existed in a steady state with continual creation of matter or whether it had begun in a Big Bang. The 1960s saw the first exploration of nearby space with satellites and spaceprobes, as well as the surprises of quasars and pulsars. The Big Bang theory of creation became dominant as new evidence surfaced. X-ray and neutrino astronomy joined optical and radio approaches. In the 1970s, theoretical physics and astronomy, always close, became even closer. There was a lot of interest in black holes, objects so massive and dense that nothing, not even light, can escape their immediate region

While Thomas Alva Edison is remembered mainly as an inventor, he also made one extremely useful scientific discovery in 1883, now known as the Edison effect. He found that electricity flows through space from a heated metal. When the electron was discovered, the Edison effect was explained as electrons boiling off the metal, like water vapor from a heated tea kettle. Sir John Ambrose Fleming in 1904 developed the first practical application of the Edison effect; he found the electrons traveled only to a positively charged anode. Since alternating current switches back and forth from negative to positive, the Edison effect could be used to turn alternating current into direct current. A device that does this is called a rectifier. Some electrical equipment, such as a radio receiver, will not work with alternating current, so rectification is needed when alternating current, which is more easily transported over long distances than direct current, is used as a power source.

The Edison effect is much more effective in a vacuum, so Fleming's device was enclosed in a partially evacuated glass tube. In America, such a device is still called a tube (in England, it is a valve). Later inventors, notably Lee De Forest, found other ways to use the Edison effect in tubes. De Forest showed how to use tubes as amplifiers, for example.

De Forest's tubes greatly improved the radio, which had previously relied on crystals for rectification. It was not clear why crystals could change alternating current to direct. When tubes came along, almost no one bothered to wonder why.

Tubes do not depend merely on the flow of electricity, as previous devices, such as the electric motors, lights, or heaters, do; instead, they depend on control of the behavior of electrons. Therefore, we speak of devices that use tubes as electronic, not merely as electrical.

Tubes made possible such electronic devices as radio and television, but they had drawbacks: heating a metal eventually causes it to boil away; the vacuum also tends to degrade; the power required for the Edison effect is fairly high; heat tends to build up in the vicinity of the tubes; and practical tubes have to be fairly large. While these problems were only a minor annoyance in building radio and television sets for the living room, they limited the size of portable sets. Also, when the first digital computers were built at the end of World War II, the many tubes needed used enough power to cause lights to dim. In addition, the computers produced immense amounts of heat that had to be got rid of, they occupied large rooms, and they failed with such frequency that it was clear that larger computers could never be built.

Although almost no one wondered why crystals worked as rectifiers, scientists at Bell Laboratories decided in the 1940s that it was worth looking into this question. One of them, William Bradford Shockley, found that some crystals worked better than others. The ones that worked best were impure crystals of germanium. By 1948, Shockley, abetted by theoretical physicists John Bardeen and Walter Brattain, was able to modify the impurities to produce crystals that were as good as rectifiers as tubes. They could also be used like a De Forest tube for amplification. The physicists called their new crystals transistors.

The transistor had none of the problems of the tube. It was small, needed no vacuum, did not wear out easily, and produced little or no heat. Soon transistor radios were the rage. The scientists won the 1956 Nobel Prize for Physics. Large commercial computers, starting with Remington-Rand's UNIVAC, could be built.

What people did not fully realize—except for a few penetrating thinkers like Richard Feynman—was that the transistor was just the beginning. The transistor works because it is possible to adjust the impurities in a crystal so that one region has an excess of electrons while another has a deficit. Electrons will flow from one region to the other just as electrons cross the space between the cathode and anode in a tube. But there is no need for heat to be used; and since this is all happening at the molecular level, not a lot of space is required. A transistor can be made very small indeed. In fact, a transistor can easily occupy just a small region on a crystal. So, you can put more than one transistor on each crystal.

And that is what scientists did. Beginning in the 1960s, ways were found to pack more and more transistors onto a given piece of crystal. Such a piece came to be called a chip. A chip is usually a fingernail-sized sliver of silicon, "doped" with impurities in a pattern that enables it to serve as a computer memory or a central processing system or a controller of fuel injection in an automobile. The chip has made possible the incredible increase in power and availability of the personal computer. It has also invaded almost all aspects of daily life, controlling everything from your microwave oven to your car.

---

as a result of their immense gravitational forces. The 1980s added space-based infrared telescopes to the available ways of knowing the universe, while the Voyager spacecraft extended our close-up knowledge of the solar system to the outer planets. The inflationary universe helped make sense of the Big Bang. A major event in 1987 was the viewing of the closest observable supernova since 1604. Various studies found planets or other objects orbiting stars. These ranged from clouds of particles to brown dwarfs—objects too large to be called planets but too small to become stars.

Space travel is an entirely new technology whose first real successes occurred during this period, although it was foreshadowed by rocket research before World War II. Since 1957, when the first artificial satellite was put into orbit about Earth, space technology has been most useful as a scientific tool when vehicles without people aboard were used. Small spaceprobes reached Venus, the moon, and Mars. Others passed by Mercury, Jupiter, Saturn, Uranus, and Comet Halley (with Neptune about to be added to the list). Satellites detected belts of radiation around Earth, the solar wind, and giant magnetic fields in space.

At the same time, much of the public's interest was in

In the history of technology, the development of computers is unique. No other technical device underwent such rapid development after its invention.

For a long time scientists have been fascinated by machines that would be able to perform calculations. During the seventeenth century, Blaise Pascal invented a machine with geared wheels that could add and subtract. Gottfried Wilhelm Leibniz, a few years later, developed one that could add, subtract, multiply, and divide. These machines, and others like them, required that a human operator direct the operation at every stage.

In the nineteenth century, the Englishman Charles Babbage designed, but never completed, a working model of a different kind of computing machine, called the Analytical Engine. It had an important feature of the present-day modern computer: it was designed so that it would perform mathematical operations from a set of instructions or a "program" supplied to the machine. The machine would "read" the instructions from perforated cards similar to those that controlled the looms that had been developed by Joseph-Marie Jacquard about 20 years earlier. The Analytical Engine was to be equipped with a memory, which Babbage called the "store," and a central processor, which he called the "mill." A long sequence of different operations could be performed with no human intervention after the punched cards were fed in.

One reason the Analytical Engine was never completed is that it operated mechanically and technicians could not produce sufficiently accurate parts for mechanical operation. The invention of the vacuum tube, however, led to a revolution in technology during the 1920s. During the 1930s scientists investigated how the vacuum tube could be used to replace the mechanical gears and levers in the calculators of the day, descendants of Pascal's and Leibniz's machines.

Although it is unclear who really can be considered to be the "inventor" of the modern computer, it is now generally accepted that the theoretical physicist John V. Atanasoff and his assistant Clifford Berry in 1942 built the first operational computer that used vacuum tubes to perform mathematical operations. It was called the Atanasoff Berry Computer, or ABC. The ABC used binary numbers that were stored in capacitors mounted on a rotating drum; it could perform operations with 16-digit binary numbers, which correspond to eight-digit decimal numbers.

Around the same time, several other scientists and engineers started devising and building electronic computing machines. As was the case for many inventions, the development of computers was accelerated because of defense needs. During World War II, the Ballistic Research Laboratory in Aberdeen, MD, needed a faster, nonmechanical computer to keep up with the calculations of trajectories of projectiles that were required for the firing tables for gunners. Construction of the computer, known as the ENIAC (electronic numerical integrator and computer), began in 1943 and was completed early in 1945. The computer contained about 18,000 vacuum tubes, whose heating filaments remained switched on to reduce tube failures. Unlike Atanasoff's machine, the ENIAC was a decimal machine, but it was an important improvement because it was programmable and could function as a general purpose machine. The programming, however, consisted of changing the wiring of the machine by plugging cables into plugboards and setting thousands of switches. Setting up the machine for a new calculation was a long and arduous task.

The completion of ENIAC's successor, the EDVAC (electronic discrete variable computer), was delayed until 1952 because of legal battles concerning the patent. It incorporated many of the ideas of the mathematician John von Neumann. An important innovation was that programs could be stored in memory. Contrary to the ENIAC, which had several specialized processor units, the EDVAC had a central processor and a random-access read-write memory. It processed binary numbers serially and its functioning was based on Boolean logic.

In England, the development of electronic computers also started because of the war effort. During World War II, Germany used an electromechanical coding machine called "Enigma" that produced coded naval messages believed to be unbreakable. In Bletchley Park, a secret installation, the English built a series of electronic machines, named "Colossus," that could decipher the German secret messages. Ten versions were built, starting in 1943. The experience gained with the Colossus enabled the English engineers, under the leadership of the mathematician Alan Turing, to complete the ACE (automatic computing engine) in 1950. Turing wrote for that machine the first programs in the modern sense. They consisted of partially numerical and partially alphabetical code. A larger machine, the Mark 1, was completed in Manchester in 1948. It became the first computer operating on a program fully stored in memory.

During the 1950s and 1960s, the development of computers took off swiftly. The main hardware improvements were the introduction of the transistor to replace the vacuum tube and the introduction of integrated circuits to replace the circuits made up of discrete components such as transistors, resistors, and capacitors. The memories of computers also evolved from bulky cathode-ray tubes and delay lines via magnetic core memories to solid-state memories. Punch cards, like those planned for the Analytical Engine, and punch paper tape were first used for input and output and mass storage. These were eventually replaced by magnetic tape and magnetic disks. The development in hardware was such that the present-day desktop computer surpasses in memory, speed, and power the large mainframe computers from the 1950s and early 1960s.

The software also underwent an evolution from simple machine-code programs to the present-day high-level languages. Progress in the development of high-level languages and artificial intelligence suggests that computers programmable in plain English will be a reality in the near future.

programs that put people in space. From the first flights in 1961, the adventure (and occasional disaster) of human space travel has been the focus of much of the space programs of both the Soviet Union and the United States. The US program first focused on putting people on the moon. It succeeded in 1969, although the project was abandoned after a few additional trips. Later the US program concentrated on the development of a system in which most of a vehicle could be used in the way that an airplane is. The US space shuttle program was closed down for a couple of years as a result of the *Challenger* accident, in which the vehicle exploded and killed all seven aboard. The Soviet Union so far has had a manned program directed toward the establishment of a permanent station in orbit around Earth. As a result, many Soviet cosmonauts have spent months at a time in space. The United States is planning to develop a space station as well, while the Soviet Union is beginning to develop a version of the space shuttle.

Space technology has had a number of direct benefits for people around the world. Communications satellites have helped make Earth a "global village." Weather satellites have given us the almost credible five-day forecast and have informed scientists about changes in the atmosphere. Resources satellites detect changes in forests and crops and locate mineral deposits. Satellites used for location of exact positions on the surface of Earth not only help ships and planes, but can even be used to find lost or injured individuals in a wilderness. Less appealing to many is the knowledge that the military forces of both the United States and the Soviet Union use satellites for spying and have vast numbers of space-going rockets armed with fusion bombs to use as weapons.

Biology. It can be argued that biology made the most dramatic progress of any science after World War II, especially biology at the level of the behavior of individual molecules. There were also dramatic gains in understanding the behavior of animals, especially primates, that emerged from careful studies in the wild. Mathematics was employed extensively by ecologists to describe the statistical behavior of communities of animals and to explain altruism and other traits in the controversial new discipline of sociobiology. The most notable development in biology, however, was the rise of genetic engineering.

A little more than 50 years after biologists became aware of Gregor Mendel's laws of heredity, James Watson and Francis Crick found the key to how these laws work. About the same time, scientists experimenting with bacteria and viruses that prey upon them discovered ways to transfer genetic material from one organism to another. The combination of new understanding and new techniques created the sudden growth of genetic engineering (see also Medicine below). Equally important, genetic information could now be used to learn about how proteins are built and what they do.

Chemistry. Although of little immediate practical use, the most dramatic story in chemistry was success in forming compounds with some of the noble gases, heretofore believed not to enter any form of molecule. Chemists are still working on making compounds with the remaining noble gases, with compounds of helium as the main goal. There is even some promise that the noble-gas compounds may have a role in the production of powerful lasers.

To many people chemistry gradually changed from friend to enemy in the postwar period. As chemists developed new herbicides and pesticides, they helped create the immense production of food and fibers with greatly decreased labor; but as fish kills and occasional problems with cancer among some who worked with the chemicals impinged upon people's consciousness, many felt it necessary to reduce the dependence on such chemicals. Similarly, chemists developed additives that prolonged food shelf life, made it cheaper, or improved its color; but again people came to view such additives with suspicion as part of a broad-based concern about the health risks of chemicals. More subtly, the chemistry that produced improved forms of Teflon also yielded the spray propellants and refrigeration gases that seem to be destroying the protective layer of ozone in the atmosphere.

Earth science. The period after World War II saw a revolution in our basic understanding of Earth's crust. In the 1950s and 1960s, geologists developed the idea that the crust of Earth was broken into a number of large plates that move relative to each other. This theory explained such features of Earth as the locations of many earthquakes and volcanoes, of most mountain ranges, and of trenches and rifts on the floor of the ocean. Although some geologists did not accept the theory at first, by the late 1980s there was firm evidence that the plates were moving in the predicted patterns.

Paleontology, although a study of once living creatures, is usually treated as part of earth science. A major controversy in paleontology arose in the 1980s following the discovery of a worldwide layer of iridium precisely at the boundary between the Cretaceous and Tertiary periods, a time of numerous extinctions.

Meteorology is another part of earth science. While weather forecasts improved as a result of data from satellites and radar as well as better computers and computer models, in 1961 Edward Lorenz established that very small changes in initial conditions result in large changes in the weather. Consequently, medium-term weather prediction (more than a week and less than a year) probably can never be certain. Scientists studying long-term changes, however, became concerned that gases released by human activity were causing major changes in the atmosphere. One such change, the greenhouse effect, may cause temperatures to climb worldwide as a result of carbon dioxide and other gases that trap heat in the atmosphere the way that the glass walls and ceiling of a

Genetic engineering consists of a set of methods to change the genes of an organism so that proteins produced by that organism differ in type or quantity from those produced by a "wild" organism—one that has not been altered. Although people have been causing such changes in organisms since domestication began about 10,000 years ago, the phrase *genetic engineering* conventionally refers to a set of techniques that only became possible in 1973.

It is easy to think that genetic engineering was the inevitable outcome of the Watson-Crick explanation of the mechanism of heredity in 1953, but that is not really the case. While knowledge of DNA structure and the genetic code is essential to genetic engineering, the path that led to practical results was separate from the path that led to understanding. Along the way, of course, both paths interacted constantly. The path that led to practical applications could not have been predicted in advance, as almost all of the important discoveries were previously unsuspected by any scientist.

The start of the path toward genetic engineering occurred in 1952 when Joshua Lederberg discovered that bacteria, like some protists, conjugate to exchange genetic material. This behavior, much like sex in multicellular organisms, led Lederberg to perceive that there are two populations of bacteria, which he called M and F. The F population contains a body that he called a *plasmid.* After conjugation, the F bacterium passes the plasmid on to the M bacterium, with which it has conjugated. (It would appear that Lederberg had his sexes backward, but that is not essential to what follows.) This discovery was completely surprising.

The next year William Hayes established that the plasmid consists of genetic material. By then it was clear that genes were DNA; therefore, plasmids were rings of DNA floating free of the main DNA in the chromosome of a bacterium.

About the same time, an apparently unrelated situation became a major problem. Both the sulfa drugs and the antibiotics of the late 1930s and 1940s were, in the 1950s, beginning not to work as well. Many bacteria were becoming resistant to these drugs. Epidemics, especially in hospitals, could no longer be controlled. Many scientists studied the problem. In 1959, Japanese scientists discovered that the genes for drug resistance were carried on plasmids, and therefore passed from bacterium to bacterium. Within a given bacterium, the plasmids multiplied, so there were plenty of copies to pass around. Putting a few drug-resistant bacteria into a colony of bacteria that showed no resistance resulted in short order in a colony that was completely resistant.

It is not surprising that some bacteria had plasmids that protected against these drugs. Antibiotics are natural substances found in the environment, so some bacteria have evolved defenses against them. The increase in the amount of antibiotics caused by human intervention led to the resistant bacteria passing the plasmids around to larger populations.

In the meantime, another line of research was also leading toward genetic engineering. Starting right after World War II, a number of biologists made an intensive study of viruses that infect bacteria, which are collectively called bacteriophages, or just phages. This line of research demonstrated that genes are DNA and not a protein, as had previously been suspected. As early as 1946 Max Delbrück and Alfred Hershey independently showed that the genes from different phages could spontaneously combine. Werner Arber studied the mutation process in phages in detail. In the process he discovered that bacteria resist phages by splitting the phage DNA with enzymes. Subsequent recombination of split genes was a consequence of this. By 1968, Arber had located the enzymes produced by bacteria that split DNA at specific locations. The split ends are "sticky;" that is, different genes that have been split at the same location by one of these restriction enzymes, as they came to be called, will recombine when placed together in the absence of the enzyme. The resulting product is called recombinant DNA.

The following year, 1969, Jonathan Beckwith and coworkers became the first to isolate a single gene. It was a bacterial gene for a part of the metabolism of sugar.

In 1973 Stanley H. Cohen and Herbert W. Brown combined the restriction enzymes with plasmids with isolation of specific genes to introduce genetic engineering. They cut a chunk out of a plasmid found in the bacterium *Escherichia coli* and inserted into the opening a gene from a different bacterium. Then they put the plasmid back into the bacteria *E. coli,* where copies were made and transferred to other bacteria. Within months other scientists repeated the trick, inserting genes from fruit flies and frogs into *E. coli.*

Not all scientists thought this was a good thing. In July 1974, Paul Berg and other biologists met under the auspices of the US National Academy of Sciences to draw up guidelines that would prohibit certain kinds of genetic engineering.

Since 1974, the tension between those who are rapidly advancing genetic engineering and those who worry about where it is going has continued. By the 1980s, the genetic engineers were producing useful products from bacteria and yeasts, including human growth hormone, human insulin, and a vaccine for hepatitis B. All these are made in tanks in controlled environments and have ceased to evoke much resistance from scientists or the public. There has been more resistance to experiments in which genetically engineered bacteria are released in the environment, although a number of small-scale releases have so far not resulted in known environmental damage.

In one area, the progress of genetic engineering has been frustratingly slow. From the beginning of the new technique, there has been hope that it could be used to cure human genetic diseases. So far, that has not proved possible. On the other hand, some of the techniques used in genetic engineering have made it possible to find genes that are markers for a number of diseases.

greenhouse trap heat. Another change is the thinning or loss of the ozone layer that protects life from excessive ultraviolet radiation; this change is caused by gases that catalyze ozone, an oxygen molecule of three atoms, back into ordinary oxygen, whose molecules have two atoms.

Mathematics. Although much of mathematics after World War II became so abstract that nonmathematicians had great difficulty following the results, some important proofs of long outstanding problems were produced. These include the proof of the long-standing conjecture that only four colors are needed to color any map. Another remarkable proof of a well-known conjecture, Mordell's conjecture, was made by Gerd Faltings in 1983.

For another conjecture, Paul Cohen showed in 1963 that Cantor's continuum hypothesis concerning infinite sets is neither consistent nor inconsistent with the generally accepted basis of mathematics. Cantor had discovered late in the nineteenth century that there are many different-sized infinities. His continuum hypothesis conjectured that two particular infinities constructed by different processes are the same size. Cohen's work (which was built on a 1940 proof by Kurt Gödel) showed that one can choose whether the two infinities will be equal or not equal. At the same time, Cohen (again building on Gödel's earlier work) showed that a widely used axiom in set theory has a similar position in mathematics.

More important than proofs of individual theorems was the development of fruitful new concepts in mathematics that could then be used to solve a range of problems. Most important of these is probably catastrophe theory, originally put forward by René Thom. Instead of dealing with smooth processes, such as continuous acceleration, catastrophe theory treats events in which "the straw on the camel's back" causes a complete change of state. Closely related to catastrophe theory is the theory of strange attractors, sets connected to unstable functions that have stable values near one or two points. Another fruitful new idea is fractal theory, originally put forward by Benoit Mandelbrot. A fractal is a figure that is self-similar at varying scales.

Medicine. The interaction of medicine and biology, especially molecular biology, became so close in recent years that it is sometimes difficult to tell where one ends and the other begins. On the other hand many innovations are pure medicine with little or no input from molecular biology. These include organ transplants; endoscopy and related techniques (for example, angioplasty, in which a tube inserted into an artery carries a balloon or laser used to flatten or remove arterial plaque); amniocentesis and other methods of diagnosing and treating the fetus; ways of handling fertilized human eggs to produce viable children; artificial substitutes for skin; cochlear implants to aid hearing; kidney dialysis; ultrasound scanning; and a variety of new vaccines.

The new technology of genetic engineering yielded various direct applications, including the production of certain proteins needed by some ill people (human insulin, human growth hormone, interleukin 2, and T.P.A., a blood clot dissolver); artificial vaccines, such as a safer and cheaper vaccine for hepatitis B; improved tissue matching capabilities for transplants; and the ability to locate gene markers for such genetic disorders as Huntington's disease or Duchenne muscular dystrophy.

Just as the discoveries of the endocrine system and of vitamins in the earlier part of the twentieth century revolutionized the understanding of disease, a new understanding of the immune system that developed after 1945 also produced major changes in our understanding of how some diseases are caused and how they may be avoided. The discovery of blood types, artificial immunity with killed or weakened germs, and allergic shock in the late nineteenth and early twentieth centuries suggested that some substances or processes in the body become activated when exposed to "foreign" proteins. After the war, Frank Burnet and Peter Medawar showed that the reaction against foreign proteins is developed largely after birth. The study of the immune system after this unraveled an increasingly complex network of interacting cells and proteins. Many diseases, including some types of arthritis and diabetes, were found to be caused when the immune system attacks a part of the self. Ironically, just as studies of the immune system began to reveal how it works, AIDS, the disease that destroys the immune system, surfaced.

A related development that has major implications for the future and many immediate applications was the invention of monoclonal antibodies. These molecules can be targeted on one specific receptor—part of a molecule. The monoclonal antibodies detect the presence of their target receptor and attach themselves to it.

Physics. World War II ended with two nuclear fission bombs ("atomic bombs") used as weapons, preceded by a single test earlier in 1945. Physics clearly had an important place in military matters, and funding for large projects in physics has continued, especially in nuclear and particle physics. More recently, governments have realized how important materials science is to them and the economy; this branch of physics has produced the transistor and its descendants, the laser, and the promise of high-temperature superconductivity. Funding for physics has contributed to a remarkable period of growth since the war.

Discovery of the Lamb shift in 1947, just before a major conference of leading physicists at Shelter Island, NY, led to the solution of mathematical problems that had surfaced in the study of atoms and subatomic particles. The mathematical theory that resulted is quantum electrodynamics (QED), often called the most accurate theory in physics.

Just as QED was being developed, however, scientists studying cosmic rays began to find new subatomic particles that did not behave as predicted—for one thing, they did not decay into other particles as fast as theory called for. The reason for this strange behavior was worked into theory and christened "strangeness," the beginning of a trend in physics toward whimsical names, from quarks and flavors to WIMPs (weakly interacting massive particles). Strangeness and the newly discovered strange particles were fit into a classification scheme, somewhat like the earlier periodic table, called the eightfold way. The eightfold way, like the periodic table, enabled prediction of previously undiscovered entities. When the predicted omega-minus particle was found in 1964, the eightfold way was viewed as confirmed.

The eightfold way was developed by Murray Gell-Mann on the basis of abstract mathematics, not on the basis of some physical understanding of why it should work. The same year that the eightfold way was confirmed, Gell-Mann introduced his physical explanation, called the quark model. The quark model explained most particles as combinations of other particles, called quarks, that have fractional charges (omitted from the quark model were electrons, neutrinos, and their relatives). The quark model was even more successful than the eightfold way, although its first confirmed prediction did not occur until 1977.

The strange particles of 1947 also started another major line of thinking. Again, the problem was that some particles decayed in ways that violated prevailing theories. Frank Yang and T.D. Lee concluded that this decay problem could be explained if nature distinguished between right and left in certain instances. A team of experimenters headed by Madame Chien-Shiung Wu showed that the Yang-Lee idea was correct.

The success of Gell-Mann in using a mathematical theory to reach the quark model inspired others to work with similar mathematical ideas—specifically a branch of mathematics called group theory—to explain the particles left out of the quark model. Starting in 1967, this was accomplished to some degree in the electroweak theory, a theory that showed that at high energies two forces that are distinct at the energy levels normally experienced become a single force. The theory was verified by the discovery of predicted new particles in 1983. With that success came many imitators who wanted to use group theory to unify the electroweak force with one of the two remaining forces, called the strong force, into a grand unified theory, or GUT. While promising, the GUTs have remained somewhat unconvincing. Along the way, theorists have turned to other mathematical approaches. The concept of supergravity, which began to be developed in 1976, was combined with such ideas as the following: particles may not be particles but very skinny strings and loops; the universe may not be four dimensional as in Einstein's theories, but 10 or 11 dimensional.

Technology. There are two very different kinds of technology. One is specific to certain problems—for ex-

ample, Velcro. The other is a generalized new technology that affects many different fields of endeavor—such as the laser. A few technologies are somewhere in between, such as liquid-crystal displays. The period after World War II is particularly rich in all varieties of new technology.

The specific technologies are the ones that perhaps change everyday life the most. Before 1946 cars did not have tubeless or radial-ply tires, power steering or power brakes, or fuel injection, not to mention some of the advanced features available in models today. Remember carbon paper? Now we have copying machines. Think of life without long-playing records, compact discs, or audio or video cassettes; battery-powered watches; microwave ovens; fax documentation; permanent press! These advances change the texture of everyday life.

Besides the laser, two intertwined technologies that started just at the end of the war stand out for the contributions they have made in many fields and the promise of vast changes in the future. These are solid-state electronic devices and digital computers.

The digital computer came first, but would never have been much more than a minor influence on life without the solid-state devices. Early computers with vacuum tubes used too much energy and were "down" most of the time. The transistor, which came along a couple of years after the war, solved both problems. The reliable transistor's descendants made computers cheaper and smaller at an astonishing rate. Today, a thousand dollars will buy a 4 kg (9 lb) computer that has 32 times the memory and speed of computers that cost millions of dollars and filled air-conditioned rooms in the 1940s. Furthermore, the 1940s computer was programmed by hard wiring and its storage memory was a host of magnetic core devices. It was not until 1956 that a computer language was developed; the keyboard for entering data came 11 years later; and it was another three years before data could be stored on floppy disks.

Although the greatest impact of solid-state electronics technology was in computers, solid-state devices, generally called microprocessors, appeared as parts of various other technologies with increasing frequency. The most obvious use has been in miniaturizing radios and television sets. But the ubiquitous microprocessors handle more of an automobile's operation each year. They show up in household appliances as well.

Another new technology that must be mentioned is genetic engineering. Its advances are covered above as biology or medicine. It is not likely, however, that this technology will continue to be limited to those fields (and agriculture). Genetic engineering may someday affect daily life more than solid-state electronics.

New technologies will also continue to change our lives in ways we cannot predict. Somewhere in the mid-1980s, a new Xerox or Velcro or even a new transistor was probably invented. We just don't know about it yet.

| 1946 | The first meeting of the United Nations takes place | Ruth Benedict's *The chrysanthemum and the sword*, financed by the US government during World War II, is an effort to explain the Japanese culture to the West | Work at Jodrell Bank begins on Oct 10 with collecting the radar reflections from the Giacobinid meteor trails

A V-2 rocket carries a spectrograph to a height of 55 km (34 mi) for study of the sun

Radio astronomy begins in Australia with solar studies by a team led by E.G. Bowen

The radio source Cygnus A is discovered by Hey, Parsons, and Phillips

Sir James Hopwood Jeans, English mathematician and astronomer, d Dorking, Surrey, Sep 17

Martin Ryle (b England, Sep 27, 1918) identifies the radio source Cygnus A, the first known radio galaxy to be identified | Max Delbrück (b Berlin, Germany, Sep 4, 1906) and Alfred Day Hershey (b Owosso, MI, Dec 4, 1908) independently discover that the genetic material from different viruses can be combined to form a new type of virus

Hermann J. Muller of the United States wins the Nobel Prize for Physiology or Medicine for his discovery of X-ray mutation of genes

Linus Pauling suggests that enzymes work by lowering the energy barrier of a reaction through binding to a transitional state as the atoms in a compound move from position to position about their central core; this is later established for many enzymes | Gilbert Newton Lewis, American chemist, d Berkeley, CA, Mar 23

Alfred Stock, German chemist, d Karlsruhe, Baden-Württemberg, Aug 12

James B. Sumner, John H. Northrop, and Wendell Stanley of the United States share the Nobel Prize for Chemistry, Sumner for the first crystallization of an enzyme, Northrop for his crystallization of enzymes, and Stanley for the crystallization of a virus |
| 1947 | | Two shepherd boys discover the Dead Sea Scrolls, Jewish religious documents from around the time of Christ, in a cave near Khirbet Qumran | Soviet astronomer Victor Amazaspovich Ambartsumian (b Tiflis, Georgia, Sep 18, 1908) shows that the Milky Way galaxy contains O stars (the hottest and earliest | Czechoslovakian-American biochemists Carl Ferdinand Cori (b Prague, Dec 15, 1896) and Gerty T. Cori (b Prague, Aug 15, 1896) and Bernardo A. Houssay of | Johannes Nicolaus Bronsted, Danish chemist, d Copenhagen, Dec 17

Moses Gomberg, Russian-American chemist, d Ann Arbor, MI, Feb 12 |

Physicist Vincent Joseph Schaefer (b Schenectady, NY, Jul 4, 1906) discovers in Jul by accident that carbon dioxide causes supercooled water vapor to turn into snow; on Nov 13 he conducts a large-scale experiment by dropping dry ice into a cloud from an airplane, causing the first artificial snowstorm

The first synchro-cyclotron is built at the University of California at Berkeley; by Nov it produces alpha particles with an energy of 380 million electron volts (MeV)

Percy W. Bridgman of the United States wins the Nobel Prize for Physics for his discovery of the laws of high-pressure physics

Paul Langevin, French physicist, d Paris, Dec 19

Abraham Pais and C. Moller introduce the term *lepton* to describe lightweight particles not affected by the strong force; today six are recognized, the electron, muon, and tauon along with their associated neutrinos

George Rochester and C.C. Butler discover the first "V particle" in Oct while detecting cosmic rays in a cloud chamber; seven months later they find another; they are not understood until 1953, when the concept of strangeness is introduced—V particles are strange particles

John Logie Baird, Scottish inventor of a precursor to television and of a device that can find objects in the dark, d Jun 14

1946

Arthur Burks, Herman Goldstine, and John von Neumann's *Preliminary discussion of the logical design of an electronic computing instrument* contribute to the theory of the digital computer

Willard Frank Libby (b Grand Valley, CO, Dec 17, 1908) introduces the radioactive carbon-14 method of dating ancient objects

On Dec 24, the first Soviet nuclear reactor goes into effect under the direction of Igor Vasilevich Kurchatov, b Sim (USSR), Jan 12, 1903

John William Mauchly (b Cincinnati, OH, Aug 30, 1907) and John Prosper Eckert complete ENIAC, the first all-purpose, all-electronic computer; it does not use binary numerals, but has its vacuum tubes arranged to display decimal numerals; it draws so much electricity that it causes the lights in a nearby town to dim each time it is used

John von Neumann starts computer research at the Institute for Advanced Study in Princeton

Victor Moritz Goldschmidt, Swiss-Norwegian geochemist, d Oslo, Norway, Mar 20

Alfred North Whitehead, English mathematician and philosopher, d Cambridge, MA, Dec 30

Chloramphenicol, a powerful antibiotic whose use is now restricted because of dangerous side effects, is discovered

Willis Lamb discovers that two states of the hydrogen atom differ in frequency, a result that contradicts accepted theory; this discovery of the Lamb shift leads to the

The first microwave cooker goes on sale in the United States

1947

The first airplane flies at supersonic speed in the United States

1947

cont

type of stars) along with T associations, groups of young stars

Daniel-Henri Chalonge and Vladimir Kourganoff analyze the solar spectrum to study the absorption spectrum of the negative hydrogen ion present in the solar atmosphere

Lyman Spitzer, Jr. (b Toledo, OH, Jun 26, 1914) speculates that astronomers might put telescopes of various kinds in orbit around Earth on artificial satellites

Argentina share the Nobel Prize for Physiology or Medicine, the Coris for their work on the metabolism of glycogen and Houssay for pituitary study

Karl von Frisch discovers that bees use the polarization of light for orientation

Sir Frederick Gowland Hopkins, English biochemist, d Cambridge, May 16

Fritz Albert Lipmann discovers the compound coenzyme A, which is involved in the control of the energy level in cells

Edward Lawrie Tatum (b Boulder, CO, Dec 14, 1909) and Joshua S. Lederberg discover genetic recombination in the bacterium *Escherichia coli*

Baron Alexander Robertus Todd (b Glasgow, Scotland, Oct 2, 1907) synthesizes both adenosine diphosphate (ADP) and adenosine triphosphate (ATP), the compounds used by cells to handle energy

Sir Robert Robinson of England wins the Nobel Prize for Chemistry for his study of plant alkaloids

development of quantum electrodynamics (QED)

Physicists working with the General Electric Company's electron synchrotron particle accelerator observe synchrotron radiation for the first time; this is electromagnetic radiation produced as moving charged particles change paths; at first an unwanted side effect, by the 1980s it is a valued source of X rays

The Berkeley proton linear accelerator, under the direction of Luis Walter Alvarez, is able to produce free protons with an energy of 32 million electron volts (MeV), a record at the time

Sir Edward Appleton of England wins the Nobel Prize for Physics for his discovery of the ionic layer in the atmosphere

A conference at Shelter Island, off Long Island, NY, Jun 2-4, includes reports on the Lamb shift, hyperfine anomalies, the two-meson hypothesis, and electron-neutron interaction; physicists at the meeting and just after it, starting with Hans Bethe on the train ride home, create quantum electrodynamics

Hungarian-British physicist Dennis Gabor (b Budapest, Jun 5, 1900) develops the basic concept of holography, although the technique does not become truly practical until after the invention of the laser

The tubeless tire is introduced by Goodyear in the United States

Henry Ford, American industrialist, d Dearborn, MI, Apr 7

1947

cont

1947

cont

Creating elements

The ancient Greeks developed theories based upon four "elements"—earth, air, fire, and water—although these were not thought of as elements in the modern sense of the word. The Greek elements were more like qualities—salt tended toward earth with a little bit of water and fire in it, for example.

The modern conception of element was first proposed by seventeenth-century chemist Robert Boyle, who said that an element is a chemical that cannot be broken down into another substance. At that time, however, it was hardly clear which substances were elements and which were not. Some of what we recognize as elements were known—carbon, copper, gold, iron, lead, mercury, silver, sulfur, tin, and zinc. Other substances that we now know are compounds, such as water and potash, or even mixtures, such as air, were mistakenly believed to be elements. Boyle just missed being the first person to discover a new element according to his definition, but he was anticipated by Hennig Brand, who discovered phosphorus some five to ten years before Boyle's independent discovery.

In the eighteenth century many new elements were discovered and recognized as such. These include barium, beryllium, chlorine, chromium, cobalt, hydrogen, magnesium, manganese, molybdenum, nickel, nitrogen, oxygen, tellurium, titanium, tungsten, and zirconium. Even more were discovered in the first part of the nineteenth century, so when Dmitri Mendeléev set about developing the periodic table in 1869, he had 63 of the 90 naturally occurring elements to work with. (It is often thought that 92 elements occur naturally, but two of the elements below uranium in the periodic table are not found in nature.) Mendeléev's periodic table not only made some sense out of the growing number of elements, it also indicated what the properties of as yet undiscovered elements might be. When new elements were discovered, it was found that they fit right into the blanks in Mendeléev's table (occasionally with a minor adjustment in the scheme).

In 1914, Henry Moseley discovered why Mendeléev's periodic table worked, and improved upon it. Moseley found that the differences in elements came from the different number of protons in each atom of the element—although Moseley would not have put it quite that way. He was looking at the number of electrons, which is the same as the number of protons for un-ionized atoms. With Moseley's work it could be seen that there were just seven "holes" left in the periodic table (below uranium). Five of these were filled by the discovery of francium, hafnium, promethium, protoactinium, and rhenium, all of which are found on Earth, although most of them are quite rare.

But that was not the whole story. Even before all of these elements were found, the first "new" element, technetium, was created in 1937. Emilio Segrè thought that since the element with atomic number 43 had not been discovered, it might be created. He got some molybdenum (atomic number 42) that had been bombarded with deutrons (particles consisting of a proton bound to a neutron) from Ernest Lawrence. Careful chemistry revealed the traces of technetium in the sample. Some of the deutrons had stuck to the molybdenum nucleus, providing the extra proton that lifted the atom to element 43. In 1940, Segrè helped complete the list of 92 elements as part of the team that produced element 85, astatine.

1948

Margaret Mead's *Male and female: A study of the sexes in a changing world* claims that many parts of Western culture are a result of child-rearing practices in the West	The 5-m (200-in.) Hale reflecting telescope at Palomar, CA, is completed Hermann Bondi (b Vienna, Austria, Nov 1, 1919), Thomas Gold (b Vienna, May 22, 1920), and Fred Hoyle (b Bingley, England, Jun 24, 1915) advance the steady-state theory of the universe in which continual creation of matter powers expansion; later, this theory loses most of its	Paul Müller of Switzerland wins the Nobel Prize for Physiology or Medicine for his discovery of the effect of DDT on insects	Arne Tiselius of Sweden wins the Nobel Prize for Chemistry for his research on serum proteins

Creating elements (continued)

Around the same time, however, scientists found that they could go beyond the known 92 elements in the periodic table. Edwin McMillan was exposing uranium to accelerated neutrons in 1940. Some of the neutrons stuck to the uranium nucleus and then underwent beta decay (changing from neutrons to protons by emitting an electron). On June 8 McMillan and Philip Abelson announced the discovery of element 93, which they named neptunium (because Neptune is the planet beyond Uranus). Later that year they showed that the same process also produced element 94, plutonium, exhausting the supply of planets beyond Uranus as names. Later elements would be named for places and persons.

McMillan moved on to other things—radar and nuclear weapons; World War II had started. After the war, a team headed by Glenn T. Seaborg resumed the work of creating new elements. With amazing regularity the team announced one new element after another: americium, curium, berkelium, californium, einsteinium, and fermium—elements 95 to 100. All of these were created by adding protons to the nuclei of previously existing elements.

Subsequently, various groups extended the list to 109. When elements 108 and 109 were created in 1984 and 1982 (109 was made first), only a tiny number of atoms of each were made. The new elements were recognized by their decay patterns, since all of the elements above uranium are radioactive. Many are so unstable that they last only a few seconds after creation. Element 108 only lasted two-thousandths of a second.

There is hope for stable elements above 109. Nuclear theory predicts that element 110 will be much more stable than 109, for example. In the meantime, only one of the elements above uranium has won a place in society. Plutonium is a terrific element for making nuclear weapons.

1947 cont

Philipp Eduard Anton von Lenard, Hungarian-German physicist, d Messelhausen, Baden-Württemberg, W Germany, May 20

R. Marshak and Hans Bethe, independently of S. Sakata and T. Inoue, rediscover the two-meson theory: the meson found in cosmic rays (now known as the muon) is not the same as the meson predicted by Yukawa (now known as the pion)

Louis Carl Heinrich Paschen, German physicist, d Potsdam, E Germany, Feb 25

Max Karl Ernst Ludwig Planck, German physicist, d Göttingen, W Germany, Oct 3

Cecil Frank Powell (b Tonbridge, England, Dec 5, 1903) and coworkers discover the pion, which is the particle predicted in 1935 by Hideki Yukawa; for about 10 years, it was believed that the muon was the Yukawa meson, although theoreticians in 1942 and 1946 independently concluded that there must be two mesons

---

1948

John Franklin Enders (b West Hartford, CT, Feb 10, 1897), Thomas Huckle Weller (b Ann Arbor, MI, Jun 15, 1915), and Frederick C. Robbins (b Auburn, AL, Aug 25, 1916) learn how to grow mumps viruses in chick tissue using penicillin to prevent bacterial contamination; the same technique works for the polio virus

Patrick Blackett of England wins the Nobel Prize for Physics for his discoveries with and improvements on the Wilson cloud chamber

Richard Feynman (b New York, NY, May 11, 1918), independently of Julian Seymour Schwinger (b New York, NY, Feb 12, 1918) and Shin'ichiro Tomonaga, develops renormalizable quantum electrodynamics

The atomic clock is introduced

Manchester University's Mark I prototype, a stored-program electronic computer, starts operating

**1948
cont**

A lot out of nothing (part two)

The existence of an actual vacuum was a subject for debate among scientists from Aristotle into the twentieth century. Since light, magnetic fields, and heat all travel though a vacuum, *something* must be there. Borrowing a word from Aristotle, scientists described various kinds of "aethers" that exist in even the hardest vacuum and that pervade space. Maxwell's theory of electromagnetism reduced these different types to just one, called the ether. Various experiments were developed to detect this ether, of which the most famous was the Michelson-Morley experiment, which failed to find it (*see* "Does the ether exist?" p 366). Finally, in 1905, Einstein banished the ether by means of special relativity and allowed the true vacuum to exist.

But not for long. The Heisenberg uncertainty principle of 1927 led particle physicists to predict that particles would arise spontaneously from the vacuum, so long as they disappeared before violating the uncertainty principle. The quantum vacuum is a very active place, with all sorts of particles appearing and disappearing. Careful experiments have demonstrated that the quantum theorists are correct in this interpretation of a vacuum. For example, two metal plates in a vacuum that are narrowly separated will be pushed together. This is because more energetic particles can appear briefly outside the plates than can appear within the narrow separation.

Furthermore, starting in 1980 with the theory of the inflationary universe, particle physicists have told us that the entire universe was created as a "false vacuum," a quantum vacuum that has more energy in its nothingness than it should. The decay of that particular vacuum to an ordinary quantum vacuum produced all the mass in the universe and started the Big Bang.

followers to the competing Big Bang theory

Gerard P. Kuiper discovers Miranda, at the time the smallest of the five known satellites of Uranus

Bernard Lyot invents the photoelectric polarimeter

A group led by Karl August Folkers (b Decatur, IL, Sep 1, 1906) works out the structure of streptomycin

A group at Merck & Company headed by Karl August Folkers discovers that certain bacteria require vitamin B_{12} for growth, enabling them to use the bacteria as indicators so that they can isolate the pure vitamin

Philip Showalter Hench (b Pittsburgh, PA, Feb, 28, 1896) discovers that cortisone can be used to treat rheumatoid arthritis

(QED); Tomonaga was actually first, but his 1943 work was not known outside of Japan

George Gamow, Ralph Alpher, and Robert Herman develop the Big Bang theory of the origin of the universe

German-American physicist Marie Goeppert-Mayer (b Kattowitz, Poland, Jun 28, 1906) suggests that protons and neutrons in the atomic nucleus occupy shells similar to those occupied by electrons, an idea independently formulated in 1949 by Johannes Hans Daniel Jenson (b Hamburg, Germany, Jun 25, 1907)

C.M.G. Lattes and E. Gardner detect the first artificially produced pions (then called pi mesons) at Berkeley, CA, in Feb

After a walk in the woods with his dog, Swiss engineer George deMestral steals an idea from the cockleburs in his socks and the dog's coat and invents the fastener Velcro

Hungarian-American physicist Peter Mark Goldmark (b Budapest, Dec 2, 1906) develops the first long-playing record in the United States

Edwin Land invents a camera and film system that develops pictures inside the camera in about a minute

Louis Lumière, French inventor who, with his brother, developed the cinematograph, an important advance over Edison's motion-picture system, d Jun 6

English-American physicist William Bradford Shockley (b London, Feb 13, 1910), American physicist Walter Houser Brattain (b Amoy, China, Feb 10, 1902), and American physicist John Bardeen (b Madison, WI, May 23, 1908) discover the transistor, a tiny device that works like a vacuum tube but uses less power

Cybernetics by Norbert Wiener (b Columbia, MO, Nov 26, 1894) is an exhaustive mathematical analysis of the theory of feedback and automated processes

Orville Wright, American inventor, d Dayton, OH, Jan 30

1949

GENERAL

The German Democratic Republic and the German Federal Republic are established, splitting the German state into east and west

The North Atlantic Treaty Organization (NATO) is created as a common defense among Western nations against the Soviet Union

ASTRONOMY & SPACE

The radio source in the Crab nebula is identified

Walter Baade discovers the asteroid Icarus

Ralph Belknap Baldwin formulates his meteor theory of the lunar surface

J.G. Bolton discovers a galactic radio source that is a supernova remnant

William Albert Hiltner and Fohn Scoville Hall discover polarization of light by interstellar matter

Gerard P. Kuiper discovers Nereid, the smaller of Neptune's two moons

Fred Lawrence Whipple (b Red Oak, IA, Nov 5, 1906) suggests that comets are "dirty snowballs" consisting of ice or ammonia ice and rock dust

A rocket testing ground is established at Cape Canaveral, FL

By adding a small rocket to the top of a captured German V-2 rocket, the first rocket with more than one stage is created ; launched from White Sands, NM, by American scientists, it is able to reach a height of 400 km (240 mi), well beyond the atmosphere

BIOLOGY

Cole and Marmont invent the "voltage clamp" for controlling the cell membrane potential

Félix Hubert D'Herelle, Canadian-French bacteriologist, d Paris, Feb 22

Donald O. Hebb's *The organization of behavior* proposes that when synapses "fire" at the same time, they build closer connections, resulting in a neural network that contains memory

Frank Macfarlane Burnet (b Traralgon, Australia, Sep 3, 1899) proposes that immune responses to other tissue develop as a person grows and are not inborn

CHEMISTRY

Derek Harold Richard Barton (b Gravesend, England, Sep 8, 1918) starts his work with complex organic molecules, such as steroids and terpenes, in which he demonstrates that shape is important to the properties of the molecule and even a small chemical change can cause a major change in properties

Friedrich Karl Rudolf Bergius, German chemist, d Buenos Aires, Argentina, Mar 30

André-Louis Debierne, French chemist, d Paris, Aug William Francis Giauque of the United States wins the Nobel Prize for Chemistry for his study of low-temperature chemistry

English biochemist Dorothy Crowfoot Hodgkin (b Cairo, Egypt, May 12, 1910) is the first to use an electronic computer to work out the structure of an organic chemical, penicillin

Frederic Stanley Kipping, English chemist, d Criccieth, Caernarvonshire, Wales, May 1

Leonor Michaelis, German-American chemist, d New York, NY, Oct 9

1950

GENERAL

On Jun 25, troops from North Korea invade South Korea, precipitating United Nations intervention; this becomes known as the Korean War

ASTRONOMY & SPACE

Funds for building the great Jodrell Bank radiotelescope are obtained by Sir Alfred Charles Barnard Lovell (b Oldland Common, England, Aug 31, 1913)

BIOLOGY

Embryo transplants for cattle are first performed

Philip S. Hench and Edward C. Kendall of the United States and Tadeusz Reichstein of Switzerland win the Nobel Prize for Physiology or Medicine for the

CHEMISTRY

Otto Diels and Kurt Alder of Germany win the Nobel Prize for Chemistry for their synthesis of organic compounds of the diene group

Sir Walter Norman Haworth, English chemist, d Birmingham, Mar 19

	Claude Shannon publishes his work on information theory, a general approach to all kinds of information handled electronically, including symbolic logic	X rays from a synchrotron are used for the first time in medical diagnosis and treatment Walter Rudolf Hess of Switzerland and Antonio E. Moniz of Portugal win the Nobel Prize for Physiology or Medicine, Hess for study of middle brain function and Moniz for prefrontal lobotomy Schack August Steenberg Krogh, Danish physiologist, d Copenhagen, Sep 13	Leo James Rainwater (b Council, ID, Dec 9, 1917) begins to work on the idea that the nucleus of an atom might not be spherical, which he later helps to establish Hideki Yukawa of Japan wins the Nobel Prize for Physics for his prediction of the pion	In the United States, General Mills and Pillsbury begin marketing prepared cake mixes EDSAC (Electronic Delay Storage Automatic Computer) at Cambridge University goes into operation BINAC (Binary Automatic Computer), the first electronic stored-program computer in the United States, goes into operation in Aug	**1949**
In Apr John von Neumann, working with a team of meteorologists and ENIAC—one of the first computers—makes the first computerized 24-hour weather predictions		George Richards Minot, American physician, d Brookline, MA, Feb 25 Robert Wallace Wilkins (b Chattanooga, TN, Dec 4, 1906) introduces reserpine for the treatment of high blood pressure,	German-French physicist Alfred Kastler, b Guebwiller (France), May 3, 1902, develops optical pumping, a system of using light or radio waves to excite atoms, which then emit electromagnetic waves that can be studied to reveal the atomic	Diner's Club introduces the first charge card, a prototype of the credit card Commercial color television begins in the United States	**1950**

1950 cont

ASTRONOMY & SPACE

Jan Hendrik Oort proposes that a great cloud of material orbiting the sun far beyond the orbit of Pluto is the cause of comets; material from this Oort cloud, as it comes to be known, is dislodged from time to time and falls toward the sun, where we perceive it as a comet

Karl Guthe Jansky, American pioneer of radio astronomy, d Red Bank, NJ, Feb 14

BIOLOGY

discovery of cortisone and other hormones of the adrenal cortex and their functions

Frederick William Twort, English bacteriologist, d Camberley, Mar 20

CHEMISTRY

The artificial sweetener cyclamate is introduced

1951

ANTHROPOLOGY/ ARCHAEOLOGY

Robert Broom, Scottish-South African paleontologist, d Pretoria, South Africa, Apr 6

ASTRONOMY & SPACE

The Lyman-alpha line, which is the resonance line of hydrogen, is discovered in the solar spectrum using measurements taken from rockets

Dutch-American astronomer Dirk Brouwer (b Rotterdam, Netherlands, Sep 1, 1902) becomes the first astronomer to use a computer to calculate planetary orbits, using data collected since 1653 and predicting the orbits or the five outer planets though 2060

Shin Hirayama proposes that asteroids are grouped in families

Rudolph Minkowski and Walter Baade discover the first optical objects that can be related to a known radio source

BIOLOGY

James Frederick Bonner (b Ansley, NE, Sep 1, 1910) shows that mitochondria carry out oxidative phosphorylation

Rachel Carson's *The sea around us* is a popular book describing the oceans

Fritz Albert Lipmann discovers acetylcoenzyme A, an essential part of the chemistry of the body, especially important in breaking down carbohydrates, fats, and portions of proteins to obtain energy for the cell

Otto Fritz Meyerhof, German-American biochemist, d Philadelphia, PA, Oct 6

Linus Pauling and Corey determine the alpha-helix structure of proteins

Nikolaas Tinbergen's *The study of instinct* is an important study of animal behavior

CHEMISTRY

Ernst Otto Fischer (b Müchen-Sollen, Germany, Nov 10, 1918) begins his research into the structure of the newly discovered ferrocene, eventually showing that it is a new type of metal-organic compound with an atom of iron sandwiched between two carbon rings

William Draper Harkins, American chemist, d Chicago, IL, Mar 7

Edwin M. McMillan and Glenn T. Seaborg of the United States win the Nobel Prize for Chemistry for their discovery of plutonium and research on transuranium elements

1950 cont

following the practice of using the drug in the form of snakeroot in India

structure; this is an important precursor to the laser

Edward Arthur Milne, English physicist, d Dublin, Ireland, Sep 21

Hantaro Nagaoka, Japanese physicist, d Tokyo, Dec 11

Cecil Powell of England wins the Nobel Prize for Physics for his photographic study of atomic nuclei

Pakistani-British physicist Abdus Salam, b Jhang Maghiana (Pakistan), Jan 29, 1926, proves that Yukawa's pion theory is renormalizable

---

1951

Antabuse, a drug that prevents alcoholics from drinking, is introduced

US surgeon John Gibbon develops the heart-lung machine

Bernardo Alberto Houssay's *Human physiology* is published

Rueben Kahn introduces a "universal reaction" blood test for detecting several disorders at an early stage

Max Theiler of South Africa wins the Nobel Prize for Physiology or Medicine for developing the 17-D yellow fever vaccine

Aage Niels Bohr (b Copenhagen, Denmark, Jun 19, 1922) and Ben Roy Mottelson (b Chicago, IL, Jul 9, 1926) work out the mathematical details of a theory of the atomic nucleus originally developed by Leo Rainwater; they show that the nucleus is not necessarily spherical and provide details useful in nuclear fusion

Sir John Cockcroft of England and Ernest T.S. Walton win the Nobel Prize for Physics for their discovery of the transmutation of atomic nuclei in accelerated particles

Erwin Wilhelm Mueller develops the field ion microscope

Arnold Johannes Sommerfeld, German physicist, d Munich, Bavaria, W Germany, Apr 26

"Three-dimensional" motion pictures are shown, although viewers must wear special polarizing glasses

Chrysler introduces power steering

The United Kingdom introduces the "zebra" street crossing, an important contribution to pedestrian safety; when there is a pedestrian on the striped crossing, cars going both ways must halt

Transcontinental television in the United States is inaugurated

John William Mauchly and John Prosper Eckert build UNIVAC I, the first electronic computer to be commercially available and the first to store data on magnetic tape; it is sold by Remington Rand

1951
cont

ASTRONOMY & SPACE

German-American astronomer Rudolph Leo B. Minkowski (b Strassburg, Germany, May 28, 1895) discovers the asteroid Geographos, which has an orbit that takes it very close to Earth

Jan Hendrik Oort, C.A. Muller, and Hendrik Christoffel Van de Hulst construct a map of the Milky Way galaxy from Doppler shifts observed in the 21-cm (8.3-in.) emission line of hydrogen

Charles Dillon Perrine, American-Argentinian astronomer, d Villa General Mitre, Argentina, Jul 21

American physicist Edward Mills Purcell (b Taylorville, IL, Aug 20, 1912) is among the first to observe the 21-cm (8.3-in.) line caused by hydrogen atoms in space that had been predicted by van de Hulst

Eugen Karl Rabe determines that the solar parallax is 498.69 seconds

BIOLOGY

Nikolay Vavilov's *The origin, variation, immunity, and breeding of cultivated plants*, published eight years after his death, describes his work in finding the evolutionary basis of the immunity of various strains of wheat to disease and his efforts to breed improved varieties

Robert Burns Woodward (b Boston, MA, Apr 10, 1917) synthesizes the steroids cortisone and cholesterol

1952

GENERAL

In Nov Kenneth P. Oakley announces that tests on the jawbone of the so-called "Piltdown man" reveal it to be a fake, an ape jawbone that had been doctored with a chemical to make it look ancient

Michael Ventris deciphers Linear B, one of the ancient languages of Crete

ASTRONOMY & SPACE

Walter Baade announces an error in the Cepheid luminosity scale, based on his discovery that Cepheid variables in population I stars have different absolute magnitudes than Cepheid variables belonging to population II, showing that the galaxies are about twice as remote as had been thought

Adriaan Blaauw shows that the expansion of the zeta-Persei cluster started 1.3 million years ago, proving that stars are continuously created in the Milky Way galaxy

BIOLOGY

A calf is produced using frozen semen

Alan Lloyd Hodgkin (b Banbury, England, Feb 5, 1914) and Andrew Fielding Huxley (b London, Nov 22, 1917) formulate the modern theory of excitation of nerves, based on changes in the sodium and potassium ions in nerve cells; as a nerve impulse passes, sodium ions flood the cell and potassium ions are pumped out

Joshua Lederberg (b Montclair, NY, May 23, 1925) discovers that viruses that attack bacteria can

CHEMISTRY

Vladimir Nikolaevich Ipatieff, Russian-American chemist, d Chicago, IL, Nov 29

Archer J.P. Martin and Richard L.M. Synge of England win the Nobel Prize for Chemistry for their separation of elements by paper chromatography

Glenn T. Seaborg discovers einsteinium, the artificial element with an atomic number of 99, in the debris of the first thermonuclear explosion

Nevil Vincent Sidgwick, English chemist, d Oxford, UK, Mar 15

Canadian-American physicist Walter Henry Zinn (b Kitchner, Ontario, Dec 10, 1906) develops an experimental breeder reactor that is built near Idaho Falls, IN

1951

cont

The volcano Mount Parícutin in Mexico, which did not exist before 1943, stops erupting, leaving a cone 2 km (6265 ft) high

James Alfred Van Allen (b Mount Pleasant, IA, Sep 7, 1914) develops the rockoon, a rocket launched from a balloon, to study the physics of the upper atmosphere

A polio epidemic in the United States strikes, affecting 47,665 persons

British doctor Douglas Bevis develops amniocentesis, a method of examining the genetic heritage of a fetus while it is still in the womb

Jean Dausset (b Toulouse, France, Oct 10, 1916) discovers that people who have had repeated blood transfusions eventually develop antibodies to the blood being transfused, an observation later used in typing tissue for organ transplants

The first accident at a nuclear reactor occurs at Chalk River in Canada, where a technician's error causes the nuclear core to explode

The first breeder reactor, which produces plutonium at the same time as it produces energy from uranium, is built by the US Atomic Energy Commission

The Cosmotron at Brookhaven, NY, a new synchrotron particle accelerator, accelerates protons beyond 1 billion electron volts (1 GeV)

The CBS television network uses a UNIVAC computer to predict the results of the US presidential election; UNIVAC's first prediction of a landslide is right on the mark but not believed by its operators; they quickly reprogram it so that it incorrectly predicts a close contest

The first commercial product using transistors instead of vacuum tubes is introduced; it is a hearing aid

Sony develops the pocket-sized transistor radio

1952

1952

cont

ASTRONOMY & SPACE

Georg H. Herbig reports his observation of Herbig-Haro objects (regions where stars are formed)

Bernard Ferdinand Lyot, French astronomer, d on a train near Cairo, Egypt, Apr 2

Paul Willard Merrill discovers technetium in S-type supergiants; technetium has a short half-life, showing that these stars have been formed recently

Forest Ray Moulton, American astronomer, d Wilmette, IL, Dec 7

Martin Schwarzschild studies the evolution of stars by studying spectrum-luminosity graphs of stars in globular clusters

BIOLOGY

transmit genetic material from one bacterium to another, a significant step leading to genetic engineering

Lederberg introduces the term *plasmid* to describe the bacterial structures he has discovered that contain extrachromosomal genetic material

Alan Turing suggests that interactions between the strengths of various chemicals, which he calls *morphogens*, determine the biological makeup of a given species or an individual; a later example of this idea is that two morphogens determine the stripes on a zebra or the spots on a leopard

Eugene Aserinsky, while studying sleep patterns in his eight-year-old son Armond with an electro-encephalograph, discovers periods of rapid eye movement (REM) in normal sleep that are later found to be associated with vivid dreams

CHEMISTRY

Chaim Weismann, Russian-British-Israeli chemist, d Rehovoth, Israel, Nov 9

1953

GENERAL

Elizabeth II is crowned queen of the UK and British Commonwealth on Jun 2

Sir Edmund Percival Hillary and Tenzing Norgay reach the summit of Mount Everest, the world's highest mountain, on May 29

ASTRONOMY & SPACE

Harold Delos Babcock measures the magnetic field over the surface of the sun

Edwin Powell Hubble, American astronomer, d San Marino, CA, Sep 28

Milton La Salle Humason discovers a galaxy with a recession speed of 60,000

BIOLOGY

Vincent Du Vigneaud works out the exact order of the amino acids in the small protein molecule of the hormone oxytocin

Rosalind Elsie Franklin (b London, Jul 25, 1920) and British physicist Maurice Hugh Frederick Wilkins (b

CHEMISTRY

Karl Ziegler (b Helsa, Germany, Nov 26, 1898) develops the first catalyst that combines monomers into a polymer in a regular fashion, producing a much stronger and more resistant polyethylene

				1952
	The world's first sex-change operation is performed on George Jorgenson, who becomes known to the world as Christine	Felix Bloch and Edward Mills Purcell of the United States win the Nobel Prize for Physics for their measurement of the magnetic moment of a neutron		cont

Jonas Edward Salk (b New York, NY, Oct 28, 1914) develops a killed-virus vaccine against polio; it is used for mass inoculations starting in 1954 and successfully prevents the disease, although it is later superseded by a live-virus vaccine developed by Albert Sabin

Sir Charles Frank predicts that, although icosahedral crystals are impossible, regions of icosahedral symmetry will form in liquids that are cooled below their freezing points; later work demonstrates that this happens when metals are cooled so fast that they have the structure of glasses; these are called metallic glasses

Sir Charles Scott Sherrington, English neurologist, d Eastbourne, Sussex, Mar 4

Selman A. Waksman of the United States wins the Nobel Prize for Physiology or Medicine for the discovery of streptomycin

Donald Glaser (b Cleveland, OH, Sep 21, 1926) observes cosmic ray tracks in his bubble chamber, which consists of a small glass bulb filled with ether—inspired by watching the bubbles rise in a glass of beer

Robert Wallace Wilkins discovers that reserpine is a tranquilizer, the first one found; he had been using it to treat high blood pressure

A group led by Hungarian-American physicist Edward Teller (b Budapest, Jan 15, 1908) develops the first thermonuclear device, known as the H bomb; the first such bomb, which works by nuclear fusion, is exploded at the Eniwetok Atoll in the South Pacific on Nov 6

				1953
Maurice Ewing announces that there is a great rift running down the middle of the Mid-Atlantic Ridge	John H. Gibbon, Jr., uses his heart-lung machine to keep Cecelia Bavolek alive while operating successfully on her heart; it is the first use of the machine on a human being	Murray Gell-Mann (b New York, NY, Sep 15, 1929) introduces in Aug a system of assigning isospin to particles that leads to the concept of strangeness	Michelin of France and Pirelli of Italy introduce radial-ply tires	

American physicist Charles Hard Townes (b Greenville, SC, Jul 28, 1915) develops the maser, the precursor of the laser

The first bubble-chamber pictures of subatomic interactions are published by Donald Glaser

1953 cont

A cease-fire is signed in the Korean War on Jul 27, restoring the boundaries that existed before the war; 54,000 Americans had died in the war

Joseph Stalin (Yosif Vissarionovich Djugashvili) dies in the USSR after 30 years as its leader

km/sec (37,000 mi/sec), one-fifth the speed of light

Roger C. Jennison, while a graduate student, develops a way to use very long baseline interferometry with radiotelescopes, dramatically improving their resolution; as he is a graduate student, the idea is not picked up; it is rediscovered in 1975 by Alan E.E. Rogers

Herbert Rollo Morgan, Albert Edward Whitford, and Arthur Dodd Code show that O associations of stars are in the arms of the Milky Way galaxy

Herbert Rollo Morgan's *Catalogue fondamental* (N30) lists 5268 fundamental stars (reference stars used for determining positions of celestial objects)

Neymann and Scott discover the existence of superclusters (clusters of clusters) of galaxies

I.S. Shklovskii explains the radio emission from the Crab nebula as being caused by synchrotron radiation

Pongaroa, New Zealand, Dec 15, 1916) make X-ray studies of deoxyribonucleic acid (DNA) that allow James Watson and Francis Crick to determine its structure

William Hayes discovers that plasmids can be used to transfer introduced genetic markers from one bacterium to another

Mabel N. and Lowell E. Hokin (b Chicago, IL, Sep 20, 1924) discover that acetylcholine causes cells of the pancreas to increase their uptake of phosphorus groups that are incorporated into the cell membrane; this is an important step toward learning how cells communicate with each other

Alfred C. Kinsey (b Hoboken, NJ, Jun 23, 1894) and coworkers produce a landmark study of sex practices of US women; among the conclusions: almost half have sexual relations before marriage, a quarter are unfaithful afterward, and a quarter of those unmarried have had a homosexual relationship

Fritz A. Lipmann of the United States and Hans Adolf Krebs of England share the Nobel Prize for Physiology or Medicine, Lipmann for the discovery of coenzyme A and Krebs for the discovery of the citric acid cycle

Austrian-British biochemist Max Ferdinand Perutz (b Vienna, May 19, 1914) introduces the technique of adding an atom of a heavy element, such as gold, to

Upon hearing of Karl Ziegler's discovery of a way to catalyze the reactions that produce plastics with metal ions, Giulio Natta (b Imperia, Italy, Feb 26, 1903) uses the idea to develop the first isotactic polymers

Hermann Staudinger of Germany wins the Nobel Prize for Chemistry for his study of polymers

Evarts A. Graham and Ernest L. Wydner demonstrate that tars from tobacco smoke cause cancer in mice

Frederick Sanger (b Rendcombe, England, Aug 13, 1918) becomes the first to determine the structure of a protein, insulin

Robert Andrews Millikan, American physicist, d Pasadena, CA, Dec 19

T. Nakano and Kasuhiko Nishijima introduce the concept of the strangeness quantum number independently of Murray Gell-Mann

Frits Zernike of the Netherlands wins the Nobel Prize for Physics for his development of the phase-contrast microscope

Plate tectonics

In 1912, Alfred Wegener, a German meteorologist, announced his belief in the idea of continental drift. Wegener was not the first to notice that the coastlines of South America and Africa seem to fit together as if they had once been part of a single continent that had broken apart. But he pursued the idea more vigorously than others had, gathering evidence that fossils and rock strata were the same on the two matching coasts. Most geologists rejected his evidence, however, because he could not suggest a workable way for the continents to move though the solid rock of Earth's crust. However, the theory was too appealing to be forgotten. After World War II, geology textbooks generally mentioned Wegener's theory of continental drift but also continued to cast doubt on its validity.

Beginning in the early 1950s, better equipment for measuring the residue of Earth's magnetism that is found in rocks, improved methods for measuring the age of rocks, and a program of exploration of the floor of the oceans all combined to vindicate partially continental drift. The new magnetic studies showed that rocks of the same age on different continents indicate that in the past the magnetic poles seemed to be in two places at the same time. The alternative explanation, which was eventually accepted, was that the continents had changed position with regard to each other.

It had been known since 1929 that the magnetic field of Earth reversed itself every few hundred thousand years. This had been observed in successive strata, one above the other, on the continents. In the early 1960s, F.J. Vine and D.H. Matthews demonstrated that on the floor of the ocean these reversals do not occur in different strata vertically but instead lay in similar strata that are side by side. In effect, from a magnetic point of view, the rocks of the ocean floor are striped. The youngest stripes are immediately adjacent to a formation in the middle of the ocean, a valley with mountains on either side that is known as a rift valley. As one travels farther from the mid-ocean rift, the rocks are older. This tends to confirm a 1960 theory of Harry Hess that new ocean crust is being formed at the rift. According to Hess's theory, the oldest crust should be sinking into deep trenches in the ocean floor, such as the one off the Philippine islands. This part of the theory, which came to be known as sea-floor spreading, was also confirmed by measuring the ages of rocks dredged from the ocean bottom.

Geologists synthesized continental drift and sea-floor spreading into a single theory, plate tectonics (*tectonics* means movements of Earth's crust). According to plate tectonics, the crust of Earth is broken into a number of large plates. Some of these plates contain only ocean crust, while others contain both continental crust and ocean crust. For example, one plate consists of most of the Pacific Ocean, while another contains North America and the western half of the Atlantic Ocean. These plates move through a semiliquid region of Earth just below the crust itself. Where the plates bump into each other, mountain ranges form. Most earthquakes and volcanoes are at the plate boundaries.

The theory was very successful at predicting many features of Earth's crust, but some geologists were not satisfied until the slow motion of the plates could be measured—typically about 2 cm (1 in.) per year. This was accomplished in the 1980s using satellites, lasers, and the positions of very distant galaxies.

1953

cont

BIOLOGY

an organic molecule to improve X-ray diffraction of the molecule

James L.B. Smith, who examined the first coelacanth known to science in 1938, acquires the second specimen of the supposedly extinct fish as a result of advertising a reward; he discovers that the elusive fish is common around the Comoro Islands, where it has been known to local fisherman all along

James Dewey Watson (b Chicago, IL, Apr 6, 1928) and Francis H. Crick (b Northhampton, England, Jun 8, 1918) develop the double-helix model for deoxyribonucleic acid, or DNA, which explains how the giant molecule is capable of transmitting heredity in living organisms

---

1954

GENERAL

France gives up its occupation of Indochina

The US Supreme Court rules against schools that are segregated by race

Aikichi Kaboyama, a Japanese fisher, is killed by fallout from a US test of a thermonuclear bomb in the Pacific

ANTHROPOLOGY/ARCHAEOLOGY

Kamal el Malakh and other archaeologists find two nearly identical chambers near the base of the Great Pyramid of Khufu in Egypt; excavating one of them, while leaving the other untouched for future work, they uncover the remains of a 43-m (142-ft) cedar boat that is probably intended to carry the dead pharaoh west

ASTRONOMY & SPACE

F.T. Haddock, C.H. Mayer, and R. Sloanaker discover radio emissions from ionized hydrogen in the Orion nebula and the Omega nebula

BIOLOGY

A mining engineer discovers seeds of the arctic lupine that have been preserved frozen in rodent burrows in the Yukon Territory of Canada since the end of the last ice age

Albert Francis Blakeslee, American botanist, d Northampton, MA, Nov 16

Vincent Du Vigneaud synthesizes oxytocin, the first hormone ever to be synthesized

John F. Enders, Thomas H. Weller, and Frederick C. Robbins win the Nobel Prize for Physiology or Medicine for their discovery of a method for cultivating viruses in tissue culture

CHEMISTRY

Otto Paul Hermann Diels, German chemist, d Kiel, Mar 7

Linus C. Pauling of the United States wins the Nobel Prize for Chemistry for his work on chemical bonds

					1954

Elso Sterrenberg Barghoorn (b New York, NY, Jun 30, 1915) and Stanley A. Tyler discover tiny fossils in the 2-billion-year-old Gunflint chert of the Canadian shield; the fossils resemble modern monerans, such as bacteria and blue-green algae, and are one and a half billion years older than any previously known

Alan Mathison Turing, English mathematician, d (probably of his own hand as a result of persecution for homosexuality) Wilmslow, Cheshire, Jun 7

Chlorpromazine (Thorazine) is introduced for the treatment of mental disorders

The Bevatron particle accelerator at the Radiation Laboratory in Berkeley (renamed the Lawrence Berkeley Laboratory in 1958) is completed; it can accelerate uranium atoms to 6.5 billion electron volts (GeV)

The *Centre Européen de Recherche Nucléaire* (European Center for Nuclear Research), known as CERN, is founded on Sep 29

It is discovered in Nov that the neutral K particle and its antiparticle decay in a way that seems to violate the laws of physics

Luis Walter Alvarez develops his liquid-hydrogen bubble

The Soviet Union builds the first small nuclear reactor for peacetime use

TV dinners are introduced in the United States

The first atomic-powered submarine, the *Nautilus*, is commissioned

Edwin Howard Armstrong, American electrical engineer, d New York, NY, Feb 1

Bell Telephone scientists Chapin, Fuller, and Peterson develop the photovoltaic cell, which can produce electric power from sunlight

Kotaro Honda, Japanese metallurgist, d Tokyo, Feb 12

1954 cont

Measuring with waves

Possibly the most accurate form of measure is based on the interaction of one wave with another. The principles involved are the same whether the waves are of water, light, or sound. If two waves are exactly the same, they can be adjusted so that they will cancel each other or increase each other. If the crests of one version of the wave are superimposed on the crests of another version, the waves will add. If they are light waves, the result will be brighter light. If the crests of one version of the wave are superimposed on the troughs of the other, the waves will cancel. If they are light waves, this will produce darkness. If the waves are not exactly the same, superimposing them will produce bands of brightness and darkness, called an interference pattern. Using an interference pattern to make measurements is called interferometry.

Among the first to use interferometry in measurement was Albert Michelson in 1881. By 1887 he and Edward Morley had used the idea to conduct one of the most famous experiments in science. They established through experimental data that the velocity of light is not affected by the movement of Earth through space. This result helped make people accept Einstein's theory of relativity, which included the concept that the velocity of light in a vacuum is a universal constant.

Astronomers, starting with Martin Ryle in the 1960s, use interferometry with radiotelescopes to create giant, fictitious telescopes that work better than real ones. Two or more telescopes acting together as a single telescope can be used to obtain better resolution than any one telescope. At first, Ryle used just two radiotelescopes that were not very far apart. By 1980, the Very Large Array in Socorro, NM, was using 27 radiotelescopes linked by computer to obtain the same resolution obtainable from a single radiotelescope 27.4 km (17 mi) in diameter. Today, with interferometry, a group of telescopes all over Earth can be linked by satellite and computer to give the equivalent of a single telescope

J. Hin Tjio and Albert Levan show that humans have 46 chromosomes rather than 48, as was believed previously

1955

GENERAL

The Warsaw Pact links the Soviet Union and the nations of Eastern Europe into an alliance against the West

ANTHROPOLOGY/ARCHAEOLOGY

Sir Arthur Keith, influential British anthropologist, d Jan 8

ASTRONOMY & SPACE

The 76.2-m (250-ft) radio "dish" at Jodrell Bank, England, is completed

Kenneth Lynn Franklin (b Alameda, CA, Mar 25, 1923) and Burke detect radio emissions from Jupiter

Jan Hendrik Oort measures the polarization of light coming from the Crab Nebula, confirming that it is synchrotron radiation

BIOLOGY

James Frederick Bonner and Paul Ts'o isolate mitochondria from cells

Sir Alexander Fleming, Scottish bacteriologist, d London, Mar 11

German biochemist Heinz Fraenkel-Conrat, b Breslau (Wroclaw, Poland) Jul 29, 1910, shows that viruses consist of a noninfective protein coat and a mildly infective nucleic-acid core

CHEMISTRY

Vincent Du Vigneaud of the United States wins the Nobel Prize for Chemistry for his synthesis of polypeptide hormones

Measuring with waves (continued)

with the diameter of the planet. Astronomers are even looking forward to space-based radiotelescopes in the future that would expand the present limitation of the size of Earth. For example, a radiotelescope on the moon could work with those on Earth or in a space station to form the equivalent of an antenna with a diameter the distance from Earth to the moon.

Today, interferometry is the king of measurement systems. Interferometry is so sensitive that it can detect the movement of an object that moves only a centimeter in hundreds of years. It can even be used to determine the rotation of distant stars. Using lasers as the source of light, interferometry can detect tiny movements in the crust of Earth that precede earthquakes.

The interferometer is also the instrument of choice for determining small distances. It can detect tiny variations in thickness of a lens, for example. Using lasers or radar and interferometry combined, space vehicles guide themselves with extreme precision as they cross the vast distances between planets.

1954 cont

PHYSICS

chamber and observes tracks caused by subatomic particles

Max Born of England and Walther Bothe of Germany share the Nobel Prize for Physics, Born for his work in quantum mechanics and Bothe for his work in cosmic radiation

Enrico Fermi, Italian-American physicist, d Chicago, IL, Nov 28

Fritz Wolfgang London, German-American physicist, d Durham, NC, Mar 30

In Nov, Abraham Pais introduces the name *baryon* to describe particles that are affected by the strong force

Chinese-American physicist Chen Ning (Frank) Yang (b Hofei, China, Sep 22, 1922) and Robert Mills in Jun establish the basis for modern quantum field theory by developing the mathematical characteristics of fields needed to allow symmetry to produce new particles; these are called Yang-Mills gauge-invariant fields

TECHNOLOGY

Auguste Lumière, French chemist and inventor who, with his brother Louis, invented the cinematograph, d Apr 10

1955

MATHEMATICS

Henri Cartan and Samuel Eilenberg develop homological algebra, a cross between abstract algebra and algebraic topology that helps unify mathematics

Karl Prachar proves that there are many numbers that are one less than a prime number that have a large number of prime factors

MEDICINE

Oswald Theodore Avery, Canadian-American physician, d Nashville, TN, Feb 20

Sir Edward Mellanby, whose work with vitamin A paved the way for the discovery of vitamin D, d Jan 30

Antonio Caetano de Abreu Moniz, Portuguese surgeon, d Lisbon, Dec 13

PHYSICS

It is discovered that two varieties of K mesons exist; they decay according to different modes, but are otherwise apparently the same particle; one particle is termed tau and the other theta, so the problem becomes known as the tau-theta puzzle

Luis Walter Alvarez designs a semiautomatic measuring machine for analyzing particle tracks on bubble-chamber photographs; the

TECHNOLOGY

In the United States, deep freezers capable of freezing fresh food go on sale

The first artificial diamonds for industrial use are produced in the United States

Velcro is patented

Christopher Cockerell develops the first practical hovercraft

The first optical fibers are produced by Narinder Kapary in London

1955

cont

Martin Ryle constructs the first radio interferometer, a device that increases the ability of radiotelescopes to resolve radio sources; with interferometry, radiotelescopes become as good as light telescopes at resolving light sources

The US Vanguard project for launching artificial satellites is announced

Mahlon Bush Hoagland (b Boston, MA, Oct 5, 1921), working with protein synthesis, establishes that transfer RNA combines with specific amino acids; these acids are later combined by messenger RNA, following instructions from DNA

Spanish-American biochemist Severo Ochoa (b Luarca, Spain, Sep 24, 1905) develops enzymes that cause nucleic acid bases to form RNA, essentially creating synthetic RNA

Henry Clapp Sherman, American biochemist, d Rensselaer, NY, Oct 7

James Batcheller Sumner, American biochemist, d Buffalo, NY, Aug 12

Hugo Theorell of Sweden wins the Nobel Prize for Physiology or Medicine for his study of oxidation enzymes

Lasers

In 1916 and 1917 Albert Einstein continued his study of the physics of light. Among other things, Einstein showed that molecules that had been suitably energized would emit light of a single color, or monochromatic light. After World War II, uses were found for this effect. The effort started in 1951 when Charles Townes wanted to produce stronger microwaves. These could only be produced by a very small vibrator. As Townes was waiting on a park bench for a restaurant to open, it occurred to him that ammonia molecules were about the right size to vibrate at the needed speed. He did some quick calculations on the back of an envelope and concluded that he could pump energy into ammonia gas and the molecules would emit microwaves. He soon built the first device that produced microwave amplification by stimulated emission of radiation; he named this the maser after the initials of the process.

Various people, including Townes, thought that the same principle could be used to amplify light, although the technical problems were more difficult. Eventually, the patent office gave credit for developing light amplification by stimulated emission of radiation to Columbia University graduate student Gordon Gould, who conceived of a laser on November 11, 1957. Despite this patent, it is generally believed that the first effective laser was built by Theodore Maiman in May, 1960.

Thereafter, lasers of various types were used in fields from entertainment to weaponry. Because lasers produce high-energy light at specific wavelengths that does not diverge as rapidly as ordinary light, they can be used to transfer energy to a specified location with great accuracy. Consequently, one of the first uses of lasers has been in cutting and welding, both in heavy industry and medical practice. Indeed, one of the places that a person is most likely to encounter a laser today is in a hospital, where surgeons use lasers as scalpels and ophthalmologists use them to weld damaged retinas in place. Military planners also hope to use the laser's cutting and burning power in weapons. Others are working on devices to use laser power to ignite the fusion of hydrogen for a power source for the future.

Because laser light diverges very slowly, a laser beam can be used to determine how level a surface is. Farmers have used lasers to make certain that fields are level; this helps in protecting fields from erosion.

Lasers are also part of the ongoing optical revolution, in which electronic devices are replaced by photonic devices. A photonic device uses photons instead of electrons. Lasers are an excellent source of photons for many applications. Although the principal photonic devices now in use are long-distance fiber-optic networks, many scientists predict the development of photonic computers.

Lasers have many scientific applications as well as technical ones. Among the more exotic is using intersecting laser beams to halt atoms in midspace. Among the more common is the use of laser interferometry to locate exact positions on Earth; for example, lasers can detect the slight movement of the crust that may precede earthquakes.

machine, called Frankenstein, is built a year later by Hugh Bradner and Jack Franck

Owen Chamberlain (b San Francisco, CA, Jul 10, 1920) and Emilio Segrè succeed in producing antiprotons

Clyde Lorrain Cowan, Jr. (b Detroit, MI, Dec 6, 1919) and Frederick Reines (b Paterson, NJ, Mar 16, 1918) observe neutrinos for the first time (they had been predicted by Wolfgang Pauli in 1930)

Albert Einstein, German-Swiss-American physicist, d Princeton, NJ, Apr 18

Willis Lamb, Jr., and Polykarp Kusch of the United States share the Nobel Prize for Physics, Lamb for his measurement of the hydrogen spectrum and Kusch for his study of the magnetic momentum of the electron

The field ion microscope, developed by American physicist Erwin Wilhelm Mueller, becomes the first instrument that can picture individual atoms; it accomplishes this by releasing ions from the tip of a needle cooled nearly to absolute zero; the ions strike a fluorescent screen, producing an image of the needle tip

Robert Williams Wood, American physicist who specialized in the study of optics, d Aug 11

1956

GENERAL

Citizens and workers in Hungary rise up against Soviet domination, but the rebellion is crushed when Soviet troops invade on Nov 4

Israeli, British, and French forces invade Egypt to prevent nationalization of the Suez Canal

The United States explodes its largest hydrogen bomb in a test at Bikini Atoll in the Pacific Ocean; islanders from Bikini are first evacuated to other islands, but are told that they can return to Bikini when the tests are over

ANTHROPOLOGY/ARCHAEOLOGY

William Clouser Boyd (b Dearborn, MO, Mar 4, 1903), assisted by his wife Lyle, after studies of blood groups extending back into the 1930s, releases a list of 13 "races" of *Homo sapiens* based on blood groups; one surprise is the Basques appear to be the remaining part of an early European race

ASTRONOMY & SPACE

Walter Sydney Adams, American astronomer, d Pasadena, CA, May 11

Herbert Friedman (b New York, NY, Jun 21, 1916) reports that solar flares are X-ray sources

Meghnad Saha, Indian astronomer who showed that temperature is as important as elemental composition in determining a star's spectrum, d Feb 16

Robert Julius Trumpler, Swiss-American astronomer, d Oakland, CA, Sep 10

BIOLOGY

Benjamin Minge Duggar, American botanist, d New Haven, CT, Sep 10

Arthur Kornberg (b Brooklyn, NY, Mar 3, 1918) synthesizes DNA by the action of enzymes on nucleic-acid bases, or nucleotides; the DNA is not biologically active, however

Rumanian-American physiologist George Emil Palade (b Iasi, Rumania, Nov 19, 1912) discovers that small bodies within the cell, now called ribosomes, are mostly ribonucleic acid (RNA); it is soon found that the ribosomes are the locations in the cell where proteins are manufactured

Earl Wilbur Sutherland, Jr. (b Burlingame, KS, Nov 19, 1915) isolates cyclic AMP (adenosine monophosphate), an intermediate in the cell's energy cycle and also important in many other reactions

CHEMISTRY

Sir Cyril Hinshelwood of England and Soviet physical chemist Nikolai Semenov (b Saratov, Apr 15, 1896) share the Nobel Prize for Chemistry for their parallel work on the kinetics of chemical chain reactions

Dorothy Crowfoot Hodgkin uses an electronic computer to work out the structure of vitamin B_{12}

Choh Hao Li and coworkers determine that adrenocorticotrophic hormone (ACTH) consists of 39 amino acids in a specific order

Choh Hao Li and coworkers isolate human growth hormone

Frederick Soddy, English chemist, d Brighton, Sussex, Sep 22

Bruce Charles Heezen, American oceanographer and geologist (b Vinton, IA, Apr 11, 1924) and Maurice Ewing discover the Mid-Oceanic Ridge, a globe-girdling formation of mountains and rifts

Werner Forssman (b Berlin, Germany, Aug 29, 1904), Dickinson Richards (b Orange, NJ, Oct 30, 1895), and French-American physiologist André F. Cournand (b Paris, Sep 4, 1895) win the Nobel Prize for Physiology or Medicine for their use of the catheter for study of the interior of the heart and circulatory system

Bruno Kirsh discovers peculiar dense bodies in the heart cells of guinea pigs; the function of these bodies is unknown at this time; much later they are discovered to release hormones that help regulate the circulatory system

Birth control pills are used in a large-scale test conducted by John Rock and Gregory Pincus in Puerto Rico

Robert Mearns Yerkes, American psychologist, d New Haven, CT, Feb 3

Heinrich Barkhausen, German physicist who discovered that iron emits sound when subjected to a steadily increasing magnetic field (the Barkhausen effect), d Dresden, Feb 20

Nicolaas Bloembergen lays the theoretical foundations for the construction of a solid-state maser (the microwave analogue of a laser) that will produce microwaves continuously instead of intermittently

B. Cook, G.R. Lambertson, O. Picconi, and W.A. Wentzel discover the antineutron

The daughter of Pierre and Marie Curie, French physicist Irène Joliot-Curie, d Paris, of leukemia probably caused by working with radioactive materials, Mar 17

Willem Hendrik Keesom, Dutch physicist who explored the properties of liquid helium and even produced solid helium, d Oegstgeest, Mar 24

K. Lande, E.T. Booth, J. Impeduglia, and Leon M. Lederman obtain the first evidence for a long-lived neutral K particle

William Shockley, Walter H. Brattain, and John Bardeen of the United States win the Nobel Prize for Physics for their studies on semiconductors and the invention of the electronic transistor

Sir Franz Eugen Francis Simon, German-British physicist who

The Lip company in France produces the first commercial watch to run on electric batteries

The first transatlantic telephone cable is put into operation on Sep 25

John Backus and a team at IBM invent FORTRAN, the first computer programming language; previously, computer programs had to be installed in machine language

Christopher Hinton (b Tisbury, England, May 12, 1901) opens Calder Hall in England, the first large-scale nuclear power plant designed for peaceful purposes

John McCarthy develops Lisp, the computer language of artificial intelligence

Stanislaw Ulam programs a computer to play chess on a 6×6 board; the program, called MANIAC I, becomes the first computer program to beat a human in a game

1956

cont

1957 The Common Market is established in Europe

Richard Evelyn Byrd, American explorer, d Boston, MA, Mar 11

Henry Norris Russell, American astronomer, d Princeton, NJ, Feb 18

Martin Ryle suggests that changes in brightness of Seyfert galaxies are caused by the ejection of blobs of matter that travel at nearly the speed of light

The first artificial satellite, Sputnik I, is launched by the Soviet Union on Oct 4

The Soviet Union launches its second satellite; this one contains a live dog

Gerty Theresa Radnitz Cori, Czech-American biochemist, d St. Louis, MO, Oct 26

G.E. Hutchinson defines the ecological niche as an abstract hypervolume in a space with axes for each of the environmental and biological variables that affect the organism whose niche it is; that is, the niche is a region both in space and in possible behaviors that the organism occupies

Irving Langmuir, American chemist, d Falmouth, MA, Aug 16

Sir Alexander Todd of England wins the Nobel Prize for Chemistry for his study of nucleic acids

Paul Walden, Russian-German chemist, d Gammertingen, Germany, Jan 24

Heinrich Otto Wieland, German chemist, d Munich, Bavaria, Aug 5

					1956

developed magnetic methods to attain temperatures close to absolute zero, d Oxford, UK, Oct 31

Chen Ning (Frank) Yang and Tsung-Dao Lee (b Shanghai, China, Nov 24, 1926) in May, in a conversation at the White Rose Cafe in New York City, realize that conservation of parity has never been checked for the weak force; by Jun they have a paper proposing experiments that could be used to determine whether or not parity is conserved

cont

1957

EARTH SCIENCE	MATHEMATICS	MEDICINE	PHYSICS	TECHNOLOGY
Frederick Alexander Lindemann (Viscount Cherwell), English scientist who first showed that the upper stratosphere is warmer than the lower layers of the atmosphere, d Jul 3	John von Neumann, Hungarian-American mathematician, d Washington, DC, Feb 8	West German workers manufacturing the herbicide 2, 4, 5-T develop a skin disease later named chloracne; their cases lead to the first recognition that a dioxin frequently contaminates such herbicides	The Columbia University Physics Department on Jan 15 announces that experiments conducted by Chien-Shiung Wu based on theoretical work by Chen Ning Yang and T.D. Lee show that parity is not conserved for the weak interactions (specifically, the decay of cobalt-60 differentiates left from right)	Columbia University physics graduate student Gordon Gould on Nov 11 has the idea that will translate into the laser; Gould does not, however, apply for a patent until 1959; by then others have also begun work on lasers; Gould's patent claims are not to be accepted until after 1986

The high-speed, painless dental drill is developed in the United States

Maurice S. Bartlett concludes that between 4000 and 5000 cases of measles are needed to keep an epidemic alive; this implies that measles cannot survive in an isolated population of less than 250,000 persons

Daniel Bovet of Italy wins the Nobel Prize for Physiology or Medicine for his discovery of antihistamines and work with curare

Alick Isaacs (b Jul 17, 1921) and Jean Lindenmann discover interferons, natural substances produced by the body that fight viruses

John Bardeen, Leon N. Cooper (b New York, NY, Feb 28, 1930), and John R. Schrieffer (b Oak Park, IL, May 31, 1931) formulate the "BCS theory," which explains superconductivity by assuming the existence of coupled electrons, called Cooper pairs, that cannot undergo scattering by collisions with atoms in the conductor

Walther Wilhelm Georg Bothe, German physicist who first thought of using a pair of geiger counters to detect individual cosmic rays, d Heidelberg, W Germany, Feb 8

God is left-handed

After World War II, physicists in the United States and elsewhere were able to return to basic research. One group used balloons and other devices to lift photographic emulsions high in the atmosphere, where they recorded cosmic-ray tracks. Cosmic rays are the particles that bombard Earth from space or the particles that these primary particles knock out of the molecules that they encounter along the way in the atmosphere. Collisions between a cosmic ray and a molecule produce higher energies than the particle accelerators ("atom smashers") of the day were able to achieve. As Albert Einstein had shown in 1905, energy can be changed into mass. In fact, the result of these collisions at high energies is the production of new particles of masses greater than are observed in ordinary matter.

The new particles often behaved in very strange ways. In fact, Murray Gell-Mann, in the early 1950s, began to call one particular property *strangeness,* a name that has stuck. In physics, it is possible to specify completely whether a particle is "strange" or not. Among the strangest of the strange were particles that have undergone numerous name changes. These particles are the neutral K (kappa) mesons, also called kaons. In the 1950s, physicists called one variety of K meson *tau* and another *theta* (today *tau* refers to an entirely different particle). The problem with tau and theta was that every bit of evidence suggested that they should be exactly the same particle; but when, like all heavy particles, they decayed into lighter particles, tau decayed into three particles and theta decayed into two. Furthermore, the products of the two decays implied that the original particles had to be different in an essential property called parity. The fundamental wisdom of the time was that two otherwise identical particles could not have unequal parities.

Parity itself is a simple concept that is based on an either/or situation. The most common representations are in terms of numbers, usually +1 and −1, or in terms of right and left. Since these two seemingly dissimilar ideas are mathematically identical, physicists often use the numbers +1 and −1, even when the actual physical operation is closer to right and left. Parity for particles is an example of this. The tau has a parity of −1 while the theta has a parity of +1. The experiment that finally explained "the tau-theta paradox," however, rested on the difference between right and left.

By 1956, this problem was a focal session for a physics conference in Rochester, NY. Physicists Martin Block and Richard Feynman were roommates at the meeting. Block suggested a radical idea to Feynman, which Feynman passed on to the conferees at the session on the tau-theta problem. Block's idea was that there is some fundamental difference between right and left. This was radical because the idea that right and left are essentially the same, called *the law of conservation of parity,* was one of the pillars of theoretical physics at the time. Frank Yang, one of the speakers at that session, and T.D. Lee met a month later at the White Rose Cafe near Columbia University and decided to work out the implications of Block's idea. In another month or so Yang and Lee were able to point to specific experiments that would resolve the question. Since they were theoreticians, others performed the experiments. By the end of the year, experiments conducted under the leadership of Chien-Shiung Wu had shown that parity is not conserved in interactions that involve the weak interaction, one of the fundamental forces of nature. Specifically, cobalt 60, whose atoms decay as a result of the weak interaction, preferentially decays toward the left, not the right.

The announcement of these results in January of 1957 forced a major revision in the ideas of physicists. It was a revolution whose implications are not completely resolved today. Wolfgang Pauli captured the idea best in a phrase calling God "a weak left-hander." He initially used those words to dismiss the results, and did not accept the fall of parity until he read papers produced by physicists who confirmed Wu's result in various other experiments.

This lack of symmetry in a world dominated by symmetry may be necessary for the universe to exist at all. In most reactions, matter and its exact opposite, antimatter, are produced in equal amounts. Yet the existing universe, as far as we can tell, contains matter and not antimatter. Matter and antimatter interact to destroy each other whenever they meet. If equal amounts of matter and antimatter were produced in the creation of the universe, this reaction would have "uncreated" it. Physicists think that some asymmetrical reaction at the beginning of time must have prevented this from happening, or we would not be here to think about it.

Polish-American microbiologist Albert Bruce Sabin (b Bialystok, Poland, Aug 26, 1906) develops a polio vaccine based upon live, weakened viruses

Leo Esaki (b Osaka, Japan, Mar 12, 1925) discovers that electrons are able to tunnel from one region of a semiconductor to another, bypassing what would ordinarily be a barrier and causing resistance to decrease rather than increase with increasing current, as would be expected

Tsung-Dao Lee and Chen Ning Yang of China win the Nobel Prize for Physics for their discovery of violations of the law of conservation of parity

Gerhart Lüders and Wolfgang Pauli prove the CPT theorem: transformation of a particle into an antiparticle leaves the laws of physics intact as long as charge, parity, and time are all reversed (that is, positive becomes negative, right becomes left, and past becomes future, or vice-versa)

In Nov, Julian Schwinger makes the first suggestion that weak interactions are mediated by charged vector bosons, later called W particles; the W particles are finally detected in 1983

Johannes Stark, Nazi physicist who discovered that an electric field splits spectral lines, d Traunstein, Bavaria, W Germany, Jun 21

Sir John Sealy Edward Townsend, Irish-English physicist who was among several physicists to determine the charge of the electron in 1897, d Feb 16

1958

A "red event" is observed in the lunar crater Alphonsus by N.A. Kozyrev from the Crimean Astrophysical Observatory in the Soviet Union; although thought at the time to be volcanic activity, this explanation is rejected after astronauts visit the moon

Venus is observed at radio wavelengths by Mayer

American physicist Eugene Norman Parker (b Houghton, MI, Jun 10, 1927) demonstrates that there is a "solar wind" of particles thrown out by the sun; it is this solar wind that makes comets' tails point away from the sun

Wernher von Braun's team launches the first American satellite to reach a successful orbit around Earth

The orbit of the Vanguard satellite, launched on Mar 17, is used by John Aloysius O'Keefe (b Lynn, MA, Oct 13, 1916) to show that Earth is slightly pear-shaped with a bulge of about 15 m (50 ft) in the Southern Hemisphere

George Wells Beadle, Edward Lawrie Tatum, and Joshua Lederberg of the United States share the Nobel Prize for Physiology or Medicine, Beadle and Tatum for relating genes to enzymes and Lederberg for genetic recombination

Ilya Darevsky discovers lizard species in Soviet Armenia that consist entirely of females who reproduce parthenogenetically—without fertilization by males; this is the first known example of an all-female vertebrate species

Harris, Michael, and Scott demonstrate the direct effect of hormones on the central nervous system via the posterior hypothalamus

Kurt Alder, German chemist, d Cologne, Jun 20

Rosalind Elsie Franklin, English physical chemist, d London, Apr 16; her untimely death is generally thought to have robbed her of a Nobel Prize for her X-ray diffraction studies of DNA that enable Watson and Crick to determine its structure

Friedrich Adolf Paneth, German-British chemist, d Mainz, W Germany, Sep 17

Frederick Sanger of England wins the Nobel Prize for Chemistry for his discovery of the structure of insulin

Bachus and Herzenberg show the existence of a "dynamo" inside Earth that creates the magnetic field

When radiation counters aboard US satellites Explorer I and III mysteriously fail, James Van Allen suspects the cause is high levels of radiation; he designs a counter for Explorer IV, launched Jul 26, to measure this; the result is the discovery of the Van Allen radiation belt that surrounds Earth

Bifocal contact lenses are introduced

Ian Donald of Scotland is the first to use ultrasound to examine unborn children

John Broadus Watson, American psychologist, d New York, NY, Sep 25

Physicists show that charge conjugation is not preserved in some interactions; that is, when positive charges are replaced by negative charges and vice versa, a different reaction occurs; this is closely related to the breakdown of conservation of parity observed in 1957

Pavel A. Cherenkov, Ilya M. Frank, b St. Petersburg (Leningrad, USSR), Oct 23, 1908, and Igor E. Tamm of the Soviet Union win the Nobel Prize for Physics for their discovery of the principle of light emission by electrically charged particles moving faster than the speed of light in a medium

Clinton Joseph Davisson, American physicist who proved experimentally that electrons are waves as well as particles, d Charlottesville, VA, Feb 1

Frédéric Joliot-Curie, French physicist who with his wife, Irène, discovered artificial radioactivity and later aided the Allies in developing the nuclear bomb (while remaining in France to work with the Resistance), d Paris, Aug 14

Ernest Orlando Lawrence, American physicist who developed the cyclotron for particle acceleration, d Palo Alto, CA

Wolfgang Pauli, Austrian-American physicist whose exclusion principle is the atomic basis of chemistry, d Zurich, Switzerland, Dec 14

The United States opens its first experimental nuclear reactor for the generation of electric power at Shippingport, PA

Alex Bernstein and Michael Roberts develop a chess program that runs on an IBM 704 and plays like a fair amateur

Charles Franklin Kettering, American inventor, d Loudonville, OH, Nov 25

	GENERAL	ANTHROPOLOGY/ ARCHAEOLOGY	ASTRONOMY & SPACE	BIOLOGY	CHEMISTRY
1959	The Antarctic Treaty is signed; signatories, including the US and the USSR, promise to keep the continent free from military activities and to use it for scientific research	English anthropologist Louis Seymour Bazett Leakey (b Kabete, Kenya, Aug 7, 1903) finds fossil remains of an early hominid from about 1,750,000 years ago that he names *Zinjanthropus*; later anthropologists usually refer to the species as *Australopithecus robustus* or *A. boisei* Derek J. Prince recognizes that a corroded mechanism discovered in the Mediterranean in 1900 is a primitive computer built about 65 BC and used to calculate the positions of planets	Radar contact is made with the sun by Eshleman at the Stanford Research Institute in the United States N.S. Kardashev predicts that spectral lines are emitted from ionized gases in space; these lines are subsequently discovered The Soviet Union launches Lunik I on Jan 2; aimed at the moon, it misses and goes into orbit around the sun, where it is widely hailed as the first artifical planet and renamed Mechta (Dream) The Soviet spaceprobe Lunik II becomes the first human-produced object to reach the moon's surface when it crashlands on Sep 14 The Soviet Union launches Lunik III on Oct 4; its path takes it past the moon in such a way that it can return photographs of the far side to Earth, giving the first ever views of the far side of the moon	Japanese scientists discover that resistance to antibodies in the bacterium *Shigella dysenteriae* is carried from one bacterium to another by plasmids, small circles of DNA apart from the chromosomes that are found in most bacteria *The molecular basis of evolution* by Christian Boehmer Anfinsen (b Monessen, PA, Mar 26, 1916) is based on his work with the structure and function of enzymes G.E. Hutchinson's "Homage to Santa Rosalia, or why are there so many kinds of animals?" points out that two different species cannot occupy the same ecological niche and also that similar species with similar niches occur in sizes that are about 1.3 times apart in all linear dimensions Severo Ochoa and Arthur Kornberg of the United States win the Nobel Prize for Physiology or Medicine for the artificial production of nucleic acids with enzymes	Jaroslav Heyrovský of Czechoslovakia wins the Nobel Prize for Chemistry for his polarography for electrochemical analysis Adolf Windaus, German chemist, d Göttingen, Jun 9
1960	The International Bureau of Weights and Measures defines the standard meter as a certain multiple of the wavelength of the light emitted when krypton gas is heated	Sir Charles Leonard Woolley, English archaeologist, d London, Feb 20	Walter Baade, German-American astronomer, d Göttingen, Germany, Jun 25 Frank Donal Drake (b Chicago, IL, May 28, 1930) runs project Ozma, a 400-hour attempt to find extraterrestrial life in the universe; none is found Sir Harold Spencer Jones, English astronomer, d London, Nov 3	Roy Chapman Andrews, American zoologist, d Carmel, CA, Mar 11 James Frederick Bonner discovers that chromosomes synthesize RNA Sir Macfarlane Burnet of Australia and English biologist Peter Brian Medawar (b Rio de Janeiro, Brazil, Feb, 1915) win the Nobel Prize for Physiology or Medicine for their study of immune reactions to tissue transplants	Georges Claude, French chemist, d Saint-Cloud, Seine-et-Oise, May 23 Lyman Creighton Craig (b Palmyra, IA, Jun 12, 1906) succeeds in purifying the hormone parathormone from the parathyroid gland Willard F. Libby of the United States wins the Nobel Prize for Chemistry for the invention of radiocarbon dating

1959

The US Weather Bureau institutes the temperature-humidity index (THI) as a way of judging how uncomfortable a hot summer day is

Murray Llewellyn Barr discovers that one of the X chromosomes of the female forms a dark spot when a cell not undergoing mitosis is stained; the dark spot is called a Barr body; apparently the other X chromosome is diffuse throughout the cell, while the Barr body is coiled up tightly

C.E. Ford shows that Turner's syndrome is a sex-chromosome defect

P.A. Jacobs and J.A Strong show that Klinefelter's syndrome is caused by a sex-chromosome defect

A team headed by Luis Walter Alvarez at Berkeley completes a 500-liter, 183-cm (132-gal, 72-in.) bubble chamber filled with liquid hydrogen for detecting subatomic particles produced by the Cosmotron particle accelerator

Sir Owen Richardson, English physicist who worked out the theory of the Edison effect, providing the basis for various types of vacuum tubes, d Alton, Hampshire, Feb 15

Emilio Segrè and Owen Chamberlain of the United States win the Nobel Prize for Physics for their demonstration of the existence of the antiproton

Charles Thomson Rees Wilson, Scottish physicist who invented the cloud chamber for detecting tracks of subatomic particles, d Carlops, Peeblesshire, UK, Nov 15

The first commercial Xerox copier is introduced

The St. Lawrence Seaway is opened, connecting the St. Lawrence River and the Great Lakes

De Beers of the Union of South Africa manufactures the first artificial diamond

Grace Murray Hopper invents COBOL, a computer language designed for business uses

1960

Geothermal power is produced for the first time in the United States at The Geysers, near San Francisco

The United States launches TIROS 1, the first weather satellite

Harry Hammond Hess (b New York, NY, May 24, 1906) develops the theory of sea-floor spreading; new crust is created at mid-ocean ridges; it moves toward deep-sea trenches, where it descends back into the mantle

Luis Walter Alvarez announces his discovery in bubble chambers of very short-lived particles, which are termed resonances

Donald Glaser of the United States wins the Nobel Prize for Physics for developing the bubble chamber for subatomic study

Igor Vasilevich Kurchatov, Soviet physicist who directed the Soviet

Astroturf, a sort of artificial grass, is used to cover the playing field at the Astrodome in Houston, TX

Theodore Harold Maimen (b Los Angeles, CA, Jul 11, 1927) develops the first laser in May using a ruby cylinder

Echo, the first passive communications satellite, is launched on Aug 12 as a result of efforts by John Robinson Pierce (b Des

1960 cont		The aperture synthesis method is developed by Martin Ryle and Anthony Hewish Allan Rex Sandage (b Iowa City, IA, Jun 18, 1926) discovers a starlike object of sixteenth magnitude that emits radio waves; this object, the 3C48, is later identified as a quasar	John Cowdery Kendrew (b Oxford, England, Mar 24, 1917) locates all of the atoms in the organic molecule myoglobin, a close relative to hemoglobin Jacques Monod (b Paris, Feb 9, 1910) and François Jacob (b Nancy, France, Jun 17, 1920) mail their proof that messenger RNA exists to the *Journal of Molecular Biology* on Dec 24 Kenneth Norris and John Prescott establish that the bottlenose dolphin uses echolocation (essentially sonar) to locate objects in water, just as bats use echolocation to locate objects in air Using the technique of adding a heavy atom, such as gold, to an organic molecule, Max Perutz determines the atom-by-atom structure of hemoglobin	Walter Karl Friedrich Noddack, German chemist, d Bamberg, Bavaria, Dec 7
1961	Louis S.B. Leakey and Mary Leakey find the first fossil remains of *Homo habilis*, or "Handy Man"	C. Roger Lynds detects radio waves from planetary nebulas at the National Radio Astronomy Observatory in Greenbank, WV The Soviet Union attempts a Venus probe, but contact with the probe is lost Soviet cosmonaut Yuri Gagarin (b near Gzhatsk, USSR, Mar 9, 1934) becomes the first human being to orbit Earth on Apr 12 in his 1.8-hour mission in *Vostok I* Alan B. Shepard, Jr., becomes the first US astronaut in space as his Mercury 3 capsule	William Thomas Astbury, English physical biochemist, d Leeds, Yorkshire, Jun 4 Hungarian-American physicist Georg von Békésy (b Budapest, Jun 3, 1899) wins the Nobel Prize for Physiology or Medicine for his study of auditory mechanisms; he establishes a new theory of hearing based on the waves formed by fluid in the cochlea; this replaces a theory put forward by Hermann von Helmholtz John Joseph Bittner, American biologist, d Minneapolis, MN, Dec 14	Melvin Calvin of the United States wins the Nobel Prize for Chemistry for his work on the chemistry of photosynthesis The artifical element 103, lawrencium, is produced by a team led by Albert Ghiorso; it is named in honor of Ernest O. Lawrence, founder of the Lawrence Radiation Laboratory, where the new element was created Morris William Travers, English chemist, d Stroud, Gloucestershire, Aug 25

1960 cont

Jacques Piccard and US naval lieutenant Don Walsh descend in the bathyscaphe *Trieste* to the bottom of the Challenger Deep, 10,900 m (35,800 ft) below the surface of the Pacific Ocean

programs for nuclear weapons and nuclear power, d Moscow, Feb 7

Max Theodor Felix von Laue, German physicist who discovered that crystals can be used as diffraction gratings for X rays, d Berlin, Apr 23

Rudolf Ludwig Mössbauer (b Munich, Germany, Jan 31, 1929) discovers that it is possible to make gamma rays with a very narrow wavelength and to measure very accurately any change in wavelength; almost immediately the Mössbauer effect is used to confirm Einstein's general relativity theory

Pound and Rebka use the Mössbauer effect to measure the change in frequency of gamma rays (gravitational red shift) between the foot and top of a water tower, confirming the theory of general relativity

Moines, IA, Mar 27, 1910), who believes in the future of communications via satellite

1961

The *Cuss I*, a converted oil-drilling rig, is used to drill a core from the ocean bottom at a depth of 3.5 km (2.2 mi); it succeeds in drilling through 183 m (600 ft) of sediment to reach the basalt floor of the ocean

Edward N. Lorenz accidentally finds the first mathematical system with chaotic behavior in a computer model of how the atmosphere behaves—small changes in initial conditions lead to chaos, making weather prediction inherently difficult; his work attracts attention in

It is discovered that the immune system is partly mediated by the thymus gland; some white blood cells must develop in the thymus after birth to become active

Judah Folkman suggests that a tumor might release a substance, which he calls tumor angiogenesis factor, that promotes the growth of blood vessels

Arnold Lucius Gesell, American psychologist, d New Haven, CT, May 29

Jeffry Goldstone (b Manchester, England, Sep 3, 1933) publishes what comes to be known as the Goldstone theorem on Jan 1; it predicts that symmetry breaking will produce particles of zero mass; for other reasons, all particles of zero mass are already known, so the Goldstone particles apparently cannot exist; later Philip Anderson solves this problem

Percy Williams Bridgman, American physicist, d Randolph, NH, Aug 20

Lee De Forest, American inventor, d Hollywood, CA, Jun 30

1961
cont

ASTRONOMY & SPACE

Freedom 7 completes a 15-minute suborbital flight on May 5

Virgil I. Grissom flies his *Liberty Bell 7* capsule on a 16-minute suborbital flight to become the second American in space on Jul 21

Soviet cosmonaut G. Titov orbits earth 17 times in 25.6 hours on Aug 6

BIOLOGY

Jules-Jean-Baptiste-Vincent Bordet, Belgian bacteriologist, d Brussels, Apr 6

Robert William Holley (b Urbana, IL, Jan 28, 1922) produces purified samples of three different varieties of transfer-RNA, the molecules that actually produce proteins from the codes preserved in DNA

Howard M. Dintis establishes that transfer-RNA, a form of ribonucleic acid, puts down one amino acid after another, starting at one end and proceeding to the other, to produce a given protein

James V. McConnell reports that planaria (flatworms) that eat flatworms that have learned their way through a simple maze learn the maze more quickly than planaria that have not cannibalized learned worms; this result is widely disputed

Marshall Warren Nirenberg (b New York, NY, Apr 10, 1927) works out the first "letter" of the genetic code when he shows that UUU (three uridylic acid bases in a row) in RNA codes for the amino acid phenylalanine

1962

GENERAL

The Cuban missile crisis occurs as the US blockades Cuba from Soviet ships bringing guided missiles; the USSR backs down and missiles already in Cuba are dismantled; in return, the US promises not to invade the island

ASTRONOMY & SPACE

The Sugar Grove radiotelescope fiasco occurs; the US attempt to build a 183-m (600-ft) fully steerable dish radiotelescope, begun in 1959, is discontinued after expenditures of $96 million

The first radar contact with Mercury is made by the radiotelescope at Arecibo, Puerto Rico

BIOLOGY

Charles William Beebe, American naturalist, d Simla Research Station, near Arima, Trinidad, Jun 4

Rachel Carson's *Silent spring* alarms environmentalists and makes the general public aware of the introduction of chemicals into the ecosystem

CHEMISTRY

Niel Bartlett (b Newcastle, England, Sep 15, 1932) demonstrates that compounds can be formed that include the noble gases by preparing xenon platinum hexafluoride; previously it was believed that the noble gases, such as xenon, could not combine with any other atoms to form molecules

				1961 cont

other fields, leading to a new branch of mathematics

Frank L. Horsfall, Jr., announces that all forms of cancer result from changes in the DNA of cells

Carl Gustav Jung, Swiss psychiatrist, d Küsnacht, near Zurich, Jun 6

Jack Lippes introduces an inert plastic IUD (intrauterine device) for birth control

Otto Loewi, German-American physiologist, d New York, NY, Dec 25

Murray Gell-Mann and, independently, Yu'val Ne'eman, D. Speiser and J. Tarski introduce what Gell-Mann christens "the eightfold way" to classify heavy sub-atomic particles; now known by its group-theoretic designation as SU(3), it is the beginning of the quark theory and a part of the standard model

Robert Hofstadter (b New York, NY, Feb 5, 1915) discovers that protons and neutrons have a structure; they are made up of a central core that is positive and surrounded by two shells of mesons, with one of the shells in the neutron being negative enough to make the particle neutral

Robert Hofstadter of the United States and Rudolf Mössbauer of Germany share the Nobel Prize for Physics, Hofstadter for his measurement of nucleons and Mössbauer for his work on gamma rays

Erwin Schrödinger, Austrian physicist, d Alpbach, Jan 4

1962

Harry H. Hess suggests that convection currents in Earth's mantle may be the cause of sea-floor spreading

Tibor Rado discovers a function connected to whether or not a Turing machine halts that is not computable; proof of the existence of noncomputable functions had showed that such functions exist, but did not provide examples

Lasers are used in eye surgery for the first time

Niels Henrik David Bohr, Danish physicist, d Copenhagen, Nov 18

William Weber Coblentz, American physicist, d Washington, DC, Sep 15

Arthur Holly Compton, American physicist, d Berkeley, CA, Mar 15

Unimation in the United States markets the world's first industrial robot

The Aviation Supply Office in Philadelphia introduces time-sharing on its computer for inventory control

The nuclear ship *Savannah* begins sea trials in Mar

1962

cont

The Wade-Giles system of transliterating Chinese characters into the Roman alphabet is replaced by the Pin-yin system

Andrew Ellicott Douglass, American astronomer, d Tucson, AZ, Mar 20

Riccardo Giacconi, H. Gursky, F.R. Paolini, and Bruno B. Rossi report the discovery in Scorpio of the first known X-ray source outside the solar system

Cyril Hazard determines that the radio source 2C273 consists of two components by observing its occultation by the moon; the source will be identified as a quasar a year later

Peter van de Kamp announces that an analysis of over 2000 photographs of Barnard's star shows that its motion wobbles in a periodic fashion; he interprets this as meaning that a planet orbits the star; later data suggest that the wobble may be caused by a defect in the telescope used

The Soviet Union launches the first attempted Mars probe, but contact with the probe is lost

The US spaceprobe Mariner 2 becomes the first object made by humans to voyage to another planet when it reaches the vicinity of Venus

Ariel I, launched from Cape Canaveral, FL, is the first British-built satellite

M. Scott Carpenter completes three orbits of Earth in his Mercury space capsule *Aurora 7* on Jun 24

John H. Glenn, Jr. is the first American to orbit Earth in his Mercury 6 space capsule *Friendship 7* on Feb 20

Francis H.C. Crick and Maurice Wilkins of England and James D. Watson of the United States win the Nobel Prize for Physiology or Medicine for their determination of the molecular structure of DNA

Conrad Arnold Elvehjem, American biochemist, d Madison, WI, Jul 27

Sir Ronald Aylmer Fisher, English biologist, d Adelaide, Australia, Jul 29

French-Australian physician Jacques F.A.P. Miller (b Nice, France, Apr 2, 1931) determines that the thymus is part of the immune system by removing it from newborn mice, who then fail to develop white blood cells (leukocytes) or lymph nodes and accept grafts they would normally reject

Erich Tschermak von Seysenegg, Austrian botanist, d Vienna, Oct 11

Emile Zuckerkandl and Linus Pauling propose that changes in genetic material over time can be used as a "biological clock" to determine how long ago one species separated from another

John C. Kendrew and Max F. Perutz of England win the Nobel Prize for Chemistry for their X-ray study of the structure of hemoproteins

A group headed by G. Danby, using the cut-up battleship *Missouri* as shielding for their experiment at Brookhaven, establishes that there are two types of neutrino—one associated with the electron (previously known) and another associated with the muon; today we also believe there is a third type, the tauon neutrino

Lev D. Landau of the Soviet Union wins the Nobel Prize for Physics for his work on superfluidity in liquid helium

L.B. Okun introduces the term *hadron* to represent collectively particles that are affected by the strong force; for example, protons, neutrons, pions, and kaons are hadrons, while the leptons, such as electrons, muons, and neutrinos, are not

Auguste Piccard, Swiss physicist, d Lausanne, Mar 24

Telstar, the first active communications satellite, is launched on Jul 10; it relays the first trans-Atlantic television pictures

1962

cont

1962

cont

Walter M. Schirra orbits Earth six times on Oct 3 in his Mercury spacecraft *Sigma 7*

1963

US President John F. Kennedy is assassinated

Ice age skeletons found in the Romito cave near Cosenza, Italy, include that of a dwarf; later analysis suggests that the 17-year-old dwarf was accepted by the hunter-gatherers despite the fact that he could not have contributed much to the community

Lascaux cave, home of the most famous cave paintings, dating back 17,000 years, is closed to the public to prevent damage to the art caused by the increased humidity from people's breath

Herbert Friedman directs the launch of a satellite to study X rays from space

Friedman and his group at the US Naval Research Laboratory identify Sco X-1 as a strong X-ray source in Scorpio

Seth Barnes Nicholson, American astronomer, d Los Angeles, CA, Jul 2

Allan Sandage discovers that the galaxy M82 is undergoing a massive explosion at its center, one that is believed to have been in progress for 1.5 million years

Dutch-American astronomer Maarten Schmidt (b Groningen, Netherlands, Dec 28, 1929) reports his discovery of the extraordinarily large red shift of the object 3C273; the red shift corresponds to a recession velocity of 47,400 km (29,400 mi) per second, and constitutes the first recognition of a quasar

Otto Struve, Russian-American astronomer, d Berkeley, CA, Apr 6

Syncom 2, launched on Feb 14, becomes the first artificial satellite to go into a geosynchronous orbit; that is, an orbit that is matched to Earth's rotation so that the satellite stays directly above one location on Earth's surface

Sir John C. Eccles of Australia and Alan Lloyd Hodgkin and Andrew F. Huxley of England win the Nobel Prize for Physiology or Medicine for their study of the mechanism of transmission of neural impulses

David Keilin, Russian-British biochemist, d Cambridge, England, Feb 27

Carl Sagan (b New York, NY, Nov 9, 1934) detects adenosine triphosphate (ATP), an important biochemical used in storing energy, in a mixture of chemicals that reflects what are thought to be the early conditions on Earth

Giulio Natta of Italy and Karl Ziegler of Germany win the Nobel Prize for Chemistry for their synthesis of polymers for plastics

1963

F.J. Vine and D.H. Matthews show that the residual magnetism of the floor of the Indian Ocean changes polarity in a periodic fashion, convincing evidence that sea-floor spreading has occurred; since Earth's magnetic field reverses from time to time, the magnetic "stripes" also indicate the rate of spreading

Paul J. Cohen shows that Cantor's continuum hypothesis is independent of the axioms of set theory; this means that there are at least two types of mathematics possible—one that says that the continuum hypothesis is true and one that says it is false

Herbert Spencer Gasser, American physiologist, d New York, NY, May 11

Nicola Cabibbo develops a theory of weak interactions that indirectly leads to the electroweak theory that is accepted today

Theodore von Karman, Hungarian-American physicist, d Aachen, W Germany, May 7

Eugene Paul Wigner and Maria Goeppert Mayer of the United States and J. Hans D. Jensen of Germany share the Nobel Prize for Physics, Wigner for the mechanics of proton-neutron interaction and Mayer and Jensen for their theory of nucleic structure

Friction welding is invented

The cassette for recording and playing back sound is introduced by Philips of the Netherlands

Semiconductor diodes that use electron tunneling go on sale, only six years after the tunneling effect was discovered by Leo Esaki

Quasars

Many scientific discoveries have put an idea scientists had—a theory—on a stronger footing. Examples are the discovery of the cosmic background radiation which turned out to be a confirmation of the Big-Bang theory, and the discovery of pulsars, which made reality of the idea of neutron stars. The discovery of quasars, however, only caused bewilderment among astronomers. The consequence of this discovery was that one either had to question the validity of the yardstick of the astronomer, the red shift, or to agree that there are processes out there for which we have no explanation at all.

During the 1950s, radio astronomers discovered a number of very compact radio sources. Because radio telescopes at that time could not pinpoint celestial objects very accurately, it was difficult to find these objects with a telescope. One of these compact sources, known as 3C273, was occulted by the moon in 1962 and its exact position could be established. Photos taken with the 5-m Hale (200-in.) telescope at Mt. Palomar showed a star-like object at that position. However, its spectrum was unusual: it contained absorption lines that could not be identified. It, and later others like it, was called a quasi-stellar radio source, or quasar for short.

In 1963 Maarten Schmidt discovered that the absorption lines in the spectrum of 3C273 were common ones, but shifted towards the red end of the spectrum by an extraordinary amount. During the following years astronomers discovered a large number of quasars with large red shifts.

It was commonly held among astronomers that the red shifts observed in the spectra of extragalactic systems are Doppler shifts caused by velocity these systems have because they participate in the expansion of the universe (see "The size of the universe," p 440). Red shifts caused by the expansion of the universe are called cosmological red shifts. If the red shifts of quasars are cosmological, the quasars must be at extraordinary distances, making them the farthest objects ever observed in the telescope. Because they can be observed over such distances, their energy output must be enormous.

Not all astronomers, however, believe that quasars have cosmological red shifts. The American astronomer Halton Arp has discovered a number of systems consisting of a quasar and a galaxy that seem to be physically connected but exhibit very different red shifts in their spectra. He therefore argues that some still unknown mechanism other than the expansion of the universe must cause these red shifts. Most astronomers believe that quasars have cosmological red shifts and that the systems discovered by Arp only appear to be connected, and that they, in reality are at very different distances from Earth.

1963
cont

ASTRONOMY & SPACE

L. Gordon Cooper completes 22 orbits of Earth in his 34-hour flight in his Mercury spacecraft *Faith 7* on May 15

For the first time, two spacecraft with people aboard (Nikolayev and Popovich, of the Soviet Union) are in orbit simultaneously

Valentina Tereshkova-Nikolayeva of the Soviet Union becomes the first woman in space, making 48 orbits in 78 hours on Jun 16

1964

GENERAL

The Aswan Dam on the Nile is completed, resulting in profound ecological changes, including increased schistosomiasis, decreased numbers of sardines in the eastern Mediterranean, and increased erosion of the Nile delta

The US Congress passes the Gulf of Tonkin Resolution, giving President Lyndon Baines Johnson the power to "take all necessary measures to ... prevent further aggression;" this is generally viewed as the start of the Vietnam War between the United States and North Vietnam

ASTRONOMY & SPACE

The International Year of the Quiet Sun begins on Jan 1 and will continue to Dec 31, 1965

US spaceprobe Ranger 7 takes the first good close-range photographs of the moon; 4316 pictures in all are taken by the six television cameras on board and relayed to Earth

The 213-cm (84-in.) reflecting telescope at Kitt Peak, AZ, goes into operation in Sep

Robert H. Dicke revives a suggestion made previously by George Gamow that the Big Bang would have left behind recognizable background radio-wave radiation

Herbert Friedman and his group at the US Naval Research Laboratory discover X rays from the Crab nebula with detectors mounted on rockets

Harold Weaver and his coworkers discover the radio emission of OH radicals in certain galactic regions; the emission is caused by "cosmic masers"

BIOLOGY

The International Rice Research Institute starts the "green revolution" with new strains of rice that have double the yield of previous strains if they are given sufficient fertilizer

German-American biochemist Konrad Emil Bloch (b Neisse, Germany, Jan 21, 1912) and Feodor Lynen (b Munich, Germany, Apr 6, 1911) win the Nobel Prize for Physiology or Medicine for their work on cholesterol and fatty acid metabolism

Gerhard Domagk, German biochemist, d Burberg, Württemburg-Baden, West Germany, Apr 24

John Burdon Haldane, English-Indian geneticist, d Bhubaneswar, India, Dec 1

W.D. Hamilton's "The genetical evolution of social behavior" shows that the "altruistic" behavior of social bees and ants can be explained; female workers pass along more of their

CHEMISTRY

Dorothy M.C. Hodgkin of England wins the Nobel Prize for Chemistry for her analysis of the structure of vitamin B_{12}

Hans Karl von Euler-Chelpin, German-Swedish chemist, d Stockholm, Sweden, Nov 7

The submersibles *Alvin* and *Aluminaut* complete preliminary tests before going into full service	Norbert Weiner, American mathematician, d Stockholm, Sweden, Mar 18	Home kidney dialysis is introduced in the United Kingdom and the United States	Raymond Y. Chiao, Boris P. Stoicheff, and Charles H. Townes discover stimulated Brillouin scattering, in which a low intensity beam of light passes through a transparent substance but a light of high enough intensity is reflected from the same substance	The Verrazano Bridge, the longest suspension bridge in the world in its time, opens to traffic in New York City	1964

The submersibles *Alvin* and *Aluminaut* complete preliminary tests before going into full service

A great earthquake strikes Alaska on Mar 27

Bids to drill the (later aborted) Project Mohole are opened on Jul 12; the Mohole was to extend to the Mohorovicic discontinuity between the crust and the mantle of Earth

Norbert Weiner, American mathematician, d Stockholm, Sweden, Mar 18

Home kidney dialysis is introduced in the United Kingdom and the United States

Baruch S. Blumberg (b New York, NY, Jul 28, 1925) discovers the "Australian antigen," which is the key to the development of a vaccine for hepatitis B

Raymond Y. Chiao, Boris P. Stoicheff, and Charles H. Townes discover stimulated Brillouin scattering, in which a low intensity beam of light passes through a transparent substance but a light of high enough intensity is reflected from the same substance

James W. Cronin (b Chicago, IL, Sep 29, 1931), Val L. Fitch (b Merriman, NE, Mar 10, 1923), and coworkers discover that the neutral K meson breaks the rule that combines conservation of charge conjugation and parity (CP conservation); this also implies that time reversal is not invariant

James Franck, German-American physicist, d Göttingen, May 21

Murray Gell-Mann's paper "A schematic model of baryons and mesons" introduces in Feb the concept of quarks, particles of fractional charge that make up all baryons and mesons

The Verrazano Bridge, the longest suspension bridge in the world in its time, opens to traffic in New York City

Permanent press clothing is introduced

Containerships are introduced, simplifying international trade

1964

cont

Ecology and sociobiology

Ecology is the science of the relationship between organisms and their environment. It may be dated to as early as 1789 or earlier. The word itself was coined in 1866 (it means "study of the house"). But ecology really came into its own after World War II.

One reason for this is certainly the environmental movement that began seriously in 1962 with Rachel Carson's book *The silent spring* and intensified in the 1970s. The popularity of environmental causes attracted people to ecology, sometimes in the mistaken belief that ecology is the science of avoiding pollution.

Another reason for recent gains in the ecology movement was the postwar development of the computer. Ecology is very much concerned with numbers and mathematical models, since it treats a great many individuals interacting with each other and with their physical surroundings all at once. Prewar models of how populations behave were computed by hand or with mechanical calculators. By the mid-1970s, an ecologist could feed a formula into a computer and see how a population stabilized, became periodic, or exploded depending on the choice of a single parameter. With such a tool in hand, much of the work in ecology during the 1960s and 1970s concentrated on finding how competition among species or individuals for niches—ways of living—affected population.

The realization in 1964 that apparently altruistic traits in species could be explained by the mathematics of Mendelian heredity led to sociobiology as a controversial offshoot of ecology. Calculations showed that among the social bees and ants a worker passes along more of her genetic heritage when the queen has all the children than when each worker has some of them (bee and ant workers are always female). Generalizing from this special example, Edward O. Wilson and others argued that many different types of social behavior in animals can be explained as mechanisms designed to transmit genes. People who believe that altruism in humans originates from nonbiological considerations fought against the precepts of sociobiology. Although the issues are not resolved, it is clear that sociobiological ideas continue to influence much of the research on animal behavior.

genes by encouraging reproduction by the queen rather than reproducing themselves; this concept is later extended to become sociobiology

Mark Richmond and Eric Johnston show that the same plasmid that carries resistance to penicillin in staphylococci also permits the bacteria that possess the plasmid to survive a mercury-based disinfectant; other workers show that the plasmid confers resistance to a wide variety of normally toxic metals

Thomas F. Roth and Keith R. Porter discover the first cell receptors embedded in the cell membrane of an egg cell; they identify the coated pits that collect the precursor to yolk from the environment and the coated vesicles that carry the yolk into the interior of the egg cell

Charles Yanofsky and Sydney Brenner prove that the order of nucleotides in DNA is precisely collinear with the order of amino acids in proteins, an important step in the elucidation of protein synthesis

The discovery of the omega-minus particle clinches the case for Murray Gell-Mann's eightfold way of classifying subatomic particles

Sheldon Glashow (b New York, NY, Dec 5, 1932) and James D. Bjorken in Aug introduce a fourth quark species carrying "charm;" no one pays much attention to it until early in 1970, when Glashow, studying the problems with the Cabibbo theory, realizes that charm will solve these problems

Victor Francis Hess, Austrian-American physicist, d Mount Vernon, NY, Dec 17

Peter Higgs suggests a mechanism of spontaneous symmetry breaking, which comes to be known as the Higgs boson; its importance is that the Higgs boson gives enough mass to particles that otherwise would surely have been detected (all particles with low mass are thought to be known); in fact, the W particles found in 1983 seem to have mass from the Higgs mechanism

Leo Szilard, Hungarian-American physicist, d La Jolla, CA, May 30

Charles H. Townes of the United States, Nikolay G. Basov (b Leningrad, USSR, Dec 14, 1922), and Soviet physicist Aleksandr Prokhorov (b Australia, Jul 11, 1916) win the Nobel Prize for Physics for the development of maser and laser principles in quantum mechanics

1964

cont

1965

A chain reaction from a faulty relay in a Canadian hydroelectric plant knocks out power for 14 hours in New York City and in much of the northeastern United States and southern Canada on Nov 9

In *Stonehenge decoded* Gerald Stanley Hawkins (b Norfork, England, Apr 20, 1928) argues that the neolithic site once served as a cross between an observatory and a computer, enabling the builders to compute the times of various astronomical events

The first cosmic maser is found; cosmic masers are regions in clouds of interstellar gas that are stimulated to produce radiation by starlight; the radiation produced is of the coherent type most familiar today as the light from a laser

Astronomers at the Arecibo Ionospheric Observatory discover that the rotation of Venus is retrograde; that is, Venus rotates in a different direction from the other planets, so on Venus the sun rises in the west and sets in the east; the period is approximately 247 days

Astronomers at the Arecibo Ionospheric Observatory in Puerto Rico discover that Mercury rotates on its axis with a period of about 59 days; previously, scientists believed that Mercury keeps one face toward the sun as the moon does toward Earth

Bertil Lindblad, Swedish astronomer, d Stockholm, Jun 26

German-American physicist Arno Allan Penzias (b Munich, Apr 26, 1933) and Robert Woodrow Wilson (b Houston, TX, Jan 10, 1936) accidentally find the radio-wave remnants of the Big Bang while trying to refine radio equipment; their discovery convinces most astronomers that the Big Bang theory is correct

It is discovered in May that injections of synthetic progestogen and estrogen can prevent ovulation for a full month, leading to the hope of developing a once a month birth control pill

Alec D. Bangham and coworkers show that phospholipid bilayers form closed spheres in water; this discovery is of theoretical importance because such bilayers are typical of the cell membrane; later such phospholipid spheres will be used in practical applications, such as delivering drugs to a specific target

William J. Dreyer and J. Claude Bennet propose that antibodies can match so many different antigens because they have many variable genes and a single constant gene; this concept is later proved correct, although much more complex than in Dreyer and Bennet's original proposal

Robert Holley in Mar works out the complete structure of a molecule of transfer-RNA, the molecule that builds proteins based on instructions from DNA

François Jacob, André Lwoff (b Aulnay-le-Château, France, May 8, 1902) and Jacques Monod of France win the Nobel Prize for Physiology or Medicine for their studies on and discoveries of the regulatory activities of genes

Paul Hermann Müller, Swiss chemist, d Basel, Oct 13

Hermann Staudinger, German chemist, d Freiburg-im-Breisgau, Sep 8

Robert Runnels Williams, American chemist, d Summit, NJ, Oct 2

Robert B. Woodward of the United States wins the Nobel Prize for Chemistry for the synthesis of organic compounds

| | | | Stephan von Molnar (b Leipzig, Germany, Jun 26, 1935) investigates the magnetic properties of europium | | 1964 cont |

EARTH SCIENCE	MATHEMATICS	MEDICINE	PHYSICS	TECHNOLOGY	1965
The US Navy's *Sealab II* project concludes after 45 days; during the project, people live in the sea off the coast of California at a depth of 62 m (205 ft)	Hugh C. Williams, R.A. German, and C.R. Zarnke, using a computer, work out the first complete solution to Archimedes' "cattle of the sun" problem, posed about 2000 years earlier; it is a number of 206,545 digits	A vaccine against measles becomes available	Sir Edward Victor Appleton, English physicist, d Edinburgh, Scotland, Apr 21	Joseph Giordmaine and Robert Miller develop the continuously tunable laser	

The US Navy's *Sealab II* project concludes after 45 days; during the project, people live in the sea off the coast of California at a depth of 62 m (205 ft)

Melvin Calvin and coworkers discover the breakdown products of chlorophyll in the Soudan shale of Minnesota, a rock formation that is between 2.5 and 2.9 billion years old

Jacques-Yves Cousteau's team of aquanauts surfaces after spending 23 days at a depth of 100 m (330 ft) in the Mediterranean

Arthur Holmes, English geologist, d London, Sep 20

J. Tuzo Wilson suggests that faults perpendicular to the mid-ocean rifts develop as the sea floor spreads; such transform faults are found to be characteristic of sea-floor spreading

Hugh C. Williams, R.A. German, and C.R. Zarnke, using a computer, work out the first complete solution to Archimedes' "cattle of the sun" problem, posed about 2000 years earlier; it is a number of 206,545 digits

A vaccine against measles becomes available

Soft contact lenses are invented

Morris E. Davis reports that estrogen therapy prevents atherosclerosis and osteoporosis in postmenopausal women

Joseph Erlanger, American physiologist, d St. Louis, MO, Dec 5

Harry Harlow demonstrates that monkeys reared in total isolation show great emotional impairment for the rest of their lives

Philip Showalter Hench, American physician, d Ocho Rios, Jamaica, Mar 30

Sir Edward Victor Appleton, English physicist, d Edinburgh, Scotland, Apr 21

Moo-Young Han and Yoichiro Nambu introduce the concept for quarks that is later named color

Julian S. Schwinger and Richard P. Feynman of the United States and Shin'ichiro Tomonaga of Japan win the Nobel Prize for Physics for their research into the basic principles of quantum electro-dynamics

The SLAC (Stanford Linear Accelerator Center) particle accelerator goes into operation

Joseph Giordmaine and Robert Miller develop the continuously tunable laser

John Kemeny and Thomas Kurtz develop BASIC (beginners all-purpose symbolic instruction code), a computer language for beginners; it becomes the main programming language used by owners of personal computers, although most commercial programs for personal computers are in more sophisticated languages

1965

cont

Frederick Reines and J.P.F Sellshop detect neutrinos from cosmic rays deep in a South African gold mine, the start of neutrino astronomy

Allan Sandage finds quasars that do not emit radio waves (radio quiet quasars)

Aleksei A. Leonov of the Soviet Union conducts the first spacewalk, or trip outside a satellite wearing only a spacesuit, taking a 20-minute excursion on Mar 18

Virgil I. Grissom and John W. Young are America's first two-man space crew, orbiting Earth three times in their Gemini spacecraft on Mar 23

The Early Bird communications satellite goes into synchronous orbit (an orbit that maintains the satellite above the same region of Earth's surface at all times) in Apr

James A. McDivitt and Edward H. White orbit Earth 62 times in their Gemini spacecraft launched on Jun 3; they perform the first extravehicular activity using a personal propulsion unit

Mariner IV reaches the neighborhood of Mars on Jul 15, passing within 12,000 km (7500 mi) of the planet

The Russian space-probe Zond III photographs areas on the opposite of the moon that had not been photographed by the probe Lunik III in 1959

W.A. Jones, Morton Beroza, and Martin Jacobson develop artificial sex attractants, or pheromones, for cockroaches and other insects

Hans Ris and Walter Plaut discover that chloroplasts in algae have their own DNA

Sol Spiegelman synthesizes a self-reproducing virus

Exploring the planets

If you are old enough, and especially if you were interested in astronomy or science fiction in your youth, you once knew a lot about the solar system that was simply not true. Among these fictions are that Mercury always keeps the same face to the sun, that Venus is covered with water, that there are canals on Mars, that Jupiter has nine moons, that Saturn is the only planet with rings, and so forth. Today people know that these ideas are wrong because we have been there—not in person but with various sensors.

For example, in 1965 radar was used to find the correct rotation of both Mercury and Venus. Not only did we learn that Mercury doesn't keep the same face to the sun, but we also learned that Venus rotates backwards, with the sun rising very slowly in the west and setting in the east, 247 of our days later.

Another technological product of the war, however, the giant liquid-fueled rocket, was the main key to our new knowledge of the solar system. These rockets provided the power to send probes to virtually all the planets.

Venus was the first planet to be visited by probes. Because of the gravitational pull of the sun, Venus and Mercury are "downhill" from Earth, while the other planets are "uphill." The first US spaceprobe aimed toward Venus, Mariner I, landed at the bottom of the Atlantic Ocean because of a programming error, but Mariner II made it to the Venus neighborhood on December 14, 1962. Sensitive radio detectors revealed that Venus was far too hot to have liquid water. In 1966 the Soviet Union actually struck Venus with a probe, but it failed to transmit any information. The next year the Soviets had a more successful landing, while the US Mariner V flew past at about the same time. Both revealed that in addition to the heat, the atmosphere was very dense. Two more Soviet craft landed in 1969 and another in 1972. These were all hard landings. In 1975, however, the USSR Venera 9 and 10 soft landed and returned pictures, showing Venus to be a bleak place. In 1978, the United States returned to Venus exploration with a Venus orbiter and another spacecraft that launched probes into the atmosphere. In 1982, the Soviets landed two more spacecraft that returned pictures and other data. As a result of this series of explorations, it has been possible to compile maps of Venus and to determine a great deal about its atmosphere. There is still not enough data to be certain of the answer to such questions as whether Venus undergoes plate tectonics, which are greatly of interest to geologists on Earth.

Mars has long been of special interest because of the possibility that it may harbor living organisms, which still has not been completely ruled out, although most evidence so far is negative. In the early 1960s, both the Soviet Union and the United States launched several failed missions to Mars. Finally on July 14, 1965, the US spaceprobe Mariner IV passed close to Mars. In 1969, two more US spaceprobes made flybys of the red planet. But 1971 was a banner year that saw two Soviet and one US spacecraft in orbit about Mars, transmitting data, including television pictures, back to Earth. At first it seemed that nothing would be learned from orbiting Mars, since it was impossible to see much through the atmosphere. Later it was found that the orbiting probes had arrived during a giant dust storm. When the dust died down, after several months, it became possible to map sections of Mars and to begin to study its geology—or "marsology," since geology is the study of Earth's rocks, landforms, and so forth.

In 1974, the Soviets made the first attempt to land on Mars, but the spaceprobe crashed onto the planet. In 1976, however, two US spacecraft, Viking 1 and 2, landed on Mars. They did not find life, but Viking 1 broadcast data about its vicinity until May, 1983, long after its expected "life."

On March 2, 1971, the United States launched Pioneer 10 on a mission designed to go beyond Mars. It was one of the most successful missions ever. In 1973, Pioneer 10 flew by Jupiter, providing data on its magnetic field and its moons. Then it just kept going. On June 13, 1983, Pioneer 10 passed beyond the orbit of the then farthest planet, Neptune. (Pluto, which is usually the farthest planet from the sun, has such an irregular orbit that for about 25 years out of its 248-year trip around the sun it is inside the orbit of Neptune.) Pioneer 10 kept going and is still broadcasting information about the solar wind from the farthest point any spaceprobe has reached. Because it is still in the solar wind, purists do not consider that Pioneer 10 has yet left the solar system, however.

In 1974, the US Mariner 10, after a glance at Venus, became the first spaceprobe to reach the vicinity of Mercury. It was also the second and third, as it went into orbit about the sun, and flew past Mercury more than once, exploring the planet from space each time.

Another Pioneer, Pioneer 11, was the first to reach Saturn, in 1979, revealing evidence of more moons than had previously been known.

But the most successful trip of all was that of Voyager 2. Together with Voyager 1, Voyager 2 flew by Jupiter in 1979 and by Saturn in 1980 and 1981, but then they parted company. While Voyager 1 headed for empty space, Voyager 2 continued on to Uranus, which it reached in 1986. Neptune is scheduled for 1989. Like the Pioneer mission, the Voyagers located additional moons. In passing Saturn, they provided spectacular close-up views of the famous rings. Voyager 2 also produced close-ups of the rings of Uranus, which had been discovered from Earth by more conventional means.

1965
cont

ASTRONOMY & SPACE

L. Gordon Cooper and Charles Conrad, Jr., begin their 190-hour, 120-orbit space mission on Aug 21 to demonstrate the feasibility of a lunar mission

Frank Borman and James A. Lovell, Jr., are launched on Dec 4 for a 13-day space mission in their *Gemini 7* capsule; they perform the first space rendezvous with Walter M. Schirra and Thomas P. Stafford, who are launched on Dec 15 in *Gemini 6*

1966

ANTHROPOLOGY/ARCHAEOLOGY

Robert Ardrey's *The territorial imperative* argues that human beings, like songbirds and other animals, are naturally territorial

Clifford Evans and Betty J. Maggers claim that pottery from 3000 BC in South America is essentially the same as contemporaneous pottery from the Jomon culture of Japan

ASTRONOMY & SPACE

Dirk Brouwer, Dutch-American astronomer, d New Haven, CT, Jan 3

Herbert Friedman, Edward T. Byram, and Talbot A. Chubb discover a powerful X-ray source in the constellation Cygnus, apparently coming from the galaxy Cygnus-A

Abbé Georges Lemaître, Belgian astronomer, d Louvain, Jun 20

Martin Rees discovers quasars whose components seem to separate from each other at velocities greater than the velocity of light; the phenomenon will later be explained as caused by the light heading at a small angle toward the observer, giving a illusion of greater speed

Allan Sandage identifies Sco X-1, a bright X-ray star that mysteriously varies by as much as 250 percent within a few days

BIOLOGY

Konrad Lorenz's *On aggresssion* argues that only human beings intentionally kill members of their own species

Sol Spiegelman and Iciro Haruna discover an enzyme that allows RNA molecules to duplicate themselves

CHEMISTRY

Georg von Hevesy, Hungarian-Danish-Swedish chemist, d Freiburg-im-Breisgau, Germany, Jul 5

Robert Mulliken of the United States wins the Nobel Prize for Chemistry for his study of atomic bonds in molecules

				1966
The 15th International Congress of Mathematicians is held in Moscow	The French Academy of Medicine is the first medical group to use brain inactivity instead of heart stoppage as the clinical definition of death	The Atomic Energy Commission announces that the United States will build a 200 GeV (200 billion electron volt) particle accelerator in Weston, IL, outside of Chicago	Fuel injection for automobile engines is developed in the United Kingdom	
Luitzen Egbertus Jan Brouwer, the Dutch founder of the intuitionist school of mathematics, d Dec 2	Daniel Carleton Gajdusek (b Yonkers, NY, Sep 9, 1923) succeeds in transferring kuru, a disease of the central nervous system thought to spread by cannibalism, to chimpanzees, the first time a viral disease of the central nervous system is transferred from humans to another species	Peter Joseph Wilhelm Debye, Dutch-American physical chemist, d Ithaca, NY, Nov 2		
		Alfred Kastler of France wins the Nobel Prize for Physics for his study of atomic structure by optical pumping		
	Charles B. Huggins and Francis Peyton Rous of the United States win the Nobel Prize for Physiology or Medicine for their research into the treatment and causes of cancer	Vladimir Iosifovich Veksler, Soviet physicist, d Moscow, Sep 22		
		Frits Zernike, Dutch physicist, d Naarden, Mar 10		
	William Claire Menninger, American psychiatrist, d Sep 6			
	Harry M. Meyer, Jr., and Paul D. Parman develop a live-virus vaccine for rubella (German measles)			

1966

cont

The Soviet spaceprobe Luna IX accomplishes the first soft landing on the moon by an automatic probe on Feb 3, landing in the Ocean of Storms

The US Environmental Science Services Administration launches ESSA I, the first weather satellite capable of viewing the entire Earth

The Soviet Union's spaceprobe Venera III becomes the first object made by humans to land on another planet when it reaches Venus on Mar 1

Neil Alden Armstrong (b Wapakoneta, OH, Aug 5, 1930) and David R. Scott, in a *Gemini 8* spacecraft, accomplish the first dual launch and docking with Agena on Mar 16 and have the first Pacific Ocean splashdown

Soviet spaceprobe Luna X on Apr 4 becomes the first spacecraft to orbit the moon

Surveyor I is launched by the United States; its mission, which it successfully accomplishes on Jun 1, is to soft land on the moon in the Ocean of Storms and return pictures of the lunar surface to Earth

Thomas P. Stafford and Eugene A. Cernan are launched in their *Gemini 9A* craft on Jun 3; they complete 44 revolutions and over two hours of extravehicular activity

John W. Young and Michael Collins are launched on Jul 18 for a 43-orbit flight; they complete the first dual rendezvous of spacecraft

Margaret Sanger, American founder of what is now the Planned Parenthood Foundation of America, d Sep 6

In May Michael B. Sporn, C. Wesley Dingman, Harriette L. Phelps, and Gerald N. Wogan report that aflatoxins caused by the mold *Aspergillus flavus* growing on peanuts cause liver damage and cancer

---

New windows to the universe

Astronomy is mainly an observational science, and advances in astronomy have often been the direct result of new observational techniques. When Galileo directed his telescope toward the night sky, he made more discoveries in a year than all the astronomers before him.

The discovery that we receive radio waves from the universe was quite accidental: in the 1930s the radio engineer Karl Jansky discovered that a source of interference that disturbs shortwave communications is the center of our galaxy.

When the first radiotelescopes came into use during and after the World War II, astronomers discovered that a large number of objects, such as galaxies, the sun, and even planets, emit radio waves. One of the most important achievements of radio astronomers was the determination of the shape of our galaxy and the establishment of its spiral structure by observing the 21-cm (8-in.) line emitted by neutral hydrogen in the spiral arms.

Because of the much longer wavelengths of radio waves as compared to light waves, the first radiotelescopes were not able to pinpoint celestial objects accurately. But in recent years radio astronomers have combined radiotelescopes to reach a higher angular resolution than that of the largest optical telescopes.

Radio waves and light waves are only two domains of the electromagnetic spectrum. After World War II, astronomers started to observe the universe in other areas of the electromagnetic spectrum, such as infrared and ultraviolet light, X rays, and gamma rays. Earth's atmosphere, however, absorbs a large fraction of electromagnetic radiation in regions of interest to astronomers other than visible light and radio frequencies. Balloons and rockets have carried instruments high into the atmosphere to make observations possible. Infrared telescopes have been placed on high mountains in dry regions where absorption is less.

The technical breakthrough that completely solves the problem of atmospheric absorption was the development of artificial satellites in which instruments could be placed high above Earth's atmosphere. The Infrared Astronomical Satellite (IRAS), launched in 1983, was equipped with a 60-cm (24-in) reflecting telescope. The telescope and the infrared detectors were cooled to 2°K by liquid helium. Although Earth's atmosphere absorbs infrared radiation, the huge clouds of dust and gas inside our galaxy are transparent to infrared radiation; we can observe directly the nucleus of our galaxy in the infrared. During the 10-month-period IRAS was operational, it discovered a huge number of infrared sources, including eight infrared galaxies.

Ultraviolet observations must also be made from space, and several satellites have been launched that have made observations of the sun.

The atmosphere is entirely opaque to X rays. The first X-ray cameras used by astronomers were pinhole cameras on balloons or rockets with special film emulsions that are blackened by exposure to X rays. More recently, several dedicated X-ray satellites have been launched, using special X-ray detectors to achieve better pinpointing of X-ray sources. *Uhuru* ("freedom" in Swahili), launched in 1970, discovered 339 new X-ray sources. The Einstein Observatory (HEAO-2), put into orbit in 1978, was equipped with an X-ray telescope, a device that uses mirrors of a special geometry to focus X rays. X rays are usually emitted by highly energetic processes, such as matter falling into a black hole.

Gamma-ray photons—the most energetic photons of the electromagnetic spectrum—are emitted during nuclear reactions. Gamma rays coming from the universe were also discovered—accidentally, this time—by the American Vela system of satellites designed to monitor nuclear explosions on Earth. The sources of these gamma rays in space are not yet known, but it is suspected that this form of radiation may originate in quasars.

1966

cont

		ASTRONOMY & SPACE		
		In Aug, Soviet spaceprobe Luna XI goes into orbit around the moon		
		Charles Conrad, Jr., and Richard F. Gordon, Jr., begin a 44-orbit flight in *Gemini 11* on Sep 12		
		US spacecraft Lunar Orbiter I, designed to send back photographs of the moon from orbit about it, begins operation; although one of its cameras fails, it returns dramatic photographs of the moon's surface, including the far side, to Earth		
		A second Lunar Orbiter is launched by the United States on Nov 6; it is even more successful than the first		
		Soviet spaceprobe Luna XIII lands on the moon and returns photographs and data about soil		
		James A. Lovell, Jr., and Edwin E. Aldrin, Jr., are launched on Nov 11 on the final Gemini mission; they complete five hours of extravehicular activity		

1967

GENERAL	ANTHROPOLOGY/ ARCHAEOLOGY	ASTRONOMY & SPACE	BIOLOGY	CHEMISTRY
The International Bureau of Weights and Measures defines the second as the time that microwaves emitted by hot cesium oscillate 9,192,631,770 times	J.R. Cann, J.E. Dixon, and Colin Renfew announce that a study of obsidian from various neolithic (New Stone Age) sites establishes that complex trade routes existed in the Mediterranean basin in the period between 8000 BC and 5000 BC

Elwyn L. Simons announces his discovery of the 30 million-year-old skull of an ape, which he names *Aegyptopithecus;* it is the earliest known primate that is definitely a part of the | Virgil I. (Gus) Grissom, Edward H. White II, and Roger B. Chaffee, US astronauts, die in a fire during a ground test of an Apollo spacecraft on Jan 27

Soviet cosmonaut Vladimir M. Komarov is killed on Apr 24 during the descent of his spacecraft, *Soyuz I*; as he attempts an emergency reentry from orbit, his parachute lines become entangled, causing the craft to plunge to Earth | Charles T. Caskey, Richard E. Marshall, and Marshall W. Nirenberg show that identical forms of messenger RNA are used to produce the same amino acids in bacteria, guinea pigs, and toads, suggesting that the genetic code is a universal system used by all life forms

Casimir Funk, Polish-American biochemist, d New York, NY, Nov 20

Haldan K. Hartline (b Bloomsburg, PA, Dec 22, 1903), George | Manfred Eigen (b Bochum, Germany, May 9, 1927), Ronald G.W. Norrish (b Cambridge, England, Nov 9, 1897), and George Porter (b Stainforth, England, Dec 6, 1920) share the Nobel Prize for Chemistry for their study of high-speed chemical reactions

Jaroslav Heyrovský, Czech physical chemist, d Prague, Mar 27

Sir Cyril Norman Hinshelwood, English physical chemist, d London, Oct 9 |

Unifying the forces

Most physical theories aim at explaining apparently different phenomena by simple sets of laws. The best example is Newton's theory of gravitation. Apparently unrelated phenomena, such as the apple falling from the tree and the moon circling the Earth are explained by a unique force: gravity. Gravity is a force that all masses exert on each other. It is an extremely weak force and can only be experienced when at least one of the masses is very large, such as the Earth.

When scientists started experimenting with electricity and magnets they discovered other forces that seemed to act independently from gravitation: the magnetic force, and the force of attraction or repulsion between charged bodies. When he developed his electromagnetic theory, James Clerk Maxwell unified these forces into the electromagnetic force.

In this century, the study of the atomic nucleus has revealed two other forces: the strong force that holds the protons and neutrons in the atomic nucleus together, and the weak force, which is responsible for slow nuclear processes, such as some types of radioactive decay.

With these four forces physicists can describe the whole range of known physical processes, and therefore these forces are called fundamental forces. The gravitational and electromagnetic forces are long-range forces because they act over very long distances, while the weak and strong force only act over very short distances. However, these forces seem to act independently from each other, and physicists have been searching for a single theory that would put the four fundamental forces into a single theory. Theories that would achieve this are called unification schemes.

Albert Einstein was convinced that it would be possible to unify gravitation and electromagnetic forces. He spent much of his later years trying to find such a theory, but without success.

In 1967 Steven Weinberg, Abdus Salam and Sheldon Lee Glashow formulated a theory called "electroweak theory" that unified the weak force and the electromagnetic force. According to this theory the weak forces in atomic nuclei are mediated by three particles with even larger masses than the particles transmitting the strong force: the W^+, W^- and Z^0 particles with charges, respectively, of $+1, -1$, and zero. Experimental verification of the existence of these particles eluded high-energy physicists until 1983, when scientists at CERN (*Centre Européen de Recherche Nucléaire*, near Geneva, Switzerland), under the direction of Carlo Rubbia, discovered these particles in head-on collisions in a huge proton-antiproton collider.

Theories that would unify the electromagnetic, the strong force, and the weak force are called grand unification theories, or GUTs. Theoretical physicists have been somewhat successful in developing GUTs, but there are several competing theories and no evidence as to which one might be correct.

It is generally believed that it will be much more difficult to find a theory that unifies all the four fundamental forces. A theory that would unify gravity with the other forces—a theory sometimes called supergrand unification, or SuperGUT, still has to be worked out. A promising beginning has been made with the development of superstring theories that emerged during the 1980s.

Data from previously classified US Navy navigation satellites are made available to the general public on Jul 30

A.C. Delany, Audrey Delany, D.W. Parking, John J. Griffen, Edward D. Goldberg, and B.E.F. Reimann discover that one-third to one-half of the sediments in the mid-Atlantic Ocean consist of dust blown from Africa and Europe

Clomiphene is introduced to increase fertility; it also results in an increase in multiple births

Mammography for detecting breast cancer is introduced

A 20-year study of fluoridation in Evanston, IL, shows that dental cavities have been reduced by 58 percent as a result of adding fluorides to the water supply

Surgeon Christiaan Neelthing Barnard (b Beaufort, South

Hans A. Bethe of the United States wins the Nobel Prize for Physics for his study of the energy production of stars

Sir John Douglas Cockcroft, English physicist, d Cambridge, Sep 18

V.M. Lobashov shows that the strong nuclear force also violates parity conservation

Bern Matthias (b Frankfurt-am-Main, Germany, Jun 8, 1918) and coworkers find an

Keyboards are put in use to provide data for computers

Overseas direct dialing from New York to London and Paris begins on Mar 1 with 80 New York customers on a three-month trial; in Jun, 1966, three calls had been dialed from Philadelphia in the first demonstration of the new technique

The US Department of Agriculture starts a test project in irradiating wheat and other foods to kill insects

1967

cont

hominid line, the line that eventually leads to *Homo sapiens*

Angelos Galanopoulos suggests that a great volcanic explosion that nearly destroyed the island of Thera (also known as Santorini) gave rise to the myth of Atlantis

In Jul an amateur British astronomer discovers a bright nova, named Nova Delphini for the constellation in which it is found; at its height it has an apparent magnitude of three, which is 1 million times as bright as the faintest stars seen with the naked eye, but about 1 million times less bright than the brightest stars

Jocelyn Bell discovers the first pulsar, CP 1919, in Jul; her supervisor, Anthony Hewish (b Fowey, England, May 11, 1924), will be given a share of the 1974 Nobel Prize for this feat, but Bell will not be so recognized; because they do not know the cause, data on the pulsar are not published until Feb, 1968; at first, its regular signal is taken to be a message from extraterrestrial beings, or "little green men," but soon Thomas Gold shows that the signal is caused by a neutron star rotating very fast

Raymond Davis, Jr., installs a 100,000-gallon tank of cleaning fluid deep in the Homestake Gold Mine in South Dakota; it is designed to detect neutrinos from the interior of the sun; as the experiment progresses, it becomes apparent that fewer neutrinos are detected than expected, a major problem for solar theory

Ejnar Hertzsprung, Danish astronomer, d Roskilde, Oct 21

Frank J. Low and D. Kleinmann start the search for galaxies emitting at infrared wavelengths; they are

Wald (b New York, NY, Nov 18, 1906), and Finnish-Swedish physiologist Ragnar Arthur Granit (b Helsinki, Finland, Oct 30, 1900) win the Nobel Prize for Physiology or Medicine for their advanced discoveries in the physiology and chemistry of the human eye

Arthur Kornberg announces that he and coworkers have synthesized biologically active DNA; specifically, they used one strand of natural DNA from a virus as a template to assemble the paired strand from laboratory chemicals

The theory of island biogeography by R.H. MacArthur and Edward O. Wilson is the beginning of the equilibrium biogeography, a school of ecology that focuses on situations where systems are in balance and therefore stable

Elmer Verner McCollum, American biochemist, d Baltimore, MD, Nov 15

Hermann Joseph Muller, American biologist, d Indianapolis, IN, Apr 5

Gregory Pincus, American biologist, d Boston, MA, Aug 22

Alf E. Porsild and Charles R. Arington announce in Oct that they have grown arctic lupines from seeds that had been frozen since the last ice age, setting a new record of about 10,000 years for the length of time that plant seeds can remain viable under proper conditions

Richard Kuhn, Austrian-German chemist, d Heidelberg, Germany, Jul 31

S. Manabe and R.T. Wetherald warn that human activities that increase the amount of carbon dioxide in the air are causing a greenhouse effect that will raise global temperatures

Africa, Nov 8, 1922) performs the first partially successful human heart transplant, Dec 3; the recipient of the new heart is Louis Washkansky, who lives for 18 days

Albert M. Cohen and coworkers report that LSD (lysergic acid diethylamide), frequently taken as a drug to induce an altered state of consciousness, produces breaks in chromosomes, creating the possibility of genetic abnormalities in children of LSD users

Cleveland surgeon Rene Favaloro develops the coronary bypass operation

Michael S. Gazzinga reports that studies of people whose brain halves have been severed reveal that the two halves function independently

Alick Isaacs, Scottish virologist who discovered interferons, d Jan 26

Wolfgang Köhler, Russian-German-American psychologist, d Enfield, NH, Jun 11

Gustave J.V. Nossal proposes in Jun that antibodies work by detecting the size and shape of the antigen

George Davis Snell (b Bradford, MA, Dec 19, 1903) discovers that tissue compatibility is determined by specific genes

alloy of niobium, aluminum, and germanium that becomes superconducting at a temperature of 20.05°K (degrees above absolute value); this sets a record more than two degrees above the previous high temperature for superconductivity

J. Robert Oppenheimer, American theoretical physicist, d Princeton, NJ, Feb 18

Robert Jemison Van de Graaff, American physicist, d Boston, MA, Jan 16

Independently, Steven Weinberg (b New York, NY, May 3, 1933) Abdus Salam, and Sheldon Lee Glashow develop electroweak unification in terms of a spontaneously broken symmetry and group theory; the new theory predicts an observable transformation of particles, which physicists call the "neutral current"

The National Academy of Sciences in Jun reports that the practice of adding antibiotics to animal food, while producing greater yields, may leave traces in meat and increase drug resistance in bacteria

Gene Amdahl proposes the construction of a computer with parallel processors; such a machine could solve certain classes of problems much faster than an ordinary computer, which always works in a linear sequence

R.M. Dolby develops a method to eliminate background sound in recordings that also improves fidelity in other ways

1967

cont

1967

cont

able to report a number of them in 1970

The third US Lunar Orbiter completes the task of surveying possible landing sites on the moon for the Apollo astronauts

US spaceprobe Surveyor III soft lands on the moon on Apr 19; it digs small trenches in the lunar surface and returns photographs of the results to Earth, where the lunar soil is thought (incorrectly) to look a lot like Earth soil and therefore safe to walk upon (which is correct)

US scientists launch Lunar Orbiter IV on May 4; its mission is to map the face of the moon in more detail than can be seen from telescopes placed on Earth

US spaceprobe Surveyor V soft lands on the Sea of Tranquility Sep 10

The Soviet spaceprobe Venera 4 parachutes a capsule into the atmosphere of Venus, where high pressures and temperatures cause it to cease functioning about the time it is thought to have hit the surface of the planet; telemetry from its descent reveals that the atmosphere of Venus is mostly carbon dioxide

US spaceprobe Mariner V flies to within 4000 km (2600 mi) of Venus; it obtains data on the atmosphere, temperature, and air pressure of Venus that are roughly consistent with the Venera probe of the previous day;

George Schaller's *The deer and the tiger* describes the interactions between tigers and their prey in India

Charles Yanofsky (b New York, NY, Apr 7, 1925) and coworkers demonstrate that the sequence of codons in a gene determines the sequence of amino acids in a protein; although this relationship had been previously assumed, Yanofsky's work was the first to show that it definitely occurs

Sydney Brenner, Francis H.C. Crick, and colleagues decode another genetic codon on DNA that signals a stop message; this nearly completes the decoding of the various DNA codons

The Return of Catastrophism

In the late eighteenth and early nineteenth centuries, earth scientists such as Hutton and Lyell introduced the doctrine of uniformitarianism as an antidote to the prevailing ideas that the surface of the Earth had been shaped by number of catastrophes. At that time, the most popular catastrophe was Noah's flood, but great volcanic catastrophes were also invoked (see "Neptunism vs. Plutonism, p 234). One of the results of such catastrophes, the proponents of catastrophism claimed, was the fossil record, showing the remains of creatures that were wiped out during the catastrophes of the past. Among the exponents of catastrophism were Charles Bonnet and Georges Cuvier. Charles Lyell was their great opponent, arguing that the surface of the Earth was formed by small changes occurring over vast reaches of time, which is the basic doctrine of uniformitarianism.

Darwin took up the concept of uniformitarianism from Lyell and applied it to evolution, offering an alternative explanation for the fossil record. For the next hundred years, uniformitarianism was dominant in geology, paleontology, and evolutionary theory. A small move away from uniformitarianism was initiated in the 1970s by Niles Eldredge and Stephen Jay Gould in evolutionary biology with their theory of "punctuated equilibrium," but their new ideas, while influential, were not accepted by many uniformitarians.

In 1980, however, a chance discovery brought catastrophism to the forefront of scientific thought again. Walter Alvarez at a site in Italy was investigating the boundary between the Cretaceous and Tertiary periods, called the K-T boundary by geologists. He asked his father, the distinguished physicist Luis Alvarez, to help analyze a clay layer that appeared right at the K-T boundary. The senior Alvarez had the equipment necessary to check the clay for heavy metals, so he did. The clay turned out to have much more of the heavy metal iridium than samples of clay usually do. Examination of samples of the K-T boundary from around the world showed the same iridium enrichment.

The Alvarezes and their coworkers proposed that the iridium was caused by a catastrophe—they postulated that a large body, such as an asteroid or a comet, loaded with iridium, had crashed into the Earth. Furthermore, they claimed that dust from that impact caused the extinctions that made up the K-T boundary by blanketing the Earth with dust and soot. The resulting loss of sunlight would cool the Earth and also stop photosynthesis, removing the base of the food chain, green plants, which in turn would lead to other extinctions. The general public was particularly taken with this idea because the dinosaurs became extinct at or near the end of the Cretaceous. Here was an explanation for why the dinosaurs became extinct that had actual evidence to back it up.

While the possibility of such an impact was still being studied, a new factor developed. David M. Raup and J. John Sepkoski, Jr., analyzed mass extinctions of the type that occurred at the K-T boundary. They found that such extinctions were periodic, with a period of about 26 million years. It is difficult to think of any Earthly factor that would have such a period; however, scientists could suggest factors outside the Earth. The most prominent idea was that something periodically disturbed the cloud of comets—the Oort cloud—that fills the edge of the solar system. When this happens, some of the comets fall toward the sun. A few of them might strike the Earth, causing such catastrophes as mass extinctions. Candidates for the disturbing force included passage of the solar system through the plane of the galaxy, a companion star to the sun called Nemesis, and a planet called X.

Although much of the new catastrophism is still in dispute, the idea that started it all, the impact of an object at the K-T time boundary, has gradually gained acceptance by most geologists and many paleontologists. Whether or not Nemesis or X exists, something apparently did strike the Earth at the K-T time boundary, and it probably caused the mass extinctions at that time. One proposed mechanism is based on the idea that such an impact would cause acid rain the strength of battery acid.

A spin-off catastrophe is nuclear winter. Scientists impressed with the Alvarezes' notion that clouds of dust and soot could cool the Earth enough to cause some species to become extinct investigated whether or not the same effect might happen as a result of a nuclear war. The debate on this potential catastrophe is continuing.

1967

cont

ASTRONOMY & SPACE

later data will show, however, that the temperature and pressure are even higher

US spaceprobe Surveyor VI soft lands on Central Bay on the moon Nov 9

1968

GENERAL

Citizens in Czechoslovakia support liberalization of their communist government, but Soviet troops take over the country to prevent change

A draft treaty to stop the spread of nuclear weapons is approved in Jun by the UN General Assembly

ASTRONOMY & SPACE

Surveyor VII becomes the first space vehicle to land undamaged on the moon on Jan 9, from which it sends back 21,000 photographs of the lunar surface

Astronomers find a pulsar with a period of 0.033 seconds in the center of the Crab nebula in Taurus, the site of the supernova of Jul 4, 1054

Nine months after discovering one bright nova, amateur British astronomer George Alcock finds another almost as bright, named Nova Vulpeculae for the constellation in which it is found

Eric Becklin and Gerry Neugebauer report the observation of the galactic center at infrared frequencies

George Clark, Gordon Garmire, and William Kraushaar report the discovery, using NASA's Third Orbiting Solar Observatory, of a gamma-ray background coming from the Milky Way galaxy, especially from its center

The Soviet spacecraft Zond 5 becomes the first human-created object to travel around the moon and return to Earth

Yuri Alekseyevich Gagarin, Russian cosmonaut, d near Kirzhach, Vladimir region, Mar 27

BIOLOGY

A report from the US House of Representatives declares that Lake Erie is so polluted that it is essentially dead, and that if all pollution ceases it will take 500 years to restore it to its condition during World War II

Werner Arber (b Granichen, Switzerland, Jun 3, 1929) discovers that bacteria defend themselves against viruses by producing enzymes that cut the DNA of the virus at a particular point; these restriction enzymes later become one of the basic tools of genetic engineering

Sir Henry Hallett Dale, English biologist, d Cambridge, Jul 23

Robert Holley, Indian-American chemist Hans Gobind Khorana (b Raipur, India, Jan 9, 1922), and Marshall W. Nirenberg of the United States win the Nobel Prize for Physiology or Medicine for deciphering the genetic code that determines cell function

Mark Ptashne and, independently, Walter Gilbert and their coworkers identify the first repressor genes

David Zipser discovers the meaning of the only unknown codon of the 64 that make up the genetic

CHEMISTRY

Otto Hahn, German physical chemist, d Göttingen, Jul 28

Lars Onsager of the United States wins the Nobel Prize for Chemistry for his work on the theoretical basis of diffusion of isotopes

The *Glomar Challenger* goes into service; its Deep Sea Drilling Project will obtain thousands of cores from the ocean floor in 96 legs over the next 15 years

US Air Force scientists show that radar can be used to detect wind shifts and precipitation

Elso S. Barghoorn and his coworkers report finding remains of amino acids in rocks that are as old as 3 billion years, showing that life existed in that remote time

The US government moves the people of Bikini Island back to the atoll, claiming that radioactive contamination from the hydrogen bomb of 1956 has reached sufficiently low levels

Dentists show that dental caries (tooth decay) is caused by a substance produced by streptococcal bacteria that normally live in the mouth when these bacteria metabolize sugar

English researchers report in Apr that oral contraceptives can cause blood clots in susceptible women

Christiaan N. Barnard performs a second human heart transplant; this time the patient, Philip Blaiberg, lives 74 days with his new heart

Howard Walter Florey, Baron Florey, Australian-English pathologist, d Oxford, England, Feb 21

French-American physiologist Roger Guillemin (b Dijon, France, Jan 11, 1924) and Polish-American biochemist Andrew Victor Schally (b Vilna, USSR, Nov 30, 1926) discover in this year and the next a simple substance produced by the brain that affects the hormones produced by the pituitary gland

Luis Walter Alvarez of the United States wins the Nobel Prize for Physics for his discovery of resonance particles

George Gamow, Russian-American physicist, d Boulder, CO, Aug 19

Lev Davidovich Landau, Soviet physicist, d Moscow, Apr 1, after an incredible international effort to save his life following an automobile accident on Jan 7, 1962

Lise Meitner, Austrian-Swedish physicist, d Cambridge, England, Oct 27

Andrey Dmitriyevich Sakharov (b Moscow, USSR, May 21, 1921), who had previously led the successful effort of the Soviet Union to develop a thermonuclear bomb (also called the hydrogen, or H, bomb), speaks out in favor of nuclear arms reduction, a point of view that puts him into conflict with the Soviet government

Joseph Weber reports the detection of gravitational waves, but does not convince the scientific community of the validity of his experiment

Regular hovercraft service starts across the English Channel

The first supertankers for carrying petroleum are put into service

The luxury ocean liner *Queen Elizabeth* II is launched

The Nuclear Materials Equipment Corporation begins sterilizing bacon and potatoes with radiation from radioactive cobalt-60 as a means of preserving them

The first supersonic airliner, the Soviet Tupolev TU-144, is demonstrated on Dec 31

James L. Goddard, commissioner of the US Food and Drug Administration, refuses in Apr to permit canned ham that has been radioactively sterilized to be used for human consumption by the US Army

1968

1968

cont

Walter M. Schirra, Donn F. Eisele, and Walter Cunningham begin the first three-man Apollo mission on Oct 11; the mission lasts 260 hours and features live TV sessions with the crew

Launched on Dec 21 in the first manned *Saturn V* flight, Frank Borman, James A. Lovell, Jr., and William A. Anders orbit the moon 10 times

code; it is the three-base codon UGA (for uracil, guanine, adenine); Zipser discovers it to be a termination signal whose meaning is "stop making this protein"

1969

A satellite detects 161 X-ray sources in the sky

In Jan, Cocke, Taylor, and Disney are the first to identify a visible star associated with a pulsar, the pulsar in the Crab nebula; previously pulsars were known from radio signals only

Thomas Gold and Franco Pacini explain pulsars as rapidly rotating neutron stars that produce beams of synchrontron radiation; as these beams intersect Earth, they are detected as rapid on-and-off radio signals

Vesto Melvin Slipher, American astronomer, d Flagstaff, AZ, Nov 8

In the first space linkup of two manned vehicles, the Soviet spacecrafts *Soyuz 4* and *Soyuz 5* dock and transfer crew on Jan 14

An Apollo 9 mission launched on Mar 3 with James A. McDivitt, David Scott, and Russell Schweickart is the first test flight of all lunar hardware in Earth orbit, including the lunar module

Jonathan Beckwith (b Cambridge, MA, Dec 25, 1935) and coworkers succeed in isolating a single gene; it is a bacterial gene for a step in the metabolism of sugar

Max Delbruck, Alfred D. Hershey, and Salvador E. Luria of the United States win the Nobel Prize for Physiology or Medicine for their discoveries in the workings and reproduction of viruses

Derek H.R. Barton of England and Odd Hassel of Norway win the Nobel Prize for Chemistry for the determination of the three-dimensional shape of organic compounds

		Joe M. McCord discovers an enzyme that renders free oxygen radicals harmless			**1968** cont

| Harry Hammond Hess, American geologist, d Woods Hole, MA, Aug 25 | The new US military draft gets off to a bad start when workers use a procedure that turns out not to be random to select draftees; people born in the later months of the year turn out to be much more likely to be drafted than people born in the early months of the year | It is recognized that exercise can produce an allergic reaction in some people, causing hives, choking, low blood pressure, and other familiar symptoms of allergic shock

In Texas, Denton Cooley and Domingo Liotta replace the diseased heart of Haskell Karp with the first artificial heart to be used in a human being; Karp lives for nearly three days

Arthur R. Jensen writes in the *Harvard Educational Review* that genetic factors cause blacks to score poorly on standardized tests, not environmental factors; this conclusion is widely disputed | Murray Gell-Mann of the United States wins the Nobel Prize for Physics for his classification of elementary particles

Cecil Frank Powell, English physicist, d near Bellano, Italy, Aug 9

Otto Stern, German-American physicist, d Berkeley, CA, Aug 17

Robert Wilson founds the Fermi National Accelerator Laboratory, generally known as Fermilab, in Batavia, IL, just west of Chicago | The scanning electron microscope reaches practical use after 15 years of development

The first home yogurt maker is marketed

"Bubble memory" devices are created for use in computers; unlike conventional memory devices, bubble memory continues to remember even when the computer is turned off

Willy Ley, German-American engineer, d New York, NY, Jun 24 | **1969** |

1969

cont

In a lunar mission development flight launched on May 18 with Eugene A. Cernan, John W. Young, and Thomas P. Stafford as crew, the lunar module descends to within 15,000 m (50,000 ft) of the lunar surface

American astronaut Neil Armstrong on Jul 20 becomes the first human being to stand on the moon; crewmate Buzz Aldrin is right behind him, while the other member of the Apollo 11 crew, Michael Collins, orbits the moon

A second lunar landing mission is begun on Nov 14 with Charles Conrad, Jr., Richard F. Gordon, Jr., and Alan L. Bean aboard; over 15 hours are spent exploring the surface of the moon

---

1970

GENERAL

In the United States, the first Earth Day is celebrated on Apr 22

ANTHROPOLOGY/ ARCHAEOLOGY

In Jun, J.M. Adovasio starts his students excavating a rock shelter named Meadowcroft, about 80 km (50 mi) west of Pittsburgh, PA; unexpectedly, carbon-14 dates obtained from 1973 to 1978 show Meadowcroft was inhabited by human beings as early as 19,000 years ago, nearly 8000 years earlier than other sites in North America

ASTRONOMY & SPACE

The 100-m (328-ft) radio "dish" at Bonn, West Germany, is completed

The large reflecting telescope at Kitt Peak, AZ, is completed, and the first large telescope is erected on Mauna Kea, Hawaii, a 224-cm (88-in.) reflector

The first Chinese and Japanese artificial satellites are launched

The third lunar mission is begun on Apr 11 with James A. Lovell, Jr., Fred W. Haise, Jr., and John L. Swigert, Jr., as crew; a landing attempt on the moon is aborted because of equipment failure

BIOLOGY

Hamilton Othanel Smith (b New York, NY, Aug 23, 1931) isolates the first restriction enzyme, an enzyme that always breaks a DNA molecule at a specific junction of bases

Julius Axelrod (b New York, NY, May 30, 1912), Ulf von Euler (b Stockholm, Sweden, Feb 7, 1905), and Sir Bernard Katz (b Leipzig, Germany, Mar 26, 1911) win the Nobel Prize for Physiology or Medicine for discoveries in the chemical transmission of nerve impulses

François Jacob's *La logique du vivant* (The logic of life) describes pre-Mendelian views of heredity

CHEMISTRY

Argentine biochemist Luis F. Leloir (b Paris, Sep 6, 1906) wins the Nobel Prize for Chemistry for his discovery of sugar nucleotides and their biosynthesis of carbohydrates

Choh Hao Li synthesizes human growth hormone

Stephen A. Cook shows that a problem in formal logic, called the satisfiability problem, is exactly as difficult as any problem in a large range of problems that can be solved by a kind of Turing machine; this opens the door to finding that many problems once thought to be distinct are essentially the same problem

Bertrand Arthur William Russell, 3rd Earl Russell, English mathematician and philosopher, d Penrhyndeudraeth, Merioneth, England, Feb 2

Francis Peyton Rous, American physician who discovered that viruses can cause cancer, d New York, NY, Feb 16

Philip Edward Smith, American endocrinologist, d Florence, MA, Dec 8

Hannes Alfvén of Sweden and Louis Néel (b Lyon, France, Nov 22, 1904) share the Nobel Prize for Physics, Alfvén for work in plasma physics and Néel for work in antiferromagnetism

Max Born, German-British physicist, d Göttingen, Germany, Jan 5

Sir Chandrasekhara Venkata Raman, Indian physicist, d Bangalore, India, Nov 21

Carbon-dioxide lasers are introduced for industrial cutting and welding

The first of the "jumbo jets," the Boeing 747, goes into service across the Atlantic

The floppy disk is introduced for storing data used by computers

1970

GENERAL
ANTHROPOLOGY/
ARCHAEOLOGY
ASTRONOMY
& SPACE
BIOLOGY
CHEMISTRY

1970
cont

A team of scientists at the University of Wisconsin led by Har Gobind Khorana announce the first complete synthesis of a gene, analine-transfer RNA; previous workers had used a natural gene as a template, but in this case the gene was assembled directly from its component chemicals

Jacques Monod's *Le hasard et la nécessité* (Chance and necessity) is published

Alfred Henry Sturtevant, American geneticist, d Pasadena, CA, Apr 5

Howard M. Temin (b Philadelphia, PA, Dec 10, 1934) and David Baltimore (b New York, NY, Mar 7, 1938) discover reverse transcriptase in viruses; this enzyme causes RNA to be transcribed onto DNA, a process that is another key step in the development of genetic engineering

Otto Heinrich Warburg, German biochemist, d Berlin-Dahlem, Aug 1

1971

Irvin Shapiro (b New York, NY, Oct 10, 1929) discovers superluminal sources by using very long baseline interferometry; these are components of quasars that appear to move away from each other at speeds greater than the speed of light

Alan B. Shepard, Jr., and Edgar D. Mitchell collect 44 kg (98 lb) of moon rocks during the US Apollo 14 lunar mission, launched on Jan 31 and piloted by Stuart A. Roosa

Bloom and coworkers at the National Institute of Mental Health show that cyclic AMP is involved in signaling between neurons

Daniel Nathans (b Wilmington, DE, Oct 30, 1928) and Hamilton Smith develop various enzymes (restriction enzymes) that break DNA at specific sites

Wendell Meredith Stanley, American biochemist, d Salamanca, Spain, Jun 15

German-Canadian physicist Gerhard Herzberg (b Hamburg, Germany, Dec 25, 1904) wins the Nobel Prize for Chemistry for his study of the geometry of molecules in gases

Paul Karrer, Swiss chemist, d Zurich, Jun 18

Theodor H.E. Svedberg, Swedish chemist, d Stockholm, Feb 25

Arne Wilhelm Kaurin Tiselius, Swedish chemist, d Uppsala, Oct 29

Genetic Markers

About 3,000 human diseases are known to be caused by specific genes. For example, a person who inherits a single gene for Huntington's disease or two genes for cystic fibrosis will have that disease.

The problem is that there are many genes—perhaps as many as 2,000,000 of them for a given human being. And these 2,000,000 are different from one person to another. Locating exactly which gene is involved in a genetic disease is far from easy.

It is possible, however, to locate certain parts of a DNA molecule. Bacteria produce chemicals that cut DNA at specific sites, producing short pieces of DNA. The pieces can be sorted by length by pushing them through a gel with electricity. Then a method called Southern blotting (developed by Edward M. Southern) can be used to pick out specific pieces of DNA from the ones that are a given length.

Suppose that you know that a certain genetic disease occurs in members of a family. You do not know which gene is involved, but you know which members of the family are affected by the disease. You can compare the identifiable pieces of DNA from affected and unaffected family members. Often there is a piece of identifiable DNA that is usually inherited with the gene for the disease. In that case, the identifiable DNA is a *marker* for the gene. By suitable separation and testing of different chromosomes, a scientist can use the marker to locate the chromosome on which the suspect gene can be found.

Using these methods, Kay Davies and Robert Williamson in 1983 located the first marker for a genetic disease, the marker for Duchenne muscular dystrophy.

Knowing a marker for the gene is useful in predicting whether or not a given person carries the gene (if it is recessive or if, like the gene for Huntington's disease, it does not produce symptoms early in life), but it is not as good as being able to identify the actual gene. With persistence, however, the actual gene can sometimes be found and copies of it made—geneticists say the gene is cloned. By 1987, for example, the gene for Duchenne muscular dystrophy had been cloned. About five months later, doctors established that the defective gene failed to produce a protein that is needed in small amounts by all the skeletal muscles in the body.

A likely candidate for clinical results is the cystic fibrosis gene, which was well marked by 1988, although not located exactly. It is believed that this gene always fails in the same simple way. If the gene can be cloned and its product analyzed, there is good reason to hope for clinical applications.

As presently practiced, identifying genetic markers requires careful study of a large family, since different families may have different markers for the same gene. A major effort, however, is underway to map all the useful genetic markers on each chromosome. Such a map would cut short the effort to locate specific genes.

There is another approach, however. It is theoretically possible to map the genes themselves. This is complicated by the fact that each human being (except possibly identical twins) has a somewhat different genome—the totality of genes—from everyone else. In 1988, the US National Academy of Sciences called for a major national effort to map the genes for human beings.

An earthquake with a magnitude of 6.5 on the Richter scale strikes San Fernando, CA

The US Food and Drug Administration asks doctors to stop prescribing diethylstilbestrol (DES) to control morning sickness in expectant mothers because of evidence that the drug predisposes their daughters to cancers of the reproductive tract

In the United States it is shown that electric currents can speed the healing of fractures

In Canada, the first nuclear-power station that is cooled by ordinary water goes into service

Sir William Lawrence Bragg, Australian-English physicist, d Ipswich, UK, Jul 1

Dennis Gabor of England wins the Nobel Prize for Physics for his invention of holography

Gerardus t'Hooft proves the renormalizability of massless Yang-Mills fields and of massive

Direct telephone dialing, as opposed to operator-assisted calling, begins between parts of the United States and Europe on a regular basis

The first microprocessor, now known as the chip, is introduced by Intel in the United States

Patrick Haggerty's Texas Instruments introduces the first pocket calculator, the Pocketronic; it can add, subtract, multiply, and divide; it weighs more than a

1971

cont

		ASTRONOMY & SPACE	BIOLOGY	CHEMISTRY

Three crew members of Soviet spacecraft *Soyuz 10* dock with *Salyut 1*, the first space station, on Mar 23

David R. Scott and James B. Irwin drive the Lunar Rover on the moon's surface as part of the US Apollo 15 lunar mission, begun on Jul 26 and piloted by Alfred M. Worden

The American spacecraft Mariner 9 becomes the first human-built object to orbit another planet when it enters orbit around Mars on Nov 13; initial results are few because Mars is covered by a huge dust storm, but Mariner 9 eventually returns 7329 pictures, many of them of great clarity

Soviet spaceprobe Mars 2 goes into orbit around Mars on Nov 27, just 14 days after the US probe Mariner 9

The Soviet Union's spaceprobe Mars 3 lands on Mars on Dec 2 and broadcasts an unreadable television signal for about 20 seconds before it goes dead

Earl W. Sutherland, Jr., of the United States wins the Nobel Prize for Physiology or Medicine for the discovery of cyclic AMP

Edward O. Wilson's *The insect societies* is a fundamental work on social insects

Robert Burns Woodward synthesizes vitamin B_{12}

1972

The California State Board of Education demands that Biblical accounts of creation receive equal attention in textbooks as Darwinian theory

Louis Seymour Bazett Leakey, English anthropologist, d London, Oct 1

The first Earth-resources satellite, Landsat I, is launched

B.J. Harris finds that 60 percent of the unidentified radio sources scintillate; this is an indication that these sources are compact

Milton La Salle Humason, American astronomer, d Mendocino, CA, Jun 18

In the United States, the use of DDT is restricted to protect the environment, especially birds, whose eggshells are dangerously thinned, lowering the birds' reproductive rate

Gerald M. Edelman (b New York, NY, Jul 1, 1929) and Rodney R. Porter of England win the Nobel Prize for Physiology or Medicine for the determination of the chemical structure of antibodies

Christian Boehmer Anfinsen, Stanford Moore (b Chicago, IL, Sep 4, 1913), and William H. Stein (b New York, NY, Jun 25, 1911) share the Nobel Prize for Chemistry for their pioneering research in enzyme chemistry

In the United Kingdom, the first completely sterile hospital units are introduced to protect patients at special risk from infection

In the United Kingdom, the diamond-bladed scalpel is introduced

Sir Cyril Lodowic Burt, English psychologist, d London, Oct 10

Bernardo Alberto Houssay, Argentinian physiologist, d Buenos Aires, Sep 21

Raymond V. Damadian applies for a patent on magnetic resonance imaging (MRI, also known as NMR) as cancer detector

Yang-Mills fields with spontaneously broken symmetry

Igor Yevgenyevich Tamm, Russian physicist, d Moscow, Apr 12

kilogram (about 2.5 pounds) and costs around $150

Niklaus Wirth develops Pascal (named for Blaise Pascal, who invented the first calculator), a popular language used on home computers

1971

cont

In the United Kingdom, the first computerized axial tomography (CAT scan) imager is introduced

David S. Janowsky and coworkers determine that bipolar disorder, commonly called "manic depression," is caused by an imbalance between two types of neurotransmitters, the adrenergic and cholinergic

The large accelerator at Fermi National Accelerator Laboratory (Fermilab) in Batavia, IL, begins operations at 200 GeV (giga electron volts, or 200 billion electron volts) and reaches 400 GeV later in the year

John Bardeen, Leon N. Cooper, and John R. Schrieffer of the United States win the Nobel Prize for Physics for their theory of superconductivity without

In Germany, an experimental power station uses coal that is converted to gas before being burned

1972

1972

cont

Richard N. Manchester calculates the galactic magnetic field strength by measuring the Faraday rotation for different wavelengths of the polarization plane of radio waves emitted by pulsars

Harlow Shapley, American astronomer, d Boulder, CO, Oct 20

A Soviet spacecraft, Venera 8, soft lands on Venus

Launched on Apr 16, John W. Young, Thomas Mattingly II, and Charles M. Duke are the fifth Apollo crew to land on the moon

US spaceprobe Pioneer 10 is launched; on Jun 13, 1983, it will become the first human-created object to leave the solar system

In the last manned lunar landing, launched on Dec 7, Eugene A. Cernan, Ronald E. Evans, and Harrison Schmitt spend 44 hours on the moon's surface and return with 110 kg (243 lb) of material

Edward Calvin Kendall, American biochemist, d Princeton, NJ, May 4

Max Theiler, South African-American microbiologist, d New Haven, CT, Aug 11

1973

The United Kingdom joins the European Common Market

Gamma rays are detected from an "invisible" source, termed Geminga (meaning "does not exist" in the Milanese dialect); a faint, very hot star is detected at its location in 1987

Astronomers using radio techniques discover complex molecules, including CH, HCN and H_2O, in Comet Kohoutek

John Hershey shows that the telescope used by Peter van de Kamp to find a wobble in Barnard's star,

A calf is produced from a frozen embryo for the first time

Timothy V. Bliss finds that high-speed bursts of electricity strengthen neural networks in the brain, tending to confirm the neural-network theory of memory proposed by Donald O. Hebb in 1949

Stanley H. Cohen and Herbert W. Boyer show that DNA molecules can be cut with restriction enzymes, joined together with other enzymes,

Ernst Otto Fischer of Germany and Geoffrey Wilkinson (b England, Jul 14, 1921) win the Nobel Prize for Chemistry for their work on the chemistry of ferrocene

Karl Ziegler, German chemist, d Muhlheim-Ruhr, Aug 12

electrical resistance at a temperature of absolute zero

Georg von Békésy, Hungarian-American physicist, d Hawaii, Jun 13

In Sep, Murray Gell-Mann presents the beginnings of quantum chromodynamics (QCD), the theory that links quarks and color forces, along with three different flavors of quarks

Marie Goeppert-Mayer, German-American physicist, d San Diego, CA, Feb 20

Boris Ya. Zel'dovich and coworkers use the phenomenon of stimulated Brillouin scattering to create a beam of time-reversed light; such a beam can be transmitted through a distorting medium and be concentrated instead of distorted

---

Charles H. Bennett demonstrates that it is possible to build a computer without the known components that cause a loss of energy

Michael S. Brown and Joseph L. Goldstein discover the receptor in the membrane of human body cells that captures low-density lipoproteins and removes them from the bloodstream, an important step in understanding how atherosclerosis (arterial plaque) develops

Computerized axial tomography, better known as the CAT scan or the CT scan, is introduced, developed by South African-American

Physicists suspect the existence of "dark pulse" solitons

Scientists at Bell Labs develop a tunable, continuous-wave laser

Lev Andreevich Artsimovich, Soviet physicist, d USSR, Mar 1

Leo Esaki of Japan, Norwegian-American physicist Ivar Glaever (b Bergen, Norway, Apr 5, 1929), and Brian Josephson (b Cardiff, England, Jan 4, 1940) win the Nobel Prize for

The push-through tab on soft-drink and beer cans is introduced

The ENIAC patent is invalidated, crediting John Vincent Atanasoff as the originator of the modern computer

IBM produces the first Josephson junction that might be used as an electronic switching device; it has a switching rated 1,000 times

1973

cont

indicating the presence of a planet, also causes wobbles in the motions of 12 other stars, throwing doubt on the hypothesis of a planetary system at Barnard's star; other astronomers also fail to detect wobble

W. Klebesadel, Ian B. Strong, and R.A. Olson report the discovery by the Vela satellite of gamma-ray bursts originating in outer space; the satellite was searching for gamma-ray bursts caused by nuclear explosions on Earth

Gerard Peter Kuiper, Dutch-American astronomer, d Mexico City, Dec 23

The first *Skylab* is launched on May 25 by a Saturn rocket; for 28 days a three-man crew headed by Charles Conrad establishes orbit and conducts medical and other experiments

The second *Skylab* mission is launched on Jul 29 and lasts 59 days; the crew performs systems and operational tests and thermal shield deployment

The third *Skylab* mission is launched on Nov 16; in an 84-day mission, the three-man crew obtains medical data for extending space flights

and reproduced by inserting them into the bacterium *Escherichia coli;* this is the beginning of genetic engineering

J.M. Diamond's "Distributional ecology of New Guinea birds" claims that similar birds occur in a gradation in which each larger bird is about 1.3 times as large as its smaller relatives in all dimensions; later work shows that this ratio also applies to inanimate objects such as skillets and violins

Karl von Frisch and Konrad Lorenz of Germany and Nikolaas Tinbergen of the Netherlands win the Nobel Prize for Physiology or Medicine for their study of individual and social behavior patterns of animal species

Artturi Ilmari Virtanen, Finnish biochemist, d Helsinki, Finland, Nov 11

Selman Abraham Waksman, Russian-American microbiologist, d Hyannis, MA, Aug 16

1974 | A revolution installs a democratic regime in Portugal

US President Richard Milhous Nixon resigns on Aug 8 as a result of the cover-up of the Watergate break-in

A team led by Don Johanson and Maurice Taieb discover "Lucy," 40 percent of the skeleton of an early hominid that is more than 3 million years old; Lucy is a representative of a previously undiscovered species, *Australopithecus afarensis*

Charles Kowal (b Buffalo, NY, Nov 8, 1940) discovers Leda, the thirteenth known moon of Jupiter

Allan Sandage, Jerome Kristian, and Basil Katem search the optical counterparts of 47 radio sources with the Hale reflector on

Gary K. Beauchamp, Kunio Yamazake, and Edward A. Boyse observe that mice are able to distinguish between other mice with the same genes for an immunity structure called the Major Histocompatibility Complex; later it is determined

Paul J. Flory (b Sterling, IL, Jun 19, 1910) wins the Nobel Prize for Chemistry for his study of long-chain molecules

physicist Allan MacLeod Cormack (b Johannesburg, Feb 23, 1924) and English physicist Godfrey N. Hounsfield

Walter Rudolf Hess, Swiss physiologist, d Zurich, Aug 12

Scientists in the United Kingdom introduce the nuclear magnetic resonance (NMR or, more recently, MRI for magnetic resonance imager) scanner for medical diagnosis

Physics for their theories on superconductors and semiconductors important to microelectronics

Paul Musset and coworkers at CERN discover neutral currents in neutrino reactions; these had been predicted by the electroweak theory and are viewed as a partial confirmation of the theory

Jogesh C. Pati and Abdus Salam suggest theoretical reasons why protons might decay into other particles

David Politzer and, independently, David Gross and Frank Wilczek show that quarks must exhibit asymptotic freedom— that is, the farther apart quarks are, the more they attract each other; when they are touching, they ignore each other

Edward P. Tryon suggests that the entire universe may have been created from absolutely nothing as a result of the probabilistic laws of quantum mechanics; an allowable fluctuation in a quantum vacuum could result in the creation of energy

Sir Robert Watson-Watt, Scottish physicist, d Inverness, Dec 6

that of the fastest semiconductor device available at that time, but required cooling to near absolute zero

---

William Maurice Ewing, American geologist, d Galveston, TX, May 4

F. Sherwood Rowland and Mario Molina warn that chlorofluorocarbons (also called Freons), commonly used as spray propellants and in

Theodore P. Baker, John Gill, and Robert M. Solovay show that two classes of problems are equivalent with one set of assumptions and not equivalent with another set of assumptions, even if both sets of assumptions are consistent

Earl Wilbur Sutherland, Jr., American physician and pharmacologist, d Miami, FL, Mar 9

The US FDA bans the use of chloroform in drugs and cosmetics because of suspected carcinogenity

Optical pulses less than a trillionth of a second long are produced by lasers at Bell Labs

Patrick Maynard Stuart Blackett, Baron Blackett, English physicist, d London, Jul 13

Radio-Electronics publishes an article describing the construction of a "personal computer"

Hewlett Packard introduces the programmable pocket calculator

1974

cont

Mount Palomar; for 21 of these radio sources they fail to find optical counterparts

Fritz Zwicky, Swiss astronomer, d Pasadena, CA Feb 8

A Soviet spaceprobe lands on Mars

that this is accomplished by the sense of smell

Genetic engineering, in which the genes of one organism are inserted in another, is viewed with alarm by a committee of 139 scientists from the US National Academy of Sciences and 18 other nations led by Paul Berg (b New York, NY, Jun 30, 1926); in Jul the committee calls for a halt in specified research until the implications are better understood

Albert Claude and George E. Palade of the United States and Belgian cytologist C. René De Duve (b Thames Ditton, England, Oct 2, 1917) share the Nobel Prize for Physiology or Medicine for the advancement of cell biology, electron microscopy, and structural knowledge of cells

Alan E. Jacob and Robert W. Hedges discover transposans, genetic elements in bacteria (later discovered in other organisms) that can move easily from one plasmid to another or that can move from the chromosome to a plasmid or to another chromosome

Lars Terenius and Ageneta Wahlstrom discover that certain small peptides naturally produced in the body act upon the opiate receptors in the brain

refrigeration, may be destroying the ozone layer in the atmosphere; the ozone layer protects against excessive ultraviolet radiation; such radiation can cause skin cancer

Pierre Deligne proves Weil's conjecture, using algebraic geometry

Roger Penrose discovers a way to tile a plane using two different rhombuses so that the ratio of fat rhombuses to thin ones is an irrational number, specifically the golden mean; such a Penrose tiling has no cell that contains an integral number of rhombuses

Satyendranath Bose, Indian physicist, d Calcutta, Feb 4

Sir James Chadwick, English physicist, d Cambridge, Jul 24

Howard M. Georgi and Sheldon Lee Glashow develop the first of the grand unified theories (or GUTs); these account for the strong, weak, and electromagnetic forces as variants of a single force that is broken into parts when the universe begins to cool down after the Big Bang

Anthony Hewish and Martin Ryle of England win the Nobel Prize for Physics, Hewish for his discovery of pulsars and Ryle for his improvements in radiotelescopy

Independently Burton Richter (b New York, NY, Mar 22, 1931) and Samuel Chao Chung Ting (b Ann Arbor, MI, Jan 27, 1936) discover a new subatomic particle which Burton calls the psi particle and Ting calls the J particle; it is important in that it tends to confirm the charm theory; today the particle is called the J/psi

Vannevar Bush, American electrical engineer, d Belmont, MA, Jun 28

Charles Augustus Lindbergh, American aviator, d Kipahulu, Hawaii, Aug 26

1974

cont

1975

GENERAL

The last American troops are evacuated from South Vietnam in Apr, ending the Vietnam War, during which 57,000 American lives were lost

ASTRONOMY & SPACE

Charles Kowal discovers a fourteenth moon of Jupiter

Alan E.E. Rogers rediscovers the forgotten idea of Roger C. Jennison for improving the resolution of radiotelescopes by very long baseline interferometry; this time the idea is applied within a year with great success, obtaining a resolution of 0.01 arc-second

Vera C. Rubin (b Philadelphia, PA, Jul 23, 1928) and W. Kent discover that the Milky Way galaxy has a proper motion of about 500 km (300 mi) per sec by measuring its speed with respect to distant galaxies, forming a reference frame

The first pictures from the surface of Venus are received, from the Russian probes *Venera 9* and *Venera 10*

The first cooperative US-Soviet space mission, the Apollo Soyuz Test Project, is launched on Jul 15; a three-man Apollo spacecraft docks with the two-man Soviet *Soyuz 19*

BIOLOGY

David Baltimore and Howard M. Temin of the United States and Italian-American virologist Renato Dulbecco (b Cantanzaro, Italy, Feb 22, 1914) win the Nobel Prize for Physiology or Medicine, Baltimore and Temin for their discovery of reverse transcriptase and Dulbecco for the interaction between viruses and host cells

Theodosius Dobzhansky, Russian-American geneticist, d Davis, CA, Dec 18

John Hughes, independently of Lars Terenius and Ageneta Wahlstrom, discovers enkephalin, a peptide that acts upon the opiate receptors in the brain, producing effects similar to morphine

Robert Michel suggests that turnover of cell membranes is connected to calcium signals between cells; subsequent research establishes that this is the case

Edward Lawrie Tatum, American biochemist, d New York, NY, Nov 5

The discovery of how to produce monoclonal antibodies is announced in the United Kingdom by César Milstein; these antibodies are selected from a single lineage (clone) and all bind with the same chosen protein, or antigen

CHEMISTRY

John Warcup Cornforth (b Sydney, Australia, Sep 7, 1917) and Yugoslav-Swiss chemist Vladimir Prelog (b Sarajevo, Jul 23, 1906) win the Nobel Prize for Chemistry, Cornforth for his work on the structure of enzyme-substrate combinations and Prelog for his study of asymmetric compounds

Sir Robert Robinson, English chemist, d Great Missenden, England, Feb 8

Jacob Aall Bonnevie Bjerknes, Norwegian-American meteorologist, d Los Angeles, CA, Jul 7

Mitchell J. Feigenbaum discovers Feigenbaum's number (represented by a lower case sigma and approximately 4.669201609103), the ratio toward which the consecutive differences of iterated functions tends

The tau lepton, or tauon, is discovered in Aug; the tauon is a "fat muon" just as a muon is a "fat electron;" that is, the muon has all the properties of the electron, but has 200 times the mass; similarly, the tauon is like an electron, but has about 3500 times the mass

William David Coolidge, American physicist, d Schenectady, NY, Feb 3

John Ray Dunning, American physicist, d Key Biscayne, FL, Aug 25

Gustav Ludwig Hertz, German physicist, d Berlin, Oct 30

James Rainwater of the United States and Ben Mottelson and Aage Bohr of Denmark win the Nobel Prize for Physics for their studies proving the asymmetrical structure of the atomic nucleus

Sir George Paget Thomson, English physicist, d Cambridge, Sep 10

The first liquid-crystal displays for pocket calculators and digital clocks are marketed in the United Kingdom

The first personal computer, the Altair 8800, is introduced in kit form in the United States; it has 256 bytes of memory

1975

1976

Astronomy & Space

The 6-m (236-in.) reflecting telescope at Mount Semirodriki in the Soviet Union is completed

Astronomers discover that Pluto is at least partially covered with frozen methane

Tom Kibble theorizes that as the universe cooled after the Big Bang, various flaws would have formed cosmic strings, long, thin bundles of the original energy of the universe that would persist even today

Donald Howard Menzel, American astronomer, d Dec 14

Rudolph Leo B. Minkowski, German-American astronomer, d Berkeley, CA, Jan 4

Rupert Wildt, German-American astronomer, d Orleans, MA, Jan 9

US spaceprobes Vikings 1 and 2 soft land on Mars and begin sending back direct pictures and other information from the surface of the planet; on Jun 19 Viking 1 is the first spacecraft ever to soft land on a planet other than Earth.

Biology

Genentech, the first commercial company aimed at developing products through genetic engineering, is established in south San Francisco, CA

Baruch S. Blumberg and D. Carleton Gajdusek of the United States win the Nobel Prize for Physiology or Medicine for their identification of hepatitis antigen and slow viruses

Carl Peter Henrik Dam, Danish biochemist, d Copenhagen, Apr 18

Czechoslovakian-American biochemist Martin F. Gellert (b Prague, Jun 5, 1929) and coworkers discover the enzyme gyrase causes DNA to form supercoils, the natural state of most DNA in the cell; although DNA is naturally coiled into a double helix, that helix is further coiled to form a larger helix called a supercoil

Har Gobind Khorana and coworkers announce construction of a functional synthetic gene, complete with regulatory mechanisms

Trofim Denisovich Lysenko, Soviet biologist, d Kiev, Ukraine, Nov 20

Jacques Lucien Monod, French biochemist, d Cannes, May 31

Susumu Tonegawa reports his experiment showing that the genes responsible for producing antibodies move physically close together on a chromosome and produce the large variety of antibodies by combination

Chemistry

The Givaudan-La Roche Icmesa pesticide plant near Seveso, Italy, accidentally releases a cloud of poisonous gas that spreads dioxins over 730 h (1800 a), killing dogs and farm animals, but causing no human fatalities, although people are evacuated from the area until it can be cleaned up

William N. Lipscomb (b Cleveland, OH, Dec 9, 1919) wins the Nobel Prize for Chemistry for his study of bonding in boranes

Lars Onsager, Norwegian-American chemist, d Coral Gables, FL, Oct 5

Leopold Stephen Ružička, Croatian-Swiss chemist, d Zurich, Switzerland, Sep 26

1977

John B. Corliss and Robert D. Ballard, aboard the submersible *Alvin*, discover deep-sea ocean vents near the Galápagos Islands; surrounding the hot spring is a specialized community of sulfur-eating bacteria, giant clams, and tube worms

Bruce Charles Heezen, American oceanographer and geologist, d near Reykjanes, Iceland, Jun 21

Oskar Morgenstern, German-American economist, d Princeton, NJ, Jul 26

Ronald L. Rivest, Adi Shamir, and Leonard M. Adleman suggest that a cryptographic system based on the difficulty of factoring large numbers would be feasible; such systems are known as public-key codes because the means of encoding can be made public without losing security

Bradley Efron develops the "bootstrap" method for statistics, a more accurate means of analysis that depends on the use of high-speed computers

Two homosexual men in New York City are diagnosed as having the then rare cancer Kaposi's sarcoma; in retrospect, they are probably the earliest known AIDS victims in New York; the disease will not be officially recognized until 1981

Dutch scientists discover that incinerator wastes are contaminated by dioxins, potentially dangerous chemicals that have been linked to cancer in animal studies

The last recorded case of smallpox found in the wild is in Somalia; after this the smallpox virus is believed to be extinct in the wild, although the virus is retained for research purposes in several laboratories

Edgar Douglas Baron Adrian, English physiologist, d London, Aug 4

Andreas R. Gruentzig invents balloon angioplasty, a method for unclogging diseased arteries

Archibald Vivian Hill, English physiologist, d Cambridge, Jun 3

Philip Warren Anderson (b Indianapolis, IN, Dec 13, 1923), John Hasbrouck Van Vleck (b Middletown, CT, Mar 13, 1899), and Nevill Francis Mott (b Leeds, England, Sep 30, 1905) share the Nobel Prize for Physics, Mott and Anderson for work on amorphous semiconductors, Van Vleck for work on the magnetic properties of atoms

Peter Carl Goldmark, Hungarian-American physicist, d Westchester County, NY, Dec 7

Leon Lederman discovers the upsilon particle, which confirms the quark theory of baryons (heavy subatomic particles, such as protons and neutrons)

Erwin Wilhelm Mueller, German-American physicist, d Washington, DC, May 17

The Apple II, the first personal computer available in assembled form and the first to be truly successful, is introduced

A fiber optic system is used for large-scale trials in a telephone system for the first time

Wernher Magnus Maximilian von Braun, German-American rocket engineer, d Alexandria, VA, Jun 16

1978

Chlorofluorocarbons are banned as spray propellants in the United States on the grounds that they damage the ozone layer in the atmosphere, the layer that screens out most of the harmful ultraviolet radiation from the sun

Kurt Gödel, Austrian-American mathematician, d Princeton, NJ, Jan 14

US medical personnel remove the islanders from Bikini Atoll in the Pacific when it is discovered that radioactive cesium-137 is being taken up from the soil by plants and entering the diet of the islanders; the people of Bikini had returned in 1968 after radiation from fusion-bomb tests was pronounced safe

Laser cooling of trapped positive ions is demonstrated for the first time

The first wiggler, a device that increases the brightness of synchrotron radiation, is installed in the SPEAR storage ring at the Stanford Synchrotron Radiation Laboratory

Samuel Abraham Goudsmit, Dutch-American physicist, d Reno, NV, Dec 4

Apple brings out the first disk drive for use with personal computers

1978 cont

ASTRONOMY & SPACE

James W. Christy (b Milwaukee, WI, Sep 15, 1938) and Robert S. Harrington discover Charon, the only known satellite of Pluto, on Jun 22

The Seasat satellite carries an imaging radar system that is used for analyzing currents and ice flow in the ocean

Two Pioneer space-probes are launched toward Venus; they represent the first American attempts to put vehicles into orbit around the planet and to land capsules there

BIOLOGY

Stephen C. Harrison reports the first high-resolution structure of an intact virus, the tomato bushy stunt virus (TBSV)

Daniel Nathans and Hamilton O. Smith of the United States and Werner Arber of Switzerland win the Nobel Prize for Physiology or Medicine for their discovery and use of restriction enzymes for DNA

1979

GENERAL

US President James Earl (Jimmy) Carter persuades Egypt and Israel to end their 30-year war

ANTHROPOLOGY/ARCHAEOLOGY

Lothar Haselberger, while visiting the Temple of Apollo at Didyma, Turkey, discovers the almost complete plans for the temple inscribed in stone on some of the temple walls; the temple was planned in 334 BC to be the largest built, although after 600 years it was still not complete

ASTRONOMY & SPACE

The High-Energy Astrophysical Observatory (HEAO 3) picks up gamma rays from Cygnus X-1

In Mar Voyager 1 discovers a ring around Jupiter

Skylab, an early US space station that had been abandoned, falls into the atmosphere as a result of an intense solar wind caused by increased sunspot activity; it breaks into pieces that land in western Australia, causing no damage

BIOLOGY

Sir Walter Bodner suggests a way to use a combination of chemical studies of DNA with restriction enzymes and studies of heredity in large families to locate gene markers for specific traits

Ernst Boris Chain, German-English biochemist, d Ireland, Aug 11

Feodor Lynen, German biochemist, d Munich, Aug 6

CHEMISTRY

English-American chemist Herbert C. Brown (b London, May 22, 1912) and George Wittig (b Berlin, Germany, Jun 16, 1897) win the Nobel Prize for Chemistry, Brown for his study of boron-containing organic compounds and Wittig for his study of phosphorus-containing organic compounds

Giulio Natta, Italian chemist, d Bergamo, May 2

Robert Burns Woodward, American chemist, d Cambridge, MA, Jul 8

Charles Herbert Best, American-Canadian physiologist, d Toronto, Ontario, Mar 31

David Botstein, Ronald W. Davis, and Mark H. Skolnick propose at a conference in Apr that DNA sequencing can be used to develop gene markers for various genetic diseases

The first human baby conceived outside the body—called a "test-tube baby"—is born to Lesley Brown in the United Kingdom

Alison A. Paul and D.A.T. Southgate determine the amount of fiber in various foods, leading the way to finding the possible connection between fiber and cancer of the colon

Wolfgang Pauli traps neutrons in a magnetic storage ring and succeeds in measuring the average lifetime of a neutron, which is about 15 minutes

Arno A. Penzias and Robert W. Wilson of the United States and Pyotr L. Kapitsa of the Soviet Union share the Nobel Prize for Physics, the Americans for their discovery of microwave radiation from the Big Bang and Kapitsa for work in low-temperature physics

Karl Manne Georg Siegbahn, Swedish physicist, d Stockholm, Sep 26

1978

cont

A severe anthrax epidemic breaks out in Sverdlovsk, USSR, killing 64 people; US experts think that it is the result of an accident at a germ-warfare facility, but Soviet doctors claim it results from contaminated meat; evidence released in 1988 tends to confirm the Soviet claims

Allan MacLeod Cormack of the United States and Godfrey N. Hounsfield of the United Kingdom win the Nobel Prize for Physiology or Medicine for their invention of computed axial tomography

Werner Forssman, German surgeon, d Schopfheim, W Germany, Jun 1

Four groups at the Deutsches Elektronen Synchrotron (DESY) in Hamburg observe the gluon, a particle that carries the strong force

Otto Robert Frisch, Austrian-British physicist, d Sep 22

Dennis Gabor, Hungarian-British physicist, d London, Feb 9

Shin'ichiro Tomonaga, Japanese physicist, d Tokyo, Jul 8

Steven Weinberg and Sheldon L. Glashow of the United States and Abdus Salam of Pakistan win the Nobel Prize for Physics for discovering the link between electromagnetism and the weak force of radioactive decay

Visicalc introduces the first spreadsheet program for personal computers; this enables personal computer users to develop business applications without learning to program a computer

The airplane *Gossamer Albatross* is the first human-powered aircraft to cross the English Channel

The nuclear reactor at Unit 2 of Three Mile Island loses its water buffer through operator error on Mar 28; a small amount of radioactive material escapes the containment dome, but the reactor itself undergoes a partial meltdown; no one is injured

Jean Ichbiah and coworkers develops ADA, a computer language named for Lady Ada Lovelace

1979

1979

cont

1980

GENERAL

Joseph Banks Rhine, American parapsychologist (investigator of extrasensory perception), d Hillsborough, NC, Feb 20

ANTHROPOLOGY/ARCHAEOLOGY

Evidence from airplane and satellite observations show that Maya cities were agricultural centers surrounded by extensive canals

ASTRONOMY & SPACE

The VLA—Very Large Array—radio telescope in Socorro, NM, with resolution equivalent to a single dish 27.4 km (17 mi) in diameter, begins operation

The US Solar Maximum Mission satellite is launched on Valentine's Day to study the solar maximum and the sun's output of radiation; the satellite experiences failure in the attitude control system in Nov and is unable to complete its mission

The quasar 3C273 is identified as an emitter of gamma rays

Uwe Fink and other astronomers report the discovery of a thin atmosphere on Pluto

Timothy J. Pearson and coworkers track a glowing blob ejected by the quasar 3C273 from Jul, 1977 to Jul, 1980 and observe that it is apparently traveling at 9.6 times the speed of light; but they conclude that this apparent motion reflects a speed of about 0.995 the speed of light toward Earth

Alistair Walker observes the eclipse of a star by Charon, the moon of Pluto

Susan Wyckoff of Arizona State University and Peter Wehinger of the Max Planck Institute discover a nebulous region around one of

BIOLOGY

American geneticists discover hypervariable regions in genes (short sequences of DNA that are repeated over and over in the same chromosome)

The US Supreme Court rules that a microbe developed by General Electric for oil cleanup can be patented

David Burpee, American horticulturist, d Jun 24

Louise Clarke and John A. Carbon clone the gene that is the centromere of the yeast chromosome, the part of a chromosome to which spindle fibers attach in meiosis (asexual cell division)

A team headed by Martin Cline succeeds in transferring a gene from one mouse to another and having the gene function

Richard Brooke Roberts, American biophysicist, d Apr 4

William Howard Stein, American biochemist, d New York, NY, Feb 2

The successful production of human interferon in bacteria is announced by Charles Weissmann of the University of Geneva

CHEMISTRY

Paul Berg, Walter Gilbert (b Boston, MA, Mar 21, 1932), and Frederick Sanger of England share the Nobel Prize for Chemistry, Berg for the development of recombinant DNA and Gilbert and Sanger for the development of methods to map the structure of DNA

Willard Frank Libby, American chemist, d Los Angeles, CA, Sep 8

(the daughter of Lord Byron who, legend has it, was the first computer programmer); it is to be used by the US armed services

1979 cont

On May 18, Mount St. Helens in Washington erupts at 8:32 PST, killing 61 people and devastating a large region around the volcano

Walter Alvarez, working with Luis Alvarez, Frank Asaro, and Helen Michel, discovers a thin layer of clay at the Cretaceous-Tertiary boundary, a time marked by mass extinctions, including the extinction of the dinosaurs; the clay is enriched with the heavy metal iridium, leading the team to speculate that a giant body from space collided with Earth, causing the layer (later found to be worldwide) and the extinctions

Leonard M. Adleman and Robert S. Rumely develop a new and improved test for prime numbers

Robert Griess, Jr., constructs a finite simple group known as the monster because of its size; with this group, the proof of the classification of all the finite simple proofs, running thousands of pages and starting in 1830, is complete

Dornier Medical Systems of Munich, W Germany, develops the lithotripter, a machine that uses sound waves to break up kidney stones while the stones are still in the kidney

Scientists at the New York Blood Center develop a successful experimental vaccine against hepatitis B

George D. Snell, Venezuelan-American geneticist Baruj Benacerraf (b Caracas, Oct 29, 1920), and Jean Dausset of France win the Nobel Prize for Physiology or Medicine for their discovery of the role of antigens in organ transplants

Several groups of researchers announce that neutrinos may have a tiny mass; this mass may represent the "missing mass" believed to hold galaxies together

The first undulator, a device for increasing the power of synchrotron radiation, is installed in the SPEAR storage ring used at the Stanford Synchrotron Radiation Laboratory

James W. Cronin and Val L. Fitch of the United States win the Nobel Prize for Physics for their studies showing that charge parity and time symmetry could be violated

Alan H. Guth proposes a cosmological model of the birth of the universe, called the inflationary universe, in which the universe expands rapidly for a short time before the Big Bang

Bern Teo Matthias, German-American physicist, d La Jolla, CA, Oct 27

Rice and his group demonstrate a miniature Nd:YAG rod laser pumped by a diode laser

Heinrich Rohrer (b Switzerland, Jun 9, 1933) and Gerd Binnig invent the scanning tunneling microscope, which uses electron tunneling from a sharp point that is moved over a

Walter Robert Dornberger, German rocket engineer, d Jun 26

Sir Vincent Ferranti, British electrical engineer, d May 20

John William Mauchly, American engineer, d Jan 8

1980

1980
cont

ASTRONOMY & SPACE

the first discovered quasars, the 3C273, showing that the quasar may be in the center of a galaxy

The US scientific satellite Magsat (for magnetic field satellite) crashes to Earth in Jun, 1980, after completing its eight-month mission to survey Earth's magnetic fields

Voyager 1 flies by Saturn on Nov 12, and Voyager 2 flies by on Aug 27, 1981; both passages provide much information about the planet, its moons, and ring system; Voyager 1 finds the thirteenth and fourteenth moons of Saturn

1981

ANTHROPOLOGY/ARCHAEOLOGY

J. Desmond Clark and Timothy D. White find two fossil bones from a humanlike creature who lived 4 million years ago in the Awash River valley in Ethiopia

David Pilbeam determines that *Ramapithecus* is probably an ancestor of the orangutan, not the first and earliest hominid, ancestor to apes, australopithecines, and humans

ASTRONOMY & SPACE

Stephen Boughn and coworkers discover that the cosmic background radiation is not entirely isotropic; there are variations of 0.3 percent in directions 90 degrees apart

Joseph P. Cassinelli and others at the University of Wisconsin in Madison discover the most massive star known: R136a; it is 100 times brighter than the sun and has 2500 times the sun's mass

Harold J. Reitsema and coworkers observe that just before Neptune is to pass in front of a star, something gets in the way of the starlight; later, William B. Hubbard suggests that the cause is a discontinuous ring around the planet

Hyron Spinrad (b New York, NY, Feb 17, 1934) and John Stauffer discover two galaxies at 10 billion

BIOLOGY

Chinese scientists become the first to clone successfully a fish, a golden carp

Scientists from Ohio University in Athens are the first to transfer genes from one animal to another by transferring genes from other species into mice

The black-footed ferret, thought to be extinct, is discovered living in small numbers in prairie-dog towns in Wyoming

George Washington Corner, American embryologist, d Sep 28

Max Delbrück, German-American microbiologist, d Pasadena, CA, Mar 9

A.H. Harcourt and coworkers determine that statistical differences in average body weight and testes weight can be used to predict monogamy, polygyny,

CHEMISTRY

Aspartame, an artificial sweetener, is introduced in the United States

Kenichi Fukui of Japan and Roald Hoffmann of the United States win the Nobel Prize for Chemistry for the application of laws of quantum mechanics to chemical reactions

Eugene Glueckauf, German chemist, d Sep 15

Adam Heller, Barry Miller, and Ferdinand A. Thiel announce a liquid junction cell that converts 11.5 percent of solar energy into electricity

Harold Clayton Urey, American chemist, d La Jolla, CA, Jan 5

surface to detect the distance of the point from the surface; this method is sensitive enough to provide pictures of individual atoms in a surface

John Hasbrouck Van Vleck, American physicist, d Cambridge, MA, Oct 27

Klaus von Klitzing discovers the quantized Hall effect, in which changes in resistance in a plate kept in a magnetic field at temperatures near absolute zero occur in discrete steps instead of continuously; it is one of the few examples of quantum behavior that is directly observable

The US Centers for Disease Control recognizes acquired immune deficiency syndrome (AIDS) for the first time

The genetic code for the hepatitis B surface antigen is found, opening up the possibility of a bioengineered vaccine

The US Food and Drug Administration on Nov 16 approves a vaccine for hepatitis B made from human blood that had been developed by scientists at Merck Institute for Therapeutic Research in Pennsylvania

Adolpho de Bold, Harald Sonnenberg, and coworkers demonstrate that ground-up rat atria (heart chambers) reduce blood pressure in rats, suggesting that the rat atrium produces a hormone; the hormone is produced by bodies in the cells first observed by Bruno Kirsh in 1956

Nicolaas Bloembergen and Arthur L. Schawlow of the United States and Kai M. Siegbahn of Sweden win the Nobel Prize for Physics for their advances in technological applications of lasers for the study of matter

Harvey Fletcher, American physicist, d Jul 23

A.D. Linde (and independently Andreas Albrecht and Paul J. Steinhardt) develops a theory of the origins of the universe, called the new inflationary universe, that greatly improves on the ideas of Alan Guth put forward in 1980

Hideki Yukawa, Japanese physicist, d Kyoto, Sep 8

The world's longest suspension bridge opens over the Humber estuary in the United Kingdom; it is 1410 m (4626 ft) long

Solar One, the world's largest solar-power station, generating up to 10 megawatts of electricity, is completed

The IBM personal computer, using what is to become an industry-standard disk operating system (DOS), is introduced

1981

cont

light-years, the most distant galaxies known at the time

John Stocke announces his discovery of "narrow-line" quasars (quasars whose spectra consist of narrow emission lines)

The first flight of STS-1, the Space Transportation System's *Columbia*, is begun on Apr 12 with John W. Young and Robert L. Crippen as crew; the concept of a reusable space shuttle is proven to be practical

In the first reuse of a spacecraft, *Columbia* is launched on Nov 12; the mission ends early because of loss of a fuel cell

or promiscuity in primates; species with large testes for their body size are promiscuous

Sir Hans Adolf Krebs, German-British biochemist, d Oxford, UK, Nov 22

Charles G. Sibley and Jon E. Ahlquist use studies of DNA to determine the evolutionary history of such flightless birds as the ostrich and the emu, suggesting various revisions in the relationships between species

Roger W. Sperry, David H. Hubel, and Torsten N. Wiesel of the United States win the Nobel Prize for Physiology or Medicine for their studies on the organization and local functions of brain areas

---

1982

From the bones discovered at Herculaneum, the sister city of Pompeii, anthropologists are able to find out physical details about the Romans, who lived at Herculaneum in 79 AD

After 17 years of work, the *Mary Rose*, a sixteenth-century warship, is pulled out of the water in Portsmouth, England, and found to be full of Tudor artifacts

Soviet spacecraft Venera 13 and Venera 14 make successful soft landings on Venus

The third flight of *Columbia* is launched on Mar 22 and lasts eight days; its payload includes space science experiments

The fourth and final developmental mission of space shuttle *Columbia* is launched on Jun 27 and lasts seven days

The fifth flight of space shuttle *Columbia*, launched on Nov 11, is its first operational mission; the first deployment of a satellite from the shuttle is accomplished

A gene from one mammal functions for the first time in another mammal as the gene for rat growth hormone is transferred to mice; some of the mice grow to double normal size because of the additional hormones they produce

René Jules Dubos, French microbiologist, d Feb 20

François Jacob's *The possible and the actual* helps clarify the the content of evolutionary theory

William S. Mason and Jesse Summers show that the hepatitis B virus replicates with the aid of reverse transcriptase; this is surprising since the virus is not an RNA virus; later they show that this occurs

West German scientists announce on Aug 29 the creation of a single atom of element 109

William Francis Giauque, American chemist, d Mar 28

Aaron Klug of South Africa wins the Nobel Prize for Chemistry for developments in electron microscopy and study of acid-protein complexes

Forrest H. Nielson and Curtiss D. Hunt announce their discovery that boron plays an important role in the development of bone and sex hormones

Ruth and Victor Nussenzweig of New York University apply for a patent on a malaria vaccine

---

The volcano El Chichón in Mexico erupts in Apr, sending dust and gases high into the stratosphere, where they remain for about three years and reduce the amount of sunlight reaching Earth's surface; it is not established whether this produces any appreciable cooling

Ronald N. Bracewell introduces an algorithm that replaces the Fourier transform; it is a fast version of the Hartley technique and is known as the Hartley transform or Hartley-Bracewell algorithm

Mike Freedman develops methods that show when two manifolds in four dimensions can be continuously translated from one to another, providing a way of classifying some four-dimensional spaces

The US Food and Drug Administration grants approval to Eli Lilly & Company to market human insulin produced by bacteria, the first commercial product of genetic engineering

Swedish physicians attempt to cure Parkinson's disease with dopamine generated by the patients' own adrenal glands; however, implanting tissues from the gland in the brain produces temporary or no improvement

A team of doctors led by William DeVries implants the first Jarvik 7 artificial heart on Dec 2; the patient, Barney Clark, lives 112 days

Margulies Lazar, American physician, d Mar 7

Blas Cabrera announces the detection of a single magnetic monopole, a particle predicted by many new theories, but not previously observed; the same experimental design fails to detect a second event and other scientists cannot locate the monopole; the existence of the monopole is still in doubt in 1988

Merle Anthony Tuve, American physicist, d May 20

Kenneth G. Wilson of the United States wins the Nobel Prize for Physics for his theory of phase transitions

Compact-disk players are introduced

Compaq brings out the first "clone" of the IBM personal computer, a computer that uses the same operating system as the IBM PC and that has other elements in common so that most IBM programs can be used

Nikolay Petrovich Kamanin, Soviet electrical engineer, d Mar 14

Paul Kollsman, American aeronautical engineer, d Sep 26

Nikolay Alekseyevich Pilyugin, Soviet electrical engineer, d Aug 2

Vladimir Zworykin, Russian-born electrical engineer, d Jul 29

1982

cont

ASTRONOMY & SPACE

A team of seven astronomers discovers that our galaxy, the galaxies of the local group, and other components of the local supercluster of galaxies move toward the "Great Attractor," a point in the direction of the Southern Cross.

BIOLOGY

because one chomosome of two in the virus consists of a double helix with one strand of RNA and one of DNA

Stanford Moore, American biochemist, d Aug 23

Hugo Theorell, Swedish biochemist, d Aug 15

John R. Vane of the United States and Sune K. Bergstrom and Bengt I. Samuelsson of Sweden win the Nobel Prize for Physiology or Medicine for their studies on formation and function of prostaglandins, hormonelike substances that combat disease

---

1983

GENERAL

The International Bureau of Weights and Measures on Oct 20 redefines the meter to be the distance that light travels in 1/299,792,458 of a second

ANTHROPOLOGY/ARCHAEOLOGY

In Kenya, Dr. Maeve Leakey finds a fossil jaw tentatively dated between 16 and 18 million years ago, and identified as *Sivapithecus*

Many discoveries are made concerning the Roman baths in the ancient city of Bath, England, owing to the underground excavations of the inner precinct of the Temple of Sulis Minerva, completed in Jan

Scientists develop a method for dating ancient objects based on chemical changes in obsidian

ASTRONOMY & SPACE

IRAS, a satellite designed to detect infrared radiation from objects in space, is launched on Jan 25; in the course of its mission, it will discover evidence of planet formation around stars outside the solar system; it completes its mission in Nov

Bart J. Bok, Dutch astronomer, d Aug 5

The second US space shuttle, *Challenger,* is successfully launched on Apr 4; a TDRS tracking satellite is deployed and the crew carries out the first extravehicular activity (EVA)

The second *Challenger* flight launched on Jun 18, carries the first five-person crew and the first American woman in space, Sally K. Ride; the Remote Manipulator Structure is used to deploy and retrieve a satellite

BIOLOGY

The first field test of a bacteria with artificially altered genetic structure is stopped by a lawsuit; the test will be delayed for several years as a result

Albert Claude, Belgian biologist, d May 22

G.L. Collingridge shows that chemically blocking certain receptors in the brain, called NMDA receptors because they are detected by N-methyl D-aspartate, stops the long-term potentiation of nerve cells that is believed to be the mechanism for storing memories

Walther J. Gehring and coworkers discover the homeobox, a common sequence of genes in a wide variety of organisms that directs development of the organism; all of the organisms containing the homeobox—

CHEMISTRY

Chemists at Duke University produce an artificial gill from hemoglobin immobilized in plastic

Joel H. Hildebrand, American chemist, d Apr 30

Henry Taube of the United States wins the Nobel Prize for Chemistry for new discoveries in the basic mechanism of chemical reactions

Aspartame is approved for use as an artificial sweetener in soft drinks

Hans Hugo Selye, Austrian endocrinologist who pioneered studies on stress, d Oct 16

Wolf Szmumess, Polish-born epidemiologist, d Jun 6

Patrick Walsh devises a new operation for prostate cancer that is generally successful in avoiding impotence, a common side effect of previous surgical treatments for the disease

---

Francis W. Reichelderfer, American meteorologist, d Jan 26

Simon Donaldson discovers that some four-dimensional spaces are not smooth

Ivan Vinogradov, Soviet mathematician, d Mar 20

Scientists show that a protein produced by genetically engineered yeast can protect chimpanzees against hepatitis B

The immunosuppressant cyclosporine is approved by the US Food and Drug Administration, making transplants of organs much safer than they previously had been

William C. Boyd, American immunologist, d Feb 19

John Buster and Maria Bustillo of the Harbor-UCLA Medical Center in Torrance, CA perform the first successful human embryo transfers

Marc Cantin, Jacques Genest, and co-workers in Jun locate the hormone produced by the heart that regulates blood pressure; it is synthesized two months later by a group headed by Ruth F. Nutt

Felix Bloch, Swiss-American physicist, d Sep 10

William A. Fowler and Subrahmanyan Chandrasekhar of the United States win the Nobel Prize for Physics for their investigations into the aging and ultimate collapse of stars

A team headed by Carlo Rubbia at CERN discovers the charged W particles and the neutral Z particle, predicted carriers of the weak force according to the electroweak theory that unifies the weak force with electric charge; this, along with the earlier discovery of neutral currents, confirms the electroweak theory

A team of German and American scientists develops a "wet" solar cell that has an energy conversion efficiency of 9.5 percent

Apple's Lisa brings the mouse and pull-down menus to the personal computer; a mouse is a device that moves the cursor on the screen as a result of moving the mouse on a hard surface; pressing a button on the mouse sends a command to the computer, depending on where the cursor is located

IBM's PC-XT is the first personal computer with a hard-disk drive built into it; it is a memory device capable of storing 10 megabits of information

Robert Buckminster Fuller, American inventor, d Jul 1

1983 cont

ASTRONOMY & SPACE

The third *Challenger* flight, on Aug 30, carries the first black American into space, Guion Bluford, Jr.; a weather/communications satellite is launched for India

Space shuttle *Columbia* carries the European Skylab on its Nov 28 mission; the six-person crew performs numerous experiments

BIOLOGY

insects, annelid worms, vertebrates, and so forth—are segmented, suggesting that the homeobox directs segmentation

Barbara McClintock of the United States wins the Nobel Prize for Physiology or Medicine for her discovery of mobile genes in the chromosomes of corn

Andrew W. Murray and Jack W. Szostak, working with yeast, create the first artificial chromosome

R.B., a retired postal worker from California, d after five years of amnesia apparently caused by a loss of blood to his brain; an autopsy reveals that the cause of the amnesia is a small damaged region in the hippocampus, the part of the brain now believed to be involved in memory

Solomon Spiegelman, American microbiologist, d Jan 21

Ulf von Euler, Swedish biochemist, d Mar 10

1984

ANTHROPOLOGY/ ARCHAEOLOGY

During an expedition led by Andrew Hill, a fossilized jawbone, believed to be about 5 million years old, is found in Kenya; it appears to be from *Australopithecus afarensis*, the oldest known representative of the hominid line

Charles G. Sibley and Jon E. Ahlquist use studies of DNA to explore the relationships among humans and the great apes; they conclude that humans are more closely related to chimpanzees than either humans or chimpanzees are to other

ASTRONOMY & SPACE

J. C. Bhattachayya and his group at the Indian Institute of Astrohysics in Bangalore and Uttar Pradesh State Observatory at Naini Tal discover two more rings of Saturn during the occultation of a star by the ring system

BIOLOGY

Archie Carr (b Mobile, AL, 1910) discovers where some sea turtles live between the time they hatch and their reappearance as young adults; the young turtles make their way to beds of floating seaweed, where they live on other small animals that make their homes in the seaweed patches

Kenneth S. Cole, American biophysicist, d Apr 18

CHEMISTRY

West German scientists announce the creation of three atoms of element 108

Eric Block and Saleem Ahmad at the State University of New York at Albany synthesize ajoene, a compound found in garlic and believed to function as a blood thinner

R. Bruce Merrifield of the United States wins the Nobel Prize for Chemistry for his discovery of a method for creating peptides and proteins

1983

cont

Fernand Daffos is the first doctor to use fetal blood taken by a needle through the umbilical cord for diagnosis of disease in the fetus

Kay Davies and Robert Williamson find a gene marker that indicates Duchenne muscular dystrophy

James F. Gusella finds a gene marker for Huntington's disease

H. Keffer Hartline, American physician, d Mar 17

Barry J. Marshall and J. Robin Warren discover that a bacterium, *Campylobacter pylori*, causes gastritis and may cause ulcers of the stomach and small intestine

---

1984

Soviet researchers drill to a depth of 12 km (7.5 mi) and reach Earth's lower crust in the Kola Hole

Carbon dioxide from Lake Monoun in Cameroon or some other volcanic gas kills 37 people

The American Heart Association lists smoking as a risk factor for strokes for the first time

Workers at the New York State Department of Health develop a genetically modified vaccinia virus that protects animals against hepatitis B, herpes simplex, and influenza

The first clinical trials of a vaccine against hepatitis B produced by yeast that has been given genes for a surface molecule of the hepatitis virus start on Jun 1

The particle accelerator at Fermilab reaches 800 billion electron volts (GeV)

Pyotr L. Kapitsa, Soviet physicist, d Apr 8

Carlo Rubbia of Italy and Simon van der Meer of the Netherlands win the Nobel Prize for Physics for their discovery of the W and Z particles

Carlo Rubbia at CERN announces that his group has discovered evidence for the top quark, the heaviest of the

Optical disks for the storage of computer data are introduced

IBM introduces a megabit RAM memory chip with four times the memory of earlier chips

Apple brings Lisa technology down to an affordable price with its instantly popular Macintosh computer

IBM's PC AT is the first personal computer to use a new chip to expand speed and memory in an existing personal-computer architecture

1984

cont

great apes, and that humans and apes diverged about 5 to 6 million years ago

A storm in the Mediterranean uncovers the underwater site of an 8000-year-old settlement, now known as Altit-Yam, off the coastal town of Atlit, Israel

Richard E.W. Adams uncovers in a jungle in Guatemala the first unlooted Maya tomb found since the early 1960s; it had been sealed effectively in the fifth century AD

Michael G.L. Baillie and coworkers establish an unbroken tree-ring chronology based on Irish oak trees that extends from the present to 7272 years earlier

On Aug 1, peat cutter Andy Mould discovers Lindow Man, a 2200-year-old body preserved in a peat bog; later researchers believe that Lindow Man was a Druid who was killed in a ritual involving a burned barley cake called a bannock; the priest who got the burned piece of bannock was sacrificed to the Celtic gods

A project to produce a Sumerian dictionary, to be edited by Ake R. Sjoberg, is started; it will take many years and is expected to run about 16,000 entries in more than 22 volumes

Many previously unknown tombs turn up at the excavations by Sayed Tewfik and coworkers at the Necropolis at Sakkara in Egypt; these tombs belong to high officials of Ramses II, perhaps the most powerful pharaoh

Donald W. McCarthy and Frank J. Low discover a cool companion to the star called Van Biesbroeck 8, at a distance of 8 light-years from Earth; this discovery is later disputed

Bradford A. Smith of the University of Arizona and Richard J. Terrile of NASA photograph a planetary system forming around the star Beta Pictoris

On the fourth *Challenger* flight, launched on Feb 3, jet-propelled backpacks carry two astronauts on the first untethered spacewalks; two satellites are lost

The fifth *Challenger* flight is launched on Apr 7; the Long Duration Exposure Facility is deployed for experiments in space durability

Alec Jeffreys discovers the technique of genetic fingerprinting: the identification of certain core sequences of DNA unique to each person; this method can not only be used for identifying individuals but also for establishing family relationships

Steen A. Willadsen sucessfully clones sheep, producing genetically identical sheep by separating an embryo into separate cells and introducing the cell nucleus into sheep ova that have had their nuclei removed; the altered eggs are then implanted in female sheep for development into fetuses and birth

Allan Wilson and Russell Higuchi of the University of California at Berkeley become the first to clone genes from an extinct species; they clone genes from a preserved skin of a quagga, a form of zebra that had been extinct for a hundred years

The hormone produced by the heart, discovered by Adolpho de Bold in 1981, is isolated and analyzed by de Bold and several other teams; one result is that it is variously named auriculin, atriopeptin, and cardionatrin, leading to its simply being called the "heart hormone"

Surgeon William H. Clewall of the University of Colorado Health Sciences Center at Denver performs the first successful surgery on a fetus before birth

César Milstein of Britain, Georges J.F. Köhler of West Germany, and Niels K. Jerne of Denmark share the Nobel Prize for Physiology or Medicine, Milstein and Köhler for their research on monoclonal antibodies and Jerne for his studies of the immune system

quarks, and for monojets, particles predicted by supersymmetry theories; neither discovery is confirmed as of 1988

Dany Shechtman, Ilan Blech, Denis Gratias, and John W. Cahn, workers at the US National Bureau of Standards, find the first quasicrystal, a material that appears to be a crystal but which has a pattern of repetition that violates the rules for crystals; the alloy of aluminum and manganese shows fivefold symmetry

Strings: Reality in 10 or 11 dimensions?

When people first began thinking about quarks, a persistent question was why isolated quarks had not been observed. One idea was that quarks might be the ends of strings. If a particle was a string and the quarks were just the ends of the string, then one could understand why one never found one quark without the other.

A string is essentially a 1-dimensional object in a space of 4 dimensions (counting time as a dimension). Physicists turned to topology, the mathematics of knots and surfaces, to find what the implications might be of using strings instead of particles in their calculations. Surprisingly, strings simplified calculations.

At about this same time, physicists working with other mathematical theories of particles and forces rediscovered an idea from the 1920s. In 1919, Theodor Kaluza had found a way to derive electromagnetism from Einstein's theory of general relativity, but Kaluza's derivation requires 5 dimensions and we only observe 4. In 1926 Oskar Klein developed an explanation of why the fifth dimension is invisible—it curls up into a tiny circle—but very few people were impressed. In the late 1970s, however, Eugene Cremmer and Bernard Julia

found that another theory that had been discarded because it was so complicated became very simple if the universe has 11 dimensions with 7 of them curled up in the way Klein suggested.

The string theorists were not to be left out of this promising idea. Already they had incorporated another theory, called supersymmetry, into string theory, resulting in superstring theory. Now they tried putting their superstrings into spaces with more dimensions. The complicated mathematics of superstrings shows that they work best in Kaluza-Klein spaces of 9 or 10 dimensions, however, not 11.

Theoretical "particle" physicists at this point have worked themselves into this dilemma: In many ways it makes sense for the universe to be a Kaluza-Klein space of 11 dimensions, but the calculations are difficult and peculiar. Superstring theory makes the calculations easy and normal, but does not seem to describe the universe. Still, many physicists, and not just the ones working on superstring theory, think that it may finally be recognized as the ultimate reality.

1985

GENERAL	ANTHROPOLOGY/ ARCHAEOLOGY	ASTRONOMY & SPACE	BIOLOGY	CHEMISTRY
		Construction of the world's largest telescope, the Keck, begins on Mauna Kea in Hawaii; the Keck telescope's mirror will be 10 m (33 ft) in diameter, but it will not be cast as a single piece	A semipalmated sandpiper, banded and released four days earlier in Plymouth Beach, MA, is shot in Guyana, 4500 km (2800 mi) away; the flight sets a record for long distance over water by a shorebird	Information about lanxides, crosses between ceramics and metals, is released for the first time when the US Defense Department declassifies the three-year-old subject
		James R. Houck and his team at Cornell University report the discovery of eight infrared galaxies located by the infrared astronomical satellite IRAS	Charles G. Sibley and Jon E. Ahlquist use DNA studies to revise the evolutionary history of perching birds and songbirds from Australia and New Guinea	Herbert A. Hauptman and Jerome Karle of the United States win the Nobel Prize for Chemistry for their work on equations to determine the structure of molecules
		Mark Morris discovers at the center of the Milky Way a number of string-shaped radio sources; termed "threads," they are believed to be candidates for low-energy cosmic strings		
		Edward F. Tedesco detects the eclipses of Pluto by its moon Charon and determines the diameter of Pluto to be less than 3000 km (1900 mi)		
		Neil Turok suggests that groups of galaxies called Abell clusters have formed as a result of cosmic strings; long, thin remants of the original energy of the Big Bang that have persisted to the present		
		The first crew change in space occurs; a crew replaces another crew on board Soviet satellite *Salyut 7* in order to do repairs		

1986

GENERAL	ANTHROPOLOGY/ ARCHAEOLOGY	ASTRONOMY & SPACE	BIOLOGY	CHEMISTRY
	Tim White and Donald Johanson locate 302 pieces of a female *Homo habilis*, now known as OH62, that is 1.8 million years old; it is the first time that bones from the limbs of *H.*	The space shuttle *Challenger* blows apart 73 seconds after launch on Jan 28, killing six astronauts and teacher S. Christa McAuliffe	All the known black-footed ferrets in the wild are captured and put into a breeding program after canine distemper kills most of the known population	Americans Dudley R. Herschback (b San Jose, CA) and Yuan T. Lee, inventors of the crossed-beam molecular technique, and Canadian John C. Polanyi, inventor of chemiluminescence as

EARTH SCIENCE	MATHEMATICS	MEDICINE	PHYSICS	TECHNOLOGY	
The volcano Nevada del Ruiz in northern Colombia erupts, melting snow and ice on its summit; the resulting torrent of mud sweeps into the town of Amero, killing about 21,000 of the 22,500 residents (and about 4000 people at other locations near the volcano) Peter A. Rona discovers the first deep-sea ocean vents, or hot springs, on the Mid-Atlantic Ridge in the Atlantic Ocean; previous vents had been found in the Red Sea and the Pacific Ocean The British Antarctic survey detects a hole in the ozone layer over Antarctica, as ozone levels fall to their lowest in Sep; satellite records confirm that the hole had been forming for several years	Robert Gompf discovers that there is an infinity of nonsmooth four-dimensional spaces Hugh C. Williams and Harvey Dunbar determine that the number formed by writing 1031 ones in a row is prime	The US Food and Drug Administration approves the implantable defibrillator, a mechanical device used to prevent the heart from beating too rapidly or from fibrillating (quivering instead of beating) Scientists find a gene marker for polycystic kidney disease on chromosome 16 Scientists find a gene marker for cystic fibrosis on chromosome 7 The US Food and Drug Administration in Oct approves marketing growth hormone manufactured by bacteria, the second drug produced by genetic engineering (after human insulin) to be sold in the United States Lasers are used in the United States for the first time to clean out clogged arteries Michael S. Brown and Joseph L. Goldstein of the United States win the Nobel Prize for Physiology or Medicine for their studies in the regulation of cholesterol metabolism Bert L. Vallee and coworkers find the tumor angiogenesis factor first predicted by Judah Folkman in 1961; it stimulates the growth of new blood vessels and is renamed angiogenin	The Tevatron particle accelerator at Fermilab in Batavia, IL, begins operation A nuclear X-ray laser explodes underground, producing X-ray radiation that is 1 million times brighter than that obtained in earlier tests Klaus von Klitzing of West Germany wins the Nobel Prize for Physics for his discovery of the quantized Hall effect	AT&T Bell Laboratories achieves the equivalent of sending 300,000 simultaneous telephone conversations or 200 high-resolution television channels at once over a single optical fiber The "synfuels" concept in the United States (designed to develop alternative energy sources based on coal or oil shales) loses nearly all of its funding, reflecting a worldwide petroleum glut	**1985**
In Aug oceanographers discover a vast plume of hot water rising off the Juan de Fuca Ridge in the Pacific; the plume is thought to have been caused by a vent in the ocean's floor	Ramachandran Balasubramanian, Jean-Marc Deshouillers, and François Dress show that every natural number is the sum of at most 19 fourth powers, a conjecture first put forth	The US Food and Drug Administration approves OKT3, the first monoclonal antibody to be approved for therapeutic use in humans; it aids in organ transplants	In Oct teams of physicists from the University of Washington, the University of Hamburg, and the US National Bureau of Standards independently observe for the first time in-	Chernobyl nuclear reactor number 4, near Kiev, USSR, explodes at 1:23 AM local time on Apr 26, leading to a catastrophic release of radioactivity that kills dozens of people	**1986**

1986

cont

habilis have been found; OH62 was short, only about 1 meter (3 ft) tall and her body was more apelike than expected

A fisherman wading in the Acula River near Veracruz, Mexico, discovers a stone stele in the riverbed; when raised, it is found to be covered with a previously unknown form of writing, dating from the first century AD; the writing is similar to glyphs used by the Maya and other Central American groups

Harold L. Dibble invents a system based on the electronic theodolite, a surveying device that records the exact three-dimensional position of artifacts as they are found in a "dig;" combined with a computer, entire levels can be shown or graphed, saving archaeologists hours of drudgery

The US spaceprobe Voyager 2 passes close to Uranus and its moons at a speed of more than 51,000 km (32,000 mi) per hour, taking pictures and measurements that are radioed to Earth; 10 more satellites are discovered, as well as much basic information about the system

Mir (Peace), a more modern version of the *Salyut 7*, is launched by the Soviet Union, becoming the first permanently manned space station

The US Department of Agriculture grants the Biologics Corporation of Omaha, NE, the world's first license to market a living organism produced by genetic engineering, a virus used as a vaccine to prevent a herpes disease in swine

On May 29, Agrracetus, a biotechnology company in Wisconsin, conducts the first field trials of genetically engineered organisms, genetically altered tobacco

Arthur Ashkin and his colleagues trap individual living organisms using the radiation pressure of a laser, allowing new methods for observing and manipulating them

David Baltimore and coworkers report that installing foreign genes in a mouse causes the mouse's own genes to produce proteins closely related to those produced by the foreign genes; later evidence suggests that this study may be flawed, partly because it is difficult to replicate

Rita Levi-Montalcini of Italy and Stanley Cohen of the United States win the Nobel Prize for Physiology or Medicine for their studies of the mechanisms of cell and organ growth

Richard Morris, Fary Lynch, Michel Baudry, and coworkers show that by blocking NMDA receptors in the brain, they can interfere with learning in rats; NMDA receptors, when stimulated by an electrical charge

a way of studying chemical reactions and bonds, win the Nobel Prize for Chemistry

where two tectonic plates are moving apart

A miner in the La Toca amber mine in the Dominican Republic finds a complete frog, 35 to 40 million years old, preserved as a fossil in amber

Lake Nyos in the Cameroon emits a strange gas that kills about 1750 people and much of the livestock around the lake; later investigators decide that the gas was largely carbon dioxide released in an underwater eruption of the volcano that formed the lake or caused by turnover of the lake waters

by Edward Waring in 1770; Waring also thought that every natural number is the sum of at most nine cubes, an idea that is still unproven

The US Food and Drug Administration approves a hepatitis B vaccine made by yeast, the first vaccine to be approved for humans that is produced by genetic engineering

Louis Kunkel and coworkers discover the gene that is defective in Duchenne muscular dystrophy, a common, fatal form of the disease

dividual quantum jumps in individual atoms

Steven Chu, John Bjorkholm, Alex Cable, and Arthur Ashkin trap individual atoms using the radiation pressure of a laser

Ephraim Fishbach discovers a fifth fundamental force, the hypercharge, which acts counter to gravity and is detected as a repulsion force between test masses; this discovery is not accepted by many physicists

K. Alex Müller and J. Georg Bednorz discover an oxide combination that is superconducting at 30° Kelvin (30 degrees above absolute zero), the highest known temperature for superconductivity and a breakthrough that leads to other materials that are superconducting at much higher temperatures

Heinrich Rohrer of Switzerland and Gerd Binnig and Ernst Ruska of West Germany share the Nobel Prize for Physics, Ruska for designing the first electron microscope and Rohrer and Binnig for designing the scanning tunneling microscope

within a few weeks; it forces the mass evacuation of all families within 30 km (18.6 mi) for an indefinite period

Compaq leaps past IBM by introducing computers using an advanced 32-bit chip, the Intel 80386

A bicycle designed by Gardner Martin and powered by Fred Marckham on May 11 sets the human-powered land speed record of 105.37 km (65.48 mi) per hour

Dick Rutan and Jeana Yeager pilot the airplane *Voyager* around the world without refueling in a nine-day trip that starts on Dec 14

1986

cont

1986

cont

and the neuro-transmitter glutamate, cause calcium ions to enter nerve cells

A team from MIT led by Robert A. Weinberg announces on Oct 16 the discovery of a gene that can suppress the cancer retinoblastoma; it is the first gene known to inhibit growth

---

1987

The US Supreme Court rejects the equal-time concept for the teaching of creationism as a science

In May, workers digging an irrigation ditch near Wenatchee, WA, find six Clovis spearpoints; Peter Mehringer starts a study of the site, dating it at about 11,500 years old; it is the first undisturbed Clovis site known; most anthropologists believe the Clovis people were the first humans in the Americas

A team of scientists coordinated by Farouk el-Baz drills into a chamber at the base of the Great Pyramid at Giza, Egypt, to assess the air and the condition of a boat believed to be buried in the chamber; the hope for obtaining ancient Egyptian air is not realized but the boat is located

Susan Pevonak, a student volunteer working for Dale Croes at a Pacific Coast Indian site near Sekiu, WA, finds a microlith attached to a wooden handle; it is the first of the small stone blades, which are known from many sites around the world, ever found attached and it provides insights into how the tools were used

David Stuart, a student at Princeton University working

Astronomers observe an object, 3C326.1, at radio frequencies; it is believed to be a galaxy in the process of formation

Bruce Campbell, Gordon Walker, and Stephenson Yang announce the discovery of planet-size bodies orbiting Gamma Cephei and Epsilon Eridani, with possible planet-size bodies around five other nearby solar-type stars

Jules Halpern discovers a very hot star at Geminga, the location of an invisible gamma-ray source

C. Roger Lynds and Vahe Petrosian locate in Jan two enormous arcs of light, each larger than a single galaxy; by Nov they have established that one of the arcs, in Abell 370, is the image of a far-distant, unseen galaxy that has been distorted by a gravitational lens; the other is probably caused the same way

Francesco Paresce and Christopher Burrows discover a disk of particles around the star Beta Pictoris (at a distance of 53 light-years); it is believed to be a disk of proto-planets, objects that eventually will crash into each other to form planets

Chinese scientists insert genes for human growth hormone into goldfish and loach; this results in these fish growing up to four times as fast as normal

The last wild California condor is trapped and placed in a breeding program in a California zoo

The US Patent and Trademark Office extends patent protection to all animals

The world's first test of free-living genetically engineered bacteria is conducted, when scientists on Apr 24 spray strawberry fields with bacteria that have had an ice-forming gene deleted

Archie Carr, American marine biologist, d Wewa Pond near Micanopy, FL, May 21

Hans Fricke uses a submersible to observe coelacanths, the famous "living-fossil" fish, in their natural habitat in the deep Indian Ocean; instead of crawling on the ocean floor as some scientists expected, they swim slowly and frequently perform headstands or swim upside down

Herbert Naarmann and N. Theophilou develop a form of polyacetylene that is doped with iodine and that is in some ways a better conductor of electricity than copper

Charles J. Pedersen and Donald J. Cram of the United States and Jean-Marie Lehn of France win the Nobel Prize for Chemistry for their work in making complicated molecules that perform the same functions as natural proteins

1987

Kevin Aulenback discovers the second known cache of dinosaur eggs that contain fossilized unhatched dinosaurs, probably duck-billed dinosaurs, near Milk River in Alberta, Canada, on Jun 24

In an experiment headed by Thomas McEvilly, an array of motion detectors, seismometers, and acoustic sensors are lowered into the mile-deep Varian oil well near Parkfield, CA; the experiment, part of the Parkfield Prediction Experiment, is intended to monitor changes along the San Andreas fault

Wade Miller discovers a fossilized dinosaur egg that contains the oldest known embryo of any kind, probably the embryo of an allosaur from about 150 million years ago; the embryo is detected by X rays of the egg and is less than 2 cm (1 in.) long

James Ryan reports in May that the US National Aeronautics and Space Administration has definitely detected the motion of Earth's crustal plates using very long baseline interferometry and radio noise from distant quasars

Clifford Taubes discovers that the infinity of nonsmooth four-dimensional spaces is uncountable (an infinity is countable if each element in it can be matched to one of the counting numbers; for example, the rational numbers are countable, but the real numbers are uncountable)

Sir Walter Bodmer, Ellen Solomon, H.J.R. Bussey, and A.J. Jeffreys announce that they have found a marker for a gene that causes cancer of the colon

Murray B. Bornstein reports on a two-year study of the drug Cop 1, which he finds is successful in slowing the early stages of multiple sclerosis

Kevin P. Campbell and Roberto Coronado announce in Dec that they have located the protein used to regulate the passage of calcium into and out of muscle cells, the key step in muscle contraction or relaxation; the new protein is named the calcium release channel

A research team led by Louis M. Kunkel announces the discovery of the protein dystrophin; absence of this protein in some human males is suspected to cause Duchenne muscular dystrophy

Ignacio Navarro Madrazo announces that implanting cells from a person's adrenal gland in the brain can cure or alleviate Parkinson's disease; earlier experiments along the same lines had been

Georg Bednorz of West Germany and K. Alex Müller of Switzerland win the Nobel Prize for Physics for their discovery of superconductivity in a material at higher temperature than any previously known

A team led by Ching-Wu Chu becomes the first group to make a material that is superconducting at the temperature of liquid nitrogen, $-196°C$ ($-321°F$)

Dieter Kroekel, Naomi Halas, Giampiero Giuliani, and Daniel Grischkowsky of IBM produce a "dark-pulse" soliton, a standing wave that propagates without spreading through an optical fiber and that consists of a short interruption of a light pulse

Michael K. Moe, Alan A. Hahn, and Steve R. Elliot observe the double beta decay of selenium-82, predicted since 1935 but previously unobserved; they also establish that selenium-82 has the longest half-life ever measured—1.1×10^{20} years

Apple's Macintosh II and Macintosh SE become the most powerful personal computers available

The Numerical Aerodynamic Simulation Facility, an advanced supercomputer devoted to simulation and capable of a top speed of 1,720,000,000 computations a second, starts operations on Mar 9

IBM brings out the Personal System/2 group of personal computers, based on 3.5 in. disk drives, hard disks, enhanced graphics, and access to a new operating system that enables interconnections between computers

A lift-slab building under construction in Bridgeport, CT, collapses on Apr 23; the technique, in which slabs of flooring are poured at ground level and then raised, comes under suspicion as 28 workers are killed in the disaster

For the first time a crime suspect is convicted on the basis of genetic fingerprinting in the United Kingdom

1987

cont

for William L. Fash, discovers a new find at Copán, the most worked-over of Mayan cities; Stuart uncovers a cache that includes jade pieces, a flint knife thought to have been used in human sacrifice, and instruments for ritual bloodletting

Harvey Weiss announces the discovery of 1100 Akkadian clay tablets, the largest cache of cuneiform tablets found since 1933; the tablets, dating from about 1740 to 1725 BC, are from northern Mesopotamia at a place called Tell Leilan; the tablets had been excavated in Sep and Oct, 1986

Shortly past midnight on Feb 24, Ian Shelton observes the nearest observable supernova to Earth since 1604; the supernova in the Large Magellanic Cloud now known as Supernova 1987A was formerly the star Sk69.202; neutrinos from the explosion reached Earth on Feb 23

R. Brent Tully's paper announcing his discovery of the Pisces-Cetus Supercluster Complex, the largest structure discovered thus far in the universe, appears in the *Astrophysical Journal*

Benjamin Zuckerman and Eric E. Becklin report that they have found a brown dwarf in orbit around Giclas 29-38, about 50 light-years from Earth; the brown dwarf is manifested as excess infrared radiation coming from the vicinity of Giclas 29-30, a white dwarf

Yuri V. Romanenko, Soviet cosmonaut, returns to Earth from the *Mir* space station after 326 days in space, a new record

François Jacob's *La statue intérieure* (The statue within) describes his growth as a scientist and the discovery, with Jacques Monod, of the operon theory of gene regulation and messenger RNA

John Larson shows that NMDA receptors in the brain, believed to be part of the mechanism for storing memory, are most effective when stimulated at the same rate as the theta-rhythm component of brain waves; the theta rhythm is known to appear in rats when they are actively exploring their environment

Orange Band, the last dusky seaside sparrow, dies in captivity of old age, completing the extinction of the species; some of the species' genetic heritage has been preserved, however, by a program of crossbreeding the last five dusky seaside males with females of the closely related Scott's seaside sparrow

David C. Page and colleagues announce their discovery of the gene that initiates maleness in mammals; it appears to be a single gene on the Y chromosome that starts the sequence that leads to the development of testes instead of ovaries

Jack Stromminger, Don Wiley, and their coworkers determine the three-dimensional structure of the major histocompatibility protein, an essential part of the immune system; part of the structure appears to be a pocket that holds and binds an antigen

Scientists flying into the stratosphere over the Antarctic continent in Aug and Sep confirm that the ozone "hole" that forms at that time exists and is probably caused by chlorine compounds, most likely chlorofluorocarbons used in industry and as spray propellants

unsuccessful, but Madrazo changes the location of the implant

Michael Zasloff announces in Aug that he has discovered potent new forms of antibiotics, which he terms magainins, in the skin of the African clawed frog

Superconductors

One of the most important breakthroughs in experimental physics was the discovery by the Dutch physicist Heike Kamerlingh Onnes in 1911 that electrical resistance in mercury vanishes when it is cooled to temperatures close to absolute zero ($-273°C$, $-460°F$, or $0°K$). This phenomenon is known as superconductivity. Soon it was found that other metals and alloys also become superconducting at very low temperatures. For example, if one cools a ring of lead below $7.22°K$ and induces a current in it, this current will keep flowing indefinitely.

The temperature at which a metal becomes superconducting is called the critical temperature. The higher the critical temperature of a metal or alloy, the easier it can be used in technical applications. Superconductors are currently used in large and powerful magnets, mainly in particle accelerators and the magnetic-resonance imaging machines used in medicine. Because these magnets require cooling with liquid helium, they are complicated and expensive. There are also experimental superconducting electric motors. Superconductors repel magnetic lines of force, so some plans for magnetically levitated trains also call for superconducting substances.

It was only in 1957, more than 45 years after the discovery of superconductivity that a theory explaining the phenomenon was articulated. According to the BCS theory, named after the American physicists John Bardeen, Leon N. Cooper, and John Robert Schrieffer, the superconducting current is carried by electrons linked together through lattice vibrations (phonons) that cannot dissipate energy through scattering, the usual cause of electrical resistance in conductors.

The breakthrough of Kamerlingh Onnes was matched by the discovery in 1986 and 1987 of materials that become superconducting at much higher temperatures, a discovery that many scientists think may open the way to the eventual development of superconductors that work without any refrigeration. Two researchers at IBM, K. Alex Müller and Georg Bednorz, first reported the development of materials that become superconducting at substantially higher temperatures than liquid helium, which boils at $4.2°K$. The new materials are ceramics that become superconducting at temperatures between $90°$ and $120°$ K, above the boiling point of liquid nitrogen, which is inexpensive and easy to maintain.

The new materials are obtained by sintering techniques common to the production of ceramic materials. The different components are brought together in powdered form and heated. A solid-state reaction forms the final compound. The compounds are very brittle and are not easily formed into films or wires suitable for technical applications. A number of such compounds are now known. The most widely investigated ones consist of a mixture of yttrium, barium, copper, and oxygen.

By 1988 no satisfactory theory existed to explain the mechanism of superconductivity in ceramic superconductors, but theorists believe that pairs of charge carriers, either electrons or holes, are linked together by magnetic and electronic interactions.

1987 cont

Susumu Tonegawa of the United States wins the Nobel Prize for Physiology or Medicine for his studies of antibodies and the immune system

---

1988

Signatories to the 1959 Antarctic Treaty agree on rules for opening the continent to economic exploitation in the form of mining minerals or drilling for oil; all such activities are to be closely monitored to prevent adverse environmental impact

French and Israeli scientists announce that fossils found in a cave in Israel are the 92,000-year-old remains of modern-type *Homo sapiens,* more than doubling the length of time that modern humans are known to have existed

Andrea Caradini and coworkers discover in Jun a wall on the Palatine hill in Rome that dates from the seventh century BC, tending to confirm legendary accounts of the foundation of Rome

Tom D. Dillehay and Michael B. Collins report that charcoal dating of artifacts found at Monte Verde in southern Chile indicates that people have been present in the Americas for at least 33,000 years

Farouk el-Baz reports that the chamber examined by remote sensing in 1987 at the foot of the Great Pyramid contains hot, humid air that is a danger to the 4600-year-old boat that is in the chamber

Scientists in an airplane traveling at 12,500 m (41,000 ft) directly observe the atmosphere of Pluto; previously the atmosphere had been suspected from indirect evidence, but there was considerable controversy over its density

Simon J. Lilly reports that he has located a fully formed galaxy that is 12 billion light-years away, and therefore 12 billion years old, indicating galaxy formation at an early period in the history of the universe

James C. Bednarz announces that he has discovered cooperative hunting behavior in family units of Harris's hawks, a hawk species of the New Mexico desert

Henry A. Erlich and coworkers announce in Apr that they have developed a method that can identify a person—or at least rule out a given individual—from the DNA in a single hair; in this method, the DNA is amplified by causing the small portion actually located to form a polymer so that the portion repeats

Ya-Ming Hou and Paul Schimmel discover how to decode part of the genetic code found in transfer RNA, the part of the protein-building mechanism directed by DNA that does the actual assembly of amino acids into proteins

The US Patent and Trade Office issues patent No. 4,736,866 to Harvard Medical School for a mouse developed by Philip Leder and Timothy A. Stewart by genetic engineering; it is the first US patent issued for a vertebrate

Sewall Wright, American geneticist and evolutionary theorist, d Madison, WI, Mar 3

Nikolaas Tinbergen, Dutch-British ethologist, d Oxford, England, Dec 21

Chemists estimate that there are 10 million specific chemical compounds that are recorded; each year, 400,000 new compounds are described

Biochemist Charles Glen King, who first isolated vitamin C, d West Chester, PA, Jan 24

British scientists who have been monitoring wave height off of Land's End since 1962 report that the average wave height has increased from 2.3 m (7.4 ft) to 2.7 m (9.0 ft)

The US Senate ratifies an international treaty intended to reduce the use of chloro-fluorocarbons, which have been implicated in destroying the protective ozone layer in the atmosphere, making the United States the first nation to ratify the treaty

Silvio Micali and coworkers claim to have found a method for generating purely random numbers that is based on the difficulty of factoring large numbers that are the product of two large primes

Yoichi Miyoaka develops an outline of a proof for Fermat's last theorem (that for natural numbers x, y, z, and n, the equation $x^n + y^n = z^n$ has no solutions when n is greater than 2); later it is found that there are flaws in the proof

In July, Michael Atiyah discovers that the basic equation of modern topological knot theory—the Jones polynomial—is mathematically the same as quantum gauge theory, an essential part of the standard model of particle physics

Italian scientists report that follow-up studies of the people exposed to dioxins in the 1976 industrial accident near Seveso, Italy, show no increase in birth defects

The US Food and Drug Administration approves alpha interferon as a treatment for genital warts

William Castelli reports that the Framingham Heart Study has identified enlargement of the left ventricle as a major risk factor for strokes and heart attacks as well as for congestive heart failure; common causes of an enlarged left ventricle are untreated high blood pressure and overweight

Graham Colditz and coworkers announce that a study of 120,000 nurses reveals that women who smoke half a pack of cigarettes a day are twice as likely to have strokes as non-smokers, while women who smoke two packs a day are six times as likely

Elias J. Corey, Myung-choi Kang, Manoj C. Desai, Arun K. Ghosh, Ioannis N. Houpis, and Wei-guo Su announce the synthesis of ginkgolide B, the chemical thought to be the active ingredient in many herbal remedies based on ginkgo leaves; ginkgolide B fights asthmas and other

In Jan scientists from Japan's National Research Institute for Metals develop a new warm-temperature superconductor based on bismuth, bringing the number of types of warm-temperature superconductors to three

Scientists at the University of Arkansas discover a fourth type of warm-temperature superconductor; based on thallium, the new type quickly sets a high-temperature record for superconductivity of 125°K (−148°C or −234°F)

Richard Philips Feynman d Los Angeles, CA, Feb 15

Paul French and Roy Taylor announce a laser that produces X-ray pulses at a wavelength of 248 nanometers and that lasts only 65 femtoseconds, an important step toward the development of useful X-ray lasers

James Van House and Arthur Rich publish the first image from a positron transmission microscope, a microscope that uses positrons, antiparticles to the electron, instead of electrons to produce images

Roger Peoppel demonstrates the "Meissner motor," an electric motor based on high-temperature superconductors

John L. Gustafson, Gary R. Montry, Robert E. Benner, and coworkers find a way to rewrite problems for computer parallel processing that speeds their solution by a factor of 1000; previously an increase in speed by a factor of 100 was thought to be the limit of this method

The human-powered aircraft *Daedalus 88*, piloted by Kanellos Kanellopoulos, flies from Crete to the shore of Santorini, where it breaks up just offshore in heavy breezes; the flight of 3 hours 54 minutes covers 119 km (74 mi) and sets new distance and time records for human-powered flight

Francis C. Moon and Rishi Raj use a warm-temperature superconductor to build an almost frictionless high-speed bearing

Roland Winston supervises a test of a new mirror system that concentrates sunlight to 60,000 times its normal intensity on Earth; it is believed that the system will have applications in the development of new types of lasers and possibly in developing new materials

1988
cont

Missing mass

In the 1920s scientists discovered that the universe was expanding. Alexander Friedmann had proved that Einstein's general theory of relativity required that the universe should expand, and Edwin Hubble had observed that it actually did expand. Friedmann's equations, however, predicted three possible types of expansion. In two of them, the universe expands forever in slightly different ways. In the third, the expansion gradually slows, stops, and finally contracts. Such a universe is called "closed." The parameter that makes the difference between the three situations is the amount of mass in the universe. Too much mass and the universe collapses. Too little mass and the expansion becomes runaway, virtually blowing the universe apart. When, like the Baby Bear's porridge, the expansion is just right, the expansion continues forever, but gradually slows down, keeping the universe much as we know it today. This is called the "flat universe."

Astronomers since World War II have been working on the problem of determining the amount of mass. Most astronomers prefer a universe that collapses or is flat. But when the amount of mass is calculated from known observable galaxies and dust, it is far too small. If these calculations are correct, the universe will undergo runaway expansion.

There may be mass in the universe that we cannot observe. Indeed, we know that there is because Vera Rubin has shown that galaxies rotate too quickly near their edges. According to the laws of gravity and rotation, if the edge we see of a galaxy were the real edge, the rotational speed would be reduced the farther away a part of the galaxy is from the center. Since observations show that this is not the case, Rubin determined that there must be great clouds of unobserved mass in which each galaxy is embedded. She could even calculate how much unobserved mass there is.

But the newly found mass in galaxies still is not enough by a large margin to close or at least flatten the universe. The inflationary universe of Alan Guth, first conceived in 1979 and improved by other physicists since, made the task a little easier because their predicted universe needed slightly less mass to close or flatten it—but still a lot more than anyone could find.

Since then, various astronomers and physicists have postulated many entities that could provide the missing mass. Astronomers have contributed brown dwarfs (objects in size and energy between a planet and a star) and black holes. But physicists have provided most of the new ideas.

For example, the universe is presumed to be filled with nearly undetectable, massless neutrinos. Give each neutrino just a little mass and there will be plenty to close the universe. Measurements have been ambiguous. There are theories in which neutrinos must have some mass and there are theories in which that is impossible. No one knows for sure which are true.

If not neutrinos, then perhaps some undiscovered particle could provide the missing mass. Particles predicted by various theories include magnetic monopoles, photinos, axions, gravitinos, and winos. Most of these are grouped together as WIMPS—weakly interacting massive particles—which means that they are hard to detect if present and also hard to produce in particle accelerators (since the more massive a particle is, the more energy it takes to create it).

A hot idea from physics in 1988 is the cosmic string. A cosmic string is not a particle, but an incredibly long, skinny bit of leftover energy from the Big Bang. Even a few inches of a cosmic string has a mass similar to that of the Earth—and some of the cosmic strings are expected to reach from one end of the universe to another. One of the nice things about cosmic strings as a way to supply the missing mass is that they are in principle observable. There was even a possible sighting of one in 1988.

allergies by suppressing the immune system

Eric Courchesne and coworkers establish that autistic children develop brain abnormalities in the womb or shortly after birth, tending to confirm findings that autism is caused by a decrease in Purkinje cells, which normally relay inhibitory signals from one part of the brain to another

André F. Cournand, French-American researcher into chest diseases, d Great Barrington, MA, Feb 19

Rudolf Jaenisch and coworkers announce on Mar 10 that they have succeeded in implanting the gene for a hereditary disease of humans in mice; this is believed to open the way to the study of such diseases and to improved treatment

Sung-Hou Kim, Susumu Nishimura, and Eiko Ohtsuka determine the complete physical structure of the protein produced by the oncogene (cancer-causing gene) c-H-ras; it is believed that this may lead to clinical applications

Louis Kunkel, Eric P. Hoffman, and coworkers announce on May 25 that they have discovered that the protein dystrophin is completely or almost completely absent in cases of Duchenne muscular dystrophy, making it possible for the first time to diagnose the disease accurately in its early stages

Ernst August Friedrich Ruska, inventor of the electron microscope, d West Berlin, W Germany, May 30

OVERVIEW

The coming era

Astonishingly, much that will happen in science in the future is known today. One result of the rise of "big science" (*see* p 490) is that considerable planning and preliminary work are required. The lead time for "big" projects is measured in years or even decades.

Big vs. small science

Every one of the 1990s big-science projects—the Hubble space telescope, the U.S. space station, the Human Genome Project, the superconducting supercollider (SSC), and the Earth Observing System (Eos)—has provoked controversy. The main question is whether the high cost of big science erodes financial support for the rest of the scientific community. This controversy was heightened by the partial failure of the Hubble space telescope.

A major achievement of small science was the discovery in 1986 of high-temperature superconductors (*see* "Superconductors," p 603). Many physicists feel that if small science received more funding, similar important discoveries would be made. While particle physicists get giant, expensive new toys with which to explore the mysteries of the universe, condensed-matter physicists, whose work generally has more practical value to humankind, lack government support. Similarly, earth scientists who are exploring the interior of the earth wish for some of the $30 billion allotted to Eos to explore the atmosphere and oceans from space.

Ethical challenges face 1990s scientists

In the 1970s scientists began to learn how to manipulate genes successfully. By 1974 scientists were worrying publicly about the possibility of harm to humankind caused by the intentional or unintentional release of new organisms created by recombinant DNA techniques. By the 1980s the creation of new organisms led to lawsuits that tried to prevent the release of genetically modified bacteria. As the 1990s began, the first large-scale introduction of a genetically engineered hormone into agriculture and the first well-planned attempts to alter a human's genetic makeup to cure a hereditary disease started. Such experiments and introductions raised ethical questions that had not previously required serious answers.

Scientists also found the ethical structure of scientific culture under attack. Some scientists have committed fraud in an attempt to enhance their own reputations. This has proven to be a tough situation to handle, in part because it is often difficult to tell fraud from sloppy but honest work and also because science is supposed to be self-policing. Congressional hearings at the start of the 1990s could lead to outside regulation of science, which most scientists view with considerable alarm.

Ethical problems cropped up in many other areas. Research on human embryos and in vitro fertilization became the target of people opposed to abortion. The Human Genome Project came under attack because it could open the way to eugenic measures. Animal protection groups mounted strong attacks, sometimes using violence, against laboratories and scientists practicing animal experimentation.

Perhaps such ethical concerns are not special to our era. In the recent past, scientists grappled with whether they should work on weapons of mass destruction, or even on any project that might be used in war. The concerns of the 1990s may simply reflect a change in the forefront of science from nuclear physics and space to biology and medicine.

Major advances

Anthropology and archaeology. Fossil finds in 1989 pushed back estimates of the origin of anatomically modern human beings: some finds in the Near East were dated 100,000 to 145,000 years old. These modern-type humans were also living in Europe at least 40,000 years ago, about 5,000 years earlier than previously believed. One of the main challenges that faces anthropology in the 1990s is to clarify the relationship among "archaic" *Homo sapiens* from Africa and the Near East, the Neanderthals from

Europe and the Near East, and the undoubtedly modern humans of the past 40,000 years in Europe.

Important new techniques are also making scientific analysis more useful. Scientists routinely create present-day versions of stone tools, smashed or worn bones, and abandoned sites. By comparing these new versions with the fossil record, anthropologists have developed new insights into the lives of our ancestors and their cousins. Microscopic examination of ancient bones shows patterns that modern re-creations can help identify as scrape marks from stone tools, crunching by hyenas' jaws, or splitting by rocks. As a result of such studies, for example, it is now widely believed that *Homo habilis* was a scavenger instead of a hunter: patterns that appear to be caused by cutting meat from bones with stone tools are superimposed on patterns that seem to have been caused by the teeth of carnivores. If *H. habilis* had been the hunter, the superimposition should have been reversed.

New techniques of genetic analysis may become even more important. Studies of mitochondrial DNA from present-day humans have indicated an origin of all modern humans in Africa some 200,000 years ago, which some anthropologists find consistent with most recent fossil evidence, although many disagree. For example, one study of human teeth suggests an origin for *H. sapiens* in southeast Asia. Increasingly clever techniques are enabling researchers to clone and multiply tiny amounts of DNA from fossils. Perhaps the 1990s will see the first reconstruction of DNA from fossils of early humans, or even human ancestors!

In archaeology, new finds show that many sites are yet to be thoroughly excavated or investigated. Even the well-known ancient Mesopotamian city of Nippur revealed in 1990 a massive new temple, dedicated to the goddess of medicine.

The ability of scholars to read Mayan writing, or glyphs, continues to contribute to a new understanding of the rise and fall of the Mayan civilization. In 1959, Heinrich Berlin became the first to recognize that some Mayan glyphs were the names of ceremonial centers. The following year, Titiana Proskouriakoff learned to read the glyphs for the names of Mayan rulers. By the time the government of Honduras started a major new excavation project at Copán, in 1975, more than half the Mayan glyphs had been translated. As a result, by 1990 the history of Copán was beginning to become as clear and well-known as that of the ancient Egyptian dynasties. The 1990s will no doubt see similar detailed knowledge come to light for other Mayan centers, perhaps finally leading to a clear understanding of their strange and wonderful civilization and its precipitous decline in Central America.

Astronomy and space. Astronomy and space in 1989 and 1990 were marked by several successes, unexpected findings, and false starts.

The discovery of a form of aberration called spherical coma in the large lens of the Hubble space telescope was the biggest disappointment. Astronomers had expected that the instrument would answer many questions about the distance of galaxies, galaxy formation, quasars, and the size of the universe. Answers to these questions seem unlikely until after 1993, when the defect can be remedied by a shuttle mission. The aberration occurred because the large mirror had been built using an optical device that had a one-millimeter error.

An unqualified success came on August 24, 1989, when Voyager 2 completed the last planetary fly-by of its mission. As the space probe passed Neptune it recorded three complete rings, six previously undiscovered moons, a significant magnetic field tilted alarmingly with respect to the planet's axis, unusual topography at Triton (Neptune's largest satellite), and violent weather on the planet itself. As the two Voyagers headed out toward the edge of the solar system, Voyager 1 looked back on February 13, 1990, and took "a family portrait." Although the sun looked almost like a star and the six planets captured were only tiny dots, it was the first image ever of our home system from space.

Another 1990 success—almost too much of one—was the Cosmic Background Explorer (COBE), a satellite that examined background radiation thought to come from the Big Bang, the explosion that formed the expanding universe. COBE found that the background radiation was the same in all directions, as closely as it could be measured. This result makes it even more difficult for cosmologists to explain why the universe is "lumpy"—divided into regions of comparatively high density (galaxies) and regions of very low density (intergalactic space).

On the definitely unsuccessful side was the news from the Soviet Phobos mission to Mars. A mistaken computer command sent from the ground knocked out Phobos 1 long before it reached Mars, while a computer failure aboard ship caused Phobos 2 to lose touch after sending 17 minutes of garbled data from the vicinity of the planet.

A major task of astronomy in the 1990s will be to resolve the problem of how galaxies have gotten to be the way they now are. Martha P. Haynes and Riccardo Giovanelli think they have discovered a new galaxy forming nearby, which does not fit with the accepted theory that all the galaxies formed shortly after the Big Bang. Not everyone agrees. Some think it is just a galaxy that has a lot of extra gas, not one forming out of the gas. An ongoing project by John Huchra and Margaret Geller, five years old in 1990, is to map the distribution of galaxies in the universe. Although only 0.0000001 percent complete, the map has already revealed more detail than expected, notably a bubblelike structure common to the distribution of galaxies and a feature, referred to as the Great Wall, that is the largest known structure in the universe. The project is expected to continue through the 1990s and may well produce other surprises.

Another mystery whose solution is expected in the 1990s involves production of neutrinos by the sun. The number of neutrinos from the sun predicted by theorists is nearly four times the number detected in an 18-year study from 1968 to 1986 at the Homestake Gold Mine in Lead, South Dakota. The Japanese used a different technique starting in 1986 and got the same result as the Homestake experiment. Several new neutrino detectors around the world will begin to operate early in the 1990s, using a variety of different detection methods. A Soviet-American experiment called SAGE (Soviet-American Gallium Experiment) operating with liquid gallium under a mountain in the Caucasus should, in theory, be able to detect more neutrinos than the Homestake and Japanese experiments, but its first 60 days of operation in 1990 turned up none at all. Kurt Wolfsburg and coworkers are going to look for neutrino traces in a molybdenum mine in Colorado; another liquid-gallium experiment called GALLEX was supposed to start in 1990 in Italy; and later in the decade a heavy-water detector called SNO (Sudbury Neutrino Observatory) will start up in Sudbury, Ontario. Each of these experiments detects a slightly different range of neutrinos, which may help resolve the problem.

Unlike most other sciences, astronomy studies some events that can be reliably predicted to happen. Such predictions date back to the first scientist, Thales (see "The first known date," p 26). In the 1990s one predict-

able event will be on July 11, 1991—the best total eclipse of the sun of the decade. Try to be in Hawaii or Mexico to get a good look! Thirteen years later, you might want to catch one of the rare transits of Venus (see "The transit of Venus," pp 214–15).

Biology. One of the most remarkable success stories of the 1980s is continuing into (and probably through) the 1990s. Since 1963 a group of workers in many different laboratories have been following up on Sydney Brenner's visionary idea of learning everything there is to know about a small, not too complicated organism. Brenner's choice, *Caenorhabditis elegans*, is a millimeter-long nematode that is built from only 959 cells. The Worm Project, as the intensive study of C. *elegans* soon became known, managed to determine the developmental history of each of the 959 cells—how the original zygote split, how cells migrated, which cells died along the way, and so forth—in just 20 years. Three years later, in 1986, they had worked out the entire nervous system of the worm. By then, other workers on the Worm Project were already three years into the next phase, producing a map of the worm's genome, or genetic structure. This was 95 percent complete in 1990, with only minor gaps. Workers on the project all had specially developed sheets of filter paper that could be used to identify virtually any of the worm's genes—or any gene found in another organism that was also found in C. *elegans.* By 1990, the Worm Project had moved on to its next goal—mapping each of the 300-million base pairs in the genome. This map, which they hope to complete by the year 2000, is expected to help pave the way for the Human Genome Project (see "The Human Genome Project," p 614).

Although one thinks of the major work in the biology of the 1990s as taking place at the molecular level, there can be surprises at the whole-organism level as well. In 1990 whole colonies of apparently novel organisms were found living near hot vents at the bottom of Lake Baikal in the Soviet Union. Other new species no doubt also remain to be discovered, not only new insects and protozoans, but even new species of mammals—even primates. The end of the 1980s turned up two new lemurs on Madagascar and a new monkey in Brazil. Maybe the 1990s will finally see the discovery of one of those large, hairy, humanlike primates said to exist in various remote regions of the world—but maybe not.

Chemistry. Two electrochemists sparked a major scientific controversy on March 23, 1989, when they announced that they could perform nuclear fusion at room temperature ("cold fusion") with simple apparatus available to most laboratories. Physicists almost immediately proclaimed it impossible. A year or so later it appeared that the physicists were right and the chemists wrong, but it was still not clear what was happening with the chemists' apparatus. Claims and counterclaims, and even possible fraud, clouded the story from the beginning. One certainly hopes that the cold-fusion issue will be completely resolved by the end of the decade.

Chemistry often seems like the only mature science—the one in which the basic theory is completely worked out and future chemists need only fill in the details and apply the theory. Perhaps this concept ought to be viewed as a warning signal—after all, that description of a mature science was commonly believed to apply to physics late in the nineteenth century. Perhaps theoretical chemistry has some major surprises coming in the 1990s.

Earth science. Climate has emerged as a principal focus of earth science of the 1990s. Efforts are being made to explain the course of events that precipitate ice ages or the warming that follows an ice age. While a new method of dating based upon growth of corals, developed by Richard Fairbanks, has tended to confirm the Milankovitch theory that ice ages are related to the cycle of changes in the relationship between the earth and the sun, the exact sequence of events is still being worked out. In 1990, Edouard Bard reported that the coral dates move the peak of the last ice age back from 18,000 years ago to 21,500 years ago, which is much closer to the 23,000 years predicted by the Milankovitch theory. The climate problem has now taken on practical importance because of the need to determine how the release of greenhouse gases into the atmosphere is changing climates.

The controversy, which started in 1980, over whether Earth's collision with a large object caused mass extinction 65 million years ago seemed to be winding down by 1990. New evidence accumulated throughout the 1980s that some kind of collision had occurred, and by 1990, the location of the collision, in the Caribbean Sea, had been tentatively identified. The relationship between this collision and the extinction of the dinosaurs was still in question, however.

Meanwhile, the dinosaurs themselves continued to receive new interpretations, a process that will no doubt go on in the 1990s as well. New technology, such as CAT scans of fossil bones, and new finds in the field (and in museum back rooms) both fuel and help resolve the controversies over the life-styles of the dinosaurs and their relationship to birds, mammals, and reptiles. For example, the 1989 discovery of hollow spaces in some dinosaur bones suggests a closer relationship to birds, while other studies suggest that dinosaur bones may be those of warm-blooded creatures. Feeding behavior of the familiar *Tyrannosaurus rex* was reinterpreted when better fossils of the small front limbs showed that they were far from being the weak appendages commonly pictured. Their strength suggests that they were a major aid in holding prey and raising it to the many-toothed mouth, where it could be disposed of.

Another major fossil find of 1989 was the discovery in Egypt, by Philip D. Gingerich, B. Holly Smith, and Elwyn L. Simons, of a whale with hind legs. The whale, a serpentlike creature of some 50 million years ago, lived in the Tethys Sea. The legs were too small to be used in locomotion on land or in the water, but evidently had some other use—perhaps as an aid in reproduction. This whale and its ancestors had already been completely marine for 10 million years, so the discovery of the fossil leg bones was something of a surprise.

Mathematics. Mathematicians have long been interested in problems associated with factoring numbers (for example, 51 as 3×17). A breakthrough came in 1971 when, for the first time, a number 40 digits long was factored. For the next ten years or so it was generally believed that factoring a 50-digit number was practically impossible. It was this belief that caused Ronald L. Rivest, Adi Shamir, and Leonard M. Adleman to base their idea for an encryption system on factoring large numbers. By the mid-1980s it seemed necessary to use numbers of about 100 digits to keep the codes safe from the factorers. But in 1988, the team of Arjen K. Lenstra and Mark S. Manasse, with the help of the computers of a dozen collaborators, factored the first 100-digit number. After that, cryptographers began to use codes based on numbers of about 150 digits.

In 1989 John Pollard and Hendrik Lenstra, Jr., the brother of Arjen, developed a shortcut for factoring some types of numbers that were purposely constructed as being difficult. One of the numbers in this class was at the top of the current mathematicians' Ten Most Wanted List—a real list of hard-to-factor numbers compiled annually by a group of mathematicians. Arjen Lenstra and Manasse used the same organizational skills they had employed to factor the 100-digit number in 1988, but this time they used about 100 computers. By June 1990, they had solved the problem and factored the number, one with 155 digits. This record is almost certain to be broken in the 1990s. In the meantime, cryptographers are beginning to use 200-digit numbers, but even that may not be enough to outwit the clever mathematicians and their arrays of computers.

In a related development, scientists at the Amdahl Corporation Key Computer Laboratories in Fresno, California, showed that the number $(391{,}581 \times 2^{216{,}123}) - 1$ is a prime with 65,087 digits—the largest prime known. Over 2000 years ago, Euclid proved that there could be no largest prime, so this record will certainly be broken, and very likely in the 1990s.

Medicine. A number of successes in 1989 and 1990 that resulted from our increased ability to handle genes presage what most people expect to be many major advances in the 1990s. In 1989 the main gene that causes cystic fibrosis was located on the chromosome and cloned, which has already led to preliminary prenatal tests for the gene. In

1990 the discovery of the gene responsible for neurofibromatosis quickly led to the determination of its protein product, thought to be an important tumor suppressor. Genetic tests were developed in 1989 that indicate how susceptible a person might be to developing type I diabetes. On July 31, 1990, the Recombinant DNA Advisory Committee approved the first real gene therapies for humans: for adenosine deaminase deficiency, a genetic disease that destroys the immune system, and for a form of cancer.

The availability of human growth hormone produced by genetic engineering resulted in a possible application that goes beyond helping young people grow taller. A study in 1990 revealed that the same hormone can cause older people to regain lost muscle and enlarge atrophying organs. Although the full implications were not known when this overview was written, the study suggests that in many ways genetically engineered hormones or other biochemicals may become the wonder drugs of the 1990s, attacking a whole host of problems previously thought to be beyond the range of medical science.

Physics. In the 1990s various new particle accelerators will begin to operate (see "High-energy physics in the 1990s," p 613). Nevertheless, as was true in the 1980s, many of the fundamental advances, especially those that affect our lives directly, will be made in the branches of physics that constitute materials science. For example, compare the detection of the W and Z particles in 1983 with the discovery of high-temperature superconductors in 1986 in a huge particle generator at CERN in Geneva. While the W and Z discovery was a marvel of big science that confirmed an important theory, the superconductivity discovery was classical "small science"—although at a big laboratory—that revealed the need for new theoretical physics. Furthermore, high-temperature superconductivity may have practical applications in the relatively near future. However, several obstacles still impede the successful application of these materials in magnets and other electrical devices. One of the problems is the brittleness of these materials. They are not easily shaped into wires that can be wound to form magnets. But a more important problem is that high-temperature superconductors cannot carry large currents as conventional superconductors can. One of the reasons for this low current carrying capacity is the granular structure of the ceramic superconductors. The electric currents pass from one grain to another with difficulty. This problem has partially been solved by the creation of superconducting films, which do not have a granular structure and which may soon find applications in electronics.

Another, more serious problem came to light in late 1988. If a high-temperature superconductor is placed in a magnetic field, the field penetrates the superconductor and arranges itself in an ordered three-dimensional structure called the flux lattice. This flux lattice consists of intermeshing stringlike areas, called fluxoids, in which the magnetic field is concentrated. In 1988 several scientists reported the discovery that these fluxoids interfere with the superconducting charge carriers by being moved about by them, causing the superconductor to lose its zero resistance. Several groups have succeeded at diminishing the influence of the flux lattice by artificially creating crystal defects in the superconductor. These crystal defects stabilize the flux lattice so that its influence on the charge carriers becomes small. Several teams reported current carrying capacities that were comparable with conventional superconductors.

Technology. Transistors are the tiny electronic devices that provide the brains and the memory in computers. Around 1965 Gordon Moore and Robert Noyce, cofounders of the Intel Corporation and pioneers in integrated circuits ("chips"), predicted that the number of transistors that could be placed on a chip would double every year and a half. If anything, that prediction has turned out to be too conservative. By 1990, the industry-standard Intel chip contained approximately half a million transistors in about the same space that a single transistor occupied when semiconductor devices first became commercially available in the 1950s.

Intel has announced its vision of what they hope will be the industry-standard microprocessor chip of the year 2000. They expect to be producing a chip that contains 100 million transistors. By using an internal clock that would be more than six times as fast as the 1990 chips, and by programming the chip to process four different sets of instructions at a time, the new chip could handle as many as two billion instructions per second.

The technology behind computers of the 1990s could easily be quite different from the silicon-based technology predicted by Intel. A major development of the 1980s was the vast improvement in artificial diamonds over the first such diamond (produced in 1959). Especially important was a technique producing thin films of diamond. Diamond may become an essential part of future computer circuits because of its ability to transmit heat efficiently. The difficulty in ridding computers of excess heat, caused by movement of electrons, has previously been thought to be a barrier to further miniaturization of electronic devices.

But diamonds may also be important in other ways in future computer circuits, since the computers may be based upon light instead of electrons. Diamonds are extremely transparent to light as well as to heat. The basic advantages light has over electrons in speed and lack of interference suggest strongly that someday optical computers will outperform electronic ones. The first rudimentary optical computer was demonstrated in 1990 at Bell Labs, but it was far from being a working technology.

Physics is generally considered the most fundamental science, and the most fundamental branch in physics is high-energy physics, also known as particle physics. It is a field in which scientists try to find answers to such questions as what are the basic units of matter and what holds matter together? Lately particle physics has begun also to ask where matter comes from in the first place. To answer these questions, scientists have to probe deep inside the atom, even inside the particles that make up atoms. Scientists probe matter by observing how matter interacts with particles.

One way to probe atoms is by looking at what happens when photons, the particles of elecromagnetic energy, interact with matter. This was first done with photons of visible light, which have energies of a few electron volts (eV). This method, called spectroscopy, was originally developed as early as 1859, before anyone had any idea of why it worked. Niels Bohr in 1913 explained the relationship between electromagnetic radiation and electrons in atoms that causes spectroscopy to work. Building on Bohr's work, it could be observed that higher-energy particles would ineract with deeper layers of the atom. X-ray photons have higher energies than light, a few thousands of eV. This energy is sufficient to probe the inner structure of the atom. To probe the nucleus, scientists need particles with even higher energies, millions of electron volts. The proton was discovered in 1911 with naturally occurring particles from radioactive elements, but the energy of the particles was low. Starting in 1932, physicists began building machines that could give higher energies to particles.

Particles with high energies not only can probe the nuclei of atoms, but also can create particles that were not originally in existence. This occurs according to Einstein's equation of the equivalence of energy and matter, $E = mc^2$. A very large energy, E, is needed for even a small amount of matter, m, because of the large size of the speed of light, c, squared. To obtain convenient size units of measurement, physicists usually use the energy equivalent for a small particle instead of its actual mass.

Today physicists use huge machines called particle accelerators that accelerate particles to energies of trillions of electron volts. Some accelerators shoot particles at fixed targets. Physicists then observe with several types of detectors which particles are created in nuclear reactions. A more efficient method to transform the energy of accelerated particles into new particles is to use head-on collisions. This is especially effective when particles can collide with their own near twins, the antiparticles. Such a collision turns all of the mass in both particles to energy. The energy quickly forms into mass again, but frequently into particles that were not present originally. For example, in 1983, the huge proton-antiproton collider at CERN produced the W^+, W^-, and Z^0 particles. The existence of these particles had been predicted by the standard model, the most commonly accepted theory for the fundamental structure of matter, but the particles had not previously been observed, since they are formed only at high energies and very quickly break apart into smaller particles. Their creation at CERN tended to verify the truth of the standard model.

According to the standard model of physics that was developed in the 1970s, matter is made up of quarks and of electronlike particles called leptons. Other particles carry the four fundamental forces (*see* "Unifying the forces," p 555). Several questions pertaining to the standard theory are not yet answered, however. No quarks have yet been observed directly, although there are hints indicating their existence. The discovery of one of them would give the standard model a much stronger footing. Scientists believe that the top quark is the most likely candidate to be observed if collision energies in colliders would be increased.

In 1983 the High-Energy Physics Advisory Panel of the Department of Energy proposed the construction of a new huge collider, the superconducting supercollider, known as the SSC. As the idea evolved, the instrument would consist of two rings of pipe placed in a tunnel 53 miles across. In each pipe, beams of protons would be accelerated to energies of 20 TeV (20 billion billion eV) each. The protons would be kept in a circular path by very powerful, superconducting magnets. The protons would travel in opposite directions and meet at certain points head on, producing collisions with an energy of 40 TeV.

Scientists expect that during the collisions, some particles with masses of 2 to 4 TeV would be created. This would be sufficient for the production of the top quark and the Higgs boson, another particle also predicted by the standard model.

In 1988 the site for the construction of the SSC was announced: Waxahachie in Texas. According to plans in 1990, the collider would start operating in the late 1990s. The cost of the SSC was estimated at $8 billion.

Both the top quark and the Higgs boson are also theoretically within the reach of an existing device, the large electron-positron collider (LEP). The collider, built by CERN, started operation on July 14, 1989. It is the most powerful electron-positron collider built to date. In a ring in a 27.6-kilometer tunnel, electrons and their antiparticles, positrons, collide with an energy of about 110 MeV (110 million eV), creating about a million Z^0 particles each year. The first important experimental result of the LEP is the demonstration that elementary particles fall into three and only three families, as predicted by the standard model. The proof is that the Z^0 particles disintegrate into three kinds of neutrinos, each belonging to one kind of family. Hopes of physicists on the project are high that this instrument will discover the top quark as a decay product of Z^0 particles. Physicists backing the SSC prefer to think that this will not happen at the energies obtainable in the LEP.

In 1984 Robert Sinsheimer, Chancellor of the University of California at Santa Cruz, was inspired by contemplating the high cost of major telescopes to think of a truly expensive project in biology. His notion was to make a map of each of the approximately 3 billion bases that form the DNA of the 46 human chromosomes. Recall that DNA, the ultimate stuff of heredity, is essentially a coded message that uses four different chemical building blocks, called nucleotides or bases, to compose genes. The four bases, thymine, cytosine, adenine, and guanine, are conveniently represented by their initial letters. The ultimate goal that Sinsheimer had in mind would be the unraveling of the entire code of genetic information: a string of the letters T, G, C, and A that would fill 510 volumes of a standard-sized encyclopedia. The next year he arranged a workshop to present his idea.

Sinsheimer was not the only one looking for a big project. The U.S. Department of Energy (DOE) was supporting national laboratories that were losing their funding. Sinsheimer's idea seemed to be just what the DOE was looking for, even though it had no obvious relationship to energy. The tenuous link claimed by DOE was that they had previously funded studies of radiation damage to genes. DOE sponsored a second conference on the idea, which led to others. The Office of Technology Assessment, the National Research Council, and, somewhat belatedly, the National Institutes of Health (NIH) got into the act. Along the way the idea gained a name, the Human Genome Project—the genome is what biologists call all the genes of an individual organism —even though it had no formal existence. And, in 1988, it gained a leader when James D. Watson, who with Francis Crick in 1953 had first unraveled the double helix that is DNA, agreed to head the NIH version of the project. This effort started formally on January 3, 1989. The DOE continued to keep its hand in, but with Watson at the forefront, the NIH was finally establishing leadership.

The NIH and DOE sponsored a conference in the summer of 1989 to figure out how to get the job done. They developed a plan that would start in 1991, run for 15 years, and cost about a buck per nucleotide. In case you forgot, that's $3 billion. They also developed a strategy for doing the work, and put it out for public comment. They got plenty.

One basic idea was to begin by developing a map of the chromosomes, where the genes are located, by identifying a marker every 2 million bases on the average. This was to be an interim goal that would be accomplished by 1996. A marker is any short section of DNA that can be recognized by a related section whose sequence is known. Because of the way the two parts of the double helix join, each segment has one and only one complement. Given that complement, the segment can be recognized. Gene "banks" already exist that carry complements to known sequences of DNA.

The problem with this interim goal was that most specialists in the field thought it could not be accomplished in the way it was planned. So they convened a separate meeting in March 1990 to work out a new plan. The new plan, which was endorsed by the project, was based on spending two years finding "index" markers—common and useful sequences that come along about once every 10 to 15 million bases. An index map not only would be of immediate use in developing the more closely spaced map, but also would be helpful to anyone looking for a particular gene.

Internationally this was another success of big science, since the Europeans enthusiastically joined the project and the Japanese reluctantly fell in line as well. The project also has been marked by a high degree of cooperation among research groups throughout the genetics community. Work has been parceled out so that geneticists are cooperating, instead of competing in the way that Genome Project leader Watson described so vividly in his book about his own major discovery, *The Double Helix.* Each of the main groups has its own chromosome or set of chromosomes to study. Researchers also agreed on common languages and tools to make sharing data easier. Late in 1990, however, the project still had unresolved questions on exactly how data would be shared and also how the money would be allocated, so this spirit of cooperation may not last the entire project. Many think that the only way to continue the early and needed emphasis on sharing data will be to institute tough rules that all researchers have to follow. Another result of massive cooperation is that the peer-review system is falling apart; since so many scientists in the same field are working on the project, there are few available for reviewing the others' grants and papers.

The Genome Project has faced serious opposition from without the genetics community as well, some of it well organized. Some socially conscious scientists and nonscientists fear that the knowledge gained will be used to discriminate against people with "bad genes."

Among the many spin-offs expected of the Genome Project will be its usefulness in developing better "evolutionary clocks." Analysis in the genetic drift of sections of mitochondrial DNA or parts of the nuclear DNA have been used to determine the evolutionary history of species. In the case of human beings, the indications as of 1990 are that the human line split from the African great apes about five million years ago and that the ancestor of all modern humans lived in Africa 200,000 years ago. While the first of these conclusions is no longer considered controversial, the second is still the subject of intense debate. Knowing the details of the genome should allow the biochemists who specialize in this field to refine their work considerably, as well as to develop new results.

A major "big-science" project of the 1990s is expected to be the new American space station. So far the United States has lagged behind the Soviet Union in space station development. The Soviet Union put a primitive space station into earth orbit as early as April 19, 1971. In 1986, the Soviets started a new space station program with *Mir*, the first station that can be expanded by adding modules. The first U.S. space station, *Skylab*, was launched on May 25, 1973. The new U.S. station, *Freedom*, is planned as a joint effort of several nations—the United States, Canada, the European Community, and Japan. In 1990 *Freedom* was expected to be complete by 1998.

One unexpected problem with *Mir* and *Freedom* is maintenance. Critics of *Freedom* claim that so much time will have to be spent by astronauts maintaining the station that the whole plan will fail. Although much simpler than *Freedom* is expected to be, *Mir* has already developed a host of minor maintenance problems. The most dramatic so far came on July 18, 1990, when Anatoly Solovyov and Aleksandr Balandin, after a seven-hour spacewalk devoted to repairs on the spacecraft that brought them to *Mir*, found that they could not enter the module in which they had been living because they could not shut its airlock hatch. They were able to stay in another part of the station until July 26, when they finally managed to close the hatch and repressurize.

In addition to the space stations, there are a number of scientific space missions planned for the 1990s and beyond:

1990   The U.S. space probe Magellan began its mission to map Venus on August 10.

After visiting Venus in February for a gravity assist and a quick look, the U.S.–German space probe Galileo—launched October 18, 1989—was expected to return in December for the first of its two passes near Earth.

Late in the year the United States was expected to launch the space probe Ulysses toward the sun in a combined mission with the European Space Agency.

1991   In August, the United States plans to orbit the Extreme Ultraviolet Explorer (EUVE), an ultraviolet telescope.

Galileo—if its fuel supply looks good at this point—will fly by the asteroid Gaspara on October 29.

1992   The European Space Agency's Giotto space probe, somewhat worse for wear and quite blind from its encounter with Comet Halley in 1986, is going to try it again, this time with Comet Grigg-Skjellerup. If all goes well, the rendezvous will be July 10.

In September, the United States plans to launch Mars Observer, a space probe that is expected to map the red planet from orbit during 1993 and 1994. The mission is a cooperative effort with Germany, Austria, France, the Soviet Union, and the United Kingdom.

In December, Galileo once again swings past Earth for its last gravity assist before heading toward its final destination of Jupiter.

1993   In May the European Space Agency plans to orbit the Infrared Space Observatory (ISO).

Depending upon fuel supply, Galileo will fly by the asteroid Ida on August 28.

In August, Mars Observer is expected to enter its orbit about Mars.

1994   The Soviet Union plans to try again for Mars, just five years after computer failures wrecked its last attempt. The probe Columbus/Mars 1994 will also include a French experiment in which a balloon will be used to sample Martian soil at various points around the planet.

1995   In August, the United States plans to launch space probe Comet Rendezvous and Asteroid Flyby (CRAF).

On December 7, Galileo finally will arrive at Jupiter. Its eleven planned orbits are scheduled to photograph the moons and the ring system, study the magnetic field, and drop a probe into the Jovian atmosphere.

1996   In April, the United States and the European Space Agency plan to launch the Cassini/Huygens mission to Saturn. Cassini is the main probe that will orbit Saturn, starting in 2002, and Huygens is a mission into the Saturnian atmosphere.

The United States, the European Space Agency, and Japan plan to orbit the Earth Observing System (Eos), a massive effort to study Earth from four satellites. Because of its complexity, this is considered the major satellite program of the decade.

The Soviet Union's Mars program is supposed to take another step with the Mars Rover Mission, which will land a vehicle on Mars that will travel about the planet for a planned five years.

The Soviet Union and France plan to launch the space probe Vesta, which is scheduled to visit between ten and twenty asteroids and a couple of comets during a period of seven years.

1998   Space probe CRAF is expected to fly by asteroid Hamburga.

In December, the United States plans to launch the Mars Sample Return Mission, expected to do what its name implies. Since this is the only launch window for such a mission for a number of years, the Soviet Union plans to do the same thing. Current thinking in 1990 is that by 1998 the two missions will have been combined.

2000   CRAF will begin a three-year mission of "tailing" Comet Kopff August 14, getting close-up shots of how the comet forms a tail as it nears the sun.

2001   There is one planet that has not been visited at all —Pluto. Conditions will be ripe for launching a probe toward Pluto in 2001. Whether anyone will do so was not known in 1990.

NAME INDEX

Entries are located in two ways. Those in the Timetables are indicated by the year and the first two letters of the subject, such as 1988 AN (anthropology and archaeology). Those in overviews or essays are indicated by a page number, preceded by p or by pp. The same system is used in the subject index, which follows. In this index, a middle name in *italics* is the name by which a scientist is commonly known.

Heraclitus of Ephesus: 480–471 BC PH, p 32
Herbig, Georg H.: 1952 AS
Herman, Robert: 1948 PH
Hermite, Charles: 1873 MA, 1901 MA
Hero: 60–69 TE, 100–109 PH, p 25
Herodotus of Halicarnassus: 450–441 BC GE, 430–421 BC GE, pp 24, 87
Herod the Great: 10–1 BC TE
Herophilus: 300–291 BC LI, p 24
Héroult, Paul: 1886 TE, 1914 TE
Herring, W. Conyers: 1939 PH
Herschbach, Dudley R.: 1986 CH
Herschel, Caroline: 1797 AS, 1848 AS
Herschel, Sir John: 1820 AS, 1825 AS, 1834 AS, 1847 AS, 1864 AS, 1871 AS
Herschel, Sir William: 1773 AS, 1781 AS, 1782 AS, 1783 AS, 1785 AS, 1786 AS, 1789 AS, 1790 AS, 1795 AS, 1800 AS, 1800 PH, 1802 AS, 1811 AS, 1822 AS, pp 190, 290, 388
Hershey, Alfred Day: 1946 BI, 1969 BI
Hershey, John: 1973 AS
Hertz, Gustav: 1914 PH, 1925 PH
Hertz, Heinrich: 1888 PH, 1892 PH, 1894 PH, p 335
Hertzsprung, Ejnar: 1905 AS, 1908 AS, 1911 AS, 1913 AS, 1967 AS
Herzberg, Gerhard: 1929 PH, 1971 CH
Hess, Germain Henri: 1840 CH, 1876 CH
Hess, Harry Hammond: 1960 EA, 1962 EA, 1969 EA
Hess, Victor Francis: 1912 AS, 1936 PH, 1964 PH
Hess, Walter Rudolf: 1881 ME, 1949 ME, 1973 ME
Hevelius, Johannes: 1641 AS, 1647 AS, 1687 AS
Hevesy, György: 1913 CH, 1923 CH, 1934 BI, 1943 CH, 1966 CH
Hewish, Anthony: 1960 AS, 1974 PH
Hewitt, Peter C.: 1901 TE
Hey, M. H.: 1942 AS, 1946 AS
Heydenberg, N.: 1936 PH
Heymans, Corneille: 1938 BI
Heyrovský, Jaroslav: 1918 CH, 1959 CH, 1967 CH
Hickman, Henry: 1824 ME
Higgs, Peter: 1964 PH
Higuchi, Russell: 1984 BI
Hilbert, David: 1897 MA, 1899 MA, 1900 MA, 1904 MA, 1943 MA, pp 383, 462
Hill, Andrew: 1984 AN
Hill, Archibald: 1913 BI, 1922 BI, 1977 ME
Hill, William: 1911 ME
Hillary, Sir Edmund: 1953 GE
Hillier, James: 1937 PH
Hiltner, William Albert: 1949 AS
Hinshelwood, Sir Cyril: 1928 CH, 1956 CH, 1967 CH
Hinton, Christopher: 1956 TE
Hippacos of Metapontum: 450–441 BC MA
Hippalus: 100–91 BC GE
Hipparchus: 160–141 BC AS, 140–131 BC MA, 130–121 BC AS, pp 22, 46, 190
Hippasus: 420–411 BC MA
Hippias of Elis: 470–461 BC MA, p 23
Hippocrates of Chios: p 23
Hippocrates of Cos: 460–451 BC LI, 400–391 BC LI, 370–361 BC LI, p 23
Hirayama, Shin: 1951 AS
Hire, J. N. de la: 1716 TE
Hisinger, Wilhelm: 1803 EA, 1852 EA
Hitler, Adolf: 1933 GE
Hittorf, Johann Wilhelm: 1853 PH, 1914 CH
Hitzig, Julius Eduard: 1870 BI
Hoagland, Mahlon: 1955 BI
Hobbes, Thomas: 1651 GE, 1679 GE
Hodgkin, Alan: 1952 BI, 1963 BI
Hodgkin, Dorothy: 1949 CH, 1956 CH, 1964 CH

Hodgkin, Thomas: 1832 ME
Hoe, Richard March: 1847 TE, 1886 TE
Hoffman, Eric: 1905 ME
Hoffmann, Friedrich: 1718 ME, 1742 PH
Hoffmann, Roald: 1981 CH
Hofmann, Albert: 1938 CH, 1943 CH
Hofmann, August Wilhelm von: 1818 CH, 1858 CH, 1892 CH, p 276
Hofmeister, Wilhelm Friedrich: 1847 BI, 1851 BI, 1877 BI
Hofstadter, Robert: 1961 PH
Hökfelt, Thomas: 1977 BI
Hokin, Lowell E. and Mabel N.: 1953 BI
Hollerith, Herman: 1890 TE, 1896 TE, p 279
Holley, Robert: 1961 BI, 1965 BI, 1968 BI
Holmes, Arthur: 1890 EA, 1965 EA
Holmes, Oliver Wendell: 1843 ME, 1894 ME
Holwarda, Phocyclides: 1638 AS
Homberg, Wilhelm: 1702 CH
Honda, Kotaro: 1916 CH, 1954 TE
Honnecourt, Villard de: p 78
Honold, G: 1902 TE
Hooke, Robert: 1658 TE, 1664 AS, 1665 PH, 1674 PH, 1676 PH, 1676 TE, 1679 PH, 1703 PH, 1705 PH, pp 164, 192, 306
Hopkins, Sir Frederick: 1900 BI, 1906 BI, 1921 ME, 1929 ME, 1947 BI
Hoppe-Seyler, Ernst Felix: 1862 BI, 1871 BI, 1875 BI, 1895 BI
Hopper, Grace Murray: 1959 TE
Horner, W. G.: 1819 MA
Horner, William E.: 1829 ME
Horrocks, Jeremiah: 1639 AS, 1641 AS, p 214
Horsfall, Frank: 1961 ME
Horsley, Victor: 1883 ME
Horton, J. W.: 1928 TE
Hou, Ya-Ming: 1988 BI
Houck, James R.: 1985 AS
Hounsfield, Godfrey: 1973 ME, 1979 ME
House, James Van: 1988 PH
Houssay, Bernardo: 1887 ME, 1947 BI, 1951 ME, 1971 ME
Houtgast, Jaap: 1940 AS
Howe, Elias: 1846 TE, 1867 TE
Hoyle, Sir Fred: 1948 AS
Hsi Han: 300–309 LI
Hubble, Edwin Powell: 1924 AS, 1926 AS, 1928 AS, 1929 AS, 1953 AS, p 382, 440
Hubel, David H.: 1981 BI
Huchra, John: p 610
Hufeland, C. W.: 1796 ME
Huggins, Charles Branton: 1941 ME, 1966 ME
Huggins, Sir William: 1863 AS, 1864 AS, 1868 AS, 1876 AS, 1879 AS, 1910 AS
Hughes, John: 1975 BI
Hull, Albert W.: 1921 TE
Humason, Milton: 1953 AS, 1972 AS
Humboldt, Friedrich *Alexander von*: 1799 GE, 1799 EA, 1827 AS, 1829 EA, 1851 AS, 1859 BI, p 269
Humboldt, Wilhelm von: 1799 AS
Hume, David: 1739 GE, 1776 GE
Hunt, Curtiss D.: 1981 ME
Hunter, John: 1771 ME, 1778 ME, 1793 ME
Hunter, William: 1774 ME
Huntsman, Benjamin: 1742 TE, 1751 TE
Hutchinson, G. E.: 1957 BI, 1959 BI
Hutchinson, Miller: 1902 TE
Hutton, James: 1785 EA, 1797 EA, pp 193, 234, 404
Huxley, Andrew: 1952 BI, 1963 BI
Huxley, Julian: 1932 BI
Huxley, Thomas Henry: 1860 BI, 1863 GE, 1895 BI
Huygens, Christiaan: 1656 AS, 1656 TE, 1657 MA, 1659 AS, 1659 TE, 1664 PH, 1666 GE, 1673 PH, 1673 TE, 1678 PH, 1695 PH, pp 131, 148, 164
Hyatt, John Wesley: 1869 TE, 1920 TE, p 276
Hypatia: 370–379 MA, 410–419 MA, p 22

Ibn ash-Shatir, 1370–1379 AS
Ibn Battuta: 1320–1329 GE
Ibn Yunus: 1000–1009 AS
Ichbiah, Jean: 1979 TE
Imhotep: 3000–2901 BC LI, p 2
Ingenhousz, Jan: 1779 BI, 1799 ME
Innes, Robert: 1915 AS, 1933 AS
Inoue, T.: 1942 PH, 1947 PH
Ipatieff, Vladimir: 1900 CH, 1952 CH
Irwin, James B.: 1971 AS
Isaacs, Alick: 1957 ME, 1967 ME
Isadore of Miletus: 530–539 TE
Isidore of Seville: 630–639 GE
Ivanov, Ilya Ivanovich: 1901 BI, 1932 BI
Ivanovsky, Dmitri: 1892 ME, 1920 BI
I-Xing: 720–729 TE

Jablochkoff, Paul: 1876 TE
Jackson, Charles: 1841 ME, 1844 ME, 1880 CH
Jackson, William: 1874 AN
Jacob, Alan: 1974 BI
Jacob, François: 1960 BI, 1965 BI, 1970 BI, 1982 BI, 1987 BI
Jacobi, Carl Gustav: 1804 MA, 1825 MA, 1851 MA
Jacobs, P. A.: 1959 ME
Jacobson, Martin: 1965 BI
Jacquard, Joseph-Marie: 1801 TE, 1805 TE, 1834 TE
Jaenisch, Rudolf: 1988 ME
Jagger, Thomas A.: 1912 EA
James, Hubert: 1933 CH
James, William: 1890 ME, 1907 GE, 1910 ME
Janowsky, David: 1972 ME
Jansky, Karl: 1931 AS, 1950 AS
Janssen, Pierre: 1868 AS, 1907 AS
Janssen, Zacherias: 1590 PH, 1609 TE
Jarchus, Solomon: 1150–1159 GE
Jeanne D'Arc: 1420–1429 GE, 1430–1439 GE
Jeans, Sir James Hopwood: 1904 PH, 1906 PH, 1908 PH, 1917 AS, 1946 AS
Jefferson, Thomas: 1743 GE, 1826 GE, p 260
Jeffreys, Alec: 1984 BI, 1987 ME
Jeffreys, Sir Harold: 1940 EA
Jenner, Edward: 1796 ME, 1797 ME, 1823 ME
Jennison, Roger: 1953 AS, 1975 AS
Jensen, Arthur R.: 1969 ME
Jensen, Johannes Hans D.: 1948 PH, 1963 PH
Jerne, Niels K.: 1984 ME
Jia Xien: 1100–1109 MA
Johannsen, Wilhelm: 1909 ME, 1917 BI, 1927 BI
Johanson, Don: 1974 AN, 1986 AN
John XXII, Pope: 1310–1319 PH
Johnston, Eric: 1964 BI
Joliot-Curie, Frédéric: 1934 PH, 1935 CH, 1939 PH, 1958 PH
Joliot-Curie, Irène: 1931 PH, 1934 PH, 1935 CH, 1939 PH, 1956 PH
Jones, F. Wood: 1919 AN
Jones, Sir Harold Spencer: 1931 AS, 1960 AS
Jones, William: 1706 MA
Jones, W. A.: 1965 BI
Jordan, Camille: 1870 MA
Jordan, F. W.: 1919 TE
Jordanus Nemorarius: 1220–1229 MA, PH
Jorgenson, Christine: 1952 ME
Josephson, Brian: 1973 PH
Jouffroy, Marquis de: 1781 TE
Joule, James Prescott: 1843 PH, 1846 PH, 1847 PH, 1852 PH, 1889 PH, p 320
Joy, Alfred Harrison: 1945 AS
Julia, Bernard: p 595
Jung, Carl Gustav: 1917 GE, 1961 ME
Junghans, S.: 1943 TE
Jussieu, Adrien de: 1797 BI
Jussieu, Antoine Laurent de: 1774 BI, 1789 BI, 1836 BI
Justinian: 530–539 TE, p 58

Prigogine, Ilya: 1977 CH
Prince, Derek J.: 1959 AN
Pringsheim, Nathanael: 1823 BI, 1894 BI
Pritchard, Charles, 1885 AS, 1803 AS
Proclus: 580–571 BC MA, 340–331 BC MA, 460–
469 MA, 480–489 MA, p 272
Proctor, Richard: 1873 AS, 1888 AS
Prokhorov, Aleksandr: 1964 PH
Proskouriakoff, Titiana: p 609
Protagoras: 480–471 BC PH, 420–411 BC GE
Proust, Joseph-Louis: 1799 CH, 1808 CH, 1826
CH, p 338
Prout, William: 1815 CH
Przhevalsky, Nikolay: 1839 GE, 1888 GE
Pseudo-Democritus: 100 CH
Ptashne, Mark: 1968 BI
Ptolemy: 140–149 AS, 150–159 PH, 170–179
AS, 820–829 AS, pp 22, 24, 46
Pullman, George M.: 1872 TE
Punnet, R. C.: 1911 BI, 1915 BI
Pupin, Michael I.: 1896 ME, 1935 PH
Purcell, Edward: 1951 AS, 1952 PH
Purkinje, Jan Evangelista: 1835 BI, 1839 ME,
1839 BI, 1869 ME
Pylarini, Giacomo: 1701 ME
Pythagoras: 580–571 BC MA, 530–521 BC MA,
510–501 BC GE, 500–491 BC MA, pp 20,
22, 24, 36
Pytheas: 330–321 BC PH

Quetelet, Lambert Adolphe Jacques: 1831 MA,
1835 BI, 1841 MA, 1874 AS
Quick, Armand: 1932 ME
Qutb al-Din al Shirazi: 1280–1289 AS

R.B.: 1983 BI
Rabe, Eugen Karl: 1951 AS
Rabi, Isidor Isaac: 1937 PH, 1944 PH
Racine, Jean-Baptiste: 1699 GE
Rado, Tibor: 1962 MA
Rahn, Johann Heinrich: 1659 MA
Rainwater, Leo James: 1949 PH, 1951 PH, 1975
PH
Raleigh, Sir Walter, 1586 GE
Raman, Sir Chandrasekhara Venkata: 1928 PH,
1930 PH, 1970 PH
Ramanujan, Srinivasa: 1913 MA, 1920 MA
Rammazzini, Bernardino: 1700 ME, p 149
Ramón y Cajal, Santiago: 1893 BI, 1904 BI, 1906
BI, 1934 BI
Ramsay, Sir William: 1892 CH, 1895 CH, 1898
CH, 1903 PH, 1904 CH, 1916 CH
Ramsden, Jesse: 1768 PH
Ramses II: 1250–1201 BC GE
Ramus, Peter (Pierre de la Ramée): 1543 PH,
1572 GE
Rankine, William John Macquorn: 1853 PH,
1859 TE, 1872 TE
Raoult, François-Marie: 1886 CH, 1901 CH
Raup, David M.: p 559
Rawlinson, Sir Henry Creswicke: 1802 GE, 1837
AN, 1895 AN, p 272
Ray, John: 1667 BI, 1682 BI, 1686 BI, 1691 BI,
1693 BI, 1705 BI, p 275
Rayleigh, John William Strutt, Lord: 1882 PH,
1885 EA, 1892 CH, 1904 PH, 1919 PH,
p 380
Realdo, Colombo: 1559 LI
Réaumur, René Antoine Ferchault de: 1720 TE,
1722 TE, 1730 PH, 1734 BI, 1752 BI, 1757
PH
Reber, Grote: 1937 AS, 1942 AS
Recorde, Robert: 1542 MA, 1551 MA, 1557
MA
Redfield, William: 1847 GE, 1857 EA
Redi, Francesco: 1668 BI, 1671 BI, 1697 ME
Reed, Ezekiel: 1786 TE
Reed, Walter: 1900 ME, 1901 ME, 1902 ME
Rees, Martin: 1966 AS
Regan, Johann: 1913 BI

Regiomantanus (Johannes Müller): 1436 AS,
1464 MA, 1471 AS, 1472 AS, 1474 AS,
1476 AS
Regius, Hudalrichus: 1536 MA
Regnault, Henri-Victor: 1852 CH, 1878 CH
Reich, Ferdinand: 1824 EA, 1863 CH, 1882 EA
Reichelderfer, Francis W.: 1895 EA, 1983 EA
Reichenbach, Karl: 1833 CH
Reichstein, Tadeusz: 1933 BI, 1936 BI
Reid, Harry Fielding: 1859 EA, 1944 EA
Reines, Frederick: 1955 PH, 1965 AS
Reinhold, Erasmus: 1551 AS, 1553 MA
Reitsema, Harold J.: 1981 AS
Remak, Robert: 1845 ME, 1865 ME
Remsen, Ira: 1846 CH, 1927 CH
Renault, Louis: 1902 TE
Renfew, Colin: 1967 AN
Rennie, John: 1761 TE, 1821 TE
Retzius, Anders Adolf: 1842 BI, 1860 ME
Réveillon, Jean-Baptiste: p 232
Rhazes (Abu-Bakr Muhammad ibn Zakariyya Ar-
Razi): 840–849 LI, 880–909 PH, 930–939
LI
Rheticus, Georg Joachim: 1540 AS, 1576 MA,
1596 MA
Rhind, A. Henry: 1858 MA
Rhine, Joseph Banks: 1980 GE
Riccati, Jacopo Francesco: 1676 MA, 1754 MA
Ricci, Michelangelo: 1619 MA, 1682 MA
Riccioli, Giovanni Battista: 1651 AS, 1671 AS
Richards, Dickinson: 1956 ME
Richards, Theodore William: 1913 CH, 1914
CH, 1928 CH
Richardson, Lewis Fry: 1922 EA
Richardson, Sir Owen Willans: 1902 PH, 1911
PH, 1914 PH, 1928 PH, 1933 PH, 1959 PH
Richer, Jean: 1672 PH, 1679 PH, 1696 AS
Richet, Charles Robert: 1913 ME, 1935 ME
Richmann, George Wilhelm: 1753 PH
Richmond, Mark: 1964 BI
Richter, Burton: 1974 PH, 1976 PH
Richter, Charles Francis: 1935 EA
Richter, Hieronymus Theodor: 1824 EA, 1898
EA
Richter, Jeremias: 1791 CH, 1807 CH
Ricketts, Howard Taylor: 1910 ME (two entries)
Rickover, Hyman George: 1900 TE
Riemann, Georg Friedrich Bernhard: 1851 MA,
1854 MA (two entries), 1861 MA, 1866
MA, p 273
Righi, Augusto: 1920 PH
Ringer, Sydney: 1883 ME, 1910 ME
Rinio, Benedetto: 1410–1419 LI
Riquet, Pierre-Paul: 1666 TE
Ris, Hans: 1965 BI
Ritter, Johann Wilhelm: 1801 PH, 1810 PH,
p 388
Rivest, Ronald L.: 1977 MA, p 611
Robbins, Frederick C.: 1954 BI
Robert of Chester: 1140–1149 MA, p 60
Roberts, Michael: 1958 TE
Roberts, Richard Brooke: 1939 PH, 1980 BI
Roberts, Richard J.: 1977 BI
Roberval, Giles Personne de: 1634 MA, 1675
MA
Robinet, Jean-Baptiste: 1761 BI
Robins, Benjamin: 1742 MA
Robinson, Sir Robert: 1925 CH, 1947 CH, 1975
CH
Robison, John: 1769 PH
Roche, Edouard-Albert: 1849 AS, 1883 AS
Rochester, George: 1946 PH
Rock, John: 1956 ME
Roebling, John Augustus: 1883 TE
Roebuck, John: 1746 TE, 1762 TE, 1794 TE
Roentgen, Wilhelm Konrad: 1895 PH, 1901 PH,
1923 PH, p 388
Roger II: 1140–1149 LI
Rogers, Alan: 1975 AS
Roggeveen, Jakob: 1722 GE

Rohrer, Heinrich: 1980 PH, 1986 PH
Rokitansky, Karl: 1804 ME, 1878 ME
Rolle, Michel: 1691 MA
Romanenko, Yuri V.: 1987 AS
Romanes, George John: 1883 BI
Rome de Lisle, Jean B. L.: 1772 PH, 1783 EA
Römer, Ole: 1675 PH, 1700 AS, 1704 AS, 1706
AS, 1710 AS, p 160
Rona, Peter A.: 1985 EA
Rooke, Lawrence: p 146
Roosevelt, Franklin Delano: 1941 GE
Roozeboom, Hendrik Willem Bakhuis: 1854 CH,
1907 CH
Rorschach, Hermann: 1921 ME, 1922 ME
Rose, William Cumming: 1935 BI, 1937 BI
Ross, Sir James Clark: 1831 EA, 1862 GE
Ross, Sir Ronald: 1897 ME, 1902 ME, 1904 ME,
1932 ME
Ross, T.: 1938 TE
Rossby, Carl-Gustaf Arvid: 1932 EA
Rosse, Earl of (William Parsons): 1845 AS, 1855
AS, 1868 AS, 1867 AS
Rossi, Bruno Benedetto: 1930 AS, 1931 AS,
1962 AS
Roth, Thomas F.: 1964 BI
Rouelle, Hilaire-Marin: 1773 BI, 1779 BI
Rous, Francis Peyton: 1910 ME, 1966 ME, 1970
ME
Rousseau, Jean-Jacques: 1761 GE
Roux, Pierre: 1853 BI, 1933 BI
Rowland, F. Sherwood: 1974 EA
Rowland, Henry Augustus: 1882 PH, 1886 AS,
1897 AS, 1901 PH
Royds, T. D.: 1908 PH
Rubbia, Carlo: 1983 PH, 1984 PH (two entries)
Rubel, W.: 1904 TE
Rubin, Vera: 1975 AS, p 606
Rubner, Max: 1884 ME, 1894 BI, 1932 ME
Rubruck, Willhelm von: 1250–1259 GE
Rudbeck, Olof: 1652 LI, 1702 BI
Rudolff, Christoff: 1525 MA, 1545 MA
Ruffini, Paolo: 1799 MA, 1824 MA, p 296
Ruhmkörff, Heinrich Daniel: 1855 PH, 1877 PH
Rumford, Count (Benjamin Thompson): 1798
PH, 1814 PH, p 193
Rümker, Carl Ludwig Christian: 1822 AS
Runge, Friedlieb Ferdinand: 1834 CH
Rush, Benjamin: 1786 ME, 1812 ME, 1813 ME
Ruska, Ernst August Friedrich: 1931 PH, 1933
PH, 1986 PH, 1988 PH
Russell, A.S.: 1913 PH
Russell, Bertrand Arthur William, Earl: 1902
MA, 1903 MA, 1910 MA, 1919 MA, 1970
MA, p 462
Russell, Henry Norris: 1905 AS, 1913 AS, 1928
AS, 1935 AS, 1957 AS
Russell, Sidney: 1912 TE
Rutherford, Daniel: 1722 CH, 1794 PH, 1819
CH
Rutherford, Ernest (Baron Rutherford of Nelson):
1899 PH (two entries), 1901 PH, 1902 PH,
1903 PH (three entries), 1904 PH, 1906 PH,
(two entries), 1908 CH, 1908 PH, 1909 PH,
1911 PH, 1914 PH, 1917 PH, 1919 PH,
1937 PH, pp 380, 385, 389, 393
Rutherford, William: 1886 BI
Rutherfurd, Lewis Morris: 1816 AS, 1892 AS
Ružička, Leopold Stephan: 1935 BI, 1939 CH,
1976 CH
Ryan, James: 1987 EA
Rydberg, Johannes Robert: 1854 PH, 1919 PH
Ryle, Sir Martin: 1946 AS, 1955 AS, 1957 AS,
1960 AS, 1974 PH

Sabatier, Paul: 1897 CH, 1912 CH, 1941 CH
Sabin, Albert: 1952 ME, 1957 ME
Sabine, Sir Edward: 1852 EA, 1883 PH
Sabine, Wallace: 1868 PH, 1919 PH
Saccheri, Girolamo: 1733 MA (two entries) pp
272–273

Sachs, Julius von: 1865 BI, 1897 BI
Sacrobosco (John of Halifax): 1250–1259 MA
Sagan, Carl: 1963 BI
Saha, Meghnad: 1921 AS, 1956 AS
Sainte-Claire Deville, Henri: 1855 TE, 1881 CH
Sakata, Shoichi: 1942 PH, 1947 PH
Sakel, Manfred J.: 1929 ME
Sakharov, Andrey: 1968 PH
Salam, Abdus: 1950 PH, 1967 PH, 1973 PH,
    1979 PH, p 555
Salk, Jonas: 1952 ME
Salomen, A.: 1913 ME
Samuelsson, Bengt: 1982 BI
Sanctorius, Sanctorius: 1603 LI, 1612 PH, 1614
    LI, 1636 LI
Sandage, Allan: 1960 AS, 1963 AS, 1965 AS,
    1966 AS, 1974 AS
Sanger, Frederick: 1953 ME, 1958 CH, 1980 CH
Sanger, Margaret: 1917 ME, 1966 ME
Sappho: 600–591 BC GE
Sargon of Akkad: 2400–2301 BC TE
Sauria, Charles: 1830 TE
Saussure, Horace Bénédict de: 1766 PH, 1779
    EA, 1783 PH, 1799 PH, p 192
Saussure, Nicholas de: 1804 BI
Sautuola, Maria: 1879 AN
Sauveur, Joseph: 1701 PH
Savage, Thomas: 1847 BI
Savary, Félix: 1827 AS
Savery, Thomas: 1698 TE, 1702 TE, 1715 TE,
    p 192
Scaliger, Julius: 1557 PH
Scaliger, Joseph Justus: 1583 TE, 1609 GE
Scarpa, Antonio: 1772 ME, 1832 ME
Schaeberle, John Martin: 1896 AS, 1924 AS
Schaefer, Vincent: 1946 EA
Schafer, J. P.: 1931 AS
Schallenberger, Oliver: 1888 TE
Schaller, George: 1967 BI
Schally, Andrew V., 1977 BI
Schaudinn, Fritz: 1905 ME, 1906 BI
Schawlow, Arthur L.: 1981 PH
Scheele, Karl Wilhelm: 1765 CH, 1770 CH,
    1772 CH, 1774 CH, 1776 CH, 1777 CH,
    1780 CH, 1781 CH, 1782 CH, 1783 CH,
    1784 CH, 1786 CH
Scheiner, Christoph: 1611 AS, 1650 AS, p 134
Scheuchzer, Johann Jakob: 1725 BI, 1733 BI
Schiaparelli, Giovanni: 1866 AS, 1877 AS, 1910
    AS
Schick, Bela: 1913 ME
Schick, Joseph: 1928 TE
Schickardt, Wilhelm: 1623 MA
Schimmel, Paul: 1988 BI
Schimper, Wilhelm Philipp: 1874 EA
Schirra, Walter: 1962 AS, 1965 AS, 1968 AS
Schlack P.: 1938 TE
Schleiden, Matthias Jakob: 1838 BI, 1842 BI,
    1881 BI, pp 306–307
Schliemann, Heinrich: 1873 AN, 1878 AN,
    1890 AN, p 272
Schmidt, Bernhard: 1930 AS, 1935 AS
Schmidt, J. F. J.: 1878 AS
Schmidt, Maarten: 1963 AS, p 541
Schnirelman, L. G.: 1930 MA
Schoeffer, Peter: 1425 TE, 1502 TE
Schoenheimer, Rudolf: 1935 BI, 1941 BI
Schönbein, Christian: 1840 CH, 1845 CH, 1868
    CH, p 276
Schöner, Johannes: 1477 TE, 1547 TE
Schönlein, Johann: 1829 ME
Schooten, Frans van: 1649 MA, 1660 MA
Schrieffer, John R.: 1957 PH, 1972 PH, p 603
Schrödinger, Erwin: 1926 PH, 1961 PH, p 381
Schröter, Johann: 1800 AS
Schultes, Johann: 1655 LI
Schultze, Max: 1825 ME, 1874 ME
Schumacher, Heinrich: 1822 AS
Schurer, Christoph: 1540 TE
Schuster, Arthur: 1890 PH

Schwabe, Heinrich: 1826 AS, 1843 AS, 1875 AS
Schwann, Theodor: 1836 BI, 1838 BI, 1839 BI,
    1882 BI, pp 306–307
Schwartz, David: 1895 TE
Schwartz, Joel: 1988 ME
Schwarzschild, Karl: 1906 AS, 1917 AS
Schwarzschild, Martin: 1952 AS
Schweickart, Russell: 1969 AS
Schweigger, Johann: 1820 PH, 1857 PH
Schwinger, Julian: 1948 PH, 1957 PH, 1965 PH
Scopes, John T.: 1925 BI
Scott, David: 1969 AS, 1971 AS
Scott, Robert: 1868 GE, 1912 GE
Seaborg, Glenn T.: 1941 CH, 1944 CH, 1951
    CH, 1952 CH, p 505
Sebokht, Bishop Severus: 660–669 MA
Secchi, Pietro Angelo: 1867 AS, 1868 AS, 1878
    AS
Sedgwick, Adam: 1835 EA, 1839 EA, 1873 EA,
    1879 EA
See, Thomas: 1892 AS
Seebeck, Thomas: 1821 PH, 1831 PH
Sefström, Nils: 1831 CH, 1845 CH
Segrè, Emilio: 1937 CH, 1955 PH, 1959 PH
Seleucus, 190–181 BC AS
Selye, Hans: 1907 BI, 1982 ME
Semenov, Nikolai: 1956 CH
Semiramis, Queen: 2200–2101 BC TE
Semmelweiss, Ignaz: 1847 ME, 1865 ME
Seneca, Lucius Annaeus: 10–1 BC GE, 50–59
    GE, 60–69 GE
Senefelder, Aloys: 1798 TE
Sennacherib: 600–591 BC TE
Sepkoski, J. John, Jr.: p 559
Sertürner, Friedrich: 1805 BI, 1841 CH
Servetus, Michael: 1553 LI
Severin, Christian (Longomontanus): 1562 AS,
    1647 AS
Shakespeare, William: 1600 GE, 1616 GE
Shamir, Adi: 1977 MA, p 611
Shanks, William: 1853 MA, 1882 MA, p 360
Shannon, Claude: 1938 TE, 1949 MA
Shapiro, Irvin: 1971 AS
Shapley, Harlow: 1914 AS, 1918 AS, 1972 AS
Sharp, Abraham: 1717 MA, p 360
Sharp, Phillip: 1977 BI
Shaw, George: 1799 BI
Shaw, Louis: 1927 ME
Shechtman, Dany: 1984 PH
Shelton, Ian: 1987 AS
Shemin, David: 1911 BI
Shen Kua: 1070–1089 PH, TE
Shepard, Alan: 1961 AS, 1971 AS
Sherman, Henry: 1875 BI, 1955 BI
Sherrington, Sir Charles: 1906 ME, 1932 BI,
    1952 ME
Shih Lu: 220–211 BC TE
Shih Shen: 300–391 BC AS
Shih Tsung: 950–959 TE
Shklovskii, I. S.: 1953 AS
Shockley, William: 1948 TE, 1956 PH, p 494
Short, James: 1740 AS
Shovel, Sir Cloudesley: 1707 GE
Shrapnel, Henry: 1784 TE, 1842 TE
Sibley, Charles G.; 1981 BI, 1984 AN, 1985 BI
Siebold, Karl Theodor Ernst von: 1846 BI, 1885
    BI, p 307
Siegbahn, Karl: 1924 PH, 1978 PH, 1981 PH
Siemens, Friedrich: 1904 TE
Siemens, Sir William: 1844 TE, 1856 TE, 1883
    TE
Siemens, W.: 1903 TE
Siemens, Ernst Werner von: 1842 TE, 1892 TE
Sikorski, Igor: 1913 TE, 1939 TE
Silliman, Benjamin: 1818 GE, 1864 CH
Simon de Monfort: 1260–1269 GE
Simon, Sir Franz: 1893 PH, 1956 PH
Simond, Paul-Louis: 1898 ME
Simons, Elwyn L.: 1965 AN, 1967 AN, p 611
Simpson, Sir James: 1846 ME, 1870 ME

Singer, Isaac Merrit: 1851 TE, 1875 TE
Sinsheimer, Robert: p 614
Sitter, Willem de: 1917 AS, 1934 AS
Sjoberg, Ake R.: 1984 AN
Skinner, B. F.: 1938 ME
Skobeltzyn, Dmitri: 1929 PH
Skolnick, Mark H.: 1978 ME
Slipher, Vesto: 1912 AS, 1914 AS, 1920 AS,
    1925 AS, 1969 AS, p 440
Sloane, Sir Hans: 1676 GE
Small, William: p 216
Smeaton, John: 1759 TE, 1765 TE, 1771 TE,
    1792 TE
Smellie, William: 1752 ME
Smith, B. Holly: p 611
Smith, Edwin: 1862 AN, p 2
Smith, Hamilton: 1970 BI, 1971 BI, 1978 BI
Smith, James, L. B.: 1953 BI
Smith, Philip: 1884 ME, 1970 ME
Smith, Theobald: 1892 ME, 1934 ME
Smith, William: 1794 EA, 1815 EA, 1816 EA,
    1839 EA
Smithson, James: 1846 GE
Snell, George Davis: 1967 ME, 1980 ME
Snell, Willebrord: 1617 MA, 1621 PH, 1626 MA
Snow, John: 1854 ME, 1858 ME
Socrates: 470–461 BC GE, 400–391 BC GE, p 20
Soddy, Frederick: 1901 PH, 1902 PH, 1903 PH,
    1911 PH, 1912 PH, 1913 PH (two entries),
    1921 CH, 1956 CH
Solomon, Ellen: 1987 BI
Solon of Athens: 600–591 BC GE
Solovay, Robert: 1974 MA
Solovyov, Anatoly: p 615
Solvay, Ernest: 1861 CH, 1922 CH
Sommeiller, Germain: 1870 TE
Somerville, Mary Fairfax: 1830 AS, 1832 MA
Sommerfeld, Arnold: 1916 PH, 1928 PH, 1951
    PH, p 381
Sonnenberg, Harald: 1981 ME
Sophocles: 560–541 BC GE
Sorby, Henry: 1847 BI
Sorensen, Soren: 1909 CH, 1939 CH
Sosigenes: 90–81 BC AS, 50–41 BC TE
Soubeiran, Eugène: 1831 CH
Soufflot, Jacques: 1780 TE
Southern, Edward M.: p 567
Southgate, D. A. T.: 1978 ME
Spallanzani, Lazzaro: 1765 BI, 1767 BI, 1768 BI,
    1773 ME, 1779 BI, 1780 BI, 1785 BI, 1799
    BI
Spedding, Frank: 1942 CH
Speiser, D.: 1961 PH
Spemann, Hans: 1924 BI, 1935 BI, 1941 BI
Spence, Peter: 1845 CH
Spencer, Herbert: 1879 GE, 1884 GE, 1903 GE
Sperry, Elmer: 1911 TE, 1930 TE
Sperry, Roger: 1981 BI
Spiegelman, Solomon: 1965 BI, 1966 BI, 1983 BI
Spies, Tom: 1938 ME
Spinrad, Hyron: 1981 AS
Spittler, Adolf: 1897 TE
Spitzer, Lyman: 1947 AS
Sporn, Michael: 1966 ME
Sprague, Frank: 1884 TE
Sprengel, Christian: 1793 BI, 1816 BI
Squier, E. George: 1846 AN, 1850 AN, 1877
    AN, p 272
Ssuma Ch'ien: 90–81 BC TE
Stafford, Thomas: 1965 AS, 1966 AS
Stahl, Georg Ernst: 1697 CH, 1707 ME, 1723
    CH, 1734 CH
Stalin, Joseph: 1953 GE
Stanley, Wendell M.: 1935 BI, 1936 BI, 1946
    CH, 1971 BI
Stanley, William: 1885 TE
Stark, Johannes: 1913 PH, 1919 PH, 1957 PH
Starling, Ernest: 1902 BI, 1905 BI, 1912 ME,
    1927 ME
Stas, Jean Servais: 1865 CH, 1891 CH

Staudinger, Hermann: 1926 CH, 1953 CH, 1965 CH

Staudt, Karl Christian von: 1847 MA

Stauffer, John: 1981 AS

Stearn, W. H.: 1903 CH

Steenbock, Harry: 1924 BI

Stefan, Josef: 1879 PH, 1893 PH

Stein, William: 1972 CH, 1980 BI

Steiner, Jakob: 1796 MA

Steinhardt, Paul J:, 1981 PH

Steinitz, Ernst: 1910 MA

Steinmetz, Charles: 1865 TE, 1910 TE, 1923 TE

Steno, Nicolaus: 1669 GE, 1675 BI, 1686 BI, p 87

Stephen of Antioch: 1120–1129 LI

Stephens, John Lloyd: 1839 AN, 1841 AN, p 272

Stephenson, George: 1814 TE, 1825 TE, 1848 TE

Stern, Otto: 1924 PH, 1933 PH, 1943 PH, 1969 PH

Stevens, John: 1802 TE

Stevenson, C.E.: 1937 PH

Stevinus, Simon: 1585 MA, 1586 PH, 1620 PH, p 119

Stewart, Balfour: 1882 EA, 1887 PH

Stewart, Timothy: 1988 BI

Stifel, Michael: 1544 MA, 1567 MA

Stiles, Charles W.: 1902 ME

Stirling, James: 1738 MA

Stock, Alfred: 1909 CH, 1946 CH

Stocke, John: 1981 AS

Stoicheff, Boris P.: 1964 PH

Stokes, Sir George: 1851 PH, 1903 MA

Stoll, Arthur: 1938 CH

Stonehill, D.: 1961 PH

Stoney, George *Johnstone*: 1871 PH, 1874 PH, 1911 PH

Strabo: 10–1 BC PH, 20–29 GE, p 54

Strasburger, Eduard: 1875 BI, 1882 BI, 1894 BI, 1912 BI

Strassman, Fritz: 1902 CH, 1939 PH

Strato: 340–331 BC PH, 270–261 BC PH

Street, J. C.: 1937 PH

Strohmeyer, Friedrich: 1817 CH, 1835 CH

Strömgren, Bengt: 1935 AS

Strominger, Jack: 1987 BI

Strong, Ian B.: 1973 AS

Strong, J. A.: 1959 ME

Strowger, Almon Brown: 1905 TE

Strutt, Jedediah: 1758 TE, 1797 TE

Strutt, John. *See* Rayleigh, Lord

Struve, Friedrich Georg von: 1837 AS, 1839 AS, 1840 AS, 1864 AS

Struve, Otto: 1897 AS, 1963 AS

Stuart, David: 1987 AN

Stumpf, Carl: 1883 ME

Sturgeon, William: 1823 PH

Sturtevant, Alfred: 1911 BI, 1915 BI, 1970 BI

Suess, Eduard: 1885 EA, 1914 EA

Suger, Abbé: 1120–1129 TE

Suleiman: 840–849 GE

Sullivan, W. N.: 1941 CH

Summers, Jesse: 1982 BI

Sumner, James B.: 1926 BI, 1946 CH, 1955 BI

Sundmann, Karl Fritiof: 1913 PH

Susrata: 500–491 BC LI

Süssmilch, Johann: 1761 MA

Su Sung: 1090–1099 TE

Sutherland, Earl Wilbur, Jr.: 1956 BI, 1971 BI, 1974 ME

Sutton, Walter: 1902 BI, 1903 BI

Svedberg, Theodor: 1923 CH, 1926 CH, 1971 CH

Swammerdam, Jan: 1658 LI, 1669 BI, 1680 BI, 1737 BI, p 149

Swan, Sir Joseph Wilson: 1879 TE, 1914 PH

Swedenborg, Emanuel: 1734 CH, 1734 TE, 1772 GE

Swigert, John L., Jr.: 1970 AS

Sydenham, Thomas: 1624 LI, 1689 ME

Sylvester, James Joseph: 1877 MA, 1878 MA, 1897 MA

Sylvius, Franciscus: 1614 LI, 1672 ME

Synge, Richard L. M.: 1944 CH, 1952 CH

Szent-Györgyi, Albert: 1928 BI, 1935 BI, 1937 BI

Szilard, Leo: 1939 PH, 1964 PH, p 480

Szostak, Jack W.: 1983 BI

Tachenius, Otto: 1699 BI

Taieb, Maurice: 1974 AN

Tainter, Charles S.: 1885 TE

Takamine, Jokichi: 1901 BI, 1922 CH

Talbot, William Henry Fox: 1839 TE, 1877 TE

Tamm, Igor: 1937 PH, 1971 PH

Tarski, J.: 1961 PH

Tartaglia (Fontana), Niccoló: 1499 MA, 1512 MA, 1535 MA, 1537 TE, 1541 MA, 1557 MA, pp 73, 107

Tasman, Abel: p 24

Tatum, Edward Lawrie: 1947 BI, 1975 BI

Taube, Henry: 1983 CH

Taubes, Clifford: 1987 MA

Taylor, Brook: 1715 MA, 1719 MA, 1731 MA

Taylor, Frank B.: 1910 EA

Taylor, Frederick Winslow: 1856 TE, 1915 TE

Taylor, Geoffrey Ingram: 1921 PH

Taylor, Roy: 1988 PH

Tedesco, Edward F.: 1985 AS

Teisserenc de Bort, Léon: 1902 EA, 1913 EA

Telesio, Bernardino: 1509 GE, 1588 GE

Telford, Thomas: 1825 TE, 1834 TE

Teller, Edward: 1952 PH

Temin, Howard: 1970 BI, 1975 BI

Tennant, Charles: 1799 CH

Tennant, Smithson: 1803 CH, 1815 CH

Terenius, Lars: 1974 BI, 1975 BI

Tereshkova-Nikolayeva, Valentina: 1963 AS

Terrile, Richard: 1984 AS

Tesla, Nikola: 1884 TE, 1888 TE, 1891 TE, 1943 TE

Tewfik, Sayed: 1984 AN

Thabit ibn Qurra: 830–839 MA, 880–909 MA

Thaddeus of Florence: 10–19 LI

Thales of Miletus: 580–571 BC AS, 580–571 BC MA, 580–571 BC PH, 560–541 BC AS, pp 21–22, 24–26

Thaer, A.D.: 1804 BI

Theaetetus: 420–411 BC MA, 370–371 BC MA, p 25

Theiler, Max: 1937 ME, 1951 ME, 1972 BI

Thénard, Louis-Jacques: 1818 CH, 1857 CH

Theodoric the Great: 510–529 GE

Theodoric of Freibourg: 1300–1309 PH, 1310–1319 PH

Theodorus of Cyrene: 400–391 BC MA

Theodorus of Samos: 530–521 BC TE

Theodosius: 390–399 GE

Theon of Alexandria: 390–399 MA

Theon of Smyrna: 120–129 MA

Theophrastus of Eresus: 30–291 BC LI, 290–281 BC LI, pp 23, 87

Theorell, Axel *Hugo* Theodor: 1932 BI, 1935 BI, 1955 BI, 1982 BI

Thiel, Ferdinand: 1981 CH

Thiout, Antoine: 1725 TE

Thölde, Johan: 1604 PH

Thom, René: p 498

Thomas, Llewellyn: 1926 PH

Thomas, Sidney: 1875 TE

Thompson, Benjamin. *See* Rumford, Count

Thompson, D'Arcy: 1915 BI

Thompson, John T.: 1920 TE

Thomsen, Christian Jorgensen: 1834 AN, 1865 AN

Thomsen, Hans Peter: 1853 CH, 1909 CH

Thomson, Sir Charles *Wyville*: 1868 BI, 1872 EA, 1882 BI

Thomson, Elihu: 1886 TE, 1937 TE

Thomson, Sir George Paget: 1927 PH, 1937 PH, 1975 PH

Thomson, James: 1849 PH

Thomson, Sir Joseph John (J. J.): 1894 PH, 1881 PH, 1897 PH, 1899 PH, 1899 PH, 1903 PH, 1904 PH, 1906 PH, (two entries), 1910 PH, 1940 PH, pp 388, 393

Thomson, Robert William: 1845 TE, 1873 TE

Thomson, Thomas: 1802 CH, 1807 CH

t'Hooft, Gerardus: 1971 PH

Thorndike, Edwin Lee: 1926 ME

Thunberg, Karl: 1828 BI

Thucydides: 480–471 BC LI, 410–401 BC GE

Thutmosis III: 1500–1451 BC AS

Timoni, Emanuel: 1713 ME

Tinbergen, Nikolaas: 1907 BI, 1951 BI, 1988 BI

Ting, Samuel Chao Chung: 1974 PH, 1976 PH

Tiselius, Arne Wilhelm: 1930 CH, 1948 CH, 1971 CH

Tissandier, Albert: 1883 TE

Tissandier, Gaston: 1883 TE

Titchener, Edward Bradford: 1901 BI, 1909 ME

Titius, Johann Daniel: 1766 AS, 1796 AS

Titov, G.: 1961 AS

Tjio, J. Hin: 1954 BI

Todd, Baron Alexander Robertus: 1947 BI, 1957 CH, p 492

Tolman, Edward Chace: 1932 ME

Tombaugh, Clyde: 1930 AS, p 291

Tomonaga, Shin'ichiro: 1943 PH, 1948 PH, 1965 PH, 1979 PH

Tonegawa, Susumu: 1976 BI, 1987 BI

Topham, F.: 1903 CH

Torello, Camillo: 1566 TE

Torrey, John: 1796 BI

Torricelli, Evangelista: 1640 PH, 1643 TE, 1647 PH, p 142

Toscanelli, Paolo: 1390–1399 LI, 1482 LI

Tournefort, Joseph de: 1700 BI, 1708 BI

Townes, Charles Hard: 1953 PH, 1964 PH, p 523

Townsend, Sir John: 1957 PH

Trajan (Marcus Ulpius Trajanus): 110–119 GE

Travers, Morris William: 1898 CH, 1961 CH

Trembley, Abraham: 1710 BI, 1742 BI, 1744 BI, 1784 BI

Tretz, J. F.: 1869 TE

Trevithick, Richard: 1800 TE, 1801 TE, 1804 TE, 1808 TE, 1833 TE

Trumpler, Robert Julius, 1930 AS, 1956 AS

Trutfetter, Jodocus: 1514 PH

Tryon, Edward P.: 1973 PH

Tsai Lun: 100–109 TE, 110–119 TE

Tschermak von Seysenegg, Erich: 1900 BI, 1962 BI

Tschirikov, Ilich: 1741 EA

Tschirnhausen, Count Ehrenfried von: 1708 TE, p 396

Tseng Kung-Liang: 1040–1049 TE, 1070–1089 TE

Tsiolkovsky, Konstantin E.: 1895 AS, 1895 TE, 1903 TE, 1935 PH

Ts'o, Paul: 1955 BI

Tsu Ch'ung-Chih: 460–469 MA

Tsu Keng: 460–469 MA

Tsvett, Mikhail Semenovich: 1906 CH, 1919 BI

Tu Shih: 20–39 TE

Tull, Jethro: 1701 TE, 1731 BI, 1741 BI

Tully, R. Brent: 1987 AS

Tulp, Nicolaas: 1641 LI

Tunnstall, Cuthbert: 1522 MA

Tuohy, Kevin: 1887 ME

Turing, Alan Mathison: 1935 MA, 1936 MA, 1937 MA, 1943 TE, 1952 BI, 1954 MA, p 495

Turok, Neil: 1985 AS

Tuve, Merle Anthony: 1901 PH, 1936 PH, 1982 PH

Twort, Frederick William: 1915 BI, 1916 BI, 1950 BI

SUBJECT INDEX

639